A MATRIX HANDBOOK
FOR STATISTICIANS

THE WILEY BICENTENNIAL–KNOWLEDGE FOR GENERATIONS

Each generation has its unique needs and aspirations. When Charles Wiley first opened his small printing shop in lower Manhattan in 1807, it was a generation of boundless potential searching for an identity. And we were there, helping to define a new American literary tradition. Over half a century later, in the midst of the Second Industrial Revolution, it was a generation focused on building the future. Once again, we were there, supplying the critical scientific, technical, and engineering knowledge that helped frame the world. Throughout the 20th Century, and into the new millennium, nations began to reach out beyond their own borders and a new international community was born. Wiley was there, expanding its operations around the world to enable a global exchange of ideas, opinions, and know-how.

For 200 years, Wiley has been an integral part of each generation's journey, enabling the flow of information and understanding necessary to meet their needs and fulfill their aspirations. Today, bold new technologies are changing the way we live and learn. Wiley will be there, providing you the must-have knowledge you need to imagine new worlds, new possibilities, and new opportunities.

Generations come and go, but you can always count on Wiley to provide you the knowledge you need, when and where you need it!

WILLIAM J. PESCE
PRESIDENT AND CHIEF EXECUTIVE OFFICER

PETER BOOTH WILEY
CHAIRMAN OF THE BOARD

A MATRIX HANDBOOK FOR STATISTICIANS

George A. F. Seber
Department of Statistics
University of Auckland
Auckland, New Zealand

BICENTENNIAL 1807 **WILEY** 2007 **BICENTENNIAL**

WILEY-INTERSCIENCE
A John Wiley & Sons, Inc., Publication

Published by John Wiley & Sons, Inc., Hoboken, New Jersey.
Published simultaneously in Canada.

For general information on our other products and services or for technical support, please contact our
Customer Care Department within the United States at (800) 762-2974, outside the United States at
(317) 572-3993 or fax (317) 572-4002.

Wiley also publishes its books in a variety of electronic formats. Some content that appears in print may
not be available in electronic format. For information about Wiley products, visit our web site at
www.wiley.com.

Wiley Bicentennial Logo: Richard J. Pacifico

Library of Congress Cataloging-in-Publication Data:

Seber, G. A. F. (George Arthur Frederick), 1938–
  A matrix handbook for statisticians / George A.F. Seber.
    p.; cm.
  Includes bibliographical references and index.
  ISBN 978-0-471-74869-4 (cloth)
1. Matrices. 2. Statistics. I. Title.
  QA188.S43   2007
  512.9'434—dc22   2007024691

Printed in the United States of America.

10 9 8 7 6 5 4 3 2 1

CONTENTS

PREFACE

This book has had a long gestation period; I began writing notes for it in 1984 as a partial distraction when my first wife was fighting a terminal illness. Although I continued to collect material on and off over the years, I turned my attention to writing in other fields instead. However, in my recent "retirement", I finally decided to bring the book to birth as I believe even more strongly now of the need for such a book. Vectors and matrices are used extensively throughout statistics, as evidenced by appendices in many books (including some of my own), in published research papers, and in the extensive bibliography of Puntanen et al. [1998]. In fact, C. R. Rao [1973a] devoted his first chapter to the topic in his pioneering book, which many of my generation have found to be a very useful source. In recent years, a number of helpful books relating matrices to statistics have appeared on the scene that generally assume no knowledge of matrices and build up the subject gradually. My aim was not to write such a how-to-do-it book, but simply to provide an extensive list of results that people could look up – very much like a dictionary or encyclopedia. I therefore assume that the reader already has a basic working knowledge of vectors and matrices. Alhough the book title suggests a statistical orientation, I hope that the book's wide scope will make it useful to people in other disciplines as well.

In writing this book, I faced a number of challenges. The first was what to include. It was a bit like writing a dictionary. When do you stop adding material; I guess when other things in life become more important! The temptation was to begin including almost every conceiveble matrix result I could find on the grounds that one day they might all be useful in statistical research! After all, the history of science tells us that mathematical theory usually precedes applications. However, this is not practical and my selection is therefore somewhat personal and reflects my own general knowledge, or lack of it! Also, my selection is tempered by my ability to access certain books and journals, so overall there is a fair dose of randomness in the selection process. To help me keep my feet on the ground and keep my focus on statistics, I have listed, where possible, some references to statistical applications of the theory. Clearly, readers will spot some gaps and I apologize in advance for leaving out any of your favorite results or topics. Please let me know about them (e-mail: seber@stat.auckland.ac.nz). A helpful source of matrix definitions is the free encyclopedia, wikipedia at http://en.wikipedia.org.

My second challenge was what to do about proofs. When I first started this project, I began deriving and collecting proofs but soon realized that the proofs would make the book too big, given that I wanted the book to be reasonably comprehensive. I therefore decided to give only references to proofs at the end of each section or subsection. Most of the time I have been able to refer to book sources, with the occasional journal article referenced, and I have tried to give more than one reference for a result when I could. Although there are many excellent matrix books that I could have used for proofs, I often found in consulting a book that a particular result that I wanted was missing or perhaps assigned to the exercises, which often didn't have outline solutions. To avoid casting my net too widely, I have therefore tended to quote from books that are more encyclopedic in nature. Occasionally, there are lesser known results that are simply quoted without proof in the source that I have used, and I then use the words "Quoted by ..."; the reader will need to consult that source for further references to proofs. Some of my references are to exercises, and I have endeavored to choose sources that have at least outline solutions (e.g., Rao and Bhimasankaram [2000] and Seber [1984]) or perhaps some hints (e.g., Horn and Johnson [1985, 1991]); several books have solutions manuals (e.g., Harville [2001] and Meyer [2000b]). Sometimes I haven't been able to locate the proof of a fairly of straightforward result, and I have found it quicker to give an outline proof that I hope is sufficient for the reader.

In relation to proofs, there is one other matter I needed to deal with. Initially, I wanted to give the original references to important results, but found this too difficult for several reasons. Firstly, there is the sheer volume of results, combined with my limited access to older documents. Secondly, there is often controversy about the original authors. However, I have included some names of original authors where they seem to be well established. We also need to bear in mind Stigler's maxim, simply stated, that "no scientific discovery is named after its original discoverer." (Stigler [1999: 277]). It should be noted that there are also statistical proofs of some matrix results (cf. Rao [2000]).

The third challenge I faced was choosing the order of the topics. Because this book is not meant to be a teach-yourself matrix book, I did not have to follow a "logical" order determined by the proofs. Instead, I was able to collect like results together for an easier look-up. In fact, many topics overlap, so that a logical order is not completely possible. A disadvantage of such an approach is that concepts are sometimes mentioned before they are defined. I don't believe this will cause any difficulties because the cross-referencing and the index will, hopefully, be sufficiently detailed for definitions to be readily located.

My fourth challenge was deciding what level of generality I should use. Some authors use a general field for elements of matrices, while others work in a framework of complex matrices, because most results for real matrices follow as a special case.

Most books with the word "statistics" in the title deal with real matrices only. Although the complex approach would seem the most logical, I am aware that I am writing mainly for the research statistician, many of whom are not involved with complex matrices. I have therefore used a mixed approach with the choice depending on the topic and the proofs available in the literature. Sometimes I append the words "real case" or "complex case" to a reference to inform the reader about the nature of the proof referenced. Frequently, proofs relating to real matrices can be readily extended with little change to those for the complex case.

In a book of this size, it has not been possible to check the correctness of all the results quoted. However, where a result appears in more than one reference, one would have confidence in its accuracy. My aim has been been to try and faithfully reproduce the results. As we know with data, there is always a percentage that is either wrong or incorrectly transcribed. This book won't be any different. If you do find a typo, I would be grateful if you could e-mail me so that I can compile a list of errata for distribution.

With regard to contents, after some notation in Chapter 1, Chapter 2 focuses on vector spaces and their properties, especially on orthogonal complements and column spaces of matrices. Inner products, orthogonal projections, metrics, and convexity then take up most of the balance of the chapter. Results relating to the rank of a matrix take up all of Chapter 3, while Chapter 4 deals with important matrix functions such as inverse, transpose, trace, determinant, and norm. As complex matrices are sometimes left out of books, I have devoted Chapter 5 to some properties of complex matrices and then considered Hermitian matrices and some of their close relatives.

Chapter 6 is devoted to eigenvalues and eigenvectors, singular values, and (briefly) antieigenvalues. Because of the increasing usefulness of generalized inverses, Chapter 7 deals with various types of generalized inverses and their properties. Chapter 8 is a bit of a potpourri; it is a collection of various kinds of special matrices, except for those specifically highlighted in later chapters such as non-negative matrices in Chapter 9 and positive and non-negative definite matrices in Chapter 10. Some special products and operators are considered in Chapter 11, including (a) the Kronecker, Hadamard, and Rao–Khatri products and (b) operators such as the vec, vech, and vec-permutation (commutation) operators. One could fill several books with inequalities so that in Chapter 12 I have included just a selection of results that might have some connection with statistics. The solution of linear equations is the topic of Chapter 13, while Chapters 14 and 15 deal with partitioned matrices and matrices with a pattern.

A wide variety of factorizations and decompositions of matrices are given in Chapter 16, and in Chapter 17 and 18 we have the related topics of differentiation and Jacobians. Following limits and sequences of matrices in Chapter 19, the next three chapters involve random variables - random vectors (Chapter 20), random matrices (Chapter 21), and probability inequalities (Chapter 22). A less familiar topic, namely majorization, is considered in Chapter 23, followed by aspects of optimization in the last chapter, Chapter 24.

I want to express my thanks to a number of people who have provided me with preprints, reprints, reference material and answered my queries. These include Harold Henderson, Nye John, Simo Puntanen, Jim Schott, George Styan, Gary Tee, Goëtz Trenkler, and Yongge Tian. I am sorry if I have forgotten anyone because of the length of time since I began this project. My thanks also go to several anonymous referees who provided helpful input on an earlier draft of the book, and to the Wiley team for their encouragement and support. Finally, special thanks go to my wife Jean for her patient support throughout this project.

GEORGE A. F. SEBER

Auckland, New Zealand
Setember 2007

CHAPTER 1

NOTATION

1.1 GENERAL DEFINITIONS

Vectors and matrices are denoted by boldface letters $\mathbf{a}$ and $\mathbf{A}$, respectively, and scalars are denoted by italics. Thus $\mathbf{a} = (a_i)$ is a vector with ith element a_i and $\mathbf{A} = (a_{ij})$ is a matrix with i, jth elements a_{ij}. I maintain this notation even with random variables, because using uppercase for random variables and lowercase for their values can cause confusion with vectors and matrices. In Chapters 20 and 21, which focus on random variables, we endeavor to help the reader by using the latter half of the alphabet $u, v, \ldots, z$ for random variables and the rest of the alphabet for constants.

Let $\mathbf{A}$ be an $n_1 \times n_2$ matrix. Then any $m_1 \times m_2$ matrix $\mathbf{B}$ formed by deleting any $n_1 - m_1$ rows and $n_2 - m_2$ columns of $\mathbf{A}$ is called a *submatrix* of $\mathbf{A}$. It can also be regarded as the intersection of m_1 rows and m_2 columns of $\mathbf{A}$. I shall define $\mathbf{A}$ to be a submatrix of itself, and when this is not the case I refer to a submatrix that is not $\mathbf{A}$ as a *proper submatrix* of $\mathbf{A}$. When $m_1 = m_2 = m$, the square matrix $\mathbf{B}$ is called a *principal submatrix* and it is said to be of *order m*. Its determinant, $\det(\mathbf{B})$, is called an *mth-order minor* of $\mathbf{A}$. When $\mathbf{B}$ consists of the intersection of the same numbered rows and columns (e.g., the first, second, and fourth), the minor is called a *principal minor*. If $\mathbf{B}$ consists of the intersection of the first m rows and the first m columns of $\mathbf{A}$, then it is called a *leading principal submatrix* and its determinant is called a *leading principal m-th order* minor.

A Matrix Handbook for Statisticians. By George A. F. Seber
Copyright © 2008 John Wiley & Sons, Inc.

Many matrix results hold when the elements of the matrices belong to a general field $\mathcal{F}$ of scalars. For most practitioners, this means that the elements can be real or complex, so we shall use $\mathbb{F}$ to denote either the real numbers $\mathbb{R}$ or the complex numbers $\mathbb{C}$. The expression $\mathbb{F}^n$ will denote the n-dimensional counterpart.

If $\mathbf{A}$ is complex, it can be expressed in the form $\mathbf{A} = \mathbf{B} + i\mathbf{C}$, where $\mathbf{B}$ and $\mathbf{C}$ are real matrices, and its *complex conjugate* is $\overline{\mathbf{A}} = \mathbf{B} - i\mathbf{C}$. We call $\mathbf{A}' = (a_{ji})$ the *transpose* of $\mathbf{A}$ and define the *conjugate transpose* of $\mathbf{A}$ to be $\mathbf{A}^* = \overline{\mathbf{A}}'$. In practice, we can often transfer results from real to complex matrices, and vice versa, by simply interchanging $'$ and $*$.

When adding or multiplying matrices together, we will assume that the sizes of the matrices are such that these operations can be carried out. We make this assumption by saying that the matrices are *conformable*. If there is any ambiguity we shall denote an $m \times n$ matrix $\mathbf{A}$ by $\mathbf{A}_{m \times n}$. A matrix partitioned into blocks is called a block matrix.

If x and y are random variables, then the symbols $\mathrm{E}(y)$, $\mathrm{var}(y)$, $\mathrm{cov}(x, y)$, and $\mathrm{E}(x \mid y)$ represent expectation, variance, covariance, and conditional expectation, respectively.

Before we give a list of all the symbols used we mention some univariate statistical distributions.

1.2 SOME CONTINUOUS UNIVARIATE DISTRIBUTIONS

We assume that the reader is familiar with the normal, chi-square, t, F, gamma, and beta univariate distributions. Multivariate vector versions of the normal and t distributions are given in Sections 20.5.1 and 20.8.1, respectively, and matrix versions of the gamma and beta are found in Section 21.9. As some noncentral distributions are referred to in the statistical chapters, we define two univariate distributions below.

1.1. (Noncentral Chi-square Distribution) The random variable x with probability density function

$$f(x) = \frac{1}{2^{\nu/2}} e^{-x^2/2} x^{(\nu/2)-1} \sum_{i=1}^{\infty} e^{-\delta/2} \left(\frac{\delta}{4} \right)^i \frac{1}{i, \Gamma(\frac{1}{2}\nu + i)} x^i$$

is called the *noncentral chi-square distribution* with ν degrees of freedom and non-centrality parameter δ, and we write $x \sim \chi_\nu^2(\delta)$.

(a) When $\delta = 0$, the above density reduces to the (central) chi-square distribution, which is denoted by χ_ν^2.

(b) The noncentral chi-square can be defined as the distribution of the sum of the squares of independent univariate normal variables y_i $(i = 1, 2, \ldots, n)$ with variances 1 and respective means μ_i. Thus if $\mathbf{y} \sim N_d(\boldsymbol{\mu}, \mathbf{I}_d)$, the multivariate normal distribution, then $x = \mathbf{y}'\mathbf{y} \sim \chi_d^2(\delta)$, where $\delta = \boldsymbol{\mu}'\boldsymbol{\mu}$ (Anderson [2003: 81–82]).

(c) $\mathrm{E}(x) = \nu + \delta$.

Since $\delta > 0$, some authors set $\delta = \tau^2$, say. Others use $\delta/2$, which, because of (c), is not so memorable.

1.2. (Noncentral F-Distribution) If $x \sim \chi^2_m(\delta)$, $y \sim \chi^2_n$, and x and y are statistically independent, then $F = (x/m)/(y/n)$ is said to have a noncentral F-distribution with m and n degrees of freedom, and noncentrality parameter δ. We write $F \sim F_{m,n}(\delta)$. For a derivation of this distribution see Anderson [2003: 185]. When $\delta = 0$, we use the usual notation $F_{m,n}$ for the F-distribution.

1.3 GLOSSARY OF NOTATION

Scalars

$\mathcal{F}$	field of scalars
$\mathbb{R}$	real numbers
$\mathbb{C}$	complex numbers
$\mathbb{F}$	$\mathbb{R}$ or $\mathbb{C}$
$z = x + iy$	a complex number
$\overline{z} = x - iy$	complex conjugate of z
$|z| = (x^2 + y^2)^{1/2}$	modulus of z

Vector Spaces

$\mathbb{F}^n$	n-dimensional coordinate space
$\mathbb{R}^n$	$\mathbb{F}^n$ with $\mathbb{F} = \mathbb{R}$
$\mathbb{C}^n$	$\mathbb{F}^n$ with $\mathbb{F} = \mathbb{C}$
$\mathcal{C}(\mathbf{A})$	column space of $\mathbf{A}$, the space spanned by the columns of $\mathbf{A}$
$\mathcal{C}(\mathbf{A}')$	row space of $\mathbf{A}$
$\mathcal{N}(\mathbf{A})$	$\{\mathbf{x} : \mathbf{A}\mathbf{x} = \mathbf{0}\}$, null space (kernel) of $\mathbf{A}$
$\mathcal{S}(A)$	span of the set A, the vector space of all linear combinations of vectors in A
$\dim \mathcal{V}$	dimension of the vector space $\mathcal{V}$
$\mathcal{V}^\perp$	the orthogonal complement of $\mathcal{V}$
$\mathbf{x} \in \mathcal{V}$	$\mathbf{x}$ is an element of $\mathcal{V}$
$\mathcal{V} \subseteq \mathcal{W}$	$\mathcal{V}$ is a subset of $\mathcal{W}$
$\mathcal{V} \subset \mathcal{W}$	$\mathcal{V}$ is a proper subset of $\mathcal{W}$ (i.e., $\mathcal{V} \neq \mathcal{W}$)
$\mathcal{V} \cap \mathcal{W}$	intersection, $\{\mathbf{x} : \mathbf{x} \in \mathcal{V} \text{ and } \mathbf{x} \in \mathcal{W}\}$
$\mathcal{V} \cup \mathcal{W}$	union, $\{\mathbf{x} : \mathbf{x} \in \mathcal{V} \text{ and/or } \mathbf{x} \in \mathcal{W}\}$
$\mathcal{V} + \mathcal{W}$	sum, $\{\mathbf{x} + \mathbf{y} : \mathbf{x} \in \mathcal{V}, \mathbf{y} \in \mathcal{W}\}$
$\mathcal{V} \oplus \mathcal{W}$	direct sum, $\{\mathbf{x} + \mathbf{y} : \mathbf{x} \in \mathcal{V}, \mathbf{y} \in \mathcal{W}; \mathcal{V} \cap \mathcal{W} = \mathbf{0}\}$
$\langle\,,\,\rangle$	an inner product defined on a vector space
$\mathbf{x} \perp \mathbf{y}$	$\mathbf{x}$ is perpendicular to $\mathbf{y}$ (i.e., $\langle \mathbf{x}, \mathbf{y} \rangle = 0$)

Complex Matrix

$\mathbf{A} = \mathbf{B} + i\mathbf{C}$	complex matrix, with $\mathbf{B}$ and $\mathbf{C}$ real
$\overline{\mathbf{A}} = (\bar{a}_{ij}) = \mathbf{B} - i\mathbf{C}$	complex conjugate of $\mathbf{A}$
$\mathbf{A}^* = \overline{\mathbf{A}}' = (\bar{a}_{ji})$	conjugate transpose of $\mathbf{A}$
$\mathbf{A} = \mathbf{A}^*$	$\mathbf{A}$ is a Hermitian matrix
$\mathbf{A} = -\mathbf{A}^*$	$\mathbf{A}$ is a skew-Hermitian matrix
$\mathbf{A}\mathbf{A}^* = \mathbf{A}^*\mathbf{A}$	$\mathbf{A}$ is a normal matrix

Special Symbols

sup	supremum
inf	infemum
max	maximum
min	minimum
$\rightarrow$	tends to
$\Rightarrow$	implies
$\propto$	proportional to
$\mathbf{1}_n$	the $n \times 1$ vector with unit elements
$\mathbf{I}_n$	the $n \times n$ identity matrix
$\mathbf{0}$	a vector or matrix of zeros
$\mathrm{diag}(\mathbf{d})$	$n \times n$ matrix with diagonal elements $\mathbf{d}' = (d_1, \ldots, d_n)$, and zeros elsewhere
$\mathrm{diag}(d_1, d_2, \ldots, d_n)$	same as above
$\mathrm{diag}\,\mathbf{A}$	diagonal matrix ; same diagonal elements as $\mathbf{A}$
$\mathbf{A} \geq \mathbf{0}$	the elements of $\mathbf{A}$ are all non-negative
$\mathbf{A} > \mathbf{0}$	the elements of $\mathbf{A}$ are all positive
$\mathbf{A} \succeq \mathbf{0}$, n.n.d	$\mathbf{A}$ is non-negative definite $(\mathbf{x}'\mathbf{A}\mathbf{x} \geq 0)$
$\mathbf{A} \succeq \mathbf{B}, \mathbf{B} \preceq \mathbf{A}$	$\mathbf{A} - \mathbf{B} \succeq \mathbf{0}$
$\mathbf{A} \succ \mathbf{0}$, p.d.	$\mathbf{A}$ is positive definite $(\mathbf{x}'\mathbf{A}\mathbf{x} > 0$ for $\mathbf{x} \neq \mathbf{0})$
$\mathbf{A} \succ \mathbf{B}, \mathbf{B} \prec \mathbf{A}$	$\mathbf{A} - \mathbf{B} \succ \mathbf{0}$
$\mathbf{x} \ll \mathbf{y}$	$\mathbf{x}$ is (strongly) majorized by $\mathbf{y}$
$\mathbf{x} \ll_w \mathbf{y}$	$\mathbf{x}$ is weakly submajorized by $\mathbf{y}$
$\mathbf{x} \ll^w \mathbf{y}$	$\mathbf{x}$ is weakly supermajorized by $\mathbf{y}$
$\mathbf{A}' = (a_{ji})$	the transpose of $\mathbf{A}$
$\mathbf{A}^{-1}$	inverse of $\mathbf{A}$ when $\mathbf{A}$ is nonsingular
$\mathbf{A}^-$	weak inverse of $\mathbf{A}$ satisfying $\mathbf{A}\mathbf{A}^-\mathbf{A} = \mathbf{A}$
$\mathbf{A}^+$	Moore-Penrose inverse of $\mathbf{A}$
$\mathrm{trace}\,\mathbf{A}$	sum of the diagonal elements of a square matrix $\mathbf{A}$
$\det \mathbf{A}$	determinant of a square matrix $\mathbf{A}$
$\mathrm{rank}\,\mathbf{A}$	rank of $\mathbf{A}$
$\mathrm{per}\,\mathbf{A}$	permanent of a square matrix $\mathbf{A}$
$\mathrm{mod}(\mathbf{A})$	modulus of $\mathbf{A} = (a_{ij})$, given by $(|a_{ij}|)$
$\mathrm{Pf}(\mathbf{A})$	pfaffian of $\mathbf{A}$
$\rho(\mathbf{A})$	spectral radius of a square matrix $\mathbf{A}$
$\kappa_v(\mathbf{A})$	condition number of an $m \times n$ matrix, $v = 1, 2, \infty$

$\langle \mathbf{x}, \mathbf{y} \rangle$	inner product of $\mathbf{x}$ and $\mathbf{y}$
$\|\mathbf{x}\|$	a norm of vector $\mathbf{x}$ $(= \langle \mathbf{x}, \mathbf{x} \rangle^{1/2})$
$\|\mathbf{x}\|_2$	length of $\mathbf{x}$ $(= (\mathbf{x}^* \mathbf{x})^{1/2})$
$\|\mathbf{x}\|_p$	L_p vector norm of $\mathbf{x}$ $(= \sum_{i=1}^{n} |x_i|^p)^{1/p})$
$\|\mathbf{x}\|_\infty$	L_∞ vector norm of $\mathbf{x}$ $(= \max_i |x_i|)$
$\|\mathbf{A}\|_p$	a generalized matrix norm of $m \times n$ $\mathbf{A}$ $(= \sum_{i=1}^{m} \sum_{j=1}^{n} |a_{ij}|^p)^{1/p}, \; p \geq 1)$
$\|\mathbf{A}\|_F$	Frobenius norm of matrix $\mathbf{A}$ $(= (\sum_i \sum_j |a_{ij}|^2)^{1/2})$
$\|\mathbf{A}\|_{v,in}$	generalized matrix norm for $m \times n$ matrix $\mathbf{A}$ induced by a vector norm $\| \cdot \|_v$
$\|\mathbf{A}\|_{ui}$	unitarily invariant norm of $m \times n$ matrix $\mathbf{A}$
$\|\mathbf{A}\|_{oi}$	orthogonally invariant norm of $m \times n$ matrix $\mathbf{A}$
$\|\|\mathbf{A}\|\|$	matrix norm of square matrix $\mathbf{A}$
$\|\|\mathbf{A}\|\|_{v,in}$	matrix norm for a square matrix $\mathbf{A}$ induced by a vector norm $\| \cdot \|_v$
$\mathbf{A}_{m \times n}$	$m \times n$ matrix
$(\mathbf{A}, \mathbf{B})$	matrix partitioned by two matrices $\mathbf{A}$ and $\mathbf{B}$
$(\mathbf{a}_1, \ldots, \mathbf{a}_n)$	matrix partitioned by column vectors $\mathbf{a}_1, \ldots, \mathbf{a}_n$
$\mathbf{A} \otimes \mathbf{B}$	Kronecker product of $\mathbf{A}$ and $\mathbf{B}$
$\mathbf{A} \circ \mathbf{B}$	Hadamard (Schur) product of $\mathbf{A}$ and $\mathbf{B}$
$\mathbf{A} \odot \mathbf{B}$	Rao–Khatri product of $\mathbf{A}$ and $\mathbf{B}$
vec $\mathbf{A}_{m \times n}$	$mn \times 1$ vector formed by writing the columns of $\mathbf{A}$ one below the other
vech $\mathbf{A}_{m \times m}$	$\frac{1}{2}m(m+1) \times 1$ vector formed by writing the columns of the lower triangle of $\mathbf{A}$ (including the diagonal elements) one below the other
$\mathbf{I}_{(m,n)}$ or $\mathbf{K}_{nm}$	vec-permutation (commutation) matrix
$\mathbf{G}_n$ or $\mathbf{D}_n$	duplication matrix
$\mathbf{P}_n$ or $\mathbf{N}_n$	symmetrizer matrix
$\lambda(\mathbf{A})$	eigenvalue of a square matrix $\mathbf{A}$
$\sigma(\mathbf{B})$	singular value of any matrix $\mathbf{B}$

CHAPTER 2

VECTORS, VECTOR SPACES, AND CONVEXITY

Vector spaces and subspaces play an important role in statistics, the key ones being orthogonal complements as well as the column and row spaces of matrices. Projections onto vector subspaces occur in topics like least squares, where orthogonality is defined in terms of an inner product. Convex sets and functions arise in the development of inequalities and optimization. Other topics such as metric spaces and coordinate geometry are also included in this chapter. A helpful reference for vector spaces and their properties is Kollo and von Rosen [2005: section 1.2].

2.1 VECTOR SPACES

2.1.1 Definitions

Definition 2.1. If S and T are subsets of some space V, then $S \cap T$ is called the *intersection* of S and T and is the set of all vectors in V common to both S and T. The *sum* of S and T, written $S + T$, is the set of all vectors in V that are a sum of a vector in S and a vector in T. Thus

$$W = S + T = \{\mathbf{w} : \mathbf{w} = \mathbf{s} + \mathbf{t}, \mathbf{s} \in S \text{ and } \mathbf{t} \in T\}.$$

(In most applications S and T are vector subspaces, defined below.)

Definition 2.2. A *vector space* $\mathcal{U}$ over a field $\mathcal{F}$ is a set of elements $\{\mathbf{u}\}$ called vectors and a set $\mathcal{F}$ of elements called scalars with four binary operations ($+$, $\cdot$, $\star$, and $\circ$) that satisfy the following axioms.

A Matrix Handbook for Statisticians. By George A. F. Seber
Copyright © 2008 John Wiley & Sons, Inc.

(1) $\mathcal{F}$ is a field with regard to the operations $+$ and $\cdot$.

(2) For all $\mathbf{u}$ and $\mathbf{v}$ in $\mathcal{U}$ we have the following:

(i) $\mathbf{u} \star \mathbf{v} \in \mathcal{U}$.

(ii) $\mathbf{u} \star \mathbf{v} = \mathbf{v} \star \mathbf{u}$.

(iii) $(\mathbf{u} \star \mathbf{v}) \star \mathbf{w} = \mathbf{u} \star (\mathbf{v} \star \mathbf{w})$ for all $\mathbf{w} \in \mathcal{U}$.

(iv) There is a vector $\mathbf{0} \in \mathcal{U}$, called the *zero vector*, such that $\mathbf{u} \star \mathbf{0} = \mathbf{u}$ for all $\mathbf{u} \in \mathcal{U}$.

(v) For each $\mathbf{u} \in \mathcal{U}$ there exists a vector $-\mathbf{u} \in \mathcal{U}$ such that $\mathbf{u} \star -\mathbf{u} = \mathbf{0}$.

(3) For all α and β in $\mathcal{F}$ and all $\mathbf{u}$ and $\mathbf{v}$ in $\mathcal{U}$ we have:

(i) $\alpha \circ \mathbf{u} \in V$.

(ii) There exists an element in $\mathcal{F}$ called the *unit element* such that $1 \circ \mathbf{u} = \mathbf{u}$.

(iii) $(\alpha + \beta) \circ \mathbf{u} = (\alpha \circ \mathbf{u}) \star (\beta \circ \mathbf{u})$.

(iv) $\alpha \circ (\mathbf{u} \star \mathbf{v}) = (\alpha \circ \mathbf{u}) \star (\alpha \circ \mathbf{v})$.

(v) $(\alpha \cdot \beta) \circ \mathbf{u} = \alpha \circ (\beta \circ \mathbf{u})$.

We note from (2) that $\mathcal{U}$ is an abelian group under "$\star$". Also, we can replace "$\star$" by "$+$" and remove "$\cdot$" and "$\circ$" wihout any ambiguity. Thus (iv) and (v) of (3) above can be written as $\alpha(\mathbf{u} + \mathbf{v}) = \alpha\mathbf{u} + \alpha\mathbf{v}$ and $(\alpha\beta)\mathbf{u} = \alpha(\beta\mathbf{u})$, which we shall do in what follows.

Normally $\mathcal{F} = \mathbb{F}$, where $\mathbb{F}$ denotes either $\mathbb{R}$ or $\mathbb{C}$. However, one field that has been useful in the construction of experimental designs such as orthogonal Latin squares, for example, is a finite field consisting of a finite number of elements. A finite field is known as a *Galois field*. The number of elements in any Galois field is p^m, where p is a prime number and m is a positive integer. For a brief discussion see Rao and Rao [1998: 6–10].

If $\mathcal{F}$ is a finite field, then a vector space $\mathcal{U}$ over $\mathcal{F}$ can be used to obtain a finite projective geometry with a finite set of elements or "points" S and a collection of subsets of S or "lines." By identifying a block with a "line" and a treatment with a "point," one can use the projective geometry to construct balanced incomplete block designs—as, for example, described by Rao and Rao [1998: 48–49].

For general, less abstract, references on this topic see Friedberg et al. [2003], Lay [2003], and Rao and Bhimasankaram [2000].

Definition 2.3. A subset $\mathcal{V}$ of a vector space $\mathcal{U}$ that is also a vector space is called a *subspace* of $\mathcal{U}$.

2.1. $\mathcal{V}$ is a vector subspace if and only if $\alpha\mathbf{u} + \beta\mathbf{v} \in \mathcal{V}$ for all $\mathbf{u}$ and $\mathbf{v}$ in $\mathcal{V}$ and all α and β in $\mathcal{F}$. Setting $\alpha = \beta = 0$, we see that $\mathbf{0}$, the zero vector in $\mathcal{U}$, must belong to every vector subspace.

2.2. The set $\mathcal{V}$ of all $m \times n$ matrices over $\mathcal{F}$, along with the usual operations of addition and scalar multiplication, is a vector space. If $m = n$, the subset $\mathcal{A}$ of all symmetric matrices is a vector subspace of $\mathcal{V}$.

Proofs.   Section 2.1.1.

2.1. Rao and Bhimasankaram [2000: 23].

2.2. Harville [1997: chapters 3 and 4].

2.1.2 Quadratic Subspaces

Quadratic subspaces arise in certain inferential problems such as the estimation of variance components (Rao and Rao [1998: chapter 13]). They also arise in testing multivariate linear hypotheses when the variance–covariance matrix has a certain structure or pattern (Rogers and Young [1978: 204] and Seeley [1971]). Klein [2004] considers their use in the design of mixture experiments.

Definition 2.4. Suppose $\mathcal{B}$ is a subspace of $\mathcal{A}$, where $\mathcal{A}$ is the set of all $n \times n$ real symmetric matrices. If $\mathbf{B} \in \mathcal{B}$ implies that $\mathbf{B}^2 \in \mathcal{B}$, then $\mathcal{B}$ is called a *quadratic subspace* of $\mathcal{A}$.

2.3. If $\mathbf{A}_1$ and $\mathbf{A}_2$ are real symmetric idempotent matrices (i.e., $\mathbf{A}_i^2 = \mathbf{A}_i$) with $\mathbf{A}_1\mathbf{A}_2 = \mathbf{0}$, and $\mathcal{A}$ is the set of all real symmetric $n \times n$ matrices, then

$$\mathcal{B} = \{\alpha_1\mathbf{A}_1 + \alpha_2\mathbf{A}_2 : \alpha_1 \text{ and } \alpha_2 \text{ real}\},$$

is a quadratic subspace of $\mathcal{A}$.

2.4. If $\mathcal{B}$ is a quadratic subspace of $\mathcal{A}$, then the following hold.

(a) If $\mathbf{A} \in \mathcal{B}$, then the Moore–Penrose inverse $\mathbf{A}^+ \in \mathcal{B}$.

(b) If $\mathbf{A} \in \mathcal{B}$, then $\mathbf{A}\mathbf{A}^+ \in \mathcal{B}$.

(c) There exists a basis of $\mathcal{B}$ consisting of idempotent matrices.

2.5. The following statements are equivalent.

(1) $\mathcal{B}$ is a quadratic subspace of $\mathcal{A}$.

(2) If $\mathbf{A}, \mathbf{B} \in \mathcal{B}$, then $(\mathbf{A} + \mathbf{B})^2 \in \mathcal{B}$.

(3) If $\mathbf{A}, \mathbf{B} \in \mathcal{B}$, then $\mathbf{A}\mathbf{B} + \mathbf{B}\mathbf{A} \in \mathcal{B}$.

(4) If $\mathbf{A} \in \mathcal{B}$, then $\mathbf{A}^k \in \mathcal{B}$ for $k = 1, 2, \ldots$.

2.6. Let $\mathcal{B}$ be a quadratic subspace of $\mathcal{A}$. Then:

(a) If $\mathbf{A}, \mathbf{B} \in \mathcal{B}$, then $\mathbf{A}\mathbf{B}\mathbf{A} \in \mathcal{B}$.

(b) Let $\mathbf{A} \in \mathcal{B}$ be fixed and let $\mathcal{C} = \{\mathbf{A}\mathbf{B}\mathbf{A} : \mathbf{B} \in \mathcal{B}\}$. Then $\mathcal{C}$ is a quadratic subspace of $\mathcal{B}$.

(c) If $\mathbf{A}, \mathbf{B}, \mathbf{C} \in \mathcal{B}$, then $\mathbf{A}\mathbf{B}\mathbf{C} + \mathbf{C}\mathbf{B}\mathbf{A} \in \mathcal{B}$.

Proofs. Section 2.1.2.

2.3. This follows from the definition and noting that $\mathbf{A}_2\mathbf{A}_1 = \mathbf{0}$.

2.3 to 2.6. Rao and Rao [1998: 434–436, 440].

2.1.3 Sums and Intersections of Subspaces

Definition 2.5. Let $\mathcal{V}$ and $\mathcal{W}$ be vector subspaces of a vector space $\mathcal{U}$. As with sets, we define $\mathcal{V} + \mathcal{W}$ to be the *sum* of the two vector subspaces. If $\mathcal{V} \cap \mathcal{W} = \mathbf{0}$ (some authors use $\{\mathbf{0}\}$), we say that $\mathcal{V}$ and $\mathcal{W}$ are *disjoint vector subspaces* (Harville [1997] uses the term "essentially disjoint"). Note that this differs from the notion of disjoint sets, namely $\mathcal{V} \cap \mathcal{W} = \phi$, which we will not need. When $\mathcal{V}$ and $\mathcal{W}$ are disjoint, we refer to the sum as a *direct sum* and write $\mathcal{V} \oplus \mathcal{W}$. Also $\mathcal{V} \cap \mathcal{W}$ is called the *intersection* of $\mathcal{V}$ and $\mathcal{W}$.

The ordered pair $(\cap, \subseteq)$ forms a lattice of subspaces so that lattice theory can be used to determine properties relating to the sum and intersection of subspaces. Kollo and von Rosen [2006: section 1.2] give detailed lists of such properties, and some of these are given below.

2.7. Let $\mathcal{A}$, $\mathcal{B}$, and $\mathcal{C}$ be vector subspaces of $\mathcal{U}$.

(a) $\mathcal{A} \cap \mathcal{B}$ and $\mathcal{A} + \mathcal{B}$ are vector subspaces. However, $\mathcal{A} \cup \mathcal{B}$ need not be a vector space. Here $\mathcal{A} \cap \mathcal{B}$ is the smallest subspace containing $\mathcal{A}$ and $\mathcal{B}$, and $\mathcal{A} + \mathcal{B}$ is the largest. Also $\mathcal{A} + \mathcal{B}$ is the smallest subspace containing $\mathcal{A} \cup \mathcal{B}$. By smallest subspace we mean one with the smallest dimension.

(b) If $\mathcal{U} = \mathcal{A} \oplus \mathcal{B}$, then every $\mathbf{u} \in \mathcal{U}$ can be expressed uniquely in the form $\mathbf{u} = \mathbf{a} + \mathbf{b}$, where $\mathbf{a} \in \mathcal{A}$ and $\mathbf{b} \in \mathcal{B}$.

(c) $\mathcal{A} \cap (\mathcal{A} + \mathcal{B}) = \mathcal{A} + (\mathcal{A} \cap \mathcal{B}) = \mathcal{A}$.

(d) (Distributive)

    (i) $\mathcal{A} \cap (\mathcal{B} + \mathcal{C}) \supseteq (\mathcal{A} \cap \mathcal{B}) + (\mathcal{A} \cap \mathcal{C})$.

    (ii) $\mathcal{A} + (\mathcal{B} \cap \mathcal{C}) \subseteq (\mathcal{A} + \mathcal{B}) \cap (\mathcal{A} + \mathcal{C})$.

(e) In the following results we can interchange $+$ and $\cap$.

    (i) $[\mathcal{A} \cap (\mathcal{B} + \mathcal{C})] + \mathcal{B} = [(\mathcal{A} + \mathcal{B}) \cap \mathcal{C}] + \mathcal{B}$.

    (ii) $\mathcal{A} \cap [\mathcal{B} + (\mathcal{A} \cap \mathcal{C})] = (\mathcal{A} \cap \mathcal{B}) + (\mathcal{A} \cap \mathcal{C})$.

    (iii) $\mathcal{A} \cap (\mathcal{B} + \mathcal{C}) = \mathcal{A} \cap [\mathcal{B} \cap (\mathcal{A} + \mathcal{C})] + \mathcal{C}$.

    (iv) $(\mathcal{A} \cap \mathcal{B}) + (\mathcal{A} \cap \mathcal{C}) + (\mathcal{B} \cap \mathcal{C}) = [\mathcal{A} + (\mathcal{B} \cap \mathcal{C})] \cap [\mathcal{B} + (\mathcal{A} \cap \mathcal{C})]$.

    (v) $\mathcal{A} \cap \mathcal{B} = [(\mathcal{A} \cap \mathcal{B}) + (\mathcal{A} \cap \mathcal{C})] \cap [(\mathcal{A} \cap \mathcal{B}) + (\mathcal{B} \cap \mathcal{C})]$.

Proofs. Section 2.1.3.

    2.7a. Schott [2005: 68].

    2.7b. Assume $\mathbf{u} = \mathbf{a}_1 + \mathbf{b}_1$ so that $\mathbf{a} - \mathbf{a}_1 = -(\mathbf{b} - \mathbf{b}_1)$, with the two vectors being in disjoint subspaces; hence $\mathbf{a} = \mathbf{a}_1$ and $\mathbf{b} = \mathbf{b}_1$.

    2.7c–e. Kollo and von Rosen [2006: section 1.2].

    2.7d. Harville [2001: 163, exercise 4].

2.1.4 Span and Basis

Definition 2.6. We can always construct a vector space $\mathcal{U}$ from $\mathcal{F}$, called an *n-tuple space*, by defining $\mathbf{u} = (u_1, u_2, \ldots, u_n)'$, where each $u_i \in \mathcal{F}$.

In practice, $\mathcal{F}$ is usually $\mathbb{F}$ and $\mathcal{U}$ is $\mathbb{F}^n$. This will generally be the case in this book, unless indicated otherwise. However, one useful exception is the vector space consisting of all $m \times n$ matrices with elements in $\mathcal{F}$.

Definition 2.7. Given a subset A of a vector space $\mathcal{V}$, we define the *span* of A, denoted by $\mathcal{S}(A)$, to be the set of all vectors obtained by taking all linear combinations of vectors in A. We say that A is a *generating set* of $\mathcal{S}(A)$.

2.8. Let A and B be subsets of a vector space. Then:

(a) $\mathcal{S}(A)$ is a vector space (even though A may not be).

(b) $A \subseteq \mathcal{S}(A)$. Also $\mathcal{S}(A)$ is the smallest subspace of $\mathcal{V}$ containing A in the sense that every subspace of $\mathcal{V}$ containing A also contains $\mathcal{S}(A)$.

(c) A is a vector space if and only if $A = \mathcal{S}(A)$.

(d) $\mathcal{S}[\mathcal{S}(A)] = \mathcal{S}(A)$.

(e) If $A \subseteq B$, then $\mathcal{S}(A) \subseteq \mathcal{S}(B)$.

(f) $\mathcal{S}(A) \cup \mathcal{S}(B) \subseteq \mathcal{S}(A \cup B)$.

(g) $\mathcal{S}(A \cap B) \subseteq \mathcal{S}(A) \cap \mathcal{S}(B)$.

Definition 2.8. A set of vectors $\mathbf{v}_i$ $(i = 1, 2, \ldots, r)$ in a vector space are *linearly independent* if $\sum_{i=1}^r a_i \mathbf{v}_i = \mathbf{0}$ implies that $a_1 = a_2 = \cdots = a_r = 0$. A set of vectors that are not linearly independent are said to be *linearly dependent*. For further properties of linearly independent sets see Rao and Bhimasankaram [2000: chapter 1].

The term "vector" here and in the following definitions is quite general and simply refers to an element of a vector space. For example, it could be an $m \times n$ matrix in the vector space of all such matrices; Harville [1997: chapters 3 and 4] takes this approach.

Definition 2.9. A set of vectors $\mathbf{v}_i$ $(i = 1, 2, \ldots, r)$ *span* a vector space $\mathcal{V}$ if the elements of $\mathcal{V}$ consist of all linear combinations of the vectors (i.e., if $\mathbf{v} \in \mathcal{V}$, then $\mathbf{v} = a_1 \mathbf{v}_1 + \cdots + a_r \mathbf{v}_r$). The set of vectors is called a *generating set* of $\mathcal{V}$. If the vectors are also linearly independent, then the $\mathbf{v}_i$ form a *basis* for $\mathcal{V}$.

2.9. Every vector space has a basis. (This follows from Zorn's lemma, which can be used to prove the existence of a maximal linearly independent set of vectors, i.e., a basis.)

Definition 2.10. All bases contain the same number of vectors so that this number is defined to be the dimension of $\mathcal{V}$.

2.10. Let $\mathcal{V}$ be a subspace of $\mathcal{U}$. Then:

(a) Every linearly independent set of vectors in $\mathcal{V}$ can be extended to a basis of $\mathcal{U}$.

(b) Every generating set of $\mathcal{V}$ contains a basis of $\mathcal{V}$.

2.11. If $\mathcal{V}$ and $\mathcal{W}$ are vector subspaces of $\mathcal{U}$, then:

(a) If $\mathcal{V} \subseteq \mathcal{W}$ and $\dim \mathcal{V} = \dim \mathcal{W}$, then $\mathcal{V} = \mathcal{W}$.

(b) If $\mathcal{V} \subseteq \mathcal{W}$ and $\mathcal{W} \subseteq \mathcal{V}$, then $\mathcal{V} = \mathcal{W}$. This is the usual method for proving the equality of two vector subspaces.

(c) $\dim(\mathcal{V} + \mathcal{W}) = \dim(\mathcal{V}) + \dim(\mathcal{W}) - \dim(\mathcal{V} \cap \mathcal{W})$.

2.12. If the columns of $\mathbf{A} = (\mathbf{a}_1, \ldots, \mathbf{a}_r)$ and the columns of $\mathbf{B} = (\mathbf{b}_1, \ldots, \mathbf{b}_r)$ both form a basis for a vector subspace of $\mathbb{F}^n$, then $\mathbf{A} = \mathbf{BR}$, where $\mathbf{R} = (r_{ij})$ is $r \times r$ and nonsingular.

Proofs.   Section 2.1.4.

2.8. Rao and Bhimasankaram [2000: 25–28].

2.9. Halmos [1958].

2.10. Rao and Bhimasankaram [2000: 39].

2.11a–b. Proofs are straightforward.

2.11c. Meyer [2000a: 205] and Rao and Bhimasankaram [2000: 48].

2.12. Firstly, $\mathbf{a}_j = \sum_i \mathbf{b}_i r_{ij}$ so that $\mathbf{A} = \mathbf{BR}$. Now assume rank $\mathbf{R} < r$; then rank $\mathbf{A} \leq \min\{\text{rank } \mathbf{B}, \text{rank } \mathbf{R}\} < r$ by (3.12), which is a contradiction.

2.1.5   Isomorphism

Definition 2.11. Let $\mathcal{V}_1$ and $\mathcal{V}_2$ be two vector spaces over the same field $\mathcal{F}$. Then a map (function) ϕ from $\mathcal{V}_1$ to $\mathcal{V}_2$ is said to be an *isomorphism* if the following hold.

(1) ϕ is a bijection (i.e., ϕ is one-to-one and onto).

(2) $\phi(\mathbf{u} + \mathbf{v}) = \phi(\mathbf{u}) + \phi(\mathbf{v})$ for all $\mathbf{u}, \mathbf{v} \in \mathcal{V}_1$.

(3) $\phi(\alpha\mathbf{u}) = \alpha\phi(\mathbf{u})$ for all $\alpha \in \mathcal{F}$ and $\mathbf{u} \in \mathcal{V}_1$.

$\mathcal{V}_1$ is said to be *isomorphic* to $\mathcal{V}_2$ if there is an isomorphism from $\mathcal{V}_1$ to $\mathcal{V}_2$.

2.13. Two vector spaces over a field $\mathcal{F}$ are isomorphic if and only if they have the same dimension.

Proofs.   Section 2.1.5.

2.13. Rao and Bhimasankaram [2000: 59].

2.2   INNER PRODUCTS

2.2.1   Definition and Properties

The concept of an inner product is an important one in statistics as it leads to ideas of length, angle, and distance between two points.

***Definition* 2.12.** Let $\mathcal{V}$ be a vector space over $\mathbb{F}$ (i.e., $\mathbb{R}$ or $\mathbb{C}$), and let $\mathbf{x}$, $\mathbf{y}$, and $\mathbf{z}$ be any vectors in $\mathcal{V}$. An inner product $\langle \cdot, \cdot \rangle$ defined on $\mathcal{V}$ is a function $\langle \mathbf{x}, \mathbf{y} \rangle$ of two vectors $\mathbf{x}, \mathbf{y} \in \mathcal{V}$ satisfying the following conditions:

(1) $\langle \mathbf{x}, \mathbf{y} \rangle = \overline{\langle \mathbf{y}, \mathbf{x} \rangle}$, the complex conjugate of $\langle \mathbf{y}, \mathbf{x} \rangle$.

(2) $\langle \mathbf{x}, \mathbf{x} \rangle \geq 0$; $\langle \mathbf{x}, \mathbf{x} \rangle = 0$ implies that $\mathbf{x} = \mathbf{0}$.

(3) $\langle \alpha \mathbf{x}, \mathbf{y} \rangle = \alpha \langle \mathbf{x}, \mathbf{y} \rangle$, where α is a scalar in $\mathbb{F}$.

(4) $\langle \mathbf{x} + \mathbf{y}, \mathbf{z} \rangle = \langle \mathbf{x}, \mathbf{z} \rangle + \langle \mathbf{y}, \mathbf{z} \rangle$.

When $\mathcal{V}$ is over $\mathbb{R}$, (1) becomes $\langle \mathbf{x}, \mathbf{y} \rangle = \langle \mathbf{y}, \mathbf{x} \rangle$, a symmetry condition. Inner products can also be defined on infinite-dimensional spaces such as a Hilbert space. A vector space together with an inner product is called an *inner product space*. A complex inner product space is also called a *unitary space*, and a real inner product space is called a *Euclidean space*.

The *norm* or *length* of $\mathbf{x}$, denoted by $\|\mathbf{x}\|$, is defined to be the positive square root of $\langle \mathbf{x}, \mathbf{x} \rangle$. We say that $\mathbf{x}$ has *unit length* if $\|\mathbf{x}\| = 1$. More general norms, which are not associated with an inner product, are discussed in Section 4.6.

We can define the angle θ between $\mathbf{x}$ and $\mathbf{y}$ by

$$\cos \theta = \langle \mathbf{x}, \mathbf{y} \rangle / (\|\mathbf{x}\| \|\mathbf{y}\|).$$

The *distance* between $\mathbf{x}$ and $\mathbf{y}$ is defined to be $d(\mathbf{x}, \mathbf{y}) = \|\mathbf{x} - \mathbf{y}\|$ and has the properties of a metric (Section 2.4). Usually, $\mathcal{V} = \mathbb{R}^n$ and $\langle \mathbf{x}, \mathbf{y} \rangle = \mathbf{x}'\mathbf{y}$ in defining angle and distance.

Suppose (2) above is replaced by the weaker condition

(2') $\langle \mathbf{x}, \mathbf{x} \rangle \geq 0$. (It is now possible that $\langle \mathbf{x}, \mathbf{x} \rangle = 0$, but $\mathbf{x} \neq \mathbf{0}$.)

We then have what is called a *semi-inner product* (quasi-inner product) and a corresponding *seminorm*. We write $\langle \mathbf{x}, \mathbf{y} \rangle_s$ for a semi-inner product.

2.14. For any inner product the following hold:

(a) $\langle \mathbf{x}, \alpha \mathbf{y} + \beta \mathbf{z} \rangle = \bar{\alpha} \langle \mathbf{x}, \mathbf{y} \rangle + \bar{\beta} \langle \mathbf{x}, \mathbf{z} \rangle$.

(b) $\langle \mathbf{x}, \mathbf{0} \rangle = \langle \mathbf{0}, \mathbf{x} \rangle = 0$.

(c) $\langle \alpha \mathbf{x}, \beta \mathbf{y} \rangle = \alpha \langle \mathbf{x}, \beta \mathbf{y} \rangle = \alpha \bar{\beta} \langle \mathbf{x}, \mathbf{y} \rangle$.

2.15. The following hold for any norm associated with an inner product.

(a) $\|\mathbf{x} + \mathbf{y}\| \leq \|\mathbf{x}\| + \|\mathbf{y}\|$ (triangle inequality).

(b) $\|\mathbf{x} - \mathbf{y}\| + \|\mathbf{y}\| \geq \|\mathbf{x}\|$.

(c) $\|x + y\|^2 + \|x - y\|^2 = 2\|x\|^2 + 2\|y\|^2$ (parallelogram law).

(d) $\|x + y\|^2 = \|x\|^2 + \|y\|^2$ if $\langle x, y \rangle = 0$ (Pythagoras theorem).

(e) $\langle x, y \rangle + \langle y, x \rangle \leq 2\|x\| \cdot \|y\|$.

2.16. (Semi-Inner Product) The following hold for any semi-inner product $\langle \cdot , \cdot \rangle_s$ on a vector space $\mathcal{V}$.

(a) $\langle 0, 0 \rangle_s = 0$

(b) $\|x + y\|_s \leq \|x\|_s + \|y\|_s$.

(c) $\mathcal{N} = \{x \in \mathcal{V} : \|x\|_s = 0\}$ is a subspace of $\mathcal{V}$.

2.17. (Schwarz Inequality) Given an inner product space, we have for all x and y

$$\langle x, y \rangle^2 \leq \langle x, x \rangle \langle y, y \rangle,$$

or

$$|\langle x, y \rangle| \leq \|x\| \cdot \|y\|,$$

with equality if either x or y is zero or $x = ky$ for some scalar k. We can obtain various inequalities from the above by changing the inner product space (cf. Section 12.1).

2.18. Given an inner product space and *unit* vectors u, v, and w, then

$$\sqrt{1 - |\langle u, v \rangle|^2} \leq \sqrt{1 - |\langle u, w \rangle|^2} + \sqrt{1 - |\langle w, v \rangle|^2}.$$

Equality holds if and only if w is a multiple of u or of v.

2.19. Some inner products are as follows.

(a) If $\mathcal{V} = \mathbb{R}^n$, then common inner products are:

(1) $\langle x, y \rangle = y'x = \sum_{i=1}^{n} x_i y_i$ $(= x'y)$. If $x = y$, we denote the norm by $\|x\|_2$, the so-called *Euclidean norm*.

The *minimal angle between two vector subspaces* $\mathcal{V}$ and $\mathcal{W}$ in $\mathbb{R}^n$ is given by

$$\cos \theta_{\min} = \max_{x \in \mathcal{V}, y \in \mathcal{W}} \frac{(x'y)^2}{\|x\|_2 \cdot \|y\|_2}.$$

For some properties see Meyer [2000a: section 5.15].

(2) $\langle x, y \rangle = y'Ax$ $(= x'Ay)$, where A is a positive definite matrix.

(b) If $\mathcal{V} = \mathbb{C}^n$, then we can use $\langle x, y \rangle = y^*x = \sum_{i=1}^{n} x_i \bar{y}_i$.

(c) Every inner product defined on $\mathbb{C}^n$ can be expressed in the form $\langle x, y \rangle = y^*Ax = \sum_i \sum_j a_{ij} x_i \bar{y}_j$, where $A = (a_{ij})$ is a Hermitian positive definite matrix. This follows by setting $\langle e_i, e_j \rangle = a_{ij}$ for all i, j, where e_i is the ith column of I_n. If we have a semi-inner product, then A is Hermitian non-negative definite. (This result is proved in Drygas [1970: 29], where symmetric means Hermitian.)

2.20. Let $\mathcal{V}$ be the set of all $m \times n$ real matrices, and in scalar multiplication all scalars belong to $\mathbb{R}$. Then:

(a) $\mathcal{V}$ is vector space.

(b) If we define $\langle \mathbf{A}, \mathbf{B} \rangle = \mathrm{trace}(\mathbf{A}'\mathbf{B})$, then $\langle \, , \rangle$ is an inner product.

(c) The corresponding norm is $(\langle \mathbf{A}, \mathbf{A} \rangle)^{1/2} = (\sum_{i=1}^{m} \sum_{j=1}^{n} a_{ij}^2)^{1/2}$. This is the so-called *Frobenius norm* $\|\mathbf{A}\|_F$ (cf. Definition 4.16 below (4.7)).

Proofs.   Section 2.2.1.

2.14. Rao and Bhimasankaram [2000: 251–252].

2.15. We begin with the Schwarz inequality $|\langle \mathbf{x}, \mathbf{y} \rangle| = |\langle \mathbf{y}, \mathbf{x} \rangle| \leq \|\mathbf{x}\| \cdot \|\mathbf{y}\|$ of (2.17). Then, since $\langle \mathbf{x}, \mathbf{y} \rangle + \langle \mathbf{y}, \mathbf{x} \rangle$ is real,

$$\langle \mathbf{x}, \mathbf{y} \rangle + \langle \mathbf{y}, \mathbf{x} \rangle \leq |\langle \mathbf{x}, \mathbf{y} \rangle + \langle \mathbf{y}, \mathbf{x} \rangle| \leq |\langle \mathbf{x}, \mathbf{y} \rangle| + |\langle \mathbf{y}, \mathbf{x} \rangle| \leq 2\|\mathbf{x}\| \cdot \mathbf{y}\|,$$

which proves (e). We obtain (a) by writing $\|\mathbf{x} + \mathbf{y}\|^2 = \langle \mathbf{x} + \mathbf{y}, \mathbf{x} + \mathbf{y} \rangle$ and using (e); the rest are straightforward. See also Rao and Rao [1998: 54].

2.16. Rao and Rao [1998: 77].

2.17. There are a variety of proofs (e.g., Schott [2005: 36] and Ben-Israel and Greville [2003: 7]). The inequality also holds for quasi-inner (semi-inner) products (Harville [1997: 255]).

2.18. Zhang [1999: 155].

2.20. Harville [1997: chapter 4] uses this approach.

2.2.2   Functionals

Definition **2.13.** A function f defined on a vector space $\mathcal{V}$ over a field $\mathbb{F}$ and taking values in $\mathbb{F}$ is said to be a *linear functional* if

$$f(\alpha_1 \mathbf{x}_1 + \alpha_2 \mathbf{x}_2) = \alpha_1 f(\mathbf{x}_1) + \alpha_2 f(\mathbf{x}_2)$$

for every $\mathbf{x}_1, \mathbf{x}_2 \in \mathcal{V}$ and every $\alpha_1, \alpha_2 \in \mathbb{F}$. For a discussion of linear functionals and the related concept of a *dual space* see Rao and Rao [1998: section 1.7].

2.21. (Riesz)   Let $\mathcal{V}$ be an an inner product space with inner product $\langle \, , \rangle$, and let f be a linear functional on $\mathcal{V}$.

(a) There exists a unique vector $\mathbf{z} \in \mathcal{V}$ such that

$$f(\mathbf{x}) = \langle \mathbf{x}, \mathbf{z} \rangle \text{ for every } \mathbf{x} \in \mathcal{V}.$$

(b) Here $\mathbf{z}$ is given by $\mathbf{z} = \overline{f(\mathbf{u})}\, \mathbf{u}$, where $\mathbf{u}$ is any vector of unit length in $\mathcal{V}^{\perp}$.

Proofs.   Section 2.2.2.

2.21. Rao and Rao [1998: 71].

2.2.3   Orthogonality

***Definition* 2.14.** Let $\mathcal{U}$ be a vector space over $\mathbb{F}$ with an inner product $\langle,\rangle$, so that we have an inner product space. We say that $\mathbf{x}$ is *perpendicular* to $\mathbf{y}$, and we write $\mathbf{x} \perp \mathbf{y}$, if $\langle \mathbf{x}, \mathbf{y} \rangle = 0$.

2.22. A set of vectors that are mutually orthogonal—that is, are pairwise orthogonal for every pair—are linearly independent.

***Definition* 2.15.** A basis whose vectors are mutually orthogonal with unit length is called an *orthonormal basis*. An orthonormal basis of an inner product space always exists and it can be constructed from any basis by the Gram–Schmidt orthogonalization process of (2.30).

2.23. Let $\mathcal{V}$ and $\mathcal{W}$ be vector subspaces of a vector space $\mathcal{U}$ such that $\mathcal{V} \subseteq \mathcal{W}$. Any orthonormal basis for $\mathcal{V}$ can be enlarged to form an orthonormal basis for $\mathcal{W}$.

***Definition* 2.16.** Let $\mathcal{U}$ be a vector space over $\mathbb{F}$ with an inner product $\langle,\rangle$, and let $\mathcal{V}$ be a subset or subspace of $\mathcal{U}$. Then the *orthogonal complement* of $\mathcal{V}$ with respect to $\mathcal{U}$ is defined to be

$$\mathcal{V}^{\perp} = \{ \mathbf{x} : \langle \mathbf{x}, \mathbf{y} \rangle = 0 \text{ for all } \mathbf{y} \in \mathcal{V} \}.$$

If $\mathcal{V}$ and $\mathcal{W}$ are two vector subspaces, we say that $\mathcal{V} \perp \mathcal{W}$ if $\langle \mathbf{x}, \mathbf{y} \rangle = 0$ for all $\mathbf{x} \in \mathcal{V}$ and $\mathbf{y} \in \mathcal{W}$.

2.24. Suppose $\dim \mathcal{U} = n$ and $\boldsymbol{\alpha}_1, \boldsymbol{\alpha}_2, \ldots, \boldsymbol{\alpha}_n$ is an orthonormal basis of $\mathcal{U}$. If $\boldsymbol{\alpha}_1, \ldots, \boldsymbol{\alpha}_r$ $(r < n)$ is an orthonormal basis for a vector subspace $\mathcal{V}$ of $\mathcal{U}$, then $\boldsymbol{\alpha}_{r+1}, \ldots, \boldsymbol{\alpha}_n$ is an orthornormal basis for $\mathcal{V}^{\perp}$.

2.25. If S and T are subsets or subspaces of $\mathcal{U}$, then we have the following results:

(a) $S^{\perp}$ is a vector space.

(b) $S \subseteq (S^{\perp})^{\perp}$ with equality if and only if S is a vector space.

(c) If S and T both contain $\mathbf{0}$, then $(S + T)^{\perp} = S^{\perp} \cap T^{\perp}$.

2.26. If $\mathcal{V}$ is a vector subspace of $\mathcal{U}$, a vector space over $\mathbb{F}$, then:

(a) $\mathcal{V}^{\perp}$ is a vector subspace of $\mathcal{U}$, by (2.25a) above.

(b) $(\mathcal{V}^{\perp})^{\perp} = \mathcal{V}$.

(c) $\mathcal{V} \oplus \mathcal{V}^{\perp} = \mathcal{U}$. In fact every $\mathbf{u} \in \mathcal{U}$ can be expressed uniquely in the form $\mathbf{u} = \mathbf{x} + \mathbf{y}$, where $\mathbf{x} \in \mathcal{V}$ and $\mathbf{y} \in \mathcal{V}^{\perp}$.

(d) $\dim(\mathcal{V}) + \dim(\mathcal{V}^{\perp}) = \dim(\mathcal{U})$.

2.27. If $\mathcal{V}$ and $\mathcal{W}$ are vector subspaces of $\mathcal{U}$, then:

(a) $\mathcal{V} \subseteq \mathcal{W}$ if and only if $\mathcal{V} \perp \mathcal{W}^{\perp}$.

(b) $\mathcal{V} \subseteq \mathcal{W}$ if and only if $\mathcal{W}^{\perp} \subseteq \mathcal{V}^{\perp}$.

(c) $(\mathcal{V} \cap \mathcal{W})^{\perp} = \mathcal{V}^{\perp} + \mathcal{W}^{\perp}$ and $(\mathcal{V} + \mathcal{W})^{\perp} = \mathcal{V}^{\perp} \cap \mathcal{W}^{\perp}$.

For more general results see Kollo and von Rosen [2005: section 1.2].

Definition 2.17. Let $\mathcal{V}$ and $\mathcal{W}$ be vector subspaces of $\mathcal{U}$, a vector space over $\mathbb{F}$, and suppose that $\mathcal{V} \subseteq \mathcal{W}$. Then the set of all vectors in $\mathcal{W}$ that are perpendicular to $\mathcal{V}$ form a vector space called the *orthogonal complement* of $\mathcal{V}$ with respect to $\mathcal{W}$, and is denoted by $\mathcal{V}^{\perp} \cap \mathcal{W}$. Thus

$$\mathcal{V}^{\perp} \cap \mathcal{W} = \{\mathbf{w} : \mathbf{w} \in \mathcal{W}, \langle \mathbf{w}, \mathbf{v} \rangle = \mathbf{0} \text{ for every } \mathbf{v} \in \mathcal{V}\}.$$

2.28. Let $\mathcal{V} \subseteq \mathcal{W}$. Then

(a) (i) $\dim(\mathcal{V}^{\perp} \cap \mathcal{W}) = \dim(\mathcal{W}) - \dim(\mathcal{V})$.

    (ii) $\mathcal{W} = \mathcal{V} \oplus (\mathcal{V}^{\perp} \cap \mathcal{W})$.

(b) From (a)(ii) we have $\mathcal{U} = \mathcal{W} \oplus \mathcal{W}^{\perp} = \mathcal{V} \oplus (\mathcal{V}^{\perp} \cap \mathcal{W}) \oplus \mathcal{W}^{\perp}$.
The above can be regarded as an orthogonal decomposition of $\mathcal{U}$ into three orthogonal subspaces. Using this, vectors can be added to any orthonormal basis of $\mathcal{V}$ to form an orthonormal basis of $\mathcal{W}$, which can then be extended to form an orthonormal basis of $\mathcal{U}$.

2.29. Let $\mathcal{A}$, $\mathcal{B}$, and $\mathcal{C}$ be vector subspaces of $\mathcal{U}$. If $\mathcal{B} \perp \mathcal{C}$ and $\mathcal{A} \perp \mathcal{C}$, then
$$\mathcal{A} \cap (\mathcal{B} \oplus \mathcal{C}) = \mathcal{A} \cap \mathcal{B}.$$

2.30. (Classical Gram–Schmidt Algorithm) Given a basis $\mathbf{x}_1, \mathbf{x}_2, \ldots, \mathbf{x}_n$ of an inner product space, there exists an orthonormal basis $\mathbf{q}_1, \mathbf{q}_2, \ldots, \mathbf{q}_n$ given by $\mathbf{q}_1 = \mathbf{x}_1/\|\mathbf{x}_1\|$, $\mathbf{q}_j = \mathbf{w}_j/\|\mathbf{w}_j\|$ $(j = 2, \ldots, n)$, where

$$\mathbf{w}_j = \mathbf{x}_j - \langle \mathbf{x}_j, \mathbf{q}_1 \rangle \mathbf{q}_1 - \langle \mathbf{x}_j, \mathbf{q}_2 \rangle \mathbf{q}_2 - \cdots - \langle \mathbf{x}_j, \mathbf{q}_{j-1} \rangle \mathbf{q}_{j-1}.$$

This expression gives the algorithm for computing the basis. If we require an orthogonal basis only without the square roots involved with the normalizing, we can use $\mathbf{w}_1 = \mathbf{x}_1$ and, for $j = 2, 3, \ldots, n$,

$$\mathbf{w}_j = \mathbf{x}_j - \frac{\langle \mathbf{x}_j, \mathbf{w}_1 \rangle \mathbf{w}_1}{\langle \mathbf{w}_1, \mathbf{w}_1 \rangle} - \cdots - \frac{\langle \mathbf{x}_j, \mathbf{w}_{j-1} \rangle \mathbf{w}_{j-1}}{\langle \mathbf{w}_{j-1}, \mathbf{w}_{j-1} \rangle}.$$

Also the vectors can be replaced by matrices using a suitable inner product such as $\langle \mathbf{A}, \mathbf{B} \rangle = \operatorname{trace}(\mathbf{A}'\mathbf{B})$.

2.31. Since, from (2.9), every vector space has a basis, it follows from the above algorithm that every inner product space has an orthonormal basis.

2.32. Let $\{\boldsymbol{\alpha}_1, \boldsymbol{\alpha}_2, \ldots, \boldsymbol{\alpha}_n\}$ be an orthonormal basis of $\mathcal{V}$, and let $\mathbf{x}, \mathbf{y} \in \mathcal{V}$ be any vectors. Then, for an inner product space:

(a) $\mathbf{x} = \langle \mathbf{x}, \boldsymbol{\alpha}_1 \rangle \boldsymbol{\alpha}_1 + \langle \mathbf{x}, \boldsymbol{\alpha}_2 \rangle \boldsymbol{\alpha}_2 + \cdots + \langle \mathbf{x}, \boldsymbol{\alpha}_n \rangle \boldsymbol{\alpha}_n$.

(b) (Parseval's identity) $\langle \mathbf{x}, \mathbf{y} \rangle = \sum_{i=1}^{n} \langle \mathbf{x}, \boldsymbol{\alpha}_i \rangle \langle \boldsymbol{\alpha}_i, \mathbf{y} \rangle$.

Conversely, if this equation holds for any $\mathbf{x}$ and $\mathbf{y}$, then $\boldsymbol{\alpha}_1, \ldots, \boldsymbol{\alpha}_n$ is an orthonormal basis for $\mathcal{V}$.

(c) Setting $\mathbf{x} = \mathbf{y}$ in (b) we have

$$\|\mathbf{x}\|^2 = |\langle \mathbf{x}, \boldsymbol{\alpha}_1 \rangle|^2 + |\langle \mathbf{x}, \boldsymbol{\alpha}_2 \rangle|^2 + \cdots + |\langle \mathbf{x}, \boldsymbol{\alpha}_n \rangle|^2.$$

(d) (Bessel's inequality) $\sum_{i=1}^{k}\langle \mathbf{x}, \boldsymbol{\alpha}_i \rangle \leq \|\mathbf{x}\|^2$ for each $k \leq n$.

Equality occurs if and only if $\mathbf{x}$ belongs to the space spanned by the $\boldsymbol{\alpha}_i$.

Proofs.   Section 2.2.3.

2.24. Schott [2005: 54].

2.25a. If $\mathbf{x}_1, \mathbf{x}_2 \in S^{\perp}$, then $\langle \mathbf{x}_i, \mathbf{y} \rangle = 0$ for all $\mathbf{y} \in S$ and $\langle \alpha_1\mathbf{x}_1 + \alpha_2\mathbf{x}_2, \mathbf{y} \rangle = \alpha_1\langle \mathbf{x}_1, \mathbf{y} \rangle + \alpha_2\langle \mathbf{x}_2, \mathbf{y} \rangle = 0$, i.e., $\alpha_1\mathbf{x}_1 + \alpha_2\mathbf{x}_2 \in S^{\perp}$.

2.25b. If $\mathbf{x} \in S$, then $\langle \mathbf{x}, \mathbf{y} \rangle = 0$ for all $\mathbf{y} \in S^{\perp}$ and $\mathbf{x} \in (S^{\perp})^{\perp}$. By (a), $(S^{\perp})^{\perp}$ is a vector space even if S is not; then use (2.26b).

2.25c. If $\mathbf{x}$ belongs to the left-hand side (LHS), then $\langle \mathbf{x}, \mathbf{s} + \mathbf{t} \rangle = \langle \mathbf{x}, \mathbf{s} \rangle + \langle \mathbf{x}, \mathbf{t} \rangle = 0$ for all $\mathbf{s} \in S$ and all $\mathbf{t} \in T$. Setting $\mathbf{s} = \mathbf{0}$, then $\langle \mathbf{x}, \mathbf{t} \rangle = 0$; similarly, $\langle \mathbf{x}, \mathbf{s} \rangle = 0$ and $LHS \subseteq RHS$. The argument reverses.

2.26. Rao and Rao [1998: 62–63].

2.27a–b. Harville [1997: 172].

2.27c. Harville [2001: 162, exercise 3] and Rao and Bhimasankaram [2000: 267].

2.28a(i). Follows from (2.26d) with $\mathcal{U} = \mathcal{W}$.

2.28a(ii). If $\mathbf{x} \in RHS$, then $\mathbf{x} = \mathbf{y} + \mathbf{z}$ where $\mathbf{y} \in \mathcal{V} \subseteq \mathcal{W}$ and $\mathbf{z} \in \mathcal{W}$ so that $\mathbf{x} \in \mathcal{W}$ and $RHS \subseteq LHS$. Then use (i) to show $\dim(RHS) = \dim(LHS)$.

2.29. Kollo and von Rosen [2005: 29].

2.30. Rao and Bhimasankaram [2000: 262] and Seber and Lee [2003: 338–339]. For matrices see Harville [1997: 63–64].

2.32a–c. Rao and Rao [1998: 59–61].

2.32d. Rao [1973a: 10].

2.2.4   Column and Null Spaces

Definition **2.18.** If $\mathbf{A}$ is a matrix (real or complex), then the space spanned by the columns of $\mathbf{A}$ is called the *column space* of $\mathbf{A}$, and is denoted by $\mathcal{C}(\mathbf{A})$. (Some authors, including myself in the past, call this the *range space* of $\mathbf{A}$ and write $\mathcal{R}(\mathbf{A})$.) The corresponding *row space* of $\mathbf{A}$ is $\mathcal{C}(\mathbf{A}')$, which some authors write as $\mathcal{R}(\mathbf{A})$; hence my choice of notation to avoid this confusion. The *null space* or *kernel*, $\mathcal{N}(\mathbf{A})$ of $\mathbf{A}$, is defined as follows:

$$\mathcal{N}(\mathbf{A}) = \{\mathbf{x} : \mathbf{A}\mathbf{x} = \mathbf{0}\}.$$

The following results are all expressed in terms of complex matrices, but they clearly hold for real matrices as well.

2.33. From the definition of a vector subspace we find that $\mathcal{C}(\mathbf{A})$ and $\mathcal{N}(\mathbf{A})$ are both vector subspaces.

2.34. Let $\mathbf{A}$ and $\mathbf{B}$ both have n columns. If any one of the following conditions holds, then all three hold:

(1) $\mathcal{C}(\mathbf{A}') \subseteq \mathcal{C}(\mathbf{B}')$.

(2) $\mathcal{N}(\mathbf{B}) \subseteq \mathcal{N}(\mathbf{A})$.

(3) $\mathbf{A}(\mathbf{I}_n - \mathbf{B}^-\mathbf{B}) = \mathbf{0}$.

If (3) holds for a particular weak inverse $\mathbf{B}^-$, then (3) holds for any weak inverse $\mathbf{B}^-$.

2.35. Let $\mathbf{A}$ be any complex matrix.

(a) $\mathcal{N}(\mathbf{A}^*\mathbf{A}) = \mathcal{N}(\mathbf{A})$.

(b) $\mathcal{C}(\mathbf{A}\mathbf{A}^*) = \mathcal{C}(\mathbf{A})$.

(c) Two more results follow from (a) and (b) by interchanging $\mathbf{A}$ and $\mathbf{A}^*$.

In most applications $\mathbf{A}$ is real so that $\mathbf{A}^* = \mathbf{A}'$.

2.36. $\mathcal{N}(\mathbf{A}) \subseteq \mathcal{C}(\mathbf{I} - \mathbf{A})$ and $\mathcal{N}(\mathbf{I} - \mathbf{A}) \subseteq \mathcal{C}(\mathbf{A})$.

2.37. If $\mathbf{x} \perp \mathbf{y}$ when $\langle \mathbf{x}, \mathbf{y} \rangle = \mathbf{x}^*\mathbf{y} = 0$, and $\mathbf{A}$ is an $m \times n$ complex matrix, then $\mathcal{N}(\mathbf{A}) = \{\mathcal{C}(\mathbf{A}^*)\}^\perp$. We therefore have an orthogonal decomposition

$$\mathcal{N}(\mathbf{A}) \oplus \mathcal{C}(\mathbf{A}^*) = \mathbb{F}^n \quad \text{and} \quad \dim \mathcal{N}(\mathbf{A}) + \dim \mathcal{C}(\mathbf{A}^*) = n.$$

We get a further result by interchanging the roles of $\mathbf{A}$ and $\mathbf{A}^*$. Note that $\dim[\mathcal{C}(\mathbf{A}^*)] = \operatorname{rank} \mathbf{A}^* = \operatorname{rank} \mathbf{A}$, by (3.3c).

2.38. If $\mathbf{A}$ is $m \times n$ and $\mathbf{B}$ is $m \times p$, then $\mathcal{C}(\mathbf{B}) \subseteq \mathcal{C}(\mathbf{A})$ if and only if there exists an $n \times p$ matrix $\mathbf{R}$ such that $\mathbf{AR} = \mathbf{B}$. Furthermore, if $p = n$, $\mathcal{C}(\mathbf{A}) = \mathcal{C}(\mathbf{B})$ if and only if there exists such a nonsingular $\mathbf{R}$. Similar results are available for row spaces by simply taking transposes. Thus if $\mathbf{C}$ is $q \times n$, then $\mathcal{C}(\mathbf{C}') \subseteq \mathcal{C}(\mathbf{A}')$ if and only if there exists a $q \times m$ matrix $\mathbf{S}$ such that $\mathbf{SA} = \mathbf{C}$.

2.39. The following hold for conformable matrices:

(a) If $\mathcal{C}(\mathbf{A}) \subseteq \mathcal{C}(\mathbf{B})$, then $\mathcal{C}(\mathbf{A}'\mathbf{B}) = \mathcal{C}(\mathbf{A}')$.

(b) $\mathcal{C}(\mathbf{B}_1) \subseteq \mathcal{C}(\mathbf{B}_2)$ implies that $\mathcal{C}(\mathbf{A}'\mathbf{B}_1) \subseteq \mathcal{C}(\mathbf{A}'\mathbf{B}_2)$.

(c) $\mathcal{C}(\mathbf{B}_1) = \mathcal{C}(\mathbf{B}_2)$ implies that $\mathcal{C}(\mathbf{A}'\mathbf{B}_1) = \mathcal{C}(\mathbf{A}'\mathbf{B}_2)$.

(d) If $\mathcal{C}(\mathbf{A} + \mathbf{BE}) \subseteq \mathcal{C}(\mathbf{B})$ for some conformable $\mathbf{E}$, then $\mathcal{C}(\mathbf{A}) \subseteq \mathcal{C}(\mathbf{B})$.

(e) If $\mathcal{C}(\mathbf{A}) \subseteq \mathcal{C}(\mathbf{B})$, then $\mathcal{C}(\mathbf{A} + \mathbf{BE}) \subseteq \mathcal{C}(\mathbf{B})$ for any conformable $\mathbf{E}$.

Proofs.   Section 2.2.4.

2.34. Scott and Styan [1985: 210].

2.35. Meyer [2000a: 212–213].

2.36. Note that $\mathbf{Bx} = \mathbf{0}$ if and only if $\mathbf{x} = (\mathbf{I} - \mathbf{B})\mathbf{x}$. Set $\mathbf{B} = \mathbf{A}$ and $\mathbf{B} = \mathbf{I} - \mathbf{A}$.

2.37. Ben-Israel and Greville [2003: 12], Rao and Bhimasankaram [2000: 269], and Seber and Lee [2003: 477, real case].

2.38. Graybill [1983: 90] and Harville [1997: 30].

2.39. Quoted by Kollo and von Rosen [2005: 49]. For (a) we first have $\mathcal{C}(\mathbf{A'B}) \subseteq \mathcal{C}(\mathbf{A'})$. Then, from (2.35), $\mathbf{A'x} = \mathbf{A'Ay} = \mathbf{A'BRy} \in \mathcal{C}(\mathbf{A'B})$, by (2.38), i.e., $\mathcal{C}(\mathbf{A'}) \subseteq \mathcal{C}(\mathbf{A'B})$. The rest are straightforward.

2.3 PROJECTIONS

Definition 2.19. A square matrix $\mathbf{P}$ such that $\mathbf{P}^2 = \mathbf{P}$ is said to be *idempotent*. In this section we focus on the geometrical properties of such matrices, which are used extensively in statistics. Algebraic properties are considered in Section 8.6.

2.3.1 General Projections

Definition 2.20. Let the vector space $\mathcal{U}$ be the direct sum of two vector spaces $\mathcal{V}_1$ and $\mathcal{V}_2$ so that $\mathcal{U} = \mathcal{V}_1 \oplus \mathcal{V}_2$ (i.e., $\mathcal{V}_1 \cap \mathcal{V}_2 = \mathbf{0}$). Then every vector $\mathbf{v} \in \mathcal{V}$ has a unique decomposition $\mathbf{v} = \mathbf{v}_1 + \mathbf{v}_2$, where $\mathbf{v}_i \in \mathcal{V}_i$ $(i = 1, 2)$. The transformation $\mathbf{v} \to \mathbf{v}_1$ is called the *projection of* $\mathbf{v}$ *on* $\mathcal{V}_1$ *along* $\mathcal{V}_2$. Here uniqueness follows by assuming another decomposition $\mathbf{v} = \mathbf{w}_1 + \mathbf{w}_2$ so that $\mathbf{v}_1 - \mathbf{w}_1 = -(\mathbf{v}_2 - \mathbf{w}_2)$, which implies $\mathbf{v}_i = \mathbf{w}_i$ for $i = 1, 2$, otherwise $\mathcal{V}_1 \cap \mathcal{V}_2 \neq \mathbf{0}$. Usually $\mathcal{U} = \mathbb{F}^n$, and the following hold if $\mathbb{F}$ is $\mathbb{R}$ or $\mathbb{C}$.

2.40. The above projection on $\mathcal{V}_1$ along $\mathcal{V}_2$ can be represented by an $n \times n$ matrix $\mathbf{P}$ called a *projector* or *projection matrix* so that $\mathbf{Pv} = \mathbf{v}_1$. Also $\mathbf{P}$ is unique and idempotent.

2.41. Using the above notation, $\mathbf{v} = \mathbf{Pv} + (\mathbf{I}_n - \mathbf{P})\mathbf{v} = \mathbf{v}_1 + \mathbf{v}_2$, so that $\mathbf{v}_2 = (\mathbf{I}_n - \mathbf{P})\mathbf{v}$ is the projection of $\mathbf{v}$ on $\mathcal{V}_2$ along $\mathcal{V}_1$. Here $\mathbf{P}$ and $\mathbf{I}_n - \mathbf{P}$ are unique and idempotent, and

$$\mathbf{P}(\mathbf{I}_n - \mathbf{P}) = \mathbf{0}.$$

2.42. Using the above notation, we can identify $\mathcal{V}_1$ and $\mathcal{V}_2$ as follows:

(a) $\mathcal{C}(\mathbf{P}) = \mathcal{V}_1$.

(b) $\mathcal{C}(\mathbf{I}_n - \mathbf{P}) = \mathcal{V}_2$.

(c) If $\mathbf{P}$ is idempotent, then from (8.61) we obtain

$$\mathcal{C}(\mathbf{P}) \oplus \mathcal{N}(\mathbf{P}) = \mathcal{V}_1 \oplus \mathcal{V}_2.$$

2.43. Using the notation of (2.42), suppose that $\mathcal{V}_1 = \mathcal{C}(\mathbf{A})$, where $\mathbf{A}$ is $n \times n$ of rank r. Let $\mathbf{A} = \mathbf{R}_{n \times r} \mathbf{C}_{r \times n}$ be a full-rank factorization of $\mathbf{A}$ (cf. 3.5). Then

$$\mathbf{P} = \mathbf{R}(\mathbf{CR})^{-1}\mathbf{C}$$

is the projection onto $\mathcal{V}_1$ along $\mathcal{V}_2$.

Proofs.   Section 2.3.1.

2.40. Assume two projectors $\mathbf{P}_i$ $(i = 1, 2)$, then $(\mathbf{P}_1 - \mathbf{P}_2)\mathbf{v} = \mathbf{v}_1 - \mathbf{v}_1 = \mathbf{0}$ for all $\mathbf{v}$ so that $\mathbf{P}_1 = \mathbf{P}_2$. Now $\mathbf{v}_1 = \mathbf{v}_1 + \mathbf{0}$ is the unique decomposition of $\mathbf{v}_1$ so that $\mathbf{P}^2\mathbf{v} = \mathbf{P}(\mathbf{P}\mathbf{v}) = \mathbf{P}\mathbf{v}_1 = \mathbf{v}_1 = \mathbf{P}\mathbf{v}$ for all $\mathbf{v}$ so that $\mathbf{P}^2 = \mathbf{P}$.

2.41. Rao and Rao [1998: 240–241]. Multiply the first equation by $\mathbf{P}$ to prove $\mathbf{P}(\mathbf{I}_n - \mathbf{P}) = \mathbf{0}$.

2.42a. $\mathcal{C}(\mathbf{P}) \subseteq \mathcal{V}_1$ as $\mathbf{P}$ projects onto $\mathcal{V}_1$. Conversely, if $\mathbf{v}_1 \in \mathcal{V}_1$, then $\mathbf{P}\mathbf{v}_1 = \mathbf{v}_1$, and $\mathcal{V}_1 \subseteq \mathcal{C}(\mathbf{P})$; (b) is similar.

2.43. Meyer [2000a: 634].

2.3.2   Orthogonal Projections

***Definition* 2.21.** Suppose $\mathcal{U}$ has an inner product $\langle,\rangle$, and let $\mathcal{V}$ be a vector subspace with orthogonal complement $\mathcal{V}^\perp$, namely

$$\mathcal{V}^\perp = \{\mathbf{x} : \langle \mathbf{x}, \mathbf{y} \rangle = 0, \text{ for every } \mathbf{y} \in \mathcal{V}\}.$$

Then $\mathcal{U} = \mathcal{V} \oplus \mathcal{V}^\perp$ so that every $\mathbf{v} \in \mathcal{U}$ can be expressed uniquely in the form $\mathbf{v} = \mathbf{v}_1 + \mathbf{v}_2$, where $\mathbf{v}_1 \in \mathcal{V}$ and $\mathbf{v}_2 \in \mathcal{V}^\perp$. The vectors $\mathbf{v}_1$ and $\mathbf{v}_2$ are called the *orthogonal projections* of $\mathbf{v}$ onto $\mathcal{V}$ and $\mathcal{V}^\perp$, respectively (we shall omit the words "along $\mathcal{V}^\perp$" and "along $\mathcal{V}$", respectively). Orthogonal projections will, of course, share the same properties as general projections. If $\mathcal{V} = \mathcal{C}(\mathbf{A})$, we shall denote the orthogonal projection $\mathbf{P}_\mathcal{V}$ onto $\mathcal{V}$ by $\mathbf{P}_\mathbf{A}$. In what follows we assume that $\mathcal{U} = \mathbb{F}^n$.

2.44. Using the above notation, $\mathbf{v}_1 = \mathbf{P}_\mathcal{V}\mathbf{v}$ and $\mathbf{v}_2 = (\mathbf{I}_n - \mathbf{P}_\mathcal{V})\mathbf{v}$, where $\mathbf{P}_\mathcal{V}$ and $\mathbf{I}_n - \mathbf{P}_\mathcal{V}$ are unique idempotent matrices. The matrix $\mathbf{P}_\mathcal{V}$ is said to be the *orthogonal projector* or *orthogonal projection matrix* of $\mathbb{F}^n$ onto $\mathcal{V}$, while $\mathbf{P}_{\mathcal{V}^\perp} = \mathbf{I}_n - \mathbf{P}_\mathcal{V}$ is the orthogonal projector of $\mathbb{F}^n$ onto $\mathcal{V}^\perp$. As we shall see below, the definition of orthogonality depends on the definition of $\langle \mathbf{x}, \mathbf{y} \rangle$.

2.45. If $\mathbb{F}^n = \mathbb{R}^n$ and $\langle \mathbf{x}, \mathbf{y} \rangle = \mathbf{x}'\mathbf{y}$, then from the orthogonality we have

$$\mathbf{P}_\mathcal{V}'(\mathbf{I} - \mathbf{P}_\mathcal{V}) = \mathbf{0},$$

and $\mathbf{P}_\mathcal{V}$ is symmetric as well as being idempotent.

2.46. Let $\mathbb{F}^n = \mathbb{C}^n$ and define $\langle \mathbf{x}, \mathbf{y} \rangle = \mathbf{y}^*\mathbf{A}\mathbf{x}$, where $\mathbf{A}$ is a Hermitian positive definite matrix. Note that $\mathbf{x} \perp \mathbf{y}$ if $\mathbf{y}^*\mathbf{A}\mathbf{x} = 0$ (cf. 2.19c).

(a) Let $\mathbf{P}_\mathcal{V}$ be the orthogonal projection matrix that projects onto $\mathcal{V}$. Then $\mathbf{P}_\mathcal{V}^2 = \mathbf{P}_\mathcal{V}$ and $\mathbf{A}\mathbf{P}_\mathcal{V}$ is Hermitian, that is,

$$\mathbf{A}\mathbf{P}_\mathcal{V} = \mathbf{P}_\mathcal{V}^*\mathbf{A}.$$

(Note that $\mathbf{P}_\mathcal{V}$ is generally not Hermitian. However, if $\mathbf{A} = \mathbf{I}_n$, then $\mathbf{P}_\mathcal{V}$ is Hermitian.)

(b) $\mathcal{C}(\mathbf{P}_\mathcal{V}) = \mathcal{V}$ and $\mathcal{C}(\mathbf{I}_n - \mathbf{P}_\mathcal{V}) = \mathcal{V}^\perp$ (from 2.42). Also

$$\mathbf{P}_\mathcal{V}^*\mathbf{A}(\mathbf{I}_n - \mathbf{P}_\mathcal{V}) = \mathbf{A}\mathbf{P}_\mathcal{V}(\mathbf{I}_n - \mathbf{P}_\mathcal{V}) = \mathbf{0}.$$

(c) Let $\mathcal{V} = \mathcal{C}(\mathbf{X})$. Then
$$\mathbf{P}_{\mathcal{V}} = \mathbf{X}(\mathbf{X}^*\mathbf{A}\mathbf{X})^-\mathbf{X}^*\mathbf{A},$$
which is unique for any weak inverse $(\mathbf{X}^*\mathbf{A}\mathbf{X})^-$ and therefore invariant. Also $\mathbf{P}_{\mathcal{V}^\perp} = \mathbf{I}_n - \mathbf{P}_{\mathcal{V}}$.

(d) If $\mathcal{V} = \mathcal{C}(\mathbf{X})$, then $\mathbf{P}_{\mathcal{V}}\mathbf{X} = \mathbf{X}$.

2.47. Of particular interest is a special case of (2.46) above, namely $\langle \mathbf{x}, \mathbf{y} \rangle = \mathbf{x}'\mathbf{V}^{-1}\mathbf{y}$, where $\mathbf{V}$ is positive definite and $\mathbf{x}, \mathbf{y} \in \mathbb{R}^n$. Because of its statistical importance in a variety of nonlinear models including nonlinear regression (e.g., generalized or weighted least squares) and multinomial models, $\langle \mathbf{x}, \mathbf{y} \rangle$ has been called the *weighted inner product space* (Wei [1997]). We now list some special cases of the previous general theory. Let $\mathbf{X}$ be $n \times p$ of rank p and $\mathcal{V} = \mathcal{C}(\mathbf{X})$. Then:

(a) $\mathbf{P}_{\mathcal{V}} = \mathbf{X}(\mathbf{X}'\mathbf{V}^{-1}\mathbf{X})^-\mathbf{X}'\mathbf{V}^{-1}$, which implies $\mathbf{P}_{\mathcal{V}}^2 = \mathbf{P}_{\mathcal{V}}$ and $\mathbf{P}_{\mathcal{V}}'\mathbf{V}^{-1} = \mathbf{V}^{-1}\mathbf{P}_{\mathcal{V}}$. Here $(\mathbf{X}'\mathbf{V}^{-1}\mathbf{X})^-$ is any weak inverse of $\mathbf{X}'\mathbf{V}^{-1}\mathbf{X}$. Further properties of $\mathbf{P}_{\mathcal{V}}$ (with $\mathbf{V}^{-1}$ replaced by $\mathbf{V}$) are given by Harville [2001: 106–112].

(b) If the columns of $\mathbf{Q}$ and $\mathbf{N}$ are respectively orthonormal bases of $\mathcal{V}$ and $\mathcal{V}^\perp$, then $\mathbf{P}_{\mathcal{V}} = \mathbf{Q}\mathbf{Q}'\mathbf{V}^{-1}$ and $\mathbf{P}_{\mathcal{V}^\perp} = \mathbf{N}\mathbf{N}'\mathbf{V}^{-1}$, where $\mathbf{P}_{\mathcal{V}} + \mathbf{P}_{\mathcal{V}^\perp} = \mathbf{I}_n$.

(c) From (b), $\mathbf{Q}'\mathbf{V}^{-1}\mathbf{N} = \mathbf{0}$.

We can set $\mathbf{V} = \mathbf{I}$ is the above to get the unweighted case.

2.48. Let $\mathbf{V}$ be an $n \times n$ positive definite matrix, $\mathbf{G}$ an $n \times g$ matrix of rank g $(g \leq n)$, and $\mathbf{F}$ an $n \times f$ matrix $(f = n - g)$ of rank f such that $\mathbf{G}'\mathbf{F} = \mathbf{0}$. Then
$$\mathbf{V}\mathbf{F}(\mathbf{F}'\mathbf{V}\mathbf{F})^{-1}\mathbf{F}' + \mathbf{G}(\mathbf{G}'\mathbf{V}^{-1}\mathbf{G})^{-1}\mathbf{G}'\mathbf{V}^{-1} = \mathbf{I}_n.$$

2.49. Let $\mathbb{F}^n = \mathbb{C}^n$, $\mathbf{v} \in \mathbb{C}^n$, and define $\langle \mathbf{x}, \mathbf{y} \rangle = \mathbf{x}^*\mathbf{y}$, i.e., $\mathbf{A} = \mathbf{I}_n$ in (2.46). Then:

(a) $\mathbf{P}_{\mathcal{V}}$ is an orthogonal projection matrix on some vector space if and only if $\mathbf{P}_{\mathcal{V}}$ is idempotent and Hermitian.

(b) From (2.42) we have $\mathcal{V} = \mathcal{C}(\mathbf{P}_{\mathcal{V}})$.

(c) Let $\mathbf{T} = (\mathbf{t}_1, \mathbf{t}_2, \ldots, \mathbf{t}_p)$, where the columns $\mathbf{t}_i$ of $\mathbf{T}$ form an orthonormal basis for $\mathcal{V}$. Then $\mathbf{P}_{\mathcal{V}} = \mathbf{T}\mathbf{T}^*$, and the projection of $\mathbf{v}$ onto $\mathcal{V}$ is $\mathbf{v}_1 = \mathbf{T}\mathbf{T}^*\mathbf{v} = \sum_{i=1}^r (\mathbf{t}_i^*\mathbf{v})\mathbf{t}_i$.

(d) If $\mathcal{V} = \mathcal{C}(\mathbf{X})$, then $\mathbf{P}_{\mathcal{V}} = \mathbf{X}(\mathbf{X}^*\mathbf{X})^-\mathbf{X}^* = \mathbf{X}\mathbf{X}^+$, where $(\mathbf{X}^*\mathbf{X})^-$ is a weak inverse of $\mathbf{X}^*\mathbf{X}$ and $\mathbf{X}^+$ is the Moore–Penrose inverse of $\mathbf{X}$. When the columns of $\mathbf{X}$ are linearly independent, $\mathbf{P}_{\mathcal{V}} = \mathbf{X}(\mathbf{X}^*\mathbf{X})^{-1}\mathbf{X}^*$.

(e) Let $\mathcal{V} = \mathcal{N}(\mathbf{A})$, the null space of $\mathbf{A}$. Then, since $\mathcal{V}^\perp = \mathcal{C}(\mathbf{A}^*)$ (by 2.37), $\mathbf{P}_{\mathcal{V}} = \mathbf{I}_n - \mathbf{A}^*(\mathbf{A}\mathbf{A}^*)^-\mathbf{A}$.

(f) If $\mathbb{F}^n = \mathbb{R}^n$, then the previous results hold by replacing $*$ by $'$ and replacing Hermitian by real symmetric. For example, if $\mathcal{V} = \mathcal{C}(\mathbf{A})$, then $\mathbf{P}_{\mathcal{V}} = \mathbf{A}(\mathbf{A}'\mathbf{A})^-\mathbf{A}'$. Furthermore, $\mathbf{x}'\mathbf{P}_{\mathcal{V}}\mathbf{x} = \mathbf{x}\mathbf{P}_{\mathcal{V}}'\mathbf{P}_{\mathcal{V}}\mathbf{x} = \mathbf{y}'\mathbf{y} \geq 0$, so that $\mathbf{P}_{\mathcal{V}}$ is non-negative definite. This result is used frequently in this book.

2.50. Let $\mathbf{A}$ be an $n \times m$ real matrix and $\mathbf{B}$ an $n \times p$ real matrix. Assuming that $\langle \mathbf{x}, \mathbf{y} \rangle = \mathbf{x}'\mathbf{y}$, let $\mathbf{P_D}$ denote the orthogonal projection onto $\mathcal{C}(\mathbf{D})$ for any matrix $\mathbf{D}$.

(a) $\mathcal{C}(\mathbf{A}) \cap \mathcal{C}(\mathbf{B}) = \mathcal{C}[\mathbf{A}(\mathbf{I}_m - \mathbf{P}_\mathcal{V})]$, where $\mathcal{V} = \mathcal{C}[\mathbf{A}'(\mathbf{I} - \mathbf{P_B})]$.

(b) $\mathcal{C}(\mathbf{A}, \mathbf{B}) = \mathcal{C}(\mathbf{A}) \oplus \mathcal{C}[(\mathbf{I} - \mathbf{P_A})\mathbf{B}]$.

(c) From (b) we have $\mathbf{P}_{(\mathbf{A},\mathbf{B})} = \mathbf{P_A} + \mathbf{P}_{(\mathbf{I}-\mathbf{P_A})\mathbf{B}}$.

(d) $\mathcal{C}(\mathbf{A}) \subseteq \mathcal{C}(\mathbf{B})$ if and only if $\mathbf{P_B} - \mathbf{P_A}$ is non-negative definite, and $\mathcal{C}(\mathbf{A}) \subset \mathcal{C}(\mathbf{B})$ if and only if $\mathbf{P_B} - \mathbf{P_A}$ is positive definite.

The above results are particularly useful in partitioned linear models.

2.51. (Some Subspace Properties) Let ω, Ω, and $\mathcal{V}$ be vector subspaces in $\mathbb{R}^n$ with $\omega \subset \Omega$, and let $\mathbf{P}_\omega$ and $\mathbf{P}_\Omega$ be the respective orthogonal projectors onto ω and Ω with respect to the inner product $\langle \mathbf{x}, \mathbf{y} \rangle = \mathbf{x}'\mathbf{y}$ defined on $\mathbb{R}^n$. Thus $\mathbf{P}_\omega$ and $\mathbf{P}_\Omega$ are symmetric and idempotent. The following results hold (see also (2.53c)).

(a) $\mathbf{P}_\Omega \mathbf{P}_\omega = \mathbf{P}_\omega \mathbf{P}_\Omega = \mathbf{P}_\omega$.

(b) $\mathbf{P}_{\omega^\perp \cap \Omega} = \mathbf{P}_\Omega - \mathbf{P}_\omega$.

(c) $\mathbf{A}\mathbf{P}_\Omega \mathbf{A}'$ is nonsingular if and only if the rows of $\mathbf{A}$ are linearly independent and $\mathcal{C}(\mathbf{A}') \cap \Omega^\perp = \mathbf{0}$.

(d) If $\omega = \Omega \cap \mathcal{N}(\mathbf{A})$, where $\mathcal{N}(\mathbf{A})$ is the null space of $\mathbf{A}$, then:

    (i) $\omega^\perp \cap \Omega = \mathcal{C}(\mathbf{P}_\Omega \mathbf{A}')$.

    (ii) $\mathbf{P}_{\omega^\perp \cap \Omega} = \mathbf{P}_\Omega \mathbf{A}'(\mathbf{A}\mathbf{P}_\Omega \mathbf{A}')^- \mathbf{A}\mathbf{P}_\Omega$, where $(\mathbf{A}\mathbf{P}_\Omega \mathbf{A}')^-$ is any weak inverse of $\mathbf{A}\mathbf{P}_\Omega \mathbf{A}'$.

(e) Let $\Omega = \mathcal{C}(\mathbf{X}) = \mathcal{C}(\mathbf{X}_1, \mathbf{X}_2)$, where the columns of $n \times p$ $\mathbf{X}$ are linearly independent, and let $\omega = \mathcal{C}(\mathbf{X}_1)$, where $\dim(\omega) = r$.

    (i) We have from (c), with $\mathcal{V} = \omega^\perp$ and $\mathbf{P}_\omega = \mathbf{X}_1(\mathbf{X}_1'\mathbf{X}_1)^{-1}\mathbf{X}_1'$ $(= \mathbf{P}_1$, say$)$, that $\mathbf{X}_2'(\mathbf{I}_n - \mathbf{P}_1)\mathbf{X}_2$ is nonsingular.

    (ii) $\omega = \Omega \cap \mathcal{N}[\mathbf{X}_2'(\mathbf{I}_n - \mathbf{P}_1)]$.

    (iii) It follows from (b) and (d)(ii)) that

$$\mathbf{P}_\Omega - \mathbf{P}_\omega = (\mathbf{I}_n - \mathbf{P}_1)\mathbf{X}_2[\mathbf{X}_2'(\mathbf{I}_n - \mathbf{P}_1)\mathbf{X}_2]^{-1}\mathbf{X}_2'(\mathbf{I}_n - \mathbf{P}_1).$$

By interchanging the subscripts 1 and 2, a further result can be obtained.

Note that (a)–(d) are used in testing a linear hypothesis for a linear regression model (e.g., Seber [1977: sections 3.9.3 and 4.5] and Seber and Lee [2003: theorems 4.1 and 4.3]); (e) is related to subset regression (see Seber and Wild [1989: Appendix D] for a summary).

2.52. If Ω and ω_i $(i = 1, 2, \ldots, k)$ are vector subspaces of $\mathbb{R}^n$ satisfying $\omega_i \subset \Omega$, with inner product $\langle \mathbf{x}, \mathbf{y} \rangle = \mathbf{x}'\mathbf{y}$, then the following results are equivalent:

(1) $\mathbf{P}_{\omega_1 \cap \omega_2 \cap \cdots \cap \omega_i} - \mathbf{P}_{\omega_1 \cap \omega_2 \cap \cdots \cap \omega_{i-1}} = \mathbf{P}_\Omega - \mathbf{P}_{\omega_i}$ for $i = 1, 2, \ldots, k$.

(2) $\omega_i^{\perp} \cap \Omega \perp \omega_j^{\perp} \cap \Omega$ for all $i, j = 1, 2, \ldots, k; \quad i \neq j.$

(3) $\omega_i^{\perp} \cap \Omega \subset \omega_j$ for all $i, j = 1, 2, \ldots, k; \quad i \neq j.$

The above results are useful in testing a sequence of nested hypotheses in an analysis of variance, when there are equal numbers of observations per cell (balanced designs) leading to an underlying orthogonal structure (cf. Darroch and Silvey [1963], Seber [1980: section 6.2], and Seber and Lee [2003: 203]).

2.53. Let ω_1 and ω_2 be vector subspaces of $\mathbb{R}^n$ with inner product $\langle \mathbf{x}, \mathbf{y} \rangle = \mathbf{x}'\mathbf{y}$.

(a) $\mathbf{P} = \mathbf{P}_{\omega_1} + \mathbf{P}_{\omega_2}$ is an orthogonal projector if and only if $\omega_1 \perp \omega_2$, in which case $\mathbf{P}_{\omega_1} + \mathbf{P}_{\omega_2} = \mathbf{P}_\omega$, where $\omega = \omega_1 \oplus \omega_2$.

(b) If $\omega_1 = \mathcal{C}(\mathbf{A})$ and $\omega_2 = \mathcal{C}(\mathbf{B})$ in (a), then $\omega_1 \oplus \omega_2 = \mathcal{C}(\mathbf{A}, \mathbf{B})$.

(c) The following statements are equivalent:

(1) $\mathbf{P}_{\omega_1} - \mathbf{P}_{\omega_2}$ is an orthogonal projection matrix.

(2) $\|\mathbf{P}_{\omega_1}\mathbf{x}\|_2 \geq \|\mathbf{P}_{\omega_2}\mathbf{x}\|_2$ for all $\mathbf{x} \in \mathbb{R}^n$.

(3) $\mathbf{P}_{\omega_1}\mathbf{P}_{\omega_2} = \mathbf{P}_{\omega_2}$.

(4) $\mathbf{P}_{\omega_2}\mathbf{P}_{\omega_1} = \mathbf{P}_{\omega_2}$.

(5) $\omega_2 \subset \omega_1$.

(d) $\mathbf{P}_{\omega_1 \cap \omega_2} = 2\mathbf{P}_{\omega_1}(\mathbf{P}_{\omega_1} + \mathbf{P}_{\omega_2})^+\mathbf{P}_{\omega_2} = 2\mathbf{P}_{\omega_2}(\mathbf{P}_{\omega_1} + \mathbf{P}_{\omega_2})^+\mathbf{P}_{\omega_1}$. Here $\mathbf{B}^+$ denotes the Moore–Penrose inverse of $\mathbf{B}$.

The above results hold for $\mathbb{C}^n$ if $\langle \mathbf{x}, \mathbf{y} \rangle = \mathbf{y}^*\mathbf{x}$ and $'$ is replaced by $*$.

***Definition* 2.22.** (Centering) Let $\mathbf{a} = (a_i)$ be an $n \times 1$ real vector, and let $\bar{a} = \sum_{i=1}^{n} a_i/n$. We say that the $\mathbf{a}$ is *centered* when we transform a_i to $b_i = a_i - \bar{a}$.

If we have the $n \times p$ matrix $\mathbf{A} = (\mathbf{a}_1, \mathbf{a}_2, \ldots \mathbf{a}_n)' = (\mathbf{a}^{(1)}, \mathbf{a}^{(2)}, \ldots, \mathbf{a}^{(p)})$ and $\bar{\mathbf{a}} = n^{-1}\sum_{i=1}^{n} \mathbf{a}_i$, then we say that $\mathbf{A}$ is *row centered* if we transform it to the matrix $\mathbf{B} = (\mathbf{a}_1 - \bar{\mathbf{a}}, \mathbf{a}_2 - \bar{\mathbf{a}}, \ldots, \mathbf{a}_n - \bar{\mathbf{a}})'$.

If $\bar{\mathbf{a}}^{(col)} = \sum_{j=1}^{p} \mathbf{a}^{(j)}/p$, then we say that $\mathbf{A}$ is *column centered* if we form the matrix $\mathbf{C} = (\mathbf{a}^{(1)} - \bar{\mathbf{a}}^{(col)}, \mathbf{a}^{(2)} - \bar{\mathbf{a}}^{(col)}, \ldots, \mathbf{a}^{(p)} - \bar{\mathbf{a}}^{(col)})$.

We say that $\mathbf{A}$ is *double-centered* if we apply both row and column centering.

2.54. Using the above notation, we have the following results:

(a) We can write $\bar{a} = \mathbf{1}_n'\mathbf{a}/n$ so that $(\bar{a}) = n^{-1}\mathbf{1}_n\mathbf{1}_n'\mathbf{a} = \mathbf{P}_{\mathbf{1}_n}\mathbf{a}$, where $\mathbf{P}_{\mathbf{1}_n} = n^{-1}\mathbf{1}_n\mathbf{1}_n'$ represents the orthogonal projection of $\mathbb{R}^n$ onto $\mathbf{1}_n$. Furthermore, $\mathbf{b} = \mathbf{a} - (\bar{a}) = (\mathbf{I}_n - \mathbf{P}_{\mathbf{1}_n})\mathbf{a}$, where $\mathbf{I}_n - \mathbf{P}_{\mathbf{1}_n}$ represents an orthogonal projection perpendicular to $\mathbf{1}_n$; this projection matrix is called a *centering matrix*.

(b) $\bar{\mathbf{a}} = \mathbf{A}'\mathbf{1}_n/n$ and $\mathbf{B} = \mathbf{A} - \mathbf{1}_n\bar{\mathbf{a}}' = (\mathbf{I}_n - \mathbf{P}_{\mathbf{1}_n})\mathbf{A}$.

(c) $\bar{\mathbf{a}}^{(col)} = \mathbf{A}\mathbf{1}_p/p$ and $\mathbf{C} = \mathbf{A}(\mathbf{I}_p - \mathbf{P}_{\mathbf{1}_p})$.

(d) When $\mathbf{A}$ is double centered we obtain $\mathbf{D} = (\mathbf{I}_n - \mathbf{P}_{\mathbf{1}_n})\mathbf{A}(\mathbf{I}_p - \mathbf{P}_{\mathbf{1}_p})$, where $d_{ij} = a_{ij} - \bar{a}_{i.} - \bar{a}_{.j} - \bar{a}_{..}$, $\bar{a}_{i.} = \sum_j a_{ij}/p$, $\bar{a}_{.j} = \sum_i a_{ij}/n$, and $\bar{a}_{..} = \sum_i \sum_j a_{ij}/(np)$.

Centering is used extensively in statistics, for example linear regression (Seber and Lee [2003: section 3.11.1 and section 11.7 for computing algorithms]) and principal component analysis, and double centering is used in classical metric scaling, in principal component analysis (Jolliffe [1992: section 14.2.3]), and in the singular-spectrum analysis (SAS) of times series, where it is applied to trajectory matrices (Golyandina et al. [2001: section 4.4, 272]).

Proofs.   Section 2.3.2.

2.46. Rao [1973a: 47].

2.47. Wei [1997: 185–187].

2.48. Seber [1984: 536].

2.49. Seber and Lee [2003: Appendices B1 and B2, real case].

2.50a. Quoted by Rao and Mitra [1971: 118, exercise 7a].

2.50b–d. Sengupta and Jammalamadaka [2003: 39, 47]; (c) uses (2.44).

2.51a–d(i). Seber and Lee [2003: Appendix B3, 477-478, real case] and Seber [1984: Appendix B3, 535, real case].

2.51d(ii). If $\mathbf{x} \in \mathcal{C}(\mathbf{X}_1) = \omega$, then $\mathbf{P}_1\mathbf{x} = \mathbf{x}$, $\mathbf{X}_2'(\mathbf{I}_n - \mathbf{P}_1)\mathbf{x} = \mathbf{0}$, and $\mathbf{x} \in \mathcal{N}(\mathbf{X}_2'(\mathbf{I}_n - \mathbf{P}_1))$. Conversely, if $\mathbf{x} = \mathbf{X}_1\boldsymbol{\alpha}_1 + \mathbf{X}_2\boldsymbol{\alpha}_2 \in \Omega$ and $\mathbf{0} = \mathbf{X}_2'(\mathbf{I}_n - \mathbf{P}_1)\mathbf{x} = \mathbf{X}_2'(\mathbf{I}_n - \mathbf{P}_1)\mathbf{X}_2\boldsymbol{\alpha}_2$ (since $\mathbf{P}_1\mathbf{X}_1 = \mathbf{X}_1$), then $\boldsymbol{\alpha}_2 = \mathbf{0}$ (by (i)) and $\mathbf{x} \in \mathcal{C}(\mathbf{X}_1)$.

2.52. Seber [1980: section 6.2].

2.53a. $\mathbf{P}$ is clearly symmetric and idempotent if and only $\mathbf{P}_{\omega_1}\mathbf{P}_{\omega_2} = -\mathbf{P}_{\omega_2}\mathbf{P}_{\omega_1}$. Multiplying on the left by $\mathbf{P}_{\omega_2}$ shows that $\mathbf{P}_{\omega_1}\mathbf{P}_{\omega_2}$ is symmetric and therefore $\mathbf{P}_{\omega_1}\mathbf{P}_{\omega_2} = \mathbf{0}$. Furthermore, since $\mathbf{P}_{\omega_i}$ is idempotent, we have from (2.35)

$$\mathcal{C}(\mathbf{P}_{\omega_1} + \mathbf{P}_{\omega_2}) = \mathcal{C}\left[(\mathbf{P}_{\omega_1}, \mathbf{P}_{\omega_2})\begin{pmatrix}\mathbf{P}_{\omega_1}\\\mathbf{P}_{\omega_2}\end{pmatrix}\right] = \mathcal{C}(\mathbf{P}_{\omega_1}, \mathbf{P}_{\omega_2}) = \omega_1 \oplus \omega_2.$$

2.53b. $\mathbf{A}'\mathbf{B} = \mathbf{0}$ implies that $\mathbf{P}_\mathbf{A}\mathbf{P}_\mathbf{B} = \mathbf{0}$.

2.53c. Quoted, less generally, by Isotalo et al. [2005a: 61]. The proofs are straightforward. For (2), note that for a symmetric idempotent matrix, $\mathbf{x}'\mathbf{A}\mathbf{x} = \mathbf{x}'\mathbf{A}'\mathbf{A}\mathbf{x} = \|\mathbf{A}\mathbf{x}\|_2^2$.

2.53d. Anderson and Duffin [1969] and Meyer [2000a: 441].

2.4   METRIC SPACES

Definition 2.23. Let S be a subset of $\mathbb{R}^n$. By a *metric* for S we mean a real-valued function $d(\cdot, \cdot)$ on $S \times S$ such that:

(a) $d(\mathbf{x}, \mathbf{y}) \geq 0$ for all $\mathbf{x}, \mathbf{y} \in S$ with equality if and only if $\mathbf{x} = \mathbf{y}$ (d is positive definite).

(b) $d(\mathbf{x}, \mathbf{y}) = d(\mathbf{y}, \mathbf{x})$ for all $\mathbf{x}, \mathbf{y} \in S$ (d is symmetric).

(c) $d(\mathbf{x}, \mathbf{y}) \leq d(\mathbf{x}, \mathbf{z}) + d(\mathbf{y}, \mathbf{z})$ for all $\mathbf{x}, \mathbf{y}, \mathbf{z} \in S$ (triangle inequality).

If we replace (c) by the stronger condition

(c') $d(\mathbf{x}, \mathbf{y}) \leq \max[d(\mathbf{x}, \mathbf{z}), d(\mathbf{y}, \mathbf{z})]$,

d is called an *ultrametric*. Note that (c') implies (c).

Definition 2.24. A *metric space* is a pair (S, d) consisting of a set S and a metric d for S.

2.55. If d is a metric, then so are d_1, d_2, and d_3, where

$$
\begin{aligned}
d_1(\mathbf{x}, \mathbf{y}) &= d(\mathbf{x}, \mathbf{y})/(1 + d(\mathbf{x}, \mathbf{y})), \\
d_2(\mathbf{x}, \mathbf{y}) &= \sqrt{d(\mathbf{x}, \mathbf{y})}, \\
d_3(\mathbf{x}, \mathbf{y}) &= kd(\mathbf{x}, \mathbf{y}) \quad (k > 0).
\end{aligned}
$$

2.56. If d is a metric, then $D(\mathbf{x}, \mathbf{y}) = [d(\mathbf{x}, \mathbf{y})]^2$ is not necessarily a metric.

2.57. (Canberra metric)   If $\mathbf{x}$ and $\mathbf{y}$ have positive elements, then the function

$$
d(\mathbf{x}, \mathbf{y}) = \frac{1}{n} \sum_{j=1}^{n} \frac{|x_j - y_j|}{x_j + y_j}
$$

is a metric.

2.58. (Minkowski Metrics) The function Δ_p is a metric, where

$$
\Delta_p(\mathbf{x}, \mathbf{y}) = \left(\sum_{i=1}^{n} |x_i - y_i|^p \right)^{1/p}, \quad p > 0.
$$

The most common ones are $p = 1$ (the *city block metric*) and $p = 2$ (the *Euclidean metric*). Various scaled versions of Δ_1 have also been used.

2.59. $\Delta_\infty(\mathbf{x}, \mathbf{y}) = \sup_{1 \leq i \leq n} |x_i - y_i|$, for all $\mathbf{x}$ and $\mathbf{y}$, is a metric.

Definition 2.25. The *Mahalanobis distance* is defined to be

$$
d(\mathbf{x}, \mathbf{y}) = \{(\mathbf{x} - \mathbf{y})' \mathbf{A} (\mathbf{x} - \mathbf{y})\}^{1/2},
$$

where $\mathbf{A}$ is positive definite. Here d is a metric. The *Mahalanobis angle* θ between $\mathbf{x}$ and $\mathbf{y}$ subtended at the origin is defined by

$$
\cos \theta = \frac{\mathbf{x}' \mathbf{A} \mathbf{y}}{(\mathbf{x}' \mathbf{A} \mathbf{x})^{1/2} (\mathbf{y}' \mathbf{A} \mathbf{y})^{1/2}}.
$$

Definition 2.26. A sequence of points $\{\mathbf{x}_i\}$ in S for a metric space (S, d) is called a *Cauchy sequence* if, for every $\epsilon > 0$, there exists a positive integer N such the $d(\mathbf{x}_i, \mathbf{x}_j) < \epsilon$ for all $i, j > N$.

A sequence $\{\mathbf{x}_i\}$ *converges* to a point $\mathbf{x}$ if, for every $\epsilon > 0$, there exists a positive integer N such that $d(\mathbf{x}, \mathbf{x}_i) < \epsilon$ for all $i > N$.

A metric space is said to be *complete* if every Cauchy sequence converges to a point in S.

Definition 2.27. Let f be a mapping of a metric space (S, d) into itself. We call f a *contraction* if there exists a constant c with $0 < c \leq 1$ such that

$$d(f(\mathbf{x}), f(\mathbf{y})) \leq cd(\mathbf{x}, \mathbf{y}), \quad \text{for all} \quad \mathbf{x}, \mathbf{y} \in S.$$

If $0 < c < 1$, we say that f is a *strict contraction*. If $f(\mathbf{x}) = \mathbf{x}$, then $\mathbf{x}$ is referred to as a *fixed point* of f.

2.60. (Contraction Mapping Theorem) Let f be a strict contraction of a complete metric space into itself. Then f has one and only one fixed point and, for any point $\mathbf{y} \in S$, the sequence

$$\mathbf{y}, f(\mathbf{y}), f^2(\mathbf{y}), f^3(\mathbf{y}), \dots,$$

where $f^r(\mathbf{y}) = f(f^{r-1}(\mathbf{y}))$, converges to the fixed point.

2.61. Let (S, d) be a metric space with $S = \mathbb{C}^n$ and $d(\mathbf{x}, \mathbf{y}) = \|\mathbf{x} - \mathbf{y}\|_2$. A matrix $\mathbf{A}$ is a contraction, that is

$$\|\mathbf{A}\mathbf{x} - \mathbf{A}\mathbf{y}\|_2 \leq c\|\mathbf{x} - \mathbf{y}\|_2 \quad \text{for} \quad 0 < c \leq 1,$$

if and only if $\sigma_{\max}(\mathbf{A}) \leq 1$, where $\sigma_{\max}(\mathbf{A})$ is the maximum singular value of $\mathbf{A}$. Further necessary and sufficient conditions for a matrix to be a contraction are given by Zhang [1999: section 5.4].

Proofs. Section 2.4.

2.55–2.57. Seber [1984: : 392, exercises 7.4–7.6, see the solutions].

2.58. Seber [1984: 352]. Use Minkowski's inequalities (12.17b) and $x_i - z_i = x_i - y_i + y_i - z_i$ to prove the triangle inequality.

2.59. Use the properties of sup.

2.60–2.61. Zhang [1999: 143–144].

2.5 CONVEX SETS AND FUNCTIONS

Definition 2.28. A subset C of $\mathbb{R}^n$ is called *convex* if, for any two points $\mathbf{x}_1, \mathbf{x}_2 \in C$, the line segment joining $\mathbf{x}_1$ and $\mathbf{x}_2$ is contained in C, that is,

$$\alpha\mathbf{x}_1 + (1 - \alpha)\mathbf{x}_2 \in C \quad \text{for} \quad 0 \leq \alpha \leq 1.$$

We will list some properties of convex sets below. For a more comprehensive discussion see Berkovitz [2002], Kelly and Weiss [1979], Lay [1982], and Rockafellar [1970].

2.62. If C_1 and C_2 are convex sets in $\mathbb{R}^n$, then:

(a) $C_1 \cap C_2$ is convex.

(b) $C_1 + C_2$ is convex.

(c) $C_1 \cup C_2$ need not be convex.

These results clearly hold for any finite number of convex sets. The result (a) also holds for a countably infinite number of convex sets.

2.63. Given any set $A \in \mathbb{R}^n$, the set C_A of points generated by taking the *convex combination* of every finite set of points $\mathbf{x}_i$ in A, namely

$$\alpha_1 \mathbf{x}_1 + \alpha_2 \mathbf{x}_2 + \cdots + \alpha_k \mathbf{x}_k \quad (\text{each } \alpha_i \geq 0 \text{ and } \sum_i \alpha_i = 1)$$

is a convex set containing A. The set C_A is the smallest convex set containing A and is called the *convex hull* of A. It is also the intersection of all convex sets containing A.

Definition 2.29. Given A a set in $\mathbb{R}^n$, we define $\mathbf{x}$ to be an *inner (interior) point* of A if there is an *open sphere* with center $\mathbf{x}$ that is a subset of A; that is, there exists $\delta > 0$ such that

$$S_\delta = \{\mathbf{y} : \mathbf{y} \in \mathbb{R}^n, (\mathbf{y} - \mathbf{x})'(\mathbf{y} - \mathbf{x}) < \delta\} \subseteq A.$$

A *boundary point* $\mathbf{x}$ of A (not necessarily belonging to A) is such that every open sphere with center $\mathbf{x}$ contains points both in A and in A^c, the complement of A with respect to $\mathbb{R}^n$.

A point $\mathbf{x}$ is a *limit (accumulation) point* if, for every $\delta > 0$, S_δ contains at least one point of S distinct from $\mathbf{x}$.

The *closure* of set A is obtained by adding to it all its boundary points not already in it, and is denoted by $\overline{A}$. It can also be obtained by adding to S all its limit points.

The set A is *closed* if $A = \overline{A}$, while the set is *open* if A^c, the complement of A, is closed. For any set A, $\overline{A}$ is the smallest closed set containing A.

An *exterior* point of A is a point in $\overline{A}^c$. A point $\mathbf{x} \in A$ is an *extreme point* of A if there are no distinct points $\mathbf{x}_1$ and $\mathbf{x}_2$ in A such that $\mathbf{x} = \alpha \mathbf{x}_1 + (1 - \alpha)\mathbf{x}_2$ for some α $(0 < \alpha < 1)$.

A set A is *bounded* if it is contained in an open sphere S_δ for some $\delta > 0$.

A set which is closed and bounded is said to be *compact*. For some properties of open and closed sets see Magnus and Neudecker [1999: 66–69].

The above results generalize to more general spaces using a more general distance metric other than $\|\mathbf{x} - \mathbf{y}\|_2$.

2.64. Let C be a convex set.

(a) The closure $\overline{C}$ is convex.

(b) C and $\overline{C}$ have the same inner, boundary, and exterior points.

(c) Let $\mathbf{x}$ be an inner point and $\mathbf{y}$ a boundary point of C. Then the points $\alpha \mathbf{x} + (1 - \alpha)\mathbf{y}$ are inner points of C for $0 < \alpha \leq 1$ and exterior points of C for $\alpha > 1$

(d) If T is an open subset of $\mathbb{R}^n$ and $T \subseteq \overline{C}$, then $T \subseteq C$.

2.65. (Separation theorems)

(a) Let C be a closed convex subset and suppose $\mathbf{0} \notin C$. Then there exists a vector $\mathbf{a}$ such that $\mathbf{a}'\mathbf{x} > 0$ for all $\mathbf{x} \in C$.

(b) Let C be a convex set and $\mathbf{y}$ an exterior point. Then there exists a unit vector $\mathbf{u}$ (i.e., $\|\mathbf{u}\|_2 = 1$) such that

$$\inf_{\mathbf{x} \in C} \mathbf{u}'\mathbf{x} > \mathbf{u}'\mathbf{y}.$$

(c) Let C be a convex set and $\mathbf{y}$ a point not in C, or a boundary point if in C. Then there exists a supporting plane through $\mathbf{y}$; that is, there exists a nonzero vector $\mathbf{a} \neq \mathbf{0}$ such that $\mathbf{a}'\mathbf{x} \geq \mathbf{a}'\mathbf{y}$ for all $\mathbf{x} \in C$, or equivalently $\inf_{\mathbf{x} \in C} \mathbf{a}'\mathbf{x} = \mathbf{a}'\mathbf{y}$, if $\mathbf{y}$ is a boundary point.

(d) Let C_1 and C_2 be convex sets with no inner point in common. Then there exists a hyperplane $\mathbf{a}'\mathbf{x} = b$ separating the two sets; that is, there exists a vector $\mathbf{a}$ and a scalar b such that $\mathbf{a}'\mathbf{x} \geq b$ for all $\mathbf{x} \in C_1$ and $\mathbf{a}'\mathbf{y} \leq b$ for all $\mathbf{y} \in C_2$. This also implies that $\mathbf{a}'\mathbf{x}_1 \geq \mathbf{a}'\mathbf{x}_2$ for all $\mathbf{x}_1 \in C_1$ and all $\mathbf{x}_2 \in C_2$.

If C_1 and C_2 are also closed, we have strict separation so that there exist $\mathbf{a}$ and b such that $\mathbf{a}'\mathbf{x} > b$ for $\mathbf{x} \in C_1$ and $\mathbf{a}'\mathbf{y} < b$ for $\mathbf{y} \in C_2$.

(e) Let C be a convex subset, symmetric about $\mathbf{0}$, so that if $\mathbf{x} \in C$, then $-\mathbf{x} \in C$ also. Let $f(\mathbf{x}) \geq 0$ be a function for which (i) $f(\mathbf{x}) = f(-\mathbf{x})$, (ii) $C_\alpha = \{\mathbf{x} : f(\mathbf{x}) \geq \alpha\}$ is convex for any positive α, and (iii) $\int_C f(\mathbf{x})\,d\mathbf{x} < \infty$. Then

$$\int_C f(\mathbf{x} + c\mathbf{y})\,d\mathbf{x} \geq \int_C f(\mathbf{x} + \mathbf{y})\,d\mathbf{x},$$

for all $0 \leq c \leq 1$ and $\mathbf{y} \in \mathbb{R}^n$.

2.66. (Convex Hull) If C_A is the convex hull of a subset $A \in \mathbb{R}^n$, then every point of A can be expressed as a convex combination of at most $n + 1$ points in A.

2.67. (Extreme Points) If C is a closed bounded convex set, it is spanned by its extreme points; that is, every point in C can be expressed as a linear combination of its extreme points. Also C has extreme points in every supporting hyperplane.

Definition 2.30. A real valued function f is *convex* in an interval I of $\mathbb{R}$ if

$$f[\alpha x + (1 - \alpha)y] \leq \alpha f(x) + (1 - \alpha)f(y), \quad \text{all } \alpha \quad \text{such that} \quad 0 < \alpha < 1,$$

for all $x, y \in I$ $(x \neq y)$. The function f is said to be *strictly convex* if $\leq$ is replaced by $<$ above.
We say that f is (strictly) *concave* if $-f$ is (strictly) convex. A linear function is both convex and concave. A similar definition applies if x is replaced by a vector or matrix.

A vector convex function is defined along the same lines. We say that $\mathbf{f}$ is *convex* if

$$\mathbf{f}(\alpha \mathbf{x} + (1 - \alpha)\mathbf{y}) \leq \alpha \mathbf{f}(\mathbf{x}) + (1 - \alpha)\mathbf{f}(\mathbf{y})$$

for every α such that $0 \leq \alpha \leq 1$ and $\mathbf{x}, \mathbf{y} \in \mathbb{R}^n$; $\mathbf{f}$ is *concave* if $-\mathbf{f}$ is convex. Here $\mathbf{a} \leq \mathbf{b}$ means $a_i \leq b_i$ for all i.

2.68. The following functions are convex.

(a) $-\log x \; (x > 0)$.

(b) $x^p, \, p > 1 \; (x > 0)$.

They can be used to establish a number of well-known inequalities (e.g., Horn and Johnson [1985: 535–536]).

2.69. The function

$$f(\mathbf{A}) = \log \det \mathbf{A}$$

is a strictly concave function on the convex set of Hermitian positive definite matrices.

2.70. Every convex and every concave function is continuous on its interior. However, a convex function may have a discontinuity at a boundary point and may not be differentiable at an interior point.

2.71. Every increasing convex (respectively concave) function of a convex (respectively concave) function is convex (respectively concave). Every strictly increasing convex (respectively concave) function of a strictly convex (respectively concave) function is strictly convex (respectively concave).

2.72. (Weirstrass's Theorem) Let S be a compact subset of a real or complex vector space. If $f : S \to \mathbb{R}$ is a continuous function, then there exist points $\mathbf{x}_{\min}, \mathbf{x}_{\max} \in S$ such that

$$f(\mathbf{x}_{\min}) \le f(\mathbf{x}) \le f(\mathbf{x}_{\max}) \quad \text{for all } \mathbf{x} \in S.$$

Definition **2.31.** The *numerical range (field of values)* of an $n \times n$ complex matrix **A** is

$$\{\mathbf{x}^* \mathbf{A} \mathbf{x} : \|\mathbf{x}\| = 1, \mathbf{x} \in \mathbb{C}^n\}.$$

2.73. (Toeplitz–Hausdorff) The numerical range of an $n \times n$ complex matrix is a convex compact subset of $\mathbb{C}^n$. For further properties of a field of values see Gustafson and Rao [1997] and Horn and Johnson [1991].

Proofs. Section 2.5.

2.62. Schott [2005: 71].

2.64a–c. Quoted by Rao [1973a: 51].

2.64d. Schott [2005: 72].

2.65a. Schott [2005: 71].

2.65b. Rao [1973a: 51].

2.65c–d. Rao [1973a: 52] and Schott [2005: 73].

2.65e. Anderson [1955], and quoted by Schott [2005: 74].

2.66–2.67. Quoted by Rao [1973a: 53].

2.69. Horn and Johnson [1985: 466–467].

2.70–2.71. Magnus and Neudecker [1999: 76].

2.73. Horn and Johnson [1991: 8] and Zhang [1999: 88–89].

2.6   COORDINATE GEOMETRY

Occasionally one may need some results from coordinate geometry. Some of these are listed below for easy reference.

2.6.1   Hyperplanes and Lines

2.74. The equation of a hyperplane passing through the points $\mathbf{x}_1, \mathbf{x}_2, \ldots, \mathbf{x}_n$ in $\mathbb{R}^n$ can be expressed in the form

$$\det \begin{pmatrix} 1 & 1 & 1 & \cdots & 1 \\ \mathbf{x} & \mathbf{x}_1 & \mathbf{x}_2 & \cdots & \mathbf{x}_n \end{pmatrix} = 0.$$

2.75. Given the points $\mathbf{x}_1 = (a_1, b_1, c_1)'$ and $\mathbf{x}_2 = (a_2, b_2, c_2)'$ in $\mathbb{R}^3$, then the equation of the line through the points is

$$\frac{x - a_1}{a_1 - a_2} = \frac{y - b_1}{b_1 - b_2} = \frac{z - c_1}{c_1 - c_2}.$$

If the two points are A and B, then $a_1 - a_2 = AB \cos\theta_1$, and so on, so that we can replace the denominators of the above line by the direction cosines $\cos\theta_i$ of the line with respect to each axis. Then $\cos\theta_1^2 + \cos\theta_2^2 + \cos\theta_3^2 = 1$. This result clearly generalizes to two points in $\mathbb{R}^n$.

2.76. Given the plane $ax + by + cz + d = 0$ in $\mathbb{R}^3$, a normal vector to the plane is given by $(a, b, c)'$, and the perpendicular distance of the point $\mathbf{x}_1 = (x_1, y_1, z_1)'$ from the plane is

$$\frac{|ax_1 + by_1 + cz_1 + d|}{\sqrt{a^2 + b^2 + c^2}}.$$

This result clearly generalises to $\mathbb{R}^n$. Given the plane $\mathbf{a}'\mathbf{x} + d = 0$, the distance of $\mathbf{x}_1$ from the plane is $(|\mathbf{a}'\mathbf{x}_1 + d|)/\|\mathbf{a}\|_2$.

2.77. Given $0 < \alpha < 1$, then $\mathbf{z} = (1 - \alpha)\mathbf{x} + \alpha\mathbf{y}$ divides the line segment joining $\mathbf{x}$ and $\mathbf{y}$ in the proportion $\alpha : (1 - \alpha)$.

Proofs. Section 2.6.1.

2.77. Abadir and Magnus [2005: 6].

2.6.2   Quadratics

2.78. If $\mathbf{A}$ is an $n \times n$ symmetric indefinite matrix (i.e., has both positive and negative eigenvalues), then $(\mathbf{x} - \mathbf{a})'\mathbf{A}(\mathbf{x} - \mathbf{a}) \leq c$ with $c > 0$ is a *hyperboloid* with center $\mathbf{a}$.

2.79. If $\mathbf{A}$ is an $n \times n$ positive definite matrix, then $(\mathbf{x} - \mathbf{a})'\mathbf{A}(\mathbf{x} - \mathbf{a}) \leq c$ with $c > 0$ is an *ellipsoid* with center $\mathbf{a}$. By shifting the origin to $\mathbf{a}$ and rotating the ellipsoid, the latter can be expressed in a standard form $\sum_{i=1}^{n} \lambda_i z_i^2 \leq c$ with $\lambda_i > 0$ $(i = 1, 2, \ldots, n)$, where the λ_i are the eigenvalues of $\mathbf{A}$. Setting all the z_is equal to zero except z_j, we see that the lengths of the semi-major axes are $b_j = \sqrt{c/\lambda_j}$ for $j = 1, 2, \ldots, n$, and the volume of the ellipsoid is

$$
\begin{aligned}
v &= \frac{\pi^{n/2}}{\Gamma(\frac{n}{2} + 1)} \prod_{j=1}^{n} b_j \\
&= \frac{\pi^{n/2} c^{n/2}}{\Gamma(\frac{n}{2} + 1)(\det \mathbf{A})^{1/2}},
\end{aligned}
$$

by (6.17c). Such a volume arises in finding the constant associated with various elliptical multivariate distributions such as the multivariate normal and the multivariate t-distributions (cf. Chapter 20).

2.80. (Quadrics) If $\mathbf{x} \in \mathbb{R}^n$, then a general quadric is $Q \equiv 0$, where $Q = \mathbf{x}'\mathbf{A}\mathbf{x} + 2\mathbf{b}'\mathbf{x} + c$ and $\mathbf{A}$ is an $n \times n$ symmetric matrix. Let $\mathbf{x}_1$ and $\mathbf{x}_2$ be two points in $\mathbb{R}^n$ that we denote by P_1 and P_2, respectively. From (2.77), the coordinates of the point P dividing the line $P_1 P_2$ in the ratio $\mu : 1$ is given by $(1 + \mu)^{-1}(\mathbf{x}_1 + \mu\mathbf{x}_2)$. Let $Q_{ij} = \mathbf{x}_i'\mathbf{A}\mathbf{x}_j + \mathbf{b}'\mathbf{x}_i + \mathbf{b}'\mathbf{x}_j + c$.

(a) Substituting for P we find that P lies on the quadric if

$$
\mu^2 Q_{22} + 2\mu Q_{12} + Q_{11} = 0.
$$

This is a quadratic in μ so that an arbitrary line meets a quadric in two points.

(b) (Tangent Plane) If P_1 lies on $Q = 0$, then $Q_{11} = 0$ and one root μ is zero. If $P_1 P_2$ is a tangent, then the other root must also be zero; that is, the sum of the roots is zero and $Q_{12} = 0$. As P_2 varies subject to $Q_{12} = 0$, P_2 lies on $Q_1 = 0$, so that

$$
\mathbf{x}_1'\mathbf{A}\mathbf{x} + \mathbf{b}'(\mathbf{x}_1 + \mathbf{x}) + c = 0,
$$

is the tangent plane at $\mathbf{x}_1$.

(c) (Tangent Cone) Suppose P_1 and P_2 are not on $Q = 0$, but $P_1 P_2$ touches the quadric so that the equation in μ has equal roots, i.e., $Q_{11}Q_{22} = Q_{12}^2$. Therefore as P_2 varies subject to this condition, we trace out the tangent cone from P_1, namely,

$$
Q_{11}Q = Q_1^2.
$$

(d) (Envelope) Suppose $Q = \mathbf{x}'\mathbf{A}\mathbf{x} - 1 \equiv 0$, where $\mathbf{A}$ is nonsingular, is a central quadric (i.e., $\mathbf{b} = \mathbf{0}$). Then using (c), $\mathbf{a}'\mathbf{x} = 1$ touches the quadric if $\mathbf{a}'\mathbf{A}^{-1}\mathbf{a} = 1$. As $\mathbf{a}$ varies, $\mathbf{a}'\mathbf{A}^{-1}\mathbf{a} = 1$ is the envelope equation.

2.6.3 Areas and Volumes

2.81. In two dimensions the area of a triangle with vertices $(x_i, y_i)'$, $i = 1, 2, 3$ is $\frac{1}{2}|\Delta|$, where

$$
\Delta = \det \begin{pmatrix} 1 & x_1 & y_1 \\ 1 & x_2 & y_2 \\ 1 & x_3 & y_3 \end{pmatrix}.
$$

The three points are collinear if and only if $\Delta = 0$.

2.82. If $\mathbf{V} = (\mathbf{v}_1, \mathbf{v}_2, \ldots, \mathbf{v}_p)$, where the $\mathbf{v}_i$ are vectors in $\mathbb{R}^n$, then the square of the two-dimensional volume of the parallelotope with $\mathbf{v}_1, \ldots, \mathbf{v}_p$ as principal edges is $\det(\mathbf{V}'\mathbf{V})$. A 2-dimensional parallelotope is a parallelogram; in this case we get the square of the area. When $p = 3$ we have the conventional parallelopiped. For statistical applications see Anderson [2003: section 7.5].

2.83. From (2.74), the four points $(x_i, y_i, z_i)'$, $i = 1, 2, 3, 4$, in three dimensions are coplanar if and only if

$$\det \begin{pmatrix} 1 & 1 & 1 & 1 \\ x_1 & x_2 & x_3 & x_4 \\ y_1 & y_2 & y_3 & y_4 \\ z_1 & z_2 & z_3 & z_4 \end{pmatrix} = 0.$$

Proofs.   Section 2.6.3.

2.81. Cullen [1997: 121].

2.82. Anderson [2003: 266]. For the area of a parallelogram see Basilevsky [1983: 64].

CHAPTER 3

RANK

The concept of rank undergirds much of matrix theory. In statistics it is frequently linked to the concept of degrees of freedom. Both equalities and inequalities are considered in this chapter, and partitioned matrices play an important role.

3.1 SOME GENERAL PROPERTIES

All the matrices in this section are defined over a general field $\mathcal{F}$, unless otherwise stated.

Definition 3.1. The *rank*, denoted by rank $\mathbf{A}$ $(= r$, say$)$, of a matrix $\mathbf{A}$ is $\dim \mathcal{C}(\mathbf{A})$, the dimension of the column space of $\mathbf{A}$. Here r is also called the *column rank* of $\mathbf{A}$. The *row rank* is $\dim \mathcal{C}(\mathbf{A}')$. If $\mathbf{A}$ is $m \times n$ of rank m (respectively n), then $\mathbf{A}$ is said to have *full row (respectively column) rank*. An $n \times n$ matrix $\mathbf{A}$ is said to be *nonsingular* if rank $\mathbf{A} = n$.

As noted in Section 2.2.4, an associated vector space of $\mathcal{C}(\mathbf{A})$ is the null space $\mathcal{N}(\mathbf{A})$, and its dimension is called the *nullity*.

3.1. rank $\mathbf{A}' = $ rank $\mathbf{A} = r$ so that the row rank equals the column rank.

3.2. Let $\mathbf{A}$ be an $m \times n$ matrix of rank r $(r \leq \min\{m, n\})$.

(a) $\mathbf{A}$ has r linearly independent columns and r linearly independent rows.

(b) There exists an $r \times r$ nonzero principal minor. When $r < \min\{m, n\}$, all principal minors of larger order than r are zero.

A Matrix Handbook for Statisticians. By George A. F. Seber
Copyright © 2008 John Wiley & Sons, Inc.

(c) If $\mathbf{B}$ is $m \times p$ and $\mathcal{C}(\mathbf{B}) \subseteq \mathcal{C}(\mathbf{A})$, then rank $\mathbf{B} \leq$ rank $\mathbf{A}$.

3.3. Let $\mathbf{A}$ be an $m \times n$ matrix over $\mathbb{F}$.

(a) rank $\mathbf{A}$ + nullity $\mathbf{A}$ = number of columns of $\mathbf{A}$.

(b) Suppose $\mathbf{A}$ is real, then

$$\operatorname{rank}(\mathbf{A}'\mathbf{A}) = \operatorname{rank}(\mathbf{A}\mathbf{A}') = \operatorname{rank} \mathbf{A}.$$

(c) Suppose $\mathbf{A}$ is complex, then:

   (i) rank $\mathbf{A}$ = rank $\overline{\mathbf{A}}$.

   (ii) Since rank $\overline{\mathbf{A}}$ = rank $\overline{\mathbf{A}}'$ by (3.1), we have rank $\mathbf{A}$ = rank $\mathbf{A}^*$.

   (iii) rank $\mathbf{A}$ = rank$(\mathbf{A}\mathbf{A}^*)$ = rank$(\mathbf{A}^*\mathbf{A})$.

Thus, combining the above,

$$\operatorname{rank} \mathbf{A} = \operatorname{rank} \overline{\mathbf{A}} = \operatorname{rank} \mathbf{A}^* = \operatorname{rank}(\mathbf{A}\mathbf{A}^*) = \operatorname{rank}(\mathbf{A}^*\mathbf{A}).$$

(d) If $\mathbf{A}$ is complex, it is not necessarily true that rank $\mathbf{A}'\mathbf{A}$ = rank $\mathbf{A}$.

3.4. We consider two special cases of rank.

(a) If rank $\mathbf{A} = 0$, then $\mathbf{A} = \mathbf{0}$. This is a simple but key result that can be used to prove the equality of two matrices.

(b) If rank $\mathbf{A} = 1$, then there exist nonzero $\mathbf{a}$ and $\mathbf{b}$ such that $\mathbf{A} = \mathbf{ab}'$.

3.5. (Full-Rank Factorization) Any $m \times n$ real or complex matrix $\mathbf{A}$ of rank r $(r > 0)$ can be expressed in the form $\mathbf{A}_{m \times n} = \mathbf{C}_{m \times r}\mathbf{R}_{r \times n}$, where $\mathbf{C}$ and $\mathbf{R}$ have (full) rank r. We call this a *full-rank factorization*. The columns of $\mathbf{C}$ may be an arbitrary basis of $\mathcal{C}(\mathbf{A})$, and then $\mathbf{R}$ is uniquely determined, or else the rows of $\mathbf{R}$ may be an arbitrary basis of $\mathcal{C}(\mathbf{A}')$, and then $\mathbf{C}$ is uniquely determined. Note that $\mathbf{C}$ has a left inverse, namely $(\mathbf{C}'\mathbf{C})^{-1}\mathbf{C}'$, and $\mathbf{R}$ has a right inverse, $\mathbf{R}'(\mathbf{R}\mathbf{R}')^{-1}$. Two full-rank factorizations can be obtained from the singular value decomposition of $\mathbf{A}$ (cf. 16.34e).

3.6. If $\mathbf{A}$ and $\mathbf{B}$ are $m \times n$ matrices, then rank $\mathbf{A}$ = rank $\mathbf{B}$ if and only if there exist a nonsingular $m \times m$ matrix $\mathbf{C}$ and an $n \times n$ nonsingular matrix $\mathbf{D}$ such that $\mathbf{A} = \mathbf{CBD}$.

3.7. If $\mathcal{C}(\mathbf{B}) = \mathcal{C}(\mathbf{C})$, then rank$(\mathbf{AB})$ = rank$(\mathbf{AC})$ for all $\mathbf{A}$.

3.8. If $\mathbf{V}$ is Hermitian non-negative definite, then $\mathbf{V} = \mathbf{RR}^*$ (by 10.10) and rank$(\mathbf{AV})$ = rank$(\mathbf{AR})$ for all $\mathbf{A}$.

Proofs. Section 3.1.

   3.1. Abadir and Magnus [2005: 77–78].

   3.2. (a) and (c) follow from the definition; for (b) see Meyer [2000a: 215].

   3.3a. Follows from (2.37) and (c)(ii) below. See also Seber and Lee [2003: 458].

3.3b. Abadir and Magnus [2005: 81] and Meyer [2000a: 212].

3.3c(i). Rao and Bhimasankaram [2000: 145].

3.3c(iii). Ben-Israel and Greville [2003: 46] and Meyer [2000a: 212].

3.3d. For a counter example consider $\mathbf{A} = (1, i)'(1, 1)$.

3.4b. Abadir and Magnus [2005: 80].

3.5. Ben-Israel and Greville [2003: 26], Marsaglia and Styan [1974a: theorem 1], and Searle [1982: 175].

3.6. If rank $\mathbf{A} =$ rank $\mathbf{B} = r$, then by (16.33a) $\mathbf{A}$ and $\mathbf{B}$ are equivalent to the same diagonal matrix. The converse follows from (3.14a).

3.7. Follows from $\mathcal{C}(\mathbf{AB}) = \mathcal{C}(\mathbf{AC})$.

3.8. By (10.10), $\mathbf{V} = \mathbf{RR}^*$ and from (2.35) we have $\mathcal{C}(\mathbf{V}) = \mathcal{C}(\mathbf{R})$. The result follows from (3.7).

3.2 MATRIX PRODUCTS

All the matrices in this section are real or complex.

3.9. Given conformable matrices $\mathbf{A}$ and $\mathbf{B}$, we have the following.

(a) rank$(\mathbf{BA}) =$ rank $\mathbf{A}$ if $\mathbf{B}$ has full row rank.

(b) rank$(\mathbf{AC}) =$ rank $\mathbf{A}$ if $\mathbf{C}$ has full column rank.

(c) rank$(\mathbf{A}'\mathbf{AB}) =$ rank$(\mathbf{AB}) =$ rank$(\mathbf{ABB}')$.

3.10. Let $\mathbf{A}$ and $\mathbf{B}$ be $m \times n$ and $n \times p$ matrices, respectively. Then:

(a) rank$(\mathbf{AB}) =$ rank $\mathbf{B} - \dim\{[\mathcal{N}(\mathbf{A})]^{\perp} \cap \mathcal{C}(\mathbf{B})\}$.

(b) rank$(\mathbf{AB}) =$ rank $\mathbf{A} - \dim\{\mathcal{C}(\mathbf{A}') \cap [\mathcal{N}(\mathbf{B})]^{\perp}\}$.

The above results immediately give us conditions for rank$(\mathbf{AB}) =$ rank $\mathbf{A}$ and rank$(\mathbf{AB}) =$ rank $\mathbf{B}$. Other conditions are given in (3.13c) and (3.13d) below.

3.11. Let $\mathbf{A}$ be a square matrix. If rank$(\mathbf{A}^m) =$ rank$(\mathbf{A}^{m+1})$, then rank$(\mathbf{A}^m) =$ rank$(\mathbf{A}^n)$ for all $n \geq m$.

3.12. rank$(\mathbf{AB}) \leq \min\{$rank $\mathbf{A},$ rank $\mathbf{B}\}$.

3.13. Let $\mathbf{A}$ have n columns and $\mathbf{B}$ have n rows. Let $\mathbf{A}^-$ and $\mathbf{B}^-$ be any weak inverses of $\mathbf{A}$ and $\mathbf{B}$, respectively. Then:

$$
\begin{aligned}
\operatorname{rank}\begin{pmatrix} \mathbf{0} & \mathbf{A} \\ \mathbf{B} & \mathbf{I}_n \end{pmatrix} &= \operatorname{rank}\mathbf{A} + \operatorname{rank}(\mathbf{B}, \mathbf{I}_n - \mathbf{A}^-\mathbf{A}) \\
&= \operatorname{rank}\begin{pmatrix} \mathbf{A} \\ \mathbf{I}_n - \mathbf{BB}^- \end{pmatrix} + \operatorname{rank}\mathbf{B} \\
&= \operatorname{rank}\mathbf{A} + \operatorname{rank}\mathbf{B} + \operatorname{rank}[(\mathbf{I}_n - \mathbf{BB}^-)(\mathbf{I}_n - \mathbf{A}^-\mathbf{A})] \\
&= n + \operatorname{rank}(\mathbf{AB}).
\end{aligned}
$$

We can deduce the following.

(a) $\mathrm{rank}(\mathbf{B}, \mathbf{I}_n - \mathbf{A}^-\mathbf{A}) = \mathrm{rank}\,\mathbf{B} + \mathrm{rank}[(\mathbf{I}_n - \mathbf{BB}^-)(\mathbf{I}_n - \mathbf{A}^-\mathbf{A})]$.

(b) $\mathrm{rank}\begin{pmatrix} \mathbf{A} \\ \mathbf{I}_n - \mathbf{BB}^- \end{pmatrix} = \mathrm{rank}\,\mathbf{A} + \mathrm{rank}[(\mathbf{I}_n - \mathbf{BB}^-)(\mathbf{I}_n - \mathbf{A}^-\mathbf{A})]$.

(c) $\mathrm{rank}(\mathbf{AB}) = \mathrm{rank}\,\mathbf{A}$ if and only if $(\mathbf{B}, \mathbf{I}_n - \mathbf{A}^-\mathbf{A})$ has full row rank n.

(d) $\mathrm{rank}(\mathbf{AB}) = \mathrm{rank}\,\mathbf{B}$ if and only if $\begin{pmatrix} \mathbf{A} \\ \mathbf{I}_n - \mathbf{BB}^- \end{pmatrix}$ has full column rank n.

(e) (Sylvester)
$$\mathrm{rank}(\mathbf{AB}) \geq \mathrm{rank}\,\mathbf{A} + \mathrm{rank}\,\mathbf{B} - n,$$

with equality if and only if $(\mathbf{I}_n - \mathbf{BB}^-)(\mathbf{I}_n - \mathbf{A}^-\mathbf{A}) = \mathbf{0}$. This result also follows from the Frobenius inequality (3.18b) by setting $\mathbf{B} = \mathbf{I}_n$. If $\mathbf{AB} = \mathbf{0}$, $\mathrm{rank}\,\mathbf{A} + \mathrm{rank}\,\mathbf{B} \leq n$.

3.14. Let $\mathbf{A}$ be any matrix.

(a) If $\mathbf{P}$ and $\mathbf{Q}$ are any conformable nonsingular matrices,
$$\mathrm{rank}(\mathbf{PAQ}) = \mathrm{rank}\,\mathbf{A}.$$

(b) If $\mathbf{C}$ has full column rank and $\mathbf{R}$ has full row rank, then
$$\mathrm{rank}\,\mathbf{A} = \mathrm{rank}(\mathbf{CA}) = \mathrm{rank}(\mathbf{AR})$$

3.15. If $\mathbf{A}$ is $p \times q$ of rank q and $\mathbf{B}$ is $q \times r$ of rank r, then $\mathbf{AB}$ is $p \times r$ of rank r.

3.16. If $\mathrm{rank}(\mathbf{AB}) = \mathrm{rank}\,\mathbf{A}$, then $\mathcal{C}(\mathbf{AB}) = \mathcal{C}(\mathbf{A})$.

3.17. Suppose that the following products of matrices exist. Then:

(a) $\mathrm{rank}(\mathbf{XA}) = \mathrm{rank}\,\mathbf{A}$ implies $\mathrm{rank}(\mathbf{XAF}) = \mathrm{rank}(\mathbf{AF})$ for every $\mathbf{F}$.

(b) $\mathrm{rank}(\mathbf{AY}) = \mathrm{rank}\,\mathbf{A}$ implies $\mathrm{rank}(\mathbf{KAY}) = \mathrm{rank}(\mathbf{KA})$ for every $\mathbf{K}$.

3.18. Let $\mathbf{A}$, $\mathbf{B}$, and $\mathbf{C}$ be conformable matrices, and let $(\mathbf{AB})^-$ and $(\mathbf{BC})^-$ be any weak inverses. Then:

(a)
$$\mathrm{rank}\begin{pmatrix} \mathbf{0} & \mathbf{AB} \\ \mathbf{BC} & \mathbf{B} \end{pmatrix} = \mathrm{rank}(\mathbf{AB}) + \mathrm{rank}(\mathbf{BC}) + \mathrm{rank}\,\mathbf{L}$$
$$= \mathrm{rank}\,\mathbf{B} + \mathrm{rank}(\mathbf{ABC}),$$

where $\mathbf{L} = [\mathbf{I} - \mathbf{BC}(\mathbf{BC})^-]\mathbf{B}[\mathbf{I} - (\mathbf{AB})^-(\mathbf{AB})]$.

(b) (Frobenius Inequality) From (a) we have
$$\mathrm{rank}(\mathbf{ABC}) \geq \mathrm{rank}(\mathbf{AB}) + \mathrm{rank}(\mathbf{BC}) - \mathrm{rank}\,\mathbf{B},$$

with equality if and only if $\mathbf{L} = \mathbf{0}$.

3.19. Let $\mathbf{V}$ be a non-negative definite $n \times n$ matrix, and let $\mathbf{X}$ be an $n \times p$ matrix. Then the following statements are equivalent.

(1) $\operatorname{rank}(\mathbf{X}'\mathbf{V}\mathbf{X}) = \operatorname{rank}\mathbf{X}$.

(2) $\operatorname{rank}(\mathbf{X}'\mathbf{V}^{+}\mathbf{X}) = \operatorname{rank}\mathbf{X}$, where $\mathbf{V}^{+}$ is the Moore–Penrose inverse of $\mathbf{V}$.

(3) $\mathcal{C}(\mathbf{X}'\mathbf{V}\mathbf{X}) = \mathcal{C}(\mathbf{X}')$.

(4) $\mathcal{C}(\mathbf{X}) \cap [\mathcal{C}(\mathbf{V})]^{\perp} = \mathbf{0}$.

Also $\operatorname{rank}(\mathbf{X}'\mathbf{V}\mathbf{X}) = p$ if and only if $\operatorname{rank}\mathbf{X} = p$ and $\mathcal{C}(\mathbf{X}) \cap [\mathcal{C}(\mathbf{V})]^{\perp} = \mathbf{0}$.

Proofs. Section 3.2.

3.9. Abadir and Magnus [2005: 82, 85].

3.10. Rao and Rao [1998: 133].

3.11. Meyer [2000a: 394] and Ben-Israel and Greville [2003: 155]; see also Section 3.8.

3.12. Abadir and Magnus [2005: 81] and Meyer [2000a: 211].

3.13. Marsaglia and Styan [1974a: theorem 6].

3.14a. By (3.12), $\operatorname{rank}\mathbf{A} = \operatorname{rank}(\mathbf{P}^{-1}\mathbf{P}\mathbf{A}) \le \operatorname{rank}(\mathbf{P}\mathbf{A}) \le \operatorname{rank}\mathbf{A}$.

3.14b. Marsaglia and Styan [1974a: theorem 2].

3.15. Let $\mathbf{A}\mathbf{B}\mathbf{x} = \mathbf{0}$. Then using (3.5), we can take a left inverse of $\mathbf{A}$ and then a left inverse of $\mathbf{B}$ to get $\mathbf{x} = \mathbf{0}$ so that the columns of $\mathbf{A}\mathbf{B}$ are linearly independent.

3.16. Graybill [1983: 89].

3.17. Harville [2001: 27, exercise 1], Marsaglia and Styan [1974a: theorem 2], and Rao and Rao [1998: 133].

3.18. Harville [1997: 396] and Marsaglia and Styan [1974a: theorem 7].

3.19. Isotalo et al. [2005b: 17].

3.3 MATRIX CANCELLATION RULES

3.20. Let $\mathbf{A}$ be any $m \times n$ matrix over $\mathcal{F}$.

(a) If $\mathbf{C}$ has full column rank and $\mathbf{R}$ has full row rank, then using left and right inverses, respectively, we have that $\mathbf{C}\mathbf{A} = \mathbf{C}\mathbf{B}$ implies $\mathbf{A} = \mathbf{B}$ and $\mathbf{P}\mathbf{R} = \mathbf{Q}\mathbf{R}$ implies $\mathbf{P} = \mathbf{Q}$.

(b) If $\operatorname{rank}(\mathbf{X}\mathbf{A}) = \operatorname{rank}\mathbf{A}$, then $\mathbf{X}\mathbf{A}\mathbf{G} = \mathbf{X}\mathbf{A}\mathbf{H}$ implies $\mathbf{A}\mathbf{G} = \mathbf{A}\mathbf{H}$.

(c) If $\operatorname{rank}(\mathbf{A}\mathbf{Y}) = \operatorname{rank}\mathbf{A}$, then $\mathbf{L}\mathbf{A}\mathbf{Y} = \mathbf{M}\mathbf{A}\mathbf{Y}$ implies $\mathbf{L}\mathbf{A} = \mathbf{M}\mathbf{A}$.

3.21. The following are useful for deriving some cancellation rules when the real or complex matrix $\mathbf{A}$ can be a function of other matrices.

(a) $\mathbf{A}^*\mathbf{A} = \mathbf{0}$ implies that $\mathbf{A} = \mathbf{0}$.

(b) $\text{trace}(\mathbf{A}^*\mathbf{A}) = 0$ implies that $\mathbf{A} = \mathbf{0}$.

We can interchange $\mathbf{A}$ and $\mathbf{A}^*$.

3.22. For real or complex matrices we have the following results.

(a) If $\mathbf{PXX}^* = \mathbf{QXX}^*$, then from (3.21a) above and

$$(\mathbf{PXX}^* - \mathbf{QXX}^*)(\mathbf{P}^* - \mathbf{Q}^*) \equiv (\mathbf{PX} - \mathbf{QX})(\mathbf{PX} - \mathbf{QX})^*,$$

we have $\mathbf{PX} = \mathbf{QX}$. We can also replace $\mathbf{X}$ by $\mathbf{X}^*$.

(b) $\mathbf{X}^*\mathbf{XAYY}^* = \mathbf{0}$ implies $\mathbf{XAY} = \mathbf{0}$. Special cases follow by setting $\mathbf{X}$ or $\mathbf{Y}$ equal to the identity matrix.

(c) $\mathbf{A}'\mathbf{AB} = \mathbf{A}'\mathbf{C}$ if and only if $\mathbf{AB} = \mathbf{AA}^+\mathbf{C}$, where $\mathbf{A}^+$ is the Moore–Penrose inverse of $\mathbf{A}$.

Proofs. Section 3.3

3.20. Harville [2001: 27, exercise 1] and Marsaglia and Styan [1974a: theorem 2].

3.21. Searle [1982: 62–63, real case]; (a) implies (b).

3.22a. Searle [1982: 63, real case].

3.22b. Use (3.21b) and $\text{trace}[(\mathbf{XAY})(\mathbf{XAY})^*] = \text{trace}(\mathbf{X}^*\mathbf{XAYY}^*\mathbf{A}^*) = 0$.

3.22c. Magnus and Neudecker [1999: 34].

3.4   MATRIX SUMS

3.23. Let $\mathbf{A}$ and $\mathbf{B}$ be any $m \times n$ matrices over $\mathcal{F}$, and let $(\mathbf{A}, \mathbf{B})^-$ and $\left(\begin{smallmatrix}\mathbf{A}\\\mathbf{B}\end{smallmatrix}\right)^-$ be any weak inverses. Define

$$\mathbf{M} = \left[\mathbf{I}_{2m} - \begin{pmatrix}\mathbf{A}\\\mathbf{B}\end{pmatrix}\begin{pmatrix}\mathbf{A}\\\mathbf{B}\end{pmatrix}^- \right] \begin{pmatrix} \mathbf{A} & \mathbf{0} \\ \mathbf{0} & \mathbf{B} \end{pmatrix} \left[\mathbf{I}_{2n} - (\mathbf{A}, \mathbf{B})^-(\mathbf{A}, \mathbf{B}) \right].$$

(a) From (2.11c), taking transposes and noting that $\text{rank}\,\mathbf{C} = \text{rank}\,\mathbf{C}'$ for any $\mathbf{C}$, we have the following results.

$$\dim[\mathcal{C}(\mathbf{A}) \cap \mathcal{C}(\mathbf{B})] \;=\; \text{rank}\,\mathbf{A} + \text{rank}\,\mathbf{B} - \text{rank}(\mathbf{A}, \mathbf{B}) \quad (= c \text{ say}),$$

$$\dim[\mathcal{C}(\mathbf{A}') \cap \mathcal{C}(\mathbf{B}')] \;=\; \text{rank}\,\mathbf{A} + \text{rank}\,\mathbf{B} - \text{rank}\begin{pmatrix}\mathbf{A}\\\mathbf{B}\end{pmatrix} \quad (= d \text{ say}).$$

(b) $0 \le \text{rank}\,\mathbf{M} \le \min\{c, d\}$. Hence $c = 0$ or $d = 0$ implies $\mathbf{M} = \mathbf{0}$.

(c)

$$\text{rank} \begin{pmatrix} \mathbf{0} & \mathbf{A} & \mathbf{B} \\ \mathbf{A} & \mathbf{A} & \mathbf{0} \\ \mathbf{B} & \mathbf{0} & \mathbf{B} \end{pmatrix} = \text{rank}(\mathbf{A}, \mathbf{B}) + \text{rank}\begin{pmatrix} \mathbf{A} \\ \mathbf{B} \end{pmatrix} + \text{rank}\, \mathbf{M}$$

$$= \text{rank}\,\mathbf{A} + \text{rank}\,\mathbf{B} + \text{rank}(\mathbf{A} + \mathbf{B}).$$

From the above we have the following:

(i) $\text{rank}(\mathbf{A}, \mathbf{B}) + \text{rank}\begin{pmatrix} \mathbf{A} \\ \mathbf{B} \end{pmatrix} + \text{rank}\,\mathbf{M} = \text{rank}\,\mathbf{A} + \text{rank}\,\mathbf{B} + \text{rank}(\mathbf{A} + \mathbf{B})$.

(ii) $\text{rank}\,\mathbf{A} + \text{rank}\,\mathbf{B} - c - d \leq \text{rank}(\mathbf{A} + \mathbf{B}) \leq \text{rank}\,\mathbf{A} + \text{rank}\,\mathbf{B} - \max\{c, d\}$.
Equality on the left occurs if and only if $\mathbf{M} = \mathbf{0}$, and equality on the right occurs if and only if $\text{rank}\,\mathbf{M} = \min\{c, d\}$.

(iii) $\text{rank}(\mathbf{A} + \mathbf{B}) \geq \text{rank}(\mathbf{A}, \mathbf{B}) + \text{rank}\begin{pmatrix} \mathbf{A} \\ \mathbf{B} \end{pmatrix} - \text{rank}\,\mathbf{A} - \text{rank}\,\mathbf{B}$,
with equality if and only if $\mathbf{M} = \mathbf{0}$.

(d) The following hold for any $\mathbf{A}$ and $\mathbf{B}$ of the same size.

(i) $\text{rank}(\mathbf{A} + \mathbf{B}) \leq \text{rank}(\mathbf{A}, \mathbf{B}) \leq \text{rank}\,\mathbf{A} + \text{rank}\,\mathbf{B}$.

(ii) $\text{rank}(\mathbf{A} + \mathbf{B}) \leq \text{rank}\begin{pmatrix} \mathbf{A} \\ \mathbf{B} \end{pmatrix} \leq \text{rank}\,\mathbf{A} + \text{rank}\,\mathbf{B}$.

(e) From the above :

(i) $\text{rank}(\mathbf{A} + \mathbf{B}) = \text{rank}(\mathbf{A}, \mathbf{B})$ if and only $d = \text{rank}\,\mathbf{M}$.

(ii) $\text{rank}(\mathbf{A} + \mathbf{B}) = \text{rank}\begin{pmatrix} \mathbf{A} \\ \mathbf{B} \end{pmatrix}$ if and only if $c = \text{rank}\,\mathbf{M}$.

(iii) $\text{rank}(\mathbf{A}, \mathbf{B}) = \text{rank}\,\mathbf{A} + \text{rank}\,\mathbf{B}$ if and only if $c = 0$.
When $c = 0$, $\mathbf{M} = \mathbf{0}$ and $\text{rank}\,\mathbf{M} = 0 = c$ so that

$$\text{rank}(\mathbf{A} + \mathbf{B}) = \text{rank}\begin{pmatrix} \mathbf{A} \\ \mathbf{B} \end{pmatrix}.$$

(iv) $\text{rank}\begin{pmatrix} \mathbf{A} \\ \mathbf{B} \end{pmatrix} = \text{rank}\,\mathbf{A} + \text{rank}\,\mathbf{B}$ if and only if $d = 0$.
When $d = 0$, $\mathbf{M} = \mathbf{0}$ and $\text{rank}\,\mathbf{M} = 0 = d$ so that

$$\text{rank}(\mathbf{A} + \mathbf{B}) = \text{rank}(\mathbf{A}, \mathbf{B}).$$

(v) $\text{rank}(\mathbf{A} + \mathbf{B}) = \text{rank}\,\mathbf{A} + \text{rank}\,\mathbf{B}$ if and only if $c = d = 0$.

3.24. If $\mathbf{A}$ and $\mathbf{B}$ are $n \times n$ matrices over $\mathcal{F}$, then, since $\mathbf{AB} - \mathbf{I}_n = (\mathbf{A} - \mathbf{I}_n)\mathbf{B} + \mathbf{B} - \mathbf{I}_n$, we have from (3.23e(iii)) and (3.12) that

$$\text{rank}(\mathbf{AB} - \mathbf{I}_n) \leq \text{rank}(\mathbf{A} - \mathbf{I}_n) + \text{rank}(\mathbf{B} - \mathbf{I}_n).$$

Definition 3.2. Suppose that $\mathbf{A} = \sum_{i=1}^{k} \mathbf{A}_i$, where each matrix is $m \times n$. We say that we have *rank additivity* if $\text{rank}\,\mathbf{A} = \sum_{i=1}^{k} \text{rank}\,\mathbf{A}_i$.

3.25. Let $\mathbf{A}$ and $\mathbf{B}$ be nonnull $m \times n$ matrices over $\mathcal{F}$ with respective ranks r and s. If any one of the following conditions hold, then they all hold.

(1) $\text{rank}(\mathbf{A} + \mathbf{B}) = \text{rank}\,\mathbf{A} + \text{rank}\,\mathbf{B}$ (i.e., rank additivity).

(2) There exist nonsingular matrices **F** and **G** such that

$$\mathbf{A} = \mathbf{F} \begin{pmatrix} \mathbf{I}_r & 0 & 0 \\ 0 & 0 & 0 \\ 0 & 0 & 0 \end{pmatrix} \mathbf{G} \quad \text{and} \quad \mathbf{B} = \mathbf{F} \begin{pmatrix} 0 & 0 & 0 \\ 0 & \mathbf{I}_s & 0 \\ 0 & 0 & 0 \end{pmatrix} \mathbf{G}.$$

The above matrices are partitioned in the same way, and the bordering zero matrices are of appropriate size; some of the latter matrices are absent if $\mathbf{A} + \mathbf{B}$ has full rank.

(3) $\dim[\mathcal{C}(\mathbf{A}) \cap \mathcal{C}(\mathbf{B})] = \dim[\mathcal{C}(\mathbf{A}') \cap \mathcal{C}(\mathbf{B}')] = 0$; that is, $c = d = 0$ in (3.23a).

(4) $\operatorname{rank}(\mathbf{A}, \mathbf{B}) = \operatorname{rank}\left(\begin{smallmatrix} \mathbf{A} \\ \mathbf{B} \end{smallmatrix}\right) = \operatorname{rank} \mathbf{A} + \operatorname{rank} \mathbf{B}$.

(5) $\left(\begin{smallmatrix} \mathbf{A} \\ \mathbf{B} \end{smallmatrix}\right)(\mathbf{A} + \mathbf{B})^{-}(\mathbf{A}, \mathbf{B}) = \begin{pmatrix} \mathbf{A} & 0 \\ 0 & \mathbf{B} \end{pmatrix}$.

(6) $\operatorname{rank} \mathbf{A} = \operatorname{rank}[\mathbf{A}(\mathbf{I}_n - \mathbf{B}^{-}\mathbf{B})] = \operatorname{rank}[(\mathbf{I}_m - \mathbf{B}\mathbf{B}^{-})\mathbf{A}]$.

(7) $\operatorname{rank} \mathbf{B} = \operatorname{rank}[(\mathbf{B}(\mathbf{I}_n - \mathbf{A}^{-}\mathbf{A})] = \operatorname{rank}[(\mathbf{I}_m - \mathbf{A}\mathbf{A}^{-})\mathbf{B}]$.

(8) $\operatorname{rank} \mathbf{A} = \operatorname{rank}[\mathbf{A}(\mathbf{I}_n - \mathbf{B}^{-}\mathbf{B})]$ and $\operatorname{rank} \mathbf{B} = \operatorname{rank}[(\mathbf{I}_m - \mathbf{A}\mathbf{A}^{-})\mathbf{B}]$.

(9) $\operatorname{rank} \mathbf{A} = \operatorname{rank}[(\mathbf{I}_m - \mathbf{B}\mathbf{B}^{-})\mathbf{A}]$ and $\operatorname{rank} \mathbf{B} = \operatorname{rank}[\mathbf{B}(\mathbf{I}_n - \mathbf{A}^{-}\mathbf{A})]$.

Here $\mathbf{A}^{-}$, $\mathbf{B}^{-}$, and $(\mathbf{A} + \mathbf{B})^{-}$ are any choices of weak inverses. If any one of (5) to (9) hold for a particular pair of weak inverses, then they hold for any pair.

3.26. Suppose that $\mathbf{A} = \sum_{i=1}^{k} \mathbf{A}_i$, where matrices are all $m \times n$ over $\mathcal{F}$. We now give a number of results about rank additivity. As idempotent matrices play a role in this theory, the reader should also refer to Section 8.6.1.

(a) The following three conditions are equivalent.

    (1) We have rank additivity.

$$(2) \ \operatorname{rank}(\mathbf{A}_1, \mathbf{A}_2, \ldots, \mathbf{A}_k) = \operatorname{rank} \begin{pmatrix} \mathbf{A}_1 \\ \mathbf{A}_2 \\ \vdots \\ \mathbf{A}_k \end{pmatrix} = \sum_{i=1}^{k} \operatorname{rank} \mathbf{A}_i.$$

    (3) $\mathbf{A}_i \mathbf{A}^{-} \mathbf{A}_i = \mathbf{A}_i$ and $\mathbf{A}_i \mathbf{A}^{-} \mathbf{A}_j = 0$ for all i, j $(i \neq j)$ where $\mathbf{A}^{-}$ is any choice of a weak inverse of $\mathbf{A}$.

If (3) holds for a particular weak inverse, then all three conditions hold for any weak inverse.

(b) Suppose $\mathbf{A}$ is idempotent (i.e., $\mathbf{A}^2 = \mathbf{A}$). Then the following three conditions are equivalent.

    (1) We have rank additivity.

$$(2) \ \operatorname{rank}(\mathbf{A}_1, \mathbf{A}_2, \ldots, \mathbf{A}_k) = \operatorname{rank} \begin{pmatrix} \mathbf{A}_1 \\ \mathbf{A}_2 \\ \vdots \\ \mathbf{A}_k \end{pmatrix} = \sum_{i=1}^{k} \operatorname{rank} \mathbf{A}_i.$$

(3) $\mathbf{A}_i^2 = \mathbf{A}_i$ and $\mathbf{A}_i\mathbf{A}_j = \mathbf{0}$ for all i, j $(i \neq j)$.

(c) If $r_i = \operatorname{rank}\mathbf{A}_i$ $(i = 1, 2, \ldots, k)$, we have rank additivity if and only if there are nonsingular matrices $\mathbf{F}$ and $\mathbf{G}$ such that

$$\mathbf{A}_1 = \mathbf{F}\begin{pmatrix} \mathbf{I}_{r_1} & \mathbf{0} & \cdots & \mathbf{0} & \mathbf{0} \\ \mathbf{0} & \mathbf{0} & \cdots & \mathbf{0} & \mathbf{0} \\ \vdots & \vdots & & \vdots & \vdots \\ \mathbf{0} & \mathbf{0} & \cdots & \mathbf{0} & \mathbf{0} \\ \mathbf{0} & \mathbf{0} & \cdots & \mathbf{0} & \mathbf{0} \end{pmatrix}\mathbf{G},$$

$$\mathbf{A}_2 = \mathbf{F}\begin{pmatrix} \mathbf{0} & \mathbf{0} & \cdots & \mathbf{0} & \mathbf{0} \\ \mathbf{0} & \mathbf{I}_{r_2} & \cdots & \mathbf{0} & \mathbf{0} \\ \vdots & \vdots & & \vdots & \vdots \\ \mathbf{0} & \mathbf{0} & \cdots & \mathbf{0} & \mathbf{0} \\ \mathbf{0} & \mathbf{0} & \cdots & \mathbf{0} & \mathbf{0} \end{pmatrix}\mathbf{G},$$

$$\vdots$$

$$\mathbf{A}_k = \mathbf{F}\begin{pmatrix} \mathbf{0} & \mathbf{0} & \cdots & \mathbf{0} & \mathbf{0} \\ \mathbf{0} & \mathbf{0} & \cdots & \mathbf{0} & \mathbf{0} \\ \vdots & \vdots & & \vdots & \vdots \\ \mathbf{0} & \mathbf{0} & \cdots & \mathbf{I}_{r_k} & \mathbf{0} \\ \mathbf{0} & \mathbf{0} & \cdots & \mathbf{0} & \mathbf{0} \end{pmatrix}\mathbf{G}.$$

Furthermore, if the $\mathbf{A}_i$ are real (respectively complex), there exist orthogonal (respectively unitary) $\mathbf{P}$ and $\mathbf{Q}$ and diagonal $\mathbf{D}$ such that $\mathbf{F} = \mathbf{PD}$ and $\mathbf{G} = \mathbf{Q}$, that is there exists a simultaneous singular value decomposition.

3.27. Let $\mathbf{A} = \sum_{i=1}^{k} \mathbf{A}_i$, where the matrices are real (or complex with $'$ replaced by $*$).

(a) We assume that the matrices are not necessarily square.

(i) If $\mathbf{A}_i'\mathbf{A}_j = \mathbf{0}$ and $\mathbf{A}_i\mathbf{A}_j' = \mathbf{0}$ for all i, j $(i \neq j)$, then the rank is additive.

(ii) If the rank is additive, then $\mathbf{A}_i\mathbf{A}_j' = \mathbf{0}$ for all i, j $(i \neq j)$ if and only if $\mathbf{A}_j\mathbf{A}' = \mathbf{AA}_j'$ $(j = 1, \ldots, k)$.

(b) We assume that the matrices are square.

(i) If $\operatorname{rank}(\mathbf{A}_i^2) = \operatorname{rank}\mathbf{A}_i$ and $\mathbf{A}_i\mathbf{A}_j = \mathbf{0}$ for all i, j $(i \neq j)$, then the rank is additive.

(ii) If the rank is additive, then $\mathbf{A}_i\mathbf{A}_j = \mathbf{0}$ for all i, j $(i \neq j)$ if and only if $\mathbf{A}_j\mathbf{A} = \mathbf{AA}_j$ $(j = 1, \ldots, k)$.

(c) If the $\mathbf{A}_i$ are all real symmetric or Hermitian non-negative definite matrices and

$$\left(\sum_{i=1}^{k} c_i\mathbf{A}_i\right)\mathbf{A}^-\left(\sum_{j=1}^{k} \mathbf{A}_j/c_j\right) = \mathbf{A}$$

holds for some distinct positive scalars c_i and for some choice of weak inverse $\mathbf{A}^-$, then the rank is additive.

Conversely, if the rank is additive, then the above equation holds for every choice of distinct positive $c_1, \ldots, c_k$ and for every choice of a weak inverse.

Proofs. Section 3.4.

3.23. Marsaglia and Styan [1974a: theorem 8 and corollary 8.1]; see also Harville [1997: 442–445].

3.25. Marsaglia and Styan [1974a: theorem 11]; see also Harville [1997: 445] for (3) and Harville [2001: 29, exercise 29] for (4) and (6)–(9).

3.26a. Marsaglia and Styan [1974a: theorem 13].

3.26b. Marsaglia and Styan [1974a: corollary 13.1]. Note that $\mathbf{AAA} = \mathbf{A}$ so that we can set $\mathbf{A}^- = \mathbf{A}$ in (a).

3.26c. Marsaglia and Styan [1974a: theorem 12].

3.27a. Marsaglia and Styan [1974a: theorem 14].

3.27b. Marsaglia and Styan [1974a: theorem 15].

3.26c. Marsaglia and Styan [1974a: theorem 16].

3.5 MATRIX DIFFERENCES

3.28. $\operatorname{rank}(\mathbf{A} - \mathbf{B}) \geq |\operatorname{rank}\mathbf{A} - \operatorname{rank}\mathbf{B}|$.

3.29. Let $\mathbf{A}$ and $\mathbf{B}$ be $m \times n$ matrices over $\mathcal{F}$. Then results on the rank of $\mathbf{A} + \mathbf{B}$ immediately lead to the results on the rank of $\mathbf{A} - \mathbf{B}$ by simply substituting $-\mathbf{B}$ for $\mathbf{B}$. We can also use the fact that $\operatorname{rank}(\mathbf{A} - \mathbf{B}) = \operatorname{rank}\mathbf{A} - \operatorname{rank}\mathbf{B}$ if and only if $\operatorname{rank}(\mathbf{A} - \mathbf{B}) + \operatorname{rank}\mathbf{B} = \operatorname{rank}\mathbf{A}$.

(a) Let

$$\mathbf{N} = \left[\mathbf{I}_{2m} - \begin{pmatrix}\mathbf{A}\\\mathbf{B}\end{pmatrix}\begin{pmatrix}\mathbf{A}\\\mathbf{B}\end{pmatrix}^-\right]\begin{pmatrix}\mathbf{A} & \mathbf{0}\\\mathbf{0} & -\mathbf{B}\end{pmatrix}\left[\mathbf{I}_{2n} - (\mathbf{A}, \mathbf{B})^-(\mathbf{A}, \mathbf{B})\right].$$

Then

$$\operatorname{rank}(\mathbf{A} - \mathbf{B}) = \operatorname{rank}(\mathbf{A}, \mathbf{B}) + \operatorname{rank}\begin{pmatrix}\mathbf{A}\\\mathbf{B}\end{pmatrix} + \operatorname{rank}\mathbf{N} - \operatorname{rank}\mathbf{A} - \operatorname{rank}\mathbf{B}.$$

(b) $\operatorname{rank}(\mathbf{A} - \mathbf{B}) = \operatorname{rank}\mathbf{A} - \operatorname{rank}\mathbf{B}$ if and only if $\mathbf{N} = \mathbf{0}$ and $\operatorname{rank}(\mathbf{A}, \mathbf{B}) = \operatorname{rank}\mathbf{A} = \operatorname{rank}\begin{pmatrix}\mathbf{A}\\\mathbf{B}\end{pmatrix}$. Furthermore, if just the latter equation is true, then

$$\operatorname{rank}\mathbf{N} = \operatorname{rank}(\mathbf{BA}^-\mathbf{B} - \mathbf{B}).$$

3.30. Let $\mathbf{A}$ and $\mathbf{B}$ be $m \times n$ matrices over $\mathcal{F}$. If one of the following five conditions is true, then all five are true.

(1) $\operatorname{rank}(\mathbf{A} - \mathbf{B}) = \operatorname{rank}\mathbf{A} - \operatorname{rank}\mathbf{B}$.

(2) $\mathcal{C}(\mathbf{B}) \subset \mathcal{C}(\mathbf{A})$, $\mathcal{C}(\mathbf{B}') \subset \mathcal{C}(\mathbf{A}')$ and $\mathbf{BA}^-\mathbf{B} = \mathbf{B}$.

(3) $\operatorname{rank}(\mathbf{A}, \mathbf{B}) = \operatorname{rank} \mathbf{A} = \operatorname{rank} \left(\begin{smallmatrix}\mathbf{A}\\\mathbf{B}\end{smallmatrix}\right)$ and $\mathbf{BA}^-\mathbf{B} = \mathbf{B}$.

(4) $\mathbf{AA}^-\mathbf{B} = \mathbf{BA}^-\mathbf{A} = \mathbf{BA}^-\mathbf{B} = \mathbf{B}$.

(5) $\operatorname{rank}(\mathbf{A} - \mathbf{B}) = \operatorname{rank}[\mathbf{A}(\mathbf{I}_n - \mathbf{B}^-\mathbf{B})] = \operatorname{rank}[(\mathbf{I}_m - \mathbf{BB}^-)\mathbf{A}]$, where $\mathbf{A}^-$ and $\mathbf{B}^-$ are any choices of weak inverses. If any one of (2) to (5) holds for any particular set of weak inverses, then all the conditions hold for every weak inverse.

3.31. Let $\mathbf{A}$ and $\mathbf{B}$ be $m \times n$ matrices over $\mathcal{F}$ with ranks r and s, respectively. Then:

(a) $\operatorname{rank}(\mathbf{A} - \mathbf{B}) = \operatorname{rank} \mathbf{A} - \operatorname{rank} \mathbf{B}$ if and only if there exist nonsingular matrices $\mathbf{F}$ and $\mathbf{G}$ such that

$$\mathbf{A} = \mathbf{F} \begin{pmatrix} \mathbf{I}_s & \mathbf{0} & \mathbf{0} \\ \mathbf{0} & \mathbf{I}_{r-s} & \mathbf{0} \\ \mathbf{0} & \mathbf{0} & \mathbf{0} \end{pmatrix} \mathbf{G} \quad \text{and} \quad \mathbf{B} = \mathbf{F} \begin{pmatrix} \mathbf{I}_s & \mathbf{0} & \mathbf{0} \\ \mathbf{0} & \mathbf{0} & \mathbf{0} \\ \mathbf{0} & \mathbf{0} & \mathbf{0} \end{pmatrix} \mathbf{G},$$

where the matrices are similarly partitioned and the bordering zero matrices are of appropriate size; some are absent if $\mathbf{A}$ or $\mathbf{B}$ has full rank.

(b) If $\mathbf{A}$ and $\mathbf{B}$ are real (complex) matrices, there exist orthogonal (unitary) matrices $\mathbf{P}$ and $\mathbf{Q}$ such that

$$\mathbf{A} = \mathbf{P} \begin{pmatrix} \mathbf{D}_1 & \mathbf{0} & \mathbf{0} \\ \mathbf{0} & \mathbf{D}_2 & \mathbf{0} \\ \mathbf{0} & \mathbf{0} & \mathbf{0} \end{pmatrix} \mathbf{Q} \quad \text{and} \quad \mathbf{B} = \mathbf{P} \begin{pmatrix} \mathbf{0} & \mathbf{0} & \mathbf{0} \\ \mathbf{0} & \mathbf{D}_2 & \mathbf{0} \\ \mathbf{0} & \mathbf{0} & \mathbf{0} \end{pmatrix} \mathbf{Q},$$

where $\mathbf{D}_1$ and $\mathbf{D}_2$ are nonsingular diagonal matrices.

3.32. If $\mathbf{A}$ and $\mathbf{B}$ are nonsingular $n \times n$ matrices, then from the identity $\mathbf{A} - \mathbf{B} = -\mathbf{B}(\mathbf{A}^{-1} - \mathbf{B}^{-1})\mathbf{A}$ and (3.14a),

$$\operatorname{rank}(\mathbf{A}^{-1} - \mathbf{B}^{-1}) = \operatorname{rank}(\mathbf{A} - \mathbf{B}) = \operatorname{rank}(\mathbf{B} - \mathbf{A}).$$

Furthermore, $\mathbf{A}^{-1} - \mathbf{B}^{-1}$ is nonsingular if and only if $\mathbf{B} - \mathbf{A}$ is nonsingular.

3.33. (Wedderburn–Guttman)   Let $\mathbf{A}$ be an $m \times n$ matrix of rank r, and let $\mathbf{M}$ and $\mathbf{N}$ be $m \times s$ and $n \times s$ matrices, respectively, such that $\mathbf{M}'\mathbf{AN}$ is nonsingular. Then

$$\operatorname{rank}[\mathbf{A} - \mathbf{N}(\mathbf{M}'\mathbf{AN})^{-1}\mathbf{M}'\mathbf{A}] = r - s,$$

with

$$\operatorname{rank}[\mathbf{AN}(\mathbf{M}'\mathbf{AN})^{-1}\mathbf{M}'\mathbf{A}] = \operatorname{rank}[\mathbf{M}'\mathbf{AN}] = s.$$

This theorem has been used in psychometrics.

3.34. (Idempotent Matrices)   Let $\mathbf{P}$ and $\mathbf{Q}$ be $n \times n$ idempotent matrices. Then

$$\begin{aligned} \operatorname{rank}(\mathbf{P} - \mathbf{Q}) &= \operatorname{rank} \begin{pmatrix} \mathbf{P} \\ \mathbf{Q} \end{pmatrix} + \operatorname{rank}(\mathbf{P}, \mathbf{Q}) - \operatorname{rank} \mathbf{P} - \operatorname{rank} \mathbf{Q} \\ &= \operatorname{rank}(\mathbf{P} - \mathbf{PQ}) + \operatorname{rank}(\mathbf{PQ} - \mathbf{Q}) \\ &= \operatorname{rank}(\mathbf{P} - \mathbf{QP}) + \operatorname{rank}(\mathbf{QP} - \mathbf{Q}). \end{aligned}$$

Proofs. Section 3.5.

3.28. Abadir and Magnus [2005: 81].

3.29. Marsaglia and Styan [1974a: 387–388].

3.30. Harville [2001: 200–203, exercise 30] and Marsaglia and Styan [1974a: theorem 17].

3.31. Marsaglia and Styan [1974a: theorem 18].

3.32. Harville [1997: 420].

3.33. Takane and Yanai [2005]. They also discuss the case when $\mathbf{M'AN}$ is rectangular.

3.34. Tian and Styan [2001]. They also give an extensive list of similar results including those for $\mathbf{P} + \mathbf{Q}$.

3.6   PARTITIONED AND PATTERNED MATRICES

Some partitioned matrices have already been mentioned above in passing so there will be a slight overlap with the following, which focuses exclusively on partitioned matrices.

3.35. (Column Partitions)   Let $\mathbf{A}$ be an $m \times n$ matrix and $\mathbf{B}$ be an $m \times q$ matrix, both over $\mathcal{F}$.

(a) $\mathcal{C}(\mathbf{A}) \cap \mathcal{C}[(\mathbf{I}_m - \mathbf{AA}^-)\mathbf{B}] = \mathbf{0}$, where $\mathbf{A}^-$ is any weak inverse of $\mathbf{A}$.

(b) $\operatorname{rank}(\mathbf{A}, \mathbf{B}) = \operatorname{rank}(\mathbf{A}, (\mathbf{I}_m - \mathbf{AA}^-)\mathbf{B})$.

(c)

$$
\begin{aligned}
\operatorname{rank}(\mathbf{A}, \mathbf{B}) &= \operatorname{rank}\mathbf{A} + \operatorname{rank}[(\mathbf{I} - \mathbf{AA}^-)\mathbf{B}] \\
&= \operatorname{rank}\mathbf{B} + \operatorname{rank}[(\mathbf{I} - \mathbf{BB}^-)\mathbf{A}].
\end{aligned}
$$

The second result is obtained by interchanging $\mathbf{A}$ and $\mathbf{B}$.

Note that $\mathbf{AA}^- = \mathbf{P_A}$ is idempotent, thus representing an (oblique) projection onto $\mathcal{C}(\mathbf{A})$ (by 7.2c); also $\mathbf{P_A A} = \mathbf{A}$.

3.36. (Row Partitions)   Let $\mathbf{A}$ be an $m \times n$ matrix and $\mathbf{C}$ be an $q \times n$ matrix, both over $\mathcal{F}$.

(a) $\mathbf{A}$ and $\mathbf{C}(\mathbf{I}_n - \mathbf{A}^-\mathbf{A})$ have disjoint row spaces (i.e., only have the zero vector in common).

(b) $\operatorname{rank}\begin{pmatrix} \mathbf{A} \\ \mathbf{C} \end{pmatrix} = \operatorname{rank}\begin{pmatrix} \mathbf{A} \\ \mathbf{C}(\mathbf{I}_n - \mathbf{A}^-\mathbf{A}) \end{pmatrix}.$

(c)

$$\mathrm{rank}\begin{pmatrix} \mathbf{A} \\ \mathbf{C} \end{pmatrix} = \mathrm{rank}\,\mathbf{A} + \mathrm{rank}[\mathbf{C}(\mathbf{I} - \mathbf{A}^-\mathbf{A})]$$
$$= \mathrm{rank}\,\mathbf{C} + \mathrm{rank}[\mathbf{A}(\mathbf{I} - \mathbf{C}^-\mathbf{C})].$$

Note that $(\mathbf{A}^-\mathbf{A})' = \mathbf{P_{A'}}$, where $\mathbf{P_{A'}}$ is the (oblique) projection onto $\mathcal{C}(\mathbf{A}')$ (by 7.2c).

3.37. For conformable matrices $\mathbf{A}$, $\mathbf{B}$, and $\mathbf{C}$ over $\mathcal{F}$ and for any choice of weak inverse $\mathbf{A}^-$, we have the following:

(a) $\mathrm{rank}(\mathbf{AB}, [\mathbf{I} - \mathbf{AA}^-]\mathbf{C}) = \mathrm{rank}(\mathbf{AB}) + \mathrm{rank}([\mathbf{I} - \mathbf{AA}^-]\mathbf{C})$.

(b) $\mathrm{rank}\begin{pmatrix} \mathbf{BA} \\ \mathbf{C}[\mathbf{I} - \mathbf{A}^-\mathbf{A}] \end{pmatrix} = \mathrm{rank}(\mathbf{BA}) + \mathrm{rank}(\mathbf{C}[\mathbf{I} - \mathbf{A}^-\mathbf{A}])$.

3.38. If all four matrices are conformable, we have

$$\mathrm{rank}\begin{pmatrix} \mathbf{A} & \mathbf{B} \\ \mathbf{C} & \mathbf{D} \end{pmatrix} = \mathrm{rank}\begin{pmatrix} \mathbf{B} & \mathbf{A} \\ \mathbf{D} & \mathbf{C} \end{pmatrix} = \mathrm{rank}\begin{pmatrix} \mathbf{C} & \mathbf{D} \\ \mathbf{A} & \mathbf{B} \end{pmatrix} = \mathrm{rank}\begin{pmatrix} \mathbf{D} & \mathbf{C} \\ \mathbf{B} & \mathbf{A} \end{pmatrix}.$$

3.39. The following hold for any conformable $\mathbf{A}$, $\mathbf{B}$, and $\mathbf{C}$ over $\mathcal{F}$.

(a) For all weak inverses $\mathbf{A}^-$ and $\mathbf{B}^-$,

$$\mathrm{rank}\begin{pmatrix} \mathbf{0} & \mathbf{A} \\ \mathbf{B} & \mathbf{C} \end{pmatrix} = \mathrm{rank}\,\mathbf{A} + \mathrm{rank}[\mathbf{B}, \mathbf{C}(\mathbf{I} - \mathbf{A}^-\mathbf{A})]$$
$$= \mathrm{rank}\,\mathbf{B} + \mathrm{rank}\begin{pmatrix} \mathbf{A} \\ (\mathbf{I} - \mathbf{BB}^-)\mathbf{C} \end{pmatrix}$$
$$= \mathrm{rank}\,\mathbf{A} + \mathrm{rank}\,\mathbf{B} + \mathrm{rank}[(\mathbf{I} - \mathbf{BB}^-)\mathbf{C}(\mathbf{I} - \mathbf{A}^-\mathbf{A})]$$
$$\leq \mathrm{rank}\,\mathbf{A} + \mathrm{rank}\,\mathbf{B} + \mathrm{rank}\,\mathbf{C},$$

with equality if and only if $\mathcal{C}(\mathbf{B}) \cap \mathcal{C}(\mathbf{C}) = \mathbf{0}$ and $\mathcal{C}(\mathbf{A}') \cap \mathcal{C}(\mathbf{C}') = \mathbf{0}$. If $\mathbf{B}$ or $\mathbf{C}$ (or both) is nonsingular, then the rank of the left-hand side is $\mathrm{rank}\,\mathbf{A} + \mathrm{rank}\,\mathbf{B}$.

(b)

$$\mathrm{rank}\begin{pmatrix} \mathbf{0} & \mathbf{A} \\ \mathbf{B} & \mathbf{C} \end{pmatrix} = \mathrm{rank}\,\mathbf{C} + \mathrm{rank}[\mathbf{A}(\mathbf{I} - \mathbf{C}^-\mathbf{C})] + \mathrm{rank}[(\mathbf{I} - \mathbf{CC}^-)\mathbf{B}]$$
$$+ \mathrm{rank}\,\mathbf{D},$$

for $\mathbf{D} = (\mathbf{I} - \mathbf{UU}^-)\mathbf{AC}^-\mathbf{B}(\mathbf{I} - \mathbf{V}^-\mathbf{V})$, with $\mathbf{U} = \mathbf{A}(\mathbf{I} - \mathbf{C}^-\mathbf{C})$ and $\mathbf{V} = (\mathbf{I} - \mathbf{CC}^-)\mathbf{B}$. The weak inverses may be any (possibly different) choices except that the $\mathbf{C}^-$ in the middle of $\mathbf{D}$ must be the same as that chosen in either $\mathbf{U}$ or $\mathbf{V}$.

3.40. (Generalized Schur Complement) Let $\mathbf{E}$, $\mathbf{F}$, $\mathbf{G}$, and $\mathbf{H}$ be conformable matrices over $\mathcal{F}$, and let $\mathbf{A}$ be given by

$$\mathbf{A} = \begin{pmatrix} \mathbf{E} & \mathbf{F} \\ \mathbf{G} & \mathbf{H} \end{pmatrix}.$$

(a) We have the following results.

(i)

$$\text{rank } \mathbf{A} = \text{rank } \mathbf{E} + \text{rank} \begin{pmatrix} \mathbf{0} & (\mathbf{I} - \mathbf{E}\mathbf{E}^-)\mathbf{F} \\ \mathbf{G}(\mathbf{I} - \mathbf{E}^-\mathbf{E}) & \mathbf{H} - \mathbf{G}\mathbf{E}^-\mathbf{F} \end{pmatrix}$$

holds for any three generalized inverses $\mathbf{E}^-$. Here $\mathbf{S} = \mathbf{H} - \mathbf{G}\mathbf{E}^-\mathbf{F}$ is the generalized Schur complement of $\mathbf{A}$ with repect to $\mathbf{E}$ and is written as $(\mathbf{A}/\mathbf{E})$ (cf. Section 14.1).

(ii) If $\mathbf{G} = \mathbf{0}$, then rank $\mathbf{A} \geq$ rank $\mathbf{E} +$ rank $\mathbf{H}$. The same is true if we have $\mathbf{F} = \mathbf{0}$ instead.

(iii) Let $\tilde{\mathbf{E}}$ be a particular weak inverse of $\mathbf{E}$. Then

$$\text{rank } \mathbf{A} = \text{rank } \mathbf{E} + \text{rank}(\mathbf{H} - \mathbf{G}\tilde{\mathbf{E}}\mathbf{F})$$

if and only if

$$\begin{aligned} \mathbf{U} &= -(\mathbf{I} - \mathbf{E}\mathbf{E}^-)\mathbf{F}\mathbf{S}^-\mathbf{G}(\mathbf{I} - \mathbf{E}^-\mathbf{E}) = \mathbf{0}, \\ \mathbf{V} &= (\mathbf{I} - \mathbf{E}\mathbf{E}^-)\mathbf{F}(\mathbf{I} - \mathbf{S}^-\mathbf{S}) = \mathbf{0}, \\ \mathbf{W} &= (\mathbf{I} - \mathbf{S}\mathbf{S}^-)\mathbf{G}(\mathbf{I} - \mathbf{E}^-\mathbf{E}) = \mathbf{0}, \end{aligned}$$

where $\mathbf{A}^-$ and $\mathbf{S}^-$ are any choices of weak inverses. Since, by (7.20), $(\mathbf{I} - \mathbf{E}\mathbf{E}^-)\mathbf{E} = \mathbf{0}$ and $(\mathbf{I} - \mathbf{E}^-\mathbf{E})'\mathbf{E}' = \mathbf{0}$, the above three conditions are satisfied when $\mathcal{C}(\mathbf{F}) \subseteq \mathcal{C}(\mathbf{E})$ and $\mathcal{C}(\mathbf{G}') \subseteq \mathcal{C}(\mathbf{E}')$ (Schott [2005: 265]). This is the case, for example, when $\mathbf{A}$ is non-negative definite and $\mathbf{E}$ and $\mathbf{H}$ are both square, that is, $\mathbf{G} = \mathbf{F}'$ (cf. 14.8c). The above result follows from (iv) below. Other conditions for the above to hold that relate to ranks are given by Tian [2002: 204].

From (3.38) we can interchange $\mathbf{E}$ and $\mathbf{H}$, and $\mathbf{F}$ and $\mathbf{G}$, in the above results, as we have done in (v) and (vi) below.

(iv) Using the above notation,

$$\text{rank } \mathbf{A} = \text{rank } \mathbf{E} + \text{rank } \mathbf{S} + \text{rank } \mathbf{V} + \text{rank } \mathbf{W} + \text{rank } \mathbf{Z},$$

where $\mathbf{Z} = (\mathbf{I} - \mathbf{V}\mathbf{V}^-)\mathbf{U}(\mathbf{I} - \mathbf{W}^-\mathbf{W})$ and any weak inverses can be used.

(v) If $\mathbf{E}$ is square and nonsingular, then the three conditions of (iii) are satisfied and

$$\text{rank } \mathbf{A} = \text{rank } \mathbf{E} + \text{rank}(\mathbf{H} - \mathbf{G}\mathbf{E}^{-1}\mathbf{F}).$$

(vi) If $\mathbf{H}$ is square and nonsingular, then

$$\text{rank } \mathbf{A} = \text{rank } \mathbf{H} + \text{rank}(\mathbf{E} - \mathbf{F}\mathbf{H}^{-1}\mathbf{G}).$$

(vii) With appropriate matrix substitutions we have from (v) and (vi)

$$\text{rank} \begin{pmatrix} \mathbf{I}_m & \mathbf{B} \\ \mathbf{B}' & \mathbf{I}_n \end{pmatrix} = n + \text{rank}(\mathbf{I}_m - \mathbf{B}\mathbf{B}') = m + \text{rank}(\mathbf{I}_n - \mathbf{B}'\mathbf{B}).$$

(b) $\operatorname{rank} \mathbf{A} = \operatorname{rank} \mathbf{E} + \operatorname{rank} \mathbf{X} + \operatorname{rank} \mathbf{Y} + \operatorname{rank} \mathbf{T}$, where

$$
\begin{aligned}
\mathbf{X} &= (\mathbf{I} - \mathbf{E}\mathbf{E}^-)\mathbf{F}, \\
\mathbf{Y} &= \mathbf{G}(\mathbf{I} - \mathbf{E}^-\mathbf{E}), \\
\mathbf{T} &= (\mathbf{I} - \mathbf{Y}\mathbf{Y}^-)(\mathbf{H} - \mathbf{G}\mathbf{E}^-\mathbf{F})(\mathbf{I} - \mathbf{X}^-\mathbf{X}).
\end{aligned}
$$

Any choices of weak inverses can be used.

3.41. If $\mathbf{A}$ is $m \times n$ and $\mathbf{B}$ is $n \times m$, then

$$
\operatorname{rank}(\mathbf{I}_m - \mathbf{A}\mathbf{B}) = \operatorname{rank}(\mathbf{I}_n - \mathbf{B}\mathbf{A}) + m - n.
$$

Proofs. Section 3.6.

3.35. Harville [1997: 385] and Marsaglia and Styan [1974a: theorem 5].

3.36. Marsaglia and Styan [1974a: theorem 5].

3.37. Marsaglia and Styan [1974a: theorem 4].

3.38. Interchange rows then columns.

3.39. Harville [1997: 388–389] and Marsaglia and Styan [1974a: theorem 19].

3.40a. Marsaglia and Styan [1974a: theorem 19 and corollary 19.1] with their restrictions in (i) and (iv) removed by Ouellette [1981: 228–229]; (ii) is proved by Abadir and Magnus [2005: 121–122]; (v) and (vi) are proved by Schott [2005: 265–266] and Abadir and Magnus [2005: 123].

3.40b. Marsaglia and Styan [1974a: theorem 19] with their restrictions removed by Ouellette [1981: 230].

3.41. Abadir and Magnus [2005: 124]. See also Groß [1999].

3.7   MAXIMAL AND MINIMAL RANKS

This topic presents some very powerful tools for handling matrix problems, as shown in a series of papers by Yongge Tian. For example, one can find the maximum and minimum ranks of an expression and then find conditions when these two are equal; this will give us the rank, subject to the conditions. One way of proving that two matrix expressions are equal is to prove that the rank of their difference is zero.

For some history of this topic see Tian [2000], and for some detailed results see Tian [2002].

3.42. For conformable matrices,

$$
\min_{\mathbf{X}_1, \mathbf{X}_2} \operatorname{rank}(\mathbf{A} - \mathbf{B}\mathbf{X}_1 - \mathbf{C}\mathbf{X}_2) = \operatorname{rank} \begin{pmatrix} \mathbf{A} & \mathbf{B} \\ \mathbf{C} & \mathbf{0} \end{pmatrix} - \operatorname{rank} \mathbf{B} - \operatorname{rank} \mathbf{C},
$$

where the minimization is with respect to all conformable $\mathbf{X}_1$ and $\mathbf{X}_2$.

3.43. For conformable matrices, define $p(\mathbf{X}_1, \mathbf{X}_2) = \mathbf{A} - \mathbf{B}_1\mathbf{X}_1\mathbf{C}_1 - \mathbf{B}_2\mathbf{X}_2\mathbf{C}_2$. Then:

(a)

$$
\min_{\mathbf{X}_1, \mathbf{X}_2} \operatorname{rank}[p(\mathbf{X}_1, \mathbf{X}_2)] = \operatorname{rank} \begin{pmatrix} \mathbf{A} \\ \mathbf{C}_1 \\ \mathbf{C}_2 \end{pmatrix} + \operatorname{rank}(\mathbf{A}, \mathbf{B}_1, \mathbf{B}_2)
$$

$$
+ \max \left\{ \operatorname{rank} \begin{pmatrix} \mathbf{A} & \mathbf{B}_1 \\ \mathbf{C}_2 & \mathbf{0} \end{pmatrix} - \operatorname{rank} \begin{pmatrix} \mathbf{A} & \mathbf{B}_1 & \mathbf{B}_2 \\ \mathbf{C}_2 & \mathbf{0} & \mathbf{0} \end{pmatrix} \right.
$$

$$
- \operatorname{rank} \begin{pmatrix} \mathbf{A} & \mathbf{B}_1 \\ \mathbf{C}_1 & \mathbf{0} \\ \mathbf{C}_2 & \mathbf{0} \end{pmatrix}, \quad \operatorname{rank} \begin{pmatrix} \mathbf{A} & \mathbf{B}_2 \\ \mathbf{C}_1 & \mathbf{0} \end{pmatrix}
$$

$$
\left. - \operatorname{rank} \begin{pmatrix} \mathbf{A} & \mathbf{B}_1 & \mathbf{B}_2 \\ \mathbf{C}_1 & \mathbf{0} & \mathbf{0} \end{pmatrix} - \operatorname{rank} \begin{pmatrix} \mathbf{A} & \mathbf{B}_2 \\ \mathbf{C}_1 & \mathbf{0} \\ \mathbf{C}_2 & \mathbf{0} \end{pmatrix} \right\}.
$$

(b)

$$
\max_{\mathbf{X}_1, \mathbf{X}_2} p(\mathbf{X}_1, \mathbf{X}_2) = \min \left\{ \operatorname{rank}(\mathbf{A}, \mathbf{B}_1, \mathbf{B}_2), \operatorname{rank} \begin{pmatrix} \mathbf{A} \\ \mathbf{C}_1 \\ \mathbf{C}_2 \end{pmatrix}, \right.
$$

$$
\left. \operatorname{rank} \begin{pmatrix} \mathbf{A} & \mathbf{B}_1 \\ \mathbf{C}_2 & \mathbf{0} \end{pmatrix}, \operatorname{rank} \begin{pmatrix} \mathbf{A} & \mathbf{B}_2 \\ \mathbf{C}_1 & \mathbf{0} \end{pmatrix} \right\}.
$$

3.44. (Generalized Schur Complement) Let

$$
\mathbf{M} = \begin{pmatrix} \mathbf{A} & \mathbf{B} \\ \mathbf{C} & \mathbf{D} \end{pmatrix},
$$

then we recall from (3.40) that the generalized Schur complement of $\mathbf{A}$ in $\mathbf{M}$ ($\mathbf{M}/\mathbf{A}$) is $\mathbf{S_A} = \mathbf{D} - \mathbf{C}\mathbf{A}^-\mathbf{B}$, where $\mathbf{A}^-$ is any weak inverse $\mathbf{A}$ (We have changed the notation for $\mathbf{M}$ to fit in with the proofs for the following results.) Then

$$
\max_{\mathbf{A}^-} \operatorname{rank}(\mathbf{S_A}) = \min \left\{ \operatorname{rank}(\mathbf{C}, \mathbf{D}), \operatorname{rank} \begin{pmatrix} \mathbf{B} \\ \mathbf{D} \end{pmatrix}, \operatorname{rank} \mathbf{M} - \operatorname{rank} \mathbf{A} \right\}
$$

and

$$
\min_{\mathbf{A}^-} \operatorname{rank}(\mathbf{S_A}) = \operatorname{rank} \mathbf{A} + \operatorname{rank}(\mathbf{C}, \mathbf{D}) + \operatorname{rank} \begin{pmatrix} \mathbf{B} \\ \mathbf{D} \end{pmatrix} + \operatorname{rank} \mathbf{M}
$$

$$
- \operatorname{rank} \begin{pmatrix} \mathbf{A} & \mathbf{0} & \mathbf{B} \\ \mathbf{0} & \mathbf{C} & \mathbf{D} \end{pmatrix} - \operatorname{rank} \begin{pmatrix} \mathbf{A} & \mathbf{0} \\ \mathbf{0} & \mathbf{B} \\ \mathbf{C} & \mathbf{D} \end{pmatrix}.
$$

Proofs. Section 3.7.

    3.42. Tian [2000].

3.43. Tian [2002]. He shows there is some simplification when $\mathcal{C}(\mathbf{B}_1) \subseteq \mathcal{C}(\mathbf{B}_2)$ and $\mathcal{C}(\mathbf{C}_2') \subseteq \mathcal{C}(\mathbf{C}_1')$. By equating (a) and (b), he obtains necessary and sufficient conditions for $\text{rank}[p(\mathbf{X}_1, \mathbf{X}_2)]$ to be invariant with respect to $\mathbf{X}_1$ and $\mathbf{X}_2$. He then finds similar conditions for $\mathcal{C}[p(\mathbf{X}_1, \mathbf{X}_2)]$ to be invariant.

3.44. Tian [2002: 201]. He uses the fact that $\mathbf{A}^-$ is a solution of $\mathbf{AXA} = \mathbf{A}$. He also gives necessary and sufficient conditions for the rank and column space of $\mathbf{S_A}$ to be invariant with respect to the choice of $\mathbf{A}^-$. Some rank and other properties of $\mathbf{CA}^-\mathbf{B}$, $\mathbf{A}^-\mathbf{B}$, $\mathbf{CA}^-$, and $\mathbf{A} - \mathbf{AB}^-\mathbf{A}$ are given by Tian [2002: 206–207].

3.8 MATRIX INDEX

Definition 3.3. If $\mathbf{A}$ is an $n \times n$, there exists a positive integer k $(1 \le k \le n)$ such that $\text{rank}(\mathbf{A}^k) = \text{rank}(\mathbf{A}^{k+1})$. The smallest k for which this is true is called the *index* of $\mathbf{A}$. If $\mathbf{A}$ is nonsingular, $k = 0$, where $\mathbf{A}^0 = \mathbf{I}_n$. The basis for the definition comes from the following results and (3.11).

3.45. If $\mathbf{A}$ is an $n \times n$ complex matrix, then:

(a) $\mathcal{N}(\mathbf{A}^0) \subseteq \mathcal{N}(\mathbf{A}) \subseteq \mathcal{N}(\mathbf{A}^2) \subseteq \cdots \subseteq \mathcal{N}(\mathbf{A}^k) \subseteq \mathcal{N}(\mathbf{A}^{k+1}) \subseteq \cdots$.

(b) $\mathcal{C}(\mathbf{A}^0) \supseteq \mathcal{C}(\mathbf{A}) \supseteq \mathcal{C}(\mathbf{A}^2) \supseteq \cdots \supseteq \mathcal{C}(\mathbf{A}^k) \supseteq \mathcal{C}(\mathbf{A}^{k+1}) \supseteq \cdots$.

There is equality at some point, in fact at the same value of k in both cases. What this means is that the index k is the smallest integer at which $\mathcal{C}(\mathbf{A}^k)$ stops shrinking and $\mathcal{N}(\mathbf{A}^k)$ stops growing.

3.46. Let $\mathbf{A}$ have index k.

(a) All matrices $\{\mathbf{A}^l : l \ge k\}$ have the same rank, the same column space, and the same null space.

(b) Their transposes $\{(\mathbf{A}^l)' : l \ge k\}$ have the same rank, the same column space, and the same null space.

(c) Their conjugate transposes $\{(\mathbf{A}^l)^* : l \ge k\}$ have the same rank, the same column space, and the same null space.

(d) For no l less than k do $\mathbf{A}^l$ and a higher power of $\mathbf{A}$ (or their transposes or conjugate transposes) have the same range or the same null space.

(e) For $l \ge k$

$$\mathcal{C}(\mathbf{A}^l) \cap \mathcal{N}(\mathbf{A}^l) = \mathbf{0} \quad \text{and} \quad \mathcal{C}(\mathbf{A}^l) \oplus \mathcal{N}(\mathbf{A}^l) = \mathbb{C}^n.$$

Proofs. Section 3.8

3.45–3.46. Ben-Israel and Greville [2003: 155] and Meyer [2000a: 395, real case].

CHAPTER 4

MATRIX FUNCTIONS: INVERSE, TRANSPOSE, TRACE, DETERMINANT, AND NORM

The topics considered in this chapter might be regarded as the "bread and butter" or, changing the metaphor, the working tools for someone using linear algebra in their research. I have not included rank, generalized inverse, and eigenvalues, as I have a separate chapter for each of these topics.

4.1 INVERSE

***Definition* 4.1.** An $m \times n$ matrix $\mathbf{A}$ is said to have a *right inverse* if there exists an $n \times m$ matrix $\mathbf{B}$ such that $\mathbf{AB} = \mathbf{I}_m$. It is said to have a *left inverse* if there exists an $n \times m$ matric $\mathbf{C}$ such that $\mathbf{CA} = \mathbf{I}_n$. These inverses are generally not unique.

4.1. An $m \times n$ matrix $\mathbf{A}$ has a left inverse if and only if it has full column rank (i.e., rank $\mathbf{A} = n$, $m \geq n$), and it has a right inverse if and only if it has full row rank (i.e., rank $\mathbf{A} = m$, $m \leq n$). Examples of such inverses are, respectively, $(\mathbf{A}'\mathbf{A})^{-1}\mathbf{A}'$ and $\mathbf{A}'(\mathbf{AA}')^{-1}$.

***Definition* 4.2.** If $\mathbf{A}$ is $n \times n$ and rank $\mathbf{A} = n$, then $\mathbf{A}$ is said to be *nonsingular* and has an inverse denoted by $\mathbf{A}^{-1}$ that satisfies $\mathbf{AA}^{-1} = \mathbf{A}^{-1}\mathbf{A} = \mathbf{I}_n$. An equivalent definition is that $\mathbf{A}$ is nonsingular if and only if $\det \mathbf{A} \neq 0$. A square matrix that is not nonsingular (i.e., $\det \mathbf{A} = 0$) is said to be *singular*. Inverses, both algebraic and numerical, can be computed using Matlab (Leon [2007: chapter 71], Maple (Jeffrey and Corless [2007: chapter 72]), and Mathematica (Ruskeepää [2007: chapter 73]).

A Matrix Handbook for Statisticians. By George A. F. Seber
Copyright © 2008 John Wiley & Sons, Inc.

4.2. If $\mathbf{A}$ is $n \times n$ and $\mathbf{AB} = \mathbf{I}_n$, then $\mathbf{B} = \mathbf{A}^{-1}$.

4.3. If $\mathbf{A}$ and $\mathbf{B}$ are nonsingular matrices of the same size, then $(\mathbf{AB})^{-1} = \mathbf{B}^{-1}\mathbf{A}^{-1}$.

4.4. If $\mathbf{A}$ is nonsingular and $c \neq 0$, then $(c\mathbf{A})^{-1} = c^{-1}\mathbf{A}^{-1}$.

4.5. If $\mathbf{A}$ is nonsingular, then $\mathbf{A}^{-1}$ is a continuous function of the elements of $\mathbf{A}$.

Proofs.   Section 4.1.

    4.1. Harville [1997: 80].

    4.2–4.4. Abadir and Magnus [2005: 83–84].

    4.5. Schott [2005: 199].

4.2   TRANSPOSE

Definition 4.3. If $\mathbf{A} = (a_{ij})$ is real or complex, we define $\mathbf{A}^* = (\bar{a}_{ji})$ to be the *conjugate transpose* of $\mathbf{A}$. When $\mathbf{A}$ is real, $\mathbf{A}^* = \mathbf{A}'$.

4.6. (Basic Results)

  (a) $(\mathbf{AB})^* = \mathbf{B}^*\mathbf{A}^*$.

  (b) $(\alpha\mathbf{A})^* = \bar{\alpha}\mathbf{A}^*$.

  (c) If $\mathbf{A}$ is a nonsingular matrix, then $(\mathbf{A}^{-1})^* = (\mathbf{A}^*)^{-1}$.

4.7. Suppose $\mathbf{A}$ and $\mathbf{B}$ are real matrices, where $\mathbf{A}$ is $p \times m$ and $\mathbf{B}$ is $p \times n$, with $m \leq n$. Then $\mathbf{AA}' = \mathbf{BB}'$ if and only if there exists an $m \times n$ matrix $\mathbf{H}$ with $\mathbf{HH}' = \mathbf{I}_m$ such that $\mathbf{AH} = \mathbf{B}$.

Proofs.   Section 4.2.

    4.6. Rao and Bhimasankaram [2000: 85, real case].

    4.7. Muirhead [1982: 589].

4.3   TRACE

Definition 4.4. If $\mathbf{A} = (a_{ij})$ is an $n \times n$ matrix, then the sum of the diagonal elements is called the *trace* of $\mathbf{A}$ and is denoted by trace $\mathbf{A}$. Thus

$$\text{trace } \mathbf{A} = \sum_{i=1}^{n} a_{ii} = \text{trace } \mathbf{A}'.$$

4.8. Let $\mathbf{A}$ be $m \times n$, and let $\mathbf{A}^-$ be any weak inverse of $\mathbf{A}$. Then, from (7.2d),

$$\text{trace}(\mathbf{A}^-\mathbf{A}) = \text{trace}(\mathbf{AA}^-) = \text{rank } \mathbf{A}.$$

4.9. If **A** is real and symmetric, then

$$\text{rank } \mathbf{A} \geq \frac{(\text{trace } \mathbf{A})^2}{\text{trace}(\mathbf{A}^2)}.$$

For related results see (6.21).

4.10. If **A** is an $n \times n$ real matrix with real eigenvalues and exactly t of them are nonzero, then

$$(\text{trace } \mathbf{A})^2 \leq t \, \text{trace}(\mathbf{A}^2).$$

4.11. Let **A** be an $n \times n$ real matrix.

(a) If **A** has real eigenvalues, then $(\text{trace } \mathbf{A})^2 \leq \text{rank}(\mathbf{A}) \, \text{trace}(\mathbf{A}^2)$.

(b) If **A** is symmetric, $(\text{trace } \mathbf{A})^2 = \text{rank}(\mathbf{A}) \, \text{trace}(\mathbf{A}^2)$ if and only if there is a non-negative integer k such that $\mathbf{A}^2 = k\mathbf{A}$.

(c) If **A** is symmetric, then $\mathbf{A}^2 = \mathbf{A}$ if and only if $\text{rank } \mathbf{A} = \text{trace } \mathbf{A} = \text{trace}(\mathbf{A}^2)$.

(d) $\text{trace}(\mathbf{A}'\mathbf{A}) \geq \text{trace}(\mathbf{A}^2)$, with equality if and only if **A** is symmetric.

4.12. If **A** is an $n \times n$ real or complex matrix, then **A** can be written as $\mathbf{A} = \mathbf{XY} - \mathbf{YX}$ for some $n \times n$ matrices **X** and **Y** if and only if $\text{trace } \mathbf{A} = 0$.

4.13. Let **A** be $m \times n$ and **B** be $n \times m$, both real or complex matrices.

(a) We have

$$\text{trace}(\mathbf{AB}) = \text{trace}(\mathbf{BA}) = \text{trace}(\mathbf{A}'\mathbf{B}') = \text{trace}(\mathbf{B}'\mathbf{A}')$$
$$= \sum_{i=1}^{m}\sum_{j=1}^{n} a_{ij}b_{ji} = \sum_{i=1}^{m}\sum_{j=1}^{n} a_{ij}b'_{ij}.$$

(b) If $m = n$ and either **A** or **B** is symmetric, then

$$\text{trace}(\mathbf{AB}) = \sum_{i=1}^{n}\sum_{j=1}^{n} a_{ij}b_{ij}.$$

This result is particularly useful in statistics.

4.14. Suppose **A** is $m \times n$, **B** is $n \times p$, and **C** is $p \times n$. Then

$$\text{trace}(\mathbf{ABC}) = \text{trace}(\mathbf{BCA}) = \text{trace}(\mathbf{CAB}).$$

4.15. Let **C** be an $m \times n$ real or complex matrix. Then

$$\text{trace}(\mathbf{C}^*\mathbf{C}) = \text{trace}(\mathbf{CC}^*) = \sum_{i=1}^{m}\sum_{j=1}^{n} |c_{ij}|^2.$$

Hence $\text{trace}(\mathbf{C}^*\mathbf{C}) = 0$ implies that $\mathbf{C} = \mathbf{0}$.

4.16. Suppose the $m \times n$ matrix $\mathbf{E}_{ij}$ has 1 in the i, jth position and zeros elsewhere. If **A** is $n \times m$, we have $\text{trace}(\mathbf{E}_{ij}\mathbf{A}) = a_{ji} = a'_{ij}$.

4.17. If $\mathbf{R}$ is $n \times n$ and nonsingular, $\text{trace}(\mathbf{R}^{-1}\mathbf{AR}) = \text{trace}(\mathbf{ARR}^{-1}) = \text{trace}\,\mathbf{A}$.

4.18. If $\mathbf{A}$ and $\mathbf{B}$ are real symmetric matrices, then $\text{trace}[(\mathbf{AB})^2] \leq \text{trace}(\mathbf{A}^2\mathbf{B}^2)$.

4.19. If $\mathbf{A}$ is $n \times n$, $\mathbf{B}$ is $m \times m$, and $\mathbf{A} \otimes \mathbf{B}$ is the Kronecker product, then (cf. 11.1l(ii))

$$\text{trace}(\mathbf{A} \otimes \mathbf{B}) = \text{trace}(\mathbf{A})\,\text{trace}(\mathbf{B}).$$

4.20. If $\mathbf{A}$ is $n \times n$ and $\mathbf{x}$ is $n \times 1$, then $\mathbf{x}'\mathbf{Ax} = \text{trace}(\mathbf{x}'\mathbf{Ax}) = \text{trace}(\mathbf{Axx}')$.

4.21. If $\mathbf{A}$ is $m \times n$, then $\text{trace}(\mathbf{AX}) = 0$ for every $n \times m$ $\mathbf{X}$ if and only if $\mathbf{A} = \mathbf{0}$.

4.22. If $\mathbf{A}$ is $n \times n$, then $\text{trace}(\mathbf{AX}) = 0$ for all Hermitian $\mathbf{X}$ if and only if $\mathbf{A} = \mathbf{0}$.

4.23. If $\mathbf{A}$ is an $n \times n$ Hermitian matrix and $\text{trace}\,\mathbf{A} \geq \Re\,\text{trace}(\mathbf{AU})$ for all unitary matrices $\mathbf{U}$, then $\mathbf{A}$ is non-negative definite. Here $\Re$ is the "real part."

4.24. Let $\mathbf{A}$ be an $n \times n$ matrix with singular value decomposition $\mathbf{A} = \mathbf{P\Sigma Q}^*$, where $\mathbf{P}$ and $\mathbf{Q}$ are $n \times n$ unitary matrices, $\mathbf{\Sigma} = \text{diag}(\sigma_1(\mathbf{A}), \dots, \sigma_n(\mathbf{A}))$, and the $\sigma_i(\mathbf{A})$ are the ordered singular values of $\mathbf{A}$. Let $\mathcal{U}_n$ be the collection of all $n \times n$ unitary matrices. Then

$$\max_{\mathbf{U} \in \mathcal{U}_n} \Re\,\text{trace}(\mathbf{AU}) = \sum_{i=1}^{n} \sigma_i(\mathbf{A}),$$

and the maximum is attained at $\mathbf{U}_0 = \mathbf{QP}^*$ (which need not be unique).

4.25. Let $\mathbf{A}$ be $m \times n$ and $\mathbf{B}$ be $n \times m$ matrices, and define $\mathcal{U}_n$ as in (4.24) above. Then

$$\max_{\mathbf{U} \in \mathcal{U}_n, \mathbf{V} \in \mathcal{U}_m} \Re\,\text{trace}(\mathbf{AUBV}) = \sum_{i=1}^{p} \sigma_i(\mathbf{A})\sigma_i(\mathbf{B}),$$

where $p = \min\{m, n\}$ and $\sigma(\cdot)$ is a singular value.

4.26. Let $\mathbf{A}$ and $\mathbf{B}$ be $n \times n$ non-negative definite matrices. Then:

(a) $\text{trace}\,\mathbf{A} \geq 0$ with equality if and only if $\mathbf{A} = \mathbf{0}$.

(b) $\text{trace}(\mathbf{AB}) \geq 0$ with equality if and only if $\mathbf{AB} = \mathbf{0}$.

4.27. Let $\mathbf{A}$ and $\mathbf{B}$ be $n \times n$ positive definite matrices. Then:

(a) $\text{trace}\,\mathbf{A} > 0$.

(b) $\text{trace}(\mathbf{AB}) > 0$.

Proofs. Section 4.3.

4.8. $\mathbf{AA}^-$ is a projection onto $\mathcal{C}(\mathbf{A})$ and is therefore idempotent so that its rank equals its trace.

4.9–4.10. Graybill [1983: 303–304].

4.11a–c. Graybill [1983: 305–306].

4.11d. Follows from $\text{trace}[(\mathbf{A} - \mathbf{A}')'(\mathbf{A} - \mathbf{A}') = 2[\text{trace}(\mathbf{A}'\mathbf{A}) - \text{trace}(\mathbf{A}^2)] \geq 0$.

4.12. Horn and Johnson [1991: 288].

4.13. Abadir and Magnus [2005: 31] and Rao and Bhimasankaram [2000: 92]. We can interchange the subscripts i and j.

4.14. Use (4.13a) with $\mathbf{AB}$ and $\mathbf{C}$, and so on.

4.15. trace$(\mathbf{C}^*\mathbf{C}) = \sum_i \sum_j c_{ij}^* c_{ji} = \sum_i \sum_j \bar{c}_{ji} c_{ji}$.

4.16. Use (4.13a).

4.18. Graybill [1983: 302].

4.19. Abadir and Magnus [2005: 277].

4.21. Use (4.16) for all $\mathbf{E}_{ij}$.

4.22–4.23. Rao and Rao [1998: 342–343].

4.24. Rao and Rao [1998: 347].

4.25. Rao and Rao [1998: 357].

4.26. Graybill [1983: 306-307].

4.27a. Each $a_{ii} > 0$ from (10.33b).

4.27b. We have trace$(\mathbf{AB}) = $ trace$(\mathbf{A}^{1/2}\mathbf{BA}^{1/2})$, where (cf. 10.32) $\mathbf{A}^{1/2}$ is the positive definite square root of $\mathbf{A}$, and $\mathbf{A}^{1/2}\mathbf{BA}^{1/2}$ is positive definite. Now apply (a) to trace$(\mathbf{A}^{1/2}\mathbf{BA}^{1/2})$.

4.4 DETERMINANTS

Determinants arise in many places in this book. In this section I concentrate on some basic properties, but the reader should also refer to Chapter 14 on partitioned matrices and Chapter 15 on patterned matrices. Determinants of special matrices are given in Chapter 8, and the differentiation of determinants is given in Chapters 17 and 18.

4.4.1 Introduction

***Definition* 4.5.** The determinant of a square matrix $\mathbf{A} = (a_{ij})$, denoted by $\det(\mathbf{A})$, is defined as

(1)
$$\det(\mathbf{A}) = \sum \varepsilon_{j_1 j_2 \cdots j_n} a_{1j_1} a_{2j_2} \cdots a_{nj_n},$$

where $\varepsilon_{j_1 j_2 \cdots j_n}$ is $+1$ or -1 according as $\{j_1, j_2 \cdots j_n\}$ is an even or odd number of permutations of the integers $\{1, 2, \ldots, n\}$, with the summation extending over all n such possible permutations. Thus $\varepsilon_{j_1 j_2 \cdots j_n} = (-1)^{j_1 + j_2 + \cdots + j_n}$.

(2) Another way of expressing $\det(\mathbf{A})$ is

$$\det(\mathbf{A}) = \sum_\pi \operatorname{sgn}(\pi) \prod_{i=1}^n a_{i\pi(i)}.$$

Here π is a permutation of the ordered set $\{1, 2, \ldots, n\}$ and $\pi(i)$ is the ith member of the permutation π. The summation extends over all permutations, and the function $\mathrm{sgn}(\pi)$ is $+1$ or -1 depending on whether the permutation is even or odd.

4.28. (Basic Properties) Let $\mathbf{A}$ be an $n \times n$ matrix.

(a) $\det(\mathbf{A}) = \det(\mathbf{A}')$.

(b) If two rows (columns) of $\mathbf{A}$ are equal, then $\det(\mathbf{A}) = 0$.

(c) If every element of a row (column) of $\mathbf{A}$ is zero, then $\det(\mathbf{A}) = 0$.

(d) If $\mathbf{B}$ is obtained from $\mathbf{A}$ by multiplying one row (column) of $\mathbf{A}$ by the scalar k, then $\det(\mathbf{B}) = k \det(\mathbf{A})$. In particular, $\det(c\mathbf{A}) = c^n \det \mathbf{A}$.

(e) If $\mathbf{B}$ is obtained from $\mathbf{A}$ by interchanging any two rows (columns), then $\det(\mathbf{B}) = -\det(\mathbf{A})$.

(f) Adding to one row (column) of a square matrix any multiple of another row (column) does not affect the value of the determinant.

(g) $\det(\mathbf{A})$ is a continuous function of the elements of $\mathbf{A}$.

Note that the transformations (d)–(f) can be represented by matrices (cf. Section 16.2.1).

4.29. (Row-Block Operations) Some of the properties of the previous result carry over to block multiplication. Let $\mathbf{A}$ be $m \times m$ and $\mathbf{B}$ be $n \times n$ matrices.

(a) If $\mathbf{E}$ is $m \times m$, then

$$\det \begin{pmatrix} \mathbf{EA} & \mathbf{EB} \\ \mathbf{C} & \mathbf{D} \end{pmatrix} = \det(\mathbf{E}) \cdot \det \begin{pmatrix} \mathbf{A} & \mathbf{B} \\ \mathbf{C} & \mathbf{D} \end{pmatrix}.$$

(b) If $\mathbf{E}$ is $n \times m$, then

$$\det \begin{pmatrix} \mathbf{A} & \mathbf{B} \\ \mathbf{C} + \mathbf{EA} & \mathbf{D} + \mathbf{EB} \end{pmatrix} = \det \begin{pmatrix} \mathbf{A} & \mathbf{B} \\ \mathbf{C} & \mathbf{D} \end{pmatrix}.$$

4.30. An $n \times n$ matrix may be written as $\mathbf{A} = \mathbf{XYX}^{-1}\mathbf{Y}^{-1}$ for some nonsingular $n \times n$ matrices $\mathbf{X}$ and $\mathbf{Y}$ if and only if $\det(\mathbf{A}) = 1$.

4.31. If $\mathbf{A}$ and $\mathbf{B}$ are $n \times n$ matrices, then:

(a) $\det(\mathbf{AB}) = \det(\mathbf{A}) \det(\mathbf{B})$.

(b) $\det(\mathbf{AA}') = \det(\mathbf{AA}') = (\det \mathbf{A})^2$.

(c) If $\mathbf{A}$ is nonsingular, then setting $\mathbf{B} = \mathbf{A}^{-1}$ gives us $\det(\mathbf{A}^{-1}) = (\det \mathbf{A})^{-1}$.

(d) $\det \begin{pmatrix} \mathbf{A} & \mathbf{B} \\ \mathbf{B} & \mathbf{A} \end{pmatrix} = \det(\mathbf{A} + \mathbf{B}) \det(\mathbf{A} - \mathbf{B})$.

4.32. (Craig–Sakamoto) If $\mathbf{A}$ and $\mathbf{B}$ are $n \times n$ real symmetric matrices, and c and d are positive scalars, then

$$\det(\mathbf{I}_n - s\mathbf{A} - t\mathbf{B}) = \det(\mathbf{I}_n - s\mathbf{A}) \cdot \det(\mathbf{I}_n - t\mathbf{B})$$

for all $|s| < c$ and $|t| < d$ if and only if $\mathbf{AB} = \mathbf{0}$.

4.33. Let $\mathbf{A}$ be an $m \times n$ matrix and $\mathbf{B}$ be an $n \times m$ matrix. Taking determinants of both sides of the following equivalence

$$\begin{pmatrix} \mathbf{I}_m - \mathbf{AB} & \mathbf{A} \\ \mathbf{0} & \mathbf{I}_n \end{pmatrix} \begin{pmatrix} \mathbf{I}_m & \mathbf{0} \\ \mathbf{B} & \mathbf{I}_n \end{pmatrix} = \begin{pmatrix} \mathbf{I}_m & \mathbf{0} \\ \mathbf{B} & \mathbf{I}_n - \mathbf{BA} \end{pmatrix} \begin{pmatrix} \mathbf{I}_m & \mathbf{A} \\ \mathbf{0} & \mathbf{I}_n \end{pmatrix},$$

using (14.18) and the fact that the determinant of a triangular matrix is the product of its diagonal elements, we get

$$\det(\mathbf{I}_m - \mathbf{AB}) = \det(\mathbf{I}_n - \mathbf{BA}).$$

If $n = 1$,

$$\det(\mathbf{I}_m - \mathbf{aa}') = 1 - \mathbf{a}'\mathbf{a}.$$

Proofs.   Section 4.4.1.

4.28a–f. Rao and Bhimasankaram [2000: 224–225] and Searle [1982: sections 4.3 and 4.4].

4.28g. Schott [2005: 198].

4.29. Abadir and Magnus [2005: 115].

4.30. Horn and Johnson [1991: 291].

4.31a–c. Searle [1982: 98-99].

4.31d. Abadir and Magnus [2005: 117–118].

4.32. Abadir and Magnus [2005: 181] and Harville [1997: 568–569; see also the theory there on polynomials].

4.4.2   Adjoint Matrix

Definition 4.6. Let $\mathbf{A}$ be an $n \times n$ matrix. If a submatrix $\mathbf{A}_{ij}$ is formed by deleting the ith row and the jth column of $\mathbf{A}$, then $\det(\mathbf{A}_{ij})$ is called the *minor* of a_{ij} and the signed minor $\alpha_{ij} = (-1)^{i+j} \det(\mathbf{A}_{ij})$ is called the *cofactor* of a_{ij}. The matrix $\mathrm{adj}(\mathbf{A}) = (\alpha_{ji})$ is called the *adjoint (adjugate)* of $\mathbf{A}$.

4.34. If $\mathbf{A}$ is $n \times n$, then

$$\sum_{j=1}^{n} a_{ij}\alpha_{kj} = \begin{cases} \det(\mathbf{A}), & \text{if } i = k, \\ 0, & \text{otherwise.} \end{cases}$$

4.35. If $\mathbf{A}$ is $n \times n$, then $\mathbf{A}(\mathrm{adj}\mathbf{A}) = (\mathrm{adj}\mathbf{A})\mathbf{A} = \det(\mathbf{A})\mathbf{I}_n$.

4.36. $\mathrm{adj}(\mathrm{adj}\mathbf{A}) = (\det \mathbf{A})^{n-2}\mathbf{A}$.

4.37. If $\mathbf{A}$ is nonsingular, it follows from (4.35) above that $\mathrm{adj}(\mathbf{A}) = (\det \mathbf{A})\mathbf{A}^{-1}$ and $\det(\mathrm{adj}\mathbf{A}) = \{\det(\mathbf{A})\}^{n-1}$.

4.38. The following hold.

(a) $\mathrm{adj}(a\mathbf{A}) = a^{n-1}\mathrm{adj}\mathbf{A}$.

(b) $\mathrm{adj}(\mathbf{A}') = (\mathrm{adj}\mathbf{A})'$.

(c) If $\mathbf{A}$ is nonsingular, then $\mathrm{adj}(\mathbf{A}^{-1}) = (\mathrm{adj}\mathbf{A})^{-1}$.

4.39. Let $\mathbf{A}$ be $n \times n$.

(a) Let rank $\mathbf{A} = n - 1$, then

$$\mathrm{adj}\mathbf{A} = (-1)^{k+1}d(\mathbf{A})\frac{\mathbf{x}\mathbf{y}'}{\mathbf{y}'(\mathbf{A}^{k-1})^{+}\mathbf{x}},$$

where k denotes the algebraic multiplicity of the zero eigenvalue of $\mathbf{A}$ ($1 \leq k \leq n$), $d(\mathbf{A})$ is the product of the $n - k$ non-zero eigenvalues of $\mathbf{A}$, $(\cdot)^{+}$ denotes the Moore–Penrose inverse, and $\mathbf{x}$ and $\mathbf{y}$ are $n \times 1$ vectors satisfying $\mathbf{A}\mathbf{x} = \mathbf{0}$ and $\mathbf{y}'\mathbf{A} = \mathbf{0}'$. If $k = n$, we put $d(\mathbf{A}) = 1$.

(b) If 0 is a simple eigenvalue of $\mathbf{A}$ (i.e., $k = 1$ in (a)), then rank $\mathbf{A} = n - 1$ and

$$\mathrm{adj}\mathbf{A} = d(\mathbf{A})\frac{\mathbf{x}\mathbf{y}'}{\mathbf{y}'\mathbf{x}}.$$

Here $d(\mathbf{A})$ is the product of the $n - 1$ nonzero eigenvalues and $\mathbf{x}$ and $\mathbf{y}$ are defined in (a).

(c) If rank $\mathbf{A} \leq n - 2$, then $\mathrm{adj}\mathbf{A} = \mathbf{0}$.

If rank $\mathbf{A} = n - 1$, then $\mathrm{rank}(\mathrm{adj}\mathbf{A}) = 1$.

4.40. If $\mathbf{A}$ and $\mathbf{B}$ are both $n \times n$, then $\mathrm{adj}(\mathbf{A}\mathbf{B}) = \mathrm{adj}\mathbf{A} \cdot \mathrm{adj}\mathbf{B}$.

Proofs. Section 4.4.2.

4.34. Abadir and Magnus [2005: 90] and Rao and Bhimasankaram [2000: 240].

4.35–4.37. Abadir and Magnus [2005: 95] and Rao and Bhimasankaram [2000: 241, 244].

4.38. Rao and Bhimasankaram [2000: 245, see solution to exercise 8].

4.39. Magnus and Neudecker [1999: 40-43].

4.40. Abadir and Magnus [2005: 95], Harville [2001: 77, exercise 14], and Rao and Bhimasankaram [2000: 244].

4.4.3 Compound Matrix

***Definition* 4.7.** Given the $m \times n$ matrix $\mathbf{A} = (a_{ij})$, a *compound matrix* of $\mathbf{A}$ is the array of all minors of a given size k, say, $(k \leq \min\{m, n\})$. The $M = \binom{m}{k}$ by $N = \binom{n}{k}$ matrix is denoted by $\mathbf{A}_{[k]} = (b_{\alpha\beta})$, where we write symbolically $\alpha = (i_1, i_2, \ldots, i_k)$ and $\beta = (j_1, j_2, \ldots, j_k)$. Here $b_{\alpha\beta}$ is the determinant of the submatrix obtained by selecting the intersection of the k rows $i_1, i_2, \ldots, i_k$ and the k columns $j_1, j_2, \ldots, j_k$. The MN elements of $\mathbf{A}_{[k]}$ are arranged in lexicographic order (for a numerical example see Horn and Johnson [1985: 19–20]). Compound matrices are useful for expressing a number of expansions of determinants like Sylvester's identity, the Cauchy–Binet formula and the Laplace expansion given in the next section (cf. Rao and Rao [1998: 146–154]).

4.41. (Basic Properties)

(a) Let $\mathbf{A}_{m \times p}$ and $\mathbf{B}_{p \times n}$ be real or complex matrices, then:

   (i) $(\mathbf{AB})_{[k]} = \mathbf{A}_{[k]}\mathbf{B}_{[k]}, \quad k \leq \min\{m, n, p\}$.

   (ii) $(c\mathbf{A})_{[k]} = c^k \mathbf{A}_{[k]}$.

(b) If $\mathbf{A}$ is an $m \times n$ real or complex matrix, then $(\mathbf{A}_{[k]})' = (\mathbf{A}')_{[k]}$.

(c) If $\mathbf{A}$ is a complex matrix, then $(\mathbf{A}_{[k]})^* = (\mathbf{A}^*)_{[k]}$.

(d) If $\mathbf{A}$ is a nonsingular real or complex matrix, then $(\mathbf{A}_{[k]})^{-1} = (\mathbf{A}^{-1})_{[k]}$.

Proofs. Section 4.4.3

   4.41. Rao and Rao [1998: 146–154] and quoted by Horn and Johnson [1985: 19–20].

4.4.4 Expansion of a Determinant

4.42. (Expanding by Row i or Column j) Referring to (4.34), we have

$$
\det(\mathbf{A}) = \sum_{j=1}^{n} a_{ij}\alpha_{ij} \quad (\text{row } i)
$$

$$
= \sum_{i=1}^{n} a_{ij}\alpha_{ij} \quad (\text{column } j).
$$

4.43. (Expanding by the Diagonal) Consider the $n \times n$ matrix

$$
\mathbf{B} = \mathbf{A} + \operatorname{diag}(x_1, x_2, \ldots, x_n).
$$

Then $\det(\mathbf{B})$ consists of the sum of all possible products of the x_i taken r at a time for $r = n, n - 1, \ldots, 2, 1, 0$, each product being multiplied by its complementary principal minor of order $n - r$ in $\mathbf{A}$. By complementary minor in $\mathbf{A}$ we mean the principal minor having diagonal elements other than those associated in $\mathbf{B}$ with the x's of the particular products concerned. For example,

$$
\det \begin{pmatrix} a_{11} + x_1 & a_{12} & a_{13} \\ a_{21} & a_{22} + x_2 & a_{23} \\ a_{31} & a_{32} & a_{33} + x_3 \end{pmatrix}
$$

$$
= x_1 x_2 x_3 + x_1 x_2 a_{33} + x_2 x_3 a_{11} + x_3 x_1 a_{22} + x_1 \det \begin{pmatrix} a_{22} & a_{23} \\ a_{32} & a_{33} \end{pmatrix}
$$

$$
+ x_2 \det \begin{pmatrix} a_{11} & a_{13} \\ a_{31} & a_{33} \end{pmatrix} + x_3 \det \begin{pmatrix} a_{11} & a_{12} \\ a_{21} & a_{22} \end{pmatrix} + \det(\mathbf{A}).
$$

When $x_1 = x_2 = \cdots = x_n = x$, we have $\det(\mathbf{B}) = \sum_{i=0}^{n} x^{n-i} s_i(\mathbf{A})$, where $s_i(\mathbf{A})$ is the sum of all the principal minors of order i of $\mathbf{A}$. We define $s_0(\mathbf{A}) \equiv 1$ and note that $s_n = \det(\mathbf{A})$.

We can obtain an expansion of $\det(\mathbf{A})$ by its diagonal elements by setting $a_{ii} = 0$ and $x_i = a_{ii}$ for $i = 1, 2, \ldots, n$. Such an expansion is particularly useful when many of the principal minors are zero. We note that

$$
\det(\mathbf{A} - \lambda \mathbf{I}_n) = \sum_{i=0}^{n} (-\lambda)^{n-i} s_i(\mathbf{A}),
$$

which leads to the characteristic equation $\det(\lambda \mathbf{I}_n - \mathbf{A}) = (-1)^n \det(\mathbf{A} - \lambda \mathbf{I}_n)$.

4.44. (Expanding by m Rows—Laplace Expansion) The Laplace expansion of $\det(\mathbf{A})$, where $\mathbf{A}$ is $n \times n$, can be obtained as follows. Firstly, consider any m ($m \geq 1$) rows of $\mathbf{A}$. They contain $R = \binom{n}{m}$ minors of order m. Secondly, multiply each of these minors, $\det(\mathbf{A}_r)$ say ($r = 1, 2, \ldots, R$), by the determinant of the complement of $\mathbf{A}_r$, $\det(\mathbf{B}_{n-r})$ say, and by a sign factor. Here the complementary minor of $\mathbf{A}_r$ is the $(n-m)$th-order minor derived from $\mathbf{A}$ by deleting the m rows and columns containing $\mathbf{A}_r$. The sign factor is $(-1)^{a_r}$, where a_r is the sum of the subscripts of the diagonal elements of $\mathbf{A}_r$. Then $\det(\mathbf{A})$ is the sum of such products, namely

$$
\det(\mathbf{A}) = \sum_{r=1}^{R} (-1)^{a_r} \det(\mathbf{A}_r) \det(\mathbf{B}_{n-r}).
$$

For example, expanding by rows 2 and 3 we have

$$
\det \begin{pmatrix} a_{11} & a_{12} & a_{13} & a_{14} \\ a_{21} & a_{22} & a_{23} & a_{24} \\ a_{31} & a_{32} & a_{33} & a_{34} \\ a_{41} & a_{42} & a_{43} & a_{44} \end{pmatrix}
$$

$$
= \det \begin{pmatrix} a_{21} & a_{22} \\ a_{31} & a_{32} \end{pmatrix} \cdot \det \begin{pmatrix} a_{13} & a_{14} \\ a_{43} & a_{44} \end{pmatrix} (-1)^{2+1+3+2}
$$

$$
+ \det \begin{pmatrix} a_{21} & a_{23} \\ a_{31} & a_{33} \end{pmatrix} \cdot \det \begin{pmatrix} a_{12} & a_{14} \\ a_{42} & a_{44} \end{pmatrix} (-1)^{2+1+3+3} + \cdots +
$$

$$
+ \det \begin{pmatrix} a_{23} & a_{24} \\ a_{33} & a_{34} \end{pmatrix} \cdot \det \begin{pmatrix} a_{11} & a_{12} \\ a_{41} & a_{42} \end{pmatrix} (-1)^{2+3+3+4}.
$$

Further extensions of the Laplace expansion method are available, many of which are named after their originators—for example, Cauchy, Binet–Cauchy (Harville [1997: 200–202] and Rao and Rao [1997: 149]), and Jacobi.

A number of other expansions are available. For example, if we are interested in relating minors of submatrices, we can use *Sylvester's Determinantal Identity* (Rao and Rao [1998: 151–153]). If $\mathbf{C} = \mathbf{A}_{n \times p}\mathbf{B}_{p \times n}$, we can expand det $\mathbf{C}$ in terms of the sum of products of a minor of $\mathbf{A}$ times a minor of $\mathbf{B}$ using the *Cauchy–Binet formula* (Rao and Bhimasankaram [2000: 238] and Rao and Rao [1997: 140]).

4.45. Given the skew-symmetric matrix

$$\mathbf{A} = \begin{pmatrix} 0 & a & b & c \\ -a & 0 & d & e \\ -b & -d & 0 & f \\ -c & -e & -f & 0 \end{pmatrix},$$

then $\det(\mathbf{A}) = (af - be + cd)^2$.

4.46.

$$\det \begin{pmatrix} a_{11}^2 & 2a_{11}a_{12} & a_{12}^2 \\ a_{11}a_{21} & a_{11}a_{22} + a_{12}a_{21} & a_{12}a_{22} \\ a_{21}^2 & 2a_{21}a_{22} & a_{22}^2 \end{pmatrix} = 4 \left\{ \det \begin{pmatrix} a_{11} & a_{12} \\ a_{21} & a_{22} \end{pmatrix} \right\}^2.$$

The above matrix occurs in genetics.

Proofs.   Section 4.4.4.

  4.43. Searle [1982: 106].

  4.44. Harville [1997: section 13.8] and Searle [1982: 109].

  4.46. Quoted by Searle [1982: 114].

4.5 PERMANENTS

Definition **4.8.** Let $\mathbf{A}$ be an $n \times n$ real matrix. The *permanent* of $\mathbf{A}$, denoted by per($\mathbf{A}$), is defined by

$$\text{per}(\mathbf{A}) = \sum_{\pi} \prod_{i=1}^{n} a_{i\pi(i)},$$

where π is a permutation of the ordered set $\{1, 2, \ldots, n\}$ and $\pi(i)$ is the ith member of the permutation π; the summation extends over all permutations π. This definition may be compared with the definition of a determinant. There $\prod_{i=1}^{n} a_{i\pi(i)}$ is multiplied by either $+1$ or -1 depending on whether π is an even or odd permutation. Note that per($\mathbf{A}$) can also be defined for an $m \times n$ matrix. For general references to permanents see Wanless [2007] and Minc [1978, 1987], and for an emphasis on applications in probability and statistics see Bapat [1990]. Permanents can be used to prove a number of properties shared by doubly stochastic matrices.

Let $\mathcal{I}$ and $\mathcal{J}$ be ordered subsets of $\{1, 2, \ldots, n\}$, each with $\#\mathcal{I}$ and $\#\mathcal{J}$ elements, respectively (the cardinality of each set), and define the $\#\mathcal{I} \times \#\mathcal{J}$ submatrix $\mathbf{A}_{(\mathcal{I},\mathcal{J})} = (a_{ij})_{i \in \mathcal{I}, j \in \mathcal{J}}$. We must have $\#\mathcal{I} = \#\mathcal{J}$ for $\mathrm{per}(\mathbf{A}_{(\mathcal{I},\mathcal{J})})$ to make sense.

4.47. Let $\mathbf{A}$ be an $n \times n$ matrix.

(a) $\mathrm{per}(\mathbf{A}) = \mathrm{per}(\mathbf{A}')$.

(b) $\mathrm{per}(\mathbf{A})$ admits a Laplace expansion along any row or column. Thus, if $\mathbf{A}_{ij}$ denotes the submatrix of $\mathbf{A}$ obtained by deleting the ith row and the jth column of $\mathbf{A}$, then

$$\mathrm{per}(\mathbf{A}) = \sum_{i=1}^{n} a_{ij} \mathrm{per} \mathbf{A}_{ij}.$$

4.48. $\mathrm{per}(\mathbf{A}_{(\mathcal{I},\mathcal{J})})$ does not depend on the order of elements in $\mathcal{I}$ or $\mathcal{J}$.

4.49. If $\mathbf{A}$ is an $n \times n$ non-negative matrix (i.e., $a_{ij} \geq 0$ for all i, j) and $\#\mathcal{I} + \#\mathcal{J} = n$, then

$$\mathrm{per}(\mathbf{A}) \geq \mathrm{per}(\mathbf{A}(\mathcal{I}, \mathcal{J}^c)) \times \mathrm{per}(\mathbf{A}(\mathcal{I}^c, \mathcal{J})),$$

where $\mathcal{I}^c$ is the complement of $\mathcal{I}$,, and so on.

4.50. If $\mathbf{A}$ is an $n \times n$ non-negative matrix, then $\mathrm{per}(\mathbf{A}) = 0$ if and only if there are subsets $\mathcal{I}$ and $\mathcal{J}$ such that

$$\#\mathcal{I} + \#\mathcal{J} \geq n + 1 \quad \text{and} \quad \mathbf{A}(\mathcal{I}, \mathcal{J}) = \mathbf{0}.$$

4.51. Let $\mathbf{A}$ and $\mathbf{B}$ be $n \times n$ complex matrices. Then
(a)

$$|\mathrm{per}(\mathbf{AB})|^2 \leq \mathrm{per}(\mathbf{AA}^*)\mathrm{per}(\mathbf{B}^*\mathbf{B}),$$

with equality if and only if one of the following occurs:

(1) A row of $\mathbf{A}$ or a column of $\mathbf{B}$ is $\mathbf{0}$,

(2) No row of $\mathbf{A}$ and no column of $\mathbf{B}$ consists of $\mathbf{0}$, and $\mathbf{A}^* = \mathbf{BD\Pi}$, where $\mathbf{D}$ is a diagonal matrix and $\mathbf{\Pi}$ is a permutation matrix.

(b) $|\mathrm{per}(\mathbf{A})|^2 \leq \mathrm{per}(\mathbf{AA}^*)$ and $|\mathrm{per}(\mathbf{A})|^2 \leq \mathrm{per}(\mathbf{A}^*\mathbf{A})$.

(c) If $\mathbf{A}$ is Hermitian non-negative definite, then

(i) $\mathrm{per}(\mathbf{A}) \leq n^{-1} \mathrm{trace}(\mathbf{A}^n)$.

(ii) $\det \mathbf{A} \leq \mathrm{per}(\mathbf{A})$.

Proofs.   Section 4.5.

4.47b. Use (4.44) with $r = 1$ and ignore the signs.

4.48. Follows from the definition.

4.49. Quoted by Rao and Rao [1998: 312].

4.50. Quoted by Rao and Rao [1998: 312].

4.51. Marcus and Minc [1964: 118, 120]. For (b), set $\mathbf{B} = \mathbf{I}$ then $\mathbf{A} = \mathbf{I}$ in (a).

4.6 NORMS

Norms, both for vectors and matrices, are used for measuring distance in vector spaces and for providing a measure of how close one matrix is to another. They can therefore be used for finding the best approximation of a matrix in a given class of matrices by a matrix in another class (e.g., of lower rank). They can also be used for investigating limits of matrix sequences and series. Norms, therefore, have a role to play in statistics in the areas of inequalities, optimization, matrix approximation, matrix analysis, and numerical analysis.

4.6.1 Vector Norms

***Definition* 4.9.** A *vector norm* on a real or complex vector space $\mathcal{V}$ is a real-valued function $\| \cdot \|$ satisfying the following three conditions.

(1) $\|\mathbf{x}\| \geq 0$ for all $\mathbf{x} \in \mathcal{V}$, and $\|\mathbf{x}\| = 0$ implies that $\mathbf{x} = \mathbf{0}$ (positive definite property).

(2) $\|\alpha \mathbf{x}\| = |\alpha| \cdot \|\mathbf{x}\|$ for all $\alpha \in \mathbb{F}$ and all $\mathbf{x} \in \mathcal{V}$ (scalar multiplication).

(3) $\|\mathbf{x} + \mathbf{y}\| \leq \|\mathbf{x}\| + \|\mathbf{y}\|$ (triangle inequality).

A vector norm is said to be *unitarily invariant* if $\|\mathbf{U}\mathbf{x}\| = \|\mathbf{x}\|$ for all $\mathbf{x} \in \mathbb{C}^n$ and all $n \times n$ unitary matrices $\mathbf{U}$.
   If (1) above is replaced by

(1a) $\|\mathbf{x}\| \geq 0$ for all $\mathbf{x} \in \mathcal{V}$,

then $\| \cdot \|$ is called a *vector seminorm*.

4.52. The following hold for both a norm and a seminorm for any $\mathbf{x}, \mathbf{y} \in \mathcal{V}$.

(a) $\| - \mathbf{x}\| = \|\mathbf{x}\|$.

(b) $| \|\mathbf{x}\| - \|\mathbf{y}\| | \leq \|\mathbf{x} - \mathbf{y}\|$.

(c) $\|\mathbf{x} - \mathbf{y}\| \leq \|\mathbf{x}\| + \|\mathbf{y}\|$.

4.53. Every vector norm on $\mathbb{R}^n$ or $\mathbb{C}^n$ is uniformly continuous.

4.54. For finite-dimensional real or complex vector spaces, all vector norms are equivalent in the sense that if $\| \cdot \|_\alpha$ and $\| \cdot \|_\beta$ are two vector norms, then there exist positive constants c_1 and c_2 such that

$$c_1\|\mathbf{x}\|_\alpha \leq \|\mathbf{x}\|_\beta \leq c_2\|\mathbf{x}\|_\alpha$$

for all $\mathbf{x}$ (cf. (4.56) below for some examples).

***Definition* 4.10.** If p is a real number with $p \geq 1$ and $\mathbf{x}$ is an $n \times 1$ vector, then

$$\|\mathbf{x}\|_p = \left(\sum_{i=1}^{n} |x_i|^p \right)^{1/p}$$

is a norm on $\mathbb{R}^n$ or $\mathbb{C}^n$ called the L_p *norm*. Letting $p \to \infty$, we find that

$$\|\mathbf{x}\|_\infty = \max_{1 \le i \le n} |x_i|$$

is also a norm called the L_∞ *norm*. The norms most commonly used are the L_1, L_2, and L_∞ norms. In particular, the so-called *Euclidean norm* L_2 is used to define the length of a vector in $\mathbb{R}^n$ or $\mathbb{C}^n$. The function $\|\mathbf{x}\|_p$ is not a norm for $0 < p < 1$.

4.55. The L_p norm $(p \ge 1)$ is a vector norm.

4.56. For all $\mathbf{x} \in \mathbb{C}^n$ we have:

(a) $n^{-1/2}\|\mathbf{x}\|_1 \le \|\mathbf{x}\|_2 \le \|\mathbf{x}\|_1$.

(b) $\|\mathbf{x}\|_\infty \le \|\mathbf{x}\|_2 \le n^{1/2}\|\mathbf{x}\|_\infty$.

(c) $\|\mathbf{x}\|_2 \le \|\mathbf{x}\|_1 \le n^{1/2}\|\mathbf{x}\|_2$.

(d) $\|\mathbf{x}\|_\infty \le \|\mathbf{x}\|_1 \le n\|\mathbf{x}\|_\infty$.

4.57. Every inner product induces a norm; we simply put $\|\mathbf{x}\| = \langle \mathbf{x}, \mathbf{x} \rangle^{1/2}$. However, there are norms not induced by an inner product as in (4.58c) below.

4.58. (Parallelogram Law)

(a) The norm induced by an inner product on a vector space (cf. 4.57) has the property
$$\|\mathbf{x} + \mathbf{y}\|^2 + \|\mathbf{x} - \mathbf{y}\|^2 = 2\|\mathbf{x}\|^2 + 2\|\mathbf{y}\|^2.$$

(b) Conversely, any norm on a real or complex vector space satisfying the above equation is induced by an inner product, namely
$$\langle \mathbf{x}, \mathbf{y} \rangle = \tfrac{1}{2}(\|\mathbf{x} + \mathbf{y}\|^2 - \|\mathbf{x}\|^2 - \|\mathbf{y}\|^2).$$

An alternative inner product that can be used is
$$\langle \mathbf{x}, \mathbf{y} \rangle = \tfrac{1}{4}(\|\mathbf{x} + \mathbf{y}\|^2 - \|\mathbf{x} - \mathbf{y}\|^2).$$

(c) If the parallelogram rule does not hold, then there is no inner product that induces the norm.

(d) Let $1 \le p \le \infty$. Then the L_p norm satisfies (a) if and only if $p = 2$. Thus L_1, for example, does not satisfy (a) so it cannot be induced by an inner product.

4.59. The sum of two vector (semi)norms is a vector (semi)norm, and any positive multiple of a vector (semi)norm is a also a vector (semi)norm.

4.60. Let $\| \cdot \|_\alpha$ and $\| \cdot \|_\beta$ be vector norms on a real or complex vector space $\mathcal{V}$. The function $\| \cdot \|$ defined by

$$\|\mathbf{x}\| = \max\{\|\mathbf{x}\|_\alpha, \|\mathbf{x}\|_\beta\}$$

is a vector norm.

4.61. (Continuity) If $\{\mathbf{x}_n\}$ and $\{\mathbf{y}_n\}$ are sequences of vectors in an inner product space such that $\|\mathbf{x}_n - \mathbf{x}\| \to 0$ and $\|\mathbf{y}_n - \mathbf{y}\| \to 0$ as $n \to \infty$, then

(a) $\|\mathbf{x}_n\| \to \|\mathbf{x}\|$ and $\|\mathbf{y}_n\| \to \|\mathbf{y}\|$.

(b) $\langle \mathbf{x}_n, \mathbf{y}_n \rangle \to \langle \mathbf{x}, \mathbf{y} \rangle$, if $\|\mathbf{x}\| < \infty$ and $\|\mathbf{y}\| < \infty$.

4.62. If $\| \cdot \|$ is a vector norm on $\mathbb{C}^n$ and $\mathbf{R}$ is a nonsingular $n \times n$ matrix, then $\| \cdot \|_{\mathbf{R}}$ defined by

$$\|\mathbf{x}\|_{\mathbf{R}} = \|\mathbf{R}\mathbf{x}\|, \quad \mathbf{x} \in \mathbb{C}^n$$

is also a vector norm on $\mathbb{C}^n$.

Proofs. Section 4.6.1.

4.52a. By (2) of Definition 4.9.

4.52b. Horn and Johnson [1985: 260].

4.52c. By (3) of Definition 4.9 with $\mathbf{y}$ replaced by $-\mathbf{y}$.

4.53–4.54. Horn and Johnson [1985: 271, 272].

4.55. Conditions (1) and (2) of Definition 4.9 are readily verified (cf. Gentle [1998: 71]), and (3) follows from Minkowski's inequality (12.17a).

4.56. (b)–(d) are quoted by Golub and Van Loan [1996: 53], while (a) and (b) are proved by Rao and Bhimasankaram [2000: 258; see the solution to exercise 14].

4.57. Rao and Bhimasankaram [2000: 256].

4.58a. This follows by simply expanding $\langle \mathbf{x} + \mathbf{y}, \mathbf{x} + \mathbf{y} \rangle$, and so on.

4.58b. Horn and Johnson [1985: 263, exercise 10] and Meyer [2000a: 290–292].

4.58c. Abadir and Magnus [2005: 64].

4.58d. Rao and Bhimasankaram [2000: 258; see the solution to exercise 9].

4.59–4.60. Horn and Johnson [1985: 268].

4.61. Abadir and Magnus [2005: 65].

4.62. Horn and Johnson [1985: 268].

4.6.2 Matrix Norms

Definition **4.11.** We can interpret the word "vector" as simply an element of a vector space. In this case, an $m \times n$ complex matrix $\mathbf{A} = (a_{ij})$ is simply an element of the space of $m \times n$ complex matrices. Alternatively, this space can also be identified with the vector space $\mathbb{C}^{mn}$ by arranging the entries of each $\mathbf{A}$ as an mn-tuple in some order (e.g., vec $\mathbf{A}$). When a norm applied to vec $\mathbf{A}$ satisfies the conditions of a vector norm, we call the norm a *generalized matrix norm*. Some examples follow.

4.63. (Generalized Matrix Norms)   Let $\mathbf{A}$ be an $m \times n$ matrix. Then the following are generalized matrix norms:

(a) $\|\mathbf{A}\|_\infty = \max_{1 \leq i \leq m, 1 \leq j \leq n} |a_{ij}|.$

(b) $\|\mathbf{A}\|_F = [\mathrm{trace}(\mathbf{A}^*\mathbf{A})]^{1/2} = (\sum_{i=1}^m \sum_{j=1}^n |a_{ij}|^2)^{1/2}$, the Frobenius norm. (We use a subscript F instead of $F = 2$ to avoid confusion later in dealing with matrix norms instead of generalized ones.)

(c) $\|\mathbf{A}\|_p = (\sum_{i=1}^m \sum_{j=1}^n |a_{ij}|^p)^{1/p} \ (p \geq 1).$

***Definition* 4.12.** (Induced Norms)   Given the vector norm $\|\cdot\|_v$, the generalized matrix norm *induced* by $\|\cdot\|_v$ for the $m \times n$ matrix $\mathbf{A}$ is defined by

$$\|\mathbf{A}\|_{v,in} = \sup_{\mathbf{x} \neq 0} \frac{\|\mathbf{A}\mathbf{x}\|_v}{\|\mathbf{x}\|_v} = \sup_{\mathbf{x} \neq 0} \left\| \mathbf{A}\left(\frac{\mathbf{x}}{\|\mathbf{x}\|_v}\right) \right\| = \sup_{\|\mathbf{x}\|_v = 1} \|\mathbf{A}\mathbf{x}\|_v,$$

where $\mathbf{x} \in \mathbb{C}^n$. As noted by Horn and Johnson [1985: 292], we can replace "sup" by "max" in the above definition. The most common vector norms are the L_p norms with $p \geq 1$. Note that $\|\cdot\|_F$ is not an induced norm, and is not to be confused with $\|\cdot\|_{2,in}$ (cf. 4.66b).

4.64. The induced norm $\|\mathbf{A}\|_{v,in}$ is a generalized matrix norm as condition (3) of Definition 4.9 in Section 4.6.1 holds.

4.65. If $\mathbf{A}$ is $m \times n$ and $\mathbf{B}$ is $n \times q$, then for an L_p vector norm $(p > 1)$

$$\|\mathbf{A}\mathbf{B}\|_{p,in} \leq \|\mathbf{A}\|_{p,in}\|\mathbf{B}\|_{p,in}.$$

This result does not hold for every $\|\cdot\|_{v,in}$. Golub and Van Loan [1996: 55], in quoting the above result, note that it represents a relationship between three different norms defined on $\mathbb{R}^{m \times q}$, $\mathbb{R}^{m \times n}$ and $\mathbb{R}^{n \times q}$, respectively. They also call the above norm a matrix norm, which it is for square matrices (see Definition 4.13 below).

4.66. Let $\mathbf{A}$ be $m \times n$. Then following are induced norms based on L_p vector norms.

(a) $\|\mathbf{A}\|_{1,in} = \max_{1 \leq j \leq n} \sum_{i=1}^m |a_{ij}|.$

(b) $\|\mathbf{A}\|_{2,in} = [\lambda_{\max}(\mathbf{A}^*\mathbf{A})]^{1/2} = \sigma_{\max}(\mathbf{A})$, where $\lambda_{\max}$ is the maximum eigenvalue of $\mathbf{A}^*\mathbf{A}$ and $\sigma_{\max}$ is the maximum singular value.

(c) $\|\mathbf{A}\|_{\infty,in} = \max_{1 \leq i \leq m} \sum_{j=1}^n |a_{ij}|.$

4.67. If $\mathbf{A}$ is $m \times n$ then:

(a) $\|\mathbf{A}\|_{2,in} \leq \|\mathbf{A}\|_F \leq n^{1/2}\|\mathbf{A}\|_{2,in}.$ (See also (4.82) below for square matrices.)

(b) $\max_{1 \leq i \leq m, 1 \leq j \leq n} |a_{ij}| \leq |\mathbf{A}\|_{2,in} \leq (mn)^{1/2} \max_{1 \leq i \leq m, 1 \leq j \leq n} |a_{ij}|.$

(c) $m^{-1/2}\|\mathbf{A}\|_{1,in} \leq \|\mathbf{A}\|_{2,in} \leq n^{1/2}\|\mathbf{A}\|_{1,in}.$

(d) $n^{-1/2}\|\mathbf{A}\|_{\infty,in} \leq \|\mathbf{A}\|_{2,in} \leq m^{1/2}\|\mathbf{A}\|_{\infty,in}.$

(e) $\|\mathbf{A}\|_{2,in} \leq \sqrt{\|\mathbf{A}\|_{1,in}\|\mathbf{A}\|_{\infty,in}}.$

The above bounds on the norm $\|\mathbf{A}\|_{2,in}$ are useful, because this norm is more difficult to compute than either $\|\mathbf{A}\|_{1,in}$ or $\|\mathbf{A}\|_{\infty,in}.$

***Definition* 4.13.** Let $\mathcal{V}$ be the vector space of $n \times n$ complex matrices. If $\mathbf{A} \in \mathcal{V}$, then the *matrix norm* of $\mathbf{A}$, denoted by $|||\mathbf{A}|||$, is any real-valued non-negative function of $\mathbf{A}$ satisfying the following conditions.

(1) $|||\mathbf{A}||| \geq 0$ and $|||\mathbf{A}||| = 0$ if and only if $\mathbf{A} = \mathbf{0}$.

(2) $|||c\mathbf{A}||| = |c| \cdot |||\mathbf{A}|||$, where c is any scalar and $|c|$ is its modulus.

(3) If $\mathbf{B} \in \mathcal{V}$ then $|||\mathbf{A} + \mathbf{B}||| \leq |||\mathbf{A}||| + |||\mathbf{B}|||$.

(4) If $\mathbf{C} \in \mathcal{V}$ then $|||\mathbf{AC}||| \leq |||\mathbf{A}||| \cdot |||\mathbf{C}|||$.

Note that the first three conditions are those of a generalized matrix norm (and of a vector norm), which can be applied to any $m \times n$ matrix. However, condition (4) applies to square matrices only. For a brief introduction to matrix norms see Meyer [2000a: section 5.2].

4.68. Let $||| \cdot |||$ be any matrix norm and $\mathbf{A}$ any $n \times n$ matrix.

(a) (i) $\rho(\mathbf{A}) \leq |||\mathbf{A}|||$, where ρ is the spectral radius.

   (ii) $\rho(\mathbf{A}) = \lim_{k \to \infty}(|||\mathbf{A}^k|||)^{1/k}.$

(b) (i) From $\mathbf{AI}_n = \mathbf{A}$ and Definition 4.13(4) we have $|||\mathbf{I}_n||| \geq 1$.

   (ii) Repeated use of Definition 4.13(4) give us $|||\mathbf{A}^k||| \leq (|||\mathbf{A}|||)^k$ (k a positive integer).

   (iii) Using $\mathbf{AA}^{-1} = \mathbf{I}_n$ and Definition 4.13(4) gives us $|||\mathbf{A}^{-1}||| \geq (|||\mathbf{A}|||)^{-1}.$

(c) From $\mathbf{A} = (\mathbf{A} - \mathbf{B}) + \mathbf{B}$, Definition 4.13(3), and interchanging the roles of $\mathbf{A}$ and $\mathbf{B}$, we have $|\ |||\mathbf{A}||| - |||\mathbf{B}|||\ | \leq |||\mathbf{A} - \mathbf{B}|||$.

(d) $|a_{ij}| \leq \theta|||\mathbf{A}|||$ for all i and j, where $\theta = \max_{1 \leq i \leq n, 1 \leq j \leq n} |||\mathbf{E}_{ij}|||$, and $\mathbf{E}_{ij}$ is an $n \times n$ matrix with 1 in the i,jth position and zeros elsewhere.

(e) If $|||\mathbf{A}|||_S = |||\mathbf{S}^{-1}\mathbf{AS}|||$ for all nonsingular $n \times n$ matrices $\mathbf{S}$, then $|||\mathbf{A}|||_S$ is a matrix norm.

4.69. Let $\mathbf{A}$ be an $n \times n$ matrix.

(a) $\|\mathbf{A}\|_p = (\sum_{i=1}^{n}\sum_{j=1}^{n}(|a_{ij}|^p)^{1/p}$ is a matrix norm for $1 \leq p \leq 2$. When $p = 2$ we use the notation $|||\mathbf{A}|||_F$.

(b) If $n \geq 2$, $\|\mathbf{A}\| = n(\max_{1 \leq i \leq n, 1 \leq j \leq n} |a_{ij}|)$ is a matrix norm, but $\|\mathbf{A}\|_\infty$ of (4.63a) is not, though it is a generalized matrix norm.

4.70. Result (4.69b) above can be generalized. For every generalized matrix norm $\|\mathbf{A}\|_\alpha$, where $\mathbf{A}$ is $n \times n$, there is a finite positive constant c_α which depends on the norm such that $c_\alpha\|\mathbf{A}\|_\alpha$ is a matrix norm.

4.71. An $n \times n$ matrix $\mathbf{A}$ is nonsingular if there is a matrix norm $||| \cdot |||$ such that $||| \mathbf{I}_n - \mathbf{A} ||| < 1$. In this case

$$\mathbf{A}^{-1} = \sum_{k=0}^{\infty} (\mathbf{I}_n - \mathbf{A})^k.$$

4.72. Given $\epsilon > 0$, there exists a matrix norm $||| \cdot |||$ such that

$$\rho(\mathbf{A}) \leq ||| \mathbf{A} ||| < \rho(\mathbf{A}) + \epsilon,$$

where ρ is the spectral radius (see also 4.68a).

***Definition* 4.14.** The generalized matrix norm induced by the vector norm $|| \cdot ||_v$ for the square matrix $\mathbf{A}$ is a matrix norm because it satisfies the four conditions of Definition 4.13. We call it the *induced matrix norm* and we denote it by $||| \mathbf{A} |||_{v,in}$ For further discussion of this norm see Horn and Johnson [1985: 292–295] and Rao and Rao [1998: 367–8].

4.73. For $n \times n$ matrices, the induced matrix norm has the following properties.

(a) $||| \mathbf{I}_n |||_{v,in} = 1$.

(b) $|| \mathbf{A}\mathbf{x} ||_v \leq ||| \mathbf{A} |||_{v,in} \cdot || \mathbf{x} ||_v$.

(c) $||| \alpha \mathbf{A} |||_{v,in} = |\alpha| \cdot ||| \mathbf{A} |||_{v,in}$.

(d) $||| - \mathbf{A} |||_{v,in} = ||| \mathbf{A} |||_{v,in}$.

(e) $||| \mathbf{I}_n + \mathbf{A} |||_{v,in} \leq ||| \mathbf{I}_n |||_{v,in} + ||| \mathbf{A} |||_{v,in} = 1 + ||| \mathbf{A} |||_{v,in}$

(f) $||| \mathbf{I}_n - \mathbf{A} |||_{v,in} \leq 1 + ||| \mathbf{A} |||_{v,in}$.

(g) Suppose $||| \mathbf{A} |||_{v,in} < 1$.

    (i) $\mathbf{B} = \mathbf{I}_n - \mathbf{A}$ and $\mathbf{I}_n + \mathbf{A}$ are nonsingular.

    (ii) From $\mathbf{B}(\mathbf{I}_n - \mathbf{A}) = \mathbf{I}_n$ we can take the norm of $\mathbf{B} = \mathbf{I}_n + \mathbf{B}\mathbf{A}$ using (e) and Definition 4.13(4) to get

$$\frac{1}{1 + ||| \mathbf{A} |||_{v,in}} \leq ||| (\mathbf{I}_n - \mathbf{A})^{-1} |||_{v,in} \leq \frac{1}{1 - ||| \mathbf{A} |||_{v,in}}.$$

    By replacing $\mathbf{A}$ by $-\mathbf{A}$ and using (c) above we see that the same bounds apply for $||| (\mathbf{I}_n + \mathbf{A})^{-1} |||_{v,in}$.

4.74. Let $\mathbf{A}$ be an $n \times n$ matrix. The matrix norm induced by an L_p vector norm is given by

$$||| \mathbf{A} |||_{p,in} = \max_{|| \mathbf{x} ||_p = 1} || \mathbf{A}\mathbf{x} ||_p, \quad p \geq 1.$$

Setting $p = 1, 2, \infty$, we have:

(a) $||| \mathbf{A} |||_{1,in} = \max_{1 \leq j \leq n} \sum_{i=1}^{n} |a_{ij}|$.

(b) $||| \mathbf{A} |||_{2,in} = [\lambda_{\max}(\mathbf{A}^*\mathbf{A})]^{1/2} = \sigma_{\max}(\mathbf{A})$, where $\lambda_{\max}$ is the maximum eigenvalue of $\mathbf{A}^*\mathbf{A}$ and $\sigma_{\max}$ is the maximum singular value. We note that $\lambda_{\max}$ is real and non-negative as $\mathbf{A}^*\mathbf{A}$ is Hermitian and non-negative definite. We note that $\lambda_{\max} = \rho(\mathbf{A}^*\mathbf{A})$, where ρ is the spectral radius. When $\mathbf{A}$ is Hermitian, $|||\mathbf{A}|||_{2,in} = [\rho(\mathbf{A}^2)]^{1/2}$, which reduces to $\rho(\mathbf{A})$ when $\mathbf{A}$ is also non-negative definite. For further properties of this induced norm see Meyer [2000a: 281-283].

This matrix norm is also called the *spectral matrix norm*.

(c) $|||\mathbf{A}|||_{\infty,in} = \max_{1 \le i \le n} \sum_{j=1}^{n} |a_{ij}|$.

The inequalities given in (4.67) apply to the above matrix norms by setting $m = n$.

4.75. If $\mathbf{A}$ and $\mathbf{B}$ are non-negative definite $n \times n$ matrices, then:

(a) $|||\mathbf{A}^s\mathbf{B}^s|||_{2,in} \le |||\mathbf{AB}|||_{2,in}^s$ for $0 \le s \le 1$.

(b) If $|||\mathbf{AB}|||_{2,in} \le 1$, then $|||\mathbf{A}^s\mathbf{B}^s|||_{2,in} \le 1$ for $0 \le s \le 1$.

(c) $|||\mathbf{AB}|||_{2,in}^t \le |||\mathbf{A}^t\mathbf{B}^t|||_{2,in}$ for $t \ge 1$.

4.76. Let $\| \cdot \|_\alpha$ and $\| \cdot \|_\beta$ be two given vector norms on $\mathbb{C}^n$, and let $||| \cdot |||_{\alpha,in}$ and $||| \cdot |||_{\beta,in}$ denote the respective induced matrix norms on the space $\mathcal{V}$ of $n \times n$ matrices.

(a) Define
$$R_{\alpha\beta} = \max_{\mathbf{x} \ne 0} \frac{\|\mathbf{x}\|_\alpha}{\|\mathbf{x}\|_\beta} \quad \text{and} \quad R_{\beta\alpha} = \max_{\mathbf{x} \ne 0} \frac{\|\mathbf{x}\|_\beta}{\|\mathbf{x}\|_\alpha}.$$

Then
$$\max_{\mathbf{A} \ne 0} \frac{|||\mathbf{A}|||_{\alpha,in}}{|||\mathbf{A}|||_{\beta,in}} = \max_{\mathbf{A} \ne 0} \frac{|||\mathbf{A}|||_{\beta,in}}{|||\mathbf{A}|||_{\alpha,in}} = R_{\alpha\beta}R_{\beta\alpha}.$$

(b) $|||\mathbf{A}|||_{\alpha,in} = |||\mathbf{A}|||_{\beta,in}$ for all $\mathbf{A} \in \mathcal{V}$ if and only if there is a positive constant c such that $\|\mathbf{x}\|_\alpha = c\|\mathbf{x}\|_\beta$ for all $\mathbf{x} \in \mathbb{C}$.

(c) $|||\mathbf{A}|||_{\alpha,in} \le |||\mathbf{A}|||_{\beta,in}$ for all $\mathbf{A} \in \mathcal{V}$ if and only if $\|\mathbf{A}\|_{\alpha,in} = \|\mathbf{A}\|_{\beta,in}$ for all $\mathbf{A} \in \mathcal{V}$

4.77. If $\mathbf{Q}$ is unitary (or orthogonal), then:

(a) $\|\mathbf{Q}\mathbf{x}\|_2 = \|\mathbf{x}\|_2$.

(b) $|||\mathbf{Q}|||_{2,in} = 1$.

Definition 4.15. A matrix norm $||| \cdot |||$ on the class of $n \times n$ matrices is a *minimal matrix norm* if the only matrix norm $N(\cdot)$ such that $N(\mathbf{A}) \le |||\mathbf{A}|||$ for all $\mathbf{A}$ is $N(\cdot) = ||| \cdot |||$.

4.78. Every induced norm is minimal and every minimal norm is induced.

Definition 4.16. If $\mathbf{A}$ is $m \times n$, the *Frobenius norm* is defined to be

$$\|\mathbf{A}\|_F = (\sum_i \sum_j |a_{ij}|^2)^{1/2} = [\text{trace}(\mathbf{A}^*\mathbf{A})]^{1/2} = \|\mathbf{A}^*\|_F.$$

When $m \ne n$, this norm is a generalized matrix norm, while if $m = n$, it is a matrix norm. However, it is not an induced norm. It is often refered to as the *Euclidean matrix norm* as $|||\mathbf{A}|||_F$, like $|||\mathbf{A}|||_{2,in}$, uses an L_2 vector norm. For this reason, Graybill [1983], for example, uses E, but we shall follow the general trend and use the subscript F to avoid confusion. Harville [1997] refers to the Frobenius norm as the "usual norm." Even when $m \neq n$, the following result shows that the norm satisfies a result like (4) of Definition 4.13.

4.79. If $\mathbf{A}$ is $m \times n$ and $\mathbf{B}$ is $n \times p$, then $\|\mathbf{AB}\|_F \leq \|\mathbf{A}\|_F \cdot \|\mathbf{B}\|_F$.

4.80. If $\mathbf{A}$ is $m \times n$ of rank r with singular values $\sigma_i = \sigma_i(\mathbf{A})$, then

$$\|\mathbf{A}\|_F^2 = \operatorname{trace}(\mathbf{A}^*\mathbf{A}) = \operatorname{trace}(\mathbf{A}\mathbf{A}^*) = \sum_{i=1}^{r} \sigma_i^2.$$

4.81. Given real symmetric $\mathbf{A}$ and real skew-symmetric $\mathbf{B}$, both $n \times n$, then

$$|||\mathbf{A} + \mathbf{B}|||_F^2 = |||\mathbf{A}|||_F^2 + |||\mathbf{B}|||_F^2.$$

4.82. $|||\mathbf{A}|||_{2,in} \leq |||\mathbf{A}|||_F \leq \sqrt{n}|||\mathbf{A}|||_{2,in}$.

4.83. If $\mathbf{A}$ and $\mathbf{U}$ are $n \times n$ and $\mathbf{U}$ is unitary, then

$$|||\mathbf{A}|||_{2,in} = |||\mathbf{U}\mathbf{A}|||_{2,in} = |||\mathbf{A}\mathbf{U}|||_{2,in} = |||\mathbf{U}^*\mathbf{A}\mathbf{U}|||_{2,in}.$$

The above also holds for $||| \cdot |||_F$.

Proofs. Section 4.6.2.

4.63. Rao and Rao [1998: 363]. The results follow by applying (4.55) to vec $\mathbf{A}$.

4.64. We see that $\|\mathbf{Ax} + \mathbf{Ay}\|_v \leq \|\mathbf{Ax}\|_v + \|\mathbf{Ay}\|_v$ implies that $\max \|\mathbf{Ax} + \mathbf{Ay}\|_v \leq \max\{\|\mathbf{Ax}\|_v + \|\mathbf{Ay}\|_v\} \leq \max \|\mathbf{Ax}\|_v + \max \|\mathbf{Ay}\|_v$.

4.65. Quoted by Golub and van Loan [1996: 55].

4.66. Horn and Johnson [1985: 294–295, the proofs hold for $m \neq n$] and Meyer [2000a: 281–284].

4.67. Golub and Van Loan [1996: 56–57].

4.68a(i). Horn and Johnson [1985: 297], Meyer [2000a: 497], and Rao and Rao [1998: 365].

4.68a(ii). Horn and Johnson [1985: 299], Meyer [2000a: 619], and Rao and Rao [1998: 373].

4.68b. Horn and Johnson [1985: 290].

4.68d. Rao and Rao [1998: 365].

4.68e. Horn and Johnson [1985: 296].

4.69a. Graybill [1983: 93, $p = 1$], Horn and Johnson [1985: 291, $p = 1, 2$], and Rao and Rao [1998: 374].

4.69b. Horn and Johnson [1985: 292].

4.70. Horn and Johnson [1985: 323].

4.71. Horn and Johnson [1985: 301]. See also (19.16a) using an infinite series.

4.72. Horn and Johnson [1985: 297] and Rao and Rao [1998: 372].

4.73a–c. Horn and Johnson [1985: 293] and Rao and Bhimasankaram [2000: 259; see the solution to exercise 15].

4.73d. Since $\|\mathbf{Ax}\|_v = \| - \mathbf{Ax}\|_v$.

4.73e. From $\sup(a + b) \le \sup a + \sup b$.

4.73f. Follows from (d).

4.73g. Rao and Bhimasankaram [2000: 259; see the solution to exercise 15].

4.74a. Horn and Johnson [1985: 294] and Rao and Rao [1998: 370].

4.74b. Rao and Rao [1998: 371].

4.74c. Horn and Johnson [1985: 295] and Rao and Rao [1998: 368–369].

4.75. Bhatia [1997: 255–256].

4.76. Horn and Johnson [1985: 303–305, further results are given there in section 5.6].

4.77. Gentle [2000: 73].

4.78. Horn and Johnson [1985: 306].

4.79. Harville [1997: 432].

4.80. From the singular value decomposition of $\mathbf{A}$, σ_i^2 is the ith ordered eigenvalue of $\mathbf{AA}^*$ and the trace is the sum of the eigenvalues.

4.81. Rao and Rao [1998: 390].

4.82. Since $\mathbf{A}^*\mathbf{A}$ is non-negative definite it has non-negative eigenvalues λ_i and trace($\mathbf{A}^*\mathbf{A}$) $= \sum_i \lambda_i$. The result then follows from $\lambda_{\max} \le \sum_i \lambda_i \le n\lambda_{max}$ and taking square roots.

4.83. Gentle [1998].

4.6.3 Unitarily Invariant Norms

***Definition* 4.17.** A real-valued function $\| \cdot \|$ on the vector space $\mathcal{V}$ of $m \times n$ complex matrices is said to be a *unitarily invariant (generalized matrix) norm*, and denoted by $\| \cdot \|_{ui}$, if it has the following properties. We shall drop the words "generalized matrix" below.

(1) $\|\mathbf{A}\| \ge 0$ for all $\mathbf{A} \in \mathcal{V}$ and $\|\mathbf{A}\| = 0$ if and only if $\mathbf{A} = \mathbf{0}$.

(2) $\|\alpha\mathbf{A}\| = |\alpha| \|\mathbf{A}\|$ for every $\alpha \in \mathbb{C}$ and $\mathbf{A} \in \mathcal{V}$.

(3) $\|\mathbf{A} + \mathbf{B}\| \leq \|\mathbf{A}\| + \|\mathbf{B}\|$ for all $\mathbf{A}$ and $\mathbf{B}$ in $\mathcal{V}$.

(4) $\|\mathbf{U}\mathbf{A}\mathbf{V}\| = \|\mathbf{A}\|$ for all $\mathbf{A} \in \mathcal{V}$ and unitary matrices $\mathbf{U}$ and $\mathbf{V}$ of orders $m \times m$ and $n \times n$, respectively.

Thus a generalized matrix norm, which satisfies the first three conditions, is unitarily invariant if it satisfies (4) as well. If $m = n$ and $\|\mathbf{AB}\| \leq \|\mathbf{A}\|\|\mathbf{B}\|$ for all $n \times n$ matrices, then $\|\cdot\|$ is a matrix norm that we denote by $\||\cdot\||_{ui}$.

If $\mathcal{V}$ is the space of real matrices, then we use the term *orthogonally invariant norm*.

4.84. Let $\|\cdot\|_{ui}$ be a unitarily invariant norm defined on $\mathcal{V}$, the space of $m \times n$ matrices. Then:

(a) $\|\mathbf{A}\mathbf{B}^*\|_{ui} \leq \sigma_1(\mathbf{A})\|\mathbf{B}\|_{ui}$ for all $\mathbf{A}, \mathbf{B} \in \mathcal{V}$.

(b) If $\mathbf{E}_{11}$ is the matrix with 1 in the $(1,1)$ position and zeros elsewhere, then $\|\mathbf{A}\|_{ui} \geq \sigma_1(\mathbf{A})\|\mathbf{E}_{11}\|_{ui}$ for all $\mathbf{A} \in \mathcal{V}$.

Here $\sigma_1(\mathbf{A})$ is the maximum singular value of of $\mathbf{A}$.

4.85. A unitarily invariant norm $\|\cdot\|_{ui}$ on the space $\mathcal{V}$ of $n \times n$ matrices is a matrix norm if and only if $\|\mathbf{A}\|_{ui} \geq \sigma_1(\mathbf{A})$ $(= \||\mathbf{A}\||_{2,in})$ for all $\mathbf{A} \in \mathcal{V}$. An equivalent condition from (4.84b) above is $\|\mathbf{E}_{11}\|_{ui} \geq 1$. Note that Bhatia [1997: 91] uses the sufficient condition $\|\mathbf{E}_{11}\|_{ui} = 1$ in his definition of matrix norm, which leads to a slightly different norm.

Definition 4.18. We define the term *general square root* of the $m \times n$ complex matrix $\mathbf{A}$ to be the unique non-negative definite matrix $(\mathbf{A}^*\mathbf{A})^{1/2}$ (cf. 10.8), and denote it by $|\mathbf{A}|$.

4.86. Let $\mathbf{A}$ be an $m \times n$ matrix. Since $(\mathbf{A}^*\mathbf{A})^{1/2}$ has the same singular values as $\mathbf{A}$, which are the same as those of $\mathbf{A}^*$ (cf. 16.34d), then from (4.87) below we have

$$\| \, |\mathbf{A}| \, \|_{ui} = \|\mathbf{A}\|_{ui}$$

for all unitarily invariant norms.

4.87. Let $\mathbf{A} = \mathbf{P}\mathbf{\Sigma}\mathbf{Q}^*$ be the singular value decomposition of an $m \times n$ matrix $\mathbf{A}$, where $\mathbf{\Sigma}$ is diagonal and $\mathbf{P}$ and $\mathbf{Q}$ are unitary $m \times m$ and $n \times n$ matrices, respectively. Then

$$\|\mathbf{A}\|_{ui} = \|\mathbf{P}^*\mathbf{P}\mathbf{\Sigma}\mathbf{Q}^*\mathbf{Q}\|_{ui} = \|\mathbf{\Sigma}\|_{ui},$$

which is a function of the singular values of $\mathbf{A}$. The nature of this function is discussed below.

4.88. Let $\mathbf{A}$ be an $m \times n$ matrix.

(a) The Frobenius norm $\|\mathbf{A}\|_F = (\sum_{i=1}^{m}\sum_{j=1}^{n}|a_{ij}|^2)^{1/2}$ and $\|\mathbf{A}\|_{2,in} = \sigma_{max}$ (the maximum singular value of $\mathbf{A}$) are both unitarily invariant generalized matrix norms. When $m = n$ they are both unitarily invariant matrix norms.

(b) $\|\mathbf{A}\|_{2,in}$ is the only unitarily invariant norm that is also an induced norm.

Two other classes of unitarily invariant norms that seem to be of particular interest are the Ky Fan k-norms and the Schatten p-norms (Bhatia [1997: 92] and Horn and Johnson [1985: 441, 445]).

4.89. Suppose $\mathbf{A}$ and $\mathbf{B}$ belong to the vector space $\mathcal{V}$ of $m \times n$ matrices and let $p = \min\{m, n\}$. In order that $\|\mathbf{A}\|_{ui} \leq \|\mathbf{B}\|_{in}$ for every unitarily invariant norm $\|\cdot\|_{ui}$ on $\mathcal{V}$, it is sufficient that

$$\sigma_i(\mathbf{A}) \leq \sigma_i(\mathbf{B}) \quad \text{for all} \quad i = 1, 2, \ldots, p,$$

and it is necessary and sufficient that

$$\sigma_1(\mathbf{A}) + \cdots + \sigma_i(\mathbf{A}) \leq \sigma_1(\mathbf{B}) + \cdots + \sigma_i(\mathbf{B}), \quad i = 1, 2, \ldots, p.$$

We now introduce a function, called a symmetric gauge function, which is intimately related to the unitarily invariant matrix norm. In fact, $\|\mathbf{A}\|$ is a unitarily invariant norm if and only it is a symmetric gauge function of the singular values of $\mathbf{A}$ (cf. (4.87) and (4.92)).

Definition 4.19. A real-valued function ϕ from $\mathbb{R}^n$ to $\mathbb{R}$ is said to be a *symmetric gauge function* if it has the following properties.

(1) $\phi(\mathbf{x}) > 0$ for all $\mathbf{x} \in \mathbb{R}^n$ with $\mathbf{x} \neq \mathbf{0}$.

(2) $\phi(\alpha \mathbf{x}) = |\alpha| \phi(\mathbf{x})$ for all $\mathbf{x} \in \mathbb{R}^n$ and $\alpha \in \mathbb{R}$.

(3) $\phi(\mathbf{x} + \mathbf{y}) \leq \phi(\mathbf{x}) + \phi(\mathbf{y})$ for all $\mathbf{x}$ and $\mathbf{y}$ in $\mathbb{R}^n$.

(4) $\phi(\mathbf{x}_\pi) = \phi(\mathbf{x})$ for all $\mathbf{x} \in \mathbb{R}^n$ and all permutations $\mathbf{x}_\pi$ of the elements of $\mathbf{x}$.

(5) $\phi(\mathbf{Jx}) = \phi(\mathbf{x})$ for all $\mathbf{x} \in \mathbb{R}^n$ and all diagonal matrices $\mathbf{J}$ with diagonal elements $+1$ or -1. This is equivalent to $\phi(\mathbf{x}) = \phi(\text{mod } \mathbf{x})$, where mod $\mathbf{x} = (|x_i|)$.

4.90. The following hold.

(a) From (1) to (3) above, ϕ is a vector norm on $\mathbb{R}^n$.

(b) A symmetric gauge function is continuous.

(c) The sum of two symmetric gauge functions is a symmetric gauge function.

(d) A positive multiple of a symmetric guage function is a symmetric gauge function.

(e) The L_p vector norm $(p \geq 1)$ on $\mathbb{R}^n$ is a symmetric gauge function.

For some examples of symmetric gauge functions, their properties, and some inequalities see Bhatia [1997: chapter 4].

4.91. Let ϕ be a symmetric gauge function, and let $\mathbf{x} = (x_i) \in \mathbb{R}^n$.

(a) If $\mathbf{y} = (y_i) = (p_i x_i)$, where $0 \leq p_i \leq 1$ for all i, then

$$\phi(y_1, \ldots, y_n) \leq \phi(x_1, \ldots, x_n).$$

(b) If $0 \leq x_i \leq y_i$ for all i, then

$$\phi(x_1, \ldots, x_n) \leq \phi(y_1, \ldots, y_n).$$

(c) $k(\max\limits_{1 \leq i \leq n} |x_i|) \leq \phi(x_1, \ldots, x_n) \leq k(\sum\limits_{i=1}^{n} |x_i|)$, where $k = \phi(1, 0, 0, \ldots, 0)$.

4.92. Let $\| \cdot \|_{ui}$ be any unitarily invariant norm on the vector space $\mathcal{V}$ of $m \times n$ real matrices, and let $p = \min\{m, n\}$. For each $\mathbf{x} = (x_i) \in \mathbb{R}^m$ let $\mathbf{X}_{m \times n} = \text{diag}(\mathbf{x})$ and $\phi(x_1, \ldots, x_m) = \|\mathbf{X}\|_{ui}$. Then ϕ is a symmetric gauge function. Thus from (4.87), with Σ replacing $\mathbf{X}$, we have that $\|\mathbf{A}\|_{ui} = \phi(\sigma_1(\mathbf{A}), \sigma_2(\mathbf{A}), \ldots, \sigma_p(\mathbf{A}))$, where ϕ is a symmetric gauge function and $\sigma_i(\mathbf{A})$ is the ith singular value of $\mathbf{A}$.

Conversely, if ϕ is a symmetric gauge function on $\mathbb{R}^p$, then the function defined by

$$\|\mathbf{A}\| = \phi(\sigma_1(\mathbf{A}), \sigma_2(\mathbf{A}), \ldots, \sigma_p(\mathbf{A}))$$

is a unitarily invariant norm on $\mathcal{V}$.

Unitarily invariant norms and gauge functions have been found useful in multivariate analysis in relation to monotone properties of power functions and simultaneous confidence intervals (Mudholkar [1965, 1966] and Wijsmann [1979]).

4.93. The $n \times n$ matrix $\mathbf{A}$ has the same singular values as $\mathbf{A}^*$ so that by (4.86) above, $\|\mathbf{A}\|_{ui} = \|\mathbf{A}^*\|_{ui}$ for all $\mathbf{A}$ and all unitarily invariant norms. Such a norm is called a *self-adjoint norm*.

4.94 (Ky Fan) If ϕ is a symmetric gauge function on $\mathbb{R}^p$ and $\sigma_1 \geq \sigma_2 \geq \cdots \geq \sigma_p \geq 0$ and $\sigma_1' \geq \sigma_2' \geq \cdots \geq \sigma_p' \geq 0$ are two sets of values, then

$$\phi(\sigma_1, \cdots, \sigma_p) \geq \phi(\sigma_1', \cdots, \sigma_p')$$

if and only if

$$\sigma_1 + \ldots + \sigma_i \geq \sigma_1' + \ldots + \sigma_i', \quad i = 1, 2, \ldots, p.$$

Proofs.   Section 4.6.3.

4.84. Horn and Johnson [1991: 206].

4.85. Horn and Johnson [1985: 450; 1991: 211, exercise 3].

4.87. Rao and Rao [1998: 375].

4.88a. Rao and Rao [1998: 376].

4.88b. Horn and Johnson [1985: 308].

4.89. Horn and Johnson [1985: 447] and Rao [1980: 6].

4.90. Rao and Rao [1998: 377–378].

4.91. Rao and Rao [1998: 377–378].

4.92. Horn and Johnson [1985: 438-441; 1991: 210] and Rao and Rao [1998: 378–380].

4.94. Fan [1951].

4.6.4 M, N-Invariant Norms

***Definition* 4.20.** Let $\mathbf{M}$ be a given positive definite $m \times m$ matrix and $\mathbf{N}$ a given $n \times n$ positive definite matrix. A generalized matrix norm on the space $\mathcal{V}$ of $m \times n$ matrices is said to be an *M,N-invariant norm* if, in addition to conditions (1), (2), and (3) of Definition 4.17 in the previous section, it satisfies the following condition

$$\|\mathbf{VAU}\| = \|\mathbf{A}\| \quad \text{for every} \quad \mathbf{A} \in \mathcal{V},$$

and any $m \times m$ matrix $\mathbf{V}$ and any $n \times n$ matrix $\mathbf{U}$ such that $\mathbf{V}^*\mathbf{M}\mathbf{V} = \mathbf{M}$ and $\mathbf{U}^*\mathbf{N}\mathbf{U} = \mathbf{N}$. This norm was introduced by Rao [1979, 1980] to deal with dimension-reducing techniques in multivariate analysis. When $\mathbf{M}$ and $\mathbf{N}$ are the identity matrices, the M, N-invariant norm becomes the unitarily invariant norm.

4.95. Using the above notation, let $\mathbf{M}^{1/2}$, $\mathbf{M}^{-1/2}$, $\mathbf{N}^{1/2}$ and $\mathbf{N}^{-1/2}$ be the respective positive definite square roots of $\mathbf{M}$, $\mathbf{M}^{-1}$, $\mathbf{N}$ and $\mathbf{N}^{-1}$ (cf. 10.32). Then the following hold.

(a) If $\|\mathbf{A}\|_\alpha$ is a unitarily invariant norm of $\mathbf{A}$, then $\|\mathbf{M}^{1/2}\mathbf{A}\mathbf{N}^{1/2}\|_\alpha$ is an M, N-invariant norm of $\mathbf{A}$.

(b) If $\|\mathbf{A}\|_\beta$ is an M, N-invariant norm of $\mathbf{A}$, then $\|\mathbf{M}^{-1/2}\mathbf{A}\mathbf{N}^{-1/2}\|_\beta$ is a unitarily invariant norm of $\mathbf{A}$.

Proofs.   Section 4.6.4.

> 4.95. Rao and Rao [1998: 394–395]. They also give a number of matrix approximations based on the M, N-invariant norm.

4.6.5   Computational Accuracy

An important question in computing is: How do errors both in the data and in the round-off affect the computation of expression—for example, the inverse of a nonsingular matrix? Suppose $\mathbf{A}$ is $n \times n$ and, instead of computing $\mathbf{A}^{-1}$, we actually compute $(\mathbf{A} + \delta\mathbf{A})^{-1}$. Then, assuming that a particular matrix norm of the error $\|\delta\mathbf{A}\|$ is small enough, Horn and Johnson [1985: 335–338] show that if $\|\delta\mathbf{A}\|\|\mathbf{A}^{-1}\| < 1$, then

$$\frac{\|\mathbf{A}^{-1} - (\mathbf{A} + \delta\mathbf{A})^{-1}\|}{\|\mathbf{A}^{-1}\|} \leq \frac{\kappa(\mathbf{A})}{1 - \kappa(\mathbf{A})(\|\delta\mathbf{A}\|/\|\mathbf{A}\|)} \cdot \frac{\|\delta\mathbf{A}\|}{\|\mathbf{A}\|},$$

where $\kappa(\mathbf{A}) = \|\mathbf{A}\|\|\mathbf{A}^{-1}\|$. The above expression bounds the relative error in the inverse in terms of the relative error in the data. For $\|\delta\mathbf{A}\|$ small, the right-hand side of the above expression is of the order of $\kappa(\mathbf{A})\|\delta\mathbf{A}\|/\|\mathbf{A}\|$. Therefore if $\kappa(\mathbf{A})$ is not large, the relative error in the inverse is of the same order as the relative error of the data.

One can obtain a similar result in computing an eigenvalue. For example, if $\hat{\lambda}$ is an eigenvalue of $\mathbf{A} + \delta\mathbf{A}$, where $\mathbf{A}$ is diagonalizable (e.g., symmetric) with $\mathbf{A} = \mathbf{R}\mathbf{\Lambda}\mathbf{R}^{-1}$ and $\mathbf{\Lambda} = \mathrm{diag}(\lambda_1, \ldots, \lambda_n)$, then there is some eigenvalue λ_i of $\mathbf{A}$ such that, for an appropriate matrix norm,

$$|\hat{\lambda} - \lambda_i| \leq \|\mathbf{R}\|\|\mathbf{R}^{-1}\|\|\delta\mathbf{A}\| = \kappa(\mathbf{R})\|\delta\mathbf{A}\|.$$

Horn and Johnson [1985: section 6.3] derive a number of perturbation results like the one above for different properties of $\mathbf{A}$ and $\delta\mathbf{A}$.

Finally, we look at a corresponding result in relation to solving linear equations. For example, consider

$$(\mathbf{A} + \delta\mathbf{A})(\mathbf{x} + \delta\mathbf{x}) = \mathbf{b} + \delta\mathbf{b}.$$

Duff et al. [1986: 89–90] show that if $\|\delta\mathbf{A}\|\|\mathbf{A}^{-1}\| < 1$, then

$$\frac{\|\delta\mathbf{x}\|}{\|\mathbf{x}\|} \leq \frac{\kappa(\mathbf{A})}{1 - \kappa(\mathbf{A})(\|\delta\mathbf{A}\|/\|\mathbf{A}\|)} \cdot \left(\frac{\|\delta\mathbf{A}\|}{\|\mathbf{A}\|} + \frac{\|\delta\mathbf{b}\|}{\|\mathbf{b}\|} \right).$$

In introducing $\kappa(\mathbf{A})$ in the above discussion, we have not specified the norm $\|\cdot\|$. Furthermore, in deriving the above expression it transpires that we only require the norm to be an induced one. Also, the definition of $\kappa(\mathbf{A})$ used above is only appropriate for nonsingular matrices. By choosing an appropriate norm, we now generalize the definition to include nonsingular and rectangular matrices.

Definition 4.21. The *condition number* of an $m \times n$ real matrix $\mathbf{A}$, denoted by $\kappa_2(\mathbf{A})$, is the ratio of the largest singular value to the smallest nonzero singular value. Thus,

$$\kappa_2(\mathbf{A}) = \left(\frac{\lambda_{\max}}{\lambda_{\min}} \right)^{1/2},$$

where $\lambda_{\max}$ is the largest and $\lambda_{\min}$ is the smallest nonzero eigenvalue of $\mathbf{A}'\mathbf{A}$. Unfortunately, this condition number is not easy to compute, and for further details see Gentle [1998: 115–116].

4.96. When $\mathbf{A}$ is positive definite, its eigenvalues are positive, $\mathbf{A}'\mathbf{A} = \mathbf{A}^2$, and

$$\kappa_2(\mathbf{A}) = \frac{\lambda_{\max}(\mathbf{A})}{\lambda_{\min}(\mathbf{A})}.$$

The same is true for a Hermitian positive definite $\mathbf{A}$, as we replace $\mathbf{A}'\mathbf{A}$ by $\mathbf{A}^*\mathbf{A}$. Some bounds on κ_2 are given in (6.21b).

4.97. If $\mathbf{A}$ is nonsingular, then

$$\kappa_2(\mathbf{A}) = |||\mathbf{A}|||_{2,in} \cdot |||\mathbf{A}^{-1}|||_{2,in},$$

where $|||\cdot|||_{2,in}$ is the induced matrix norm corresponding to the L_2 vector norm (cf. 4.74b).

We can also define $\kappa_1(\mathbf{A})$ and $\kappa_\infty(\mathbf{A})$ corresponding to the L_1 and L_∞ norms.

4.98. If $v = 1, 2$, or ∞, then:

(a) $\kappa_v(\mathbf{A}) = \kappa_v(\mathbf{A}^{-1})$.

(b) $\kappa_v(c\mathbf{A}) = \kappa_v(\mathbf{A})$ for $c \neq 0$.

(c) $\kappa_v(\mathbf{A}) \geq 1$.

(d) $\kappa_1(\mathbf{A}) = \kappa_\infty(\mathbf{A})$.

(e) $\kappa_2(\mathbf{A}) = \kappa_2(\mathbf{A}')$.

(f) $\kappa_2(\mathbf{A}'\mathbf{A}) = \kappa_2^2(\mathbf{A}) \geq \kappa_2(\mathbf{A})$.

Proofs.   Section 4.6.5.

   4.98. Gentle [2000: 78].

CHAPTER 5

COMPLEX, HERMITIAN, AND RELATED MATRICES

Although complex matrices have been refered to in previous chapters, it seems appropriate to have a chapter that looks more closely at complex matrices. Complex matrices arise, for example, in time series and the related topic of signal processing, and in experimenal designs. We shall initially list some general properties of complex matrices before looking at Hermitian matrices. The related matrices that are considered are the skew-Hermitian, complex symmetric, real symmetric, skew-symmetric, complex orthogonal, and normal matrices. Factorizations and decompositions for these matrices are given in Chapter 16, while results about eigenvalues and eigenvectors for these matrices are located in Chapter 6. Unitary and real orthogonal matrices are considered in greater detail in Section 8.1, and Fourier matrices are covered in Section 8.12.2. At the end of this chapter we briefly consider quaternions, which are used, for example, in nuclear physics.

5.1 COMPLEX MATRICES

Definition 5.1. Given a complex number $x = x_1 + ix_2$, where x_1 and x_2 are both real, then its *complex conjugate* is defined to be $\bar{x} = x_1 - ix_2$, and its *modulus* or *absolute value* is defined to be $|x|$, where $|x| = (x_1^2 + x_2^2)^{1/2}$. If $\mathbf{A}$ is complex, it can be expressed in the form $\mathbf{A} = \mathbf{A}_1 + i\mathbf{A}_2$, where $\mathbf{A}_1$ and $\mathbf{A}_2$ are real matrices, and its complex conjugate is $\overline{\mathbf{A}} = \mathbf{A}_1 - i\mathbf{A}_2$. We also define the *conjugate transpose* of $\mathbf{A}$ to be $\mathbf{A}^* = \overline{\mathbf{A}}'$.

A Matrix Handbook for Statisticians. By George A. F. Seber
Copyright © 2008 John Wiley & Sons, Inc.

Definition 5.2. An $n \times n$ matrix $\mathbf{A}$ is said to be a *Hermitian matrix* if $\mathbf{A}^* = \mathbf{A}$ and *skew-Hermitian (anti-Hermitian)* if $\mathbf{A} = -\mathbf{A}^*$.

A real or complex matrix $\mathbf{A}$ is symmetric if $\mathbf{A} - \mathbf{A}'$.

A *complex orthogonal* $n \times n$ matrix $\mathbf{T}$ is a complex matrix such that $\mathbf{T}'\mathbf{T} = \mathbf{I}_n$. We omit the word "complex" if $\mathbf{T}$ is real.

An $n \times n$ matrix $\mathbf{U}$ is called a *unitary* matrix if $\mathbf{U}^*\mathbf{U} = \mathbf{I}_n$.

5.1.1   Some General Results

5.1. For complex scalars x and y we have:

(a) $|xy| = |x||y|$.

(b) $|x + y| \le |x| + |y|$.

(c) $|x|^2 + |y|^2 \ge 2\Re(x\bar{y})$, where $\Re$ is the "real part."

5.2. A complex orthogonal matrix $\mathbf{T}$ need not be unitary.

5.3. (Isomorphism Between Complex and Real Matrices)   Let $\mathbf{Z} = \mathbf{Z}_1 + i\mathbf{Z}_2$ be an $n \times n$ complex matrix with $\mathbf{Z}_i$ $(i = 1, 2)$ real matrices. Let

$$\mathbf{Z}^R = \left(\begin{array}{cc} \mathbf{Z}_1 & -\mathbf{Z}_2 \\ \mathbf{Z}_2 & \mathbf{Z}_1 \end{array} \right),$$

and define $\mathbf{X}^R$ and $\mathbf{Y}^R$ in a similar fashion.

(a) If $\mathbf{Z} = \mathbf{X} + \mathbf{Y}$, then $\mathbf{Z}^R = \mathbf{X}^R + \mathbf{Y}^R$.

(b) If $\mathbf{Z} = \mathbf{XY}$, then $\mathbf{Z}^R = \mathbf{X}^R\mathbf{Y}^R$.

(c) If $\mathbf{W} = \mathbf{Z}^{-1}$, then $\mathbf{W}^R = (\mathbf{Z}^R)^{-1}$.

(d) $\det \mathbf{Z}^R = |\det \mathbf{Z}|^2$.

(e) If $\mathbf{Z}$ is Hermitian, then $\mathbf{Z}^R$ is symmetric.

(f) If $\mathbf{Z}$ is unitary, then $\mathbf{Z}^R$ is orthogonal.

(g) Suppose the eigenvalues and eigenvectors of $\mathbf{Z}$ are λ_j and $\boldsymbol{\alpha}_j = \boldsymbol{\alpha}_{1j} + i\boldsymbol{\alpha}_{2j}$, $j = 1, 2, \ldots, n$, where the $\boldsymbol{\alpha}_{rj}$ are real for $r = 1, 2$, and all j. Then those of $\mathbf{Z}^R$ are, respectively,

$$\lambda_j, \left(\begin{array}{c} \boldsymbol{\alpha}_{j1} \\ \boldsymbol{\alpha}_{j2} \end{array} \right); \quad \lambda_j, \left(\begin{array}{c} -\boldsymbol{\alpha}_{2j} \\ \boldsymbol{\alpha}_{j1} \end{array} \right), \quad j = 1, 2, \ldots, n.$$

This result could be useful for carrying out numerical computations involving complex matrices.

5.4. Let $\mathbf{x} = (x_1, x_2, \ldots, x_n)'$ be a complex vector with $|\sum_{i=1}^{n} x_i| = \sum_{i=1}^{n} |x_i|$. Then $x_i = \theta|x_i|$, $i = 1, 2, \ldots, n$, for some complex number θ satisfying $|\theta| = 1$.

5.5. If $\mathbf{A}$ is a square complex matrix and $\mathbf{x}^*\mathbf{A}\mathbf{x} = 0$ for all complex $\mathbf{x}$, then $\mathbf{A} = \mathbf{0}$. Thus if $\mathbf{x}^*\mathbf{A}\mathbf{x} = \mathbf{x}^*\mathbf{B}\mathbf{x}$ for all complex $\mathbf{x}$, then $\mathbf{A} = \mathbf{B}$. However, these results do not necessarily hold if the matrices are real and the equalities hold for all real $\mathbf{x}$.

5.6. If $\mathbf{A}$ is an $n \times n$ real or complex matrix, then there exists a nonsingular matrix $\mathbf{S}$ such that $\mathbf{SAS}^{-1}$ is symmetric. There also exists a nonsingular matrix $\mathbf{R}$ such that $\mathbf{A}' = \mathbf{RAR}^{-1}$

5.7. Let $\mathbf{A}$ be an $n \times n$ real or complex matrix. Then every product of n entries of $\mathbf{A}$ taken from distinct rows and columns equals 0 (i.e., $a_{1i_1}, a_{2i_2} \cdots a_{ni_n} = 0$), with distinct i_j, if and only if $\mathbf{A}$ contains an $r \times s$ zero submatrix, where $r + s = n + 1$.

5.8. Let $\mathbf{A} = (a_{ij})$ be an $n \times n$ real or complex matrix with eigenvalues λ_i ($i = 1, 2, \ldots, n$), then

$$\sum_{i=1}^{n} |\lambda_i|^2 \le \sum_{i=1}^{n} \sum_{j=1}^{n} |a_{ij}|^2.$$

Equality occurs if and only if $\mathbf{A}$ is normal (cf. Section 5.6).

Proofs.   Section 5.1.1.

    5.1. Abadir and Magnus [2005: 12].

    5.2. For a 2×2 counterexample see Horn and Johnson [1985: 71, exercise 8].

    5.3. Quoted by Brillinger [1975: 71] with a corrected sign change. All the results except (c) and (d) can be verified directly, while (c) amounts to showing that if $\mathbf{WZ} = \mathbf{I}$ then $\mathbf{W}^R \mathbf{Z}^R = \mathbf{I}$; (d) follows from (5.10).

    5.4. Bapat and Raghavan [1997: 19].

    5.5. Davis [1979: 61–62]. For a counter example see (5.25).

    5.6. Horn and Johnson [1985: 209–210].

    5.7. Zhang [1999: 126–127].

    5.8. Zhang [1999: 260].

5.1.2   Determinants

5.9. Let $\mathbf{A} = \mathbf{A}_1 + i\mathbf{A}_2$, where $\mathbf{A}_1$ and $\mathbf{A}_2$ are real $n \times n$ matrices. If $\det \mathbf{A} = a + ib$ and $|\cdot|$ represents the modulus, then:

(a) $\det \overline{\mathbf{A}} = a - ib$.

(b) $\det \mathbf{A}' = \det \mathbf{A}$.

(c) $|\det \mathbf{A}|^2 = |\det \overline{\mathbf{A}}|^2 = a^2 + b^2 = |\det \mathbf{A} \det \overline{\mathbf{A}}| = \det \mathbf{A} \det \overline{\mathbf{A}}$.

(d) $|\det \mathbf{A} \det \overline{\mathbf{A}}| = |\det \mathbf{A} \det \overline{\mathbf{A}}'| = |\det(\mathbf{A}\overline{\mathbf{A}}')| = |\det(\mathbf{A}\mathbf{A}^*)|$.

5.10. Let

$$\mathbf{A} = \mathbf{A}_1 + i\mathbf{A}_2, \quad \mathbf{B} = \begin{pmatrix} \mathbf{A}_1 & \mathbf{A}_2 \\ -\mathbf{A}_2 & \mathbf{A}_1 \end{pmatrix} \quad \text{and} \quad \mathbf{C} = \begin{pmatrix} \mathbf{A}_1 & -\mathbf{A}_2 \\ \mathbf{A}_2 & \mathbf{A}_1 \end{pmatrix},$$

where $\mathbf{A}_1$ and $\mathbf{A}_2$ are real matrices. Then, for $\det \mathbf{A}_1 \neq 0$,

$$\det \mathbf{B} - \det \mathbf{C} \quad \text{and} \quad |\det \mathbf{A}| = |\det \mathbf{B}|^{1/2} = |\det \mathbf{C}|^{1/2}.$$

Proofs.   Section 5.1.2.

5.9. The results (a)–(c) follow from the definition and the product rule for determinants, and (d) follows from the expansion of a determinant.

5.10. Mathai [1997: 171–172].

5.2   HERMITIAN MATRICES

5.11. An $n \times n$ matrix $\mathbf{A}$ is Hermitian if and only if one (and therefore all) of the following five conditions hold.

(1) $\mathbf{x}^* \mathbf{A} \mathbf{x}$ is real for all $\mathbf{x} \in \mathbb{C}^n$.

(2) $\mathbf{A}^2 = \mathbf{A}^* \mathbf{A}$.

(3) $\mathrm{trace}(\mathbf{A}^2) = \mathrm{trace}(\mathbf{A}^* \mathbf{A})$.

(4) $\mathbf{A}$ is normal and all the eigenvalues of $\mathbf{A}$ are real.

(5) $\mathbf{S}^* \mathbf{A} \mathbf{S}$ is Hermitian for all $n \times n$ $\mathbf{S}$.

5.12. Suppose $\mathbf{A}$ is an $n \times n$ Hermitian matrix. Then the following hold.

(a) $\mathbf{A}^k$ is Hermitian for $k = 1, 2, \ldots$.

(b) $i\mathbf{A}$ is skew-Hermitian.

(c) If $\mathbf{A}$ is nonsingular, then $\mathbf{A}^{-1}$ is Hermitian.

(d) The diagonal elements of $\mathbf{A}$ are real.

(e) The eigenvalues of $\mathbf{A}$ are real (see Section 6.1.6 for further details).

5.13. Let $\mathbf{A}$ be an $n \times n$ matrix.

(a) $\mathbf{A}$ can be expressed uniquely in the form $\mathbf{A} = \mathbf{S} + i\mathbf{T}$, where $\mathbf{S}$ and $\mathbf{T}$ are Hermitian.

(b) $\mathbf{A}$ can be expressed uniquely in the form $\mathbf{A} = \mathbf{B} + \mathbf{C}$, where $\mathbf{B}$ is Hermitian and $\mathbf{C}$ is skew-Hermitian.

5.14. (Complex Householder Matrix) If $\mathbf{A} = \mathbf{I}_n - 2\mathbf{b}\mathbf{b}^*$, where $\mathbf{b}$ is a complex $n \times 1$ vector such that $\mathbf{b}^* \mathbf{b} = 1$, then $\mathbf{A}$ is Hermitian, unitary (i.e., $\mathbf{A}^* \mathbf{A} = \mathbf{I}_n$), and *involutionary* (i.e., $\mathbf{A}^2 = \mathbf{I}_n$).

5.15. (Trace) If $\mathbf{A}$ is $n \times n$, then:

(a) $\mathrm{trace}(\mathbf{A}\mathbf{X}) = 0$ for all Hermitian matrices $\mathbf{X}$ if and only if $\mathbf{A} = \mathbf{0}$.

(b) $\mathrm{trace}(\mathbf{A}\mathbf{X})$ is real for all Hermitian $\mathbf{X}$ if and only if $\mathbf{A}$ is Hermitian.

5.16. A square matrix $\mathbf{A}$ is a product of two Hermitian matrices if and only if it is similar to $\mathbf{A}^*$.

Proofs.   Section 5.2.

> 5.11. Zhang [1999: 209] proves (1)–(3), while Horn and Johnson [1985: 170–171] prove (1), (4), and (5).
>
> 5.12. Horn and Johnson [1985: 169–170].
>
> 5.13a. Horn and Johnson [1985: 170]. We set $\mathbf{A} = \frac{1}{2}(\mathbf{A}+\mathbf{A}^*)+i[-\frac{1}{2}(\mathbf{A}-\mathbf{A}^*)]$, and assume two such representations.
>
> 5.13b. Set $\mathbf{A} = \frac{1}{2}(\mathbf{A} + \mathbf{A}^*) + \frac{1}{2}(\mathbf{A} - \mathbf{A}^*)$.
>
> 5.15. Rao and Rao [1998: 342].
>
> 5.16. Zhang [1999: 215].

5.3   SKEW-HERMITIAN MATRICES

5.17. $\mathbf{A} - \mathbf{A}^*$ is skew-Hermitian for all square matrices $\mathbf{A}$.

5.18. Let $\mathbf{A}$ be skew-Hermitian.

(a) $i\mathbf{A}$ is Hermitian.

(b) The diagonal elements of $\mathbf{A}$ are all purely imaginary.

(c) Since the eigenvalues of an Hermitian matrix are real, the eigenvalues of $\mathbf{A}$ (and therefore of a real skew-symmetric matrix) are purely imaginary or zero.

(d) $(\mathbf{I}_n + \mathbf{A})$ is nonsingular.

5.19. Suppose $\mathbf{A}$ is a skew-Hermitian $n \times n$ matrix. Then, using (5.18d) above, we have:

(a) $\mathbf{U} = (\mathbf{I}_n - \mathbf{A})(\mathbf{I}_n + \mathbf{A})^{-1} = (\mathbf{I}_n + \mathbf{A})^{-1}(\mathbf{I}_n - \mathbf{A})$ is unitary as $\mathbf{U}^*\mathbf{U} = \mathbf{I}_n$.

   This follows from $(\mathbf{I}_n - \mathbf{A})(\mathbf{I}_n + \mathbf{A}) = (\mathbf{I}_n + \mathbf{A})(\mathbf{I}_n - \mathbf{A})$.

(b) $\mathbf{U} = [2\mathbf{I}_n - (\mathbf{I}_n + \mathbf{A})](\mathbf{I}_n + \mathbf{A})^{-1}$.

(c) From (a) we see that $\mathbf{I}_n - \mathbf{A}$ and $(\mathbf{I}_n + \mathbf{A})^{-1}$ commute.

(d) $\mathbf{I}_n + \mathbf{U}$ is nonsingular because, by (b), it equals $2(\mathbf{I}_n + \mathbf{A})^{-1}$.

(e) The matrices $\mathbf{U}$ and $\mathbf{A}$ are in $(1, 1)$-correspondence on account of the pair of equations

$$\begin{aligned} \mathbf{U} &= 2(\mathbf{I}_n + \mathbf{A})^{-1} - \mathbf{I}_n, \\ \mathbf{A} &= 2(\mathbf{I}_n + \mathbf{U})^{-1} - \mathbf{I}_n. \end{aligned}$$

   Thus $\mathbf{A}$ is skew-Hermitian if and only if $\mathbf{U}$ is unitary.

(f) These results hold if $\mathbf{A}$ is (real) skew-symmetric and $\mathbf{U}$ is real orthogonal.

Apparently the above results were first applied to statistics by Hsu [1953].

Proofs. Section 5.3.

5.18a. $(i\mathbf{A})^* = \bar{i}\mathbf{A}^* = (-i)(-\mathbf{A}) = i\mathbf{A}$.

5.18b. Use (a) and (5.12d).

5.18c. Use (5.12e).

5.18d. The determinant of a matrix is the product of its eigenvalues. Also, $\lambda(\mathbf{I}_n + \mathbf{A}) = 1 + \lambda(\mathbf{A}) = 1 + ia \neq 0$, as from (c) a is real or zero.

5.4   COMPLEX SYMMETRIC MATRICES

Although real symmetric matrices play a fundamental role in statistics, we shall first consider some results that hold for both real and complex symmetric matrices. Note that real symmetric matrices are also Hermitian (Section 5.2), normal (Section 5.6), and diagonalizable (Section 16.1), so that the results in those sections also apply to symmetric matrices.

5.20. We assume that $\mathbf{A}$ is an $n \times n$ real or complex matrix.

(a) $\mathbf{A}$ is symmetric if and only if there exists an $n \times n$ matrix $\mathbf{S}$ such that $\mathbf{A} = \mathbf{SS}'$. We may choose $\mathbf{S} = \mathbf{UD}$, where $\mathbf{U}$ is unitary,

$$\mathbf{D} = \mathrm{diag}(\sqrt{\sigma_1}, \sqrt{\sigma_2}, \ldots, \sqrt{\sigma_n}),$$

and the σ_i are the singular values of $\mathbf{A}$, in which case

$$\mathrm{rank}\,\mathbf{S} = \mathrm{rank}\,\mathbf{A}.$$

(b) If $\mathbf{A}$ is symmetric, then $\mathbf{A}$ is diagonalizable (cf. Definition 16.3) if and only if it is complex orthogonally diagonalizable. Thus $\mathbf{A} = \mathbf{S\Lambda S}^{-1}$ for a diagonal matrix $\mathbf{\Lambda}$ of eigenvalues of $\mathbf{A}$ (cf. 16.17a) if and only if $\mathbf{A} = \mathbf{Q\Lambda Q}'$, where $\mathbf{Q}$ is an $n \times n$ complex orthogonal matrix (i.e., $\mathbf{Q}'\mathbf{Q} = \mathbf{I}_n$).

5.21. If $\mathbf{A}$ and $\mathbf{B}$ are real or complex symmetric $n \times n$ matrices, then there exists a nonsingular $n \times n$ matrix $\mathbf{R}$ such that $\mathbf{A} = \mathbf{RBR}'$ if and only if $\mathrm{rank}\,\mathbf{A} = \mathrm{rank}\,\mathbf{B}$.

5.22. By considering a 2×2 matrix, we see that the eigenvalues values of a complex symmetric matrix are not necessarily real.

Proofs. Section 5.4.

5.20a. Horn and Johnson [1985: 207].

5.20b. Horn and Johnson [1985: 211–212].

5.21. Horn and Johnson [1985: 225]

5.22. For a counterexample, Abadir and Magnus [2005: 175] consider

$$\begin{pmatrix} 1 & i \\ i & 1 \end{pmatrix},$$

which has eigenvalues $1 \pm i$.

5.5   REAL SKEW-SYMMETRIC MATRICES

***Definition* 5.3.** A matrix $\mathbf{A}$ is said to be *skew-symmetric* if $\mathbf{A}' = -\mathbf{A}$. Note that a complex matrix like

$$\mathbf{A} = \begin{pmatrix} 0 & a \\ -a & 0 \end{pmatrix},$$

where a is complex, is skew-symmetric. However, my focus is on real matrices as they are a special case of skew-Hermitian matrices; some of the properties in Section 5.3 will then apply for real matrices. For a factorization of a real skew-symmetric matrix see (16.46b(ii)).

5.23. The diagonal elements of a real skew-symmetric matrix are all zero.

5.24. Let $\mathbf{A}$ be an $n \times n$ real skew-symmetric matrix.

(a) From (5.18c), the eigenvalues $\lambda_i(\mathbf{A})$ of $\mathbf{A}$ are zero or purely imaginary and occur in conjugate pairs, as the characteristic polynomial has real coefficients. Hence the eigenvalues take the form $\pm i a_i$ with a_i real $(i = 1, 2, \ldots, p)$, along with $(n - 2p)$ zeros. Thus:

   (i) If n is odd, $\det(\mathbf{A}) = 0$.

   (ii) If n is even, $\det(\mathbf{A}) \geq 0$.

   (iii) $\det(\mathbf{I}_n + \mathbf{A}) = \prod_{i=1}^{n}(1 + \lambda_i(\mathbf{A})) \geq 1$ with equality if and only if $\mathbf{A} = \mathbf{0}$.

(b) Let $n = 2m$, then $\det(\mathbf{A})$ is the square of a polynomial of degree m in the matrix entries (e.g., (4.45)). The polynomial is called the *pfaffian* of $\mathbf{A}$ and is denoted by $\mathrm{Pf}(\mathbf{A})$. There are two ways of defining a pfaffian and a helpful resource is http://en.wikipedia.org/wiki/Pfaffian. We have:

   (i) $\det(\mathbf{A}) = [\mathrm{Pf}(\mathbf{A})]^2$.

   (ii) $\mathrm{Pf}(\mathbf{B}\mathbf{A}\mathbf{B}') = \det(\mathbf{B})\mathrm{Pf}(\mathbf{A})$.

   (iii) $\mathrm{Pf}(c\mathbf{A}) = c^m \mathrm{Pf}(\mathbf{A})$.

   (iv) $\mathrm{Pf}(\mathbf{A}') = (-1)^m \mathrm{Pf}(\mathbf{A})$.

   (v) For an arbitrary $m \times m$ matrix $\mathbf{C}$,

$$\mathrm{Pf}\begin{pmatrix} \mathbf{0} & \mathbf{C} \\ -\mathbf{C}' & \mathbf{0} \end{pmatrix} = (-1)^{m(m-1)/2} \det(\mathbf{C}).$$

For further references see Halton [1966b], Mehta [2004: 543-545, examples of computation] and Northcott [1984].

5.25. Let $\mathbf{A}$ be a real square matrix and $\mathbf{x}$ a real vector, then $\mathbf{x}'\mathbf{A}\mathbf{x} = 0$ for all $\mathbf{x}$ if and only if $\mathbf{A}$ is skew-symmetric.

Proofs.   Section 5.5.

   5.23. Follows from (5.18b).

   5.24a(i). The determinant of a matrix is the product of its eigenvalues.

   5.24a(ii). $(ia)(-ia) = a^2$, where a may be zero.

5.24a(iii). Use (5.18d) and $(1 + ia)(1 - ia) = 1 + a^2$.

5.24b. Quoted in http://en.wikipedia.org/wiki/Pfaffian. Depending on the definition used, several proofs are available for (i) (originally due to Cayley); for example, Parameswaran [1954], Dress and Wenzel [1995], and Halton [1966a]. Serre [2002: 22–23] proves (ii).

5.25. Davis [1979: 60–61].

5.6   NORMAL MATRICES

Definition **5.4.** A square matrix $\mathbf{A}$ is said to be *normal* if $\mathbf{AA}^* = \mathbf{A}^*\mathbf{A}$. Note that Hermitian, skew-Hermitian, and unitary matrices are all normal, as are their real counterparts.

5.26. An $n \times n$ matrix $\mathbf{A}$ with eigenvalues $\lambda_1, \lambda_2, \ldots, \lambda_n$ is normal if and only if there exists a unitary matrix $\mathbf{Q}$ such that

$$\mathbf{Q}^*\mathbf{AQ} = \mathrm{diag}(\lambda_1, \lambda_2, \ldots, \lambda_n).$$

We say that $\mathbf{A}$ is *unitarily diagonalizable*. Note that this applies to Hermitian and unitary matrices (see also (16.46)).

5.27. If $\mathcal{A}$ is a commuting family of $n \times n$ normal matrices (i.e., $\mathbf{A}_1\mathbf{A}_2 - \mathbf{A}_2\mathbf{A}_1$ for all $\mathbf{A}_1, \mathbf{A}_2 \in \mathcal{A}$), then every member of $\mathcal{A}$ is unitarily diagonalizable by the same unitary matrix.

5.28. In addition to being unitarily diagonalizable, normal matrices have many unique properties, some of which are listed below. The following statements are equivalent.

(1) $\mathbf{A}$ is normal.

(2) There exists a polynomial $p(x)$ of degree at most $n - 1$ such that $\mathbf{A}^* = p(\mathbf{A})$.

(3) The singular values of $\mathbf{A}$ are $|\lambda_1(\mathbf{A})|, |\lambda_2(\mathbf{A})|, \ldots, |\lambda_n(\mathbf{A})|$.

(4) $\mathbf{A} = \mathbf{R} + i\mathbf{S}$, where $\mathbf{R}$ and $\mathbf{S}$ are real, symmetric, and commute (i.e., $\mathbf{RS} = \mathbf{SR}$).

(5) Every eigenvector of $\mathbf{A}$ is an eigenvector of $\mathbf{A}^*$

(6) There exists a set of eigenvectors of $\mathbf{A}$ that form an orthonormal basis for $\mathbb{C}^n$.

(7) $\sum_i \sum_j |a_{ij}|^2 = \sum_i |\lambda_i(\mathbf{A})|^2$.

5.29. If $\mathbf{A}$ is normal and $p(z)$ is a polynomial, then $p(\mathbf{A})$ is normal.

5.30. An upper-triangular matrix is normal if and only if it is diagonal.

5.31. A normal matrix is unitary if and only if its eigenvalues have absolute value 1.

5.32. If $\mathbf{A}$ and $\mathbf{B}$ are normal, then so is their Kronecker product $\mathbf{A} \otimes \mathbf{B}$.

5.33. If $\mathbf{A}$ and $\mathbf{B}$ are $n \times n$ normal matrices and $\mathbf{AB} = \mathbf{BA}$, then $\mathbf{AB}$ is normal.

5.34. A normal matrix is Hermitian if and only if its eigenvalues are real, and it is skew-Hermitian if and only if its eigenvalues have zero real part.

Proofs. Section 5.6.

5.26. Horn and Johnson [1985: section 2.5] and Zhang [1999: 65–66].

5.27. Horn and Johnson [1985: 103].

5.28. For these and further properties see Horn and Johnson [1985: 100–111] and Zhang [1999: 241–242].

5.29. Horn and Johnson [1985: 110, exercise 17] and Marcus and Minc [1964: 71].

5.30. Rao and Bhimasankaram [2000: 313].

5.31. From (5.26), $\mathbf{I} = \mathbf{AA}^* = \mathbf{U\Lambda U}^* \mathbf{U\overline{\Lambda}U}^* = \mathbf{U\Lambda\overline{\Lambda}U}^*$ and $\mathbf{\Lambda\overline{\Lambda}} = \mathbf{I}$.

5.32. From (11.1f),

$$(\mathbf{A} \otimes \mathbf{B})^*(\mathbf{A} \otimes \mathbf{B}) = (\mathbf{A}^* \otimes \mathbf{B}^*)(\mathbf{A} \otimes \mathbf{B}) = \mathbf{A}^*\mathbf{A} \otimes \mathbf{B}^*\mathbf{B} = \mathbf{AA}^* \otimes \mathbf{BB}^*,$$

which, by reversing the argument, is $(\mathbf{A} \otimes \mathbf{B})(\mathbf{A} \otimes \mathbf{B})^*$.

5.33. Using (5.27), we have $\mathbf{A} = \mathbf{U\Lambda_A U}^*$, $\mathbf{B} = \mathbf{U\Lambda_B U}^*$, and $\mathbf{AB} = \mathbf{U\Lambda_A\Lambda_B U}^* = \mathbf{U\Lambda U}^*$. Then $\mathbf{AB(AB)}^* = \mathbf{U\Lambda\overline{\Lambda}U}^* = \mathbf{U\overline{\Lambda}\Lambda U}^* = (\mathbf{AB})^*\mathbf{AB}$. Also $\mathbf{A} = -\mathbf{A}^*$ if and only if $\mathbf{\Lambda} = -\overline{\mathbf{\Lambda}}$.

5.34. If $\mathbf{A} = \mathbf{U\Lambda U}^*$, then $\mathbf{A} = \mathbf{A}^*$ if and only if $\mathbf{\Lambda} = \overline{\mathbf{\Lambda}}$.

5.7 QUATERNIONS

Definition **5.5.** Just as a complex number has two components, a *quaternion number* has four components

$$q = q^{(0)} + q^{(1)}e_1 + q^{(2)}e_2 + q^{(3)}e_3 = q_0 + \mathbf{q} \cdot \mathbf{e}, \quad \text{say,}$$

where the e_i are quantities (not ordinary numbers) satisfying the symbolic rules $e_1^2 = e_2^2 = e_3^2 = -1$, $e_1e_2 = -e_2e_1 = e_3$, $e_2e_3 = -e_3e_2 = e_1$, and $e_3e_1 = -e_1e_3 = e_2$, where "1" is a particular unit identity. This 1 and the e_i can be expressed as the matrices $\mathbf{C}(1) = \mathbf{I}_2$, and the so-called *Pauli matrices*

$$\mathbf{C}(e_1) = \begin{pmatrix} i & 0 \\ 0 & -i \end{pmatrix}, \quad \mathbf{C}(e_2) = \begin{pmatrix} 0 & 1 \\ -1 & 0 \end{pmatrix}, \quad \text{and} \quad \mathbf{C}(e_3) = \begin{pmatrix} 0 & i \\ i & 0 \end{pmatrix},$$

where $i = \sqrt{-1}$. Then

$$\mathbf{C}(q) = \begin{pmatrix} q^{(0)} + iq^{(1)} & q^{(2)} + iq^{(3)} \\ -q^{(2)} + iq^{(3)} & q^{(0)} - iq^{(1)} \end{pmatrix}$$

is a matrix representation of the quaternion q. For any 2×2 complex matrix we have

$$\begin{pmatrix} a & b \\ c & d \end{pmatrix} = \tfrac{1}{2}(a+d)\mathbf{C}(1) - \frac{i}{2}(a-d)\mathbf{C}(e_1) + \frac{1}{2}(b-c)\mathbf{C}(e_2) - \frac{i}{2}(b+c)\mathbf{C}(e_3).$$

Thus $q^{(0)} = \tfrac{1}{2}(a+d)$, $q^{(1)} = \tfrac{1}{2}(a-d)$, and so on.

The $q^{(i)}$ can be real or complex. If they are all real, then we call the quaternion real, though $\mathbf{C}(q)$ isn't necessarily real. The notation for quaternions is a little different from the usual for complex numbers. For example, the *conjugate quaternion* of a complex quaternion $q = q^{(0)} + \mathbf{q} \cdot \mathbf{e}$ is

$$\bar{q} = q^{(0)} - \mathbf{q} \cdot \mathbf{e},$$

which is different from its *complex conjugate quaternion*

$$q^* = q^{(0)*} + \mathbf{q}^* \cdot \mathbf{e}.$$

A quaternion with $q^* = q$ is real. Applying both types of conjugation together, we obtain the *Hermitian conjugate*

$$q^\dagger = \bar{q}^* = q^{(0)*} - \mathbf{q}^* \cdot \mathbf{e}.$$

When $q^\dagger = q$, q is called a *Hermitian quaternion,* and it can be shown directly after some algebra that $\mathbf{C}(q)$ is a 2×2 Hermitian matrix. If $q^\dagger = -q$, then it is called an *anti-Hermitian quaternion* and the corresponding matrix $\mathbf{C}(q)$ is skew-Hermitian. For further information about quaternions see Carmeli [1983: chapters 8 and 9], Kantor and Solodovnikov [1989], Mehta [2004: 39], and, particularly, Zhang [1997]; for a geometrical perspective see Hanson [2006].

Since any $2n \times 2n$ complex matrix $\mathbf{Q}$ can be expressed in terms of n^2 blocks of 2×2 matrices, we can write $\mathbf{Q} = (q_{ij})$ for $i, j = 1, 2, \ldots, n$, where q_{ij} is a quaternion with matrix representation $\mathbf{C}(q_{ij})$. We call $\mathbf{Q}$ an $n \times n$ *quaternion matrix.* Using quaternion arithmetic, we can define certain matrix properties for quaternion matrices, namely transposition

$$(\mathbf{Q}')_{ij} = -e_2 \bar{q}_{ji} e_2,$$

Hermitian conjugation

$$(\mathbf{Q}^\dagger)_{ij} = q_{ji}^\dagger,$$

and dual

$$(\mathbf{Q}^R)_{ij} = e_2 (\mathbf{Q}')_{ij} e_2^{-1} = \bar{q}_{ji}.$$

If $\mathbf{Q} = \mathbf{Q}^R$, the matrix is said to be *self-dual.* For further matrix details see Mehta [1989].

5.35. Let $\mathbf{Q}$ be a quaternion matrix. Then:

(a) $\mathbf{Q}^R = \mathbf{Q}^\dagger$ is necessary and sufficient for the elements of $\mathbf{Q}$ to be real quaternions. When this holds we call such a matrix *quaternion real.*

(b) If $\mathbf{Q}$ is both Hermitian and self-dual, then it is also quaternion real. Furthermore, since $q_{ij}^\dagger = \bar{q}_{ij} = q_{ji}$ for all i, j, the 2×2 corresponding matrix $q_{ij}^{(0)}$

must form a real symmetric matrix, whereas $q_{ij}^{(1)}$, $q_{ij}^{(2)}$, and $q_{ij}^{(3)}$ must lead to real skew-symmetric matrices. Self-dual Hermitian quaternion matrices have an important role in nuclear physics and are related to random matrices (cf. Section 21.10). The corresponding $2n \times 2n$ Hermitian matrix is called a self-dual Hermitian matrix.

***Definition* 5.6.** Let $\mathbf{Z}_1 = \mathbf{C}(e_2) \otimes \mathbf{I}_n = \begin{pmatrix} \mathbf{0} & \mathbf{I}_n \\ -\mathbf{I}_n & \mathbf{0} \end{pmatrix}$, a $2n \times 2n$ matrix, where "$\otimes$" is the Kronecker product. A real $2n \times 2n$ matrix $\mathbf{A}$ is said to be *Hamiltonian* if $(\mathbf{Z}_1\mathbf{A})' = \mathbf{Z}_1\mathbf{A}$. Note that $\mathbf{Z}_1$ is skew-symmetric. Hamiltonian matrices are used in classical mechanics for the study of Hamiltonian dynamical systems.

5.36. $[\mathbf{C}(e_2)]^{-1} = [\mathbf{C}(e_2)]' = -\mathbf{C}(e_2)$. Also $\mathbf{C}(e_2)^2 = -\mathbf{I}_2$.

5.37. $\mathbf{Z}_1^2 = \mathbf{C}(e_2)^2 \otimes \mathbf{I}_n = -\mathbf{I}_2 \otimes \mathbf{I}_n = -\mathbf{I}_{2n}$ and $\mathbf{Z}_1^{-1} = -\mathbf{Z}_1 = \mathbf{Z}_1'$.

5.38. Let $\mathbf{A}$ be an a $2n \times 2n$ Hamiltonian matrix. Then:

(a) Since $\mathbf{Z}_1\mathbf{A}$ is symmetric, $\mathbf{Z}_1\mathbf{A} + \mathbf{A}'\mathbf{Z}_1 = \mathbf{0}$ and, by (5.37), $\mathbf{A} = -\mathbf{Z}_1^{-1}\mathbf{A}'\mathbf{Z}_1 = \mathbf{Z}_1\mathbf{A}'\mathbf{Z}_1$.

(b) $\mathbf{A}'$ is Hamiltonian.

(c) trace $\mathbf{A} = 0$.

5.39. Let

$$\mathbf{A} = \begin{pmatrix} \mathbf{C} & \mathbf{D} \\ \mathbf{E} & \mathbf{F} \end{pmatrix},$$

where all matrices are $n \times n$, $\mathbf{D}$ and $\mathbf{E}$ are symmetric, and $\mathbf{C} + \mathbf{F}' = \mathbf{0}$. Then $\mathbf{A}$ is Hamiltonian.

***Definition* 5.7.** A real or complex $2n \times 2n$ matrix $\mathbf{B}$ is said to be *symplectic* if $\mathbf{B}'\mathbf{Z}\mathbf{B} = \mathbf{Z}$, where $\mathbf{Z}$ is a nonsingular, skew-symmetric matrix. Typically, $\mathbf{Z} = \mathbf{Z}_1$, as defined above, or $\mathbf{Z} = \mathbf{Z}_2$, where

$$\mathbf{Z}_2 = \mathbf{I}_n \otimes \mathbf{C}(e_2) = \begin{pmatrix} 0 & 1 & 0 & 0 & 0 & \cdots & 0 & 0 \\ -1 & 0 & 0 & 0 & 0 & \cdots & 0 & 0 \\ 0 & 0 & 0 & 1 & 0 & \cdots & 0 & 0 \\ 0 & 0 & -1 & 0 & 0 & \cdots & 0 & 0 \\ . & . & . & . & . & \cdots & . & . \\ 0 & 0 & 0 & 0 & 0 & \cdots & -1 & 0 \end{pmatrix},$$

a matrix used in nuclear physics. In this case, $\mathbf{Z}_2$ can expressed as an $n \times n$ quaternion matrix $e_2\mathbf{I}_n$ (Mehta [2004: 38-41]).

5.40. $\mathbf{Z}_2$ has the same properties as $\mathbf{Z}_1$ in (5.37).

5.41. If $\mathbf{B}$ is symplectic, then (i) $\mathbf{B}^{-1} = \mathbf{Z}^{-1}\mathbf{B}'\mathbf{Z}$ and (ii) $\det(\mathbf{B}) = 1$.

5.42. The matrix $\mathbf{Z}_i$ $(i = 1, 2)$ is symplectic.

5.43. Let $\mathbf{H}$ be any quaternion real $2n \times 2n$ matrix. Then there exist a symplectic matrix $\mathbf{B}$ such that $\mathbf{H} = \mathbf{B}^{-1}\mathbf{D}\mathbf{B}$, where $\mathbf{D}$ is a real, scalar, and diagonal matrix. Here scalar means that $\mathbf{D} = \text{diag}(d_1, d_1, d_2, d_2, \ldots, d_n, d_n)$ so that the eigenvalues of $\mathbf{H}$ consist of equal pairs. For further extensions see Carmeli [1983: 70–71].

Proofs. Section 5.7.

5.36-5.37. These are straightforward; we use (11.1e), (11.1i), and (11.11).

5.38b. Taking transposes in (a) and using (5.37), we have $\mathbf{A}' = \mathbf{Z}_1 \mathbf{A} \mathbf{Z}_1$.

5.38c. Using (a), trace $\mathbf{A} = -\operatorname{trace}(\mathbf{Z}_1^{-1} \mathbf{A}' \mathbf{Z}_1) = -\operatorname{trace} \mathbf{A}' = -\operatorname{trace} \mathbf{A}$.

5.39. Show that $\mathbf{Z}_1 \mathbf{A}$ is symmetric.

5.40. $\mathbf{Z}_2^2 = \mathbf{I}_n \otimes \mathbf{C}(e_2)^2 = -\mathbf{I}_n \otimes \mathbf{I}_2 = -\mathbf{I}_{2n}$.

5.41. The result (i) follows from the definition by multiplying on the left by $\mathbf{Z}^{-1}$ and on the right by $\mathbf{B}^{-1}$, and (ii) follows from (5.24b(ii)).

5.42. $(\mathbf{Z}_i' \mathbf{Z}_i)\mathbf{Z}_i = (\mathbf{Z}_i^{-1} \mathbf{Z}_i)\mathbf{Z}_i = \mathbf{Z}_i$.

5.43. Carmeli [1983: 70].

CHAPTER 6

EIGENVALUES, EIGENVECTORS, AND SINGULAR VALUES

Eigenvalues, eigenvectors, and singular values play an important role in statistics, and they arise in most of the chapters in this book. In this chapter we deal with these topics in a general sense. They also occur in a number of important inequalities in this chapter, in Chapter 14, and in Chapter 23 on majorization, and they underlie many of the factorizations and decompositions in Chapter 16. For those relating to specific matrices and some patterned matrices, the reader will need to refer to the index for those matrices. This chapter closes with a a brief introduction to antieigenvalues and antieigenvectors, which are becoming of increasing interest to statisticians in recent years.

6.1 INTRODUCTION AND DEFINITIONS

Definition 6.1. Let $\mathbf{A}$ be an $n \times n$ matrix, which we assume to have elements in $\mathbb{F}$ (i.e., either $\mathbb{R}$ or $\mathbb{C}$, unless otherwise stated). The polynomial $c(\lambda) = \det(\mathbf{A} - \lambda\mathbf{I}_n)$ is called the *characteristic polynomial*. The equation $c(\lambda) = 0$ is called the *characteristic equation*, and its roots are called the *eigenvalues (characteristic roots, latent roots)* of $\mathbf{A}$. Many authors use $f(\lambda) = \det(\lambda\mathbf{I}_n - \mathbf{A}) = (-1)^n \det(\mathbf{A} - \lambda\mathbf{I}_n)$ for the characteristic polynomial, as the coefficient of λ^n is now 1. This alternative version is sometimes more convenient, so both $c(\cdot)$ and $f(\cdot)$ are used below.

Eigenvalues may be real, complex, or a mixture of both. We shall order the eigenvalues by their modulus values, i.e., $|\lambda_1| \geq |\lambda_2| \geq \cdots \geq |\lambda_n| \geq 0$. If λ_1 is unique, we shall call it the *dominant eigenvalue*. In this case there exists a unique right eigenvector $\mathbf{x}_1$ of unit length such that $\mathbf{A}\mathbf{x}_1 = \lambda_1 \mathbf{x}_1$, called the *dominant eigenvector*.

The s distinct eigenvalues are denoted by $\mu_1, \mu_2, \ldots, \mu_s$ (or $\mu_j(\mathbf{A})$, $j = 1, 2, \ldots, s$), where $|\mu_1| > |\mu_2| > \cdots > |\mu_s| \geq 0$. The set of μ_j is called the *spectrum* of $\mathbf{A}$, and $\rho(\mathbf{A}) = |\mu_1| = |\lambda_1|$ is called the *spectral radius*. We can therefore write $f(\lambda) = \prod_{j=1}^{s} (\lambda - \mu_j)^{m_j}$, where $\sum_{j=1}^{s} m_j = n$. Here m_j [or $m(\mu_j)$] is called the *algebraic multiplicity* of the eigenvalue μ_j. If $m_j = 1$, μ_j is called a *simple eigenvalue*, while if $m_j > 1$, μ_j is called a *multiple eigenvalue*.

For every μ_j there exists a nonzero solution $\mathbf{x}$ such that $\mathbf{A}\mathbf{x} = \mu_j \mathbf{x}$, and $\mathbf{x}$ is called an eigenvector associated with μ_j. The set of all such $\mathbf{x}$ together with $\mathbf{0}$, namely $\mathcal{N}(\mathbf{A} - \mu_j \mathbf{I}_n)$ the null space of $\mathbf{A} - \mu_j \mathbf{I}_n$, is called the *eigenspace* of μ_j. The dimension g_j [or $g(\mu_j)$] of this space is called the *geometric multiplicity* of μ_j. To avoid ambiguity, we shall refer to such an $\mathbf{x}$ as a *right eigenvector*. There similarly exists a nonzero $\mathbf{y}$ such that $\mathbf{y}'\mathbf{A} = \mu_j \mathbf{y}'$ called the *left eigenvector* of $\mathbf{A}$ associated with μ_j.

If $m(\mu_j) = g(\mu_j)$, then μ_j is said to be a *semisimple eigenvalue*.

6.1. (Multiplicities). Let $\mathbf{A}$ be an $n \times n$ matrix.

(a) $g(\mu_j) \leq m(\mu_j)$; that is, the geometric multiplicity is no greater than the algebraic multiplicity.

(b) $\mathrm{rank}(\mathbf{A} - \mu_j \mathbf{I}_n) = n - g(\mu_j) \geq n - m(\mu_j)$ for all j.

(c) If $m(\mu_j) = 1$ so that μ_j is a simple eigenvalue, then $g(\mu_j) = 1$ and $\mathrm{rank}(\mathbf{A} - \mu_j \mathbf{I}_n) = n - 1$. Conversely, if $\mathrm{rank}(\mathbf{A} - \mu_j \mathbf{I}_n) = n - 1$, then μ_j is an eigenvalue, but not necessarily a simple eigenvalue.

Proofs. Section 6.1

6.1a. Schott [2005: 89] and Rao and Bhimasankaram [2000: 286].

6.1b. From (3.3a), $\dim \mathcal{C}(\mathbf{B}) + \dim \mathcal{N}(\mathbf{B}) = n$ for $\mathbf{B} = \mathbf{A} - \mu_j \mathbf{I}_n$, as $\dim \mathcal{C}(\mathbf{A}^*) = \mathrm{rank}\,\mathbf{A}^* = \mathrm{rank}\,\mathbf{A} = \dim \mathcal{C}(\mathbf{A})$.

6.1c. Magnus and Neudecker [1999: 20].

6.1.1 Characteristic Polynomial

***Definition* 6.2.** (Symmetric Functions) Given a set of constants $\lambda_1, \ldots, \lambda_n$, we define the *elementary symmetric functions* as

$$S_1 = \sum_{i=1}^{n} \lambda_i,$$

$$S_2 = \sum_{i<j} \lambda_i \lambda_j,$$

$$S_r = \sum_{i_1 < i_2 < \cdots < i_r} \lambda_{i_1} \lambda_{i_2} \cdots \lambda_{i_r},$$

$$S_n = \lambda_1 \lambda_2 \cdots \lambda_n.$$

Also, let $c(\lambda) = (-1)^n(\lambda^n + a_1\lambda^{n-1} + \cdots + a_{n-1}\lambda + a_n) = (-1)^n f(\lambda)$ be the characteristic polynomial.

6.2. If the characteristic polynomial has real coefficients, then any complex eigenvalues must come in conjugate pairs.

6.3. (Cayley–Hamilton theorem) $f(\mathbf{A}) = \mathbf{0}$.

6.4. The coefficient a_r $(r = 1, 2, \ldots, n)$ is $(-1)^r$ times the sum of all the $r \times r$ principal minors of $\mathbf{A}$. These are obtained by striking out $n - r$ rows and the same numbered columns of $\mathbf{A}$ and taking the determinant of the remaining submatrix.

(a) $a_r = (-1)^r S_r$ $(r = 1, 2, \ldots, n)$ with $a_n = (-1)^n \det \mathbf{A}$.

(b) From (a), S_r is the sum of all the $r \times r$ principal minors of $\mathbf{A}$.

(c) If $t_r = \lambda_1^r + \lambda_2^r + \ldots + \lambda_n^r$ for $r = 1, 2, \ldots, n$ and we define $0 \equiv a_{n+1} \equiv a_{n+2} \equiv \ldots$, then $t_r + t_{r-1}a_1 + \ldots + t_1 a_{r-1} + r a_r = 0$ $(r = 1, 2, \ldots)$. These expressions for the a_r are known as *Newton's identities* (Hunter [1983a: 156–157]).

6.5. If $\mathbf{A}$ is $n \times n$, and μ is not an eigenvalue of $\mathbf{A}$, then $\mathbf{A} - \mu\mathbf{I}_n$ is nonsingular as its determinant is nonzero.

6.6. If $\mathbf{A}$, $\mathbf{B}$, and $\mathbf{R}$ are $n \times n$ matrices, and $\mathbf{B} = \mathbf{RAR}^{-1}$ (i.e, $\mathbf{A}$ and $\mathbf{B}$ are similar), then $\mathbf{A}$ and $\mathbf{B}$ have the same characteristic polynomial. Note that having the same eigenvalues is a necesssary but not sufficient condition for similarity.

6.7. If $\mathbf{A}$ and $\mathbf{B}$ are real $n \times n$ matrices, then the eigenvalues of

$$\mathbf{C} = \left(\begin{array}{cc} \mathbf{A} & \mathbf{B} \\ \mathbf{B} & \mathbf{A} \end{array} \right)$$

are those of $\mathbf{A} + \mathbf{B}$ and $\mathbf{A} - \mathbf{B}$.

Definition 6.3. If f is a polynomial such that $f(\mathbf{A}) = \mathbf{0}$, we say that f *annihilates* $\mathbf{A}$. A polynomial is said to be *monic* if the coefficient of the highest power is unity.

6.8. If $\mathbf{A}$ is $n \times n$, there exists a unique monic polynomial of minimum degree no greater than n that annihilates $\mathbf{A}$.

Definition 6.4. The monic polynomial $q(\lambda)$ of the least degree that annihilates $\mathbf{A}$ is called the *minimal polynomial*.

6.9. Every monic polynomial is both the minimal polynomial and the characteristic polynomial ($f(\lambda)$ version) of its companion matrix (cf. 6.14).

6.10. The minimal polynomial divides every polynomial that annihilates $\mathbf{A}$. It therefore divides the characteristic polynomial $f(\lambda)$ (by 6.3).

6.11. If $q(\lambda)$ is the minimal polynomial of $\mathbf{A}$, then λ is a root $q(\lambda) = 0$ if and only if it is an eigenvalue of $\mathbf{A}$. Thus every root of the characteristic equation is a root of $q(\lambda) = 0$.

6.12. If $\mathbf{A}$, $\mathbf{B}$, and $\mathbf{R}$ are $n \times n$ matrices, then $\mathbf{A}$ and the similar matrix $\mathbf{B} = \mathbf{RAR}^{-1}$ have the same minimal polynomial.

6.13. Let $\mathbf{A}$ be an $n \times n$ matrix, and let $\mathbf{A}^k$ be the first matrix for which the set $\{\mathbf{I}_n, \mathbf{A}, \mathbf{A}^2, \ldots, \mathbf{A}^t\}$ is linearly independent, that is, $\mathbf{A}^k = \sum_{i=0}^{k-1} a_i \mathbf{A}^i$. Then the minimum polynomial of $\mathbf{A}$ is $x_k - \sum_{i=0}^{k-1} a_i x^i$.

6.14. (Companion Matrix) Consider the polynomial $p_n(x) = x^n + a_{n-1}x^{n-1} + \cdots + a_1 x + a_0$. The matrix

$$\mathbf{A} = \begin{pmatrix} 0 & 0 & \cdots & 0 & 0 & -a_0 \\ 1 & 0 & \cdots & 0 & 0 & -a_1 \\ . & . & \cdots & . & . & . \\ 0 & 0 & \cdots & 0 & 1 & -a_{n-2} \\ 0 & 0 & \cdots & 0 & 1 & -a_{n-1} \end{pmatrix}$$

is called the *companion matrix* of the polynomial p_n (Golub and Van Loan [1996: 348] and Horn and Johnson [1985: 146]). However, variations on the above matrix are also called the companion matrix, such as the transpose of $\mathbf{A}$ (e.g., Abadir and Magnus [2005: 173–174] and Rao and Bhimasankaram [2000: 283, solution to exercise 3]). Some authors take the transpose, then move the bottom row to the top and shift the other rows down one.

If $\mathbf{A}$ is defined above, then:

(a) $\det(x\mathbf{I}_n - \mathbf{A}) = \det(x\mathbf{I}_n - \mathbf{A}') = p_n(x)$.

(b) $p_n(x)$ is also the minimal polynomial of $\mathbf{A}$.

A version of the companion matrix can be used to find various upper and lower bounds on the roots of $f_n(\lambda) = 0$, as in Horn and Johnson [1985: 316–320]. Boshnakov [2002] extends the above concept to multi-companion matrices.

Proofs.   Section 6.1.1

6.2. Abadir and Magnus [2005: 164].

6.3. Meyer [2000a: 509, 532-533] and Rao and Bhimasankaram [2000: 292].

6.4a. Basilevsky [1983:192], Horn and Johnson [1985: 41–42], and Searle [1982: 278].

6.4b. Horn and Johnson [1985: 42].

6.4c. Hunter [1983a: 156–157].

6.6. Horn and Johnson [1985: 45].

6.7. Let $(\mathbf{A} + \mathbf{B})\mathbf{u} = \lambda\mathbf{u}$ and $(\mathbf{A} - \mathbf{B})\mathbf{v} = \mu\mathbf{v}$. Then $\mathbf{C}$ has eigenvectors $(\mathbf{u}', \mathbf{u}')'$ and $(\mathbf{v}', -\mathbf{v}')'$.

6.8. Horn and Johnson [1985: 142] and Rao and Bhimasankaram [2000: 293].

6.9. Horn and Johnson [1985: 147].

6.10. Horn and Johnson [1985: 142-143] and Rao and Bhimasankaram [2000: 293].

6.11. Horn and Johnson [1985: 143].

6.12. Horn and Johnson [1985: 143] and Rao and Bhimasankaram [2000: 295].

6.13. Meyer [2000a: 643].

6.14. Meyer [2000a: 648].

6.1.2   Eigenvalues

We assume that $\mathbf{A}$ is $n \times n$.

6.15. For every j, and $\mathbf{A}$ real or complex:

(a) $\lambda_j(\mathbf{A}') = \lambda_j(\mathbf{A})$.

(b) $\lambda_j(\overline{\mathbf{A}}) = \overline{\lambda}_j(\mathbf{A})$.

(c) $\lambda_j(\mathbf{A}^*) = \overline{\lambda}_j(\mathbf{A})$.

(d) $\lambda_j(\mathbf{K}^{-1}\mathbf{A}\mathbf{K}) = \lambda_j(\mathbf{A})$ for all j.

6.16. If $\mathbf{A}$ has r nonzero eigenvalues, then:

(a) rank $\mathbf{A} \geq r$

(b) It is possible to have $r = 0$, but have rank $\mathbf{A} = n - 1$.

6.17. We have the following:

(a) If k is a positive integer, then

$$\operatorname{trace} \mathbf{A}^k = \sum_{i=1}^{n} \lambda_i^k.$$

(b) Taking $k = 1$, trace $\mathbf{A} = \sum_{i=1}^{n} \lambda_i$.

(c) $\det(\mathbf{A}) = \prod_{i=1}^{n} \lambda_i$.

6.18. $\mathbf{A}$ is nonsingular if and only if $\lambda_j(\mathbf{A}) \neq 0$ for all j (cf. 6.17c).

6.19. If $\mathbf{A}$ is triangular, then, since the determinant of the upper-triangular matrix $\mathbf{A} - \lambda\mathbf{I}_n$ is the product of its diagonal elements, the eigenvalues are the diagonal elements of $\mathbf{A}$.

6.20. If $\lambda_n = 0$ is the only zero eigenvalue, then

$$\operatorname{trace}(\operatorname{adj}\mathbf{A}) = \prod_{i=1}^{n-1} \lambda_i,$$

where adj$\mathbf{A}$ is the adjoint matrix of $\mathbf{A}$.

6.21. (Bounds Using Traces) Let $\mathbf{A}$ be an $n \times n$ real or complex matrix with real eigenvalues λ_i; for example, $\mathbf{A}$ is Hermitian or symmetric. Define

$$m = \frac{1}{n} \sum_{i=1}^{n} \lambda_i = \frac{1}{n} \text{ trace } \mathbf{A}$$

and

$$s^2 = \frac{1}{n} \left(\sum_{i=1}^{n} \lambda_i^2 \right) - m^2 = \frac{1}{n} \text{ trace}(\mathbf{A}^2) - m^2.$$

(a) Then

$$m - s(n-1)^{1/2} \leq \lambda_{\min}(\mathbf{A}) \leq m - s(n-1)^{-1/2},$$
$$m + s(n-1)^{-1/2} \leq \lambda_{\max}(\mathbf{A}) \leq m + s(n-1)^{1/2}.$$

Equality on the left (respectively right) of the first equation holds if and only if equality holds on the left (respectively right) of the second equation, if and only if the $n - 1$ largest (respectively smallest) eigenvalues are equal. When $n = 2$ we have $\lambda_{\min}(\mathbf{A}) = m - s$ and $\lambda_{\max}(\mathbf{A}) = m + s$.

(b) (Bounds on the Condition Number) Let $\mathbf{A}$ be Hermitian positive definite with condition number $\kappa_2(\mathbf{A}) = \lambda_{\max}\mathbf{A}/\lambda_{\min}(\mathbf{A})$ (cf. 4.96).

   (i) When n is even,

$$1 + \frac{2s}{m - s(n-1)^{-1/2}} \leq \kappa_2(\mathbf{A}).$$

When $n > 2$, equality holds if and only if $\mathbf{A} = c\mathbf{I}_n$, where c is a real constant.

   (ii) When n is odd, (i) holds along with

$$1 + \frac{2sn(n^2-1)^{-1/2}}{m - s(n-1)^{-1/2}} \leq \kappa_2(\mathbf{A}).$$

When $n = 3$, equality holds if and only if the two smallest eigenvalues are equal. When $n > 3$, equality holds if and only if $\mathbf{A} = c\mathbf{I}_n$.

   (iii) In general,

$$\kappa_2(\mathbf{A}) \leq 1 + \frac{(2n)^{1/2}s[m + s(n-1)^{-1/2}]^{n-1}}{\det \mathbf{A}}.$$

When $n > 2$, equality holds if and only if $\mathbf{A} = c\mathbf{I}_n$.

   (iv) If $\mathbf{A}$ is Hermitian, trace $\mathbf{A} > 0$, and $(\text{trace } \mathbf{A})^2 > (n-1)\,\text{trace}(\mathbf{A}^2)$, then $\mathbf{A}$ is positive definite, (i) holds, and

$$\kappa_2(\mathbf{A}) \leq 1 + \frac{(2n)^{1/2}s}{m - s(n-1)^{1/2}}.$$

When $n > 2$, equality holds if and only if $\mathbf{A} = c\mathbf{I}_n$.

(c) Suppose $\mathbf{A}$ has real eigenvalues with f eigenvalues of $\mathbf{A}$ positive and g negative. Let $\mathrm{trace}(\mathbf{A}^2) > 0$.

   (i) When $\mathrm{trace}\,\mathbf{A} \geq 0$, then

$$(\mathrm{trace}\,\mathbf{A})^2 / \mathrm{trace}(\mathbf{A}^2) \geq f,$$

   with equality if and only if all the positive eigenvalues are equal and all the nonpositive eigenvalues are equal.

   (ii) When $\mathrm{trace}\,\mathbf{A} \leq 0$, then

$$(\mathrm{trace}\,\mathbf{A})^2 / \mathrm{trace}(\mathbf{A}^2) \leq g,$$

   with equality if and only if all the negative eigenvalues are equal and all the non-negative eigenvalues are equal.

(d) Let $\lambda_1 \geq \lambda_2 \geq \cdots \geq \lambda_n$.

   (i) Then

$$\lambda_k - \lambda_l \leq s n^{1/2} \left(\frac{1}{k} + \frac{1}{n - l + 1} \right)^{1/2}, \quad 1 \leq k \leq l \leq n.$$

   Equality occurs if and only if

$$\begin{aligned}
\lambda_1 &= \lambda_2 = \cdots = \lambda_k, \\
\lambda_{k+1} &= \lambda_{k+2} = \cdots = \lambda_{l-1} = m, \\
\lambda_l &= \lambda_{l+1} = \cdots = \lambda_n.
\end{aligned}$$

   (ii) From (i) we have
$$\lambda_1 - \lambda_n \leq (2n)^{1/2} s.$$

   When $n > 2$, equality holds if and only if

$$\lambda_2 = \lambda_3 = \cdots = \lambda_{n-1} = \tfrac{1}{2}(\lambda_1 + \lambda_n).$$

   (iii) If $n = 2q$ is even, then
$$2s \leq \lambda_1 - \lambda_n,$$

   with equality if and only if

$$\lambda_1 = \lambda_2 = \cdots = \lambda_q \quad \text{and} \quad \lambda_{q+1} = \lambda_{q+2} = \cdots = \lambda_n.$$

   (iv) If $n = 2q \pm 1$ is odd, the previous inequality (iii) holds and

$$2sn(n^2 - 1)^{-1/2} \leq \lambda_1 - \lambda_n,$$

   with equality if and only if the conditions for the equality of (iii) hold.

   (v)

$$\frac{(\mathrm{trace}\,\mathbf{A})^2}{\mathrm{trace}(\mathbf{A}^2)} \leq n - 2 + \frac{(\lambda_1 + \lambda_n)^2}{\lambda_1^2 + \lambda_n^2}.$$

When $n > 2$, equality holds if and only if $\lambda_1 + \lambda_n \neq 0$ and

$$\lambda_2 = \lambda_3 = \cdots = \lambda_{n-1} = \frac{\lambda_1^2 + \lambda_n^2}{\lambda_1 + \lambda_n}.$$

(e) The above results can be extended to complex matrices with complex eigenvalues. For example, if $\mathbf{A}$ now has complex eigenvalues $|\lambda_1| \geq |\lambda_2| \geq \cdots \geq |\lambda_n|$ and we define

$$m = \frac{1}{n} \operatorname{trace} \mathbf{A} \quad \text{and} \quad s_a^2 = \frac{1}{n} \operatorname{trace}(\mathbf{A}^*\mathbf{A}) - |m|^2,$$

then:

(i) $|m| - s_a(n-1)^{1/2} \leq |\lambda_n| \leq [\operatorname{trace}(\mathbf{A}^*\mathbf{A})/n]^{1/2}$.

Equality holds on the left if and only if $\mathbf{A}$ is normal, $\lambda_1 = \lambda_2 = \cdots = \lambda_{n-1}$, and $\lambda_n = cm$ for some real non-negative scalar $c \leq 1$. Equality holds on the right if and only if $\mathbf{A}$ is normal and $|\lambda_1| = |\lambda_2| = \cdots = |\lambda_n|$.

(ii) $|m| \leq |\lambda_1| \leq |m| + s_a(n-1)^{1/2}$.

Equality holds on the left if and only if $\lambda_1 = \lambda_2 = \cdots = \lambda_n$. Equality holds on the right if and only if $\mathbf{A}$ is normal, $\lambda_2 = \lambda_3 = \cdots = \lambda_n$, and $\lambda_1 = cm$ for scalar $c \geq 1$.

(iii) If $\mathbf{A}$ has k nonzero eigenvalues, then

$$|\operatorname{trace} \mathbf{A}|^2 / \operatorname{trace}(\mathbf{A}^*\mathbf{A}) \leq k \leq \operatorname{rank} \mathbf{A}.$$

Equality holds on the left if and only if $\mathbf{A}$ is normal and $|\lambda_1| = |\lambda_2| = \cdots = |\lambda_k|$. Equality holds on the right if and only if $\operatorname{rank} \mathbf{A} = \operatorname{rank}(\mathbf{A}^2)$.

Further extensions are given by Wolkowicz and Styan [1980].

Additional results relating to sums of eigenvalues are given by Wolkowicz and Styan [1980]. Extensions are given by Merikoski and Virtanen [2004] and are used to give a lower bound for the Perron root of a non-negative matrix.

6.22. Let $\mathbf{A} = (a_{ij})$ and $\mathbf{B} = (b_{ij})$ be $n \times n$ matrices with eigenvalues λ_i and γ_i, respectively. Define

$$M = \max_{1 \leq i \leq n, 1 \leq j \leq n} (|a_{ij}|, |b_{ij}|) \quad \text{and} \quad \delta(\mathbf{A}, \mathbf{B}) = \frac{1}{n} \sum_{i=1}^{n} \sum_{j=1}^{n} |a_{ij} - b_{ij}|,$$

then

$$\max_{1 \leq i \leq n} \min_{1 \leq j \leq n} |\lambda_i - \gamma_j| \leq (n+2) M^{1-n^{-1}} [\delta(\mathbf{A}, \mathbf{B})]^{1/n}.$$

6.23. Let $\mathbf{A}$ and $\mathbf{B}$ be real $n \times n$ symmetric matrices with $\mathcal{C}(\mathbf{A}) \subseteq \mathcal{C}(\mathbf{B})$. Suppose that $\mathbf{B}$ is non-negative definite, and let $\mathbf{X}$ be an $n \times k$ matrix. Then:

(a) $\mathcal{C}(\mathbf{X}'\mathbf{A}\mathbf{X}) \subseteq \mathcal{C}(\mathbf{X}'\mathbf{B}\mathbf{X})$.

(b) Consider the eigenvalues of $(\mathbf{X}'\mathbf{B}\mathbf{X})^- \mathbf{X}'\mathbf{A}\mathbf{X}$ for any weak inverse $(\mathbf{X}'\mathbf{B}\mathbf{X})^-$.

(i) The eigenvalues are all real and do not depend on the choice of generalized inverse.

(ii) The eigenvalues are the generalized eigenvalues of $\mathbf{X}'\mathbf{A}\mathbf{X}$ with respect to $\mathbf{X}'\mathbf{B}\mathbf{X}$ (cf. Section 6.1.8).

(c) $\mathcal{I}n[(\mathbf{X}'\mathbf{B}\mathbf{X})^{-}\mathbf{X}'\mathbf{A}\mathbf{X})] = \mathcal{I}n(\mathbf{X}'\mathbf{A}\mathbf{X})$, where $\mathcal{I}n(\cdot)$ is the inertia.

6.24. For each i, λ_i is a continuous function of the elements of $\mathbf{A}$.

6.25. (Quadratic Inequalities) Suppose $\mathbf{A}$ is an $n \times n$ Hermitian matrix and $\mathbf{x}_i$ $(i = 1, 2, \ldots, n)$ are a set of mutually orthonormal vectors, i.e., $\mathbf{x}_i^*\mathbf{x}_j = \delta_{ij}$, then:

(a) $\sum_{i=1}^{k} \mathbf{x}_i^*\mathbf{A}\mathbf{x}_i \leq \sum_{i=1}^{k} \lambda_i(\mathbf{A})$, $\quad k = 1, 2, \ldots, n-1$.

(b) $\sum_{i=1}^{n} \mathbf{x}_i^*\mathbf{A}\mathbf{x}_i = \sum_{i=1}^{n} \lambda_i(\mathbf{A})$.

6.26. (Hirsch and Bendixson) Let $\mathbf{A} = (a_{ij})$ be an $n \times n$ complex matrix with eigenvalues λ_i, and define the Hermitian matrices $\mathbf{B} = (\mathbf{A} + \mathbf{A}^*)/2$ and $\mathbf{C} = (\mathbf{A} - \mathbf{A}^*)/(2i)$. Then:

(a) $|\lambda_i| \leq n \max\limits_{i,j} |a_{ij}|$.

(b) $|\Re e(\lambda_i)| \leq n \max\limits_{i,j} |b_{ij}|$.

(c) $|\Im m(\lambda_i)| \leq n \max\limits_{i,j} |c_{ij}|$.

Here $\Re e$ and $\Im m$ denote the "real" and "imaginary" parts, respectively. When $\mathbf{A}$ is Hermitian, the three results all reduce to (a).

6.27. (Schur) If $\mathbf{A} = (a_{ij})$ is an $n \times n$ complex matrix with eigenvalues λ_i, then

$$\sum_{i=1}^{n} |\lambda_i|^2 \leq \sum_i \sum_j |a_{ij}|^2,$$

with equality if and only if $\mathbf{A}$ is a normal matrix.

6.28. If $\mathbf{A}$ is any $n \times n$ matrix, then, given $\epsilon > 0$, there exists an $n \times n$ matrix $\mathbf{B}$ with distinct eigenvalues such that

$$\sum_{i=1}^{n} \sum_{j=1}^{n} |a_{ij} - b_{ij}| < \epsilon.$$

6.29. (Geršgorin) Let $\mathbf{A} = (a_{ij})$ be an $n \times n$ matrix, and let

$$R_i = \sum_{j=1:j\neq i}^{n} |a_{ij}|, \quad i = 1, 2, \ldots, n.$$

(a) All the eigenvalues of $\mathbf{A}$ are located in the union of n discs (called *Geršgorin discs*)

$$\cup_{i=1}^{n}\{z \in \mathbb{C} : |z - a_{ii}| \leq R_i\}.$$

Furthermore, if the union of k of these discs forms a connected region that is disjoint from the remaining $n - k$ discs, then there are precisely k eigenvalues in this region.

(b) Since $\mathbf{A}$ and $\mathbf{A}'$ have the same eigenvalues, a similar result holds with R_i replaced by

$$C_j = \sum_{i=1: i \neq j}^{n} |a_{ij}|, \quad j = 1, 2, \ldots, n$$

and the union of discs by

$$\cup_{j=1}^{n} \{ z \in \mathbb{C} : |z - a_{jj}| \leq C_j \}.$$

(c) The eigenvalues of $\mathbf{A}$ lie in the intersection of the above two regions.

(d) $\max_i |\lambda_i(\mathbf{A})| \leq \max_i \sum_{j=1}^{n} |a_{ij}|$.

Some generalizations of the above results are given by Horn and Johnson [1985: section 6.4].

6.30. (Commuting Matrices)   Let $\mathbf{A}$ be an $n \times n$ matrix with distinct eigenvalues, and let $\mathbf{B}$ be an $n \times n$ matrix that commutes with $\mathbf{A}$, that is, $\mathbf{AB} = \mathbf{BA}$. Then $\mathbf{B}$ can be expressed uniquely as a polynomial in $\mathbf{A}$ with degree no more than $n - 1$.

6.31. (Perturbations)   Suppose that a Hermitian $n \times n$ matrix with (real) eigenvalues $\lambda_1 \geq \cdots \geq \lambda_n$ is perturbed by a Hermitian matrix $\mathbf{E}$ with ranked eigenvalues ϵ_i to give $\mathbf{B} = \mathbf{A} + \mathbf{E}$ with ranked eigenvalues β_i. Then

$$\lambda_i + \epsilon_1 \geq \beta_i \geq \lambda_i + \epsilon_n, \quad i = 1, 2, \ldots, n.$$

Proofs.   Section 6.1.2

6.15. Abadir and Magnus [2005: 166–167] and Horn and Johnson [1985: 57, (a)–(c)].

6.16a. Graybill [1983: 305]), Magnus and Neudecker [1999: 19–20], and Schott [2005: 160].

6.16b. For a counter example see Abadir and Magnus [2005: 165, exercise 7.19].

6.17. Schott [2005: 91].

6.20. If $\mathbf{A}$ has nonzero eigenvalues, then $\text{trace}(\text{adj}\mathbf{A}) = \text{trace}[(\det \mathbf{A})\mathbf{A}^{-1}] = \Pi_i \lambda_i \sum_i \lambda_i^{-1}$. Let $\lambda_n \to 0$.

6.21. Wolkowicz and Styan [1980: (a), 474–476; (b) 484–485; (c) 480–481; (d) 482–483; and (e) 491, 495].

6.22. Ostrowski [1973]; see also Elsner [1982] for some other bounds.

6.23. Scott and Styan [1985: 212].

6.24. Schott [2005: 103].

6.25. Rao and Rao [1998: 383].

6.26. Marcus and Minc [1964: 141].

6.27. Marcus and Minc [1964: 142] and Zhang [1997: 241]; Tsatsomeros [2007: 14.2] also lists this and other inequalities.

6.28. Bellman [1970: 199].

6.29. Horn and Johnson [1985: 344–346] and Meyer [2000a: 498].

6.30. Zhang [1999: 59].

6.31. Meyer [2000a: 551].

6.1.3 Singular Values

Definition **6.5.** Suppose $\mathbf{B}$ is an $m \times n$ real or complex matrix of rank r, where $r \le p = \min(m,n)$. The p largest eigenvalues of $\mathbf{B}^*\mathbf{B}$, which are the same as those for $\mathbf{BB}^*$ (by 6.54c) are non-negative (by 10.10 and 10.2), as $\mathbf{B}^*\mathbf{B}$ is non-negative definite. Their positive square roots are called the *singular values* of $\mathbf{B}$. Denote these by $\sigma_1 \ge \sigma_2 \ge \ldots \ge \sigma_r > \sigma_{r+1} = \cdots = \sigma_p = 0$; we shall use the notation $\sigma_i = \sigma_i(\mathbf{B})$. (See Section 16.3 for further details and the singular value decomposition of a matrix.) Some interesting historical comments are given by Horn and Johnson [1991: section 3.0].

6.32. Suppose that $\mathbf{B}$ is an $m \times n$ matrix with singular values $\sigma_1 \ge \sigma_2 \ge \cdots \ge \sigma_p \ge 0$, and $p = \min\{m,n\}$. Let

$$\mathbf{A} = \begin{pmatrix} \mathbf{0} & \mathbf{B}^* \\ \mathbf{B} & \mathbf{0} \end{pmatrix}.$$

Then $\mathbf{A}$ is an $(m+n) \times (m+n)$ Hermitian matrix with eigenvalues

$$\sigma_1 \ge \sigma_2 \ge \cdots \ge \sigma_p \ge 0 = \cdots = 0 \ge -\sigma_p \ge -\sigma_{p-1} \ge \cdots \ge -\sigma_1,$$

with $|m - n|$ zeros in the middle.

6.33. Suppose $\mathbf{B} \in \mathcal{B}$, the set of all $m \times n$ matrices. Then, for every $\epsilon > 0$, there exists $\mathbf{B}_\epsilon \in \mathcal{B}$ with distinct singular values such that $\|\mathbf{B} - \mathbf{B}_\epsilon\| < \epsilon$, where $\|\cdot\|$ is any generalized matrix norm on $\mathcal{B}$.

6.34. Let $\mathbf{A}$ be an $n \times n$ matrix with $\lambda_i(\mathbf{A})$ and $\sigma_i(\mathbf{A})$ the ordered eigenvalues and singular values, respectively, in decreasing order of magnitude.

(a) $\prod_{i=1}^{k} |\lambda_i(\mathbf{A})| \le \prod_{i=1}^{k} \sigma_i(\mathbf{A})$ for $k = 1, 2, \ldots, n$, with equality for $k = n$.

(b) $\sum_{i=1}^{k} |\lambda_i(\mathbf{A})| \le \sum_{i=1}^{k} \sigma_i(\mathbf{A})$ for $k = 1, 2, \ldots, n$.

(c) $|\operatorname{trace} \mathbf{A}| \le \sigma_1(\mathbf{A}) + \sigma_2(\mathbf{A}) + \cdots + \sigma_n(\mathbf{A})$. Equality holds if and only if $\mathbf{A} = u\mathbf{C}$ for some non-negative definite matrix $\mathbf{C}$ and some complex scalar u with unit modulus. When equality holds, $\mathbf{A}$ is a normal matrix.

(d) $\sigma_i(\mathbf{A}) = \sigma_i(\mathbf{UAV})$ $(i = 1, 2, \ldots, n)$ for all $n \times n$ unitary matrices $\mathbf{U}$ and $\mathbf{V}$.

(e) $\lim_{n \to \infty} [\sigma_i(\mathbf{A}^n)]^{1/n} = |\lambda_i(\mathbf{A})|$ for $i = 1, 2, \ldots, n$.

6.35. If $\mathbf{A}$ is an $n \times n$ matrix and $\mathbf{H}(\mathbf{A})$ is the Hermitian matrix $\frac{1}{2}(\mathbf{A} + \mathbf{A}^*)$, then, for $i = 1, 2, \ldots, n$, the following results hold.

(a) $\sigma_i(\mathbf{A}) \geq \lambda_i(\mathbf{H}(\mathbf{A}))$.

(b) $\sigma_i(\mathbf{A}) \geq \lambda_i[\mathbf{H}(\mathbf{UAV})]$ for all $n \times n$ unitary $\mathbf{U}$ and $\mathbf{V}$.

6.36. If $\mathbf{A}$ is an $m \times n$ matrix, $p = \min\{m, n\}$, and $\sigma_i = \sigma_i(\mathbf{A})$, then:

(a) $\sum_{i=1}^{k} a_{ii}^2 \leq \sum_{i=1}^{k} \sigma_i^2$, $\quad k = 1, 2, \ldots, p$.

(b) $\sum_{i=1}^{k} a_{ii} \leq \sum_{i=1}^{k} \sigma_i$, $\quad k = 1, 2, \ldots, p$.

Equality in (a) holds if and only if the leading $k \times k$ principal submatrix of $\mathbf{A}$ is diagonal and $|a_{ii}| = \sigma_i$ $(i = 1, 2, \ldots, k)$. Equality in (b) occurs when equality in (a) holds and $a_{ii} \geq 0$ $(i = 1, 2, \ldots, k)$.

6.37. (Bilinear Inequalities) Let $\mathbf{A}$ be an $m \times n$ matrix, and let $p = \min\{m, n\}$. For $i = 1, 2, \ldots, p$, let $\mathbf{z}_i' = (\mathbf{x}_i', \mathbf{y}_i')$ be any mutually orthonormal vectors, where $\mathbf{x}_i$ is $m \times 1$ and $\mathbf{y}_i$ is $n \times 1$. Then

$$\sum_{i=1}^{k} 2\mathbf{x}_i{}^* \mathbf{A} \mathbf{y}_i \leq \sum_{i=1}^{k} \sigma_i(\mathbf{A}), \quad k = 1, 2, \ldots, p.$$

Equality is attained when $\mathbf{x}_i$ and $\mathbf{y}_i$ are, respectively, the left and right singular vectors of $\mathbf{A}$ associated with σ_i (cf. Section 16.3).

6.38. Let $\mathbf{A}$ be a real or complex square matrix with *numerical radius*

$$w(\mathbf{A}) = \sup_{\|\mathbf{x}\|=1} |\mathbf{x}^* \mathbf{A} \mathbf{x}|.$$

Then $\rho(\mathbf{A}) \leq w(\mathbf{A}) \leq \sigma_{\max} \leq 2w(\mathbf{A})$, where $\rho(\mathbf{A})$ is the spectral radius of $\mathbf{A}$.

Proofs.   Section 6.1.3

6.32. Horn and Johnson [1985: 418] and Rao and Rao [1998: 325].

6.33. Horn and Johnson [1985: 417].

6.34a. Horn and Johnson [1991: 171] and Rao and Rao [1998: 339–340].

6.34b. Horn and Johnson [1991: 176].

6.34c. Horn and Johnson [1991: 176] and Zhang [1999: 260–261].

6.34d. Horn and Johnson [1991: 146].

6.34e. Horn and Johnson [1991: 180].

6.35. Horn and Johnson [1991: 151].

6.36. Rao and Rao [1998: 385].

6.37. Rao and Rao [1998: 383–384].

6.38. Zhang [1999: 90].

6.1.4 Functions of a Matrix

6.39. If $\mathbf{Ax} = \lambda_i \mathbf{x}$ and k is a positive integer, then $\mathbf{A}^k \mathbf{x} = \lambda_i^k \mathbf{x}$, so that $\mathbf{A}^k$ has eigenvalues λ_i^k and the same eigenvectors as $\mathbf{A}$.
If $\mathbf{A}^t = \mathbf{0}$ for some positive integer t, then $\lambda_i(\mathbf{A}) = 0$ for all i.

6.40. If $\mathbf{A}$ has eigenvalues $\lambda_i(\mathbf{A})$, a polynomial $g(\mathbf{A})$ has eigenvalues $g(\lambda_i)$ ($i = 1, 2, \ldots, n$) and the same eigenvectors as $\mathbf{A}$.

6.41. If $\mathbf{A}$ is nonsingular with eigenvalues λ_j, then $\mathbf{A}^{-1}$ has eigenvalues λ_j^{-1}.

6.42. Let $\mathbf{A}$ be an $n \times n$ matrix. If $a_0, a_1, \ldots, a_m$ are real or complex numbers, and
$$\mathbf{B} = a_0 \mathbf{I}_n + a_1 \mathbf{A} + \cdots + a_m \mathbf{A}^m,$$
then the eigenvalues of $\mathbf{B}$ are
$$a_0 + a_1 \mu_j(\mathbf{A}) + a_2 \mu_j^2(\mathbf{A}) + \cdots + a_m \mu_j^m(\mathbf{A}) \quad \text{for} \quad j = 1, 2, \ldots, s,$$
where the $\mu_i(\mathbf{A})$ are the distinct eigenvalues. If $\mathbf{B} = \mathbf{0}$, then any eigenvalue λ of $\mathbf{A}$ must satisfy the equation
$$a_0 + a_1 \lambda + a_2 \lambda^2 + \cdots + a_m \lambda^m = 0.$$

Proofs.   Section 6.1.4

   6.39. Schott [2005: 90].

   6.40. Rao and Bhimasankaram [2000: 289].

   6.41. Schott [2005: 90].

   6.42. Quoted by Marcus and Minc [1964: 23].

6.1.5 Eigenvectors

6.43. Right (left) eigenvectors associated with distinct eigenvalues μ_j are linearly independent.

6.44. The eigenspace corresponding to a distinct eigenvalue μ_j, say, is a vector subspace.

6.45. Let $\mathbf{A}$ be a real or complex square matrix, and let $\mathbf{x}$ be any $n \times 1$ nonzero vector. Then there exists an eigenvector $\mathbf{y}$ of $\mathbf{A}$ belonging to the span of $\{\mathbf{x}, \mathbf{Ax}, \mathbf{A}^2\mathbf{x}, \ldots\}$.

6.46. (Left and Right Eigenvectors) Suppose $\mathbf{A}$ is a complex square matrix.

(a) If $\mathbf{Ax} = \lambda \mathbf{x}$, $\mathbf{y}^*\mathbf{A} = \mu \mathbf{y}^*$, and $\lambda \neq \mu$, then $\mathbf{x}$ is orthogonal to $\mathbf{y}$ (i.e., $\mathbf{x}^*\mathbf{y} = 0$).

(b) $\mathbf{A}^*\mathbf{y} = \bar{\mu}\mathbf{y}$.

Proofs.   Section 6.1.5

   6.43. Rao and Bhimasankaram [2000: 287].

   6.44. Schott [2005: 88].

   6.45. Rao and Bhimasnakaram [2000: 288] and Rao and Rao [1998: 184].

   6.46. Abadir and Magnus [2005: 173].

6.1.6 Hermitian Matrices

Hermitian matrices are also discussed in Sections 5.2.

6.47. Suppose $\mathbf{A}$ is an $n \times n$ Hermitian matrix. Then the following hold.

(a) The eigenvalues of $\mathbf{A}$ are real.

(b) Eigenvectors corresponding to different eigenvalues are orthogonal (with respect to the inner product $\langle \mathbf{x}, \mathbf{y} \rangle = \mathbf{x}^*\mathbf{y}$). A right eigenvalue is also a left eigenvalue, and vice versa.

(c) There is a complete set of n orthonormal eigenvectors.

(d) $\sum_{i=1}^{n} \sum_{j=1}^{n} |a_{ij}|^2 = \sum_{i=1}^{n} |\lambda_i|^2$.

(e) There exists a unitary matrix $\mathbf{U}$ (i.e., $\mathbf{U}^*\mathbf{U} = \mathbf{I}_n$) such that $\mathbf{U}^*\mathbf{A}\mathbf{U} = \mathbf{\Lambda}$, where $\mathbf{\Lambda}$ is a diagonal matrix of the eigenvalues of $\mathbf{A}$ (cf. 16.44).

(f) Since $\mathbf{A}$ is also normal, the results relating to normal matrices apply.

6.48. (Real Symmetric Matrices) If $\mathbf{A}$ is an $n \times n$ real symmetric matrix, then it is also Hermitian and all the results for Hermitian matrices in (6.47) above apply here. However, we collect some of the results below for easy reference.

(a) The eigenvalues λ_i are all real and the corresponding eigenvectors can be chosen to be real.

(b) If rank $\mathbf{A} = r$, there are r nonzero eigenvalues and $\lambda = 0$ has algebraic multiplicity $(n - r)$.

(c) Since $\mathbf{x}'\mathbf{A} = \lambda\mathbf{x}'$ if and only if $\mathbf{A}\mathbf{x} = \lambda\mathbf{x}$, right eigenvectors are also left eigenvectors.

(d) Eigenvectors corresponding to different eigenvalues are orthogonal so that the corresponding eigenspaces are orthogonal.

(e) There exist n mutually orthogonal eigenvectors.

(f) $\operatorname{rank}(\mathbf{A} - \lambda_i\mathbf{I}_n) = n - m_i$, where m_i is the algebraic multiplicity of λ_i.

(g) There exists an orthogonal matrix $\mathbf{T}$ such that (cf. Section 16.6)

$$\mathbf{T}'\mathbf{A}\mathbf{T} = \operatorname{diag}(\lambda_1, \lambda_2, \ldots, \lambda_n).$$

(h) $\sum_{i=1}^{n} \sum_{j=1}^{n} a_{ij}^2 = \operatorname{trace}(\mathbf{A}^2) = \sum_{i=1}^{n} \lambda_i^2$.

(i) If $\mathbf{x}$ is any nonzero vector, then, for some $r \geq 1$, the vector space spanned by the vectors $\mathbf{x}, \mathbf{A}\mathbf{x}, \ldots, \mathbf{A}^{r-1}\mathbf{x}$ contains an eigenvector of $\mathbf{A}$.

Proofs. Section 6.1.6

6.47. Horn and Johnson [1985: 169–172]. For the second part of (b), if $\mathbf{y}^*\mathbf{A} = \lambda\mathbf{y}^*$, then $\mathbf{A}\mathbf{y} = \mathbf{A}^*\mathbf{y} = \overline{\lambda}\mathbf{y} = \lambda\mathbf{y}$.

6.48a–h. Abadir and Magnus [2005: section 7.2] and Searle [1982: 290–291].

6.48i. Schott [2005: 96].

6.1.7 Computational Methods

6.49. (Power Method) Let $\mathbf{A}$ be an $n \times n$ real diagonalizable matrix with real eigenvalues and a dominant eigenvalue λ_1 (i.e., $|\lambda_1| > |\lambda_2| \geq \cdots \geq |\lambda_n|$). Since $\mathbf{A}$ is diagonalizable, there exist n real right eigenvectors $\mathbf{u}_1, \mathbf{u}_2, \ldots, \mathbf{u}_n$, with $\mathbf{u}_i$ corresponding to λ_i, which are scaled to have unit length and are linearly independent.

(a) Let $\mathbf{y} = \sum_{i=1}^{n} a_i \mathbf{u}_i$, where $a_1 > 0$. Set $\mathbf{y}_0 = \mathbf{y}$ and define $\mathbf{z}_k$ and $\mathbf{y}_k$ inductively by the following: $\mathbf{z}_k = \mathbf{A}\mathbf{y}_{k-1}$ and $\mathbf{y}_k = (1/\|\mathbf{z}_k\|_2)\mathbf{z}_k$ for $k = 1, 2, \ldots$. Then $\|\mathbf{z}_k\|_2 \to |\lambda_1|$ as $k \to \infty$, and $\mathbf{y}_{2m} \to \mathbf{u}_1$ as $m \to \infty$. Also $\mathbf{y}_{2m+1}$ converges to $\mathbf{u}_1$ or $-\mathbf{u}_1$ according as λ_1 is positive or negative. One can determine the sign of λ_1 by considering successive iterations. See also Golub and Van Loan [1996: 406].

(b) If $\mathbf{R}$ is any nonsingular matrix with $\mathbf{u}_1$ as the first column, then

$$\mathbf{R}^{-1}\mathbf{A}\mathbf{R} = \begin{pmatrix} \lambda_1 & \mathbf{a}' \\ \mathbf{0} & \mathbf{B} \end{pmatrix}$$

for some $\mathbf{a}$ and $\mathbf{B}$, and the eigenvalues of $\mathbf{A}$ are those of $\mathbf{B}$ together with λ_1. If $\mu \neq \lambda_1$ is an eigenvalue of $\mathbf{B}$ with $\mathbf{v}$ as a corresponding eigenvector, then, setting $b = (\mathbf{a}'\mathbf{v})/(\mu - \lambda_1)$, we find that $\mathbf{R}\binom{b}{\mathbf{v}}$ is an eigenvector of $\mathbf{A}$ corresponding to μ. This approach can be used to obtain the eigenvalues and corresponding eigenvectors if $|\lambda_1| > |\lambda_2| > \cdots > |\lambda_n|$.

(c) Suppose $|\lambda_i| > |\lambda_j|$ and let $\mathbf{v}_i$ and $\mathbf{v}_j$ be the real, left unit eigenvectors coresponding to λ_i and λ_j, respectively. Then, since $\mathbf{A}'\mathbf{v}_j = \lambda_j \mathbf{v}_j$ and $\mathbf{A}\mathbf{u}_i = \lambda_i \mathbf{u}_i$, we have:

  (i) $(\mathbf{A}\mathbf{u}_i)'\mathbf{v}_j = \mathbf{u}_i' \mathbf{A}'\mathbf{v}_j = \lambda_j \mathbf{u}_i'\mathbf{v}_j$ and $(\mathbf{A}\mathbf{u}_i)'\mathbf{v}_j = \lambda_i \mathbf{u}_i'\mathbf{v}_j$, so that $\mathbf{u}_i \perp \mathbf{v}_j$ as $\lambda_i \neq \lambda_j$.

  (ii) If $\mathbf{B} = \mathbf{A} - \lambda_i \mathbf{v}_i \mathbf{u}_i'$, then $\mathbf{B}\mathbf{v}_j = \lambda_j \mathbf{v}_j$. As in (b), this method can be also be used for finding other eigenvalues.

6.50. (Jacobi's Method) Let $\mathbf{A}$ be a real symmetric matrix. Jacobi's method is based on the spectral decomposition of $\mathbf{A}$ (cf. 16.44), and the method may be decribed broadly as follows. Let $\mathbf{Q}_k$ be an orthogonal matrix, and consider the iteration process $\mathbf{A}^{(k+1)} = \mathbf{Q}_k'\mathbf{A}^{(k)}\mathbf{Q}_k = \mathbf{P}_{k+1}'\mathbf{A}\mathbf{P}_{k+1}$, where $\mathbf{P}_{k+1} = \mathbf{Q}_1\mathbf{Q}_2\cdots\mathbf{Q}_k$ is orthogonal. The starting values are $\mathbf{A}^{(1)} = \mathbf{A}$ and $\mathbf{P}_1 = \mathbf{I}_n$. Each $\mathbf{Q}_i$ is a Givens rotation matrix that reduces a current off-diagonal element to zero, thus reducing the sum of squares of the off-diagonal elements. We then find that $\mathbf{A}^{(k)}$ tends towards a diagonal matrix so that

$$\lim_{k \to \infty} \mathbf{P}_k'\mathbf{A}\mathbf{P}_k = \mathbf{\Lambda} \quad \text{and} \quad \lim_{k \to} \mathbf{P}_k = \mathbf{P},$$

where $\mathbf{\Lambda}$ is a diagonal matrix consisting of the eigenvalues of $\mathbf{A}$, and the columns of $\mathbf{P}$ are corresponding eigenvectors. Some theory is provided by Rao and Bhimasankaram [2000: 323–324] and a good description of the method along with further computational details are given by Gentle [1998: section 4.2].

6.51. (QR Method) This seems to be the most common method, and it can be used for both symmetric and nonsymmetric matrices $\mathbf{A} = (a_{ij})$, though the symmetric case is easier, since the eigenvalues are now real. The first step is to transform $\mathbf{A}$ into upper Hessenberg form using Householder or Givens transformations. When $\mathbf{A}$ is symmetric, the upper Hessenberg form is tridiagonal. For some details see Gentle [1998: section 4.3] and Golub and Van Loan [1996: section 7.4])

Proofs.   Section 6.1.7

6.49a. Rao and Bhimasankaram [2000: 326].

6.49b. Rao and Bhimasankaram [2000: 327 and exercise 3 for a correction].

6.49c. Gentle [1998: section 4.1].

6.1.8   Generalized Eigenvalues

Definition 6.6. If $\mathbf{A}$ and $\mathbf{B}$ are $n \times n$ matrices, we say that λ is an *eigenvalue of* $\mathbf{A}$ *with respect to* $\mathbf{B}$ if there exists a nonzero $\mathbf{x}$ that does not belong to both $\mathcal{N}(\mathbf{A})$ and $\mathcal{N}(\mathbf{B})$ such that $\mathbf{A}\mathbf{x} = \lambda\mathbf{B}\mathbf{x}$. Here λ is one of the n roots of $\det(\mathbf{A} - \lambda\mathbf{B})$, and these roots are also called the *generalized eigenvalues*. As $\boldsymbol{\mu}$ varies over $\mathbb{R}$, the matrix $\mathbf{A} - \mu\mathbf{B}$ is called a *matrix pencil.*

Generalized eigenvalues are used extensively in mulitivariate analysis—for example, in dimension-reducing techniques and for hypothesis testing in multivariate analysis of variance (Chapter 21). In this regard, some computational aspects using Cholesky decompositions are discussed by Maindonald [1984: section 6.5].

6.52. Let $\mathbf{A}$ and $\mathbf{B}$ be real $n \times n$ matrices with $\mathbf{B}$ nonsingular.

(a) The generalized eigenvalues are the eigenvalues of $\mathbf{B}^{-1}\mathbf{A}$, which are the same as those of $\mathbf{A}\mathbf{B}^{-1}$.

(b) Suppose $\mathbf{A}$ is symmetric and $\mathbf{B}$ is positive definite.

   (i) The eigenvalues of $\mathbf{A}\mathbf{B}^{-1}$ are real.

   (ii) From (6.54a), $\lambda(\mathbf{B}^{-1/2}\mathbf{A}\mathbf{B}^{-1/2}) = \lambda(\mathbf{B}^{-1}\mathbf{A})$, where $\mathbf{B}^{1/2}$ is the unique positive definite square root of $\mathbf{B}$ (cf. 10.32).

(c) The $\lambda(\mathbf{B}^{-1}\mathbf{A})$ can be computed using a Schur decomposition (cf. 16.37).

For further details see Harville [1997: section 21.14], and some computational aspects of the problem are discussed by Golub and Van Loan [1996: section 7.7].

Proofs.   Section 6.1.8

6.52a. This follows from $\det(\mathbf{A} - \lambda\mathbf{B}) = 0$ if and only if $\det\mathbf{B}\det(\mathbf{B}^{-1}\mathbf{A} - \lambda\mathbf{I}) = 0$ if and only if $\det(\mathbf{A}\mathbf{B}^{-1} - \lambda\mathbf{I})\det\mathbf{B} = 0$.

6.52b. Graybill [1983: 404–405] for (i).

6.1.9 Matrix Products

6.53. If $\mathbf{A}$ and $\mathbf{B}$ are real symmetric $n \times n$ matrices, then the eigenvalues of $\mathbf{AB}$ are real if either $\mathbf{A}$ or $\mathbf{B}$ is non-negative definite.

6.54. Suppose $\mathbf{A}$ is $m \times n$ and $\mathbf{B}$ is $n \times m$ ($m \leq n$), both complex matrices.

(a) $\lambda^{n-m} \det(\lambda \mathbf{I}_m - \mathbf{AB}) = \det(\lambda \mathbf{I}_n - \mathbf{BA})$,
   and $\mathbf{AB}$ and $\mathbf{BA}$ have the same nonzero eigenvalues, counting algebraic multiplicities.

(b) If λ is a nonzero eigenvalue of $\mathbf{AB}$, then λ is an eigenvalue of $\mathbf{BA}$ with the same geometric multiplicity. Also, if $\mathbf{x}_1, \ldots, \mathbf{x}_r$ are linearly independent eigenvectors of $\mathbf{AB}$ corresponding to λ, then $\mathbf{Bx}_1, \ldots, \mathbf{Bx}_r$ are linearly independent eigenvectors of $\mathbf{BA}$ corresponding to λ.

(c) If $\mathbf{A}$ is $m \times n$, then $\mathbf{AA}^*$ and $\mathbf{A}^*\mathbf{A}$ have the same nonzero eigenvalues.

6.55. If $\mathbf{A}$ and $\mathbf{B}$ are $n \times n$ matrices and $\mathbf{A}$ is nonsingular, then $\mathbf{AB}$ and $\mathbf{BA}$ have the same eigenvalues.

6.56. (Frobenius) Let $\mathbf{A}$ and $\mathbf{B}$ be $n \times n$ matrices that commute with $\mathbf{AB} - \mathbf{BA}$. Let $f(x_1, x_2)$ be any polynomial in x_1 and x_2 with possibly complex coefficients. Then there exists an ordering of the eigenvalues of $\mathbf{A}$ and $\mathbf{B}$, namely (α_i, β_i) for $i = 1, 2, \ldots, n$, such that the eigenvalues of $f(\mathbf{A}, \mathbf{B})$ are $f(\alpha_i, \beta_i)$ for $i = 1, 2, \ldots, n$.

6.57. (Von Neumann) Let $\mathbf{A}$ be $m \times n$ and $\mathbf{B}$ be $n \times m$ matrices such that $\mathbf{AB}$ and $\mathbf{BA}$ are Hermitian non-negative definite. Let $p = \min\{m, n\}$ and $q = \max\{m, n\}$. If we define $\sigma_j(\mathbf{A}) = \sigma_j(\mathbf{B}) = 0$ for $p + 1 \leq j \leq q$, where $\sigma(\cdot)$ is a singular value, then there exists a permutation τ of $\{1, 2, \ldots, q\}$ such that

$$\text{trace}(\mathbf{AB}) = \text{trace}(\mathbf{BA}) = \sum_{i=1}^{q} \sigma_i(\mathbf{A}) \sigma_{\tau(i)}(\mathbf{B}).$$

where $\tau(i)$ is the ith element of permutation τ.

Proofs.   Section 6.1.9

6.53. Graybill [1983: 404-405].

6.54a. Rao and Bhimasankaram [2000: 282] and Zhang [1999: 51–53, four proofs].

6.54b. Rao and Bhimasankaram [2000: 287].

6.54c. This follows from (a) with $\mathbf{B} = \mathbf{A}^*$.

6.55. This follows from (6.54a) with $m = n$.

6.56. Quoted by Marcus and Minc [1964: 25].

6.57. Rao and Rao [1997: 348].

6.2 VARIATIONAL CHARACTERISTICS FOR HERMITIAN MATRICES

A common statistical problem is that of finding the maximum or minimum of a ratio of two quadratic forms subject to some linear constraints—for example, in multivariate analysis. As we shall see below, eigenvalues and eigenvectors feature prominently in the theory. We shall work mainly with the more general complex quadratics as real quadratics follow as a special case. In following up proofs of the following results, the reader should note that we rank the eigenvalues λ_i in decreasing order of magnitude, whereas some authors such as Horn and Johnson [1985] and Magnus and Neudecker [1999] do the reverse. In the latter case, we change the sign of the suffix and add $n + 1$ to get corresponding results; thus λ_i becomes λ_{n+1-i}. However, Horn and Johnson [1985: 419] do not reverse the order of the singular values, but rank them in decreasing order.

6.58. Let $\mathbf{A}$ be an $n \times n$ Hermitian matrix with (real) eigenvalues $\lambda_1 \geq \lambda_2 \geq \ldots \geq \lambda_n$ and a corresponding set of orthonormal eigenvectors $\mathbf{u}_1, \mathbf{u}_2, \cdots, \mathbf{u}_n$ (i.e., $\mathbf{u}_i^*\mathbf{u}_j = \delta_{ij}$) in $\mathbb{C}^n$ such that $\mathbf{A}\mathbf{u}_i = \lambda_i\mathbf{u}_i$. For $k = 1, 2, \ldots, n$, let $\mathbf{U}_k = (\mathbf{u}_1, \mathbf{u}_2, \ldots, \mathbf{u}_k)$ and $\mathbf{V}_k = (\mathbf{u}_k, \mathbf{u}_{k+1}, \ldots, \mathbf{u}_n)$. Define $\mathbf{U} = \mathbf{U}_n = \mathbf{V}_1$. In what follows, we assume that $\mathbf{x} \in \mathbb{C}^n$ and $\mathbf{x} \neq \mathbf{0}$. We shall give properties of the ratio $r(\mathbf{x}) = \mathbf{x}^*\mathbf{A}\mathbf{x}/\mathbf{x}^*\mathbf{x}$, which is sometimes called the *Raleigh (– Ritz) ratio* (quotient). (In what follows some authors use "sup" and "inf" instead of "max" and "min," respectively. However, these expressions are equivalent as the extrema are attained.)

The results below immediately follow for real symmetric matrices by replacing $*$ by $'$. We note that $r(\mathbf{x})$ does not depend on $\|\mathbf{x}\|_2$ so that if $\mathbf{x} \neq \mathbf{0}$ we can scale $\mathbf{x}$ to satisfy $\|\mathbf{x}\|_2 = 1$; the denominator of $r(\mathbf{x})$ becomes 1. This alternative representation will be mentioned only once below, but it holds in all the following results. For general references see Horn and Johnson [1985: 176–180], Magnus and Neudecker [1999: 203–207], Rao and Rao [1998: 332–335], Schott [2005: 104–110], and Seber [1984: 525–526].

(a) (Raleigh–Ritz Theorem)

(i) $\lambda_n \leq r(\mathbf{x}) \leq \lambda_1$.

(ii) $\max_{\|\mathbf{x}\|_2=1} \mathbf{x}^*\mathbf{A}\mathbf{x} = \max_{\mathbf{x}\neq\mathbf{0}} r(\mathbf{x}) = \lambda_1$, and the maximum occurs when $\mathbf{x} = \mathbf{u}_1$.

(iii) $\min_{\mathbf{x}\neq\mathbf{0}} r(\mathbf{x}) = \lambda_n$, and the minimum occurs when $\mathbf{x} = \mathbf{u}_n$.

(b) The following hold for $k = 2, \ldots, n - 1$.

(i) $$\max_{\mathbf{x}\neq\mathbf{0}:\mathbf{U}_{k-1}^*\mathbf{x}=\mathbf{0}} r(\mathbf{x}) = \lambda_k,$$

and the maximum is attained when $\mathbf{x} = \mathbf{u}_k$. Note that $\mathbf{U}_{k-1}^*\mathbf{x} = \mathbf{0}$ implies that $\mathbf{x} \perp \{\mathbf{u}_1, \mathbf{u}_2, \ldots, \mathbf{u}_{k-1}\}$, i.e., $\mathbf{x} \in \mathcal{S}(\mathbf{u}_k, \mathbf{u}_{k+1}, \ldots, \mathbf{u}_n)$, where $\mathcal{S}$ is the span.

(ii) $$\min_{\mathbf{x}\neq\mathbf{0}:\mathbf{V}_{k+1}^*\mathbf{x}=\mathbf{0}} r(\mathbf{x}) = \lambda_k,$$

and the minimum is attained when $\mathbf{x} = \mathbf{u}_k$. Note that $\mathbf{V}_{k+1}^*\mathbf{x} = \mathbf{0}$ implies that $\mathbf{x} \perp \{\mathbf{u}_{k+1}, \mathbf{u}_{k+2}, \ldots, \mathbf{u}_n\}$, i.e., $\mathbf{x} \in \mathcal{S}(\mathbf{u}_1, \mathbf{u}_2, \ldots, \mathbf{u}_k)$.

(c) $\min\limits_{\mathbf{C}^*\mathbf{x}=0} r(\mathbf{x}) \le \lambda_k \le \max\limits_{\mathbf{B}^*\mathbf{x}=0} r(\mathbf{x})$

   for every $n \times (k-1)$ matrix $\mathbf{B}$ and $n \times (n-k)$ matrix $\mathbf{C}$.

(d) (Courant–Fischer Min–Max Theorem) Let $\mathbf{B}$ be any $n \times (k-1)$ complex matrix. Then for $k = 2, \ldots, n$ we have the following:

   (i) $\min\limits_{\mathbf{B}} \max\limits_{\mathbf{x}\ne 0:\mathbf{B}^*\mathbf{x}=0} r(\mathbf{x}) = \lambda_k$,

   and the result is attained when $\mathbf{B} = \mathbf{U}_{k-1}$ and $\mathbf{x} = \mathbf{u}_k$.

   (ii) $\max\limits_{\mathbf{B}} \min\limits_{\mathbf{x}\ne 0:\mathbf{B}^*\mathbf{x}=0} r(\mathbf{x}) = \lambda_{n-k+1}$,

   and the result is attained when $\mathbf{B} = \mathbf{V}_{n-k+2}$ and $\mathbf{x} = \mathbf{u}_{n-k+1}$.

   Since $\mathbf{U}_{k-1}^*\mathbf{U}_{k-1} = \mathbf{V}_{n-k+2}^*\mathbf{V}_{n-k+2} = \mathbf{I}_{k-1}$, we can impose the restriction $\mathbf{B}^*\mathbf{B} = \mathbf{I}_{k-1}$ without changing the above two results. Some authors use this formulation of the Courant-Fischer theorem (e.g., Schott [2005: 108–110]). Rao [1973a: 62], Seber [1984: 525–526], and Magnus and Neudecker [1999: 205–208, with the labeling $\lambda_1 \le \cdots \le \lambda_n$] prove the above for real matrices and Horn and Johnson [1985: 176] for the complex case. The complex case follows directly from proofs for the real case by simply replacing x_i^2 by $|x_i|^2$.

   The reader should note that there is a confusing variation in the proofs depending on how the constraints are defined (in our case by $\mathbf{B}^*\mathbf{x} = \mathbf{0}$). For example, if $\mathbf{B}$ is replaced by an $n \times (n-k)$ matrix $\mathbf{C}$ in (ii), then λ_{n-k+1} is replaced by λ_k (Abadir and Magnus [2005: 346] and Schott [2005: 108]). Furthermore, if $\mathbf{C}$ is used in (i) and $\mathbf{B}$ in (ii), then λ_k now refers to the kth largest eigenvalue rather than the kth smallest (Horn and Johnson [1985: 179] and Magnus and Neudecker [1999: 207]). One can also replace $\mathbf{B}$ by a general vector space, as in Meyer [2000a: 550] and Rao and Rao [1998: 332].

(e) The min–max theorem extends to singular values by replacing $\mathbf{A}$ by $\mathbf{A}^*\mathbf{A}$, as $\sigma_i(\mathbf{A})^2 = \lambda_i(\mathbf{A}^*\mathbf{A})$, and by noting that

$$\left(\frac{\mathbf{x}^*\mathbf{A}^*\mathbf{A}\mathbf{x}}{\mathbf{x}^*\mathbf{x}}\right) = \left(\frac{\|\mathbf{A}\mathbf{x}\|_2}{\|\mathbf{x}\|_2}\right)^2,$$

   where $\|\cdot\|_2$ is the Euclidean vector norm. For example, let $\mathbf{B}$ be any $n \times (k-1)$ complex matrix. Then, for $k = 2, \ldots, n$, we have the following.

   (i) $\min\limits_{\mathbf{B}} \max\limits_{\mathbf{x}\ne 0:\mathbf{B}^*\mathbf{x}=0} \left(\frac{\|\mathbf{A}\mathbf{x}\|_2}{\|\mathbf{x}\|_2}\right) = \sigma_k$.

   (ii) $\max\limits_{\mathbf{B}} \min\limits_{\mathbf{x}\ne 0:\mathbf{B}^*\mathbf{x}=0} \left(\frac{\|\mathbf{A}\mathbf{x}\|_2}{\|\mathbf{x}\|_2}\right) = \sigma_{n-k+1}$.

(f) The min–max theorem also extends to the eigenvalues of the product of two non-negative definite matrices. For details see Mäkeläinen [1970: 33].

6.59. Let $\mathbf{A}$ be a real $n \times n$ symmetric matrix, and let $\mathbf{B}$ be any $n \times n$ positive definite matrix. Let $\gamma_1 \ge \gamma_2 \ge \cdots \ge \gamma_n$ be the eigenvalues of $\mathbf{B}^{-1}\mathbf{A}$—that is, $\gamma_i = \lambda_i(\mathbf{B}^{-1}\mathbf{A})$—with corresponding right eigenvectors $\mathbf{v}_1, \mathbf{v}_2, \ldots, \mathbf{v}_n$, all of which are real by (6.52b(i)). Then

(a) $\displaystyle \max_{\mathbf{x} \neq 0} \frac{\mathbf{x}'\mathbf{A}\mathbf{x}}{\mathbf{x}'\mathbf{B}\mathbf{x}} = \gamma_1$ and $\displaystyle \min_{\mathbf{x} \neq 0} \frac{\mathbf{x}'\mathbf{A}\mathbf{x}}{\mathbf{x}'\mathbf{B}\mathbf{x}} = \gamma_n,$

with the bounds being attained when $\mathbf{x} = \mathbf{v}_1$ and $\mathbf{x} = \mathbf{v}_n$, respectively. In particular, for any $\mathbf{a}$ we have

$$\max_{\mathbf{x} \neq 0} \frac{(\mathbf{a}'\mathbf{x})^2}{\mathbf{x}'\mathbf{B}\mathbf{x}} = \mathbf{a}'\mathbf{B}^{-1}\mathbf{a},$$

and the maximum occurs when $\mathbf{x} \propto \mathbf{B}^{-1}\mathbf{a}$. The result for γ_1 applies to hypothesis testing for multivariate linear hypotheses and to the dimension reduction technique of discriminant coordinate analysis (cf. 21.49b)).

(b) Let $\mathbf{U}_i = (\mathbf{v}_1, \ldots, \mathbf{v}_i)$ and $\mathbf{W}_i = (\mathbf{v}_i, \ldots, \mathbf{v}_n)$. Then, for $\mathbf{x} \neq 0$ and $i = 2, 3, \ldots, n-1$,

$$\max_{\mathbf{U}'_{i-1}\mathbf{B}\mathbf{x}=0} \frac{\mathbf{x}'\mathbf{A}\mathbf{x}}{\mathbf{x}'\mathbf{B}\mathbf{x}} = \gamma_i \quad \text{and} \quad \min_{\mathbf{W}'_{i+1}\mathbf{B}\mathbf{x}=0} \frac{\mathbf{x}'\mathbf{A}\mathbf{x}}{\mathbf{x}'\mathbf{B}\mathbf{x}} = \gamma_i.$$

6.60. Let $\mathbf{A}$ be a real $n \times n$ symmetric matrix, and let $\mathbf{B}$ be any $n \times n$ positive definite matrix. For $i = 1, 2, \ldots, n$, let $\mathbf{B}_i$ be any $n \times (i-1)$ matrix and $\mathbf{C}_i$ be any $n \times (n-i)$ matrix satisfying $\mathbf{B}'_i\mathbf{B}_i = \mathbf{I}_{i-1}$ and $\mathbf{C}'_i\mathbf{C}_i = \mathbf{I}_{n-i}$, respectively. Then

$$\min_{\mathbf{B}_i} \max_{\mathbf{x} \neq 0 : \mathbf{B}'_i\mathbf{x}=0} \frac{\mathbf{x}'\mathbf{A}\mathbf{x}}{\mathbf{x}'\mathbf{B}\mathbf{x}} = \lambda_i(\mathbf{B}^{-1}\mathbf{A})$$

and

$$\max_{\mathbf{C}_i} \min_{\mathbf{x} \neq 0 : \mathbf{C}'_i\mathbf{x}=0} \frac{\mathbf{x}'\mathbf{A}\mathbf{x}}{\mathbf{x}'\mathbf{B}\mathbf{x}} = \lambda_i(\mathbf{B}^{-1}\mathbf{A}),$$

where the inner min and max are over all $\mathbf{x} \neq 0$ when $i = 1$ and $i = n$, respectively. The results will hold for Hermitian matrices with $'$ replaced by $*$.

6.61. Let $\mathbf{A}$ and $\mathbf{B}$ be positive definite $n \times n$ matrices. Then

$$\max_{\mathbf{x} \neq 0, \mathbf{y} \neq 0} \left\{ \frac{(\mathbf{x}'\mathbf{L}\mathbf{y})^2}{\mathbf{x}'\mathbf{A}\mathbf{x} \cdot \mathbf{x}'\mathbf{B}\mathbf{x}} \right\} = \theta_{\max},$$

where $\theta_{\max}$ is the largest eigenvalue of $\mathbf{A}^{-1}\mathbf{L}\mathbf{B}^{-1}\mathbf{L}'$, and also of $\mathbf{B}^{-1}\mathbf{L}'\mathbf{A}^{-1}\mathbf{L}$. The maximum occurs when $\mathbf{x}$ is a right eigenvector of $\mathbf{A}^{-1}\mathbf{L}\mathbf{B}^{-1}\mathbf{L}'$ corresponding to $\theta_{\max}$, and $\mathbf{y}$ is a right eigenvector of $\mathbf{B}^{-1}\mathbf{L}'\mathbf{A}^{-1}\mathbf{L}$ corresponding to $\theta_{\max}$. This result is used, for example, in applying the union–intersection method to testing hypotheses relating to variance matrices in multivariate analysis (Seber [1984: 89]).

6.62. Let $\mathbf{A}$ be a real $m \times n$ matrix of rank r $(r \leq \min\{m, n\})$, and let $\sigma_1^2 \geq \sigma_2^2 \geq \cdots \geq \sigma_r^2 > 0$ be the nonzero eigenvalues of the symmetric matrix $\mathbf{A}\mathbf{A}'$ (and of $\mathbf{A}'\mathbf{A}$), where σ_i is the ith singular value of $\mathbf{A}$. Referring to the singular value decomposition of $\mathbf{A}$ (Section 16.3), let $\mathbf{t}_1, \mathbf{t}_2, \ldots, \mathbf{t}_r$ be the corresponding orthogonal right eigenvectors of $\mathbf{A}\mathbf{A}'$, and let $\mathbf{w}_1, \mathbf{w}_2, \ldots, \mathbf{w}_r$ be the corresponding orthogonal right eigenvectors of $\mathbf{A}'\mathbf{A}$. Define $\mathbf{T}_k = (\mathbf{t}_1, \mathbf{t}_2, \ldots, \mathbf{t}_k)$ and $\mathbf{W}_k = (\mathbf{w}_1, \mathbf{w}_2, \ldots, \mathbf{w}_k)$ $(k < r)$, and assume $\mathbf{x} \neq 0$ and $\mathbf{y} \neq 0$. Then

(a) $\displaystyle \max_{\mathbf{x} \neq 0, \mathbf{y} \neq 0} \left\{ \frac{(\mathbf{x}'\mathbf{A}\mathbf{y})^2}{\mathbf{x}'\mathbf{x} \cdot \mathbf{y}'\mathbf{y}} \right\} = \sigma_1^2.$

The maximum occurs when $\mathbf{x} = \mathbf{t}_1$ and $\mathbf{y} = \mathbf{w}_1$.

(b) $\displaystyle\max_{\mathbf{T}'_k\mathbf{x}=0,\mathbf{W}'_k\mathbf{y}=0}\left\{\frac{(\mathbf{x}'\mathbf{A}\mathbf{y})^2}{\mathbf{x}'\mathbf{x}\cdot\mathbf{y}'\mathbf{y}}\right\}=\sigma^2_{k+1},\ (k=1,2,\ldots,r-1),$

and the maximum occurs when $\mathbf{x}=\mathbf{t}_{k+1}$ and $\mathbf{y}=\mathbf{w}_{k+1}$.

The above results are sometimes expressed in a square root version—for example, $\sup\left\{\dfrac{\mathbf{x}'\mathbf{A}\mathbf{y}}{\sqrt{\mathbf{x}'\mathbf{x}\cdot\mathbf{y}'\mathbf{y}}}\right\}=\sigma_1$, and so on. Another way of expressing this result is

$$\max_{\|\mathbf{x}\|=1,\|\mathbf{y}\|=1}\mathbf{x}'\mathbf{A}\mathbf{y}=\sigma_1,$$

and the $\mathbf{t}_i$ and $\mathbf{w}_i$ are now scaled to have unit norms. The above results are used in the multivariate technique of canonical correlation analysis (Seber [1984: 259]).

6.63. (Some Matrix Extensions)   Let $\mathbf{A}$ be an $n\times n$ positive definite matrix, and let $\mathbf{X}$ be an $n\times r$ matrix of rank r. Then

$$\max_{\mathbf{X}'\mathbf{X}=\mathbf{I}_r}\det(\mathbf{X}'\mathbf{A}\mathbf{X})=\prod_{i=1}^{r}\lambda_i(\mathbf{A})\quad\text{and}\quad\min_{\mathbf{X}'\mathbf{X}=\mathbf{I}_r}\det(\mathbf{X}'\mathbf{A}\mathbf{X})=\prod_{i=1}^{r}\lambda_{n-r+i}(\mathbf{A}).$$

Proofs.   Section 6.2

6.58a. Meyer [2000a: 549] and Seber [1984: 525].

6.58b. Meyer [2000a: 549] and Seber [1984: 525, with λ_{n-k} changed to λ_k].

6.58c. Abadir and Magnus [2005: 345].

6.58e. Horn and Johnson [1985: 420], Meyer [2000a: 555], and Rao and Rao [1998: 335].

6.59a. Rao and Bhimasankaram [2000: 348–349], Schott [2005: 121], and Seber [1984: 526–527].

6.59b. Schott [2005: 121].

6.60. Schott [2005: 123].

6.61. Seber [1984: 527].

6.62. Rao and Bhimasankaram [2000: 349] and Seber [1984: 528].

6.63. Abadir and Magnus [2005: 349].

6.3   SEPARATION THEOREMS

In this section we follow our usual practice and rank the eigenvalues of an $n\times n$ matrix $\mathbf{C}$ as $\lambda_1(\mathbf{C})\geq\lambda_2(\mathbf{C})\geq\cdots\geq\lambda_n(\mathbf{C})$.

6.64. Let $\mathbf{A}$ be an $n\times n$ Hermitian or real symmetric matrix, and let $\mathbf{A}_k$ be the leading principal $k\times k$ submatrix of $\mathbf{A}$, that is, $\mathbf{A}_k=(a_{rs})$, $r,s=1,2,\ldots,k$ for $k=1,2,\ldots,n-1$; we define $\mathbf{A}_n=\mathbf{A}$. Let $\lambda_1(\mathbf{A}_k)\geq\lambda_2(\mathbf{A}_k)\geq\cdots\geq\lambda_k(\mathbf{A}_k)$ for each k (including $k=n$), and let $\sigma_1(\mathbf{A}_k)\geq\cdots\geq\sigma_k(\mathbf{A}_k)$ be the singular values.

(a) (Sturmian Separation Theorem)   From the Courant–Fisher theorem we obtain the inequality

$$\lambda_{i+1}(\mathbf{A}_{k+1}) \leq \lambda_i(\mathbf{A}_k) \leq \lambda_i(\mathbf{A}_{k+1}), \quad i = 1, 2, \ldots, k.$$

(b) (Interlacing Theorem for Eigenvalues)   From the left- and right-hand sides of (a) we get

(i)

$$\begin{aligned}
\lambda_{n-k+i}(\mathbf{A}_n) &\leq \lambda_{n-k+i-1}(\mathbf{A}_{n-1}) \leq \cdots \leq \lambda_i(\mathbf{A}_k) \\
\lambda_i(\mathbf{A}_k) &\leq \lambda_i(\mathbf{A}_{k+1}) \leq \cdots \leq \lambda_i(\mathbf{A}_n).
\end{aligned}$$

(ii) From (i) we get

$$\lambda_{n-k+i}(\mathbf{A}) \leq \lambda_i(\mathbf{A}_k) \leq \lambda_i(\mathbf{A}), \quad i = 1, 2, \ldots, k.$$

(iii) If we reverse the order of listing the above inequalities in (i), we get the alternative expression

$$\lambda_{n-i+1}(\mathbf{A}) \leq \lambda_i(\mathbf{A}_{k-i+1}) \leq \lambda_{k-i+1}(\mathbf{A}), \quad i = 1, 2, \ldots, k.$$

(c) (Interlacing Theorems for Singular Values) Let $\mathbf{A}$ be $m \times n$ with singular values $\sigma_1(\mathbf{A}) \geq \sigma_2(\mathbf{A}) \geq \cdots \geq \sigma_r(\mathbf{A})$, where $r = \min\{m, n\}$.

(i) Let $\mathbf{B}$ be a $p \times q$ submatrix of $\mathbf{A}$ with singular values $\sigma_1(\mathbf{B}) \geq \sigma_2(\mathbf{B}) \geq \cdots \geq \sigma_s(\mathbf{B})$, where $s = \min\{p, q\}$. Then

$$\sigma_i(\mathbf{A}) \geq \sigma_i(\mathbf{B}), \quad i = 1, 2, \ldots, s.$$

(ii) Assume $m \geq n$. If $\mathbf{B}$ is a submatrix obtained from $\mathbf{A}$ by deleting one of the columns, then

$$\sigma_1(\mathbf{A}) \geq \sigma_1(\mathbf{B}) \geq \sigma_2(\mathbf{A})$$
$$\geq \sigma_2(\mathbf{B}) \geq \sigma_3(\mathbf{A}) \geq \cdots \geq \sigma_{n-1}(\mathbf{A}) \geq \sigma_{n-1}(\mathbf{B}) \geq \sigma_n(\mathbf{A}).$$

(iii) Assume $m < n$. If $\mathbf{B}$ is a submatrix obtained from $\mathbf{A}$ by deleting one of the columns, then

$$\sigma_1(\mathbf{A}) \geq \sigma_1(\mathbf{B}) \geq \sigma_2(\mathbf{A}) \geq \sigma_2(\mathbf{B}) \geq \cdots \geq \sigma_m(\mathbf{A}) \geq \sigma_m(\mathbf{B}),$$

which we now combine with (ii).

(iv) Suppose we extend the definition of singular values so that $\sigma_j(\mathbf{A}) = 0$ for $j > r$. Let $\mathbf{A}_s$ be any matrix obtained from $\mathbf{A}$ by deleting a total of s rows and columns (i.e., $s - k$ rows and k columns for some $0 \leq k \leq s$), then

$$\sigma_i(\mathbf{A}) \geq \sigma_i(\mathbf{A}_s) \geq \sigma_{i+s}(\mathbf{A}), \quad i = 1, 2, \ldots, \min\{m, n\}.$$

Note that since $\sigma_i(\mathbf{A}') = \sigma(\mathbf{A})$, we can obtain the result for deleting a single row by interchanging the two cases (ii) and (iii). Also (i)–(iii) follow from (iv).

6.65. (Eigenvalue Inequalities)

(a) (Poincaré's Separation Theorems) Let $\mathbf{A}$ be be an $n \times n$ Hermitian matrix, and let $\mathbf{B}_k$ be any $n \times k$ matrix such that $\mathbf{B}_k^* \mathbf{B}_k = \mathbf{I}_k$. Then:

(i)

$$\lambda_{n-k+i}(\mathbf{A}) \leq \lambda_i(\mathbf{B}_k^* \mathbf{A} \mathbf{B}_k) \leq \lambda_i(\mathbf{A}), \quad i = 1, 2, \ldots, k.$$

The first equalities on the left are attained if and only if $\mathbf{B}_k = \mathbf{V}_k \mathbf{U}$, where $\mathbf{U}$ is unitary and the k columns of $\mathbf{V}_k$ are any set of right eigenvectors corresponding to the k smallest eigenvalues, while the second equalities on the right are attained if and only if $\mathbf{B}_k = \mathbf{W}_k \mathbf{U}$, where $\mathbf{W}_k$ has k columns consisting of any set of right eigenvectors corresponding to the largest k eigenvalues. Scott and Styan [1985: 213–214] give some historical remarks on the history of the above result and use it to obtain bounds on the distribution of chi-square statistics used in sample surveys. Such inequalities are also used for the Durbin–Watson bounds test for serial correlation in regression.

The left-hand side can also be written in the form

$$\lambda_{n-j}(\mathbf{A}) \leq \lambda_{k-j}(\mathbf{B}_k^* \mathbf{A} \mathbf{B}_k), \quad j = 0, 1, \ldots, k - 1.$$

By setting $\mathbf{B}_k = (\mathbf{I}_k, \mathbf{0})'$, we can obtain the left-hand side of (6.64b(ii)).

(ii) Summing $i = 1, \ldots, k$ in (i), we get, for $k = 1, 2, \ldots, n$,

$$\min_{\mathbf{B}_k^* \mathbf{B}_k = \mathbf{I}_k} \operatorname{trace}(\mathbf{B}_k^* \mathbf{A} \mathbf{B}_k) = \sum_{i=1}^{k} \lambda_{n-k+i}(\mathbf{A}),$$

$$\max_{\mathbf{B}_k^* \mathbf{B}_k = \mathbf{I}_k} \operatorname{trace}(\mathbf{B}_k^* \mathbf{A} \mathbf{B}_k) = \sum_{i=1}^{k} \lambda_i(\mathbf{A}).$$

The bounds are achieved by a suitable choice of $\mathbf{B}_k$.
By setting $\mathbf{B}_k = (\mathbf{I}_k, \mathbf{0})'$, we have $\lambda_n \leq a_{ii} \leq \lambda_1$ $(i = 1, 2, \ldots n)$, $\lambda_{n-1} + \lambda_n \leq a_{ii} + a_{jj} \leq \lambda_1 + \lambda_2$, $(i, j = 1, 2, \ldots, n; i \neq j)$, and so on. In particular,

$$\sum_{i=1}^{k} \lambda_{n-k+i}(\mathbf{A}) \leq \sum_{i=1}^{k} a_{ii} \leq \sum_{i=1}^{k} \lambda_i(\mathbf{A}).$$

(iii) If $\mathbf{P}$ is an $n \times n$ idempotent Hermitian matrix (i.e., $\mathbf{P}^2 = \mathbf{P}$) of rank k, then

$$\lambda_{n-k+i}(\mathbf{A}) \leq \lambda_i(\mathbf{P} \mathbf{A} \mathbf{P}) \leq \lambda_i(\mathbf{A}), \quad i = 1, 2, \ldots, k.$$

(b) Let $\mathbf{A}$ and $\mathbf{B}$ be real $n \times n$ matrices with $\mathbf{A}$ symmetric and $\mathbf{B}$ non-negative definite with Moore–Penrose inverse $\mathbf{B}^+$. Also, let $\mathbf{T}$ be an $n \times k$ matrix of rank k such that $\mathcal{C}(\mathbf{T}) \subseteq \mathcal{C}(\mathbf{B})$ and $\mathbf{T}'\mathbf{B}\mathbf{T} = \mathbf{I}_k$, and let $\lambda_i = \lambda_i(\mathbf{B}^+\mathbf{A})$. Then the following maxima and minima with respect to $\mathbf{T}$ hold.

(i) $\max\{\operatorname{trace}(\mathbf{T}'\mathbf{A}\mathbf{T})\} = \lambda_1 + \cdots + \lambda_k$.

(ii) $\min\{\operatorname{trace}(\mathbf{T}'\mathbf{A}\mathbf{T})\} = \lambda_{n-k+1} + \cdots + \lambda_n$.

(iii) $\max\{\mathrm{trace}[(\mathbf{T}'\mathbf{AT})^2]\} = \lambda_1^2 + \cdots + \lambda_k^2$.

(iv) $\min\{\mathrm{trace}[(\mathbf{T}'\mathbf{AT})^2]\} = \lambda_{n-k+1}^2 + \cdots + \lambda_n^2$.

(v) $\max\{\mathrm{trace}[(\mathbf{T}'\mathbf{AT})^{-1}]\} = \lambda_{r-k+1}^{-1} + \cdots + \lambda_r^{-1}$, for $\mathbf{A}$ positive definite and rank $\mathbf{B} = r$.

(vi) $\min\{\mathrm{trace}[(\mathbf{T}'\mathbf{AT})^{-1}]\} = \lambda_1^{-1} + \cdots + \lambda_k^{-1}$ for $\mathbf{A}$ positive definite.

The optimum values are reached when $\mathbf{T} = (\mathbf{t}_1, \ldots, \mathbf{t}_k)$, where $\mathbf{B}^{1/2}\mathbf{t}_i$ are orthonormal right eigenvectors of $(\mathbf{B}^+)^{1/2}\mathbf{A}(\mathbf{B}^+)^{1/2}$ associated with the eigenvalues λ_i $(i = 1, 2, \ldots, k)$.

(c) If $\mathbf{A}$ and $\mathbf{B}$ are $n \times n$ positive definite matrices, then:

(i) $\lambda_1(\mathbf{A}^s\mathbf{B}^s) \le \lambda_1^s(\mathbf{AB})$ for $0 \le s \le 1$.

(ii) $[\lambda_1(\mathbf{AB})]^t \le \lambda_1(\mathbf{A}^t\mathbf{B}^t)$ for $t \ge 1$

6.66. (Singular Values) Let $\mathbf{A}$ be an $m \times n$ matrix with singular values $\sigma_i(\mathbf{A})$. Let $\mathbf{B} = \mathbf{U}^*\mathbf{AV}$, where $\mathbf{U}$ and $\mathbf{V}$ are $m \times p$ and $n \times q$, respectively, such that $\mathbf{U}^*\mathbf{U} = \mathbf{I}_p$ and $\mathbf{V}^*\mathbf{V} = \mathbf{I}_q$.

(a) If $r = (m - p) + (n - q)$,

$$\sigma_{i+r}(\mathbf{A}) \le \sigma_i(\mathbf{B}) \le \sigma_i(\mathbf{A}), \quad i = 1, 2, \ldots, \min\{m, n\}.$$

(b) If $p = q = k$,

$$|\det \mathbf{B}|^2 = \det(\mathbf{BB}^*) = \prod_i \lambda_i(\mathbf{BB}^*) = \prod_i \sigma_i^2(\mathbf{B}),$$

so that

$$|\det \mathbf{B}| \le \sigma_1(\mathbf{A}) \cdots \sigma_k(\mathbf{A}).$$

(c) If $p = q = k$, we can sum in (a) and obtain, for $k = 1, 2, \ldots, \min\{m, n\}$,

$$\max_{\mathbf{U}^*\mathbf{U}=\mathbf{I}_k, \mathbf{V}^*\mathbf{V}=\mathbf{I}_k} |\mathrm{trace}\,\mathbf{B}| = \sum_{i=1}^k \sigma_i(\mathbf{A}).$$

6.67. Let $\mathbf{A}$ be an $n \times n$ real symmetric matrix, and let $\mathbf{B}$ be an $n \times n$ positive definite matrix. If $\mathbf{F}$ is any $n \times k$ matrix of rank k, then for $i = 1, 2, \ldots, k$,

$$\lambda_i[(\mathbf{F}'\mathbf{BF})^{-1}(\mathbf{F}'\mathbf{AF})] \le \lambda_i(\mathbf{B}^{-1}\mathbf{A}),$$

and

$$\max_{\mathbf{F}} \lambda_i[(\mathbf{F}'\mathbf{BF})^{-1}(\mathbf{F}'\mathbf{AF})] = \lambda_i(\mathbf{B}^{-1}\mathbf{A}).$$

6.68. Let $\mathbf{A}$ and $\mathbf{B}$ be $n \times n$ non-negative definite matrices satisfying $\mathcal{C}(\mathbf{A}) \subseteq \mathcal{C}(\mathbf{B})$, and let $\mathbf{X}$ be an $n \times k$ real matrix with

$$b = \mathrm{rank}\,\mathbf{B} \quad \text{and} \quad r = \mathrm{rank}(\mathbf{BX}).$$

Then

$$\lambda_{b-r+i}(\mathbf{B}^-\mathbf{A}) \le \lambda_i([(\mathbf{X}'\mathbf{BX})^-\mathbf{X}'\mathbf{AX}]) \le \lambda_i(\mathbf{B}^-\mathbf{A}), \quad i = 1, 2, \ldots, r.$$

In the above equation, any choices of the weak inverses $\mathbf{B}^-$ and $(\mathbf{X}'\mathbf{B}\mathbf{X})^-$ may be made. Equality occurs on the left simultaneously for all $i = 1, 2, \ldots, r$ if and only if there exists a real $n \times r$ matrix $\mathbf{Q}_0$ such that

$$\mathbf{Q}_0'\mathbf{B}\mathbf{Q}_0 = \mathbf{I}_r, \quad \mathbf{A}\mathbf{Q}_0 = \mathbf{B}\mathbf{Q}_0\boldsymbol{\Lambda}_o, \quad \text{and} \quad \mathcal{C}(\mathbf{B}\mathbf{Q}_0) = \mathcal{C}(\mathbf{B}\mathbf{X}).$$

Here $\boldsymbol{\Lambda}_0$ is an $r \times r$ diagonal matrix containing the r smallest, not necessarily zero, generalized eigenvalues of $\mathbf{A}$ with respect to $\mathbf{B}$.

Equality holds on the right simultaneously for all $i = 1, 2, \ldots, r$ if and only if there exists a real $r \times r$ matrix $\mathbf{Q}_1$ such that

$$\mathbf{Q}_1'\mathbf{B}\mathbf{Q}_1 = \mathbf{I}_r, \quad \mathbf{A}\mathbf{Q}_1 = \mathbf{B}\mathbf{Q}_1\boldsymbol{\Lambda}_1, \quad \text{and} \quad \mathcal{C}(\mathbf{B}\mathbf{Q}_1) = \mathcal{C}(\mathbf{B}\mathbf{X}).$$

Here $\boldsymbol{\Lambda}_1$ is an $r \times r$ diagonal matrix containing the r largest generalized eigenvalues of $\mathbf{A}$ with respect to $\mathbf{B}$. Scott and Styan [1985] give an application to finding distributional bounds on two standard asymptotic hypothesis tests in multiway contingency tables.

6.69. A product version of (6.65a(ii)) is as follows. If $\mathbf{A}$ is a Hermitian positive definite matrix and $\mathbf{B}$ is an $n \times k$ matrix, then

$$\min_{\mathbf{B}_k^*\mathbf{B}_k = \mathbf{I}_k} \det(\mathbf{B}_k^*\mathbf{A}\mathbf{B}_k) = \prod_{i=1}^{k} \lambda_{n-k+i}(\mathbf{A}),$$

$$\max_{\mathbf{B}_k^*\mathbf{B}_k = \mathbf{I}_k} \det(\mathbf{B}_k^*\mathbf{A}\mathbf{B}_k) = \prod_{i=1}^{k} \lambda_i(\mathbf{A}).$$

By setting $\mathbf{B}_k = (\mathbf{I}_k, \mathbf{0})'$ and defining $\mathbf{A}_k$ as in (6.64), we have

$$\prod_{i=1}^{k} \lambda_{n-k+i}(\mathbf{A}) \leq \det \mathbf{A}_k \leq \prod_{i=1}^{k} \lambda_i(\mathbf{A}).$$

Proofs.   Section 6.3

6.64a. Rao and Bhimasankaram [2000: 347–348, real symmetric case with i and k interchanged; the proof is identical for Hermitian matrices].

6.64b(ii). Rao and Rao [1998: 328, with $\mathbf{A}_k$ replaced by $\mathbf{B}$] and Zhang [1999: 222–225].

6.64b(iii). Schott [2005: 112].

6.64c(i). Rao and Rao [1998: 330].

6.64c(ii)–(iii). Horn and Johnson [1985: 419] and Rao and Rao [1998: 330].

6.64c(iv). Horn and Johnson [1991: 149], Rao and Rao [1998: 329–332], and Zhang [1999: 229].

6.65a(i). For the real case see Abadir and Magnus [2005: 347], Schott [2005: 111], and Rao and Bhimasankaram [2000: 348].

6.65a(ii). Abadir and Magnus [2005: 348–349].

6.65a(iii). Abadir and Magnus [2005: 348].

6.65b. Quoted by Rao and Rao [1998: 495].

6.65c. Quoted by Rao and Rao [1998: 495].

6.66a. Rao and Rao [1998: 338].

6.66b. Horn and Johnson [1991: 170].

6.66c. Horn and Johnson [1991: 195].

6.67. Schott [2005: 123].

6.68. Scott and Styan [1985].

6.69. Magnus and Neudecker [1999: 212, real case, with order of eigenvalues reversed] and quoted by Schott [2005: 136, exercise 3.54].

6.4 INEQUALITIES FOR MATRIX SUMS

6.70. (Eigenvalues) Let $\mathbf{A}$ and $\mathbf{B}$ be $n \times n$ Hermitian or real symmetric matrices, and let $\mathbf{C} = \mathbf{A} + \mathbf{B}$, with corresponding eigenvalues

$$\alpha_1 \geq \alpha_2 \geq \cdots \geq \alpha_n; \quad \beta_1 > \beta_2 > \cdots > \beta_n \quad \text{and} \quad \gamma_1 \geq \gamma \geq \cdots \geq \gamma_n,$$

respectively. Then:

(a)

$$\alpha_1 + \beta_1 \quad \geq \quad \gamma_1 \quad \geq \quad \begin{cases} \alpha_1 + \beta_n \\ \alpha_2 + \beta_{n-1} \\ \cdots \\ \alpha_n + \beta_1 \end{cases}$$

$$\left. \begin{array}{c} \alpha_1 + \beta_2 \\ \alpha_2 + \beta_1 \end{array} \right\} \quad \geq \quad \gamma_2 \quad \geq \quad \begin{cases} \alpha_2 + \beta_n \\ \alpha_3 + \beta_{n-1} \\ \cdots \\ \alpha_n + \beta_2 \end{cases}$$

$$\left. \begin{array}{c} \alpha_1 + \beta_3 \\ \alpha_2 + \beta_2 \\ \alpha_3 + \beta_1 \end{array} \right\} \quad \geq \quad \gamma_3 \quad \geq \quad \begin{cases} \alpha_3 + \beta_n \\ \alpha_4 + \beta_{n-1} \\ \cdots \\ \alpha_n + \beta_3 \end{cases}$$

$$\cdots \qquad \cdots \qquad \cdots$$

$$\left. \begin{array}{c} \alpha_1 + \beta_n \\ \alpha_2 + \beta_{n-1} \\ \cdots \\ \alpha_n + \beta_1 \end{array} \right\} \quad \geq \quad \gamma_n \quad \geq \quad \alpha_n + \beta_n$$

(b) It follows from (a) that

$$\gamma_i \;\leq\; \alpha_j + \beta_{i-j+1}, \quad \text{for} \quad j = 1, 2, \ldots, i; \; i = 1, 2, \ldots, n,$$
$$\text{and } \gamma_i \;\geq\; \alpha_j + \beta_{n-j+i}, \quad \text{for} \quad j = i, i+1, \ldots, n; \; i = 1, 2, \ldots, n.$$

(c) (Weyl's Theorem) From (b) we have:

(i) For $i, j \leq n$

$$\lambda_i(\mathbf{A} + \mathbf{B}) \;\leq\; \lambda_j(\mathbf{A}) + \lambda_{i-j+1}(\mathbf{B}) \quad \text{for} \quad j \leq i,$$
$$\lambda_i(\mathbf{A} + \mathbf{B}) \;\geq\; \lambda_j(\mathbf{A}) + \lambda_{n+i-j}(\mathbf{B}) \quad \text{for} \quad j \geq i.$$

(ii) If in (i) we make the subscript substitution $j = a$ and $i - j + 1 = b$ so that $i = a + b - 1$, and then relabel, we get from the first equation

$$\lambda_{a+b-1}(\mathbf{A} + \mathbf{B}) \leq \lambda_a(\mathbf{A}) + \lambda_b(\mathbf{B}), \quad a + b - 1 \leq n, \, b \geq 1.$$

(iii) Setting $j = i$ in (i) we have, for $i = 1, 2, \ldots, n$,

$$\lambda_i(\mathbf{A}) + \lambda_n(\mathbf{B}) \leq \lambda_i(\mathbf{A} + \mathbf{B}) \leq \lambda_i(\mathbf{A}) + \lambda_1(\mathbf{B}).$$

(iv) (Monotonicity of Eigenvalues) If $\mathbf{B}$ is real non-negative definite and $\mathbf{A}$ is real symmetric, then $\lambda_i(\mathbf{B}) \geq 0$ for all i and, from (iii),

$$\lambda_i(\mathbf{A}) \leq \lambda_i(\mathbf{A} + \mathbf{B}), \quad i = 1, 2, \ldots, n.$$

If $\mathbf{B}$ is positive definite, then the inequality is strict.

(d) (i) (Lidskiĭ) Let $i_1, i_2, \ldots, i_k$ be integers satisfying $1 \leq i_1 < \cdots < i_k \leq n$. Then for $k = 1, 2, \ldots, n$,

$$\sum_{j=1}^{k} \{\lambda_{i_j}(\mathbf{A}) + \lambda_{n-k+j}(\mathbf{B})\} \leq \sum_{j=1}^{k} \lambda_{i_j}(\mathbf{A} + \mathbf{B}) \leq \sum_{j=1}^{k} \{\lambda_{i_j}(\mathbf{A}) + \lambda_j(\mathbf{B})\}.$$

(ii) (Sum of the k largest eigenvalues) For $k = 1, 2, \ldots, n$,

$$\sum_{i=1}^{k} \lambda_i(\mathbf{A}) + \sum_{i=1}^{k} \lambda_{n-k+i}(\mathbf{B}) \leq \sum_{i=1}^{k} \lambda_i(\mathbf{A} + \mathbf{B}) \leq \sum_{i=1}^{k} [\lambda_i(\mathbf{A}) + \lambda_i(\mathbf{B})].$$

(e) Suppose $\mathbf{B}$ is a real symmetric matrix with rank $\mathbf{B} \leq r$ and $\mathbf{A}$ is real symmetric. For $i = 1, 2, \ldots, n - r$ we have:

$$\lambda_{i+r}(\mathbf{A}) \;\leq\; \lambda_i(\mathbf{A} + \mathbf{B})$$
$$\lambda_{i+r}(\mathbf{A} + \mathbf{B}) \;\leq\; \lambda_i(\mathbf{A}).$$

6.71. (Convexity) For any two real symmetric $n \times n$ matrices $\mathbf{A}$ and $\mathbf{B}$, and $0 \leq \alpha \leq 1$,

$$\lambda_1[\alpha \mathbf{A} + (1 - \alpha)\mathbf{B}] \;\leq\; \alpha \lambda_1(\mathbf{A}) + (1 - \alpha)\lambda_1(\mathbf{B}),$$
$$\lambda_n[\alpha \mathbf{A} + (1 - \alpha)\mathbf{B}] \;\geq\; \alpha \lambda_n(\mathbf{A}) + (1 - \alpha)\lambda_n(\mathbf{B}).$$

Hence, λ_1 is convex and λ_n is concave on the space of real symmetric matrices. Putting $\alpha = 1/2$ gives us

$$\lambda_1(\mathbf{A} + \mathbf{B}) \leq \lambda_1(\mathbf{A}) + \lambda_1(\mathbf{B}),$$
$$\lambda_n(\mathbf{A} + \mathbf{B}) \geq \lambda_n(\mathbf{A}) + \lambda_n(\mathbf{B}).$$

6.72. (Singular Values) Let $\mathbf{A}$ and $\mathbf{B}$ be $m \times n$ matrices, and let $p = \min\{m, n\}$. Then:

(a)

$$\sigma_i(\mathbf{A} + \mathbf{B}) \leq \sigma_j(\mathbf{A}) + \sigma_{i-j+1}(\mathbf{B}), \quad j = 1, 2, \ldots, i; \ i = 1, 2, \ldots, p.$$

$$\sigma_{i+j-1}(\mathbf{A} + \mathbf{B}) \leq \sigma_i(\mathbf{A}) + \sigma_j(\mathbf{B}), \quad 1 \leq i, j \leq p; i + j \leq p + 1.$$

(b) In particular,

   (i) $\sigma_1(\mathbf{A} + \mathbf{B}) \leq \sigma_1(\mathbf{A}) + \sigma_1(\mathbf{B})$.

   (ii) $\sigma_p(\mathbf{A} + \mathbf{B}) \leq \min\{\sigma_p(\mathbf{A}) + \sigma_1(\mathbf{B}), \sigma_1(\mathbf{A}) + \sigma_p(\mathbf{B})\}$.

   (iii) $\sigma_i(\mathbf{A}) + \sigma_n(\mathbf{B}) \leq \sigma_i(\mathbf{A} + \mathbf{B}) \leq \sigma_i(\mathbf{A}) + \sigma_1(\mathbf{B})$.

(c) $|\sigma_i(\mathbf{A} + \mathbf{B}) - \sigma_i(\mathbf{A})| \leq \sigma_1(\mathbf{B})$ for $i = 1, 2, \ldots, p$.

(d) $\displaystyle\sum_{i=1}^{k} \sigma_i(\mathbf{A} + \mathbf{B}) \leq \sum_{i=1}^{k}[\sigma_i(\mathbf{A}) + \sigma_i(\mathbf{B})], \quad k = 1, 2, \ldots, p.$

Proofs. Section 6.4

6.70a. Rao and Rao [1998: 322].

6.70c(i). Bhatia [1997: 62, with i and j interchanged].

6.70c(ii). Schott [2005: 114, real case].

6.70c(iii). Schott [2005: 112, real case] and Zhang [1999: 227].

6.70c(iv). Magnus and Neudecker [1999: 208–209] and Schott [2005: 119–120].

6.70d(i). Wielandt [1955] and Dümbgen [1995].

6.70d(ii). Schott [2005: 115–116].

6.70e. Schott [2005: 112–114].

6.71. Abadir and Magnus [2005: 344–345] and Magnus and Neudecker [1999: 205, λ_1 and λ_n are interchanged].

6.72a. Rao and Rao [1998: 326–327, 360] and Horn and Johnson [1991: 178, subscripts reordered].

6.72b(iii). Zhang [1999: 228].

6.72c. Horn and Johnson [1991: 178].

6.72d. Horn and Johnson [1991: 196].

6.5   INEQUALITIES FOR MATRIX DIFFERENCES

6.73. Let $\mathbf{A}$, $\mathbf{B}$, and $\mathbf{A} - \mathbf{B}$ be Hermitian non-negative definite $n \times n$ matrices with rank $\mathbf{B} \leq k$. Then

$$\lambda_i(\mathbf{A} - \mathbf{B}) \geq \lambda_{k+i}(\mathbf{A})$$

for all i $(k + i \leq n)$ with equality for all i if and only if

$$\mathbf{B} = \sum_{i=1}^{k} \lambda_i(\mathbf{A})\mathbf{u}_i\mathbf{u}_i',$$

where $\mathbf{u}_1, \mathbf{u}_2, \ldots, \mathbf{u}_k$ are the first k orthonormal eigenvectors of $\mathbf{A}$ (i.e., those corresponding to the $\lambda_i(\mathbf{A})$, $i = 1, 2, \ldots, k$).

6.74. Let $\mathbf{A}$ and $\mathbf{B}$ be $m \times n$ matrices with ranks r and s, respectively. Then:

(a)
$$\begin{aligned} \sigma_i(\mathbf{A} - \mathbf{B}) &\geq \sigma_{i+s}(\mathbf{A}), & i + s \leq r, \\ &\geq 0, & i + s > r. \end{aligned}$$

(b) The equalities in (a) are attained if and only if $s \leq r$ and

$$\mathbf{B} = \sum_{i=1}^{s} \sigma_i(\mathbf{B})\mathbf{u}_i\mathbf{u}_i^*,$$

where the singular value decomposition of $\mathbf{A}$ is $\mathbf{A} = \sum_{i=1}^{r} \sigma_i(\mathbf{A})\mathbf{u}_i\mathbf{u}_i^*$.

Proofs.   Section 6.5

6.73. Quoted by Rao and Rao [1998: 382], though the proof is similar to that of (6.74).

6.74. Rao [1980: 8–9].

6.6   INEQUALITIES FOR MATRIX PRODUCTS

6.75. Let $\mathbf{A}$ be an $n \times n$ non-negative definite matrix, and let $\mathbf{B}$ be an $n \times n$ positive definite matrix. If $i, j, k = 1, 2, \ldots, n$ such that $j + k \leq i + 1$, then:

(a) $\lambda_i(\mathbf{AB}) \leq \lambda_j(\mathbf{A})\lambda_k(\mathbf{B})$.

(b) $\lambda_{n-i+1}(\mathbf{AB}) \geq \lambda_{n-j+1}(\mathbf{A})\lambda_{n-k+1}(\mathbf{B})$.

The case when $\mathbf{A}$ is symmetric and $\mathbf{B}$ is non-negative definite is discussed in detail by Mäkeläinen [1970].

6.76. If $\mathbf{A}$ and $\mathbf{B}$ are $n \times n$ Hermitian non-negative definite matrices, then

$$\lambda_i(\mathbf{A})\lambda_n(\mathbf{B}) \leq \lambda_i(\mathbf{AB}) \leq \lambda_i(\mathbf{A})\lambda_1(\mathbf{B}), \quad i = 1, 2, \ldots, n.$$

6.77. (von Neumann) (Trace)   If $\mathbf{A}$ and $\mathbf{B}$ are $n \times n$ Hermitian matrices, then

$$\sum_{i=1}^{n} \lambda_i(\mathbf{A})\lambda_{n-i+1}(\mathbf{B}) \leq \operatorname{trace}(\mathbf{AB}) \leq \sum_{i=1}^{n} \lambda_i(\mathbf{A})\lambda_i(\mathbf{B}).$$

Equality holds on the right when $\mathbf{B} = \sum_{i=1}^{n} \lambda_i(\mathbf{B})\mathbf{u}_i\mathbf{u}_i^*$, and equality holds on the left when $\mathbf{B} = \sum_{i=1}^{n} \lambda_{n-i+1}(\mathbf{B})\mathbf{u}_i\mathbf{u}_i^*$. Here $\mathbf{u}_i$ is a right eigenvector of $\mathbf{A}$ for the eigenvalue $\lambda_i(\mathbf{A})$, $i = 1, 2, \ldots, n$.

6.78. Let $\mathbf{A}$ and $\mathbf{B}$ be $n \times n$ non-negative definite matrices. If $1 \leq i_1 < \cdots < i_k \leq n$, then

$$\prod_{j=1}^{k} \lambda_{i_j}(\mathbf{AB}) \leq \prod_{j=1}^{k} \lambda_{i_j}(\mathbf{A})\lambda_j(\mathbf{B}), \quad k = 1, 2, \ldots, n,$$

with equality for $k = n$.

6.79. (Partial Sum)   Let $\mathbf{A}$ and $\mathbf{B}$ be $n \times n$ (real) non-negative definite matrices. Then

$$\sum_{i=1}^{k} \lambda_i(\mathbf{A})\lambda_{n-i+1}(\mathbf{B}) \leq \sum_{i=1}^{k} \lambda_i(\mathbf{AB}), \quad k = 1, 2, \ldots, n.$$

6.80. Let $\mathbf{A}$ be an $m \times n$ and $\mathbf{B}$ an $n \times m$ real or complex matrices. Then

$$\sigma_i(\mathbf{A})\sigma_m(\mathbf{B}) \leq \sigma_i(\mathbf{AB}) \leq \sigma_i(\mathbf{A})\sigma_1(\mathbf{B}), \quad i = 1, 2, \ldots, m.$$

6.81. Let $\mathbf{A}$ be an $m \times n$ and $\mathbf{B}$ an $n \times m$ real or complex matrices, and let $p = \min\{m, n\}$. Then, for singular values $\sigma(\cdot)$,

$$-\sum_{i=1}^{p} \sigma_i(\mathbf{A})\sigma_i(\mathbf{B}) \leq \operatorname{trace}(\mathbf{AB}) \leq \sum_{i=1}^{p} \sigma_i(\mathbf{A})\sigma_i(\mathbf{B}).$$

Equality holds on the right when $\mathbf{B} = \sum_{i=1}^{p} \sigma_i(\mathbf{B})\mathbf{q}_i\mathbf{p}_i^*$, and equality on the left holds when $\mathbf{B} = \sum_{i=1}^{p} \sigma_i(\mathbf{B})(-\mathbf{q}_i)\mathbf{p}_i^*$, where $\mathbf{p}_i$ and $\mathbf{q}_i$ are the singular vectors of $\mathbf{A}$ for $\sigma_i(\mathbf{A})$, $i = 1, 2, \ldots, p$ (cf. Section 16.3).

6.82. (Horn)   Let $\mathbf{A}$ be an $m \times p$ and $\mathbf{B}$ an $p \times n$ real or complex matrices, and let $q = \min\{m, n, p\}$.

(a) $\prod_{j=1}^{i} \sigma_j(\mathbf{AB}) \leq \prod_{j=1}^{i} \sigma_j(\mathbf{A})\sigma_j(\mathbf{B}), \quad i = 1, 2, \ldots, q.$

   If $\mathbf{A}$ and $\mathbf{B}$ are square matrices of the same order (i.e., $m = n = p$), then equality holds in the above equation for $i = n$.

(b) $\sum_{j=1}^{j}[\sigma_j(\mathbf{AB})]^p \leq \sum_{j=1}^{i}[\sigma_j(\mathbf{A})\sigma_j(\mathbf{B})]^p$ for $i = 1, 2, \ldots, q$ and any $p > 0$.

   Horn and Johnson [1991: 177] give some extensions to functions of the singular values.

6.83. Let $\mathbf{A}$ and $\mathbf{B}$ be real $n \times n$ symmetric matrices, and let $\mathbf{T}$ be an $n \times n$ orthogonal matrix. Then

$$\max_{\mathbf{T}} \text{trace}(\mathbf{TAT'B}) = \sum_{i=1}^{n} \lambda_i(\mathbf{A})\lambda_i(\mathbf{B}) \quad \text{and}$$

$$\min_{\mathbf{T}} \text{trace}(\mathbf{TAT'B}) = \sum_{i=1}^{n} \lambda_i(\mathbf{A})\lambda_{n+1-i}(\mathbf{B}).$$

Setting

$$\mathbf{B} = \begin{pmatrix} \mathbf{I}_k & \mathbf{0} \\ \mathbf{0} & \mathbf{0} \end{pmatrix},$$

we have

$$\max_{\mathbf{R'R}=\mathbf{I}_k} \text{trace}(\mathbf{R'AR}) = \sum_{i=1}^{k} \lambda_i(\mathbf{A}).$$

6.84. Let $\mathbf{X}_i$ be an $n \times p_i$ matrix of rank p_i $(i = 1, 2)$. Then the eigenvalues of $(\mathbf{X}_2'\mathbf{X}_2)^{-1}\mathbf{X}_2'\mathbf{X}_1(\mathbf{X}_1'\mathbf{X}_1)^{-1}\mathbf{X}_1'\mathbf{X}_2$ are less than or equal to one. This result arises in the correspondence analysis of a contingency table.

6.85. If $\mathbf{A}$ and $\mathbf{B}$ are $n \times n$ real or complex matrices of which at least one is nonsingular, then

$$\lambda_{\min}(\mathbf{AA^*})\lambda_{\min}(\mathbf{BB^*}) \leq \lambda_i(\mathbf{AB})\overline{\lambda}_i(\mathbf{AB}) \leq \lambda_{\max}(\mathbf{AA^*})\lambda_{\max}(\mathbf{BB^*})$$

for all i. If $\mathbf{A}$ and $\mathbf{B}$ are both Hermitian, one is positive definite (say $\mathbf{A}$), and the other is non-negative definite, then

$$\lambda_{\min}(\mathbf{A})\lambda_{\min}(\mathbf{B}) \leq \lambda_i(\mathbf{AB}) \leq \lambda_{\max}(\mathbf{A})\lambda_{\max}(\mathbf{B}).$$

Proofs.   Section 6.6

6.75. Schott [2005: 126–127].

6.76. Zhang [1999: 227].

6.77. Rao and Rao [1998: 386].

6.78. Lidskiĭ [1950] and quoted by Schott [2005: 127].

6.79. Quoted by Schott [2005: 128; see also 137, exercise 3.57].

6.80. Zhang [1999: 228].

6.81. Rao and Rao [1998: 387].

6.82a. Horn and Johnson [1991: 172] and Rao and Rao [1998: 340–342].

6.82b. Horn and Johnson [1991: 177].

6.83. Anderson [2003: 645].

6.84. Bénasséni [2002].

6.85. Roy [1954].

6.7   ANTIEIGENVALUES AND ANTIEIGENVECTORS

If $\mathbf{A}$ is an $n \times n$ positive definite matrix, then the cosine of the angle θ between $n \times 1$ real vectors $\mathbf{x}$ and $\mathbf{A}\mathbf{x}$ is (cf. Definition 2.12 in Section 2.2.1)

$$\cos \theta = \frac{\mathbf{x}'\mathbf{A}\mathbf{x}}{\sqrt{(\mathbf{x}'\mathbf{x})(\mathbf{x}'\mathbf{A}^2\mathbf{x})}},$$

which has the value of unity when $\mathbf{x}$ is an eigenvalue of $\mathbf{A}$, that is $\mathbf{A}\mathbf{x} = \lambda\mathbf{x}$ for some λ. This raises the question of what value of $\mathbf{x}$ minimizes $\cos \theta$, or equivalently maximises the angle beween $\mathbf{x}$ and $\mathbf{A}\mathbf{x}$. This question motivates the following definitions.

Definition 6.7. Let $\lambda_1 \geq \lambda_2 \geq \cdots \geq \lambda_n > 0$ be the eigenvalues of positive definite $\mathbf{A}$ and $\mathbf{x}_1, \mathbf{x}_2, \ldots, \mathbf{x}_n$ be the corresponding right eigenvectors. Referring to the above introduction, $\cos \theta$ takes its minimum value of

$$\mu_1 = \frac{2\sqrt{\lambda_1 \lambda_n}}{\lambda_1 + \lambda_n}$$

by the Kantorovich inequality (12.2a) (with $\mathbf{x} = \mathbf{A}^{1/2}\mathbf{y}$), and the minimum is attained at

$$\mathbf{x} = \frac{\sqrt{\lambda_n}\mathbf{x}_1 \pm \sqrt{\lambda_1}\mathbf{x}_n}{\sqrt{\lambda_1 + \lambda_n}} = (\mathbf{u}_1, \mathbf{u}_2), \quad \text{say.}$$

The vectors $(\mathbf{u}_1, \mathbf{u}_2)$ are called the *first antieigenvectors* and μ_1 the *first antieigenvalue*. The angle θ is called the *angle of the operator of* $\mathbf{A}$. We then define

$$\begin{aligned} \mu_2 &= \min_{\mathbf{x} \perp \mathbf{x}_1, \mathbf{x}_n} \frac{\mathbf{x}'\mathbf{A}\mathbf{x}}{\sqrt{(\mathbf{x}'\mathbf{x})(\mathbf{x}'\mathbf{A}^2\mathbf{x})}} \\ &= \frac{2\sqrt{\lambda_2 \lambda_{n-1}}}{\lambda_2 + \lambda_{n-1}}, \end{aligned}$$

which is attained at

$$\mathbf{x} = \frac{\sqrt{\lambda_{n-1}}\mathbf{x}_2 \pm \sqrt{\lambda_2}\mathbf{x}_{n-1}}{\sqrt{\lambda_2 + \lambda_{n-1}}} = (\mathbf{u}_3, \mathbf{u}_4), \quad \text{say.}$$

We call μ_2 the second antieigenvalue of $\mathbf{A}$ and $(\mathbf{u}_3, \mathbf{u}_4)$ the second antieigenvectors. We then find the third set by minimizing $\cos \theta$ subject to $\mathbf{x} \perp \{\mathbf{x}_1, \mathbf{x}_2, \mathbf{x}_{n-1}, \mathbf{x}_n\}$, and carry on this process until we have $\mu_1 \leq \mu_2 \leq \cdots \leq \mu_r$ $(r = [p/2])$, where

$$\mu_i = \frac{2\sqrt{\lambda_i \lambda_{n-i+1}}}{\lambda_i + \lambda_{n-i+1}}$$

are the ordered antieigenvalues and $(\mathbf{u}_1, \mathbf{u}_2), (\mathbf{u}_3, \mathbf{u}_4), \ldots (\mathbf{u}_{2r-1}, \mathbf{u}_{2r})$ are the corresponding pairs of antieigenvectors. When p is odd, the antieigenvalue of order $(n+1)/2$ is unity, with the corresponding antieigenvector $\mathbf{x}_{(p+1)/2}$.

The above terminology and concepts were introduced by Gustafson [1968] under the umbrella of *operator trigonometry*. The theory was extended to arbitrary nonsingular matrices by Gustafson [2000]. He also applied the theory to the question of one measure of efficiency of the ordinary least squares estimator (OLSE) with respect to the best linear unbiased estimator (BLUE) in Gustafson [2002, 2005]. Rao [2005] also discussed this question in detail.

6.86. If $\mathbf{A}$ is an $n \times n$ positive definite matrix, then

$$\frac{\mathbf{A}^2\mathbf{x}}{\mathbf{x}'\mathbf{A}^2\mathbf{x}} - \frac{2\mathbf{A}\mathbf{x}}{\mathbf{x}'\mathbf{A}\mathbf{x}} + \mathbf{x} = \mathbf{0}$$

is called the *Euler Equation*. This equation is satisfied by all the eigenvectors $\mathbf{x}_i$ of $\mathbf{A}$, and the only other solutions are the antieigenvectors

$$\frac{\sqrt{\lambda_j}\mathbf{x}_k \pm \sqrt{\lambda_k}\mathbf{x}_j}{\sqrt{\lambda_j + \lambda_k}}.$$

This topic has links with canonical correlations (Gustafson [2005: 116]).

6.87. Let $\mathbf{A}$ be a positive definite $n \times n$ matrix. Then

$$\max_{\mathbf{x}'\mathbf{x}=1}[\mathbf{x}'\mathbf{A}\mathbf{x} - (\mathbf{x}'\mathbf{A}^{-1}\mathbf{x})^{-1}] = (\sqrt{\lambda_1} - \sqrt{\lambda_n})^2,$$

with the maximum occurring at

$$\mathbf{x} = \left(\frac{\sqrt{\lambda_1}}{\sqrt{\lambda_1} + \sqrt{\lambda_n}}\right)^{1/2}\mathbf{x}_1 \pm \left(\frac{\sqrt{\lambda_n}}{\sqrt{\lambda_1} + \sqrt{\lambda_n}}\right)^{1/2}\mathbf{x}_n,$$

where $\mathbf{x}_1$ and $\mathbf{x}_n$ are the eigenvectors corresponding to λ_1 and λ_n, the maximum and minimum eigenvalues of $\mathbf{A}$. Rao [2005: 64–65] uses the above result to define the first of another series of antieigenvalues that he calls the SM–antieigenvalues, with corresponding antieigenvectors.

Proofs.   Section 6.7

6.86. Gustafson [2002, 2005].

6.87. Shisha and Mond [1967] and Styan [1983].

CHAPTER 7

GENERALIZED INVERSES

When a matrix is not square, or square and singular, then an inverse does not exist. However, a type of inverse does exist for these matrices called a generalized inverse that functions very much like an inverse. Such inverses are very useful in statistics for finding explicit solutions for a variety of problems such as the solution of linear equations so that this chapter has close links with Chapter 13. The reader should also consult Chapter 14 on partitioned matrices. A summary of some computational aspects of generalized inverses, along with references, is given by Ben-Israel and Greville [2003: chapter 7].

7.1 DEFINITIONS

Definition 7.1. A weak inverse of an $m \times n$ matrix $\mathbf{A}$ is defined to be any $n \times m$ matrix $\mathbf{G}$ that satisfies the condition

(1) $\mathbf{AGA} = \mathbf{A}$.

Such a matrix always exists (by 7.1 below), but it is not unique. We shall write $\mathbf{G} = \mathbf{A}^-$. Many of the results below are proved by verifying, or finding conditions, that (1) is true.

Note that the name "generalized inverse" is fairly common but not universal. Other terms used include *conditional inverse* (cf. Graybill [1983: chapter 6]), *pseudoinverse*, *g-inverse*, and *weak inverse*. I shall use the term weak inverse to avoid confusion.

A Matrix Handbook for Statisticians. By George A. F. Seber
Copyright © 2008 John Wiley & Sons, Inc.

If $\mathbf{A}$ is real and $\mathbf{G}$ also satisfies

(2) $\mathbf{GAG} = \mathbf{A}$,

(3) $\mathbf{AG}$ is symmetric,

(4) $\mathbf{GA}$ is symmetric,

then we call $\mathbf{G}$ the *Moore–Penrose* inverse and write $\mathbf{G} = \mathbf{A}^{+}$. The above definition applies to complex matrices $\mathbf{A}$ if we replace "symmetric" by "Hermitian."

There are other matrices $\mathbf{G}$ that satisfy just one or more of the above four conditions, and we shall use subcripts to identify the conditions. For example, if $\mathbf{G}$ satisfies at least (1) and (2) we shall call $\mathbf{G}$ a g_{12}-inverse and write $\mathbf{G} = \mathbf{A}^{-}_{(1,2)}$. Similarly we can write $\mathbf{A}^{-} = \mathbf{A}^{-}_{(1)}$ and refer to $\mathbf{A}^{-}$ as a g_1-inverse. We shall only use the subscript notation if there is any danger of ambiguity. For one list of the various inverses see Rao and Rao [1998: 294].

We shall also define $\mathbf{A}\{i, j, \dots, p\}$ to be the set of all matrices $\mathbf{G}$ which satisfy at least the conditions $(i), (j), \dots, (p)$. Thus $\mathbf{A}^{-}_{(1,2)} \in \mathbf{A}\{1, 2\}$, $\mathbf{A}^{-} \in \mathbf{A}\{1\}$, and so on. We shall discuss these inverses later.

If $\mathbf{A}$ is square, then a generalized inverse $\mathbf{G}$ that satisfies (1), (2), and $\mathbf{AG} = \mathbf{GA}$ is called the *group inverse*, which we denote by $\mathbf{A}^{\#}$.

7.2 WEAK INVERSES

7.2.1 General Properties

Let $\mathbf{A}$ be an $m \times n$ real or complex matrix of rank r. Many of the following results can be proved by simply checking that condition (1) above holds.

7.1. (Existence) From (16.33) there exist conformable nonsingular matrices $\mathbf{B}$ and $\mathbf{C}$ such that

$$\mathbf{BAC} = \begin{pmatrix} \mathbf{I}_r & \mathbf{0} \\ \mathbf{0} & \mathbf{0} \end{pmatrix}, \quad \text{or} \quad \mathbf{A} = \mathbf{B}^{-1} \begin{pmatrix} \mathbf{I}_r & \mathbf{0} \\ \mathbf{0} & \mathbf{0} \end{pmatrix} \mathbf{C}^{-1}.$$

Then $\mathbf{A}^{-} = \mathbf{C} \begin{pmatrix} \mathbf{I}_r & \mathbf{X} \\ \mathbf{Y} & \mathbf{Z} \end{pmatrix} \mathbf{B}$ for arbitrary $\mathbf{X}, \mathbf{Y}$, and $\mathbf{Z}$ of appropriate sizes. Although a weak inverse always exists, we see that it is not unique. Another version based on the singular value decomposition is given in (7.82).

7.2. (Basic Properties)

(a) Taking the transpose of both sides of $\mathbf{AA}^{-}\mathbf{A} = \mathbf{A}$, we see that $\mathbf{A}^{-\prime}$ is a weak inverse of $\mathbf{A}'$. Although we shall write $(\mathbf{A}^{-})' = (\mathbf{A}')^{-}$, what we mean, technically, is that $\mathbf{A}^{-\prime} \in \mathbf{A}'\{1\}$. This idea underlies all the results below.

(b) For $k \neq 0$, $k^{-1}\mathbf{A}^{-}$ is a weak inverse $k\mathbf{A}$.

(c) $\mathbf{A}^{-}\mathbf{A}$ and $\mathbf{AA}^{-}$ are each idempotent. Also, since $\mathbf{P}_{\mathcal{C}(\mathbf{A})} = \mathbf{AA}^{-}$ is not generally symmetric, it represents a nonorthogonal (oblique) projection onto $\mathcal{C}(\mathbf{A})$. Similarly, $\mathbf{P}_{\mathcal{C}(\mathbf{A}')} = (\mathbf{A}^{-}\mathbf{A})' = \mathbf{A}'\mathbf{A}'^{-}$ represents an oblique projection onto $\mathcal{C}(\mathbf{A}')$.

(d) $\operatorname{rank}(\mathbf{A}\mathbf{A}^-) = \operatorname{trace}(\mathbf{A}\mathbf{A}^-) = \operatorname{trace}(\mathbf{A}^-\mathbf{A}) = \operatorname{rank}(\mathbf{A}^-\mathbf{A}) = \operatorname{rank}\mathbf{A} \le \operatorname{rank}(\mathbf{A}^-)$.

(e) $\mathcal{C}(\mathbf{A}\mathbf{A}^-) = \mathcal{C}(\mathbf{A})$, $\mathcal{N}(\mathbf{A}\mathbf{A}^-) = \mathcal{N}(\mathbf{A})$, and $\mathcal{C}[(\mathbf{A}^-\mathbf{A})^*] = \mathcal{C}(\mathbf{A}^*)$.

(f) Taking conjugate transposes of $\mathbf{A}\mathbf{A}^-\mathbf{A} = \mathbf{A}$ we get $(\mathbf{A}^*)^- = (\mathbf{A}^-)^*$.

(g) $\operatorname{rank}\mathbf{A} = m$ if and only if $\mathbf{A}\mathbf{A}^- = \mathbf{I}_m$ (i.e., $\mathbf{A}^-$ is a right inverse of $\mathbf{A}$).

(h) $\operatorname{rank}\mathbf{A} = n$ if and only if $\mathbf{A}^-\mathbf{A} = \mathbf{I}_n$ (i.e., $\mathbf{A}^-$ is a left inverse of $\mathbf{A}$).

(i) $\mathbf{A}(\mathbf{A}^*\mathbf{A})^-\mathbf{A}^*\mathbf{A} = \mathbf{A}$ and $\mathbf{A}^*\mathbf{A}(\mathbf{A}^*\mathbf{A})^-\mathbf{A}^* = \mathbf{A}^*$. This means that $(\mathbf{A}^*\mathbf{A})^-\mathbf{A}^*$ is a weak inverse of $\mathbf{A}$, and $\mathbf{A}(\mathbf{A}^*\mathbf{A})^-$ is a weak inverse of $\mathbf{A}^*$.

(j) $\mathbf{A}(\mathbf{A}^*\mathbf{A})^-\mathbf{A}^*$ is Hermitian, idempotent, and invariant for any choice of the weak inverse $(\mathbf{A}^*\mathbf{A})^-$.

(k) $\mathbf{A}^*\mathbf{A}\mathbf{G}\mathbf{A} = \mathbf{A}^*\mathbf{A}$ if and only if $\mathbf{G}$ is a weak inverse of $\mathbf{A}$.

7.3. The following conditions are equivalent.

(1) $\mathbf{G}$ is a weak inverse of $\mathbf{A}$.

(2) $\mathbf{A}\mathbf{G}$ is idempotent and $\operatorname{rank}(\mathbf{A}\mathbf{G}) = \operatorname{rank}\mathbf{A}$.

(3) $\mathbf{G}\mathbf{A}$ is idempotent and $\operatorname{rank}(\mathbf{G}\mathbf{A}) = \operatorname{rank}\mathbf{A}$.

(4) $\operatorname{rank}(\mathbf{I}_n - \mathbf{G}\mathbf{A}) = n - \operatorname{rank}\mathbf{A}$.

7.4. (Symmetric and Hermitian Matrices)

(a) A Hermitian matrix has a Hermitian weak inverse, namely $\frac{1}{2}(\mathbf{A}^- + (\mathbf{A}^-)^*)$.

(b) A Hermitian matrix $\mathbf{A}$ has a non-negative definite weak inverse if and only if $\mathbf{A}$ is non-negative definite.

7.5. (Rank of Inverse)

(a) Taking $\mathbf{X} = \mathbf{0}$ and $\mathbf{Y} = \mathbf{0}$ in (7.1) and noting that the rank is unchanged by multiplying by a nonsingular matrix, we see that

$$\operatorname{rank}(\mathbf{A}^-) = \operatorname{rank}\mathbf{A} + \operatorname{rank}\mathbf{Z}.$$

Since $\mathbf{Z}$ is arbitrary, there exists an $\mathbf{A}^-$ having any specified rank between $\operatorname{rank}\mathbf{A}$ and $\min\{m, n\}$ (Rao and Mitra [1971: 31]). In particular, we can choose $\mathbf{Z}$ such that $\mathbf{A}^-$ has full row or column rank (i.e., the rows or columns are linearly independent).

(b) $\operatorname{rank}(\mathbf{A}^-) = \operatorname{rank}\mathbf{A}$ if and only if $\mathbf{A}^-$ is also a $g_{1,2}$-inverse

7.6. (Representation of $\mathbf{A}\{1\}$) Let $\mathbf{A}^-$ be any weak inverse of $\mathbf{A}$. Then we have the following representations.

(a)   (i) $\mathbf{A}\{1\} = \{\mathbf{X} : \mathbf{X} = \mathbf{A}^- + \mathbf{H} - \mathbf{A}^-\mathbf{A}\mathbf{H}\mathbf{A}\mathbf{A}^-; \mathbf{H}$ arbitrary$\}$.

   (ii) $\mathbf{A}\{1\} = \{\mathbf{X} : \mathbf{X} = \mathbf{A}^- + (\mathbf{I} - \mathbf{A}^-\mathbf{A})\mathbf{F} + \mathbf{G}(\mathbf{I} - \mathbf{A}\mathbf{A}^-); \mathbf{F}, \mathbf{G}$ arbitrary$\}$.

(b) Let $\mathbf{A}_1^-$, $\mathbf{A}_2^-$, and $\mathbf{A}_3^-$ be any (not necessarily the same) fixed weak inverses of $\mathbf{A}$. Then $\mathbf{B}_1$ and $\mathbf{B}_2$ are also weak inverses of $\mathbf{A}$, where

$$
\begin{aligned}
\mathbf{B}_1 &= \mathbf{A}_1^- + \mathbf{F} - \mathbf{A}_2^- \mathbf{AFAA}_3^-, \\
\mathbf{B}_2 &= \mathbf{A}_1^- + (\mathbf{I} - \mathbf{A}_3^- \mathbf{A})\mathbf{F} + \mathbf{G}(\mathbf{I} - \mathbf{AA}_2^-).
\end{aligned}
$$

Here $\mathbf{F}$ and $\mathbf{G}$ are abitrary matrices of appropriate sizes. Also, any weak inverse of $\mathbf{A}$ can be written as $\mathbf{B}_1$ and as $\mathbf{B}_2$ for some matrices $\mathbf{F}$ and $\mathbf{G}$.

If we consider the special case of $\mathbf{A}_1^-$, $\mathbf{A}_2^-$, and $\mathbf{A}_3^-$ being all the same, we see that $\mathbf{B}_1$ and $\mathbf{B}_2$ reduce to (a)(i) and (a)(ii), respectively.

(c) If $\mathbf{A}$ and $\mathbf{B}$ are $m \times n$ matrices with $\mathbf{A}\{1\} = \mathbf{B}\{1\}$, that is every weak inverse of of $\mathbf{A}$ is a weak inverse of $\mathbf{B}$, and vice-versa, then $\mathbf{A} = \mathbf{B}$.

7.7. (Rank and Products)

(a) $\mathrm{rank}(\mathbf{ABC}) = \mathrm{rank}\,\mathbf{B}$ implies that $\mathbf{C}(\mathbf{ABC})^- \mathbf{A}$ is a weak inverse of $\mathbf{B}$. We can set $\mathbf{A} = \mathbf{I}$ or $\mathbf{C} = \mathbf{I}$.

(b) Let $\mathbf{V}$ be a matrix such that $\mathrm{rank}(\mathbf{A}^* \mathbf{VA}) = \mathrm{rank}\,\mathbf{A}$ (which is automatically satisfied if $\mathbf{A}$ is Hermitian positive definite), then:

  (i) $\mathbf{A}(\mathbf{A}^* \mathbf{VA})^- (\mathbf{A}^* \mathbf{VA}) = \mathbf{A}$ and $(\mathbf{A}^* \mathbf{VA})(\mathbf{A}^* \mathbf{VA})^- \mathbf{A}^* = \mathbf{A}^*$.

  (ii) $\mathbf{A}(\mathbf{A}^* \mathbf{VA})^- \mathbf{A}^*$ is invariant for any choice of $(\mathbf{A}^* \mathbf{VA})^-$ and is of the same rank as $\mathbf{A}$. If $\mathbf{A}^* \mathbf{VA}$ is Hermitian, then so is $\mathbf{A}(\mathbf{A}^* \mathbf{VA})^- \mathbf{A}^*$.

7.8. If $\mathbf{A}$ is $m \times n$ and $\mathbf{D}$ is $m \times m$, and both are of rank m, then

$$
\mathbf{D}^{-1} = \mathbf{A}(\mathbf{A}'\mathbf{DA})^- \mathbf{A}'.
$$

7.9. (Hermite Form)

(a) If $\mathbf{A}$ is $n \times n$ and $\mathbf{B}$ is nonsingular such that $\mathbf{BA} = \mathbf{H_A}$, where $\mathbf{H_A}$ is in Hermite form (Section 16.2.4), then $\mathbf{B}$ is weak inverse of $\mathbf{A}$.

(b) Let $\mathbf{A}$ be an $m \times n$ ($m > n$) matrix, and let $\mathbf{A}_0 = (\mathbf{A}, \mathbf{0}_{m \times (m-n)})$. Let $\mathbf{B}_0$ be a nonsingular matrix such that $\mathbf{B}_0 \mathbf{A}_0 = \mathbf{H}$, where $\mathbf{H}$ is in Hermite form. Suppose $\mathbf{B}_0$ is partitioned as

$$
\mathbf{B}_0 = \begin{pmatrix} \mathbf{B} \\ \mathbf{B}_1 \end{pmatrix}
$$

where $\mathbf{B}$ is $n \times m$. Then $\mathbf{B}$ is a weak inverse of $\mathbf{A}$.
A similar result holds for $m < n$.

7.10. Let $\mathbf{A}$ and $\mathbf{B}$ be $m \times n$ complex matrices. Then the following statements are equivalent:

(1) The nonzero eigenvalues of $\mathbf{B}^- \mathbf{A}$ are invariant with respect to $\mathbf{B}^-$.

(2) $\mathrm{trace}(\mathbf{B}^- \mathbf{A})$ is invariant with respect to $\mathbf{B}^-$.

(3) $\mathcal{C}(\mathbf{A}) \subseteq \mathcal{C}(\mathbf{B})$ and $\mathcal{C}(\mathbf{A}^*) \subseteq \mathcal{C}(\mathbf{B}^*)$.

Proofs. Section 7.2.1.

7.1. Ben-Israel and Greville [2003: 41] and Graybill [1983: 136].

7.2b. The result follows from the definition of a weak inverse.

7.2c. $\mathbf{A}^-\mathbf{A}\mathbf{A}^-\mathbf{A} = \mathbf{A}^-\mathbf{A}$ and $\mathbf{A}\mathbf{A}^-\mathbf{A}\mathbf{A}^- = \mathbf{A}\mathbf{A}^-$. Also $\mathbf{P}_{\mathcal{C}(\mathbf{A})}\mathbf{A} = \mathbf{A}$ and then take the transpose of $\mathbf{A}\mathbf{P}'_{\mathcal{C}(\mathbf{A}')} = \mathbf{A}$.

7.2d. Graybill [1983: 134], Rao and Bhimasnakaram [2000: 195], and Schott [2005: 203–204].

7.2e. Ben-Israel and Greville [2003: 43].

7.2g–7.2h. Ben-Israel and Greville [2003: 43] and Schott [2005: 204].

7.2i. Rao [1973a: 26] and Rao and Mitra [1971: 22].

7.2j. Rao [1973a: 26] and Rao and Rao [1998: 268–269].

7.2k. Rao and Mitra [1971: 22].

7.3. Rao and Bhimasankaram [2000: 195, (1)–(3)] and Rao and Mitra [1971: 21, 23].

7.4b. Arguing as in (7.21) for Hermitian matrices, we have $\mathbf{A} = \mathbf{U}\mathbf{\Lambda}\mathbf{U}^*$, where $\mathbf{U}$ is unitary and $\mathbf{\Lambda}$ is a diagonal matrix of non-negative eigenvalues. We then set $\mathbf{A}^- = \mathbf{U}\mathbf{\Lambda}^-\mathbf{U}^*$, which is Hermitian non-negative definite.

7.5b. Rao and Mitra [1971: 28].

7.6a(i). Rao and Mitra [1971: 26] and Schott [2005: 204].

7.6a(ii). Rao and Mitra [1971: 26].

7.6b. Graybill [1983: 137]

7.6c. Rao and Mitra [1971: 27] and Rao and Rao [1998: 277].

7.7a. Rao and Mitra [1971: 22] and Schott [2005: 205].

7.7b. Rao and Mitra [1971: 22].

7.8. This follows from (7.7a) by noting that $\operatorname{rank}(\mathbf{A}'\mathbf{D}\mathbf{A}) = \operatorname{rank}\mathbf{D}$, since $\mathbf{D}$ is nonsingular.

7.9. Graybill [1983: 132].

7.10. Baksalary and Puntanen [1990].

7.2.2   Products of Matrices

7.11. (Invariance Properties)

(a) The matrix $\mathbf{BA}^-\mathbf{C}$ is invariant for any choice of $\mathbf{A}^-$ if and only if $\mathcal{C}(\mathbf{B}') \subseteq \mathcal{C}(\mathbf{A}')$ and $\mathcal{C}(\mathbf{C}) \subseteq \mathcal{C}(\mathbf{A})$.

(b) From (a), If $\mathbf{A}$ is a real symmetric matrix, then $\mathbf{B}'\mathbf{A}^-\mathbf{B}$ is invariant for any choice of $\mathbf{A}^-$ if and only if $\mathcal{C}(\mathbf{B}) \subseteq \mathcal{C}(\mathbf{A})$.

(c) From (a), if $\mathbf{A}$ is any $n \times n$ matrix with $\mathbf{c} \subseteq \mathcal{C}(\mathbf{A})$ and $\mathbf{c} \subseteq \mathcal{C}(\mathbf{A}')$, then $\mathbf{c}'\mathbf{A}^-\mathbf{c}$ is invariant with respect to $\mathbf{A}^-$.

(d) (Regression) Let $\mathbf{X}$ be any real matrix and let $\mathbf{G} = (\mathbf{X}'\mathbf{X})^-$ be any weak inverse of $\mathbf{X}'\mathbf{X}$.

   (i) If $\mathbf{c} \subset \mathcal{C}(\mathbf{X}')$, then $\mathbf{c}'(\mathbf{X}'\mathbf{X})^-\mathbf{X}'$ is invariant for any weak inverse of $\mathbf{X}'\mathbf{X}$.

   (ii) $\mathbf{X}(\mathbf{X}'\mathbf{X})^-\mathbf{X}' = \mathbf{XX}^+$ is invariant and symmetric, being the orthogonal projector onto $\mathcal{C}(\mathbf{X})$. Here $\mathbf{X}^+$ is the Moore–Penrose inverse of $\mathbf{X}$.

   (iii) $\mathbf{G}'$ is also a weak inverse of $\mathbf{X}'\mathbf{X}$.

   (iv) $(\mathbf{X}'\mathbf{X})^-\mathbf{X}'$ is a weak inverse of $\mathbf{X}$.

(e) If $\text{rank}(\mathbf{CAB}) = \text{rank}\,\mathbf{C} = \text{rank}\,\mathbf{B}$, then $\mathbf{B}(\mathbf{CAB})^-\mathbf{C}$ is invariant for any choice of $(\mathbf{CAB})^-$.

7.12. If $\mathbf{P}$ is idempotent, then $\mathbf{P}(\mathbf{PAP})^-\mathbf{P}$ is a weak inverse of $\mathbf{PAP}$.

7.13. Noting that $\mathbf{FF}^-$ and $(\mathbf{F}^-\mathbf{F})' = \mathbf{F}'\mathbf{F}'^-$, being idempotent, represent (oblique) projections onto $\mathbf{F}$ and $\mathbf{F}'$, respectively (cf. 7.2c), we have the following for conformable matrices.

(a) $\mathbf{BA}^-\mathbf{A} = \mathbf{B}$ if and only if $\mathcal{C}(\mathbf{B}') \subseteq \mathcal{C}(\mathbf{A}')$, that is, if and only if there exists a matrix $\mathbf{D}$ such that $\mathbf{B} = \mathbf{DA}$.

(b) $\mathbf{AA}^-\mathbf{B} = \mathbf{B}$ if and only if $\mathcal{C}(\mathbf{B}) \subseteq \mathcal{C}(\mathbf{A})$, that is, if and only if there exists a matrix $\mathbf{D}$ such that $\mathbf{B} = \mathbf{AD}$.

(c) $(\mathbf{CAB})(\mathbf{CAB})^-\mathbf{C} = \mathbf{C}$ if and only if $\text{rank}(\mathbf{CAB}) = \text{rank}\,\mathbf{C}$.

(d) $\mathbf{B}(\mathbf{CAB})^-(\mathbf{CAB}) = \mathbf{B}$ if and only if $\text{rank}(\mathbf{CAB}) = \text{rank}\,\mathbf{B}$.

7.14. Let $\mathbf{A}$ be an $m \times n$ matrix, $\mathbf{B}$ be an $m \times m$ matrix, and $\mathbf{C}$ be an $n \times n$ matrix.

(a) If $\mathbf{B}$ and $\mathbf{C}$ are nonsingular, $(\mathbf{BAC})^- = \mathbf{C}^{-1}\mathbf{A}^-\mathbf{B}^{-1}$ for some weak inverse $\mathbf{A}^-$ of $\mathbf{A}$.

(b) If $\mathbf{A}$ has rank m and $\mathbf{B}$ is nonsingular, then $(\mathbf{A}'\mathbf{BA})^- = \mathbf{A}^-\mathbf{B}^{-1}\mathbf{A}'^-$.

(c) $(\mathbf{AB})^- = \mathbf{B}^-\mathbf{A}^-$ if and only if $\mathbf{P} = \mathbf{A}^-\mathbf{ABB}^-$ is idempotent.

7.15. Let $\mathbf{A}$, $\mathbf{B}$, and $\mathbf{C}$ be $m \times n$, $p \times m$, and $n \times q$ matrices, respectively. If $\mathbf{B}$ has full column rank m, and $\mathbf{C}$ has full row rank n, then

$$(\mathbf{BAC})^- = \mathbf{C}^-\mathbf{A}^-\mathbf{B}^-.$$

We can also get special cases by setting one of the matrices equal to the identity matrix.

7.16. If $(\mathbf{A}'\mathbf{A})^-$ is a weak inverse of $\mathbf{A}'\mathbf{A}$, then so is $(\mathbf{A}'\mathbf{A})^{-\prime}$.

7.17. The following hold for weak inverses $\mathbf{A}^-$ and $\mathbf{B}^-$.

(a) $(\mathbf{I} - \mathbf{A}\mathbf{A}^-)\mathbf{B}^-\mathbf{B}\mathbf{A} = -(\mathbf{I} - \mathbf{A}\mathbf{A}^-)(\mathbf{I} - \mathbf{B}^-\mathbf{B})\mathbf{A}$.

(b) $\mathbf{B}\mathbf{A}\mathbf{A}^-(\mathbf{I} - \mathbf{B}^-\mathbf{B}) = -\mathbf{B}(\mathbf{I} - \mathbf{A}\mathbf{A}^-)(\mathbf{I} - \mathbf{B}^-\mathbf{B})$.

(c) $(\mathbf{B}\mathbf{A})^- = \mathbf{A}^-\mathbf{B}^- - \mathbf{A}^-(\mathbf{I} - \mathbf{B}^-\mathbf{B})[(\mathbf{I} - \mathbf{A}\mathbf{A}^-)(\mathbf{I} - \mathbf{B}^-\mathbf{B})]^-(\mathbf{I} - \mathbf{A}\mathbf{A}^-)\mathbf{B}^-$.

(d) Let $\mathbf{A}$ be an $m \times n$ matrix and $\mathbf{B}$ be an $n \times p$ matrix. If rank $\mathbf{B} = n$, then $(\mathbf{A}\mathbf{B})^- = \mathbf{B}^-\mathbf{A}^-$.

(e) $\mathcal{C}(\mathbf{A}) \cap \mathcal{C}(\mathbf{B}) = \mathbf{0}$ if and only if $(\mathbf{A}\mathbf{A}' + \mathbf{B}\mathbf{B}')^-$ is a weak inverse of $\mathbf{A}\mathbf{A}'$.

Proofs. Section 7.2.2.

7.11a. Graybill [1983: 134–135, with the notation change $\mathbf{A}^c \rightarrow \mathbf{A}^-$ and $\mathbf{A}^- \rightarrow \mathbf{A}^+$] and Rao and Mitra [1971: 21]. An alternative proof using the idea of extremal ranks is given by Tian [2006a: 95]. He also gives necessary and sufficient conditions for rank$(\mathbf{B}\mathbf{A}^-\mathbf{C})$ to be invariant with respect to $\mathbf{A}^-$.

7.11d(i). Graybill [1983: 135].

7.11d(ii)–(iv). Searle [1982: 221–222].

7.11e. Rao and Mitra [1971: 22].

7.12. Follows from the definition of a weak inverse.

7.13a–b. Schott [2005: 205] and Rao and Mitra [1971: 21–22].

7.13c–d. Harville [2001: 106, exercise 44].

7.14a. Harville [1997: 113, lemma 9.2.4].

7.14b. From (7.2g) we have $\mathbf{A}\mathbf{A}^- = \mathbf{I}_m$, and then use the definition of weak inverse.

7.14c. Harville [2001: 51, exercise 8]. If $\mathbf{P}^2 = \mathbf{P}$, then $\mathbf{A}\mathbf{P}^2\mathbf{B} = \mathbf{A}\mathbf{P}\mathbf{B}$, which implies that $(\mathbf{A}\mathbf{B})^- = \mathbf{B}^-\mathbf{A}^-$. The converse is straightforward.

7.15. We use the fact that $\mathbf{B}^-\mathbf{B} = \mathbf{I}_m$ and $\mathbf{C}\mathbf{C}^- = \mathbf{I}_n$ from (7.2g) and (7.2h).

7.16. Schott [2005: 206–207].

7.17. Isotalo et al. [2005b: chapter 12] and (a)–(c) quoted by Searle [1982: 226]. For (d) we have $\mathbf{A}\mathbf{B}(\mathbf{B}^-\mathbf{A}^-)\mathbf{A}\mathbf{B} = \mathbf{A}\mathbf{B}$ since $\mathbf{B}\mathbf{B}^- = \mathbf{I}_n$.

7.2.3 Sums and Differences of Matrices

7.18. The following conditions are equivalent for any weak inverse $(\mathbf{A} + \mathbf{B})^-$.

(1) $\begin{pmatrix} \mathbf{A} \\ \mathbf{B} \end{pmatrix} (\mathbf{A} + \mathbf{B})^- (\mathbf{A}, \mathbf{B}) = \begin{pmatrix} \mathbf{A} & \mathbf{0} \\ \mathbf{0} & \mathbf{B} \end{pmatrix}$.

(2) $(\mathbf{A} + \mathbf{B})^-$ is a weak inverse of both $\mathbf{A}$ and $\mathbf{B}$.

(3) $\mathcal{C}(\mathbf{A}) \cap \mathcal{C}(\mathbf{B}) = \mathbf{0}$ and $\mathcal{C}(\mathbf{A}') \cap \mathcal{C}(\mathbf{B}') = \mathbf{0}$.

7.19. Let $\mathbf{A}$, $\mathbf{B}$, $\mathbf{C}$, and $\mathbf{V}$ be real conformable matrices with $\mathbf{V}$ positive definite, and $\mathcal{C}(\mathbf{C}) = \mathcal{C}(\mathbf{A}') \cap \mathcal{C}(\mathbf{B})$. Let $\mathbf{Q_B} = \mathbf{I} - \mathbf{P_B}$, where $\mathbf{P_B} = \mathbf{B}(\mathbf{B}'\mathbf{B})^- \mathbf{B}'$.

(a)

$$\mathbf{A}(\mathbf{A}'\mathbf{V}\mathbf{A})^- \mathbf{A}' - \mathbf{A}\mathbf{Q_B}(\mathbf{Q_B}\mathbf{A}'\mathbf{V}\mathbf{A}\mathbf{Q_B})^- \mathbf{Q_B}\mathbf{A}'$$
$$= \mathbf{A}(\mathbf{A}'\mathbf{V}\mathbf{A})^- \mathbf{C}[\mathbf{C}'(\mathbf{A}'\mathbf{V}\mathbf{A})^- \mathbf{C}]^- \mathbf{C}'(\mathbf{A}'\mathbf{V}\mathbf{A})^- \mathbf{A}'.$$

(b) $\mathbf{V}^{-1} - \mathbf{Q_B}(\mathbf{Q_B}\mathbf{V}\mathbf{Q_B})^- \mathbf{Q_B} = \mathbf{V}^{-1}\mathbf{B}(\mathbf{B}'\mathbf{V}^{-1}\mathbf{B})^- \mathbf{B}'\mathbf{V}^{-1}$.

$\mathbf{Q_B}$ can be replaced by a matrix with the same range. The above results are used in the theory of singular linear regression models.

Definition 7.2. Given $\mathbf{A}$ and $\mathbf{B}$ both $m \times n$ matrices, then $\mathbf{A}(\mathbf{A}+\mathbf{B})^-\mathbf{B}$ is called the *parallel sum* of $\mathbf{A}$ and $\mathbf{B}$. Some authors call $(\mathbf{A}^+ + \mathbf{B}^+)^+$ the parallel sum and, under certain conditions, the two definitions are equivalent. For properties relating to both definitions and their equivalence, see Rao and Mitra [1971: 186–192]. They also define a *parallel difference*.

7.20. For conformable matrices

(a) $\mathbf{A}\mathbf{A}'(\mathbf{A}\mathbf{A}' + \mathbf{B}\mathbf{B}')^- \mathbf{B}\mathbf{B}' = \mathbf{B}\mathbf{B}'(\mathbf{A}\mathbf{A}' + \mathbf{B}\mathbf{B}')^- \mathbf{A}\mathbf{A}'$.

(b) $[\mathbf{A}\mathbf{A}'(\mathbf{A}\mathbf{A}' + \mathbf{B}\mathbf{B}')^- \mathbf{B}\mathbf{B}']^- = (\mathbf{A}\mathbf{A}')^- + (\mathbf{B}\mathbf{B}')^-$.

Proofs. Section 7.2.3.

> 7.18. Harville [1997: 421]. We obtain (2) by multiplying out (1) to get $\mathbf{A}(\mathbf{A}+\mathbf{B})^-\mathbf{A} = \mathbf{A}$, $\mathbf{B}(\mathbf{A}+\mathbf{B})^-\mathbf{B} = \mathbf{B}$, $\mathbf{A}(\mathbf{A}+\mathbf{B})^-\mathbf{B}$, and $\mathbf{B}(\mathbf{A}+\mathbf{B})^-\mathbf{A}$.

> 7.19–7.20. Kollo and van Rosen [2005: 50].

7.2.4 Real Symmetric Matrices

Let $\mathbf{A}$ be a real symmetric matrix.

7.21. By (16.44) there exists orthogonal $\mathbf{T}$ such that

$$\mathbf{T}'\mathbf{A}\mathbf{T} = \begin{pmatrix} \mathbf{\Lambda}_r & \mathbf{0} \\ \mathbf{0} & \mathbf{0} \end{pmatrix},$$

where $\mathbf{\Lambda}_r$ is a nonsingular $r \times r$ diagonal matrix consisting of the nonzero eigenvalues of $\mathbf{A}$. Then

$$\mathbf{A}^- = \mathbf{T} \begin{pmatrix} \mathbf{\Lambda}_r^{-1} & \mathbf{X} \\ \mathbf{Y} & \mathbf{Z} \end{pmatrix} \mathbf{T}',$$

where $\mathbf{X}$, $\mathbf{Y}$, and $\mathbf{Z}$ are arbitrary. Note that $\mathbf{A}^-$ need not be symmetric. (When $\mathbf{A}$ is Hermitian, then $\mathbf{T}$ is unitary and $\mathbf{T}'$ is replaced by $\mathbf{T}^*$.)

7.22. Suppose $\mathbf{A}$ is $n \times n$, $\mathbf{P}$ is symmetric and idempotent, $\mathbf{A} + c\mathbf{P}$ is nonsingular, and $\mathbf{PA} = \mathbf{0}$.

(a) $(\mathbf{A} + c\mathbf{P})^{-1}$ is a weak inverse of both $\mathbf{A}$ and $\mathbf{P}$.

(b) In particular, if $\mathbf{A}\mathbf{1}_n = \mathbf{0}$, $\mathbf{J}_n = \mathbf{1}_n\mathbf{1}'_n$ $(=n\mathbf{P})$, and $\mathbf{A} + d\mathbf{J}_n$ is nonsingular, then $(\mathbf{A} + d\mathbf{J}_n)^{-1}$ is a weak inverse of $\mathbf{A}$. Furthermore,

$$\mathbf{A}^+ = (\mathbf{A} + d\mathbf{J}_n)^{-1} - (dn^2)^{-1}\mathbf{J}_n.$$

These results are useful in experimental designs (e.g., John and Williams [1995: 23]).

7.23. Suppose that $1 + 1 \neq 0$ in the underlying field $\mathcal{F}$. Since $(\mathbf{A}^-)'$ is a weak inverse of $\mathbf{A}'$, a symmetric weak inverse of a symmetric matrix $\mathbf{A}$ always exists, namely $\mathbf{B} = \frac{1}{2}[\mathbf{A}^- + (\mathbf{A}^-)']$.

7.24. From the definition of a weak inverse, if $\mathbf{A}$ and $\mathbf{A}^-\mathbf{A}$ are symmetric, then $(\mathbf{A}^-)^2$ is a weak inverse of $\mathbf{A}^2$.

Proofs. Section 7.2.4.

7.21. Searle [1982: 220].

7.22. John and Williams [1995: 23].

7.23. Since $\mathbf{A}^{-\prime} = \mathbf{A}'^- = \mathbf{A}^-$, we have $\mathbf{A}\mathbf{A}^{-\prime}\mathbf{A} = \mathbf{A}$ and $\mathbf{B} = \mathbf{A}^-$.

7.2.5   Decomposition Methods

7.25. (Diagonalizable Matrices) If $\mathbf{A}$ is diagonalizable of rank r, we have from (16.17) the spectral decomposition $\mathbf{A} = \sum_{i=1}^n \lambda_i\mathbf{F}_i = \sum_{i=1}^r \lambda_i\mathbf{F}_i$, where $\lambda_{r+1} = \cdots = \lambda_n = 0$. Then:

(a) $\sum_{i=1}^r \lambda_i^{-1}\mathbf{F}_i$ is a weak inverse of $\mathbf{A}$.

(b) $(\mathbf{A} + \sum_{i=r+1}^n a_i\mathbf{F}_i)^{-1}$ is a weak inverse of $\mathbf{A}$ for all nonzero real $a_{r+1}, \ldots, a_n$.

7.26. There exist permutation matrices $\mathbf{\Pi}_1$ and $\mathbf{\Pi}_2$ such that

$$\mathbf{\Pi}_1\mathbf{A}\mathbf{\Pi}_2 = \mathbf{B} = \begin{pmatrix} \mathbf{B}_{11} & \mathbf{B}_{12} \\ \mathbf{B}_{21} & \mathbf{B}_{22} \end{pmatrix},$$

where $\mathbf{B}_{11}$ is a nonsingular $r \times r$ matrix and $r = \operatorname{rank} \mathbf{A}$. Then $\mathbf{A} = \mathbf{\Pi}'_1\mathbf{B}\mathbf{\Pi}'_2$ and

$$\mathbf{B}^- = \begin{pmatrix} \mathbf{B}_{11}^{-1} & \mathbf{0} \\ \mathbf{0} & \mathbf{0} \end{pmatrix}$$

is a weak inverse of $\mathbf{B}$. Also $\mathbf{\Pi}_2\mathbf{B}^-\mathbf{\Pi}_1$ is a weak inverse of $\mathbf{A}$.

Proofs. Section 7.2.5.

7.25. Hunter [1983a: 150].

7.26. Searle [1982: 217–218].

7.3   OTHER INVERSES

In this section we assume real matrices. However, many of the results hold for complex matrices by simply replacing $'$ by $*$.

7.3.1   Reflexive (g_{12}) Inverse

Let $\mathbf{A}$ be an $m \times n$ matrix and $\mathbf{G}$ an $n \times m$ matrix. As noted at the beginning of this chapter, $\mathbf{G} = \mathbf{A}_{12}^-$ is a g_{12}-inverse of $\mathbf{A}$ if $\mathbf{AGA} = \mathbf{A}$ and $\mathbf{GAG} = \mathbf{G}$, i.e., if $\mathbf{G}$ is a weak inverse of $\mathbf{A}$ and $\mathbf{A}$ is a weak inverse of $\mathbf{G}$. Such an inverse is usually refered to as a *reflexive generalized inverse* or reflexive g-inverse.

7.27. If $\mathbf{A}$ is $m \times n$, we have from (3.5) the full-rank factorization $\mathbf{A} = \mathbf{C}_{m \times r} \mathbf{R}_{r \times n}$, where $\mathbf{C}$ and $\mathbf{R}$ have rank r. Let $\mathbf{D}$ and $\mathbf{S}$ be the left and right inverses of $\mathbf{C}$ and $\mathbf{R}$, respectively, so that $\mathbf{DC} = \mathbf{I}_r$ and $\mathbf{RS} = \mathbf{I}_r$. Then $\mathbf{SD}$ is a reflexive g-inverse of $\mathbf{A}$.

7.28. If $\mathbf{A}_1^-$ and $\mathbf{A}_2^-$ are any (possibly different) weak inverses of $\mathbf{A}$, then $\mathbf{A}_1^- \mathbf{AA}_2^-$ is a g_{12}-inverse of $\mathbf{A}$.

7.29. Every reflexive g-inverse of a matrix $\mathbf{A}$ can be expressed in the form of $\mathbf{A}^- \mathbf{AA}^-$ for some weak inverse $\mathbf{A}^-$ of $\mathbf{A}$.

7.30. A weak inverse $\mathbf{G}$ of $\mathbf{A}$ is a g_{12}-inverse if and only if $\operatorname{rank} \mathbf{G} = \operatorname{rank} \mathbf{A}$.

7.31. If $\mathbf{G}$ is a g_{12}-inverse of $\mathbf{A}$, then $\mathbf{G}'$ is a g_{12}-inverse of $\mathbf{A}'$.

7.32. (Invariance)   If $\mathbf{A}$, $\mathbf{B}$, and $\mathbf{C}$ are nonzero conformable matrices, then $\mathbf{AB}_{12}^- \mathbf{C}$ is invariant with respect to $\mathbf{B}_{12}^-$ if and only if $\mathcal{C}(\mathbf{A}') \subseteq \mathcal{C}(\mathbf{B}')$ and $\mathcal{C}(\mathbf{C}) \subseteq \mathcal{C}(\mathbf{B})$.

Proofs. Section 7.3.1.

   7.27. Rao and Rao [1998: 279].

   7.28. Harville [1997: 496, lemma 20.3.2].

   7.29. Rao and Mitra [1971: 28].

   7.30. Harville [1997: 497] and Rao and Rao [1998: 279].

   7.31. Harville [1997: 497].

   7.32. Tian [2006a: 100] proved this using his extremal rank technique. He also gave necessary and sufficient conditions for the rank to be invariant.

7.3.2   Minimum Norm (g_{14}) Inverse

The matrix $\mathbf{G}$ is a g_{14}-inverse of $\mathbf{A}$ if $\mathbf{AGA} = \mathbf{A}$ and $\mathbf{GA}$ is symmetric (or Hermitian if $\mathbf{A}$ is complex). It is usually refered to as a *minimum norm g-inverse*.

7.33. The following conditions are equivalent.

   (1) $\mathbf{G}$ is a g_{14}-inverse of $\mathbf{A}$.

(2) $\mathbf{GAA'} = \mathbf{A'}$.

(3) $\mathbf{AA'G} = \mathbf{A}$.

(4) $\mathbf{GA} = \mathbf{P}_{\mathcal{C}(\mathbf{A'})}$, where $\mathbf{P}_{\mathcal{C}(\mathbf{A'})}$, being symmetric and idempotent, represents the orthogonal projection onto $\mathcal{C}(\mathbf{A'})$. (In this case $\mathbf{GA}$ is invariant to the choice of $\mathbf{G}$.)

In the complex case we replace $'$ by $*$.

7.34. If $\mathbf{G}$ is a g_{14}-inverse, then $\mathbf{x} = \mathbf{Gy}$ minimizes $\|\mathbf{x}\|_2$ subject to $\mathbf{Ax} = \mathbf{y}$.

7.35. $\mathbf{A}\{14\} = \{\mathbf{G} : \mathbf{G} = \mathbf{A}_{14}^- + \mathbf{Z}(\mathbf{I}_n - \mathbf{AA}_{14}^-)\}$, where $\mathbf{Z}$ is an arbitrary $n \times m$ matrix.

7.36. (Product Invariance) If $\mathbf{A}$, $\mathbf{B}$, and $\mathbf{C}$ are nonzero conformable matrices, then $\mathbf{AB}_{14}^-\mathbf{C}$ is invariant with respect to $\mathbf{B}_{14}^-$ if and only if $\mathcal{C}(\mathbf{C}) \subseteq \mathcal{C}(\mathbf{B})$.

Proofs. Section 7.3.2.

> 7.33. Harville [1997: 498–499].

> 7.34. Harville [1997: 497, theorem 20.3.6] and Rao and Rao [1998: 288].

> 7.35. Ben-Israel and Greville [2003: 55].

> 7.36. Tian [2000a: 105].

7.3.3 Minimum Norm Reflexive (g_{124}) Inverse

Let $\mathbf{A}$ be an $m \times n$ matrix and $\mathbf{G}$ an $n \times m$ matrix. As noted at the beginning of this chapter, $\mathbf{G}$ is a g_{124}-inverse of $\mathbf{A}$ if $\mathbf{AGA} = \mathbf{A}$, $\mathbf{GAG} = \mathbf{G}$ and $\mathbf{GA}$ is symmetric. Since it combines a g_{12} and a g_{14} inverse, it is refered to as a *minimum norm reflexive g-inverse*.

7.37. A matrix $\mathbf{G}$ is a g_{124}-inverse of $\mathbf{A}$ if and only if $\mathbf{G} = \mathbf{A'}(\mathbf{AA'})^-$ for some weak inverse $(\mathbf{AA'})^-$ of $\mathbf{AA'}$.

7.38. If $\mathbf{G}$ is a g_{124}-inverse of $\mathbf{A}$, then $\mathcal{C}(\mathbf{G}) = \mathcal{C}(\mathbf{A'})$.

7.39. $\mathbf{A}\{124\} = \{\mathbf{G} : \mathbf{G} = \mathbf{A}_{124}^- + \mathbf{A}_{124}^-\mathbf{Z}(\mathbf{I}_n - \mathbf{AA}_{124}^-)\}$, where $\mathbf{Z}$ is an arbitrary $n \times m$ matrix.

Proofs. Section 7.3.3.

> 7.37–7.38. Harville [1997: 499].

> 7.39. Quoted by Ben-Israel and Greville [2003: 56].

7.3.4   Least Squares (g_{13}) Inverse

Let $\mathbf{A}$ be an $m \times n$ matrix and $\mathbf{G}$ an $n \times m$ matrix. As noted at the beginning of this chapter, $\mathbf{G}$ is a g_{13}-inverse of $\mathbf{A}$ if $\mathbf{AGA} = \mathbf{A}$ and $\mathbf{AG}$ is symmetric. It is usually refered to as a *least squares g-inverse* and is denoted by $\mathbf{A}_{13}^-$. In what follows, we can replace $'$ by $*$ in the complex case.

7.40. A $p \times n$ matrix $\mathbf{G}$ is a g_{13}-inverse of the $n \times p$ matrix $\mathbf{X}$ if and only if $(\mathbf{y} - \mathbf{Xb})'(\mathbf{y} - \mathbf{Xb})$ is minimized at $\mathbf{b} = \mathbf{Gy}$.

7.41. The following statements are equivalent.

(1) A matrix $\mathbf{G}$ is a g_{13}-inverse of $\mathbf{A}$.

(2) $\mathbf{A}'\mathbf{AG} = \mathbf{A}'$ or, equivalently, $\mathbf{G}'\mathbf{A}'\mathbf{A} = \mathbf{A}$.

(3) $\mathbf{AG} = \mathbf{P}_{\mathcal{C}(\mathbf{A})}$, where $\mathbf{P}_{\mathcal{C}(\mathbf{A})} = \mathbf{A}(\mathbf{A}'\mathbf{A})^-\mathbf{A}'$ represents the orthogonal projection onto $\mathcal{C}(\mathbf{A})$.

7.42. Let $\mathbf{G}$ be a g_{13}-inverse of $\mathbf{A}$. Then:

(a) $\mathbf{AG}$ is invariant to the choice of $\mathbf{G}$.

(b) $\mathcal{C}(\mathbf{G}'\mathbf{A}') = \mathcal{C}(\mathbf{A})$.

7.43. $(\mathbf{A}'\mathbf{A})^-\mathbf{A}'$ is a g_{13}-inverse of $\mathbf{A}$ for any weak inverse, $(\mathbf{A}'\mathbf{A})^-$, of $\mathbf{A}'\mathbf{A}$.

7.44. $\mathbf{A}\{13\} - \{\mathbf{G} . \mathbf{G} - \mathbf{A}_{13}^- + (\mathbf{I}_n - \mathbf{A}_{13}^-\mathbf{A})\mathbf{Z}\}$,

where $\mathbf{Z}$ is an arbitrary $n \times m$ matrix.

7.45. (Product Invariance) If $\mathbf{A}$, $\mathbf{B}$, and $\mathbf{C}$ are nonzero conformable matrices, then $\mathbf{AB}_{13}^-\mathbf{C}$ is invariant with respect to $\mathbf{B}_{13}^-$ if and only if $\mathcal{C}(\mathbf{A}') \subseteq \mathcal{C}(\mathbf{B}')$.

Proofs. Section 7.3.4.

7.40. Harville [1997: 500–501, corollary 20.3.14] and Schott [2005: 233].

7.41. Harville [1997: 500] and Rao and Rao [1998: 289-290].

7.42. Harville [1997: 501, corollary 20.3.15].

7.43. Ben-Israel and Greville [2003: 47] and Schott [2005: 207].

7.44. Quoted by Ben-Israel and Greville [2003: 55].

7.45. This result is proved by Tian [2006a] using his extremal rank method. The same condition also applies for rank invariance.

7.3.5 Least Squares Reflexive (g_{123}) Inverse

Let $\mathbf{A}$ be an $m \times n$ matrix and $\mathbf{G}$ an $n \times m$ matrix. As noted at the beginning of this chapter, $\mathbf{G}$ is a g_{123}-inverse of $\mathbf{A}$ if $\mathbf{AGA} = \mathbf{A}$, $\mathbf{GAG} = \mathbf{G}$, and $\mathbf{AG}$ is symmetric. Such an inverse is also called a *least squares reflexive g-inverse*

7.46. If $\mathbf{G}$ is a g_{123}-inverse of $\mathbf{A}$, then $\mathcal{C}(\mathbf{G}') = \mathcal{C}(\mathbf{A})$ and $\mathcal{N}(\mathbf{G}) = [\mathcal{C}(\mathbf{A})]^{\perp} = \mathcal{N}(\mathbf{A}')$.

7.47. $\mathbf{G}$ is a g_{123} inverse of $\mathbf{A}$ if and only if $\mathbf{G} = (\mathbf{A}'\mathbf{A})^{-}\mathbf{A}'$ for some weak inverse $(\mathbf{A}'\mathbf{A})^{-}$ of $\mathbf{A}'\mathbf{A}$.

7.48. $\mathbf{A}\{123\} = \{\mathbf{G} : \mathbf{G} = \mathbf{A}_{123}^{-} + (\mathbf{I}_n - \mathbf{A}_{123}^{-}\mathbf{A})\mathbf{Z}\mathbf{A}_{123}^{-}\}$, where $\mathbf{Z}$ is an arbitrary $n \times m$ matrix.

Proofs. Section 7.3.4.

> 7.46. Harville [1997: 501, lemma 20.3.16].
>
> 7.47. Harville [1997: 502, theorem 20.3.17].
>
> 7.48. Quoted by Ben-Israel and Greville [2003: 56].

7.4 MOORE–PENROSE (G_{1234}) INVERSE

7.4.1 General Properties

Let $\mathbf{A}$ be an $m \times n$ matrix and $\mathbf{G}$ an $n \times m$ matrix. If $\mathbf{G}$ satisfies all four conditions mentioned at the beginning of this chapter, then it is called the Moore–Penrose inverse of $\mathbf{A}$ and is denoted by $\mathbf{A}^{+}$. This definition was given by Penrose [1955]. For convenience, we list the four conditions for the complex case, namely: (1) $\mathbf{AGA} = \mathbf{A}$, (2) $\mathbf{GAG} = \mathbf{G}$, (3) $\mathbf{AG} = (\mathbf{AG})^{*}$, and (4) $\mathbf{GA} = (\mathbf{GA})^{*}$.

The Moore–Penrose inverse of a general matrix $\mathbf{A}$ can be obtained using a QR decomposition (16.42) or the singular value decomposition given below (cf. 7.50). For diagonalizable matrices, which includes symmetric matrices, see (16.17c).

Moore–Penrose inverses are particularly useful in experimental design. John and Williams [1995] discuss the Moore–Penrose inverse of the so-called *information matrix* of a design for a wide range of designs including the incomplete block, the connected, and the cyclic designs.

There are a number of references referring to the real case, namely Abadir and Magnus [2005: section 10.3], Graybill [1983: chapter 6, with $\mathbf{A}^{-} \to \mathbf{A}^{+}$], Harville [1997: chapter 20], Magnus and Neudecker [1999: 33, 34, 38], and Schott [2005: section 5.2]. For the complex case see Ben-Israel and Greville [2003], Campbell and Meyer [1979: chapter 1], and Rao and Mitra [1971: section 3.3 and, for some miscellaneous expansions of $\mathbf{A}^{+}$, section 3.5].

7.49. (Representation) If $\mathbf{A}$ is a complex matrix of rank r, then we have the singular value decomposition of $\mathbf{A}$, namely $\mathbf{A} = \mathbf{P}_r \mathbf{\Delta}_r \mathbf{Q}_r^{*}$ (cf. Section 16.3), where $\mathbf{P}_r$ is $m \times r$ with orthonormal columns, $\mathbf{Q}_r$ is $n \times r$ with orthonormal columns, and $\mathbf{\Delta}_r$ is an $r \times r$ diagonal matrix with positive diagonal elements. Then

$$\mathbf{A}^{+} = \mathbf{Q}_r \mathbf{\Delta}_r^{-1} \mathbf{P}_r^{*}.$$

7.50. $\mathbf{A}^+$ is unique.

7.51. Let $\mathbf{A}$ be an $m \times n$ real matrix of rank r. Then $\mathbf{A}^+$ can be computed by the following steps.

(1) Compute $\mathbf{B} = \mathbf{A}'\mathbf{A}$.

(2) Let $\mathbf{C}_1 = \mathbf{I}_n$.

(3) Compute $\mathbf{C}_{j+1} = \mathbf{I}_n(1/j)\operatorname{trace}(\mathbf{C}_j\mathbf{B}) - \mathbf{C}_j\mathbf{B}$, for $j = 1, 2, \ldots, r - 1$.

(4) Compute $\mathbf{A}^+ = r\mathbf{C}_r\mathbf{A}'/\operatorname{trace}(\mathbf{C}_r\mathbf{B})$.

Also $\mathbf{C}_{r+1}\mathbf{B} = \mathbf{0}$ and $\operatorname{trace}(\mathbf{C}_r\mathbf{B}) \neq 0$. Since $\mathbf{C}_{r+1}\mathbf{B} = \mathbf{0}$, r does not need to be known in advance. This result is mainly of historical interest, but it does give a method for small matrices. Numerically stable methods for computing $\mathbf{A}^+$ are given by Golub and Van Loan [1996].

7.52. Below we give some basic properties of the Moore–Penrose inverse of a single matrix or vector. We assume matrices and vectors are complex, unless otherwise stated. Most of the following are readily proved by showing that the four conditions are satisfied and also invoking the uniqueness of the Moore–Penrose inverse.

(a) From (7.49) and (7.50), $\mathbf{A}^+$ always exists and is unique.

(b) $\mathbf{A}^+$ is the minimum norm least-squares g-inverse of $\mathbf{A}$, i.e., for every $\mathbf{b}$ that minimizes $(\mathbf{y} - \mathbf{A}\mathbf{b})^*(\mathbf{y} - \mathbf{A}\mathbf{b})$, $\mathbf{b}^*\mathbf{b}$ is minimized when $\mathbf{b} = \mathbf{A}^+\mathbf{y}$.

(c) $\mathbf{a}^+ = (\mathbf{a}^*\mathbf{a})^{-1}\mathbf{a}^*$ and $(\mathbf{a}\mathbf{b}^*)^+ = (\mathbf{a}^*\mathbf{a})^{-1}(\mathbf{b}^*\mathbf{b})^{-1}(\mathbf{b}\mathbf{a}^*)$.

(d) If $c \neq 0$, then $(c\mathbf{A})^+ = (1/c)\mathbf{A}^+$.

(e) If $\mathbf{D} = \operatorname{diag}(d_1, d_2, \ldots, d_n)$, then $\mathbf{D}^+ = \operatorname{diag}(d_1^+, d_2^+, \ldots, d_n^+)$, where (for $i = 1, 2, \ldots, n$)

$$d_i^+ = \begin{cases} 1/d_i, & \text{if } d_i \neq 0, \\ 0, & \text{if } d_i = 0. \end{cases}$$

(f) $\mathbf{A}^+ = \mathbf{A}^{-1}$ for nonsingular $\mathbf{A}$.

(g) $\mathbf{A}^+ = \mathbf{A}^*$ if the columns of $\mathbf{A}$ are orthogonal with respect to the inner product $\langle \mathbf{x}, \mathbf{y} \rangle = \mathbf{y}^*\mathbf{x}$.

(h) $(\mathbf{A}^+)^+ = \mathbf{A}$.

(i) $(\mathbf{A}')^+ = (\mathbf{A}^+)'$, $(\overline{\mathbf{A}})^+ = \overline{(\mathbf{A}^+)}$, and $(\mathbf{A}^*)^+ = (\mathbf{A}^+)^*$.

(j) $\mathbf{A}\mathbf{A}^+ = \mathbf{A}^+\mathbf{A}$ if and only if $\mathcal{C}(\mathbf{A}) = \mathcal{C}(\mathbf{A}^*)$, i.e., it holds when $\mathbf{A}$ is Hermitian.

(k) $\operatorname{rank} \mathbf{A} = \operatorname{rank} \mathbf{A}^+ = \operatorname{rank}(\mathbf{A}\mathbf{A}^+) = \operatorname{rank}(\mathbf{A}^+\mathbf{A})$.

(l) For any $m \times n$ complex matrix $\mathbf{A}$:

    (i) $\mathcal{C}(\mathbf{A}) = \mathcal{C}(\mathbf{A}\mathbf{A}^+) = \mathcal{C}(\mathbf{A}\mathbf{A}^*)$.

    (ii) $\mathcal{C}(\mathbf{A}^+) = \mathcal{C}(\mathbf{A}^*) = \mathcal{C}(\mathbf{A}^+\mathbf{A}) = \mathcal{C}(\mathbf{A}^*\mathbf{A})$.

    (iii) $\mathcal{C}(\mathbf{I}_m - \mathbf{A}\mathbf{A}^+) = \mathcal{N}(\mathbf{A}\mathbf{A}^+) = \mathcal{N}(\mathbf{A}^*) = \mathcal{N}(\mathbf{A}^+)$.

(iv) $\mathcal{C}(\mathbf{I}_n - \mathbf{A}^+\mathbf{A}) = \mathcal{N}(\mathbf{A}^+\mathbf{A}) = \mathcal{N}(\mathbf{A})$.

(m) $\mathbf{A}^+$ need not be a continuous function of the elements of $\mathbf{A}$. Not only can $\mathbf{A}^+(t)$ be discontinuous in the sense that $\lim_{t\to 0}\mathbf{A}^+(t) \neq \mathbf{A}^+(0)$, but as $\mathbf{A}(t)$ moves closer to $\mathbf{A}(0)$, $\mathbf{A}^+(t)$ can move further away from $\mathbf{A}^+(0)$. However, $\mathbf{A}^+(t)$ is continuous on $[a,b]$ if and only if $\mathrm{rank}[\mathbf{A}(t)]$ is constant on $[a,b]$.

7.53. Let $\mathbf{A}$ be an $n \times n$ real symmetric matrix with r nonzero eigenvalues λ_1, $\lambda_2,\ldots,\lambda_r$, and let $\mathbf{\Lambda}_r = \mathrm{diag}(\lambda_1,\lambda_2,\ldots,\lambda_r)$. Then:

(a) From (16.44) there exists an orthogonal matrix $\mathbf{Q}$ such that

$$\mathbf{A} = \mathbf{Q}\,\mathrm{diag}(\lambda_1,\ldots,\lambda_r,0,\ldots,0)\mathbf{Q}'.$$

Then

$$\mathbf{A}^+ = \mathbf{Q}\begin{pmatrix} \mathbf{\Lambda}_r^{-1} & \mathbf{0} \\ \mathbf{0} & \mathbf{0} \end{pmatrix}\mathbf{Q}' = \sum_{i=1}^{r}\lambda_i^{-1}\mathbf{q}_i\mathbf{q}_i'.$$

(b) From (a) we have

$$\mathrm{trace}\,\mathbf{A}^+ = \sum_{i=1}^{r}\lambda_i^{-1}.$$

7.54. For any real matrix $\mathbf{A}$:

(a) $\mathbf{A}^+\mathbf{A}$ and $\mathbf{A}\mathbf{A}^+$ are symmetric and idempotent, and they are equal if $\mathbf{A}$ is symmetric.

(b) Since $\mathbf{A}^+\mathbf{A}$ and $\mathbf{A}\mathbf{A}^+$ are symmetric:

   (i) $\mathbf{A}'\mathbf{A}\mathbf{A}^+ = \mathbf{A}' = \mathbf{A}^+\mathbf{A}\mathbf{A}'$.
   (ii) $\mathbf{A}'\mathbf{A}^{+\prime}\mathbf{A}^+ = \mathbf{A}^+ = \mathbf{A}^+\mathbf{A}^{+\prime}\mathbf{A}'$.

(c) $(\mathbf{A}'\mathbf{A})^+ = \mathbf{A}^+\mathbf{A}^{+\prime}$ and $(\mathbf{A}\mathbf{A}')^+ = \mathbf{A}^{+\prime}\mathbf{A}^+$.

(d) $\mathbf{A}^+ = (\mathbf{A}'\mathbf{A})^+\mathbf{A}' = \mathbf{A}'(\mathbf{A}\mathbf{A}')^+$. Also:

   (i) If $\mathbf{A}$ has full column rank, then $\mathbf{A}^+ = (\mathbf{A}'\mathbf{A})^{-1}\mathbf{A}'$ and $\mathbf{A}^+\mathbf{A} = \mathbf{I}_n$.
   (ii) If $\mathbf{A}$ has full row rank, then $\mathbf{A}^+ = \mathbf{A}'(\mathbf{A}\mathbf{A}')^{-1}$ and $\mathbf{A}\mathbf{A}^+ = \mathbf{I}_m$.

(e) $(\mathbf{A}\mathbf{A}^+)^+ = \mathbf{A}\mathbf{A}^+$ and $(\mathbf{A}^+\mathbf{A})^+ = \mathbf{A}^+\mathbf{A}$.

(f) $\mathbf{A}(\mathbf{A}'\mathbf{A})^+\mathbf{A}'\mathbf{A} = \mathbf{A} = \mathbf{A}\mathbf{A}'(\mathbf{A}\mathbf{A}')^+\mathbf{A}$.

(g) If $\mathbf{V}$ is positive definite, then $(\mathbf{X}'\mathbf{V}^{-1}\mathbf{X})(\mathbf{X}'\mathbf{V}^{-1}\mathbf{X})^+\mathbf{X}' = \mathbf{X}'$.

(h) For any weak inverse $\mathbf{A}^-$, $\mathbf{A}^+ = \mathbf{A}'(\mathbf{A}\mathbf{A}')^-\mathbf{A}(\mathbf{A}'\mathbf{A})^-\mathbf{A}'$.

(i) If $\mathbf{A}^-\mathbf{A}$ is real symmetric, then it is unique and equals $\mathbf{A}^+\mathbf{A}$.

(j) If $\mathrm{rank}\,\mathbf{A} = 1$, then $\mathbf{A}^+ = [\mathrm{trace}(\mathbf{A}\mathbf{A}')]^{-1}\mathbf{A}'$.

The above results also hold for $\mathbf{A}$ complex if we replace $'$ by $*$ and symmetric by Hermitian.

7.55. Any weak inverse of $\mathbf{A}$ can be expressed as

$$\mathbf{A}^- = \mathbf{A}^+ + \mathbf{H} - \mathbf{A}^+\mathbf{A}\mathbf{H}\mathbf{A}\mathbf{A}^+,$$

for any $\mathbf{H}$ of appropriate size. This follows from (7.6a).

7.56. Let $\mathbf{A}$ be a real matrix. The following conditions are equivalent.

(1) A matrix $\mathbf{G}$ is the Moore–Penrose inverse of $\mathbf{A}$.

(2) $\mathbf{A}^*\mathbf{A}\mathbf{G} = \mathbf{A}^*$ and $\mathbf{G}^*\mathbf{G}\mathbf{A} = \mathbf{G}^*$.

(3) $\mathbf{A}\mathbf{G} = \mathbf{P_A}$ and $\mathbf{G}\mathbf{A} = \mathbf{P_G}$,

   where $\mathbf{P_A}$ and $\mathbf{P_G}$ represent orthogonal projections onto $\mathcal{C}(\mathbf{A})$ and $\mathcal{C}(\mathbf{G})$, respectively. (This was the original definition of the Moore–Penrose inverse given by Moore [1935]. The equivalence of the two definitions is proved by Campbell and Meyer [1979: 9] and Schott [2005: 181–182, real case].)

7.57. If $\mathbf{A}$ is a real normal matrix (i.e., $\mathbf{A}'\mathbf{A} = \mathbf{A}\mathbf{A}'$) with Moore–Penrose inverse $\mathbf{A}^+$, then:

(a) $\mathbf{A}^+\mathbf{A} = \mathbf{A}\mathbf{A}^+$.

(b) $(\mathbf{A}^k)^+ = (\mathbf{A}^+)^k$ for any positive integer k.

(c) If $\mathbf{A}$ is symmetric, it is normal and $\mathbf{A}\mathbf{A}^+ = \mathbf{A}^+\mathbf{A}$.

7.58. Let $\mathbf{A}$ be a real $m \times n$ matrix, and suppose that certain rows are identical (respectively zero). Then the same rows in $\mathbf{A}'^+$ and also in $\mathbf{A}\mathbf{A}^-$ are identical (respectively zero).

7.59. (Expressed as a Limit) If $\mathbf{A}$ is an $m \times n$ matrix then

$$\mathbf{A}^+ = \lim_{\delta \to 0}(\mathbf{A}'\mathbf{A} + \delta^2\mathbf{I}_n)^{-1}\mathbf{A}' = \lim_{\delta \to 0}\mathbf{A}'(\mathbf{A}\mathbf{A}' + \delta^2\mathbf{I}_m)^{-1}.$$

7.60. (Continuity) Let $\mathbf{A}$ be an $m \times n$ matrix and $\mathbf{A}_1, \mathbf{A}_2, \ldots, \ldots$ be a sequence $m \times n$ matrices such that $\mathbf{A}_k \to \mathbf{A}$ as $k \to \infty$ (cf. Definition 19.3), then

$$\mathbf{A}_k^+ \to \mathbf{A}^+ \text{ as } k \to \infty$$

if and only if an integer N exists such that

$$\operatorname{rank} \mathbf{A}_k = \operatorname{rank} \mathbf{A} \quad \text{for all } k > N.$$

7.61. Given a real matrix $\mathbf{A}$, let $\mathbf{F}_{13}$ be any g_{13}-inverse of $\mathbf{A}\mathbf{A}'$, and let $\mathbf{H}_{14}$ be any g_{14}-inverse of $\mathbf{A}'\mathbf{A}$. Then

$$\mathbf{A}^+ = \mathbf{A}'\mathbf{F}_{13} = \mathbf{H}_{14}\mathbf{A}'.$$

7.62. (Idempotent matrices) Let $\mathbf{A}$ be a real matrix.

(a) If $\mathbf{A}$ is symmetric and idempotent, then $\mathbf{A}^+ = \mathbf{A}$.

(b) $\mathbf{A}'\mathbf{A}$ is idempotent if and only if $\mathbf{A}^+ = \mathbf{A}'$.

7.63. (Non-negative Definite Matrices)

(a) Suppose $\mathbf{A}$ is an $n \times n$ (real) non-negative definite matrix of rank r. We can write $\mathbf{A} = \mathbf{R}'\mathbf{R}$, where $\mathbf{R}$ is $r \times n$ of rank r (cf. 10.10)). Then, since $(\mathbf{RAR}')^{-1} = (\mathbf{RR}')^{-2}$ and, using (7.65a) and (7.54d), we have

$$
\begin{aligned}
\mathbf{A}^+ &= \mathbf{R}'(\mathbf{RAR}')^{-1}\mathbf{R} \\
&= \mathbf{R}'(\mathbf{RR}')^{-2}\mathbf{R} \\
&= \mathbf{R}^+(\mathbf{R}^+)'.
\end{aligned}
$$

The last result also follows directly from (7.54c).

(b) It follows from (a) that if $\mathbf{A}$ is non-negative definite (respectively positive definite), then $\mathbf{A}^+$ is also non-negative definite (respectively positive definite).

7.64. (Non-negative Definite Difference) Suppose that $\mathbf{A}$, $\mathbf{B}$, and $\mathbf{A} - \mathbf{B}$ are non-negative definite matrices, then $\mathbf{B}^+ - \mathbf{A}^+$ is non-negative definite if and only if rank $\mathbf{A} = $ rank $\mathbf{B}$.

7.65. (Full-Rank Factorization) If $\mathbf{A} = \mathbf{CR}$ is a full rank decomposition of an $n \times n$ complex matrix of rank r, where $\mathbf{C}$ is $m \times r$ of rank r and $\mathbf{R}$ is $r \times n$ of rank r, (cf. 3.5), then:

(a) $\mathbf{A}^+ = \mathbf{R}^*(\mathbf{C}^*\mathbf{AR}^*)^{-1}\mathbf{C}^* = \mathbf{R}^*(\mathbf{RR}^*)^{-1}(\mathbf{C}^*\mathbf{C})^{-1}\mathbf{C}^*$.

(b) $\mathbf{A}^+ = \mathbf{R}^+\mathbf{C}^+$.

We note that (7.63) is a special case of the above results.

7.66. Let $\mathbf{A}$ be an $m \times n$ matrix, and let $\mathbf{B}$ be an $n \times m$ matrix. Then $\mathbf{B}$ is the Moore–Penrose inverse of $\mathbf{A}$ if and only if $\mathbf{B}$ is a least squares (g_{13}) inverse of $\mathbf{A}$ and $\mathbf{A}$ is a least squares inverse of $\mathbf{B}$.

Proofs. Section 7.4.1.

7.49. Abadir and Magnus [2005: 284–285] and Schott [2005: 180–181], real case only.

7.50. Schott [2005: 181].

7.51. This is quoted by Graybill [1983: 128] and proved by Penrose [1956].

7.52b. Campbell and Meyer [1979: 28–29].

7.52c h. Simply check that the four conditions are satisfied.

7.52i. Take the conjugate transpose of the four conditions.

7.52j. Abadir and Magnus [2005: 290, real case].

7.52k. Abadir and Magnus [2005: 286, real case] and Schott [2005: 184].

7.52l. Campbell and Meyer [1979: 12].

7.52m. Meyer [2000a: 424] and Campbell and Meyer [1979: 225].

7.53a. Schott [2005: 185–186].

7.54a. Follows from the definition of $\mathbf{A}^+$.

7.54b. Abadir and Magnus [2005: 287] and Graybill [1983: 112].

7.54c. Abadir and Magnus [2005: 287], Graybill [1983: 109], and Schott [2005: 183].

7.54d. Abadir and Magnus [2005: 287, 288] and Schott [2005: 183].

7.54e. Graybill [1983: 110] and Schott [2005: 183].

7.54f. Abadir and Magnus [2005: 287]. Follows from the fact that $\mathbf{A}(\mathbf{A}'\mathbf{A})^+\mathbf{A}$ is the orthogonal projection onto $\mathcal{C}(\mathbf{A})$.

7.54g. Abadir and Magnus [2005: 287].

7.54h. Searle [1982: 216].

7.54i. Both are equal to the orthogonal projection matrix, which is unique.

7.54j. Abadir and Magnus [2005: 288].

7.56. Harville [1997: 503].

7.57a. From (2.35b), $\mathcal{C}(\mathbf{A}) = \mathcal{C}(\mathbf{A}\mathbf{A}') = \mathcal{C}(\mathbf{A}'\mathbf{A}) = \mathcal{C}(\mathbf{A}')$; we then apply (7.52p(i)).

7.57b. Proof can be demonstrated for $k = 2$. Using (a), $\mathbf{A}^2(\mathbf{A}^+)^2\mathbf{A}^2 = \mathbf{A}\mathbf{A}\mathbf{A}^+\mathbf{A}^+\mathbf{A}\mathbf{A} = \mathbf{A}\mathbf{A}^+\mathbf{A}\mathbf{A}\mathbf{A}^+\mathbf{A} = \mathbf{A}^2$; then use induction.

7.58. Graybill [1983: 117–118].

7.59. Harville [1997: 508–510]. This result holds for complex matrices if we replace $'$ by $*$, as quoted by Rao and Mitra [1971: 64].

7.60. Campbell and Meyer [1979: 217] and Penrose [1955].

7.61. Harville [1997: 506].

7.62a. Abadir and Magnus [2005: 286] and Schott [2005: 185].

7.62b. Graybill [1983: 116–117].

7.63b. Harville [1997: 505] and Searle [1982: 220].

7.64. Quoted by Schott [2005: 215, exercise 5.19].

7.65a. Ben-Israel [2003: 48], Harville [1997: 494, real case], and Searle [1982: 212, real case].

7.65b. Follows from (a) and (7.54d(i)–(ii)); see also Schott [2005: 189, real case]. For some general conditions for $(\mathbf{CR})^+ = \mathbf{R}^+\mathbf{C}^+$ to hold when ranks are not specified see Ben-Israel and Greville [2003: 160–161].

7.66. Quoted by Schott [2005: 219, exercise 5.47].

7.4.2 Sums of Matrices

7.67. Let $\mathbf{U}$ and $\mathbf{V}$ be real $m \times n$ matrices. Define

$$
\begin{aligned}
\mathbf{C} &= (\mathbf{I}_m - \mathbf{U}\mathbf{U}^+)\mathbf{V}, \\
\mathbf{M} &= \{\mathbf{I}_n + (\mathbf{I}_n - \mathbf{C}^+\mathbf{C})(\mathbf{V}'\mathbf{U}^{+\prime}\mathbf{U}^+\mathbf{V}(\mathbf{I}_n - \mathbf{C}^+\mathbf{C})\}^{-1}, \text{ and} \\
\mathbf{W} &= (\mathbf{I}_n - \mathbf{C}^+\mathbf{C})(\mathbf{M}\mathbf{V}'\mathbf{U}^{+\prime}\mathbf{U}^+(\mathbf{I}_m - \mathbf{V}\mathbf{C}^+).
\end{aligned}
$$

(a) If $\mathbf{U}\mathbf{V}' = \mathbf{0}$, then

$$
(\mathbf{U} + \mathbf{V})^+ = \mathbf{U}^+ + (\mathbf{I}_n - \mathbf{U}^+\mathbf{V})(\mathbf{C}^+ + \mathbf{W}).
$$

(b) If $\mathbf{U}\mathbf{V}' = \mathbf{0}$ and $\mathbf{U}'\mathbf{V} = \mathbf{0}$, then

$$
(\mathbf{U} + \mathbf{V})^+ = \mathbf{U}^+ + \mathbf{V}^+.
$$

7.68. (Orthogonal Sum) Let $\mathbf{A} = \sum_{i=1}^{k} \mathbf{A}_i$, where the $\mathbf{A}_i$ are all real $m \times n$ matrices. If $\mathbf{A}_i\mathbf{A}'_j = \mathbf{0}$ and $\mathbf{A}'_i\mathbf{A}_j = \mathbf{0}$ for all $i, j = 1, 2, \dots, k, i \neq j$, then from (7.67b) we have

$$
\mathbf{A}^+ = \sum_{i=1}^{k} \mathbf{A}_i^+.
$$

Proofs. Section 7.4.2.

7.67a. Boullion and Odell [1971].

7.67a–b. Schott [2005: 197].

7.4.3 Products of Matrices

7.69. Suppose $\mathbf{A}$ is any $n \times n$ complex matrix.

(a) Let $\mathbf{P}$ be any $r \times n$ $(r \geq n)$ matrix with orthonormal columns (i.e., $\mathbf{P}^*\mathbf{P} = \mathbf{I}_n$) and $\mathbf{Q}$ be any $s \times n$ $(s \geq n)$ matrix with orthonormal columns (i.e., $\mathbf{Q}^*\mathbf{Q} = \mathbf{I}_n$), then

$$
(\mathbf{P}\mathbf{A}\mathbf{Q}^*)^+ = \mathbf{Q}\mathbf{A}^+\mathbf{P}^*.
$$

(b) If $\mathbf{B} = \mathbf{U}^*\mathbf{A}\mathbf{U}$ for some unitary matrix $\mathbf{U}$, then $\mathbf{B}^+ = \mathbf{U}^*\mathbf{A}^+\mathbf{U}$.

7.70. Let $\mathbf{A}$ be a real $m \times n$ matrix and $\mathbf{B}$ be a real $n \times p$ matrix. The following conditions are equivalent.

(1) $(\mathbf{A}\mathbf{B})^+ = \mathbf{B}^+\mathbf{A}^+$.

(2) $\mathbf{A}^+\mathbf{A}\mathbf{B}\mathbf{B}'\mathbf{A}' = \mathbf{B}\mathbf{B}'\mathbf{A}'$ and $\mathbf{B}\mathbf{B}^+\mathbf{A}'\mathbf{A}\mathbf{B} = \mathbf{A}'\mathbf{A}\mathbf{B}$.

(3) $\mathbf{A}^+\mathbf{A}\mathbf{B}\mathbf{B}'$ and $\mathbf{A}'\mathbf{A}\mathbf{B}\mathbf{B}^+$ are symmetric matrices.

(4) $\mathbf{A}^+\mathbf{A}\mathbf{B}\mathbf{B}'\mathbf{A}'\mathbf{A}\mathbf{B}\mathbf{B}^+ = \mathbf{B}\mathbf{B}'\mathbf{A}'\mathbf{A}$.

(5) $\mathbf{A}^+\mathbf{A}\mathbf{B} = \mathbf{B}(\mathbf{A}\mathbf{B})^+\mathbf{A}\mathbf{B}$ and $\mathbf{B}\mathbf{B}^+\mathbf{A}' = \mathbf{A}'\mathbf{A}\mathbf{B}(\mathbf{A}\mathbf{B})^+$.

7.71. Let $\mathbf{A}$ be an $m \times p$ real matrix, and let $\mathbf{B}$ be a $p \times n$ real matrix. If $\mathbf{B}_1 = \mathbf{A}^+\mathbf{AB}$ and $\mathbf{A}_1 = \mathbf{AB}_1\mathbf{B}_1^+$, then $\mathbf{AB} = \mathbf{A}_1\mathbf{B}_1$ and $(\mathbf{AB})^+ = \mathbf{B}_1^+\mathbf{A}_1^+$.

7.72. Let $\mathbf{A}$ be any $m \times n$ matrix, and let $\mathbf{K}$ be an $n \times n$ nonsingular matrix. If $\mathbf{B} = \mathbf{AK}$, then $\mathbf{BB}^+ = \mathbf{AA}^+$. It may not be true that $\mathbf{B}^+\mathbf{B} = \mathbf{A}^+\mathbf{A}$.

7.73. If $\mathbf{A}$ and $\mathbf{B}$ are conformable complex matrices, then

$$(\mathbf{AB})^+ = (\mathbf{A}^+\mathbf{AB})^+(\mathbf{ABB}^+)^+ = (\mathbf{P}_{\mathcal{C}(\mathbf{A}^*)}\mathbf{B})^+(\mathbf{AP}_{\mathcal{C}(\mathbf{B})})^+,$$

where $\mathbf{P}_{\mathcal{C}(\mathbf{B})}$ is the orthogonal projection onto $\mathcal{C}(\mathbf{B})$, and so on (cf. 7.54a).

7.74. The following hold:

(a) $\mathbf{A} = \mathbf{0}$ if and only if $\mathbf{A}^+ = \mathbf{0}$.

(b) $\mathbf{AB} = \mathbf{0}$ if and only if $\mathbf{B}^+\mathbf{A}^+ = \mathbf{0}$.

(c) $\mathbf{A}^+\mathbf{B} = \mathbf{0}$ if and only if $\mathbf{A}'\mathbf{B} = \mathbf{0}$.

(d) If $\mathbf{Ax} = \mathbf{0}$ for some vector $\mathbf{x}$, then $\mathbf{A}^+\mathbf{x} = \mathbf{0}$ also.

7.75. (Cancellation)   Suppose we have real conformable matrices.

(a) $\mathbf{A}'\mathbf{AB} = \mathbf{A}'\mathbf{C}$ if and only if $\mathbf{AB} = \mathbf{AA}^+\mathbf{C}$.

(b) If $\mathbf{B}$ has full row rank so that $\det(\mathbf{BB}') \neq 0$, then $(\mathbf{AB})(\mathbf{AB})^+ = \mathbf{AA}^+$.

7.76. Let $\mathbf{A}$ be a real $m \times n$ matrix with $n \leq m$, and let $\mathbf{B}$ be any real $n \times n$ matrix satisfying $(\mathbf{A}'\mathbf{A})^2\mathbf{B} = \mathbf{A}'\mathbf{A}$. Then $\mathbf{A}^+ = \mathbf{B}'\mathbf{A}'$.

7.77. Suppose $\mathbf{A}$ is a real symmetric matrix.

(a) $\mathbf{A}^+ = \mathbf{B}'\mathbf{AB}$, where $\mathbf{B}$ is any solution of $\mathbf{A}^2\mathbf{B} = \mathbf{A}$.

(b) $\mathbf{A}^+ = (\mathbf{AK})^2\mathbf{A}$, where $\mathbf{K}$ is any solution of $\mathbf{A}^2\mathbf{KA}^2 = \mathbf{A}^2$.

Proofs. Section 7.4.3.

7.69a. Harville [1997: 506, real case]).

7.69b. Quoted by Ben-Israel and Greville [2003: 49].

7.70. Schott [2005: 190]. For many other equivalent but more complex conditions involving the Moore–Penrose inverse of products, see Tian [2006c] and references therein. For a related paper see also Tian [2005a].

7.71. Schott [2005: 191].

7.72. Graybill [1983: 115].

7.73. Campbell and Meyer [1979: 20].

7.74. Abadir and Magnus [2005: 288].

7.75. Abadir and Magnus [2005: 291] and Magnus and Neudecker [1999: 34].

7.76–7.77. Graybill [1983: 123].

7.5 GROUP INVERSE

We recall from Definition 7.1 that $\mathbf{A}^{\#}$ is the group inverse of a square real or complex matrix $\mathbf{A}$ if it satisfies the three conditions

$$\mathbf{A}\mathbf{A}^{\#}\mathbf{A} = \mathbf{A}, \quad \mathbf{A}^{\#}\mathbf{A}\mathbf{A}^{\#} = \mathbf{A}^{\#}, \quad \text{and} \quad \mathbf{A}\mathbf{A}^{\#} = \mathbf{A}^{\#}\mathbf{A}.$$

Such an inverse is a special case of the so-called *Drazin inverse*, discussed by Ben-Israel and Greville [2003: chapter 4, section 4] and Campbell and Meyer [1979: chapters 7-9]. Group inverses are particularly useful in the theory of finite Markov chains (cf. Meyer [1975] and Noumann and Xu [2005]).

7.78. An $n \times n$ matrix $\mathbf{A}$ has a group inverse if and only if $\mathcal{C}(\mathbf{A}) \oplus \mathcal{N}(\mathbf{A}) = \mathbb{C}^{n}$. When the group inverse exists, it is unique.

7.79. A square matrix $\mathbf{A}$ has a group inverse if and only if rank $\mathbf{A} = \text{rank}(\mathbf{A}^2)$.

7.80. Let a square matrix $\mathbf{A}$ have a full-rank factorization $\mathbf{A} = \mathbf{FG}$ (cf. 3.5). Then $\mathbf{A}$ has a group inverse if and only if $\mathbf{GF}$ is nonsingular, in which case

$$\mathbf{A}^{\#} = \mathbf{F}(\mathbf{GF})^{-2}\mathbf{G}.$$

7.81. (General Properties) From the definition we have the following:

(a) If $\mathbf{A}$ is nonsingular, then $\mathbf{A}^{\#} = \mathbf{A}^{-1}$.

(b) $(\mathbf{A}^{\#})^{\#} = \mathbf{A}$.

(c) $(\mathbf{A}^{*})^{\#} = (\mathbf{A}^{\#})^{*}$.

(d) $(\mathbf{A}')^{\#} = (\mathbf{A}^{\#})'$

(e) $(\mathbf{A}^{k})^{\#} = (\mathbf{A}^{\#})^{k}$ for every positive integer k.

(f) $\mathbf{A}^{\#} = \mathbf{A}(\mathbf{A}^{3})^{-}\mathbf{A}$.

Proofs. Section 7.5.

7.78–7.81. Ben-Israel and Greville [2003: 156-158].

7.6 SOME GENERAL PROPERTIES OF INVERSES

7.82. (Representations) The following is a useful summary from Rao and Rao [1998: 295–296]) giving representations for all the inverses of an $m \times n$ real (respectively complex) matrix $\mathbf{A}$ of rank r. We begin with the singular value decomposition of $\mathbf{A}$, namely

$$\mathbf{A}_{m \times n} = \mathbf{P}_{m \times m} \begin{pmatrix} \mathbf{\Delta}_r & \mathbf{0} \\ \mathbf{0} & \mathbf{0} \end{pmatrix} \mathbf{Q}'_{n \times n},$$

where $\mathbf{P}$ is an $m \times m$ orthogonal (respectively unitary) matrix, $\mathbf{Q}$ is an $n \times n$ orthogonal (respectively unitary) matrix and $\mathbf{\Delta}_r = \text{diag}(\delta_1, \delta_2, \cdots, \delta_r)$ is an $m \times n$ matrix with $\delta_1 \geq \delta_2 \geq \cdots \geq \delta_r > 0$. For complex matrices we replace $'$ by $*$. We shall use the notation $\mathbf{G}_{i...}$ to denote the $n \times m$ $g_{i...}$-inverse of $\mathbf{A}$.

(a) $\mathbf{G}_1 = \mathbf{Q} \begin{pmatrix} \boldsymbol{\Delta}^{-1} & \mathbf{X} \\ \mathbf{Y} & \mathbf{Z} \end{pmatrix} \mathbf{P}'$, where $\mathbf{X}$, $\mathbf{Y}$ and $\mathbf{Z}$ are arbitrary.

(b) $\mathbf{G}_{12} = \mathbf{Q} \begin{pmatrix} \boldsymbol{\Delta}^{-1} & \mathbf{X} \\ \mathbf{Y} & \mathbf{Y}\boldsymbol{\Delta}\mathbf{X} \end{pmatrix} \mathbf{P}'$, where $\mathbf{X}$ and $\mathbf{Y}$ are arbitrary.

(c) $\mathbf{G}_{14} = \mathbf{Q} \begin{pmatrix} \boldsymbol{\Delta}^{-1} & \mathbf{X} \\ \mathbf{0} & \mathbf{Z} \end{pmatrix} \mathbf{P}'$, where $\mathbf{X}$ and $\mathbf{Z}$ are arbitrary.

(d) $\mathbf{G}_{124} = \mathbf{Q} \begin{pmatrix} \boldsymbol{\Delta}^{-1} & \mathbf{X} \\ \mathbf{0} & \mathbf{0} \end{pmatrix} \mathbf{P}'$, where $\mathbf{X}$ is arbitary.

(e) $\mathbf{G}_{13} = \mathbf{Q} \begin{pmatrix} \boldsymbol{\Delta}^{-1} & \mathbf{0} \\ \mathbf{Y} & \mathbf{Z} \end{pmatrix} \mathbf{P}'$, where $\mathbf{Y}$ and $\mathbf{Z}$ are arbitrary.

(f) $\mathbf{G}_{123} = \mathbf{Q} \begin{pmatrix} \boldsymbol{\Delta}^{-1} & \mathbf{0} \\ \mathbf{Y} & \mathbf{0} \end{pmatrix} \mathbf{P}'$, where $\mathbf{Y}$ is arbitrary.

(g) $\mathbf{A}^+ = \mathbf{Q} \begin{pmatrix} \boldsymbol{\Delta}^{-1} & \mathbf{0} \\ \mathbf{0} & \mathbf{0} \end{pmatrix} \mathbf{P}'$

7.83. (Matrix Bounds) If $\mathbf{A}$ is $m \times n$, then using the Löwner ordering (cf. Definition 10.1):

(a) $\quad (\mathbf{I}_m - \mathbf{A}\mathbf{G}_1)'(\mathbf{I}_m - \mathbf{A}\mathbf{G}_1) \succeq (\mathbf{I}_m - \mathbf{A}\mathbf{G}_{13})'(\mathbf{I}_m - \mathbf{A}\mathbf{G}_{13})$.

(b) $\quad (\mathbf{I}_m - \mathbf{A}\mathbf{G}_1)'(\mathbf{I}_m - \mathbf{A}\mathbf{G}_1) \succeq (\mathbf{I}_m - \mathbf{A}\mathbf{G}_{14})'(\mathbf{I}_m - \mathbf{A}\mathbf{G}_{14})$.

(c)

$$(\mathbf{I}_m - \mathbf{A}\mathbf{G}_1)'(\mathbf{I}_m - \mathbf{A}\mathbf{G}_1) \succeq (\mathbf{I}_m - \mathbf{A}\mathbf{G}_{123})'(\mathbf{I}_m - \mathbf{A}\mathbf{G}_{123})$$
$$(\mathbf{I}_m - \mathbf{A}\mathbf{G}_1)(\mathbf{I}_m - \mathbf{A}\mathbf{G}_1)' \succeq (\mathbf{I}_m - \mathbf{A}\mathbf{G}_{123})(\mathbf{I}_m - \mathbf{A}\mathbf{G}_{123})'.$$

(d)

$$(\mathbf{I}_n - \mathbf{G}_1\mathbf{A})'(\mathbf{I}_n - \mathbf{G}_1\mathbf{A}) \succeq (\mathbf{I}_n - \mathbf{G}_{124}\mathbf{A})'(\mathbf{I}_n - \mathbf{G}_{124}\mathbf{A})$$
$$(\mathbf{I}_n - \mathbf{G}_1\mathbf{A})(\mathbf{I}_n - \mathbf{G}_1\mathbf{A})' \succeq (\mathbf{I}_n - \mathbf{G}_{124}\mathbf{A})(\mathbf{I}_n - \mathbf{G}_{124}\mathbf{A})'.$$

(e)

$$(\mathbf{I}_m - \mathbf{A}\mathbf{G}_1)'(\mathbf{I}_m - \mathbf{A}\mathbf{G}_1) \succeq (\mathbf{I}_m - \mathbf{A}\mathbf{A}^+)'(\mathbf{I}_m - \mathbf{A}\mathbf{A}^+)$$
$$(\mathbf{I}_m - \mathbf{A}\mathbf{G}_1)(\mathbf{I}_m - \mathbf{A}\mathbf{G}_1)' \succeq (\mathbf{I}_m - \mathbf{A}\mathbf{A}^+)(\mathbf{I}_m - \mathbf{A}\mathbf{A}^+)'$$
$$(\mathbf{I}_n - \mathbf{G}_1\mathbf{A})'(\mathbf{I}_n - \mathbf{G}_1\mathbf{A}) \succeq (\mathbf{I}_n - \mathbf{A}^+\mathbf{A})'(\mathbf{I}_n - \mathbf{A}^+\mathbf{A})$$
$$(\mathbf{I}_n - \mathbf{G}_1\mathbf{A})(\mathbf{I}_n - \mathbf{G}_1\mathbf{A})' \succeq (\mathbf{I}_n - \mathbf{A}^+\mathbf{A})(\mathbf{I}_n - \mathbf{A}^+\mathbf{A})'.$$

(f) From (a) and (b), the first results of (c) and (d), and the first and third results of (e) we can obtain lower bounds, as in the following example. For any unitarily invariant norm $\| \cdot \|_{ui}$ on the space of all $m \times m$ matrices,

$$\min_{\mathbf{G}} \|\mathbf{I}_m - \mathbf{A}\mathbf{G}\|_{ui} = \|\mathbf{I}_m - \mathbf{A}\mathbf{G}_{13}\|_{ui},$$

where the minimum is taken over all weak inverses $\mathbf{G}_1$.

Proofs. Section 7.6.

7.83. Rao and Rao [1998: 296–299].

CHAPTER 8

SOME SPECIAL MATRICES

In this chapter we put collect together a number of matrices that have a special structure or properties. Other more general types of matrix occur elsewhere in this book such as Hermitian, symmetric, and normal matrices in Chapter 5, various non-negative matrices in Chapter 9, and non-negative definite matrices in Chapter 10.

8.1 ORTHOGONAL AND UNITARY MATRICES

Definition 8.1. An $n \times n$ matrix $\mathbf{T}$ is *orthogonal* if $\mathbf{T}'\mathbf{T} = \mathbf{I}_n$. It immediately follows by taking determinants that $\mathbf{T}$ is nonsingular, $\mathbf{T}' = \mathbf{T}^{-1}$ and $\mathbf{T}\mathbf{T}' = \mathbf{I}_n$. An $n \times n$ complex matrix is *unitary* if $\mathbf{U}^*\mathbf{U} = \mathbf{I}$, and then $\mathbf{U}^{-1} = \mathbf{U}^*$. Although an orthogonal matrix can be real or complex, we shall focus on real orthogonal matrices rather than complex orthogonal matrices in this chapter, unless otherwise stated.

8.1. A unitary matrix is also a normal matrix so that all the properties of a normal matrix apply. For example, if $\mathbf{U}$ is unitary, there exists a unitary matrix $\mathbf{V}$ such that $\mathbf{U} = \mathbf{V}\operatorname{diag}(\lambda_1, \lambda_2, \ldots, \lambda_n)\mathbf{V}^*$, where the λ_i are the eigenvalues of $\mathbf{U}$ and satisfy $|\lambda_i| = 1$ for all i (cf. 5.31). Note that if $\mathbf{U}$ is unitary, then so are $\overline{\mathbf{U}}$, $\mathbf{U}'$, and $\mathbf{U}^+$, the Moore–Penrose inverse.

8.2. An $n \times n$ complex matrix $\mathbf{A}$ is unitary if and only if $\|\mathbf{A}\mathbf{x}\|_2 = \|\mathbf{x}\|_2$ for all $\mathbf{x} \in \mathbb{C}^n$.

A Matrix Handbook for Statisticians. By George A. F. Seber
Copyright © 2008 John Wiley & Sons, Inc.

8.3. Let $\mathbf{U}$ be a unitary matrix partitioned as

$$\mathbf{U} = \begin{pmatrix} \mathbf{A} & \mathbf{B} \\ \mathbf{C} & \mathbf{D} \end{pmatrix},$$

where $\mathbf{A}$ is $m \times m$ and $\mathbf{D}$ is $n \times n$.

(a) If $m = n$, then $\mathbf{A}$ and $\mathbf{D}$ have the same singular values.

(b) If $m < n$ and the singular values of $\mathbf{A}$ are $\sigma_1, \sigma_2, \ldots, \sigma_m$, then the singular values of $\mathbf{D}$ are also $\sigma_1, \sigma_2, \ldots, \sigma_m$ together with $n - m$ values equal to 1.

(c) $\det \mathbf{A} = \det \mathbf{D}$.

8.4. (Symmetric Unitary Matrix) Let $\mathbf{U}$ be a symmetric unitary matrix, that is, $\mathbf{U}' = \mathbf{U}$. Then there exists a complex matrix $\mathbf{S}$ with the following properties.

(a) $\mathbf{S}^2 = \mathbf{U}$.

(b) $\mathbf{S}$ is unitary.

(c) $\mathbf{S}$ is symmetric.

(d) $\mathbf{S}$ commutes with every matrix that commutes with $\mathbf{U}$.

8.5. A unitary matrix is an *isometry*, that is, a linear transformation that preserves Euclidean length.

8.6. Let $\mathbf{T}$ be a real $n \times n$ orthogonal matrix and $\mathbf{U}$ a unitary matrix.

(a) Given $\langle \mathbf{x}, \mathbf{y} \rangle = \mathbf{x}'\mathbf{y}$, the columns (rows) of $\mathbf{T}$ form an orthonormal set. The same holds for $\mathbf{U}$ if we define $\langle \mathbf{x}, \mathbf{y} \rangle = \mathbf{x}^*\mathbf{y}$.

(b) $\det \mathbf{T} = \pm 1$. If $\det = 1$ then $\mathbf{T}$ represents a rotation.

(c) $|\det \mathbf{U}| = 1$, where $|\cdot|$ is the complex modulus.

(d)   (i) If λ is an eigenvalue of $\mathbf{T}$ then so is λ^{-1}.

(ii) The eigenvalues of $\mathbf{T}$ are ± 1 or occur in conjugate pairs $e^{i\theta}$ and $e^{-i\theta}$ (θ real) on the unit circle, so that all the eigenvalues have unit modulus (cf. 16.46b).

(iii) It follows from (ii) that if n is odd, then at least one eigenvalue is $+1$ or -1.

(e) The eigenvalues of $\mathbf{U}$ are $\lambda_i = e^{i\theta}$ (θ real) for all i, so that $|\lambda| = 1$.

8.7. If the $n \times n$ matrix $\mathbf{A}$ has all its eigenvalues equal to 1 in absolute value, then $\mathbf{A}$ is unitary if $\|\mathbf{A}\mathbf{x}\|_2 \leq \|\mathbf{x}\|_2$ for all $\mathbf{x} \in \mathbb{C}^n$.

8.8. Suppose $\mathbf{C}$ is an $n \times n$ real skew-symmetric matrix, that is, $\mathbf{C}' = -\mathbf{C}$ (cf. (5.19) for real matrices). Then:

(a) $\mathbf{I}_n + \mathbf{C}$ is nonsingular.

(b) $\mathbf{A} = (\mathbf{I}_n - \mathbf{C})(\mathbf{I}_n + \mathbf{C})^{-1}$ is orthogonal with $\det \mathbf{A} = 1$.

(c) $\mathbf{A} = e^{\mathbf{C}}$ is orthogonal with det $\mathbf{A} = 1$.

8.9. (Rotation Matrix in the Plane) The matrix

$$\mathbf{T}_\theta = \begin{pmatrix} \cos\theta & -\sin\theta \\ \sin\theta & \cos\theta \end{pmatrix}$$

represents a rotation in two dimensions in a counter clockwise direction through an angle θ. We note that $\mathbf{T}_\theta\mathbf{T}_\phi = \mathbf{T}_{\theta+\phi}$ and $\mathbf{T}_{-\theta} = \mathbf{T}_\theta^{-1}$. Every 2×2 orthogonal matrix with determinant equal to $+1$ can be expressed in the form of $\mathbf{T}_\theta$ for some θ.

The following matrix

$$\mathbf{S}_\theta = \begin{pmatrix} 1 & 0 \\ 0 & -1 \end{pmatrix} \begin{pmatrix} \cos\theta & -\sin\theta \\ \sin\theta & \cos\theta \end{pmatrix} = \begin{pmatrix} \cos\theta & -\sin\theta \\ -\sin\theta & -\cos\theta \end{pmatrix}$$

represents a rotation combined with a reflection in the x-axis.

A reflection matrix is symmetric; for example,

$$\mathbf{V}_\theta = \begin{pmatrix} \cos\theta & \sin\theta \\ \sin\theta & -\cos\theta \end{pmatrix}$$

represents a reflection across a line at an angle of $\theta/2$ and has a determinant of -1.

8.10. (Helmert Matrix) We have the following orthogonal matrix $\mathbf{T}$;

$$\begin{pmatrix} \frac{1}{\sqrt{n}} & \frac{1}{\sqrt{n}} & \frac{1}{\sqrt{n}} & \frac{1}{\sqrt{n}} & \cdots & \frac{1}{\sqrt{n}} & \frac{1}{\sqrt{n}} \\ \frac{1}{\sqrt{2}} & -\frac{1}{\sqrt{2}} & 0 & 0 & \cdots & 0 & 0 \\ \frac{1}{\sqrt{6}} & \frac{1}{\sqrt{6}} & -\frac{2}{\sqrt{6}} & 0 & \cdots & 0 & 0 \\ \frac{1}{\sqrt{12}} & \frac{1}{\sqrt{12}} & \frac{1}{\sqrt{12}} & -\frac{3}{\sqrt{12}} & \cdots & 0 & 0 \\ \cdot & \cdot & \cdot & & \cdots & \cdot & \cdot \\ \frac{1}{\sqrt{n(n-1)}} & \frac{1}{\sqrt{n(n-1)}} & \frac{1}{\sqrt{n(n-1)}} & \frac{1}{\sqrt{n(n-1)}} & \cdots & \frac{1}{\sqrt{n(n-1)}} & -\frac{(n-1)}{\sqrt{n(n-1)}} \end{pmatrix}.$$

This matrix has been used for proving the the statistical independence of a number of statistics.

8.11. (Householder Transformation) This $n \times n$ orthogonal matrix is defined to be $\mathbf{H}_n = \mathbf{I}_n - 2\mathbf{h}\mathbf{h}'$, where $\mathbf{h}'\mathbf{h} = 1$. Since, from (4.33), det $\mathbf{H}_n = (1 - 2\mathbf{h}'\mathbf{h}) = -1$, $\mathbf{H}_n$ represents a reflection.

Given $\mathbf{x} = (x_1, x_2, \ldots, x_n)'$ with $x_1 \neq 0$, let $y_1 = -(\text{sign } x_1)\sqrt{\mathbf{x}'\mathbf{x}}$, and define $h_1 = [\frac{1}{2}(1 - x_1/y_1)]^{1/2}$ and $h_i = -x_i/(2h_1y_1)$ for $i = 2, 3, \ldots, n$. Then $\mathbf{H}_n\mathbf{x} = (y_1, 0, \ldots, 0)'$. Similarly we can define

$$\mathbf{H} = \begin{pmatrix} 1 & \mathbf{0}' \\ \mathbf{0} & \mathbf{H}_{n-1} \end{pmatrix}$$

so that $\mathbf{H}\mathbf{x} = (y_1, y_2, 0, \ldots, 0)'$, where $y_1 = x_1$ and $\mathbf{H}$ is orthogonal. By using a succession of such transformations, a matrix can be transformed to an upper-triangular matrix. For further details see Golub and Van Loan [1996: chapter 5] and Seber and Lee [2003: 343–347].

8.12. (Givens Transformation)   This orthogonal matrix $\mathbf{G} = (g_{ij})$ takes the form of an identity matrix except for four elements: $g_{rr} = g_{ss} = \cos\theta$ and, for $r > s$, $-g_{rs} = g_{sr} = \sin\theta$. Premultiplying by $\mathbf{G}$ rotates the rth and sth rows in a clockwise direction through angle θ. An example of a 4×4 Givens matrix is

$$\mathbf{G} = \begin{pmatrix} 1 & 0 & 0 & 0 \\ 0 & \cos\theta & 0 & \sin\theta \\ 0 & 0 & 1 & 0 \\ 0 & -\sin\theta & 0 & \cos\theta \end{pmatrix}.$$

Products of such matrices can be used to transform a matrix to upper-triangular form. For further details see Golub and Van Loan [1996: 215–221] and Seber and Lee [2003: 348–352].

8.13. If $\mathbf{B}$ is a real nonsingular matrix, then $\mathbf{B}(\mathbf{B}'\mathbf{B})^{-1/2}$ is orthogonal (cf. 10.32).

Proofs. Section 8.1.

8.1. $\overline{\mathbf{U}}^*\overline{\mathbf{U}} = \mathbf{U}'\overline{\mathbf{U}} = (\mathbf{U}^*\mathbf{U})' = \mathbf{I}_n$.

8.2. $\mathbf{x}^*\mathbf{A}^*\mathbf{A}\mathbf{x} = \mathbf{x}^*\mathbf{x}$ for all $\mathbf{x}$ implies $\mathbf{A}^*\mathbf{A} = \mathbf{I}_n$.

8.3. Zhang [1999: 134].

8.4. Zhang [1999: 152–153].

8.5. $\|\mathbf{U}\mathbf{x}\|_2^2 = \mathbf{x}^*\mathbf{U}^*\mathbf{U}\mathbf{x} = \|\mathbf{x}\|_2^2$.

8.6b. Follows from $(\det\mathbf{T})^2 = \det(\mathbf{T}'\mathbf{T}) = 1$.

8.6c. Rao and Bhimasankaram [2000: 314] and Zhang [1999: 132]. Follows from $1 = \det(\mathbf{U}\mathbf{U}^*) = \det\mathbf{U}\cdot\det\overline{\mathbf{U}} = (a + ib)(a - ib) = a^2 + b^2 = |\mathbf{U}|^2$.

8.6d. For (i), $\mathbf{T}\mathbf{x} = \lambda\mathbf{x}$ implies $\lambda^{-1}\mathbf{x} = \mathbf{T}'\mathbf{x}$ and $\det(\mathbf{T} - \lambda\mathbf{I}_n) = \det(\mathbf{T} - \lambda\mathbf{I}_n) = 0$ (i.e., $\mathbf{T}$ and $\mathbf{T}'$ have the same eigenvalues).

8.6e. Zhang [1999: 132].

8.7. Zhang [1999: 133].

8.8. Abadir and Magnus [2005: 263].

8.9. $\mathbf{T}_\theta'\mathbf{T}_\theta = \mathbf{I}_2$.

8.10. $\mathbf{T}'\mathbf{T} = \mathbf{I}_n$.

8.11. $\mathbf{H}_n'\mathbf{H}_n = \mathbf{I}_n$.

8.13. Abadir and Magnus [2005: 263].

8.2   PERMUTATION MATRICES

Definition **8.2.** Let $\mathbf{\Pi}_{ij}$ be the identity matrix $\mathbf{I}_n = (\mathbf{e}_1, \mathbf{e}_2, \ldots \mathbf{e}_n)$ with its ith and jth rows interchanged. Then $\mathbf{\Pi}_{ij}^2 = \mathbf{I}_n$, so that $\mathbf{\Pi}_{ij}$ is a symmetric and orthogonal matrix. Premultiplying any matrix by $\mathbf{\Pi}_{ij}$ will interchange its ith and jth rows so that $\mathbf{\Pi}_{ij}$ is an (elementary) permutation matrix. Postmultiplying a matrix by an elementary permutation matrix will interchange two columns.

Any reordering of the rows of a matrix can be done using a sequence of elementary permutations $\mathbf{\Pi} = \mathbf{\Pi}_{i_K j_K} \cdots \mathbf{\Pi}_{i_1 j_1}$, where

$$\mathbf{\Pi}\mathbf{\Pi}' = \mathbf{\Pi}_{i_K j_K} \cdots \mathbf{\Pi}_{i_1 j_1} \mathbf{\Pi}_{i_1 j_1} \cdots \mathbf{\Pi}_{i_K j_K} = \mathbf{I}_n.$$

The orthogonal matrix $\mathbf{\Pi}$ is called a *permutation matrix*.

The permutation matrix $\mathbf{\Pi}_0 = (\mathbf{e}_n, \mathbf{e}_1, \mathbf{e}_2, \ldots, \mathbf{e}_{n-1})$, which has been called the *forward shift permutation matrix* (and also *primary permutation matrix*), is useful in the theory of circulants.

For a helpful discussion of permutations and cyclic permutations see Davis [1979] and Rao and Bhimasankaram [2000: section 6.2].

8.14. $\mathbf{\Pi}_0$ has the following properties.

(a) $\mathbf{\Pi}_0 \mathbf{A} \mathbf{\Pi}_0' = (a_{i+1,j+1})$ with $n + 1 \equiv 1$ (i.e., subscripts are taken mod n).

(b) $\mathbf{\Pi}_0^2 = (\mathbf{e}_{n-1}, \mathbf{e}_n, \mathbf{e}_1, \cdots, \mathbf{e}_{n-2})$.

(c) $\mathbf{\Pi}_0^n = \mathbf{I}_n$.

(d) $\mathbf{\Pi}_0 = \mathbf{F}^* \mathbf{\Gamma} \mathbf{F}$, where $\mathbf{\Gamma} = \text{diag}(1, \omega, \omega^2, \ldots, \omega^{n-1})$, the ω^i are the nth roots of 1, and $\mathbf{F}$ is an $n \times n$ Fourier matrix (cf. Section 8.12.2).

8.15. An $n \times n$ permutation matrix $\mathbf{\Pi}_n$ has exactly one entry in each row and column equal to 1, and zeros elsewhere. For example

$$\mathbf{\Pi}_3 = \begin{pmatrix} 0 & 1 & 0 \\ 1 & 0 & 0 \\ 0 & 0 & 1 \end{pmatrix} \quad (= (\mathbf{e}_2, \mathbf{e}_1, \mathbf{e}_3), \quad \text{say}).$$

Thus $\mathbf{\Pi}_n$ consists of $\mathbf{I}_n$ with its rows resequenced. It is also $\mathbf{I}_n$ with its columns resequenced, but not necessarily in the same sequence; that is, $\mathbf{\Pi}_n$ is not necessarily symmetric. Left multiplying an $m \times n$ matrix $\mathbf{A}$ by $\mathbf{\Pi}_m$ produces the same resequences of the rows of $\mathbf{A}$ as $\mathbf{\Pi}_m$, while right multiplying by $\mathbf{\Pi}_n$ does the same for the columns.

8.16. If $\mathbf{\Pi}_n$ is a permutation matrix, then so is $\mathbf{\Pi}_n^k$, where k is any positive integer.

8.17. If $\mathbf{A}$ is $n \times n$, then the diagonal elements of $\mathbf{\Pi}_n' \mathbf{A} \mathbf{\Pi}_n$ are the same elements (rearranged) as the diagonal elements of $\mathbf{A}$.

Definition **8.3.** If $\mathbf{A}$ is $n \times n$ and $\mathbf{\Pi}$ is a permutation matrix, then the matrix $\mathbf{\Pi}' \mathbf{A} \mathbf{\Pi} = \mathbf{\Pi}^{-1} \mathbf{A} \mathbf{\Pi}$ is said to be *permutation similar* to $\mathbf{A}$. This concept is linked to irreducibility in (8.101).

Proofs. Section 8.2.

8.14. Davis [1979: 72].

8.16–8.17. Graybill [1983: 277]

8.3 CIRCULANT, TOEPLITZ, AND RELATED MATRICES

8.3.1 Regular Circulant

Definition 8.4. An $n \times n$ real or complex matrix $\mathbf{A}$ is a *(regular) circulant* if it has the form

$$
\mathbf{A} = \begin{pmatrix}
a_0 & a_1 & a_2 & \cdots & a_{n-1} \\
a_{n-1} & a_0 & a_1 & \cdots & a_{n-2} \\
a_{n-2} & a_{n-1} & a_0 & \cdots & a_{n-3} \\
\cdot & \cdot & \cdot & \cdots & \cdot \\
a_1 & a_2 & a_3 & \cdots & a_0
\end{pmatrix},
$$

that is, all the elements are equal on the main diagonal and on each of the diagonals parallel to the main diagonal. Note that $\mathbf{A}$ is countersymmetric as it symmetric about its main counter (opposite) diagonal. Most authors omit the word "regular" from the definition, but we follow Graybill [1983] and include it so as to be able to distinguish between two types of symmetric circulant below. Thus, if we define $(j - i)$ modulo n as

$$
(j - i)|n = \begin{cases} n + j - i, & \text{when } i > j, \\ j - i, & \text{when } i \leq j, \end{cases}
$$

then $\mathbf{A}$ is a regular circulant if and only if $(j-i)|n = (s-r)|n$ implies that $a_{ij} = a_{rs}$. Alternatively, $\mathbf{A}$ is a regular circulant if and only if $a_{ij} = a_{(j-i)|n}$. Another way of defining a regular circulant is $a_{ij} = a_{1m}$, where

$$
m = \begin{cases} j + i - 1, & j \geq i, \\ n - (j - i + 1), & j < i. \end{cases}
$$

We can also use the notation $\mathbf{A} = \text{circ}(a_0, a_1, \ldots, a_{n-1})$. In applications the circulants are generally real.

   Regular circulants can arise as incident matrices of experimental designs such as the balanced incomplete block design (BIBD) and cyclic designs (e.g., Rao and Rao [1998: 513]). There are other types of circulant such as skew circulants (Davis [1979: 83] and alternating circulants (Tee [2005: 136]).

Definition 8.5. The polynomial $p(z) = a_0 + a_1 z + \cdots + a_{n-1} z^{n-1}$ is sometimes called the *representer of the circulant* and it occurs, for example, in signal processing.

8.18. The forward shift permutation matrix $\mathbf{\Pi}_0$ of (8.14) can be expressed as $\mathbf{\Pi}_0 = \text{circ}(0, 1, \ldots, 0)$.

8.19. $\text{circ}(a_0, a_1, \ldots, a_{n-1}) = p(\mathbf{\Pi}_0) = a_0 + a_1 \mathbf{\Pi}_0 + a_2 \mathbf{\Pi}_0^2 + \cdots + a_{n-1} \mathbf{\Pi}_0^{n-1}$.

8.20. The following conditions are equivalent.

   (1) $\mathbf{A}$ is an $n \times n$ regular circulant.

   (2) $\mathbf{\Pi}_0 \mathbf{A} \mathbf{\Pi}_0' = \mathbf{A}$.

   (3) $\mathbf{\Pi}_0' \mathbf{A} \mathbf{\Pi}_0 = \mathbf{A}$.

8.21. If $\mathbf{A}$ is a real regular circulant, then so is the Moore–Penrose inverse $\mathbf{A}^+$.

8.22. Let **A** be a regular circulant. Then:

(a) $\mathbf{A}^*$ is a regular circulant.

(b) $\mathbf{A}^k$ is a regular circulant, where k is a positive integer.

(c) $\mathbf{A}^{-1}$ is also a regular circulant, if **A** is nonsingular. In this case, to compute $\mathbf{A}^{-1}$ we only need to find its first row.

8.23. If **A** is a regular circulant with first row $(a_0, a_1, a_2, \ldots, a_{n-1})$, and if

$$|a_q| > \sum_{i=0:i\neq q}^{n-1} |a_i|$$

for some q, then **A** is nonsingular.

8.24. If **A** and **B** are any $n \times n$ regular circulants, then **AB** is a regular circulant and $\mathbf{AB} = \mathbf{BA}$.

8.25. If **A** is a regular circulant, then so is $\mathbf{A}^*$ (by 8.22a) and $\mathbf{A}^*\mathbf{A}$ (by 8.24), with $\mathbf{A}^*\mathbf{A} = \mathbf{A}\mathbf{A}^*$. Thus a regular circulant is a normal matrix.

8.26. Let **A** and **C** be $n \times n$ regular circulants, and suppose there exists a matrix **X** such that $\mathbf{AX} = \mathbf{C}$. Then there exists a regular circulant **B** such that $\mathbf{AB} = \mathbf{C}$.

***Definition* 8.6.** The $n \times n$ matrix $\mathbf{C}_h$ $(h = 1, 2, \ldots, n-1)$ that has $a_h = 1$ and the other $a_i = 0$ is sometimes refered to as a *basic circulant* matrix. For example, if $n = 3$,

$$\mathbf{A} = \begin{pmatrix} a_0 & a_1 & a_2 \\ a_2 & a_0 & a_1 \\ a_1 & a_2 & a_0 \end{pmatrix} \quad \text{and} \quad \mathbf{C}_1 = \begin{pmatrix} 0 & 1 & 0 \\ 0 & 0 & 1 \\ 1 & 0 & 0 \end{pmatrix}.$$

Here $\mathbf{C}_1 = \text{circ}(0, 1, 0)$. Note that, in general, $\mathbf{C}_1$ is the same as $\mathbf{\Pi}_0$ of Definition 8.2.

8.27. If $\mathbf{C}_0 = \mathbf{I}_n$, then:

(a) $\mathbf{A} = \displaystyle\sum_{h=0}^{n-1} a_h \mathbf{C}_h.$

(b) $\mathbf{C}_h = \mathbf{C}_1^h$ $(h = 1, 2, \ldots, n-1)$ and $\mathbf{C}_n = \mathbf{I}_n$ (by 8.14c).

8.28. (Eigenvalues and Eigenvectors)   Referring to (8.14d), we have the following results.

(a) The eigenvectors of $\mathbf{C}_1$ are given by

$$\boldsymbol{\gamma}_j = n^{-1/2}(1, \omega^j, \omega^{2j}, \ldots, \omega^{(n-1)j})',$$

with corresponding eigenvalues $\lambda_{1j} = \omega^j$, for $j = 0, 1, \ldots, n-1$, where $\omega = \exp(2\pi i/n) = \cos(2\pi/n) + i\sin(2\pi/n)$ and $i = \sqrt{-1}$. (Note that the λ_{1j} are the n roots of unity.)

(b) Since $\mathbf{C}_h \mathbf{x} = \mathbf{C}_1^h \mathbf{x} = \lambda_1^h \mathbf{x}$, the eigenvectors of $\mathbf{C}_h$ are still the $\boldsymbol{\gamma}_j$ with eigenvalues $\lambda_{hj} = \omega^{jh}$ $(h, j = 0, 1, \ldots, n-1)$.

(c) We now turn our attention to $\mathbf{A}$ of (8.27).

(i) $\mathbf{A}$ has eigenvectors $\boldsymbol{\gamma}_j$ with (not necessarily distinct) eigenvalues

$$\lambda_j = \sum_{h=0}^{n-1} a_h\, w^{jh} = p(\omega_j) = \sum_{h=0}^{n-1} a_h \omega_j^h, \quad j = 0, 1, \ldots, n-1,$$

where $\omega_j = w^j$, and $p(z)$ is given in Definition 8.5 (above 8.18). Note that the $\boldsymbol{\gamma}_j$ are the same for all regular circulants.

(ii) Setting $j = 0$ in (i), $\lambda_0 = a_0 + a_1 + \ldots + a_{n-1}$ is always an eigenvalue.

(iii) The eigenvectors are mutually orthogonal, that is, $\boldsymbol{\gamma}_j^* \boldsymbol{\gamma}_k = \delta_{jk}$.

(iv) If $\mathbf{F}$ is an $n \times n$ Fourier matrix (cf. Section 8.12.2), then it is unitary and $\mathbf{FAF}^* = \boldsymbol{\Lambda}$, that is, $\mathbf{AF}^* = \mathbf{F}^*\boldsymbol{\Lambda}$, where $\boldsymbol{\Lambda} = \mathrm{diag}(\lambda_1, \lambda_2, \ldots, \lambda_n)$. Thus the columns of $\mathbf{F}^*$ are a universal set of right eigenvectors for *all* regular circulants. Also $\mathbf{A} = \mathbf{F}^*\boldsymbol{\Lambda}\mathbf{F}$.

(v) A spectral decomposition of $\mathbf{A}$ is given by

$$\mathbf{A} = \sum_{j=0}^{n-1} \lambda_j \boldsymbol{\gamma}_j \boldsymbol{\gamma}_j^*.$$

Here $\boldsymbol{\gamma}_j^*$ is the complex conjugate of $\boldsymbol{\gamma}_j'$ obtained by replacing ω_j in $\boldsymbol{\gamma}_j$ by its complex conjugate

$$\bar{\omega} = \exp(-2\pi i/n) = \cos(2\pi/n) - i\sin(2\pi/n).$$

(vi) The matrix $\boldsymbol{\gamma}_j \boldsymbol{\gamma}_j^*$ is a regular circulant that can be written in the form

$$\boldsymbol{\gamma}_j \boldsymbol{\gamma}_j^* = \sum_{h=0}^{n-1} w^{-hj} \mathbf{C}_h.$$

(g) The Moore–Penrose inverse of $\mathbf{A}$, which is also a regular circulant, is given by

$$\mathbf{A}^+ = \sum \lambda_j^{-1} \boldsymbol{\gamma}_j \boldsymbol{\gamma}_j^*,$$

where the summation is over all r nonzero eigenvalues of $\mathbf{A}$, with $r = \mathrm{rank}\,\mathbf{A}$. If $r = n$ then $\mathbf{A}^+ = \mathbf{A}^{-1} = \mathbf{F}^*\boldsymbol{\Lambda}^{-1}\mathbf{F}$.

(h) $\mathbf{A}^+ = \sum_{h=0}^{n-1} \psi_h \mathbf{C}_h$, where $\psi_h = n^{-1}\sum_j \lambda_j^{-1}\omega^{-hj}$.

8.29. Let $\mathbf{A} = \mathrm{circ}(c_1, c_2, \ldots, c_n)$ be a real $n \times n$ regular circulant.

(a) If n is odd and $\lambda_0 = \sum_{i=1}^n c_i \geq 0$ (cf. 8.28f(ii)), then $\det \mathbf{A} \geq 0$.

(b) If n is even, $n = 2r + 2$, $\lambda_0 > 0$, and

$$\left| \sum_{j=1}^{r+1} c_{2j-1} \right| \geq \left| \sum_{j=1}^{r+1} c_{2j} \right|,$$

then $\det \mathbf{A} \geq 0$.

Proofs. Section 8.3.1.

8.19. Schott [2005: 330] and Zhang [1999: 107].

8.20. Schott [2005: 329]. The second result follows by multiplying on the left by $\mathbf{\Pi}_0'$ and on the right by $\mathbf{\Pi}_0$.

8.21. Graybill [1983: 249; his $\mathbf{A}^-$ is our $\mathbf{A}^+$].

8.22. Schott [2005: 330].

8.23. Graybill [1983: 253] and Schott [2005: 330–331].

8.24. Graybill [1983: 236, 238] and Schott [2005: 330–331].

8.26. Quoted by Graybill [1983: 239].

8.27. We simply multiply out the expressions.

8.28. John and Williams [1995: Appendix A7]. For (c)(iv) see Davis [1979: 72], Schott [2005: 332], and Zhang [1999: 107].

8.29. Davis [1979: 76–77].

8.3.2   Symmetric Regular Circulant

To obtain a symmetric regular circulant one writes down a regular circulant and then determines which elements are equal to achieve symmetry; for example, we have the following matrix

$$\mathbf{A} = \begin{pmatrix} a_0 & a_1 & a_2 & a_1 \\ a_1 & a_0 & a_1 & a_2 \\ a_2 & a_1 & a_0 & a_1 \\ a_1 & a_2 & a_1 & a_0 \end{pmatrix}.$$

Note that this matrix is symmetric about its main diagonal and about its counter (opposite) diagonal, so it is *doubly symmetric*. Trivial examples are $\mathbf{I}_n$ and $\mathbf{J}_n = \mathbf{1}_n \mathbf{1}_n'$. Although our focus is on real symmetric matrices, the eigenvalue theory below applies generally to real and complex matrices.

Symmetric regular circulants arise in cyclic designs as the product of the incidence matrix and its transpose (the *concurrence matrix* of John and Williams [1995: 51]). The eigenvalues are related to the so-called canonical efficiency factors. Symmetric regular circulants also arise with variance matrices, and Khattree [1996] gives seven applications.

8.30. (Some General Properties) Let $\mathbf{A}$ be an $n \times n$ symmetric regular circulant. Then:

(a) $\mathbf{A}$ has at most $[n/2] + 1$ distinct elements, where $[a]$ is the integral part of a.

(b) $\mathbf{A}'$ is a symmetric regular circulant.

(c) If $\mathbf{A}$ is nonsingular, then $\mathbf{A}^{-1}$ is a symmetric regular circulant.

8.31. Let $\mathbf{A}$ and $\mathbf{B}$ be $n \times n$ symmetric regular circulants, then

(a) $\mathbf{AB} = \mathbf{BA}$.

(b) $\mathbf{AB}$ is a symmetric regular circulant.

(c) $a\mathbf{A} + b\mathbf{B}$, where a and b are any real numbers, is a symmetric regular circulant. Hence $a\mathbf{I}_n + b\mathbf{J}_n$, where $\mathbf{J}_n = \mathbf{1}_n\mathbf{1}_n'$, is a symmetric regular circulant.

8.32. (Eigenvalues and Eigenvectors) If $\mathbf{A}$ is a symmetric regular circulant, then we have the following results.

(a) $a_h = a_{n-h}$ $(h = 1, 2, \ldots, m)$, where

$$m = \begin{cases} n/2, & n \text{ even,} \\ (n-1)/2, & n \text{ odd.} \end{cases}$$

(b) The eigenvectors are $\boldsymbol{\gamma}_j$ $(j = 0, 1, \ldots, n-1)$ of (8.28a).

(c) The eigenvalues of $\mathbf{A}$ are

$$\lambda_j = \sum_{h=0}^{n-1} a_h \cos(2\pi jh/n), \quad j = 0, 1, \ldots, n-1.$$

(d) $\lambda_j = \lambda_{n-j}$.

(e) If $n = 2m$, then $\lambda_0 = a_0 + 2(a_1 + a_2 + \cdots + a_{m-1}) + a_m$. If $n = 2m + 1$, then $\lambda_0 = a_0 + 2(a_1 + a_2 + \cdots + a_m)$.

(f) $\mathbf{A}^+ = \sum_{h=0}^{n-1} \psi_h \mathbf{C}_h$, where $\psi_h = n^{-1} \sum_j \lambda_j^{-1} \cos(2\pi jh/n)$. Here we have used (d) so that the sum is over all nonzero eigenvalues, but only choosing one of λ_j and λ_{n-j}.

Proofs. Section 8.3.2.

8.30–8.31. Graybill [1983: 242].

8.32. John and Williams [1995: appendix A7].

8.3.3 Symmetric Circulant

Definition 8.7. A matrix is a symmetric circulant if $a_{ij} = a_{(i+j-2)|n}$. An example of an $n \times n$ symmetric circulant is

$$\mathbf{A} = \begin{pmatrix} a_0 & a_1 & a_2 & \cdots & a_{n-1} \\ a_1 & a_2 & a_3 & \cdots & a_0 \\ a_2 & a_3 & a_4 & \cdots & a_1 \\ \cdot & \cdot & \cdot & \cdots & \cdot \\ a_{n-1} & a_0 & a_1 & \cdots & a_{n-2} \end{pmatrix}.$$

Note that the elements on each of the counterdiagonals are equal.

8.33. The Moore–Penrose inverse $\mathbf{A}^+$ of a symmetric circulant is a symmetric circulant.

8.34. If $\mathbf{A}$ is a symmetric circulant and is nonsingular, then $\mathbf{A}^{-1}$ is a symmetric circulant.

8.35. Let $\mathbf{A}$ be an $n \times n$ symmetric circulant with first row elements $a_0, a_1, \ldots, a_{n-1}$ and eigenvalues $\lambda_0, \lambda_1, \ldots, \lambda_{n-1}$. If $\omega_j = \omega^j$ $(j = 0, 1, \ldots, n-1; \omega^0 = 1)$ are the n roots of unity, then

$$\lambda_i^2 = \omega_i^0 \sum_{j=0}^{n-1} a_j^2 + \omega_i \sum_{j=0}^{n-1} a_j a_{(j+1)|n} + \omega_i^2 \sum_{j=0}^{n-1} a_j a_{(j+2)|n}$$
$$+ \cdots + \omega_i^{n-1} \sum_{j=0}^{n-1} a_j a_{(j+n-1)|n}.$$

Also $\lambda_0 = a_o + a_1 + \cdots + a_{n-1}$.

8.36. If $\mathbf{A}$ and $\mathbf{B}$ are $n \times n$ symmetric circulants, then $\mathbf{AB}$ is a regular circulant, but, in general, $\mathbf{AB} \neq \mathbf{BA}$.

8.37. If $\mathbf{B}$ is a regular circulant and $\mathbf{C}$ is a symmetric circulant, then $\mathbf{BC}$ and $\mathbf{CB}$ are symmetric circulants and, in general, $\mathbf{BC} \neq \mathbf{CB}$.

8.38. Combining the above two results, we have that the product of an even number of symmetric circulants is a regular circulant, and the product of an odd number of symmetric circulants is a symmetric circulant.

8.39. Let $\mathbf{A}$ be an $n \times n$ regular circulant, $\mathbf{C}$ be an $n \times n$ symmetric circulant, and suppose there is a solution $\mathbf{X}$ to the matrix equation $\mathbf{AX} = \mathbf{C}$. Then there exists an $n \times n$ symmetric circulant $\mathbf{B}$ such that $\mathbf{AB} = \mathbf{C}$.

8.40. Let $\mathbf{A}$ be a regular circulant, and let $\mathbf{B}$ be a symmetric circulant with the same first row $a_0, a_1, \ldots, a_{n-1}$. If the matrices are both $n \times n$, then

$$\det \mathbf{A} = (-1)^{[(n-1)/2]} \det \mathbf{B},$$

where $[(n-1)/2]$ is the integral part of $(n-1)/2$.

8.41. Let $\mathbf{A}$ be an $n \times n$ matrix that is both a symmetric circulant and a symmetric regular circulant. If n is odd, then $\mathbf{A}$ takes the form $a_{ij} = a$ for all i, j. If n is even, then $\mathbf{A}$ takes the form $a_{ij} = a_0$ if $i + j$ is even and $a_{ij} = a_1$ if $i + j$ is odd.

Proofs. Section 8.3.3.

8.33. Graybill [1983: 249, his $\mathbf{A}^-$ is our $\mathbf{A}^+$].

8.34. Graybill [1983: 243].

8.35. Graybill [1983: 246–247].

8.36–8.39. Graybill [1983: 244–245].

8.40–8.41. Graybill [1983: 248–249].

8.3.4   Toeplitz Matrix

Definition 8.8. An $n \times n$ matrix $\mathbf{A}$ is a *Toeplitz* matrix if all the elements on the main diagonal are equal, all the elements on each superdiagonal are equal, and all elements on each subdiagonal are equal, that is, $a_{ij} = a_{i+s,j+s}$ for all i, j, s. For example,

$$
\mathbf{A}_1 = \begin{pmatrix}
a_0 & a_1 & a_2 & a_3 & \cdots & a_{n-1} \\
a_{-1} & a_0 & a_1 & a_2 & \cdots & a_{n-2} \\
a_{-2} & a_{-1} & a_0 & a_1 & \cdots & a_{n-2} \\
. & . & . & . & \cdots & . \\
a_{-(n-1)} & a_{-(n-2)} & a_{-(n-3)} & a_{-(n-4)} & \cdots & a_0
\end{pmatrix}
$$

is a Toeplitz matrix. The general term is $a_{ij} = a_{j-i}$ for some sequence

$$
a_{-(n-1)}, a_{-(n-2)}, \cdots, a_{-1}, a_0, a_1, a_2, \ldots, a_{n-2}, a_{n-1} \in \mathbb{C}.
$$

For general references see Grenander and Szegő [1958], Böttcher and Silbermann [1999], and Widom [1965].

A symmetric Toeplitz matrix has $a_{ij} = a_{ji}$, and there are at the most n "free" elements in the matrix. In the above example these would be the first row of elements. An example of a symmetric Toeplitz matrix arises in the study of a stationary process consisting of a set of random variables $\{u_t \mid t = 1, 2, \ldots, n\}$ with $\mathrm{cov}(u_t + \tau, u_t) = \kappa(|\tau|)$. Then the variance matrix of $\mathbf{u} = (u_1, u_2, \ldots, u_n)'$ is the positive definite Toeplitz matrix

$$
\mathrm{var}(\mathbf{u}) = \begin{pmatrix}
\kappa(0) & \kappa(1) & \kappa(2) & \cdots & \kappa(n-1) \\
\kappa(1) & \kappa(0) & \kappa(1) & \cdots & \kappa(n-2) \\
. & & . & \cdots & . \\
\kappa(n-1) & \kappa(n-2) & \kappa(n-3) & \cdots & \kappa(0)
\end{pmatrix}.
$$

When $\kappa(0) = 1$, $n = p$, and $\kappa(i) = \rho_i$ for each i, the above matrix comes from the so-called *Yule–Walker* equations that arise in the study of a pth-order autoregressive $(\mathrm{AR}(p))$ time series. Algorithms for solving these and similar equations, and for inverting a symmetric positive definite Toeplitz matrix, are given by Golub and Van Loan [1996: section 4.7].

8.42. Let

$$
\mathbf{B} = \begin{pmatrix}
0 & 1 & 0 & 0 & \cdots & 0 \\
0 & 0 & 1 & 0 & \cdots & 0 \\
0 & 0 & 0 & 1 & \cdots & 0 \\
. & . & . & . & \cdots & . \\
0 & 0 & 0 & 0 & \cdots & 1 \\
0 & 0 & 0 & 0 & \cdots & 0
\end{pmatrix}
\quad \text{and} \quad
\mathbf{F} = \begin{pmatrix}
0 & 0 & \cdots & 0 & 0 & 0 \\
1 & 0 & \cdots & 0 & 0 & 0 \\
0 & 1 & \cdots & 0 & 0 & 0 \\
. & . & \cdots & . & & . \\
0 & 0 & \cdots & 1 & 0 & 0 \\
0 & 0 & \cdots & 0 & 1 & 0
\end{pmatrix}.
$$

Then $\mathbf{B}$ and $\mathbf{F}$ are Toeplitz matrices, and are sometimes referred to as *backward shift* and *forward shift* matrices because of their effect on the elements of the columns of $\mathbf{I}_n = (\mathbf{e}_1, \mathbf{e}_2, \ldots, \mathbf{e}_n)$. Then $\mathbf{A}$, defined in the above definition 8.8, satisfies

$$
\mathbf{A} = \sum_{i=0}^{n-1} a_i \mathbf{F}^i + \sum_{i=0}^{n-1} a_i \mathbf{B}^i.
$$

8.43. A regular circulant is a Toeplitz matrix, but a Toeplitz matrix is not necessarily a regular circulant, though it is sometimes approximated by a regular circulant (cf. Brillinger [1975: 73-74] and the references therein). Any symmetric regular circulant is a symmetric Toeplitz matrix.

8.44. Let $\mathbf{A}$ be a Toeplitz matrix.

(a) $\mathbf{A}'$ is also a Toeplitz matrix.

(b) Any symmetric Toeplitz is also *doubly symmetric*.

(c) If a_{ij} is defined by $a_{ij} = a_{|i-j|}$, then $\mathbf{A}$ is a symmetric Toeplitz matrix.

The case when $\mathbf{A}$ is tridiagonal is considered in (8.110).

Proofs. Section 8.3.4.

8.44. Graybill [1983: 284–287].

8.3.5 Persymmetric Matrix

Definition 8.9. An $n \times n$ matrix $\mathbf{B} = (b_{ij})$ is called *persymmetric (countersymmetric)* if $b_{ij} = b_{n-j+1,n-i+1}$ for all i, j. Such a matrix is symmetric around the counter diagonal. An example is

$$\mathbf{B} = \begin{pmatrix} a & b & c \\ d & e & b \\ f & d & a \end{pmatrix}.$$

8.45. Let

$$\mathbf{E}_n = (\mathbf{e}_n, \mathbf{e}_{n-1}, \dots, \mathbf{e}_1)$$

$$= \begin{pmatrix} 0 & 0 & \cdots & 0 & 1 \\ 0 & 0 & \cdots & 1 & 0 \\ \cdot & \cdot & \cdots & \cdot & \cdot \\ 1 & 0 & \cdots & 0 & 0 \end{pmatrix},$$

the so-called *exchange permutation matrix*. Then, if $\mathbf{B}$ is $n \times n$, it is easy to show the following.

(a) If $\mathbf{x}' = (x_1, x_2, \dots, x_n)$, then $(\mathbf{E}_n\mathbf{x})' = (x_n, x_{n-1}, \dots, x_1)$.

(b) $\mathbf{E}_n^{-1} = \mathbf{E}_n$.

(b) $\mathbf{B}$ is persymmetric if and only if $\mathbf{B} = \mathbf{E}_n\mathbf{B}'\mathbf{E}_n$.

(c) If $\mathbf{B}$ is persymmetric and nonsingular, then $\mathbf{B}^{-1}$ is persymmetric.

(d) If $\mathbf{T}$ is an $n \times n$ Toeplitz matrix, then $\mathbf{T}$ is persymmetric. The converse is not necessarily true.

8.3.6   Cross-Symmetric (Centrosymmetric) Matrix

Definition 8.10. An $m \times n$ matrix $\mathbf{A} = (a_{ij})$ is said to be *cross-symmetric (centrosymmetric)* if $a_{ij} = a_{m+1-i,n+1-j}$ for all i, j and, we call it a C-matrix. For a list of examples of such matrices in statistics and time series see Dagum and Luati [2004]. They also consider a useful transformation and its properties, called a *t*-transformation, which takes $a_{ij} \to a_{m+1-i,n+1-j}$.

Note that when $m = n$:

(1) The elements of the first column read downwards are the same as the elements read upwards in the nth column; the elements in the second column read downwards are the same as the elements read upwards in the $(n-1)$th columns; and so forth.

(2) The elements read from left to right in the first row are the same as the elements read from right to to left in the nth row; the elements in the second row read from left to right are the same as the elements read from right to left in the $(n-1)$th row; and so forth.

(3) If n is odd, then the middle row (and column) are symmetric about the diagonal element.

An example is

$$\begin{pmatrix} a & b & e \\ c & d & c \\ e & b & a \end{pmatrix}.$$

8.3.7   Block Circulant

Definition 8.11. Given an $n \times n$ regular circulant matrix with first row elements $a_0, a_2, \ldots, a_{n-1}$, we can construct an $nk \times nk$ *block circulant* matrix $\mathbf{A}$ by replacing a_j by a $k \times k$ matrix $\mathbf{A}_j$ $(j = 0, 1, \ldots, n-1)$. Thus

$$\mathbf{A} = \begin{pmatrix} \mathbf{A}_0 & \mathbf{A}_1 & \mathbf{A}_2 & \cdots & \mathbf{A}_{n-1} \\ \mathbf{A}_{n-1} & \mathbf{A}_0 & \mathbf{A}_1 & \cdots & \mathbf{A}_{n-2} \\ \mathbf{A}_{n-2} & \mathbf{A}_{n-1} & \mathbf{A}_0 & \cdots & \mathbf{A}_{n-3} \\ \cdot & \cdot & \cdot & \cdots & \cdot \\ \mathbf{A}_1 & \mathbf{A}_2 & \mathbf{A}_3 & \cdots & \mathbf{A}_0 \end{pmatrix}.$$

Note that $\mathbf{A}$ is not necessarily a regular circulant. Typically each $\mathbf{A}_j$ is also a regular circulant or Toeplitz matrix, or it may even be a block circulant with components which are also regular circulants or Toeplitz matrices. For example, in experimental designs we might encounter the symmetric block matrix ($n = 2, k = 3$)

$$\mathbf{A} = \left(\begin{array}{ccc|ccc} 4 & 0 & 0 & 0 & 2 & 2 \\ 0 & 4 & 0 & 2 & 0 & 2 \\ 0 & 0 & 4 & 2 & 2 & 0 \\ \hline 0 & 2 & 2 & 4 & 0 & 0 \\ 2 & 0 & 2 & 0 & 4 & 0 \\ 2 & 2 & 0 & 0 & 0 & 4 \end{array} \right) = \begin{pmatrix} \mathbf{A}_0 & \mathbf{A}_1 \\ \mathbf{A}_1 & \mathbf{A}_0 \end{pmatrix}.$$

Block circulants are used, for example, in n-cyclic designs (cf. John and Williams [1995: 57–58]), while block Toeplitz matrices occur in vector-valued time series.

8.46. A regular circulant of (composite) order $n = pq$, where p and q are integers, is automatically a block circulant in which each block is Toeplitz. The blocks are of order q, and the arrangement of blocks is $p \times p$ (cf. Davis [1979: 70–71]). The family of such circulants we denote by $\mathcal{B}_{p,q}$.

8.47. If $\mathbf{A} \in \mathcal{B}_{p,q}$, then we have the sum of Kronecker products

$$\mathbf{A} = \mathbf{I}_p \otimes \mathbf{A}_0 + \mathbf{\Pi}_0 \otimes \mathbf{A}_1 + \mathbf{\Pi}_0^2 \otimes \mathbf{A}_2 + \cdots + \mathbf{\Pi}_0^{p-1} \otimes \mathbf{A}_{p-1},$$

where $\mathbf{\Pi}_0$ is the forward shift permutation matrix of order p (cf. Definition 8.2 in Section 8.2), and the $\mathbf{A}_j$ are Toeplitz of order q.

8.48. If $\mathbf{A}, \mathbf{B} \in \mathcal{B}_{p,q}$ and the α_i are scalars, then $\mathbf{A}, \mathbf{A}^*, \alpha_1 \mathbf{A} + \alpha_2 \mathbf{B}, \mathbf{AB}, p(\mathbf{A}) = \sum_{i=1}^r \alpha_i \mathbf{A}^i, \mathbf{A}^+$ and $\mathbf{A}^{-1}$ (if it exists) all belong to $\mathcal{B}_{p,q}$. We can use the relationship (8.47) so that $\mathbf{AB} = \mathbf{BA}$ if $\mathbf{A}_j \mathbf{B}_k = \mathbf{B}_k \mathbf{A}_j$ for all j, k.

8.49. Let $\mathbf{C}_{h_k}$ be a basic circulant of order n_k (cf. Definition 8.6 below (8.26)), and define the $n \times n$ Kronecker product matrix

$$\mathbf{C_h} = \mathbf{C}_{h_1} \otimes \mathbf{C}_{h_2} \otimes \cdots \otimes \mathbf{C}_{h_m},$$

where $n = n_1 n_2 \cdots n_m$. Then a block circulant matrix $\mathbf{A}$ of order n can be defined by

$$\mathbf{A} = \sum_{h_1=0}^{n_1-1} \sum_{h_2=0}^{n_2-1} \cdots \sum_{h_m=0}^{n_m-1} a_{h_1 h_2 \cdots h_m} \mathbf{C_h}.$$

The eigenvalues and eigenvectors of $\mathbf{A}$ can then be readily found (cf. John and Williams [1995: 232] and Tee [2005]).

Proofs. Section 8.3.7.

    8.47. Davis [1979: 178].

    8.48. Davis [1979: 181].

8.3.8 Hankel Matrix

Definition 8.12. A *Hankel* matrix $\mathbf{A} = (a_{ij})$ has the following structure:

$$\mathbf{A} = \begin{pmatrix} a_0 & a_1 & a_2 & \cdots & a_{K-1} \\ a_1 & a_2 & a_3 & \cdots & a_K \\ a_2 & a_3 & a_4 & \cdots & a_{K+1} \\ \vdots & \vdots & \vdots & \vdots & \vdots \\ a_{L-1} & a_L & a_{L+1} & \cdots & a_{K+L-2} \end{pmatrix},$$

where $a_{ij} = a_{i+j-2}$, so that the elements are equal on each of the counterdiagonals $i + j = const$. This matrix arises, for example, from a real time series $X = (a_0, a_1, \ldots, a_{N-1})$ of length N with L the *window length* $(1 < L < N)$ and

$K = N - L + 1$; it is called the *trajectory matrix* of X. If N and L are fixed, then there is a one-to-one relationship between $\mathbf{A}$ and X (cf. Golyandina et al. [2001: 16]).

If $L = K = n$, so that $\mathbf{A}$ is $n \times n$, then the general term is given by $a_{ij} = a_{i+j-2}$ for some given sequence $a_0, a_1, \ldots, a_{2n-3}, a_{2n-2}$. In this case $\mathbf{A}$ is symmetric.

For further details about Hankel matrices and structured matrices in general see Bini et al. [2001].

8.50. Let $\mathbf{\Pi} = (\mathbf{e}_n, \mathbf{e}_{n-1}, \ldots, \mathbf{e}_1)$ be the *backward identity* permutation matrix. Then:

(a) $\mathbf{\Pi T}$ is a Hankel marix for any Toeplitz matrix $\mathbf{T}$.

(b) $\mathbf{\Pi H}$ is Toeplitz matrix for any square Hankel matrix $\mathbf{H}$.

(c) Since $\mathbf{\Pi} = \mathbf{\Pi}' = \mathbf{\Pi}^{-1}$ and square Hankel matrices are symmetric, any Toeplitz matrix is product of two symmetric matrices ($\mathbf{\Pi}$ and a Hankel matrix).

Proofs. Section 8.3.8.

8.50. Quoted by Horn and Johnson [1985: 28].

8.4   DIAGONALLY DOMINANT MATRICES

Definition 8.13. Let $\mathbf{A} = (a_{ij})$ be a real or complex $n \times n$ matrix ($n \geq 2$), and define

$$R_p = \sum_{j=1:j\neq p}^{n} |a_{pj}| \quad \text{and} \quad C_q = \sum_{i=1:i\neq q}^{n} |a_{iq}|$$

to be, respectively, the sum of the absolute values of the off-diagonal elements of the pth row of $\mathbf{A}$, and the sum of the absolute values of the off-diagonal elements of the qth column of $\mathbf{A}$. (In the above, $|x|$ denotes the modulus of x if x is not real.)

Considering first the rows, if $|a_{pp}| > R_p$, then the pth row is said to have a strictly dominant diagonal. If $|a_{pp}| \geq R_p$ for $p = 1, 2, \ldots, n$, then $\mathbf{A}$ is said to be *(row) diagonally dominant*, while if $|a_{pp}| > R_p$ for $p = 1, 2, \ldots, n$, then $\mathbf{A}$ is said to be *strictly (row) diagonally dominant*; we denote this by r.d.d. Some authors omit the word "row" and then do not refer to columns. However, there is a corresponding set of definitions for columns. For example, if $|a_{qq}| > C_q$, then the qth column of $\mathbf{A}$ is said to have a dominant diagonal, while if $|a_{qq}| > C_q$ for $q = 1, 2, \ldots, n$, then $\mathbf{A}$ is said to be *strictly column diagonally dominant* and we write c.d.d. If $\mathbf{A}$ is either r.r.d. or c.d.d., we say that $\mathbf{A}$ is d.d.

8.51. Let $\mathbf{A}$ be any $n \times n$ matrix (real or complex).

(a) If $\mathbf{\Pi}$ is any $n \times n$ permutation matrix and $\mathbf{A}$ is r.d.d. (respectively c.d.d.), then $\mathbf{\Pi}'\mathbf{A\Pi}$ is r.d.d. (respectively c.d.d.).

(b) If $\mathbf{D}$ is any $n \times n$ nonsingular diagonal matrix and $\mathbf{A}$ is r.d.d. (respectively c.d.d.), then $\mathbf{DA}$ (respectively $\mathbf{AD}$) is r.d.d. (respectively c.d.d.).

(c) If any diagonal element of $\mathbf{A}$ is zero, then $\mathbf{A}$ is neither r.d.d. nor c.d.d.

(d) If λ is any eigenvalue of $\mathbf{A}$, then $\mathbf{A} - \lambda \mathbf{I}_n$ is neither r.d.d. nor c.d.d.

(e) If $\mathbf{A}$ is r.d.d. (respectively c.d.d.), then at least one column (respectively row) must have a dominant diagonal.

(f) If $\mathbf{A}$ is a regular circulant with first row elements $a_0, a_1, \ldots, a_{n-1}$ such that

$$|a_j| > \sum_{i=0:i\neq j}^{n-1} |a_i|$$

for some j, then $\mathbf{A}$ is nonsingular.

Definition 8.14. Let $\mathbf{A}$ a matrix such that

$$|a_{ij}| > \sum_{h=1:h\neq i}^{n} |a_{hj}|.$$

Then the jth column is said to have a *dominant element*, and it is in the ith row.

8.52. (Conditions for Nonsingularity) Let $\mathbf{A}$ be $n \times n$.

(a) (Levy–Desplanques) If $\mathbf{A}$ is d.d., then $\mathbf{A}$ is nonsingular. Conversely, if $\mathbf{A}$ is singular, then $\mathbf{A}$ is neither r.d.d. nor c.d.d. This result is linked to (6.29) as 0 cannot then lie in any closed Geršgorin disc, so that 0 is not an eigenvalue.

(b) If $\mathbf{R}$ and $\mathbf{S}$ are any nonsingular $n \times n$ matrices and $\mathbf{RAS}$ is a d.d. matrix, then $\mathbf{A}$ is nonsingular. Conversely, if $\mathbf{A}$ is nonsingular, there exist nonsingular matrices $\mathbf{R}$ and $\mathbf{S}$ such that $\mathbf{RAS}$ is d.d.

(c) Suppose each row, except one (say the kth row), has a strictly dominant diagonal, and suppose the kth row is such that $0 < |a_{kk}| = R_k$. Then $\mathbf{A}$ is nonsingular. A similar theorem exists for columns.

(c) If each column of $\mathbf{A}$ has a strictly dominant element, and each row contains one of the dominant elements, then $\mathbf{A}$ is nonsingular. This result also holds if each row has a strictly dominant element and these are in distinct columns.

(d) Suppose that for one value of $j = 1, 2, \ldots, n$, say $j = t$, either of the following equations hold, namely

$$0 < |a_{tt}| < R_t \text{ and } |a_{ii}| \cdot |a_{tt}| > R_i R_t \quad \text{for } i = 1, 2, \ldots, n; \ i \neq t$$
$$0 < |a_{tt}| < C_t \text{ and } |a_{ii}| \cdot |a_{tt}| > C_i C_t \quad \text{for } i = 1, 2, \ldots, n; \ i \neq t,$$

then $\mathbf{A}$ is nonsingular.

(e) Suppose that all the elements of $\mathbf{A}$ are nonzero. If $\mathbf{A}$ is diagonally dominant (i.e., not strictly so), and $|a_{ii}| > R_i$ for at least one value of $i = 1, \ldots, n$, then $\mathbf{A}$ is nonsingular.

8.53. (Positive Determinant) Let $\mathbf{A}$ be an $n \times n$ real matrix that is d.d. and has positive diagonal elements.

(a) $\det \mathbf{A} > 0$.

(b) If $\mathbf{A}_r$ is any $r \times r$ principal submatrix, then $\det \mathbf{A}_r > 0$.

(c) If the signs of any set of off-diagonal elements are changed, then $\det \mathbf{A} > 0$.

(d) The real part of each eigenvalue of $\mathbf{A}$ is positive. Thus all real eigenvalues are positive. If, instead, the diagonal elements of $\mathbf{A}$ are all negative, then the real parts of all eigenvalues are negative.

(e) From (d) it follows that if $\mathbf{A}$ is also Hermitian and all its main diagonal elements are positive, then all the eigenvalues of $\mathbf{A}$ are real and positive.

8.54. If $\mathbf{A}$ is $n \times n$ and $\mathbf{T}'\mathbf{AT}$ is d.d. with positive diagonal elements, where $\mathbf{T}$ is any orthogonal matrix, then $\det \mathbf{A} = \det |\mathbf{T}|^2 \det \mathbf{A} = \det(\mathbf{T}'\mathbf{AT}) > 0$ (by 8.6b). Note that $\mathbf{T}$ could be a permutation matrix.

8.55. Let $\mathbf{A}$ be an $n \times n$ matrix that is d.d., and let $\mathbf{D}$ be a diagonal matrix with the same diagonal elements as $\mathbf{A}$. Then $\rho(\mathbf{B}) < 1$, where $\mathbf{B} = \mathbf{I}_n - \mathbf{D}^{-1}\mathbf{A}$ and ρ is the spectral radius of $\mathbf{B}$.

8.56. (Linear Equations) Let $\mathbf{A}$ be an $n \times n$ real matrix with positive diagonal elements and nonpositive off-diagonal elements. For each $n \times 1$ vector $\mathbf{b}$ with non-negative elements, there exists a unique vector $\mathbf{x}$ with non-negative elements that is a solution to $\mathbf{Ax} = \mathbf{b}$ if $\mathbf{A}$ is d.d.

Proofs. Section 8.4.

8.51. Graybill [1983: section 8.11, here dominant means strictly dominant, and the complex case is mentioned in the note on p. 261].

8.52. Graybill [1983: 251–256, here dominant means strictly dominant]; also Horn and Johnson [1985: 302 and 355, for (a) and (b) respectively].

8.53. Graybill [1983: 258–261; here dominant means strictly dominant].

8.54. Graybill [1983: 260].

8.55. Graybill [1983: 262].

8.56. Graybill [1983: 265].

8.5   HADAMARD MATRICES

Definition 8.15. An $n \times n$ *Hadamard matrix* $\mathbf{H}$ is a matrix with elements all ± 1 such that $\mathbf{H}'\mathbf{H} = \mathbf{HH}' = n\mathbf{I}_n$, that is, $n^{-1/2}\mathbf{H}$ is orthogonal. If all the elements of the first column are equal to $+1$, then $\mathbf{H}$ is called a *seminormalized Hadamard matrix*. If all the elements in the first row and column are equal to $+1$, then $\mathbf{H}$ is said to be *normalized*. These matrices are closely linked to balanced incomplete block designs, group divisible designs, Youden designs, 2^n factorial experiments, optimal weighing designs, and response surface methodology. For further details and applications see Agaian [1985].

8.57. We have the following properties of an $n \times n$ Hadamard matrix $\mathbf{H}$.

(a) $\mathbf{H}'$ and $n\mathbf{H}^{-1}$ are Hadamard matrices.

(b) If $\mathbf{D}_1$ and $\mathbf{D}_2$ are diagonal matrices with diagonal elements ± 1, then $\mathbf{D}_1\mathbf{H}\mathbf{D}_2$ is a Hadamard matrix. We can set $\mathbf{D}_i = \mathbf{I}_n$ for $i = 1$ or 2.

(c) n must equal 1 or 2 or be a multiple of 4.

(d) From $\det(\mathbf{H}'\mathbf{H}) = n^n$, $\det \mathbf{H} = \pm n^{n/2}$.

(e) (Hadamard) If $\mathbf{A}$ is a real $n \times n$ matrix with $|a_{ij}| \leq 1$, then $|\det \mathbf{A}| \leq n^{n/2}$. We find that the Hadamard family is the only family of matrices which attains the upper bound.

(f) If $\mathbf{H}_1$ and $\mathbf{H}_2$ are $n_1 \times n_1$ and $n_2 \times n_2$ Hadamard matrices, respectively, then $\mathbf{H}_1 \otimes \mathbf{H}_2$ (the Knonecker product) is an $n_1 n_2 \times n_1 n_2$ Hadamard matrix.

(g) Setting $n = 2$ and applying (f) repeatedly, we see that there is a $2^k \times 2^k$ Hadamard matrix for every positive integer k.

(h) If an $n \times n$ Hadamard matrix exists, then an $n \times n$ normalized Hadamard matrix exists.

8.58. In digital signal processing, Hadamard matrices, $\mathbf{H}_n$ say, are restricted to be of order 2^n given by the recursion

$$\mathbf{H}_1 = \begin{pmatrix} 1 & 1 \\ 1 & -1 \end{pmatrix}, \quad \mathbf{H}_2 = \begin{pmatrix} \mathbf{H}_1 & \mathbf{H}_1 \\ \mathbf{H}_1 & -\mathbf{H}_1 \end{pmatrix} = \mathbf{H}_1 \otimes \mathbf{H}_1,$$

$$\mathbf{H}_n = \mathbf{H}_1 \otimes \mathbf{H}_{n-1}.$$

Also $\mathbf{H}_{2^n}$ is symmetric so that $\mathbf{H}_{2^n}^2 = 2^n \mathbf{I}_{2^n}$.

8.59. Let $\mathbf{H}_n$ be defined in (8.58) above, and consider the iteration

$$\mathbf{x}_1 = \begin{pmatrix} 1 \\ -1 + \sqrt{2} \end{pmatrix} \quad \text{and} \quad \mathbf{x}_n = \begin{pmatrix} \mathbf{x}_{n-1} \\ (-1 + \sqrt{2})\mathbf{x}_{n-1} \end{pmatrix}.$$

Then $\mathbf{H}_n$ has eigenvalues $+2^{n/2}$ and $-2^{n/2}$, each of multiplicity 2^{n-1}, and an eigenvector $\mathbf{x}_n$ corresponding to the positive eigenvalue $2^{n/2}$.

8.60. If $\mathbf{H}$ is an $m \times m$ Hadamard matrix that contains $\mathbf{J}_n = \mathbf{1}_n \mathbf{1}_n'$, then $m \geq n^2$.

Proofs. Section 8.5.

8.57. Graybill [1983: section 8.14 for (a)–(c)] and Schott [2005: 334–335, for (c)–(f)].

8.59–8.60. Zhang [1999: 120–121].

8.6 IDEMPOTENT MATRICES

8.6.1 General Properties

Definition 8.16. An $n \times n$ real or complex matrix is said to be *idempotent* if $\mathbf{A}^2 = \mathbf{A}$. In Section 2.3 we called such a matrix a projection matrix. If $\mathbf{A}$ is also real and symmetric (with $\langle \mathbf{x}, \mathbf{y} \rangle = \mathbf{x}'\mathbf{y}$), or Hermitian (with $\langle \mathbf{x}, \mathbf{y} \rangle = \mathbf{y}^*\mathbf{x}$), it represents an orthogonal projection matrix. Some other geometrical properties of such matrices are given in Section 2.3. We assume below that $\mathbf{A}$ is real, unless otherwise stated, though many of the following results hold for complex matrices.

8.61. The following statements are equivalent.

(1) An $n \times n$ matrix $\mathbf{P}$ is idempotent.

(2) $\mathbf{I}_n - \mathbf{P}$ is idempotent.

(3) $\mathcal{C}(\mathbf{P}) \cap \mathcal{C}(\mathbf{I}_n - \mathbf{P}) = \mathbf{0}$.

(4) $\mathcal{C}(\mathbf{P}) = \mathcal{N}(\mathbf{I}_n - \mathbf{P})$.

(5) $\mathcal{C}(\mathbf{I}_n - \mathbf{P}) = \mathcal{N}(\mathbf{P})$.

8.62. The following statements are equivalent.

(1) $\mathbf{A}$ is an $n \times n$ idempotent matrix of rank r with Moore–Penrose inverse $\mathbf{A}^+$.

(2) There exist orthogonal projection matrices $\mathbf{R}$ and $\mathbf{S}$ such that $\mathbf{A}^+ = \mathbf{R}\mathbf{S}$.

(3) $\mathbf{A}^+\mathbf{A}' = \mathbf{A}^+$.

(3) $\mathbf{A}'\mathbf{A}^+ = \mathbf{A}^+$.

(4) $\mathbf{A} = \mathbf{B}\mathbf{C}'$, where $\mathbf{C}'\mathbf{B} = \mathbf{I}_r$ with $\mathbf{B}$ and $\mathbf{C}$ being $n \times r$ matrices.

(5) The Jordan canonical form of $\mathbf{A}$ (cf. 16.7) can be written as

$$\begin{pmatrix} \mathbf{I}_r & \mathbf{0} \\ \mathbf{0} & \mathbf{0} \end{pmatrix}.$$

(6) There exists an orthogonal matrix $\mathbf{T}$ such that

$$\mathbf{A} = \mathbf{T} \begin{pmatrix} \mathbf{I}_r & \mathbf{K} \\ \mathbf{0} & \mathbf{0} \end{pmatrix} \mathbf{T}',$$

where $\mathbf{K}$ is $r \times (n - r)$.

For further results and discussion see Trenkler [1994].

8.63. If $\mathbf{A}$ is an $n \times n$ idempotent matrix of rank r, then there exist nonsingular $\mathbf{R}$ and unitary $\mathbf{U}$ such that

$$\mathbf{R}^{-1}\mathbf{A}\mathbf{R} = \begin{pmatrix} \mathbf{I}_r & \mathbf{0} \\ \mathbf{0} & \mathbf{0} \end{pmatrix} \quad \text{and} \quad \mathbf{U}^*\mathbf{A}\mathbf{U} = \begin{pmatrix} \mathbf{I}_r & \mathbf{Q} \\ \mathbf{0} & \mathbf{0} \end{pmatrix}$$

for some $\mathbf{Q}$. If $\mathbf{A}$ is symmetric, we can replace $\mathbf{R}$ by an orthogonal matrix.

8.64. An $n \times n$ matrix $\mathbf{A}$ is idempotent if and only if rank $\mathbf{A} + \text{rank}(\mathbf{I}_n - \mathbf{A}) = n$.

8.65. Let $\mathbf{A}$ be an $n \times n$ idempotent (real or complex) matrix of rank r, then:

(a) $\mathbf{A}$ has r eigenvalues equal to 1 and $n - r$ eigenvalues equal to 0. Also, if $\mathbf{A}$ is real and symmetric, then it is idempotent if and only if each eigenvalue of $\mathbf{A}$ is 1 or 0.

(b) $\det \mathbf{A}^2 = \det \mathbf{A}$ and $\det \mathbf{A}$ is 0 or 1. If $\det \mathbf{A} = 1$ then, by (a), $\mathbf{A} = \mathbf{I}_n$.

(c) From (8.63) we have rank $\mathbf{A} = \text{trace} \, \mathbf{A} = r$.

(d) $\mathbf{I}_n - \mathbf{A}$ is idempotent and, from (c), $\text{rank}(\mathbf{I}_n - \mathbf{A}) = n - \text{rank} \, \mathbf{A}$.

(e) $\mathbf{A}$ can be expressed in the form $\mathbf{A} = \mathbf{Q}\mathbf{R}^*$, where $\mathbf{Q}$ and $\mathbf{R}$ are $n \times r$ and $\mathbf{R}^*\mathbf{Q} = \mathbf{I}_r$.

(f) There exists a Hermitian positive definite matrix $\mathbf{C}$ such that $\mathbf{A} = \mathbf{C}^{-1}\mathbf{A}^*\mathbf{C}$.

8.66. Let $\mathbf{A}$ be an $n \times n$ matrix with Moore–Penrose inverse $\mathbf{A}^+$. Then $\mathbf{A}$ is a real symmetric idempotent matrix if and only if one of the following conditions is satisfied.

(1) $\mathbf{A}'\mathbf{A} = \mathbf{A}$.

(2) $\mathbf{I} - \mathbf{A}$ is symmetric and idempotent.

(3) $\mathbf{A}$ is idempotent and $\mathbf{A}\mathbf{A}' = \mathbf{A}'\mathbf{A}$.

(4) $\mathbf{A}$ and $\mathbf{A}'\mathbf{A}$ are idempotent.

(5) $\mathbf{A}\mathbf{A}'\mathbf{A} = \mathbf{A}$ and $\mathbf{A}$ is idempotent.

(6) $\mathbf{A}'\mathbf{A}\mathbf{A}' = \mathbf{A}'$ and $\mathbf{A}$ is idempotent.

(7) $\mathbf{A}$ and $\frac{1}{2}(\mathbf{A} + \mathbf{A}')$ are idempotent.

(8) $\mathbf{A}\mathbf{A}' + \mathbf{A}'\mathbf{A} = \mathbf{A} + \mathbf{A}'$ and $\mathbf{A}$ is idempotent.

(9) $\mathbf{I}_n - 2\mathbf{A}$ is a symmetric, orthogonal matrix.

(10) $\mathbf{A}^2 = \mathbf{A}'$ and $\mathbf{A}$ is tripotent (i.e., $\mathbf{A}^3 = \mathbf{A}$).

(11) $\mathbf{A}\mathbf{A}' = \mathbf{A}'\mathbf{A}\mathbf{A}'$.

(12) $\mathbf{A}$ is idempotent and $\text{rank}(\mathbf{I}_n - \mathbf{A}'\mathbf{A}) = n - \text{rank} \, \mathbf{A}$.

(13) $\mathbf{A}$ is idempotent and $\|\mathbf{A}\mathbf{x}\|_2 \leq \|\mathbf{x}\|_2$ for all $\mathbf{x} \in \mathbb{R}^n$.

(14) $\mathbf{x}'\mathbf{A}'\mathbf{A}\mathbf{x} = \mathbf{x}'\mathbf{A}\mathbf{x}$ for all $\mathbf{x} \in \mathbb{R}^n$.

(15) $\mathbf{A}$ is idempotent and $\mathbf{x}'\mathbf{A}\mathbf{x} \geq 0$ for all $\mathbf{x} \in \mathbb{R}^n$.

(16) $\|\mathbf{y} - \mathbf{A}\mathbf{y}\|_2 \leq \|\mathbf{y} - \mathbf{x}\|_2$ for all $\mathbf{y} \in \mathbb{R}^n$ and all $\mathbf{x} \in \mathcal{C}(\mathbf{A})$.

(17) $\mathbf{A}$ is idempotent and $\mathbf{A} = \mathbf{A}^+$.

(18) $\mathbf{A}$ is idempotent and $\mathbf{A}\mathbf{A}' = \mathbf{A}\mathbf{A}^+$.

(19) $\mathbf{A}^+ = \mathbf{A}$ and $\mathbf{A}^2 = \mathbf{A}'$.

(20) $\mathbf{A}$ and $\mathbf{A}^+$ are idempotent.

(21) There exists an $n \times m$ matrix $\mathbf{B}$ such that $\mathbf{A} = \mathbf{B}\mathbf{B}^+$.

(22) $\mathbf{A} = \mathbf{A}\mathbf{A}^+$.

(23) $\mathbf{A} = \mathbf{B}(\mathbf{B}'\mathbf{B})^{-1}\mathbf{B}$ for some $n \times m$ matrix $\mathbf{B}$ of rank m.

See Trenkler [1994] for these and further results of a similar nature. He also gives necessary and sufficient conditions that a symmetric matrix is idempotent.

8.67. (Generalized Inverse and Idempotency) Let $\mathbf{A}$ be $m \times n$ with Moore–Penrose inverse $\mathbf{A}^+$. Then the following conditions are equivalent.

(1) $\mathbf{A}'\mathbf{A}$ is idempotent (i.e., is an orthogonal projection matrix, because it is symmetric).

(2) $\mathbf{A}\mathbf{A}'$ is idempotent.

(3) $\mathbf{A}\mathbf{A}'\mathbf{A} = \mathbf{A}$, that is, $\mathbf{A}'$ is a weak inverse of $\mathbf{A}$.

(4) $\mathbf{A}' = \mathbf{A}^+$.

8.68. If $\mathbf{A}$ is $m \times n$ with any weak inverse $\mathbf{A}^-$, then $\mathcal{N}(\mathbf{A}) = \mathcal{C}(\mathbf{I} - \mathbf{A}^-\mathbf{A})$.

8.69. If $(\mathbf{C}\mathbf{B})^{-1}$ exists, then $\mathbf{B}(\mathbf{C}\mathbf{B})^{-1}\mathbf{C}$ is idempotent.

8.70. Let $\mathbf{A}$ be an $n \times n$ symmetric idempotent matrix, and let $\mathbf{B}$ be an $n \times m$ matrix of rank m.

(a) If $\mathbf{A}\mathbf{B} = \mathbf{B}$ and rank $\mathbf{A}$ = rank $\mathbf{B}$, then $\mathbf{A} = \mathbf{B}(\mathbf{B}'\mathbf{B})^{-1}\mathbf{B}'$.

(b) If $\mathbf{A}\mathbf{B} = \mathbf{0}$ and rank $\mathbf{A}$ + rank $\mathbf{B}$ = n, then $\mathbf{A} = \mathbf{I}_n - \mathbf{B}(\mathbf{B}'\mathbf{B})^{-1}\mathbf{B}'$.

8.71. (Symmetric Matrix) Let $\mathbf{A}$ be an $n \times n$ symmetric idempotent matrix of rank r, where $r < n$. Then we have the following.

(a) $0 \leq a_{ii} \leq 1$ for $i = 1, 2, \ldots, n$.

(b) If $a_{ii} = 0$ or $a_{ii} = 1$, then $a_{ij} = 0$ for all j, $j \neq i$.

(c) $\mathbf{A}$ is non-negative definite.

(d) If $\mathbf{T}$ is orthogonal, then $\mathbf{P} = \mathbf{T}'\mathbf{A}\mathbf{T}$ is a symmetric idempotent matrix.

(e) If $\mathbf{R}$ is nonsingular, then $\mathbf{P} = \mathbf{R}^{-1}\mathbf{A}\mathbf{R}$ is idempotent.

(f) $\mathbf{Q} = \mathbf{I}_n - 2\mathbf{A}$ is a symmetric orthogonal matrix.

(g) We can write $\mathbf{A} = \mathbf{T}_r\mathbf{T}_r'$, where $\mathbf{T}_r'\mathbf{T}_r = \mathbf{I}_r$, and the columns of $\mathbf{T}_r$ form an orthonormal basis for $\mathcal{C}(\mathbf{A})$. This result holds if $\mathbf{A}$ is Hermitian and we replace $\mathbf{T}'$ by $\mathbf{T}^*$.

(h) If $\mathbf{V}$ is positive definite, then

$$\text{rank}(\mathbf{A}\mathbf{V}\mathbf{A}) = \text{trace}(\mathbf{A}\mathbf{V}).$$

8.72. Let $\mathbf{A}$ be a symmetric matrix that satisfies $\mathbf{A}^{k+1} = \mathbf{A}^k$ for some positive integer k. Then $\mathbf{A}$ is idempotent.

8.73. Let $\mathbf{A}_1$ and $\mathbf{A}_2$ be $n \times n$ symmetric idempotent matrices, and suppose $\mathbf{A}_1 - \mathbf{A}_2$ is non-negative definite.

(a) $\mathbf{A}_1\mathbf{A}_2 = \mathbf{A}_2\mathbf{A}_1 = \mathbf{A}_2$.

(b) $\mathbf{A}_1 - \mathbf{A}_2$ is a symmetric idempotent matrix.

8.74. Suppose $\mathbf{A}$ and $\mathbf{B}$ are $n \times n$ matrices. If $\mathbf{AB} = \mathbf{A}$ and $\mathbf{BA} = \mathbf{B}$, then $\mathbf{A}$ and $\mathbf{B}$ are both idempotent.

8.75. (Kronecker Products) Let $\mathbf{A}$ be $m \times n$ and $\mathbf{B}$ be $m \times p$ real matrices. Let $\mathbf{A} \otimes \mathbf{B}$ be their Kronecker product, and denote by $\mathbf{P_A}$, $\mathbf{P_B}$, and $\mathbf{P_{A \otimes B}}$ the symmetric idempotent matrices that project orthogonally onto $\mathcal{C}(\mathbf{A})$, $\mathcal{C}(\mathbf{B})$, and $\mathcal{C}(\mathbf{A} \otimes \mathbf{B})$. Then:

(a) $\mathbf{P_{A \otimes B}} = \mathbf{P_A} \otimes \mathbf{P_B}$.

(b) $\mathbf{P_{A \otimes I}} = \mathbf{P_A} \otimes \mathbf{I}$.

(c) If $\mathbf{Q} = \mathbf{I} - \mathbf{P}$ in each case, then

$$\mathbf{Q_{A \otimes B}} = \mathbf{Q_A} \otimes \mathbf{Q_B} + \mathbf{Q_A} \otimes \mathbf{P_B} + \mathbf{P_A} \otimes \mathbf{Q_B}.$$

8.76. Let $\mathbf{A}$ and $\mathbf{B}$ be $n \times n$ symmetric matrices, with $\mathbf{B}$ positive definite. Then $\mathbf{AB}$ is idempotent if and only if each eigenvalue of $\mathbf{AB}$ is 0 or 1.

Proofs. Section 8.6.1.

8.61(2). Follows directly from (1).

8.61(3). Harville [1997: 384, lemma 17.2.6].

8.61(4). Let $\mathbf{P}^2 = \mathbf{P}$. If $\mathbf{y} = \mathbf{Px}$ then $(\mathbf{I}_n - \mathbf{P})\mathbf{y} = (\mathbf{I}_n - \mathbf{P})\mathbf{Px} = \mathbf{0}$ and $\mathbf{y} \in \mathcal{N}(\mathbf{I}_n - \mathbf{P})$. Conversely, if $(\mathbf{I}_n - \mathbf{P})\mathbf{y} = \mathbf{0}$ then $\mathbf{y} = \mathbf{Py} \in \mathcal{C}(\mathbf{P})$.

8.61(5). Similar to (4); see Harville [1997: 146].

8.62. Trenkler [1994].

8.63. Abadir and Magnus [2005: 234] and Schott [2005: 396].

8.64. Abadir and Magnus [2005: 235], Harville [1997: 435], Rao and Rao [1998: 253], and Trenkler [1994].

8.65a. Abadir and Magnus [2005: 233] and Schott [2005: 397].

8.65e–f. Rao and Rao [1998: 251].

8.66. Trenkler [1994].

8.67. Trenkler [1994: 266].

8.68. Harville [1997: 140].

8.69. Simply square the matrix.

8.70. Abadir and Magnus [2005: 236].

8.71a–b. Schott [2005: 399].

8.71c. Follows from $\mathbf{x}'\mathbf{A}\mathbf{x} = \mathbf{x}'\mathbf{A}'\mathbf{A}\mathbf{x} = \mathbf{y}'\mathbf{y}$.

8.71d–e. We have $\mathbf{P}^2 = \mathbf{P}$.

8.71f. We show that $\mathbf{Q}'\mathbf{Q} = \mathbf{I}_n$.

8.71g. Rao and Rao [1998: 252] and Seber and Lee [2003: 475, real case].

8.71h. Harville [2001: 82, exercise 10]. We have $\text{trace}(\mathbf{A}\mathbf{V}\mathbf{A}) = \text{trace}(\mathbf{A}\mathbf{V}^{1/2}\mathbf{V}^{1/2}\mathbf{A}) = \text{trace}(\mathbf{V}^{1/2}\mathbf{A}^2\mathbf{V}^{1/2}) = \text{trace}(\mathbf{V}^{1/2}\mathbf{A}\mathbf{V}^{1/2}) = \text{trace}(\mathbf{A}\mathbf{V})$.

8.72. Schott [2005: 399].

8.73. Seber and Lee [2003: 465].

8.74. Abadir and Magnus [2005: 236].

8.75. Quoted by Rao and Rao [1998: 262].

8.76. Schott [2005: 397].

8.6.2 Sums of Idempotent Matrices and Extensions

There are many results given for sums of idempotent matrices, and these are often expressed with different conditions. We give several versions of these below, and there is some overlap. For a very general investigation of a linear combination of two projectors see Baksalary and Baksalary [2004a] and the references therein. Questions relating to the nonsingularity of such combinations of idempotent matrices, including just sums and differences, are considered by Baksalary and Baksalary [2004b] and Koliha et al. [2004]. We assume below that all matrices are real, unless otherwise stated, though some of the results hold for complex matrices as well.

8.77. If $\mathbf{A}$ and $\mathbf{B}$ are $n \times n$ idempotent matrices, then $\mathbf{A} + \mathbf{B}$ is idempotent if and only if $\mathbf{A}\mathbf{B} = \mathbf{B}\mathbf{A} = \mathbf{0}$. We generalize this result below.

8.78. (Cochran's Theorem) Suppose $\mathbf{A}_1, \mathbf{A}_2, \ldots, \mathbf{A}_k$ is a sequence of symmetric $n \times n$ matrices such that $\sum_{i=1}^{k} \mathbf{A}_i = \mathbf{I}_n$. Then the following conditions are equivalent (i.e., each one implies the other two).

(1) $\mathbf{A}_i^2 = \mathbf{A}_i$ for $i = 1, 2, \ldots, k$.

(2) $\mathbf{A}_i\mathbf{A}_j = \mathbf{0}$ for all $i, j, i \neq j$.

(3) $\sum_{i=1}^{k} \text{rank } \mathbf{A}_i = n$.

This can be derived from (8.79) below.

8.79. Let $\mathbf{A} = \sum_{i=1}^{k} \mathbf{A}_i$, where each $\mathbf{A}_i$ is a symmetric $n \times n$ matrix. Any two of the following three conditions implies the third.

(1) $\mathbf{A} = \mathbf{A}^2$.

(2) $\mathbf{A}_i^2 = \mathbf{A}_i$ for $i = 1, 2, \ldots, n$.

(3) $\mathbf{A}_i \mathbf{A}_j = \mathbf{0}$ for all i, j, $i \neq j$.

From (8.80) we can include further results involving

(4) rank $\mathbf{A} = \sum_{i=1}^{k}$ rank $\mathbf{A}_i$,

For example, any two of (1), (2), and (3) implies all four. Furthermore, (1) and (4) imply (2) and (3) (Rao [1973a: 28]). We can relate this theorem to the previous one by defining $\mathbf{A}_0 = \mathbf{I}_n - \mathbf{A}$ so that $\sum_{i=0}^{k} \mathbf{A}_i = \mathbf{I}_n$. Alternatively, we can set $\mathbf{A} = \mathbf{I}_n$.

8.80. Let $\mathbf{A}_i$ be an $n \times n$ matrix $(i = 1, 2, \ldots, k)$, and let $\mathbf{A} = \sum_{i=1}^{k} \mathbf{A}_i$.

(a) If $\mathbf{A}^2 = \mathbf{A}$, then the following conditions are equivalent.

(1) $\mathbf{A}_i \mathbf{A}_j = \mathbf{0}$, for all $i, j, i \neq j$, and rank $\mathbf{A}_i^2 =$ rank $\mathbf{A}_i$ for $i = 1, 2, \ldots, k$.

(2) $\mathbf{A}_i^2 = \mathbf{A}_i$ for $i = 1, 2, \ldots, k$.

(3) rank $\mathbf{A} = \sum_{i=1}^{k}$ rank $\mathbf{A}_i$.

If $\mathbf{A} = \mathbf{I}_n$, then the condition on $\mathbf{A}$ is automatically satisfied and rank $\mathbf{A} = n$. Furthermore, if each $\mathbf{A}_i$ is also symmetric, then $\mathbf{A}_i^2 = \mathbf{A}_i' \mathbf{A}_i$, which implies rank $\mathbf{A}_i^2 =$ rank $\mathbf{A}_i$, and condition (1) reduces to the condition

$$\mathbf{A}_i \mathbf{A}_j = \mathbf{0} \quad \text{for all} \quad i, j, \, i \neq j.$$

(b) If the $\mathbf{A}_i$ are all idempotent and $\mathbf{A}_i \mathbf{A}_j = \mathbf{0}$ for all $i, j, i \neq j$, then

(i) $\mathbf{A}^2 = \mathbf{A}$.

(ii) rank $\mathbf{A} = \sum_{i=1}^{k}$ rank $\mathbf{A}_i$.

(c) Suppose $\mathbf{V}$ is an $n \times n$ non-negative definite matrix, and let $\mathbf{R}$ be any matrix such that $\mathbf{V} = \mathbf{R}'\mathbf{R}$ (cf. 10.10). Then (a) and (b) still hold if we replace each $\mathbf{A}_i$ by $\mathbf{R}\mathbf{A}_i\mathbf{R}'$ throughout and $\mathbf{A}$ by $\mathbf{R}\mathbf{A}\mathbf{R}'$.

8.81. Let $\mathbf{A}_i$ $(i = 1, 2, \ldots, k)$ be square (not necessarily symmetric) matrices, and let $\mathbf{A} = \sum_{i=1}^{k} \mathbf{A}_i$. Consider the following conditions:

(1) $\mathbf{A}_i^2 = \mathbf{A}_i$, $i = 1, 2, \ldots, k$.

(2) $\mathbf{A}_i \mathbf{A}_j = \mathbf{0}$ for all $i \neq j$.

(3) $\mathbf{A}^2 = \mathbf{A}$.

(4) $\sum_i$ rank $\mathbf{A}_i =$ rank $\mathbf{A}$.

(5) rank$(\mathbf{A}_i^2) =$ rank $\mathbf{A}_i$, $i = 1, 2, \ldots, k$.

Then

$$
\begin{aligned}
(1),(2) &\rightarrow (3),(4),(5),\\
(1),(3) &\rightarrow (2),(4),(5),\\
(2),(3),(5) &\rightarrow (1),(4),\\
(3),(4) &\rightarrow (1),(2),(5).
\end{aligned}
$$

For references to extensions of these results see Tian and Styan [2006]. They also add a new rank subtractivity condition of the form $\mathrm{rank}(\mathbf{I}_n - \mathbf{A}) = n - \sum_{i=1}^{k} \mathrm{rank}\,\mathbf{A}_i$.

8.82. Let $\mathbf{A}_i$ $(i = 1, 2, \ldots, k)$ be $p \times q$ matrices, and let $\mathbf{A} = \sum_{i=1}^{k} \mathbf{A}_i$. Consider the following conditions:

(1) $\mathbf{A}_i \mathbf{A}^- \mathbf{A}_i = \mathbf{A}_i$, $i = 1, 2, \ldots, k$.

(2) $\mathbf{A}_i \mathbf{A}^- \mathbf{A}_j = \mathbf{0}$ for all $i, j, i \neq j$.

(3) $\mathrm{rank}(\mathbf{A}_i \mathbf{A}^- \mathbf{A}_i) = \mathrm{rank}\,\mathbf{A}_i$, $i = 1, 2, \ldots, k$.

(4) $\sum_i \mathrm{rank}\,\mathbf{A}_i = \mathrm{rank}\,\mathbf{A}$,

for some weak inverse $\mathbf{A}^-$ of $\mathbf{A}$. Then

$$
\begin{aligned}
(1) &\rightarrow (2),(3),(4),\\
(2),(3) &\rightarrow (1),(4),\\
(4) &\rightarrow (1),(2),(3).
\end{aligned}
$$

If (1) or if (2) and (3) hold for some weak inverse $\mathbf{A}^-$, then (1), (2), and (3) hold for every weak inverse $\mathbf{A}^-$.

8.83. Let $\mathbf{A}_i$ $(i = 1, 2, \ldots, k)$ be an $n \times n$ matrix, let $\mathbf{A} = \sum_{i=1}^{k} \mathbf{A}_i$, and let $\mathbf{V}$ be a non-negative definite matrix.

(a) If $\mathbf{VAVAV} = \mathbf{VAV}$, then each of the following three conditions implies the other two.

(1) $\mathbf{VA}_i \mathbf{VA}_j \mathbf{V} = \mathbf{0}$ for all $i, j, i \neq j$, and $\mathrm{rank}(\mathbf{VA}_i \mathbf{VA}_i \mathbf{V}) = \mathrm{rank}(\mathbf{VA}_i \mathbf{V})$ for $i = 1, 2, \ldots, k$.

(2) $\mathbf{VA}_i \mathbf{VA}_i \mathbf{V} = \mathbf{VA}_i \mathbf{V}$ for $i = 1, 2, \ldots, k$.

(3) $\mathrm{rank}(\mathbf{VAV}) = \sum_{i=1}^{k} \mathrm{rank}(\mathbf{VA}_i \mathbf{V})$.

When the $\mathbf{A}_i$ are symmetric, condition (1) reduces to $\mathbf{VA}_i \mathbf{VA}_j \mathbf{V} = \mathbf{0}$ for all $i, j, i \neq j$.

(b) If $\mathbf{VA}_i \mathbf{VA}_i \mathbf{V} = \mathbf{VA}_i \mathbf{V}$ for all i and $\mathbf{VA}_i \mathbf{VA}_j \mathbf{V} = \mathbf{0}$ for all $i, j, i \neq j$, then:

(i) $\mathbf{VAVAV} = \mathbf{VAV}$.

(ii) $\mathrm{rank}(\mathbf{VAV}) = \sum_{i=1}^{k} \mathrm{rank}(\mathbf{VA}_i \mathbf{V})$.

A generalization of the above results involving rectangular matrices and an arbitrary rectangular $\mathbf{V}$ is given by Tian and Styan [2006]. For related results see Tian and Styan [2005].

8.84. Let $\mathbf{A}$ be an $n \times n$ symmetric idempotent matrix, and let $\mathbf{B}$ be a non-negative definite $n \times n$ matrix. If $\mathbf{I}_n - \mathbf{A} - \mathbf{B}$ is non-negative definite, then $\mathbf{AB} = \mathbf{BA} = \mathbf{0}$.

8.85. Let $\mathbf{A}_i$ be a symmetric idempotent $n \times n$ matrix of rank r_i $(i = 1, 2, \ldots, k)$, and let $\mathbf{A}_{k+1}$ be an $n \times n$ non-negative definite matrix such that $\mathbf{I}_n = \sum_{i=1}^{k+1} \mathbf{A}_i = \mathbf{A} + \mathbf{A}_{k+1}$, say. Then:

(a) $\mathbf{A}_i \mathbf{A}_j = \mathbf{0}$ for all $i, j = 1, 2, \ldots, k+1, i \neq j$.

(b) $\mathbf{A}_{k+1}$ is symmetric and idempotent of rank $n - \sum_{i=1}^{k} r_i$.

8.86. Let $\mathbf{A} = \sum_{i=1}^{k} \mathbf{A}_i$, where each $\mathbf{A}_i$ is an $n \times n$ non-negative definite matrix $(i = 1, 2, \ldots, k)$, and let $\mathbf{A}^2 = \mathbf{A}$. If

$$\text{trace } \mathbf{A} \leq \text{trace}(\sum_{i=1}^{k} \mathbf{A}_i^2),$$

then:

(a) $\mathbf{A}_i^2 = \mathbf{A}_i$ for $i = 1, 2, \ldots, k$.

(b) $\mathbf{A}_i \mathbf{A}_j = \mathbf{0}$ for all $i, j, i \neq j$.

8.87. Let $\mathbf{A}_i$ $(i = 1, 2, \ldots, k)$ be symmetric idempotent matrices such that

$$\mathbf{A}_i \mathbf{A}_j = \mathbf{0}, \text{ all } i, j, (i \neq j), \quad \text{and} \quad \sum_{i=1}^{k} \mathbf{A}_i = \mathbf{I}_n.$$

Then, for positive α_i $(i = 1, 2, \ldots, k)$, $\sum_{i=1}^{k} \alpha_i \mathbf{A}_i$ is positive definite.

8.88. Let $\mathbf{A}_i$ $(i = 1, 2, \ldots, k)$ be symmetric idempotent matrices such that $\mathbf{A}_i \mathbf{A}_j = \mathbf{0}$ for all $i, j, i \neq j$, and let $\alpha_i, i = 0, 1, \ldots, k$, be positive scalars. Then:

(a) $\mathbf{V} = \alpha_0 \mathbf{I}_n + \sum_{i=1}^{k} \alpha_i \mathbf{A}_i$ is positive definite.

(b) $\mathbf{V}^{-1} = \beta_0 \mathbf{I}_n + \sum_{i=1}^{k} \beta_i \mathbf{A}_i$, where

$$\beta_0 = \alpha_0^{-1} \quad \text{and} \quad \beta_i = \frac{-\alpha_i}{\alpha_0(\alpha_0 + \alpha_i)}, \quad i = 1, 2, \ldots, k.$$

8.89. Let $\mathbf{A}$ be any $n \times n$ symmetric matrix of rank r with nonzero eigenvalues λ_i $(i = 1, 2, \ldots, r)$. Then, since $\mathbf{A}$ is diagonalizable, $\mathbf{A}$ can be expressed in the form (cf. 16.17)

$$\mathbf{A} = \sum_{i=1}^{r} \lambda_i \mathbf{E}_i,$$

where, for each $i = 1, 2, \ldots, r$, $\mathbf{E}_i$ is symmetric and idempotent and $\mathbf{E}_i \mathbf{E}_j = \mathbf{0}$ for all $i, j, i \neq j$. If $\mathbf{A}$ is also idempotent, then

$$\mathbf{A} = \sum_{i=1}^{r} \mathbf{E}_i.$$

8.90. Let $\mathbf{A}_i$ be an $n \times n$ symmetric idempotent matrix of rank r_i $(i = 1, 2, \ldots, k)$ such that $\sum_{i=1}^{k} \mathbf{A}_i = \mathbf{I}_n$, and let $\mathbf{C}_i$ $(i = 1, 2, \ldots, k)$ be a $p \times p$ square matrix (possibly complex). If

$$\mathbf{\Omega}_1 = \sum_{i=1}^{k}(\mathbf{A}_i \otimes \mathbf{C}_i) \quad \text{and} \quad \mathbf{\Omega}_2 = \sum_{i=1}^{k}(\mathbf{C}_i \otimes \mathbf{A}_i)$$

are $np \times np$, then:

(a) From (8.78) we have $\mathbf{A}_i \mathbf{A}_j = \mathbf{0}$ for all $i \neq j$, and $\sum_{i=1}^{k} r_i = n$.

(b) The eigenvalues of $\mathbf{\Omega}_1$ and $\mathbf{\Omega}_2$ are the eigenvalues of $\mathbf{C}_1, \ldots, \mathbf{C}_k$ with respective algebraic multiplicities $r_1, \ldots, r_k$.

(c) $\det \mathbf{\Omega}_1 = \det \mathbf{\Omega}_2 = \prod_{i=1}^{k}(\det \mathbf{C}_i)^{r_i}$.

(d) The matrices $\mathbf{\Omega}_1$ and $\mathbf{\Omega}_2$ are nonsingular if and only if all the $\mathbf{C}_i$ $(i = 1, 2, \ldots, k)$ are nonsingular, in which case

$$\mathbf{\Omega}_1^{-1} = \sum_{i=1}^{k}(\mathbf{A}_i \otimes \mathbf{C}_i^{-1}) \quad \text{and} \quad \mathbf{\Omega}_2^{-1} = \sum_{i=1}^{k}(\mathbf{C}_i^{-1} \otimes \mathbf{A}_i).$$

This result is used in multivariate error component analysis.

Proofs. Section 8.6.2

8.77. Harville [1997: 435] and Schott [2005: 398].

8.79. Graybill [1983: 421] and Schott [2005: 401].

8.80. Harville [1997: 435–438].

8.81. Anderson and Styan [1982: 3].

8.82. Anderson and Styan [1982: 4].

8.83. Harville [1997: 439].

8.84. $\mathbf{C} = \mathbf{A}(\mathbf{I}_n - \mathbf{A} - \mathbf{B})\mathbf{A} = -\mathbf{A}\mathbf{B}\mathbf{A} = \mathbf{0}$ as $\mathbf{C}$ is non-negative definite. Then $\mathbf{A}\mathbf{B}^{1/2} = \mathbf{0}$ and $\mathbf{A}\mathbf{B} = \mathbf{0}$.

8.85. Quoted by Graybill [1983: : 423]. For (a) we consider $\mathbf{I}_n - \mathbf{A}_i - \mathbf{A}_j = (\mathbf{A} - \mathbf{A}_i - \mathbf{A}_j) + \mathbf{A}_k$, which is non-negative definite for all $i, j = 1, 2, \ldots, k+1$, $(i \neq j)$, as each $\mathbf{A}_i$ is non-negative definite, and then use (8.84). For (b), $\mathbf{A}_{k+1} = \mathbf{I}_n - \mathbf{A}$, where $\mathbf{A}$ is idempotent with rank $\mathbf{A} = \text{trace } \mathbf{A}$.

8.86–8.88. Graybill [1983: 423, 425–426].

8.90. Magnus [1982: 242, 270].

8.6.3 Products of Idempotent Matrices

8.91. Every singular $n \times n$ matrix can be written as the product of idempotent matrices.

8.92. If $\mathbf{A}$ and $\mathbf{B}$ are $n \times n$ idempotent matrices, then $\mathbf{AB}$ is idempotent if $\mathbf{AB} = \mathbf{BA}$.

Proofs. Section 8.6.3.

      8.91. Ballantyne [1978].

      8.92. Schott [2005: 398].

8.7 TRIPOTENT MATRICES

Definition 8.17. An $n \times n$ matrix is said to be *tripotent* if $\mathbf{A}^3 = \mathbf{A}$. A nonsingular tripotent matrix $\mathbf{A}$ is called a *involutionary matrix* and satisfies $\mathbf{A}^2 = \mathbf{I}_n$. An idempotent matrix is also tripotent.

8.93. (General Properties) Let $\mathbf{A}$ be an $n \times n$ tripotent matrix.

(a) rank $\mathbf{A} = \text{trace}(\mathbf{A}^2)$.

(b) The eigenvalues of $\mathbf{A}$ are ± 1 or 0. If n_1 are equal to $+1$, n_2 equal to -1 and n_3 equal to 0, then:

    (i) $\frac{1}{2}\text{trace}(\mathbf{A}^2 + \mathbf{A}) = n_1$.

    (ii) $\frac{1}{2}\text{trace}(\mathbf{A}^2 - \mathbf{A}) = n_2$.

    (iii) $\text{trace}(\mathbf{I}_n - \mathbf{A}^2) = n_3$.

    (iv) $\text{trace}\,\mathbf{A} = n_1 - n_2$.

(c) $\mathbf{A}$ is equal to a weak inverse of itself if and only if $\mathbf{A}$ is tripotent.

(d) If $\mathbf{A}$ is nonsingular, then:

    (i) $\mathbf{A}^{-1} = \mathbf{A}$.

    (ii) $\mathbf{A}^2 = \mathbf{I}_n$.

    (iii) $(\mathbf{A} + \mathbf{I}_n)(\mathbf{A} - \mathbf{I}_n) = \mathbf{0}$.

(e) If $\mathbf{T}$ is orthogonal, then $\mathbf{T}'\mathbf{AT}$ is tripotent.

(f) If $\mathbf{R}$ is nonsingular, then $\mathbf{R}^{-1}\mathbf{AR}$ is tripotent.

(g) $\mathbf{A}^2$ is idempotent.

(h) $-\mathbf{A}$ is tripotent.

8.94. Let $\mathbf{A}$ and $\mathbf{B}$ be $n \times n$ matrices.

(a) If $\mathbf{A}$ is symmetric, then $\mathbf{A}$ is tripotent if and only if its eigenvalues can only take the values $+1$, -1, or 0.

(b) If $\mathbf{A}$ is symmetric, then $\mathbf{A}$ is tripotent if and only if there exists two symmetric $n \times n$ idempotent matrices $\mathbf{C}$ and $\mathbf{D}$ such that $\mathbf{A} = \mathbf{C} - \mathbf{D}$ and $\mathbf{CD} = \mathbf{0}$. These two matrices are unique with $\mathbf{C} = \frac{1}{2}(\mathbf{A}^2 + \mathbf{A})$ and $\mathbf{D} = \frac{1}{2}(\mathbf{A}^2 - \mathbf{A})$. This result has been generalized by Baksalary et al. [2002], and they give conditions for when a linear combination of an idempotent and tripotent matrices is idempotent.

(c) $\mathbf{A}$ is tripotent if and only if $\mathbf{A}^2$ is idempotent.

(d) If $\mathbf{A}$ is symmetric, then $\mathbf{A}$ is tripotent if and only if rank $\mathbf{A} = \operatorname{rank}(\mathbf{A} + \mathbf{A}^2) + \operatorname{rank}(\mathbf{A} - \mathbf{A}^2)$.

(e) If $\mathbf{A}$ and $\mathbf{B}$ are symmetric idempotent matrices and $\mathbf{AB} = \mathbf{BA}$, then $\mathbf{A} - \mathbf{B}$ is a symmetric tripotent matrix.

8.95. Let $\mathbf{A}_i$ $(i = 1, 2, \ldots, k)$ be square matrices (not necessarily symmetric), and let $\mathbf{A} = \sum_{i=1}^{k} \mathbf{A}_i$. Consider the following conditions:

(1) $\mathbf{A}_i^3 = \mathbf{A}_i$, $i = 1, 2, \ldots, k$.

(2) $\mathbf{A}_i \mathbf{A}_j = \mathbf{0}$, for all $i \neq j$.

(3) $\mathbf{A}^3 = \mathbf{A}$.

(4) $\sum_i \operatorname{rank} \mathbf{A}_i = \operatorname{rank} \mathbf{A}$.

(5) $\mathbf{A}_i \mathbf{A} = \mathbf{A}_i^2$, $i = 1, 2, \ldots, k$.

(6) $\mathbf{A}_i^2 \mathbf{A} = \mathbf{A}_i$, $i = 1, 2, \ldots, k$.

(7) $\mathbf{A}_i \mathbf{A} = \mathbf{A} \mathbf{A}_i$, $i = 1, 2, \ldots, k$.

Then (1) and (2) hold if and only if (3), (4), and (5) hold. Condition (5) may be replaced by (6) or (7). Anderson and Styan [1982] prove the above result and a similar result for symmetric matrices. An extension is also given to r-potent matrices, which have the property that $\mathbf{A}^r = \mathbf{A}$, where r is a positive integer.

Proofs. Section 8.7.

8.93. Graybill [1983: section 12.4].

8.94a–b. Graybill [1983: 432].

8.94c. If $\mathbf{A}^3 = \mathbf{A}$, then $(\mathbf{A}^2)^2 = \mathbf{A}^2$. Conversely, if $\mathbf{A}^2$ is symmetric and idempotent, its eigenvalues are 0 and there exists orthogonal $\mathbf{T}$ such that

$$\mathbf{\Lambda}^2 = \mathbf{T}'\mathbf{A}\mathbf{T}\mathbf{T}'\mathbf{A}\mathbf{T} = \begin{pmatrix} \mathbf{I}_r & \mathbf{0} \\ \mathbf{0} & \mathbf{0} \end{pmatrix},$$

and the eigenvalues of $\mathbf{A}$ are $0, \pm 1$; then use (a).

8.94d. Anderson and Styan [1982: 13].

8.94e. We multiply out $(\mathbf{A} - \mathbf{B})^3$.

8.8   IRREDUCIBLE MATRICES

Definition **8.18.** An $n \times n$ matrix $\mathbf{A}$ is said to be *reducible* if and only if, by permuting a set of rows and the corresponding set of columns, $\mathbf{A}$ can be transformed to a matrix of the form

$$\mathbf{B} = \begin{pmatrix} \mathbf{B}_{11} & \mathbf{0} \\ \mathbf{B}_{12} & \mathbf{B}_{22} \end{pmatrix},$$

or equivalently of the form

$$\mathbf{B} = \begin{pmatrix} \mathbf{B}_{11} & \mathbf{B}_{12} \\ \mathbf{0} & \mathbf{B}_{22} \end{pmatrix},$$

where $\mathbf{B}_{11}$ and $\mathbf{B}_{22}$ are square matrices, i.e., there exists a permutation matrix $\mathbf{\Pi}$ such that $\mathbf{\Pi}\mathbf{A}\mathbf{\Pi}' = \mathbf{B}$. When $n = 1$ we have $\mathbf{A} = 0$. A matrix that is not reducible is said to be *irreducible*.

8.96. Given $\mathrm{mod}(\mathbf{A}) = (|a_{ij}|)$, where $\mathbf{A}$ is a real $n \times n$ matrix (cf. Section 9.1.2), then $\mathbf{A}$ is irreducible if and only if $(\mathbf{I}_n + \mathrm{mod}(\mathbf{A}))^{n-1} > \mathbf{0}$ (i.e., every element is positive) or, equivalently, if $[\mathbf{I}_n + \mathcal{I}(\mathbf{A})]^{n-1} > \mathbf{0}$, where $\mathcal{I}(\mathbf{A})$ is the indicator matrix of $\mathbf{A}$ (each nonzero element is replaced by 1).

8.97. Let $\mathbf{A}$ be any $n \times n$ real matrix.

(a) If $\mathbf{A}$ has no zero elements, then it is irreducible.

(b) If $\mathbf{A}$ has zero diagonal elements and nonzero off-diagonal elements, then $\mathbf{A}$ is irreducible.

(c) If $\mathbf{A}$ is reducible, it must have at least $n - 1$ elements equal to zero.

(d) if $\mathbf{A}$ has at least one row (column) of zeros, then $\mathbf{A}$ is reducible.

8.98. Let $\mathbf{A}$ be an $n \times n$ irreducible real matrix and let R_i be the sum of the absolute values of the off-diagonal elements of the ith row and C_j the same for the jth column. Suppose that either $|a_{ii}| \geq R_i$ for $i = 1, \ldots, n$ with $|a_{ii}| > R_i$ for at least one value of i, or $|a_{jj}| \geq C_j$ for $j = 1, \ldots, n$ with $|a_{jj}| > C_j$ for at least one value of j. Then $\mathbf{A}$ is nonsingular.

8.99. An $n \times n$ $(n \geq 2)$ matrix $\mathbf{A} = (a_{ij})$ is reducible if $a_{ij} = 0$ for $i \in S$ and $j \notin S$ for some nonempty proper subset S of $\{1, 2, \ldots, n\}$.

8.100. The forward shift permutation matrix $\mathbf{\Pi}_0 = (\mathbf{e}_n, \mathbf{e}_1, \ldots, \mathbf{e}_{n-1})$ is irreducible.

8.101. A permutation matrix is irreducible if and only if it is permutation similar (cf. Definition 8.3 below (8.17)) to a forward shift permutation matrix.

8.102. An $n \times n$ permutation matrix is irreducible if and only if its eigenvalues are $1, \omega, \ldots, \omega^{n-1}$, where ω is the nth primitive root of unity.

Proofs. Section 8.8.

8.96. Horn and Johnson [1985: 361].

8.97–8.98. Graybill [1983: 264].

8.99. Bapat and Raghavan [1997: 2].

8.100–8.102. Zhang [1999: 124–125].

8.9   TRIANGULAR MATRICES

Definition 8.19. A matrix is *lower-triangular* if the elements above the main diagonal are all zero. The transpose of this is said to be *upper-triangular*. A triangular matrix need not be square. A *unit triangular* matrix is a triangular matrix with unit diagonal elements, and a *strictly triangular matrix* is a triangular matrix with zero diagonal elements.

8.103. (Basic Properties)

(a) The determinant of a square triangular matrix is the product of the diagonal elements.

(b) The eigenvalues of a square triangular matrix are the diagonal elements.

(c) The inverse of a nonsingular lower (respectively upper) triangular matrix is a lower (respectively upper) triangular matrix.

(d) The product of a finite number of square lower (respectively upper) triangular matrices of the same order is a lower (respectively upper) triangular matrix.

(e) The product of two square unit upper (respectively lower) triangular matrices is unit upper (respectively lower) triangular.

(e) If $\mathbf{B}$ is an $n \times n$ triangular matrix with inverse $\mathbf{C} = \mathbf{B}^{-1}$, then $b_{ii}c_{ii} = 1$ for $i = 1, 2, \ldots, n$.

(f) From (e), the inverse of a nonsingular unit triangular matrix is also unit triangular.

8.104. If $\mathbf{K}$ is a real lower (upper) triangular matrix and if $\mathbf{K}'\mathbf{K} = \mathbf{K}\mathbf{K}'$, then $\mathbf{K}$ is a diagonal matrix.

8.105. (Factorization) Let $\mathbf{A}$ be a real square matrix such that every leading principal minor (excluding $\mathbf{A}$ itself) is nonzero.

(a) Then $\mathbf{A}$ can be written as the product of a real lower-triangular matrix $\mathbf{L}$ and a real upper-triangular matrix $\mathbf{U}$, that is,

$$\mathbf{A} = \mathbf{LU}.$$

Furthermore, if each of the diagonal elements of $\mathbf{L}$ (or $\mathbf{U}$) is set equal to unity, then the two triangular matrices are unique. It should be noted that $\mathbf{A}$ does not need to be square to have such a factorization, and the reader is referred to Section 16.4 for further details.

(b) If $\mathbf{A}$ is also symmetric, then there exists a real upper-triangular matrix $\mathbf{U}$ and a diagonal matrix $\mathbf{D}$ with diagonal elements equal to ± 1 such that

$$\mathbf{A} = \mathbf{U}'\mathbf{DU}.$$

8.106. Every real square matrix $\mathbf{A}$ is similar to a triangular matrix (either upper or lower) whose diagonal elements are the eigenvalues of $\mathbf{A}$, that is, there exists a nonsingular matrix $\mathbf{R}$ (not necessarily real) such that $\mathbf{R}^{-1}\mathbf{AR} = \mathbf{K}$, where $\mathbf{K}$ is triangular (and not necessarily real). If the eigenvalues of $\mathbf{A}$ are real, then $\mathbf{R}$ and $\mathbf{K}$ are both real (cf. 16.1e).

8.107. If $\mathbf{A}$ is a real $n \times n$ matrix with real eigenvalues, then there exists an orthogonal matrix $\mathbf{T}$ such that $\mathbf{T}'\mathbf{AT}$ is upper-triangular with diagonal elements the eigenvalues of $\mathbf{A}$ (cf. 16.37b).

8.108. (Block Triangular Matrices) An upper block triangular matrix takes the form

$$\mathbf{A} = \begin{pmatrix} \mathbf{A}_{11} & \mathbf{A}_{12} & \cdots & \mathbf{A}_{1p} \\ 0 & \mathbf{A}_{22} & \cdots & \mathbf{A}_{2p} \\ . & . & \cdots & . \\ 0 & 0 & \cdots & \mathbf{A}_{pp} \end{pmatrix},$$

where the diagonal blocks are all square matrices of possibly different sizes. We have that

$$\det \mathbf{A} = \prod_{i=1}^{p} \det \mathbf{A}_{ii}.$$

Thus $\mathbf{A}$ is nonsingular if and only if all the $\mathbf{A}_{ii}$ are nonsingular. In this case $\mathbf{A}^-$ is also upper block triangular. An algorithm for computing the inverse is given by Harville [1997: 94]. Similar results apply for lower block triangular matrices, as the inverse is also lower block triangular.

Proofs. Section 8.9.

8.103a. Simply expand the determinant by the first row or column depending on whether the matrix is lower- or upper-triangular, respectively.

8.103b. Follows from (a).

8.103c. We use the identity $\mathbf{A}\mathbf{A}^{-1} = \mathbf{I}_n$.

8.103d. Prove for just two matrices first.

8.103e. Use $\mathbf{BC} = \mathbf{I}_n$.

8.104. Graybill [1983: 212].

8.105. Graybill [1983: 207, 210].

8.106. Quoted by Graybill [1983: 211–212] and proved by Rao and Bhimasankaram [2000: 288–289].

8.107. Muirhead [1982: 587].

8.10 HESSENBERG MATRICES

Definition 8.20. An $n \times n$ matrix $\mathbf{A}$ is said to be an *upper Hessenberg* matrix if all its elements below the subdiagonal are zero (i.e., $a_{ij} = 0$ for $i > j + 1$). Its transpose is called a *lower Hessenberg* matrix. Upper Hessenberg matrices play an important role in the QR decomposition (Meyer [2000a: 536–538]). Many eigenvalue algorithms reduce their input to a Hessenberg form as a first step, and the latter play a similar role in the Schur decomposition (Golub and Van Loan [1996: section 7.4). Hessenberg matrices appear elsewhere in this book.

8.11   TRIDIAGONAL MATRICES

Definition 8.21. An $n \times n$ matrix $\mathbf{A}$ is *tridiagonal* if all its elements are zero except those in the middle three diagonals, (i.e., $a_{ij} \neq 0$ if $|i - j| \leq 1$ and $a_{ij} = 0$ if $|i - j| > 1$). Tridiagonal matrices play a role in matrix decompositions and factorizations—for example, (16.43), (16.45), and (16.46b).

8.109. If $\mathbf{A} = (a_{ij})$ is tridiagonal, then expanding $c_n(\lambda) = \det(\lambda \mathbf{I}_n - \mathbf{A})$ by the last column we find that

$$
\begin{aligned}
c_0(\lambda) &= 1, \quad c_1(\lambda) = (\lambda - a_{11}) \quad \text{and} \\
c_i(\lambda) &= (\lambda - a_{ii})c_{i-1}(\lambda) - a_{i,i-1}a_{i-1,i}c_{i-2}(\lambda), \qquad i = 2, 3, \ldots, n.
\end{aligned}
$$

8.110. Suppose that the $n \times n$ tridiagonal matrix $\mathbf{A}$ is given by

$$
\mathbf{A} = \begin{pmatrix}
a & b & 0 & \cdot & 0 & 0 & 0 \\
c & a & b & \cdot & 0 & 0 & 0 \\
0 & c & a & \cdot & 0 & 0 & 0 \\
\cdot & \cdot & \cdot & \cdot & \cdot & \cdot & \cdot \\
0 & 0 & 0 & \cdot & c & a & b \\
0 & 0 & 0 & \cdot & 0 & c & a
\end{pmatrix},
$$

where a, b, and c are real or complex. This matrix is both a Toeplitz matrix and a regular circulant.

(a) Then

$$
\det \mathbf{A} = \begin{cases}
a^n & \text{if} \quad bc = 0, \\
(n+1)(a/2)^n & \text{if} \quad a^2 = 4bc, \\
(\alpha^{n+1} - \beta^{n+1})/(\alpha - \beta) & \text{if} \quad a^2 \neq 4bc,
\end{cases}
$$

where

$$
\alpha = \frac{a + \sqrt{a^2 - 4bc}}{2} \quad \text{and} \quad \beta = \frac{a - \sqrt{a^2 - 4bc}}{2}.
$$

(b) If a is real and $bc > 0$, the eigenvalues of $\mathbf{A}$ are

$$
\lambda_j = a + 2\sqrt{bc}\cos(j\pi/(n+1)), \quad j = 1, 2, \ldots, n.
$$

(c) Let $b = c$ so that $\mathbf{A}$ is symmetric. Then $\mathbf{A}$ is positive definite if and only if the eigenvalues are positive (i.e., $a + 2b\cos(j\pi/(n+1)) > 0$ for $j = 1, 2, \ldots, n$). A sufficient condition is $a > 0$ and $|b/a| \leq \frac{1}{2}$.

(d) If $\mathbf{A}$ is positive definite and $b \neq 0$, then $\mathbf{B} = \mathbf{A}^{-1}$ is given by $b_{ij} = b_{ji}$ for $i > j$ and

$$
b_{ij} = \frac{(1 - \gamma^{2n-2j+2})(\gamma^{j+i+1} - \gamma^{j-i+1})}{(b/a)(1 - \gamma^2)(1 - \gamma^{2n+2})} \quad \text{for } i \leq j,
$$

where $\gamma = (\frac{b}{2a})(\sqrt{1 - 4(b/a)^2} - 1)$.

8.111. Given $\mathbf{A}$ in (8.110), with a real and $bc > 0$, then $\mathbf{A}$ has real eigenvectors.

8.112. The tridiagonal matrix

$$
\mathbf{A} = \begin{pmatrix}
0 & 1 & 0 & \cdot & 0 & 0 & 0 \\
-c_n & 0 & 1 & \cdot & 0 & 0 & 0 \\
0 & -c_{n-1} & 0 & \cdot & 0 & 0 & 0 \\
\cdot & & \cdot & & \cdot & \cdot & \\
0 & 0 & 0 & \cdot & -c_3 & 0 & 1 \\
0 & 0 & 0 & \cdot & 0 & -c_2 & c_1
\end{pmatrix},
$$

is called the *Schwartz matrix*. It often arises in stability analysis. $\mathbf{A}$ is positive stable if and only if $c_1 c_2 \cdots c_n > 0$ (cf. Section 8.14.4).

8.113. The inverse of a symmetric $n \times n$ matrix $\mathbf{B}$ is tridiagonal if and only if for $i = 2, 3, \ldots, n$,

$$
b_{ij}/b_{1j} = \theta_i, \quad b_{1j} \neq 0 \quad \text{for} \quad i \leq j \leq n.
$$

This condition means that all the elements on and to the right of the diagonal element in the ith row of $\mathbf{B}$ have a constant relation to the corresponding elements of the first row.

If $\mathbf{B}$ satisfies the above condition, then the inverse $\mathbf{B}^{-1} = (b^{ij})$ is given by

$$
\begin{aligned}
b^{11} &= -\theta_2 (b_{12} - \theta_2 b_{11})^{-1}, \\
b^{rr} &= -\frac{b_{r-1,r+1} - \theta_{r+1} b_{1,r-1}}{(b_{r-1,r} - \theta_r b_{1,r-1})(b_{r,r+1} - \theta_{r+1} b_{1,r})} \quad \text{for } r = 2, 3, \ldots, n-1, \\
b^{nn} &= -\frac{b_{1,n-1}}{b_{1,n}(b_{n-1,n} - \theta_n b_{1,n-1})}, \\
b^{r,r-1} &= b^{r-1,r} = (b_{r-1,r} - \theta_r b_{1,r-1})^{-1} \quad \text{for } r = 2, 3, \ldots, n, \\
b^{ij} &= 0 \quad \text{for } |i - j| > 1.
\end{aligned}
$$

8.114. (Applications of the Above Result) In all of the following cases we can simply confirm that $\mathbf{B}\mathbf{B}^{-1} = \mathbf{I}$.

(a) If

$$
\mathbf{B} = \begin{pmatrix}
n & n-1 & n-2 & n-3 & \cdots & 1 \\
n-1 & 2(n-1) & 2(n-2) & 2(n-3) & \cdots & 2 \\
n-2 & 2(n-2) & 3(n-2) & 3(n-3) & \cdots & 3 \\
n-3 & 2(n-3) & 3(n-3) & 4(n-3) & \cdots & 4 \\
\cdot & \cdot & \cdot & \cdot & \cdots & \cdot \\
1 & 2 & 3 & 4 & \cdots & n
\end{pmatrix},
$$

then $\mathbf{B}^{-1}$ is tridiagonal with $b^{ii} = 2/(n+1)$ for $i = 1, 2, \ldots, n$, and $b^{i-1,i} = b^{i,i+1} = -1/(n+1)$ for $i = 2, 3, \ldots, n-1$. ($\mathbf{B}$ is the variance matrix of ordered observations from a random sample of size n from a uniform distribution.)

(b) The *autocorrelation matrix* of an AR(1) time series is the symmetric Toeplitz matrix $\sigma^2 \mathbf{B}$, where

$$
\mathbf{B} = \begin{pmatrix}
1 & \rho & \rho^2 & \rho^3 & \cdots & \rho^{n-1} \\
\rho & 1 & \rho & \rho^2 & \cdots & \rho^{n-2} \\
\rho^2 & \rho & 1 & \rho & \cdots & \rho^{n-3} \\
\cdot & \cdot & \cdot & & \cdots & \cdot \\
\rho^{n-1} & \rho^{n-2} & \rho^{n-3} & \rho^{n-4} & \cdots & 1
\end{pmatrix}.
$$

and $|\rho| < 1$, that is, $b_{ij} = \rho^{|i-j|}$. Then

$$
\mathbf{B}^{-1} = (1 - \rho^2)^{-1}
\begin{pmatrix}
1 & -\rho & 0 & \cdots & 0 & 0 \\
-\rho & 1 + \rho^2 & -\rho & \cdots & 0 & 0 \\
0 & -\rho & 1 + \rho^2 & \cdots & 0 & 0 \\
\cdot & \cdot & & \cdots & \cdot & \cdot \\
0 & 0 & 0 & \cdots & -\rho & 1
\end{pmatrix}.
$$

Also $\mathbf{B}^{-1} = (1 - \rho^2)^{-1}\mathbf{L}'\mathbf{L}$, where

$$
\mathbf{L} =
\begin{pmatrix}
\sqrt{1 - \rho^2} & 0 & 0 & \cdots & 0 & 0 & 0 \\
-\rho & 1 & 0 & \cdots & 0 & 0 & 0 \\
\cdot & \cdot & \cdot & \cdots & \cdot & \cdot & \cdot \\
0 & 0 & 0 & \cdots & -\rho & 1 & 0 \\
0 & 0 & 0 & \cdots & 0 & -\rho & 1
\end{pmatrix}.
$$

Then $\det \mathbf{L} = \sqrt{1 - \rho^2}$ and $\det \mathbf{B} = (1 - \rho^2)^{n-1}$.

(c) If

$$
\mathbf{B} = \sigma^2
\begin{pmatrix}
1 & 1 & 1 & 1 & \cdots & 1 \\
1 & 2 & 2 & 2 & \cdots & 2 \\
1 & 2 & 3 & 3 & \cdots & 3 \\
1 & 2 & 3 & 4 & \cdots & 4 \\
\cdot & \cdot & \cdot & \cdot & \cdots & \cdot \\
1 & 2 & 3 & 4 & \cdots & n
\end{pmatrix},
$$

then

$$
\mathbf{B}^{-1} = (\sigma^2)^{-1}
\begin{pmatrix}
2 & -1 & 0 & \cdots & 0 \\
-1 & 2 & -1 & \cdots & 0 \\
0 & -1 & 2 & \cdots & 0 \\
\cdot & \cdot & \cdot & \cdots & \cdot \\
0 & 0 & 0 & \cdots & -1 \\
0 & 0 & 0 & \cdots & 2
\end{pmatrix}.
$$

(d) If

$$
\mathbf{B} =
\begin{pmatrix}
a_1^2 b_1 & a_2 a_1 b_1 & \cdots & a_n a_1 b_1 \\
a_1 a_2 b_1 & a_2^2 (b_1 + b_2) & \cdots & a_n a_2 (b_1 + b_2) \\
a_1 a_3 b_1 & a_2 a_3 (b_1 + b_2) & \cdots & a_n a_3 (b_1 + b_2 + b_3) \\
\cdot & \cdot & \cdots & \cdot \\
a_1 a_n b_1 & a_2 a_n (b_1 + b_2) & \cdots & a_n^2 (b_1 + \cdots + b_n)
\end{pmatrix},
$$

where none of the a_i or b_j is zero, then

$$
\mathbf{B}^{-1} =
\begin{pmatrix}
\frac{1}{a_1^2}\left(\frac{1}{b_1} + \frac{1}{b_2}\right) & -\frac{1}{a_1 a_2 b_2} & 0 & \cdots & 0 \\
-\frac{1}{a_1 a_2 b_2} & \frac{1}{a_2^2}\left(\frac{1}{b_2} + \frac{1}{b_3}\right) & -\frac{1}{a_2 a_3 b_3} & \cdots & 0 \\
0 & -\frac{1}{a_2 a_3 b_3} & \frac{1}{a_3^2}\left(\frac{1}{b_3} + \frac{1}{b_4}\right) & \cdots & 0 \\
\cdot & \cdot & \cdot & \cdots & \cdot \\
0 & 0 & 0 & \cdots & -\frac{1}{a_{n-1} a_n b_n} \\
0 & 0 & 0 & \cdots & \frac{1}{a_n^2 b_n}
\end{pmatrix}.
$$

Also, $\det \mathbf{B} = \Pi_{i=1}^n (a_i^2 b_i)$. A special case of the above result holds for the variance matrix of order statistics for a random sample of size n from an exponential distribution by setting

$$a_1 = a_2 = \cdots = a_n = 1 \quad \text{and} \quad b_i = 1/(n-i+1)^2 \quad (i = 1, 2, \ldots, n).$$

Proofs. Section 8.11.

8.109. Cullen [1997: 311]

8.110a. Zhang [1999: 101].

8.110b–d. Graybill [1983: 284–286].

8.111. Basilevsky [1983: 221–224].

8.112. Quoted by Horn and Johnson [1991: 111, exercise 9].

8.113. Ukita [1955] and Guttman [1955].

8.114a. Quoted by Graybill [1983: 200–201] and proved by Greenberg and Sarhan [1959].

8.114b. Graybill [1983: 201].

8.114c. We check that $\mathbf{B}\mathbf{B}^{-1} = \mathbf{I}$.

8.114d. Graybill [1983: 187–188, 202] and Roy and Sarhan[1956].

8.12   VANDERMONDE AND FOURIER MATRICES

8.12.1   Vandermonde Matrix

Definition 8.22. Let $a_1, a_2, \ldots, a_n$ be a set of real numbers, and let

$$\mathbf{V} = \begin{pmatrix} 1 & 1 & 1 & \cdots & 1 \\ a_1 & a_2 & a_3 & \cdots & a_n \\ a_1^2 & a_2^2 & a_3^2 & \cdots & a_n^2 \\ \vdots & \vdots & \vdots & \vdots & \vdots \\ a_1^{n-1} & a_2^{n-1} & a_3^{n-1} & \cdots & a_n^{n-1} \end{pmatrix}.$$

Then $\mathbf{V}$ and $\mathbf{V}'$ are called $n \times n$ *Vandermonde matrices*. The matrix $\mathbf{V}'$ arises in relation to the *Lagrange interpolation polynomial* (Meyer [2000a: 186]). Note that every $k \times k$ leading principal submatrix is also a Vandermonde matrix. A helpful notation on occasion is $\mathbf{V}(a_1, a_2, \ldots, a_n)$, which we shall use below.

8.115. $\det \mathbf{V} = \displaystyle\prod_{1 \le i < j \le n} (a_j - a_i)$.

8.116. If there are r distinct a_i values, then rank $\mathbf{A} = r$.

8.117. If $\mathbf{V}_n$ is an $n \times n$ Vandermonde matrix, then

$$\det \mathbf{V}_n = (a_n - a_1)(a_n - a_2) \cdots (a_n - a_{n-1}) \det \mathbf{V}_{n-1}.$$

8.118. Let $\mathbf{V}$ be an $n \times n$ Vandermonde matrix with distinct a_i (i.e., the inverse exists), and define

$$P_i(x) = \prod_{j=1:j\neq i}^{n} (x - a_j), \text{ for } i = 1, 2, \ldots, n$$

$$= \sum_{j=1}^{n} b_{ij} x^{j-1}, \text{ say.}$$

If $\mathbf{C} = \mathbf{V}^{-1}$, then $c_{ij} = b_{ij}/P_i(a_i)$.

8.119. (Extended Vandermonde Matrix) Let

$$\mathbf{V} = \begin{pmatrix} 1 & 1 & 1 & \cdots & 1 \\ a_1 & a_2 & a_3 & \cdots & a_n \\ a_1^2 & a_2^2 & a_3^2 & \cdots & a_n^2 \\ \vdots & \vdots & \vdots & \vdots & \vdots \\ a_1^{p-1} & a_2^{p-1} & a_3^{p-1} & \cdots & a_n^{p-1} \end{pmatrix}$$

be an $p \times n$ matrix ($n \geq p$). Then rank $\mathbf{V} = \min(p, d)$, where d is the number of distinct values of a_i. Note that $\mathbf{V}'$ is the regression matrix for a $(p-1)$th-degree polynomial regression model.

Proofs. Section 8.12.1.

8.115. Graybill [1983: 266], Schott [2005: 335–336], and Zhang [1999: 111].

8.116. Suppose $a_1, \ldots, a_r$ are distinct, then the leading principal $r \times r$ sub-matrix is nonsingular (by 8.115).

8.117. Harville [1997: section 13.6].

8.118. Graybill [1983: 270]. For another formulation of this result see Zhang [1999: 114].

8.119. Graybill [1983: 269].

8.12.2 Fourier Matrix

Definition 8.23. Let $\omega = e^{2\pi i/n} = \cos(2\pi/n) + i \sin(2\pi/n)$, where $i = \sqrt{-1}$, so that $\overline{\omega} = e^{-2\pi i/n}$; also $\omega^r = \cos(2\pi r/n) + i \sin(2\pi r/n)$. Then

$$\mathbf{F} = n^{-1/2} \mathbf{V}(1, \overline{\omega}, \overline{\omega}^2, \ldots, \overline{\omega}^{n-1})$$

is defined to be a *Fourier matrix*. Since $\mathbf{F}$ is symmetric, we have

$$\mathbf{F}^* = \overline{\mathbf{F}} = \frac{1}{\sqrt{n}} \begin{pmatrix} 1 & 1 & 1 & \cdots & 1 \\ 1 & \omega & \omega^2 & \cdots & \omega^{n-1} \\ 1 & \omega^2 & \omega^4 & \cdots & \omega^{2n-2} \\ \vdots & \vdots & \vdots & \vdots & \vdots \\ 1 & \omega^{n-1} & \omega^{2n-2} & \cdots & \omega^{(n-1)(n-1)} \end{pmatrix}.$$

The (i,j)th element is $n^{-1/2}\omega^{(i-1)(j-1)}$. Note that $\omega^n = 1$, $\overline{\omega} = \omega^{-1}$, $\omega^{2n-2} = \omega^{n-2}$, $\omega^{(n-1)(n-1)} = \omega$, $\omega^{-i} = \omega^{n-i}$, $\omega^r = \cos(2\pi r/n) + i\sin(2\pi r/n)$, and we have $\sum_{j=0}^{n-1}\omega^{ij} = 0$ if $n > 1$ and i is an integer such that $0 < i < n$ ($\omega^0 = 1$). Note that Graybill [1983: 271] interchanges ω and $\overline{\omega}$ in his notation so that $\mathbf{F}$ looks like $\mathbf{F}^*$ above, although it is not. Schott [2005: 331] interchanges $\mathbf{F}$ and $\mathbf{F}^*$, while Meyer [2000a: 357] uses $\overline{\omega} = \omega^{-1}$ and omits $n^{-1/2}$ as a multiplier.

8.120. Suppose $\mathbf{F}$ is defined above.

(a) $\mathbf{F}$ and $\mathbf{F}^*$ are both symmetric.

(b) $\mathbf{F}$ is unitary, i.e., $\mathbf{F}\mathbf{F}^* = \mathbf{I}_n$.

(c) $\mathbf{F}^{-1} = \overline{\mathbf{F}}$.

(d) $\mathbf{F}^2 = \mathbf{F}^{*2} = \mathbf{\Pi}$, where $\mathbf{\Pi}$ is the $n \times n$ permutation matrix

$$\mathbf{\Pi} = (\mathbf{e}_1, \mathbf{e}_n, \mathbf{e}_{n-1}, \ldots, \mathbf{e}_2),$$

and $\mathbf{e}_i$ is the ith column of $\mathbf{I}_n$.

(e) $\mathbf{F}^4 = \mathbf{F}^{*4} = \mathbf{I}_n$.

(f) $\mathbf{F}^*$ can be written as $n^{1/2}\mathbf{F}^* = \mathbf{C} + i\mathbf{S}$, where $\mathbf{C}$ and $\mathbf{S}$ are real matrices with

$$c_{ij} = \cos[2\pi(i-1)(j-1)/n] \quad \text{and} \quad s_{ij} = \sin[2\pi(i-1)(j-1)/n].$$

Also $\mathbf{CS} = \mathbf{SC}$ so that from $n\mathbf{F}\mathbf{F}^*$ we get $\mathbf{C}^2 + \mathbf{S}^2 = n\mathbf{I}_n$.

(g) The eigenvalues of $\mathbf{F}$ are ± 1 and $\pm i$ with appropriate algebraic multiplicities.

8.121. Let $c_n(\lambda) = \det(\lambda\mathbf{I}_n - \mathbf{F}^*)$. Then

$$
\begin{array}{lll}
n &\equiv& 0(\mathrm{mod}\ 4): \quad c_n(\lambda) = (\lambda-1)^2(\lambda-i)(\lambda+1)(\lambda^4-1)^{(n/4)-1}, \\
n &\equiv& 1(\mathrm{mod}\ 4): \quad c_n(\lambda) = (\lambda-1)(\lambda^4-1)^{(1/4)(n-1)}, \\
n &\equiv& 2(\mathrm{mod}\ 4): \quad c_n(\lambda) = (\lambda^2-1)(\lambda^4-1)^{(1/4)(n-2)}, \\
n &\equiv& 3(\mathrm{mod}\ 4): \quad c_n(\lambda) = (\lambda-i)(\lambda^2-1)(\lambda^4-1)^{(1/4)(n-3)}.
\end{array}
$$

Definition 8.24. Let $\mathbf{y}$ and $\mathbf{z}$ be n-dimensional vectors, and let $\mathbf{F}$ be an $n \times n$ Fourier matrix. Then $\mathbf{y} = \mathbf{F}\mathbf{z}$ is known as the *discrete Fourier transform* of the elements of $\mathbf{z}$. Typically, $\mathbf{z} = (z(0), z(1), \ldots z(n-1))'$, a times series sequence, or else $\mathbf{z} = \mathbf{z}(t)$ ($t = 0, 1, \ldots, T-1$), where $\mathbf{z}(t)$ is vector time series. The Fourier transform can be computed using a so-called *Fast Fourier Transform Algorithm* in which one reduces the calculation of the discrete Fourier transform for a long stretch of data to the calculation of successive transforms of shorter sets of data (cf. Brillinger [1975: section 3.5] and Meyer [2000a: section 5.8]). One can also make use of the fact that a Fourier matrix of order 2^n can be expressed as Kronecker products (Davis [1979: 36–37]).

Other applications of the transform include the convolution of two time series, computing filtered values from a transfer function, the estimation of the mixing distribution of a compound distribution, and the determination of the cumulative distribution of a random variable from its characteristic function (Brillinger [1975: 67–69]).

8.122. If $\mathbf{y} = \mathbf{F}\mathbf{z}$ then $\mathbf{z} = \mathbf{F}^{-1}\mathbf{y} = \mathbf{F}^*\mathbf{z}$.

8.123. Let $p(z) = a_o + a_1 z + a_2 z^2 + \cdots + a_{n-1} z^{n-1}$ be a polynomial of degree $n - 1$. It will be determined uniquely by specifying its values $p(z)$ at n distinct points z_k $(k = 1, 2, \ldots, n)$ in the complex plane. Suppose we select these points as the n roots of unity, namely, $1, \omega, \omega^2, \ldots, \omega^{n-1}$. Then

$$n^{1/2}\mathbf{F}^* \begin{pmatrix} a_0 \\ a_1 \\ \vdots \\ a_{n-1} \end{pmatrix} = \begin{pmatrix} p(1) \\ p(\omega) \\ \vdots \\ p(\omega^{n-1}) \end{pmatrix},$$

so that

$$\begin{pmatrix} a_0 \\ a_1 \\ \vdots \\ a_{n-1} \end{pmatrix} = n^{-1/2}\mathbf{F} \begin{pmatrix} p(1) \\ p(\omega) \\ \vdots \\ p(\omega^{n-1}) \end{pmatrix}.$$

This gives a relationship between the coefficients of $p(z)$ and its values.

Proofs. Section 8.12.2.

8.120. Davis [1979: 31–37] and Graybill [1983: 272–273, with corrections in (f)].

8.121. Carlitz [1959].

8.13   ZERO–ONE *(0,1)* MATRICES

A matrix whose elements are all 0 or 1 is called a *(0,1) matrix*. I have highlighted this topic as such matrices occur widely throughout statistics. Examples of such matrices are the permutation matrices in this chapter as well as the various vec-permutation and commutation matrices in Chapter 11. There is also the so-called *incidence* matrix discussed in (8.124) below, and there are Boolean matrices, both of which occur in combinatorial and graph theory. Zero-one matrices play an important role in the solution of equations with large sparse matrices (e.g., Duff et al. [1986]).

8.124. (Incidence Matrix)

(a) (Experimental Design)   The incidence matrix for a block design has a row for each treatment and a column for each block. Thus a 1 for the (i, j)th element of the matrix tells us that the ith treatment is applied to the jth block (cf. John and Williams [1995: chapter 1]).

(b) (Non-negative Matrix) As noted by Seneta [1981: 55], many properties of a non-negative matrix $\mathbf{A}$ depend only on the positions of the positive and zero elements within the matrix, and not on the actual size of the positive elements. Also, those positions will determine the corresponding positions in all powers $\mathbf{A}^k$, with k a positive integer. This means that in the investigation of the properties of irreducibility and primitivity, the classification of indices into essential and inessential (cf. Section 9.3), and the periodicity of indices that communicate with each other, depend only on the location of the positive elements of $\mathbf{A}$. Therefore given a non-negative matrix $\mathbf{A}$ (i.e., all its elements are non-negative), then the matrix obtained by replacing each positive element by 1 is called the *incidence matrix* of $\mathbf{A}$. Particular matrices for which incidence matrices have useful applications are stochastic and Leslie matrices. A related $(0,1)$ matrix is the *indicator* matrix, whereby the nonzero elements of any matrix are replaced by 1. For a non-negative matrix, the indicator matrix is the same as the incidence matrix.

8.125. Let $\mathbf{A}$ be an $n \times n$ *(0,1)* matrix. If $\mathbf{J}_n = \mathbf{1}_n \mathbf{1}_n'$ and

$$\mathbf{A}\mathbf{A}' = k\mathbf{I}_n + \mathbf{J}_n$$

for some positive integer k, then $\mathbf{A}$ is a normal matrix, that is, $\mathbf{A}\mathbf{A}' = \mathbf{A}'\mathbf{A}$.

8.126. If $\mathbf{A}$ and $\mathbf{B}$ are $n \times n$ *(0,1)* matrices such that $\mathbf{A}\mathbf{B} = \mathbf{J}_n - \mathbf{I}_n$, then $\mathbf{A}\mathbf{B} = \mathbf{B}\mathbf{A}$.

***Definition* 8.25.** (Boolean Matrix) The *binary Boolean algebra* $\mathcal{B}$ consists of the set $\{0,1\}$, together with the usual operations of addition and multiplication (i.e., $1 + 0 = 1$, $0 + 0 = 0$, $1 \times 0 = 0$, $0 \times 0 = 0$, $1 \times 1 = 1$), except that $1 + 1 = 1$. A *Boolean matrix* is a *(0,1)* matrix over $\mathcal{B}$. Boolean matrices have some properties that differ from matrices over $\mathbb{R}$; for example, the row rank need not equal the column rank. Some properties of Boolean matrices and Boolean vector spaces are given by Bapat and Raghavan [1997: section 5.6].

8.127. If $\mathbf{T}$ is the incidence matrix of the non-negative matrix $\mathbf{A}$, then $\mathbf{T}^k$ is the incidence matrix of $\mathbf{A}^k$ when $\mathbf{T}$ is a Boolean matrix.

Proofs. Section 8.13.

   8.125. Zhang [1999: 251].

   8.126. Zhang [1999: 252–253].

   8.127. Seneta [1981: 56].

8.14 SOME MISCELLANEOUS MATRICES AND ARRAYS

8.14.1 Krylov Matrix

***Definition* 8.26.** If $\mathbf{x} \in \mathbb{R}^n$ and $\mathbf{A}$ is an $n \times n$ matrix, then the matrix

$$(\mathbf{x}, \mathbf{A}\mathbf{x}, \mathbf{A}^2\mathbf{x}, \ldots, \mathbf{A}^{n-1}\mathbf{x})$$

is called a *Krylov matrix*. This matrix arises in the so-called Lanczos method of obtaining approximations for some eigenvalues and eigenvectors, especially for large sparse matrices (Slapničar [2007: chapter 42, 8]). The column space of the Krylov matrix is called *Krylov subspace*, and it is associated with the solution of linear equations (Greenbaum [2007: section 41.1]).

8.14.2   Nilpotent and Unipotent Matrices

***Definition* 8.27.** An $n \times n$ real or complex matrix is *nilpotent* if $\mathbf{A}^k = \mathbf{0}$ for some positive integer k, and is *unipotent* if $\mathbf{A}^2 = \mathbf{I}_n$. For example,

$$\mathbf{A} = \begin{pmatrix} \mathbf{I} & \mathbf{B} \\ \mathbf{0} & -\mathbf{I} \end{pmatrix}$$

is unipotent. For a nilpotent matrix, the smallest k such that $\mathbf{A}^k = \mathbf{0}$ is called the *index of nilpotency*.

8.128. The eigenvalues of a nilpotent matrix are all zero.

8.129. Let $\mathbf{A}$ be a real or complex $n \times n$ singular matrix with matrix index k (cf. Section 3.8) such that $\text{rank}(\mathbf{A}^k) = r$. Then there exists a nonsingular matrix $\mathbf{R}$ such that

$$\mathbf{R}^{-1}\mathbf{A}\mathbf{R} = \begin{pmatrix} \mathbf{C} & \mathbf{0} \\ \mathbf{0} & \mathbf{N} \end{pmatrix},$$

where $\mathbf{C}$ is a nonsingular $r \times r$ matrix and $\mathbf{N}$ is nilpotent with k its index of nilpotency.

8.130. If $\mathbf{A}$ and $\mathbf{B}$ are nilpotent matrices, then so is $\mathbf{A} + \mathbf{B}$.

8.131. Any Jordan block $\mathbf{J}_m(\lambda)$ (cf. Definition 16.2) can written as $\mathbf{J}_m(\lambda) = \lambda\mathbf{I}_m + \mathbf{A}_m$, where $\mathbf{A}_m$ is nilpotent as $(\mathbf{A}_m)^m = \mathbf{0}$.

  More generally, a Jordan matrix can be written as $\mathbf{J} = \mathbf{D} + \mathbf{N}$, where $\mathbf{D}$ is diagonal matrix whose main diagonal is the same as that of $\mathbf{J}$, and $\mathbf{N} = \mathbf{J} - \mathbf{D}$. Here $\mathbf{N}$ is nilpotent as $\mathbf{N}^k = \mathbf{0}$, where k is the order of the largest Jordan block in $\mathbf{J}$.

8.132. Any strictly upper-triangular $n \times n$ matrix $\mathbf{A}$ is nilpotent with index of nilpotency at most n (as $\mathbf{A}^n = \mathbf{0}$). If $r_j = \text{rank}(\mathbf{A}^j)$, then $r_{j+1} < r_j$ if $r_j > 0$.

Proofs. Section 8.14.2.

  8.128. $\mathbf{A}\mathbf{x} = \lambda\mathbf{x}$ implies that $\mathbf{x} \neq \mathbf{0}$ and $\mathbf{0} = \mathbf{A}^k\mathbf{x} = \lambda^k\mathbf{x}$.

  8.129. Meyer [2000a: 397].

  8.130. $\mathbf{A}^r = \mathbf{0}$ and $\mathbf{B}^s = \mathbf{0}$ for some r and s, which imply $(\mathbf{A} + \mathbf{B})^{r+s} = \mathbf{0}$.

  8.132. Abadir and Magnus [2005: 183].

8.14.3   Payoff Matrix

***Definition* 8.28.** Suppose we have a game consisting of two players I and II. At each stage of the game, Player I chooses a strategy j with probability y_j ($j = 1, 2, \ldots, n$), where $\sum_{j=1}^{n} y_j = 1$, and Player II independently chooses a strategy i with probability x_i ($i = 1, 2, \ldots, m$), where $\sum_{i=1}^{m} x_i = 1$. Player II then pays Player I the amount a_{ij}, or if a_{ij} is negative Player I pays Player II $(-a_{ij})$. The $m \times n$ matrix $\mathbf{A} = a_{ij}$ is called a *payoff matrix*. (Some authors reverse the roles of the two players so that Player I chooses a strategy i, etc.) The *expected income* to

Player I is $\sum_{j=1}^{n} a_{ij}y_j$, and the game is called a *matrix game*. Optimal strategies exist for each player, as is proved in (8.133) below.

Let $\mathbf{x} = (x_1, x_2, \ldots, x_m)'$ and $\mathbf{y} = (y_1, y_2, \ldots, y_n)'$. If $\mathbf{x}$ has more than one nonzero element, then the stratgy is called a *mixed* strategy. If all the elements are positive (i.e., $\mathbf{x} > \mathbf{0}$), the strategy is said to be *completely mixed*. A matrix game is called a *completely mixed game* if every optimal strategy $\mathbf{x}$ for Player II and $\mathbf{y}$ for Player I are completely mixed. For further details see Bapat and Raghaven [1997: chapter 1].

8.133. (Minimax Theorem—von Neumann) Let $\mathbf{A}$ be an $m \times n$ payoff matrix. There exists a unique constant v, called the *value of the matrix game* $\mathbf{A}$, and mixed strategies $\mathbf{x}$ for Player II and $\mathbf{y}$ for Player I such that

$$\sum_{j=1}^{n} a_{ij}y_j \geq v, \quad i = 1, 2, \ldots, m, \quad \text{and} \quad \sum_{i=1}^{n} a_{ij}x_i \leq v, \quad j = 1, 2, \ldots, n.$$

The strategy $\mathbf{x}$ is called an *optimal strategy* for Player II and $\mathbf{y}$ is called an optimal strategy for Player I.

8.134. Let v be the value of a matrix game $\mathbf{A}$, and suppose some optimal strategy of Player II is completely mixed. Then, for any optimal strategy $\mathbf{y}$ of Player I, $\mathbf{A}\mathbf{y} = v\mathbf{1}$.

8.135. Let the value of the $m \times n$ matrix game $\mathbf{A}$ be zero (i.e., $v = 0$), and suppose that every optimal strategy for Player II is completely mixed. Then $m - 1 \leq \text{rank}\,\mathbf{A} \leq n - 1$. If $\text{rank}\,\mathbf{A} = m - 1$, then the optimal strategy for Player II is unique.

8.136. Let $\mathbf{A}$ be an $n \times n$ matrix with cofactors A_{ij}. If the matrix game $\mathbf{A}$ is completely mixed, then $\sum_{i=1}^{n} \sum_{j=1}^{n} A_{ij}$ is nonzero and the value v of the game is given by

$$v = \det \mathbf{A}/(\sum_{i=1}^{n} \sum_{j=1}^{n} A_{ij}).$$

Proofs. Section 8.14.3.

8.133. Parthasarathy and Raghaven [1971].

8.134–8.136. Bapat and Raghaven [1997: 10–11, 14].

8.14.4 Stable and Positive Stable Matrices

Definition **8.29.** An $n \times n$ real or complex matrix $\mathbf{A}$ is said to be *stable* if every eigenvalue of $\mathbf{A}$ has a negative real part. The matrix is said to be *positive stable* if every eigenvalue has a positive real part. These concepts are related to the long-term equilibrium of a dynamical system, and are discussed in detail by Horn and Johnson [1991: chapter 2]. For a further discussion see Meyer [2000a: section 7.4].

8.137. $\exp(\mathbf{A}t) \to \mathbf{0}$ as $t \to \infty$ if and only if $\mathbf{A}$ is stable (cf. Section 19.6).

8.138. If $\mathbf{A}$ is an $n \times n$ real or complex matrix and

$$\Re(a_{ii}) < - \sum_{j=1:j\neq i}^{n} |a_{ij}|, \quad i = 1, 2, \ldots, n,$$

where $\Re e$ is the real part, then $\mathbf{A}$ is stable.

8.139. Suppose $\mathbf{A}$ is a positive stable matrix.

(a) $\mathbf{A}^{-1}$, $\mathbf{A}^*$ and $\mathbf{A}'$ are all positive stable.

(b) $\det \mathbf{A} > 0$.

(c) $\Re e(\text{trace } \mathbf{A}) > 0$.

(d) $\det(\mathbf{A}^k) > 0$ for any positive integer k.

8.140. If $\mathbf{A}$ is Hermitian positive definite, then $\mathbf{A}$ is positive stable.

8.141. (Lyapunov's Equation) Suppose $\mathbf{X}$, $\mathbf{A}$, and $\mathbf{C}$ are all $n \times n$ matrices such that $\mathbf{XA} + \mathbf{A}^*\mathbf{X} = \mathbf{C}$ (see also (13.17c) for further details).

(a) $\mathbf{A}$ is positive stable if and only if there exists an Hermitian positive definite solution $\mathbf{X}$ such that $\mathbf{C}$ is Hermitian positive definite.

(b) Suppose $\mathbf{X}$ and $\mathbf{C}$ are Hermitian and $\mathbf{C}$ is positive definite. Then $\mathbf{A}$ is positive stable if and only if $\mathbf{X}$ is positive definite.

(c) If $\mathbf{A}$ is positive stable, then given $\mathbf{C}$, there is a unique solution $\mathbf{X}$ to Lyapunov's equation. If $\mathbf{C}$ is Hermitian, then $\mathbf{X}$ is Hermitian, while if $\mathbf{C}$ is Hermitian positive definite, then $\mathbf{X}$ is Hermitian positive definite.

(d) A special case of the above is when $\mathbf{C} = \mathbf{I}_n$, which is Hermitian and positive definite.

Horn and Johnson [1991: 96–98] give a number of generalizations of the above theory.

Proofs. Section 8.14.4.

8.137. Horn and Johnson [1991: 92].

8.138. Marcus and Minc [1964: 159].

8.139. Horn and Johnson [1991: 93].

8.140. Horn and Johnson [1991: 95].

8.141. Horn and Johnson [1991: 96–98].

8.14.5 P-Matrix

***Definition* 8.30.** An $n \times n$ real matrix $\mathbf{A}$ is called a *P-matrix* if all its $k \times k$ principal minors are positive for $k = 1, 2, \ldots, n$.

8.142. Let $\mathbf{A}$ be an $n \times n$ P-matrix. Then:

(a) $\mathbf{A}'$ is also a P-matrix.

(a) $\mathbf{DA}$ and $\mathbf{AD}$ are also P-matrices, where $\mathbf{D}$ is a diagonal matrix with positive diagonal elements.

(b) Every principal submatrix of $\mathbf{A}$ is also a P-matrix.

(c) $a_{ii} > 0$ for $i = 1, 2, \ldots, n$.

(d) If $\mathbf{\Pi}$ is any $n \times n$ permutation matrix, then $\mathbf{\Pi}'\mathbf{A}\mathbf{\Pi}$ is a P-matrix.

(e) $\mathbf{A} + \mathbf{D}$ is a P-matrix, where $\mathbf{D}$ is a diagonal matrix with non-negative diagonal elements.

8.143. Let $\mathbf{A}$ be a real $n \times n$ matrix. Each of the following conditions is necessary and sufficient for $\mathbf{A}$ to be a P-matrix.

(1) For every $n \times 1$ vector $\mathbf{x}$, there is an element in $\mathbf{x}$ (say the qth) and the corresponding element in $\mathbf{y} = \mathbf{A}\mathbf{x}$ such that $x_q y_q > 0$.

(2) For every $\mathbf{x} \neq \mathbf{0}$, there exists a diagonal matrix $\mathbf{D}$, a function of $\mathbf{x}$, with positive diagonal elements such that $\mathbf{x}'\mathbf{D}\mathbf{A}\mathbf{x} > 0$.

(3) For every $\mathbf{x} \neq \mathbf{0}$, there exists a diagonal matrix $\mathbf{D}$, a function of $\mathbf{x}$, with non-negative diagonal elements such that $\mathbf{x}'\mathbf{D}\mathbf{A}\mathbf{x} > 0$.

(4) Every real eigenvalue of $\mathbf{A}$ and of each principal submatrix of $\mathbf{A}$ is positive.

Proofs. Section 8.14.5.

8.142. These results quoted by Graybill [1983: 376] follow directly from the definition.

8.143. Graybill [1983: 377] and Horn and Johnson [1991: 120].

8.14.6 Z- and M-Matrices

***Definition* 8.31.** An $n \times n$ real matrix $\mathbf{A} = (a_{ij})$ for which $a_{ij} \leq 0$ for all i, j, $i \neq j$ is called a *Z-matrix*. Note that if $\mathbf{B}$ is an ML-matrix (cf. Definition 9.11 above (9.43)) then $\mathbf{A} = -\mathbf{B}$ is a Z-matrix, and vice versa.

A Z-matrix $\mathbf{A}$ is called a *(nonsingular) M-matrix* if it is a nonsingular Z-matrix and $\mathbf{A}^{-1} \geq \mathbf{0}$ (i.e., has non-negative elements). This was the definition introduced by Ostrowski in 1937. An equivalent definition used by Horn and Johnson [1991: 113] is that $\mathbf{A}$ is an M-matrix if it is a Z-matrix and positive stable (cf. Section 8.14.4). I have included the word "nonsingular" to avoid ambiguity as definitions vary in the literature. For example, Bapat and Raghavan [1997: section 1.5] allow an M-matrix to be singular and use a different definition. For general references relating to M-matrices see Varga [1962] and Berman and Plemmons [1994]. Non-singular M-matrices arise in game theory (Bapat and Raghavan [1997: section 1.5]).

8.144. $\mathbf{A}$ is a Z-matrix if and only if $\mathbf{A} = s\mathbf{I}_n - \mathbf{B}$ for some $\mathbf{B} \geq \mathbf{0}$ and some real s.

8.145. Let $\mathbf{A}$ be an $n \times n$ Z-matrix such that $\mathbf{A} = \mathbf{LU}$, where $\mathbf{L}$ is lower-triangular and $\mathbf{U}$ is upper riangular, both with positive diagonal elements. Then:

(a) $\mathbf{A}$ has positive leading principal minors including $\det \mathbf{A}$ itself.

(b) $\mathbf{L}$ and $\mathbf{U}$ are nonsingular.

(c) The off-diagonal elements of both $\mathbf{L}$ and $\mathbf{U}$ are nonpositive.

(d) No element of $\mathbf{L}^{-1}$ or $\mathbf{U}^{-1}$ is negative, and the diagonal elements of $\mathbf{L}^{-1}$ and $\mathbf{U}^{-1}$ are all positive.

8.146. Let $\mathbf{A}$ be an $n \times n$ Z-matrix such that each real eigenvalue of $\mathbf{A}$ is positive. Let $\mathbf{B}$ be a Z-matrix such that $\mathbf{A} \leq \mathbf{B}$ (i.e., $a_{ij} \leq b_{ij}$ for all i, j). Then:

(a) $\mathbf{A}$ and $\mathbf{B}$ are nonsingular.

(b) $\mathbf{0} \leq \mathbf{B}^{-1} \leq \mathbf{A}^{-1}$ (i.e., $\mathbf{A}^{-1} - \mathbf{B}^{-1} \geq \mathbf{0}$).

(c) Each real eigenvalue of $\mathbf{B}$ is positive.

(d) $\det \mathbf{B} \geq \det \mathbf{A} > 0$.

8.147. (Equivalence of Definitions) Let $\mathbf{A}$ be a Z-matrix. Then $\mathbf{A}$ is an M-matrix if and only if $\Re e(\lambda) > 0$ (where $\Re e$ is the real part) for all eigenvalues λ, that is, if and only if $\mathbf{A}$ is stable.

8.148. Let $\mathbf{A}$ be a Z-matrix. Then each of the following conditions is necessary and sufficient for $\mathbf{A}$ to be an M-matrix.

(1) All principal minors of $\mathbf{A}$ are positive, including $\det \mathbf{A}$; that is, $\mathbf{A}$ is a P-matrix.

(2) The leading principal minors of $\mathbf{A}$ are all positive, including $\det \mathbf{A}$.

(3) Every real eigenvalue of $\mathbf{A}$ is positive.

(4) $\mathbf{A} + t\mathbf{I}_n$ is nonsingular for all $t \geq 0$.

(5) $\mathbf{A} + \mathbf{D}$ is nonsingular for every non-negative diagonal matrix $\mathbf{D}$.

(6) There exists an $\mathbf{x} > \mathbf{0}$ such that $\mathbf{Ax} > \mathbf{0}$.

(7) $\mathbf{Ax} \geq \mathbf{0}$ implies $\mathbf{x} \geq \mathbf{0}$.

8.149. If $\mathbf{A}$ is a Z-matrix, then it is a (nonsingular) M-matrix if and only if it can be expressed in the form $\mathbf{A} = s\mathbf{I}_n - \mathbf{B}$, where $\mathbf{B} \geq \mathbf{0}$ and $s > \rho(\mathbf{B})$, with $\rho(\mathbf{B})$ being the spectral radius of $\mathbf{B}$.

8.150. If $\mathbf{A}$ is an M-matrix, then so is every principal submatrix.

8.151. If $\mathbf{A}$ is an M-matrix, it is also a P-matrix.

8.152. Let $\mathbf{A}$ and $\mathbf{B}$ be $n \times n$ Z-matrices. If $\mathbf{A}$ is an M-matrix and $\mathbf{B} \geq \mathbf{A}$ (i.e., $b_{ij} \geq a_{ij}$ for all i, j), then:

(a) $\mathbf{B}$ is an M-matrix.

(b) $\mathbf{A}^{-1} \geq \mathbf{B}^{-1} \geq \mathbf{0}$.

(c) $\det \mathbf{B} \geq \det \mathbf{A} > 0$.

(d) The matrix $\mathbf{A}$ satisfies the Hadamard inequality

$$\det \mathbf{A} \leq a_{11} a_{22} \cdots a_{nn}.$$

(e) $\mathbf{A}^{-1} \mathbf{B} \geq \mathbf{I}_n$ and $\mathbf{B} \mathbf{A}^{-1} \geq \mathbf{I}_n$.

(f) $\mathbf{B}^{-1} \mathbf{A} \leq \mathbf{I}_n$ and $\mathbf{A} \mathbf{B}^{-1} \leq \mathbf{I}_n$.

(g) $\mathbf{A} \mathbf{B}^{-1}$ and $\mathbf{B}^{-1} \mathbf{A}$ are M-matrices

8.153. Let $\mathbf{A}$ be an M-matrix. Then there exists a positive eigenvalue of $\mathbf{A}$, λ_0 say, such that the real part of any eigenvalue of $\mathbf{A}$ is greater than or equal to λ_0.

Proofs. Section 8.14.6.

8.144. Horn and Johnson [1991: 113]. Take $c_{ij} = \max\{-a_{ij}, 0\}$ and $s \geq \max_i\{a_{ii}\}$ so that $\mathbf{A} = s\mathbf{I}_n - (\mathbf{C} + s\mathbf{I}_n - \mathrm{diag}(a_{11}, \ldots, a_{nn}))$.

8.145. Graybill [1983: 380].

8.146. Graybill [1983: 380–381].

8.147. Meyer [2000a: 626].

8.148. For further equivalent conditions and details see Horn and Johnson [1991: 114–115]. Some proofs are also given by Graybill [1983: section 11.3] and Meyer [2000a: 626]. A game theoretic proof for some of the results like these are given by Bapat and Raghavan [1991: 25–28].

8.149–8.150. Horn and Johnson [1991: 113] and Meyer [2000a: 626].

8.152. Graybill [1983: 386, (a)–(g) except (d)] and Horn and Johnson [1991: 117, (a)–(d)].

8.153. Quoted by Graybill [1983: 385].

8.14.7   Three-Dimensional Arrays

In nonlinear regression models, the expected value of a random response variable y_i is usually of the form $f_i(\mathbf{x}_i; \boldsymbol{\theta})$, and this leads to looking at $\partial f_i/\partial\theta_j\partial\theta_k$, which is a 3-dimensional array. Such arrays have been used for a wide variety of models including nonlinear models (cf. Seber and Wild [1989] and Wei [1997]) and multinomial models (e.g., Seber and Nyangoma [2000] and Wei [1997: section 7.2]).

Definition 8.32. Consider the $n \times p \times p$ array $\mathcal{W} = \{(\mathbf{w}_{rs})\}$ made up of a $p \times p$ array of n-dimensional vectors $\mathbf{w}_{rs}$ $(r, s = 1, 2, \ldots p)$. If w_{irs} is the ith element of $\mathbf{w}_{rs}$, then the matrix of ith elements $\mathbf{W}_i = (w_{irs})$ is called the ith *face* of $\mathcal{W}$. We now define two types of multiplication. Firstly, if $\mathbf{B}$ and $\mathbf{C}$ are $p \times p$ matrices, then

$$\mathcal{V} = \{(\mathbf{v}_{rs})\} = \mathbf{B}\mathcal{W}\mathbf{C}$$

denotes the array with ith face $\mathbf{V}_i = \mathbf{B}\mathbf{W}_i\mathbf{C}$, i.e.,

$$\mathbf{v}_{rs} = \sum_\alpha \sum_\beta b_{r\alpha} \mathbf{w}_{\alpha\beta} c_{\beta s}.$$

Secondly, if $\mathbf{D}$ is a $q \times n$ matrix, then we define square bracket multiplication by the equation

$$[\mathbf{D}][\mathcal{W}] = \{(\mathbf{D}\mathbf{w}_{rs})\},$$

where the right-hand side is a $q \times p \times p$ array.

We can also define trace $\mathcal{W}$, a vector with ith element trace $\mathbf{W}_i$, and vec $\mathcal{W}$, which is a $p^2 \times n$ matrix with ith column vec $\mathbf{W}_i$.

8.154. Using the above notation, we have the following.

(a) $[\mathbf{I}_n][\mathcal{W}] = \mathcal{W}$.

(b) $[\alpha\mathbf{B} + \beta\mathbf{C}][\mathcal{W}] = \alpha[\mathbf{B}][\mathcal{W}] + \beta[\mathbf{C}][\mathcal{W}]$.

(c) trace$[\mathbf{B}\mathcal{W}]$ = trace$[\mathcal{W}\mathbf{B}]$.

(d) $\mathbf{B}$ trace $\mathcal{W}$ = trace$([\mathbf{B}][\mathcal{W}])$.

(e) vec $([\mathbf{B}][\mathcal{W}])$ = (vec $\mathcal{W}$)$\mathbf{B}'$.

(f) $[\mathbf{D}][\mathbf{B}\mathcal{W}\mathbf{C}] = \mathbf{B}[\mathbf{D}][\mathcal{W}]\mathbf{C}$.

(g) vec $([\mathbf{B}\mathcal{W}\mathbf{C}])$ = $(\mathbf{C}' \otimes \mathbf{B})$vec $\mathcal{W}$, where "$\otimes$" is the Kronecker product.

(h) $[\mathbf{D}\mathbf{B}][\mathcal{W}] = [\mathbf{D}][\{(\mathbf{B}\mathbf{w}_{rs})\}] = [\mathbf{D}][[\mathbf{B}][\mathcal{W}]]$.

(i) $\mathbf{a}'\mathcal{W}\mathbf{b} = \sum_u \sum_v a_u b_v \mathbf{w}_{uv}$.

(j) $[\mathbf{d}'][\mathcal{W}]$ is a matrix with (r, s)th element $\sum_i d_i w_{irs}$.

Proofs. Section 8.14.7.

8.154. Seber and Wild [1989: 692, (h)–(j)] and Wei [1997: 188–191, (a)–(j)].

CHAPTER 9

NON-NEGATIVE VECTORS AND MATRICES

Any matrix of probabilities has non-negative entries and is therefore a non-negative matrix. Consequently, such matrices play a varied role in probability and statistics. For example, they are used in genetic and population growth models, general stochastic processes, and various scaling problems. Such matrices are also encountered in the previous chapter where a number of matrices were mentioned such as permutation matrices, where the elements are zero or one. Non-negative matrices also play an important role in combinatorics (e.g., Sachkov and Tarakonov [2002]). In this chapter we look at a wide range of such matrices. For a concise reference to the subject see Rothblum [2007: chapter 9].

9.1 INTRODUCTION

Definition 9.1. A nonzero matrix $\mathbf{A} = (a_{ij})$ is said to be *non-negative (positive)* if $a_{ij} \geq 0 \, (> 0)$ for all i, j. We write $\mathbf{A} \geq \mathbf{0} \, (> \mathbf{0})$. Also we say that $\mathbf{A} \leq \mathbf{0} \, (< \mathbf{0})$ if $a_{ij} \leq 0 \, (< 0)$. The same definition applies to vectors, namely $\mathbf{a} \geq \mathbf{0}$ if $a_i \geq 0$ for all i, and $\mathbf{a} \neq \mathbf{0}$. Finally we say that $\mathbf{A} \leq \mathbf{B}$ if and only if $\mathbf{B} - \mathbf{A} \geq \mathbf{0}$.

In most applications, $\mathbf{A}$ is square. Unless stated otherwise, we shall assume that $\mathbf{A}$ is $n \times n$. Although certain aspects of the general theory of non-negative matrices extend to countably infinite matrices, we shall consider only infinite stochastic matrices (Section 9.6.3).

A Matrix Handbook for Statisticians. By George A. F. Seber
Copyright © 2008 John Wiley & Sons, Inc.

9.1. (Frobenius–König) If $\mathbf{A} \geq \mathbf{0}$, then $\text{per}(\mathbf{A}) = 0$, where $\text{per}(\mathbf{A})$ is the permanent of $\mathbf{A}$ (cf. Section 4.5), if and only if $\mathbf{A}$ has an $r \times s$ zero submatrix with $r + s = n + 1$.

Proofs. Section 9.1.

9.1. Bapat and Raghavan [1997: 62].

9.1.1   Scaling

Definition **9.2.** Let $\mathbf{A} \geq \mathbf{0}$ be an $m \times n$ matrix. The problem of scaling $\mathbf{A}$ to obtain a non-negative $m \times n$ matrix $\mathbf{B}$ with prescribed row and column sums will be called the *scaling problem*.

$\mathbf{A}$ and $\mathbf{B}$ are said to have the *same pattern* if $a_{ij} = 0$ if and only if $b_{ij} = 0$ for all i, j.

The procedure whereby we alternatively scale the rows of $\mathbf{A}$ to give the required rows sums, then scale the columns sums of the new $\mathbf{A}$ to give the required column sums (this will upset the row sums), and then continuing to repeat these two operations, we shall call the *iterative scaling algorithm*. Under certain conditions, this procedure converges to give a solution to the scaling problem. Of related interest is the doubly stochastic matrix discussed in Section 9.7.

Scaling problems arise in many contexts. For example, Bapat and Raghavan [1997: chapter 6] mention budget allocations, probability estimation in Markov chains, Leontief input–output systems, estimating cell entries in contingency tables, and transportation planning.

9.2. (Bacharach) Let $\mathbf{A} \geq \mathbf{0}$ be $m \times n$ with no zero row or column, and let $\mathcal{I}$ and $\mathcal{J}$ be subsets of $\{1, 2, \ldots, m\}$ and $\{1, 2, \ldots, n\}$, respectively, with complements $\mathcal{I}^c$ and $\mathcal{J}^c$. Let $\mathbf{x}$ and $\mathbf{y}$ be fixed $m \times 1$ and $n \times 1$ positive vectors, respectively. Then there exists an $m \times n$ matrix $\mathbf{B} \geq \mathbf{0}$ such that $a_{ij} = 0 \Rightarrow b_{ij} = 0$, with $\mathbf{B}\mathbf{1}_m = \mathbf{x} > \mathbf{0}$ and $\mathbf{1}'_m \mathbf{B} = \mathbf{y}' > \mathbf{0}'$ if and only if

$$a_{ij} = 0 \text{ for all } i \in \mathcal{I}^c, j \in \mathcal{J} \quad \text{implies} \quad \sum_{i \in \mathcal{I}^c} x_i \leq \sum_{j \in \mathcal{J}^c} y_j \text{ and } \sum_{i \in \mathcal{I}} x_i \geq \sum_{j \in \mathcal{J}} y_j.$$

Here a_{ij} is to be understood as zero if i or $j \in \phi$, as is summation over an empty set. Using a concept relating to the elements of $\mathbf{A}$ called "connectedness," Seneta [1981: 70–77] gives a number of general theorems to establish the convergence of the iterative scaling algorithm to the matrix $\mathbf{B}$ described above with prescribed row and column sums. If $\mathbf{A} > \mathbf{0}$, then $\mathbf{A}$ is connected, and the theory simplifies. A different approach to this problem is embodied in the next two results.

9.3. Let K be a nonempty, bounded polyhedron given by

$$K = \{\boldsymbol{\pi} \in \mathbb{R}^n : \boldsymbol{\pi} \geq \mathbf{0}, \ \mathbf{C}\boldsymbol{\pi} = \mathbf{b}\},$$

where $\mathbf{C} = (c_{ij})$ is an $m \times n$ matrix and $\mathbf{b} \in \mathbb{R}^m$ is a nonzero vector. Let $\mathbf{y} \in K$. Then, for any $\mathbf{x} \geq \mathbf{0}$ with the same pattern as $\mathbf{y}$, there exist $z_i > 0$ $(i = 1, 2, \ldots, m)$ and $\boldsymbol{\pi} \in K$ such that

$$\pi_j = x_j \prod_{i=1}^{m} z_i^{c_{ij}}, \quad j = 1, 2, \ldots, n.$$

Furthermore, any $\pi \in K$ of the above type is unique. Bapat and Raghavan [1997: 247] show how to use this theorem to prove the existence of a solution to the scaling problem.

9.4. Let $\mathbf{A}$ and $\mathbf{B}$ be any pair of positive $m \times n$ matrices. Then there exists a unique matrix $\mathbf{C} = \mathbf{D}_1 \mathbf{A} \mathbf{D}_2$, where $\mathbf{D}_1$ and $\mathbf{D}_2$ are diagonal matrices with positive diagonal entries, and $\mathbf{C}$ and $\mathbf{B}$ have the same row and column sums. Also the iterative scaling algorithm applied to $\mathbf{A}$ converges to $\mathbf{C}$.

Proofs. Section 9.1.1.

   9.2. Bacharach [1965] and Seneta [1981: 79, exercise 2.34].

   9.3. Bapat and Raghavan [1997: 247].

   9.4. Bapat and Raghavan [1997: 251, 260].

9.1.2   Modulus of a Matrix

Definition 9.3. The *modulus* of any a real or complex matrix $\mathbf{A} = (a_{ij})$ is the matrix of absolute values, namely $\mathrm{mod}(\mathbf{A}) = (|a_{ij}|)$. Thus $\mathrm{mod}(\mathbf{A}) \geq \mathbf{0}$. Schott [2005: 318] uses the term $\mathrm{abs}(\mathbf{A})$.

9.5. Clearly $\mathbf{A} \leq \mathrm{mod}(\mathbf{A})$.

9.6. The following are readily proved using (5.1).

   (a) If $\mathbf{A}$ and $\mathbf{B}$ are any two conformable matrices, then

$$\mathrm{mod}(\mathbf{AB}) \leq \mathrm{mod}(\mathbf{A})\mathrm{mod}(\mathbf{B}).$$

   Here $\mathbf{B}$ could also be a vector.

   (b) If $\mathbf{A}$ is square, $\mathrm{mod}(\mathbf{A}^k) \leq [\mathrm{mod}(\mathbf{A})]^k$.

   (c) If $\mathbf{A}$ and $\mathbf{B}$ are $n \times n$ matrices such that $\mathrm{mod}(\mathbf{A}) \leq \mathrm{mod}(\mathbf{B})$, then

$$||\mathrm{mod}(\mathbf{A})||_F \leq ||\mathrm{mod}(\mathbf{B})||_F,$$

   where $||\cdot||_F$ represents the Frobenius norm.

Proofs. Section 9.1.2.

   9.6. Quoted by Rao and Rao [1998: 470].

9.2   SPECTRAL RADIUS

9.2.1   General Properties

We recall that the spectral radius $\rho(\mathbf{A})$ of a square matrix $\mathbf{A}$ is the maximum of the absolute values of the eigenvalues of $\mathbf{A}$. Note that $\rho(\mathbf{A})$ need not be an eigenvalue of $\mathbf{A}$, though we note below that it can be an eigenvalue in the case of non-negative matrices. Although the emphasis is on non-negative matrices in this chapter, further results concerning the spectral radius are given in Section 4.6.2 (e.g., 4.68a) on matrix norms.

9.7. Let $\mathbf{A} = (a_{ij})$ be a complex matrix and $\mathbf{B} = (b_{ij}) \geq \mathbf{0}$ be a real matrix, both $n \times n$, such that $\mathrm{mod}(\mathbf{A}) \leq \mathbf{B}$. Then:

(a) $\rho(\mathbf{A}) \leq \rho[\mathrm{mod}(\mathbf{A})] \leq \rho(\mathbf{B})$.

(b) (Ky Fan) Every eigenvalue of $\mathbf{A}$ lies in the region

$$\cup_{i=1}^{n}\{z \in \mathbb{C} : |z - a_{ii}| \leq \rho(\mathbf{B}) - b_{ii}\}.$$

9.8. Let $\mathbf{A}$ and $\mathbf{B}$ be $n \times n$ non-negative matrices. If $\mathbf{0} \leq \mathbf{A} \leq \mathbf{B}$, then

$$\rho(\mathbf{A}) \leq \rho(\mathbf{B});$$

that is, $\rho(\cdot)$ is monotonically increasing on the set of all $n \times n$ non-negative matrices.

9.9. Let $\mathbf{A} \geq \mathbf{0}$ be an $n \times n$ matrix.

(a) If $\mathbf{C}$ is a principal submatrix of $\mathbf{A}$, then $\rho(\mathbf{C}) \leq \rho(\mathbf{A})$. In particular,

$$\max_{1 \leq i \leq n} a_{ii} \leq \rho(\mathbf{A}).$$

(b) Let r_i be the row sum of row i and c_j be the column sum of column j. Then:

(i) $\min\limits_{1 \leq i \leq n} r_i \leq \rho(\mathbf{A}) \leq \max\limits_{1 \leq i \leq n} r_i$.

(ii) $\min\limits_{1 \leq j \leq n} c_j \leq \rho(\mathbf{A}) \leq \max\limits_{1 \leq j \leq n} c_j$.

(iii) If $r_i = \alpha$ for all i, then $\rho(\mathbf{A}) = \alpha$. If $c_j = \beta$ for all j, then $\rho(\mathbf{A}) = \beta$.

(c) Let $\mathbf{x} = (x_1, x_2, \ldots, x_n)' > \mathbf{0}$. Then:

(i) $\min\limits_{1 \leq i \leq n} \dfrac{1}{x_i} \sum\limits_{j=1}^{n} a_{ij}x_j \leq \rho(\mathbf{A}) \leq \max\limits_{1 \leq i \leq n} \dfrac{1}{x_i} \sum\limits_{j=1}^{n} a_{ij}x_j$.

(ii) $\min\limits_{1 \leq j \leq n} x_j \sum\limits_{i=1}^{n} \dfrac{a_{ij}}{x_i} \leq \rho(\mathbf{A}) \leq \max\limits_{1 \leq j \leq n} x_j \sum\limits_{i=1}^{n} \dfrac{a_{ij}}{x_i}$.

(iii) If $\mathbf{A}$ has a positive right eigenvector, then

$$\rho(\mathbf{A}) = \max_{\mathbf{x}>\mathbf{0}} \min_{1 \leq i \leq n} \frac{1}{x_i} \sum_{j=1}^{n} a_{ij}x_j = \min_{\mathbf{x}>\mathbf{0}} \max_{1 \leq i \leq n} \frac{1}{x_i} \sum_{j=1}^{n} a_{ij}x_j.$$

(d) $\rho(\mathbf{I}_n + \mathbf{A}) = 1 + \rho(\mathbf{A})$.

9.10. Let $\mathbf{A} > \mathbf{0}$ with maximum and minimum row sums of R and r, respectively, and let $m = \min_{i,j} a_{ij}$. Then

$$r + m(h - 1) \leq \rho(\mathbf{A}) \leq R - m(1 - g^{-1}),$$

where

$$g = \frac{R - 2m + \sqrt{R^2 - 4m(R - r)}}{2(r - m)}, \quad h = \frac{-r + 2m + \sqrt{r^2 + 4m(R - r)}}{2m}.$$

There exist matrices for which the bounds are attained.

Proofs. Section 9.2.1.

9.7a. Horn and Johnson [1985: 491], Meyer [2000a: 619], and Rao and Rao [1998: 471].

9.7b. Horn and Johnson [1985: 501] and Marcus and Minc [1964: 152].

9.8. Horn and Johnson [1985: 491].

9.9a. Horn and Johnson [1985: 491] and Rao and Rao [1998: 471].

9.9b. Horn and Johnson [1985: 492–493], Rao and Rao [1998: 471], and Schott [2005: 318–319].

9.9c. Horn and Johnson [1985: 493], Rao and Rao [1998: 472, for (i) and (ii)], and Schott [2005: 318–319].

9.9d. Horn and Johnson [1985: 507] and Rao and Rao [1998: 475].

9.10. Marcus and Minc [1964: 155].

9.2.2   Dominant Eigenvalue

Definition 9.4. If $|\lambda_1| > |\lambda_2| \geq \cdots \geq |\lambda_n|$, then λ_1 is called the *dominant eigenvalue* of $\mathbf{A}$. We note that $|\lambda_1|$ is also the spectral radius $\rho(\mathbf{A})$ of $\mathbf{A}$.

9.11. (Perron–Frobenius Theorem for Non-negative Matrices) If $\mathbf{A} \geq \mathbf{0}$, then the following hold.

(a) $\mathbf{A}$ has a real eigenvalue $\rho \geq 0$.

(b) With ρ can be associated non-negative left and right eigenvectors (which need not be unique even when scaled to have unit length).

(c) If $\mathbf{A}$ has a positive eigenvector, then the corresponding eigenvalue is ρ; that is, if $\mathbf{A}\mathbf{x} = \lambda\mathbf{x}$ and $\mathbf{x} > \mathbf{0}$, then $\lambda = \rho$.

(d) $|\lambda| \leq \rho$ for any eigenvalue λ of $\mathbf{A}$, i.e., ρ is the spectral radius of $\mathbf{A}$.

(d) If $\mathbf{0} \leq \mathbf{B} \leq \mathbf{A}$ and β is an eigenvalue of $\mathbf{B}$, then $|\beta| \leq \rho$.

Seneta [1981: 25–26] gives a helpful history of this and related results.
    There is a corresponding theorem, originally proved by Perron in 1907 for positive matrices; for recent proofs see Bapat and Raghavan [1997: 5–6], Rao and Rao [1998: 473] and Schott [2005: section 8.8]. However, when $\mathbf{A} > \mathbf{0}$, $\mathbf{A}$ is also irreducible and primitive (see below for definitions), so that a more general theorem is therefore

given in (9.30). For completeness, we give some related results for $\mathbf{A} > \mathbf{0}$ below in (9.16) from Horn and Johnson [1985] and Schott [2005].

9.12. If $\mathbf{A} \geq \mathbf{0}$ and $s > \rho$, where ρ is defined in (9.11) above, then $(s\mathbf{I}_n - \mathbf{A})$ has an inverse and
$$(s\mathbf{I}_n - \mathbf{A})^{-1} \geq \mathbf{0}.$$

9.13. Let $\mathbf{A} \geq \mathbf{0}$ with spectral radius ρ, and let $\mathrm{adj}(\mathbf{A})$ denote the adjoint matrix. Then:

(a) $\mathbf{B}(s) = \mathrm{adj}(s\mathbf{I}_n - \mathbf{A}) \geq \mathbf{0}$ for $s > \rho$.

(b) $\frac{d\mathbf{B}(s)}{ds} \geq \mathbf{0}$ for $s > \rho$.

(c) $\mathbf{B}(\rho) \geq \mathbf{0}$.

(d) $\frac{d}{ds}(s\mathbf{I}_n - \mathbf{A})^{-1}|_{s=\rho} \geq \mathbf{0}$.

9.14. Let $\mathbf{A} \geq \mathbf{0}$ have spectral radius ρ.

(a) Suppose $\mathbf{x} > \mathbf{0}$ and $\alpha, \beta \geq 0$. Then:

   (i) If $\alpha\mathbf{x} \leq \mathbf{A}\mathbf{x} \leq \beta\mathbf{x}$, then $\alpha \leq \rho \leq \beta$.
   (ii) If $\alpha\mathbf{x} < \mathbf{A}\mathbf{x}$, then $\alpha < \rho$
   (iii) If $\mathbf{A}\mathbf{x} < \beta\mathbf{x}$, then $\rho < \beta$.

(b) If $\mathbf{x} \geq \mathbf{0}$ ($\mathbf{x} \neq \mathbf{0}$) and $\mathbf{A}\mathbf{x} \geq \alpha\mathbf{x}$ for some α, then $\rho \geq \alpha$.

9.15. Let $\mathbf{A} \geq \mathbf{0}$.

(a) $(\mathbf{I}_n - \mathbf{A})^{-1}$ exists and is non-negative if and only if there exists $\mathbf{x} \geq \mathbf{0}$ such that $\mathbf{x} > \mathbf{A}\mathbf{x}$.

(b) If each of the row sums of $\mathbf{A}$ is less than 1, then $(\mathbf{I}_n - \mathbf{A})^{-1}$ exists and is non-negative. The same is true if each of the columns sums is less than 1.

(c) Consider the equation $(\mathbf{I}_n - \mathbf{A})\mathbf{y} = \mathbf{b}$, where $\mathbf{b} \geq \mathbf{0}$. If $(\mathbf{I}_n - \mathbf{A})^{-1}$ exists and is non-negative, then there is a unique non-negative solution $\mathbf{y} = (\mathbf{I}_n - \mathbf{A})^{-1}\mathbf{b}$. This result applies, for example, to Leontief's input–output economic model.

An irreducible version of the above theorem is given in (9.36) below.

9.16. (Perron's Theorem for Positive Matrices, with Additions) Suppose $\mathbf{A} > \mathbf{0}$ with spectral radius $\rho = \rho(\mathbf{A})$. Then:

(a) ρ is positive and is an eigenvalue.

(b) There are positive right and left eigenvectors $\mathbf{A}$ corresponding to ρ.

(c) Suppose $|\lambda| = \rho$, with any corresponding eigenvector $\mathbf{x}$. Then:

   (i) $\mathbf{A}\,\mathrm{mod}(\mathbf{x}) = \rho\,\mathrm{mod}(\mathbf{x})$, where "mod" is defined in Definition 9.3 above.
   (ii) There exists an angle θ such that $e^{-i\theta}\mathbf{x} > \mathbf{0}$.

(d) The eigenvalue ρ has algebraic and geometric mutliplicities both equal to 1.

(e) If λ is an eigenvalue of $\mathbf{A}$ and $\lambda \neq \rho$, then $|\lambda| < \rho$.

(f) Suppose $\mathbf{x}$ and $\mathbf{y}$ are positive vectors such that $\mathbf{A}\mathbf{x} = \rho\mathbf{x}$, $\mathbf{y}'\mathbf{A} = \rho\mathbf{y}'$, and $\mathbf{x}'\mathbf{y} = 1$. Then:

  (i) $(\mathbf{A} - \rho\mathbf{x}\mathbf{y}')^k = \mathbf{A}^k - \rho^k\mathbf{x}\mathbf{y}'$, for $k = 1, 2, \ldots$.

  (ii) Each nonzero eigenvalue of $\mathbf{A} - \rho\mathbf{x}\mathbf{y}'$ is an eigenvalue of $\mathbf{A}$.

  (iii) ρ is not an eigenvalue of $\mathbf{A} - \rho\mathbf{x}\mathbf{y}'$.

  (iv) $\rho(\mathbf{A} - \rho\mathbf{x}\mathbf{y}') < \rho$.

  (v) $\lim_{k \to \infty} (\rho^{-1}\mathbf{A})^k = \mathbf{x}\mathbf{y}'$.

Proofs. Section 9.2.2.

9.11. Debreu and Herstein [1953], Meyer [2000a: 670, (a) and (b)], and quoted by Seneta [1981: 28, exercise 1.12]. Horn and Johnson [1985: 493] prove (c).

9.12. Bapat and Raghavan [1997: 35].

9.13. Bapat and Raghavan [1997: 37].

9.14. Horn and Johnson [1985: 493, 504].

9.15. Rao and Rao [1998: 479–480].

9.16. Horn and Johnson [1985: 495–500] and Schott [2005: 319–323].

9.3   CANONICAL FORM OF A NON-NEGATIVE MATRIX

***Definition* 9.5.** Let $\mathbf{A} = (a_{ij}) \geq \mathbf{0}$ be $n \times n$, and define $\mathbf{A}^m = (a_{ij}^{(m)})$. If $a_{ij}^{(m)} > 0$ for some positive integer m (a function of i and j), we say that i *leads to* j or i *can reach* j (or state i can reach state j in the case of a Markov chain and its transition matrix; see Definition 9.16 in Section 9.6), or j is *accessible* from i, and we write $i \to j$. If $i \to j$ and $j \to i$, we say that the i and j *communicate* and write $i \leftrightarrow j$. If $i \leftrightarrow i$, the period of index i is defined to be $d(i) = \gcd\{k : a_{ii}^{(k)} > 0\}$—that is, the greatest common divisor of those positive integers k such that $a_{ii}^{(k)} > 0$. If $d(i) > 1$, then i is said to be *periodic (cyclic)*, while if $d(i) = 1$, then i is said to be *aperiodic (acyclic)*. Clearly, if there exists at least one j such that $i \leftrightarrow j$, we must have $i \leftrightarrow i$.

The indices can be classified as essential or inessential. If $i \to j$, but $j \not\to i$ for some j, then i is called *inessential*, and an index which leads to no index at all is also called inessential; otherwise, an index is called *essential*. Essential indices can be divided into *self-communicating classes* where all the indices within the class communicate with each other, but do not communicate with any indices outside the class. Similarly, inessential indices (if any) can also be divided into self-communicating classes in which an index in a class can reach another index outside the class, but can't get back, together with a class of individuals that communicate with no index (Seneta [1981: 12]).

9.17. If $\mathbf{A} \geq \mathbf{0}$ has at least one positive entry in each row, then it possesses at least one essential class of indices.

9.18. If $\mathbf{A} = (a_{ij}) \geq \mathbf{0}$ and $i \leftrightarrow j$, then $d(i) = d(j)$.

9.19. (Canonical Form) Given $\mathbf{A} \geq \mathbf{0}$, there exists a permutation matrix $\mathbf{\Pi}$ such that

$$
\mathbf{B} = \mathbf{\Pi}' \mathbf{A} \mathbf{\Pi} = \left(
\begin{array}{cccc|c}
\mathbf{A}_1 & \mathbf{0} & \cdots & \mathbf{0} & \mathbf{0} \\
\mathbf{0} & \mathbf{A}_2 & \cdots & \mathbf{0} & \mathbf{0} \\
\cdot & \cdot & \cdots & \cdot & \cdot \\
\mathbf{0} & \mathbf{0} & \cdots & \mathbf{A}_r & \mathbf{0} \\
\hline
& \mathbf{R} & & & \mathbf{Q}
\end{array}
\right),
$$

where the $\mathbf{A}_i$ $(i = 1, 2, \ldots, r)$ correspond to the r self-communicating classes of essential indices, and $\mathbf{Q}$ corresponds to the inessential indices, with $\mathbf{R} \neq \mathbf{0}$ in general. The matrix $\mathbf{B}$ is simply $\mathbf{A}$ with the indices reordered, and $\mathbf{Q}$ has a structure similar to $\mathbf{A}$, except that there may be nonzero elements to the left of any of its diagonal blocks, that is,

$$
\mathbf{Q} = \left(
\begin{array}{cccc}
\mathbf{Q}_1 & \mathbf{0} & \cdots & \mathbf{0} \\
\mathbf{0} & \mathbf{Q}_2 & \cdots & \mathbf{0} \\
\cdot & \cdot & \cdot & \cdot \\
& \mathbf{C} & & \mathbf{Q}_s
\end{array}
\right).
$$

In practice the matrix $(\mathbf{R} \mid \mathbf{Q})$ in $\mathbf{B}$ may be missing from $\mathbf{B}$ and we could have $r = 1$. Also

$$
\mathbf{B}^k = \left(
\begin{array}{cccc|c}
\mathbf{A}_1^k & \mathbf{0} & \cdots & \mathbf{0} & \mathbf{0} \\
\mathbf{0} & \mathbf{A}_2^k & \cdots & \mathbf{0} & \mathbf{0} \\
\cdot & \cdot & \cdots & \cdot & \cdot \\
\mathbf{0} & \mathbf{0} & \cdots & \mathbf{A}_r^k & \mathbf{0} \\
\hline
& \mathbf{R}_k & & & \mathbf{Q}^k
\end{array}
\right) \text{ and } \mathbf{Q}^k = \left(
\begin{array}{cccc}
\mathbf{Q}_1^k & \mathbf{0} & \cdots & \mathbf{0} \\
\mathbf{0} & \mathbf{Q}_2^k & \cdots & \mathbf{0} \\
\cdot & \cdot & \cdot & \cdot \\
& \mathbf{C}_k & & \mathbf{Q}_s^k
\end{array}
\right).
$$

Proofs. Section 9.3.

9.17–9.19. Seneta [1981: 14–17].

9.4 IRREDUCIBLE MATRICES

9.4.1 Irreducible Non-negative Matrix

In Section 8.8 we introduced the concept of irreducibiblity for general matrices. In this section we concentrate on non-negative matrices, the major application of irreducibility, and recall the following definition.

Definition 9.6. An $n \times n$ non-negative matrix $\mathbf{A}$ is said to be *reducible* if there exist a permutation matrix $\mathbf{\Pi}$ such that

$$
\mathbf{B} = \mathbf{\Pi} \mathbf{A} \mathbf{\Pi}' = \left(
\begin{array}{cc}
\mathbf{B}_{11} & \mathbf{0} \\
\mathbf{B}_{12} & \mathbf{B}_{22}
\end{array}
\right),
$$

where $\mathbf{B}_{11}$ and $\mathbf{B}_{22}$ are square matrices. A matrix which is not reducible is said to be *irreducible*. We note that if $\mathbf{B}$ has the general canonical form (9.19), then it is reducible. Some authors use the equivalent definition

$$\mathbf{B} = \begin{pmatrix} \mathbf{B}_{11} & \mathbf{B}_{12} \\ \mathbf{0} & \mathbf{B}_{22} \end{pmatrix}.$$

An equivalent but more useful definition of irreducibility in the present context is as follows. The $n \times n$ matrix $\mathbf{A} \geq \mathbf{0}$ is *irreducible* if and only if every pair of indices in its index set communicate, that is, for every pair i, j there exists a positive integer $m \, (\leq n)$, a function of i and j, such that $a_{ij}^{(m)} > 0$. The equivalence of the two definitions is proved by Bapat and Raghavan [1997: 2–4].

An irreducible non-negative matrix is said to be *periodic (cyclic)* with period d if the period of any one (and so of each one, by (9.18)) of its indices satisfies $d > 1$, and it is said to be *aperiodic (acyclic)* if $d = 1$.

9.20. An irreducible non-negative matrix cannot have a zero row or column.

9.21. If the matrix $\mathbf{A} = (a_{ij}) \geq \mathbf{0}$ is reducible, then so is $\mathbf{A}^k$ for any positive integer k.

9.22. An $n \times n$ non-negative matrix $\mathbf{A}$ is irreducible if and only if $(\mathbf{I}_n + \mathbf{A})^{n-1} > \mathbf{0}$.

9.23. If $\mathbf{A}$ is irreducible, then so is $\mathbf{A}'$.

Definition 9.7. The matrix $\mathbf{A} \geq \mathbf{0}$ is said to be *primitive* if there exists a positive integer p such that $\mathbf{A}^p > \mathbf{0}$. (Thus if $\mathbf{A}$ is primitive, it is irreducible as $a_{ij}^{(p)} > 0$ for all i, j.) Clearly, if $\mathbf{A} > \mathbf{0}$, then $\mathbf{A}$ is primitive.

An alternative but equivalent definition is that $\mathbf{A} \geq \mathbf{0}$ is primitive if it is irreducible and it has only one eigenvalue of maximum modulus. The equivalence follows from (9.26) below.

The smallest positive integer q such that $\mathbf{A}^q > \mathbf{0}$ is called the *index of primitivity*.

9.24. If $\mathbf{A} \geq \mathbf{0}$ is primitive, then $\mathbf{A}^k$ is non-negative, irreducible, and primitive for all $k = 1, 2, \ldots$.

9.25. If $\mathbf{A} \geq \mathbf{0}$ is primitive, then $\mathbf{A}^k > \mathbf{0}$ for some integer $k \leq (n-1)n^n$.

9.26. A non-negative matrix $\mathbf{A}$ is primitive if and only if it is irreducible and aperiodic.

9.27. If $\mathbf{A} \geq \mathbf{0}$ has $a_{ii} > 0$ for all i, then $\mathbf{A}^{n-1} > \mathbf{0}$ and $\mathbf{A}$ is primitive.

9.28. (Limit Theorem for Primitive Matrices) Let $\mathbf{A}$ be an $n \times n$ primitive non-negative matrix with distinct eigenvalues $\rho, \lambda_2, \ldots, \lambda_t$ $(t \leq n)$, where $\rho > |\lambda_2| \geq |\lambda_3| \geq \cdots \geq |\lambda_t|$. In the case $|\lambda_2| = |\lambda_3|$ $(\lambda_2 \neq \lambda_3)$ we stipulate that the algebraic multiplicity m_2 of λ_2 is at least as great as that of λ_3 and of any other eigenvalues having the same modulus as λ_2. By (9.30) there exist positive vectors $\mathbf{x}$ and $\mathbf{y}$ such that $\mathbf{A}\mathbf{x} = \rho\mathbf{x}$, $\mathbf{y}'\mathbf{A} = \rho\mathbf{y}'$ and $\mathbf{x}'\mathbf{y} = 1$. We then have the following:

(a) Suppose $\lambda_2 \neq 0$.

  (i) As $k \to \infty$,

$$\mathbf{A}^k = \rho^k \mathbf{x}\mathbf{y}' + \mathbf{O}(k^s |\lambda_2|^k)$$

  elementwise, where $s = m_2 - 1$.

(ii)

$$\lim_{k \to \infty} \frac{\mathbf{A}^k}{\rho^k} = \mathbf{xy}'.$$

(b) Suppose $\lambda_2 = 0$, then for $k \geq n - 1$,

$$\mathbf{A}^k = \rho^k \mathbf{xy}'.$$

For matrix limits see Section 19.2.

Definition 9.8. An irreducible non-negative matrix that is periodic is said to be *imprimitive*. Thus irreducible matrices can be subdivided into primitive or imprimitive matrices depending on whether they are aperiodic or periodic.

9.29. The powers of an imprimitive matrix may be studied in terms of powers of primitive matrices.

9.30. (Perron–Frobenius Theorem for Irreducible Matrices) Let $\mathbf{A} \geq 0$ be an irreducible matrix. Then we have the following.

(a) $\mathbf{A}$ has a real positive eigenvalue ρ.

(b) With ρ can be associated strictly positive left and right eigenvalues.

(c) $|\lambda| \leq \rho$ for any eigenvalue λ of $\mathbf{A}$. Thus ρ is the spectral radius of $\mathbf{A}$.

(d) ρ has geometric multiplicity 1, that is, the left and right eigenvectors associated with ρ are unique to constant multiples.

(e) ρ has algebraic multiplicity 1, that is, ρ is a simple root of the characteristic equation.

(f) If $0 \leq \mathbf{B} \leq \mathbf{A}$ and β is an eigenvalue of $\mathbf{B}$, then $|\beta| \leq \rho$. Moreover, $|\beta| = \rho$ implies $\mathbf{B} = \mathbf{A}$ so that ρ increases when any element of $\mathbf{A}$ increases.

(g) (Primitive matrices) If $\mathbf{A}$ is primitive then (a)–(f) still hold except that (c) is replaced by $|\lambda| < \rho$ for any eigenvalue $\lambda \neq \rho$.

Definition 9.9. We call ρ the *Perron–Frobenius eigenvalue* of an irreducible non-negative matrix, and its corresponding positive eigenvectors are called the *Perron–Frobenius eigenvectors*. As noted above, ρ is the spectral radius.

9.31. Let $\mathbf{A} \geq 0$ be an irreducible $n \times n$ matrix with Perron–Frobenius eigenvalue ρ, and let $\mathbf{x}$ and $\mathbf{y}$ be the right and left Perron–Frobenius eigenvectors of $\mathbf{A}$ satisfying $\mathbf{x}'\mathbf{y} = 1$. Then:

(a) $\mathbf{y}'\mathbf{Ax} = \rho \leq \mathbf{x}'\mathbf{Ay}$.

(b) (Limit Theorem) If $\mathbf{L} = \mathbf{xy}'$, then

$$\lim_{N \to \infty} \frac{1}{N} \sum_{k=1}^{N} (\rho^{-1}\mathbf{A})^k = \mathbf{L}.$$

9.32. (Subinvariance Theorem and Variations) Let $\mathbf{A} \geq \mathbf{0}$ be an irreducible matrix with Perron–Frobenius eigenvalue ρ. Let $c > 0$.

(a) If $\mathbf{Ax} \leq c\mathbf{x}$ for any nonzero $\mathbf{x} \geq \mathbf{0}$, then $\rho \leq c$ and $\mathbf{x} > \mathbf{0}$. Furthermore, $\rho = c$ if and only if $\mathbf{Ax} = c\mathbf{x}$.

(b) If $\mathbf{Ax} \geq c\mathbf{x}$ for any nonzero $\mathbf{x} \geq \mathbf{0}$, then $\rho \geq c$. Also $\rho = c$ if and only if $\mathbf{Ax} = c\mathbf{x}$.

(c) If $\mathbf{Ax} \leq c\mathbf{x}$ ($\neq c\mathbf{x}$) for some $\mathbf{x} \geq \mathbf{0}$, then $\rho < c$.

(d) If $\mathbf{Ax} \geq c\mathbf{x}$ ($\neq c\mathbf{x}$) for some $\mathbf{x} \geq \mathbf{0}$, then $\rho > c$.

9.33. (Bounds on ρ) Let $\mathbf{A} \geq \mathbf{0}$ be irreducible with Perron–Frobenius eigenvalue ρ. Then (9.9b) holds with $\rho(\mathbf{A}) = \rho$. In the case of (i) and (ii), equality on one side implies equality on both sides, that is, ρ can only be equal to a maximal or minimal row (respectively column) sum if all the row (respectively column) sums are equal. The same is true for (c)(i).

9.34. Let $\mathcal{P} = \{\mathbf{x} : \mathbf{x} > \mathbf{0}\}$. Then:

(a)

$$\sup_{\mathbf{x} \in \mathcal{P}} \min_i \left\{ \frac{\sum_{j=1}^n a_{ij} x_j}{x_i} \right\} = \rho = \inf_{\mathbf{x} \in \mathcal{P}} \max_i \left\{ \frac{\sum_{j=1}^n a_{ij} x_j}{x_i} \right\}.$$

There also exists an $\mathbf{x} \in \mathcal{P}$ for which both the supremum and the infimum are attained.

(b)

$$\sup_{\mathbf{x} \in \mathcal{P}} \left\{ \inf_{\mathbf{y} \in \mathcal{P}} \frac{\mathbf{y}'\mathbf{Ax}}{\mathbf{y}'\mathbf{x}} \right\} = \rho = \inf_{\mathbf{x} \in \mathcal{P}} \left\{ \sup_{\mathbf{y} \in \mathcal{P}} \frac{\mathbf{y}'\mathbf{Ax}}{\mathbf{y}'\mathbf{x}} \right\}.$$

9.35. Let $\mathbf{A} \geq \mathbf{0}$ be irreducible with Perron–Frobenius eigenvalue ρ, and let $\mathbf{E} \geq \mathbf{0}$ ($\mathbf{E} \neq \mathbf{0}$). If $\delta > 0$, then $\mathbf{B} = \mathbf{A} + \delta\mathbf{E}$ is irreducible with a Perron–Frobenius eigenvalue that, by a suitable choice of δ, may be made equal to any positive number exceeding ρ.

9.36. Let $\mathbf{A} \geq \mathbf{0}$ be irreducible with Perron–Frobenius eigenvalue ρ.

(a) A necessary and sufficient condition for a solution $\mathbf{x}$ ($\mathbf{x} \geq \mathbf{0}, \mathbf{x} \neq \mathbf{0}$) to the equation $(s\mathbf{I}_n - \mathbf{A})\mathbf{x} = \mathbf{c}$ to exist for any $\mathbf{c} \geq \mathbf{0}$ ($\mathbf{c} \neq \mathbf{0}$) is that $s > \rho$. In this case, there is only one solution $\mathbf{x}$ that is strictly positive, and it is given by $\mathbf{x} = (s\mathbf{I}_n - \mathbf{A})^{-1}\mathbf{c}$.

(b) Of those real numbers s for which the inverse exists, $(s\mathbf{I} - \mathbf{A})^{-1} > 0$ if and only if $s > \rho$.

(c) If $s = 1$, then $\rho < 1$ if none of the row (or column) sums of $\mathbf{A}$ exceed 1, and at least one is less than 1. For applications see Leontief's input-output economic model and an extension described by Bapat and Raghavan [1997: chapter 7], Rao and Rao [1998: 477–481], and Seneta [1981: chapter 2].

9.37. Let $\mathbf{A} \geq \mathbf{0}$ be irreducible with Perron–Frobenius eigenvalue $\rho = 1$. Then the sequence $\{\mathbf{A}^k\}$ converges if and only if $\mathbf{A}$ is primitive.

9.38. If $\mathbf{A} > \mathbf{0}$ is irreducible with Perron–Frobenius eigenvalue ρ, and $\mathbf{A}^k = (a_{ij}^{(k)})$, then for each pair (i, j) the power series

$$A_{ij}(s) = \sum_{k=0}^{\infty} a_{ij}^{(k)} s^k$$

all have the same radius of convergence $R = \rho^{-1}$.

9.39. Suppose $\mathbf{A} \geq \mathbf{B} \geq \mathbf{0}$ and $\mathbf{A} \neq \mathbf{B}$. If $\mathbf{A} + \mathbf{B}$ is irreducible, then $\rho(\mathbf{A}) > \rho(\mathbf{B})$, where $\rho(\cdot)$ is the dominant eigenvalue of the appropriate matrix.

Proofs. Section 9.4.1.

   9.20. Seneta [1981: 18].

   9.21. We take powers of $\mathbf{B}$ in the definition (e.g., $\mathbf{B}^2 = \mathbf{\Pi A \Pi' \Pi A \Pi'} = \mathbf{\Pi A^2 \Pi'}$).

   9.22. Bapat and Raghavan [1997: 3], Rao and Rao [1998: 469], and Schott [2005: 324].

   9.23. This follows from either definition of irreducibility.

   9.24–9.25. Horn and Johnson [1985: 518].

   9.26. Seneta [1981: 21].

   9.27. Horn and Johnson [1985: 517].

   9.28. Seneta [1981: 9].

   9.29. Seneta [1981: 21].

   9.30. Bapat and Raghavan [1997: 17, proved the result using the theory of completely mixed matrix games], Horn and Johnson [1985: 508, for (a)–(e)], Schott [2005: 325–326, for (a)–(e)], and Seneta [1981: 22, 3–7].

   9.31a. Bapat and Raghavan [1997: 121, with $\mathbf{x}$ and $\mathbf{y}$ interchanged].

   9.31b. Horn and Johnson [1985: 524].

   9.32a. Seneta [1981: 23].

   9.32b. Quoted by Seneta [1981: 29, exercise 1.17].

   9.32c–d. Debreu and Hurstein [1953].

   9.33. Quoted by Seneta [1981: 27, exercise 1.7].

   9.34–9.35. Birkhoff and Varga [1958] and quoted by Seneta [1981: 27, exercises 1.7 and 1.8].

   9.36. Seneta [1981: 30–31].

9.37. Hunter [1983a: 170].

9.38. Quoted by Seneta [1981: 29, exercise 1.14].

9.39. Quoted by Seneta [1981: 29, exercise 1.16].

9.4.2 Periodicity

Definition **9.10.** If $\mathbf{A} \geq \mathbf{0}$ is irreducible, then by (9.18) each index i has the same period, d, say, which we call the period of $\mathbf{A}$.

9.40. Let $\mathbf{A} \geq \mathbf{0}$ be an irreducible matrix with h eigenvalues whose moduli are equal to the spectral radius ρ. We know from (9.30c) that $h \geq 1$. Then $h = d$, the period of $\mathbf{A}$ (cf. (9.41b) below).

9.41. Let $\mathbf{A} \geq \mathbf{0}$ be an $n \times n$ irreducible matrix with period d.

(a) $\mathbf{A}$ is primitive if and only if $d = 1$.

(b) If $d > 1$, there exist d distinct eigenvalues with $|\lambda| = \rho$, where ρ is the spectral radius. These eigenvalues are $\rho \exp i(2\pi k/d)$, $k = 0, 1, \ldots, d - 1$, the d roots of $\lambda^d - \rho^d = 0$.

(c) If $\lambda \neq 0$ is any eigenvalue of $\mathbf{A}$, then the numbers $\lambda \exp[i(2\pi k/d)]$, $k = 0, 1, \ldots, d - 1$, are also eigenvalues.

(d) The set of n eigenvalues when plotted as points in the complex λ-plane is invariant under a rotation of the plane through the angle $2\pi/d$.

(e) Combining (b) and (c),

$$\det(\lambda \mathbf{I}_n - \mathbf{A}) = \lambda^m (\lambda^d - \rho^d) \prod_{i=1}^{r}(\lambda^d - \lambda_i^d),$$

where $|\lambda_i| < \rho$ for $i = 1, 2, \ldots, r$ and $m = n - (r + 1)d$.

9.42. Let $\mathbf{A} \geq \mathbf{0}$ be irreducible with period d $(d > 1)$.

(a) There exists a permutation matrix $\mathbf{\Pi}$ such that

$$\mathbf{\Pi A \Pi'} = \begin{pmatrix} \mathbf{0} & \mathbf{B}_{12} & \mathbf{0} & \cdots & \mathbf{0} \\ \mathbf{0} & \mathbf{0} & \mathbf{B}_{23} & \cdots & \mathbf{0} \\ \cdot & \cdot & \cdot & \cdots & \cdot \\ \mathbf{0} & \mathbf{0} & \mathbf{0} & \cdots & \mathbf{B}_{d-1,d} \\ \mathbf{B}_{d1} & \mathbf{0} & \mathbf{0} & \cdots & \mathbf{0} \end{pmatrix} (= \mathbf{B}),$$

where the zero submatrices on the main diagonal are square. Note that $\mathbf{\Pi}$ permutes the rows, while $\mathbf{\Pi'}$ permutes the columns in the same order.

(b) Conversely, suppose $\mathbf{A} \geq \mathbf{0}$ and there exists a permutation matrix such that $\mathbf{\Pi A \Pi'} = \mathbf{B}$, as defined in (a). If $\mathbf{A}$ has no zero rows or columns and $\mathbf{B}_{12}\mathbf{B}_{23} \cdots \mathbf{B}_{d-1,d}\mathbf{B}_{d1}$ is irreducible, then $\mathbf{A}$ is irreducible.

Proofs. Section 9.4.2.

9.41a. Seneta [1981: 21].

9.41b. Horn and Johnson [1985: 510, 512] and Seneta [1981: 23].

9.41c. Seneta [1981: 24].

9.41d. Bapat and Raghavan [1997: 41–42].

9.41e. Bapat and Raghavan [1997: 43].

9.42a. Bapat and Raghavan [1997: 41–42].

9.42b. Seneta [1981: 29, exercise 1.18].

9.4.3 Non-negative and Nonpositive Off-Diagonal Elements

Definition **9.11.** An $n \times n$ real matrix $\mathbf{B} = (b_{ij})$ for which $b_{ij} \geq 0$, for all i, j $(i \neq j)$ is called an *ML-matrix*. This matrix arises in the theory of Markov processes.

9.43. If $\mathbf{B}$ is an ML-matrix, there exists a non-negative α sufficiently large so that

$$\mathbf{T} = \alpha\mathbf{I}_n + \mathbf{B} \geq \mathbf{0}.$$

Definition **9.12.** An ML-matrix $\mathbf{B}$ is said to be an *irreducible ML matrix* if $\mathbf{T} = \alpha\mathbf{I}_n + \mathbf{B} \geq \mathbf{0}$ is irreducible. By taking $\alpha > \max_i |b_{ii}|$, we can make the irreducible $\mathbf{T}$ aperiodic and primitive.

9.44. Suppose $\mathbf{B}$ is an $n \times n$ irreducible ML-matrix. Then there exists an eigenvalue τ with the following properties.

(a) τ is real.

(b) With τ are associated stricly positive left and right eigenvectors, which are unique to constant multiples.

(c) τ is greater than the real part of any other eigenvalue λ of $\mathbf{B}$, $\lambda \neq \tau$.

(d) τ is a simple root of the characteristic equation of $\mathbf{B}$.

(e) $\tau \leq 0$ if and only if there exists $\mathbf{y} \geq \mathbf{0}$ $(\mathbf{y} \neq \mathbf{0})$ such that $\mathbf{By} \leq \mathbf{0}$, in which case $\mathbf{y} > \mathbf{0}$; and $\tau < 0$ if and only if there is a strict inequality in at least one position in $\mathbf{By} \leq \mathbf{0}$.

(f) $\tau < 0$ if and only if $\Delta_i > 0$, $i = 1, 2, \ldots, n$, where Δ_i is the principal minor of $-\mathbf{B}$ formed from the first i rows and columns of $-\mathbf{B}$.

(g) $\tau < 0$ if and only if $-\mathbf{B}^{-1} > \mathbf{0}$.

9.45. An ML-matrix $\mathbf{B}$ is irreducible if and only if $e^{\mathbf{B}t} > \mathbf{0}$ for all $t > 0$ (see Section 19.6 for matrix exponentials). In this case

$$e^{\mathbf{B}t} = e^{\rho t}\mathbf{w}\mathbf{v}' + \mathbf{O}(e^{t_1 t})$$

elementwise as $t \to \infty$, where $\mathbf{w}$ and $\mathbf{v}'$ are the positive right and left eigenvectors of $\mathbf{B}$ corresponding to the dominant eigenvalue ρ of $\mathbf{B}$, normed so that $\mathbf{v}'\mathbf{w} = 1$, and having $t_1 < \rho$.

Proofs. Section 9.4.3.

9.43. Choose $\alpha = \max_i |b_{ii}|$.

9.44–9.45. Seneta [1981: 46–48].

9.4.4 Perron Matrix

***Definition* 9.13.** An $n \times n$ matrix $\mathbf{A}$ is said to be a *Perron matrix* (polynomially positive matrix) if $f(\mathbf{A}) > 0$ for some polynomial f with real coefficients. A matrix $\mathbf{A}$ is called a *power-positive matrix* if $\mathbf{A}^k > 0$ for some positive integer k.

9.46. If $\mathbf{A}$ is an irreducible ML-matrix, then it is is a Perron matrix. Also $\mathbf{B} = -\mathbf{A}$ is a Perron matrix.

9.47. If $\mathbf{A} \geq 0$ is irreducible, then $f(\mathbf{A}) = \sum_{i=1}^{n} \mathbf{A}^i > 0$ and $\mathbf{A}$ is a Perron matrix.

9.48. A power-positive matrix is a Perron matrix. Setting $k = 1$, we see that this includes positive matrices.

9.49. If $\mathbf{A}$ is a Perron matrix, then there exists an eigenvalue τ such that:

(a) τ is real.

(b) With τ can be associated strictly positive left and right eigenvectors, which are unique to constant multiples.

(c) τ is a simple root of the characteristic equation of $\mathbf{A}$.

9.50. Let $\mathbf{A}$ be a Perron matrix with τ defined above, and let adj denote an adjoint matrix.

(a) (i) $\min_i \sum_j a_{ij} \leq \tau \leq \max_i \sum_j a_{ij}$.
    (ii) $\min_j \sum_i a_{ij} \leq \tau \leq \max_j \sum_i a_{ij}$.

(b) Either $\text{adj}(\tau \mathbf{I}_n - \mathbf{A}) > 0$ or $-\text{adj}(\tau \mathbf{I}_n - \mathbf{A}) > 0$.

(c) If $\mathbf{A}\mathbf{x} \leq c\mathbf{x}$ for some nonzero $\mathbf{x} \geq 0$ and scalar c, then $c \geq \tau$; $c = \tau$ if and only if $\mathbf{A}\mathbf{x} = c\mathbf{x}$.

Proofs. Section 9.4.4.

9.46. From Definition 9.12 we see that $\mathbf{A}$ can be written in the form $\mathbf{T} - \alpha \mathbf{I}_n$, $\alpha > 0$, where $\mathbf{T}$ is non-negative and primitive, so that for some positive integer k, $(\mathbf{A} + \alpha \mathbf{I}_n)^k > 0$, which is a real polynomial.

9.47. Seneta [1981: 49].

9.48. Set $f(\mathbf{x}) = x^k$.

9.49. Bapat and Raghavan [1997: 44, proof using matrix game theory] and Seneta [1981: 49].

9.50. Seneta [1981: 52].

9.4.5 Decomposable Matrix

Definition 9.14. An square matrix $\mathbf{A}$ is called *partly decomposable* if there exist permutation matrices $\mathbf{\Pi}_1$ and $\mathbf{\Pi}_2$ such that

$$\mathbf{\Pi}_1\mathbf{A}\mathbf{\Pi}_2 = \begin{pmatrix} \mathbf{B}_{11} & \mathbf{0} \\ \mathbf{B}_{12} & \mathbf{B}_{22} \end{pmatrix},$$

where $\mathbf{B}_{11}$ and $\mathbf{B}_{22}$ are square matrices. A matrix is said to be *fully indecomposable* if it is not partly decomposable. Clearly an irreducible matrix is also fully indecomposable, but not necessarily vice versa. A major role of indecomposability is in investigating the combinatorial properties of non-negative matrices.

9.51. If $\mathbf{A} \geq \mathbf{0}$ is $n \times n$ and fully indecomposable, then $\mathbf{A}^{n-1} > \mathbf{0}$.

9.52. If $\mathbf{A}$ and $\mathbf{B}$ are non-negative $n \times n$ fully indecomposable matrices, then so is $\mathbf{AB}$. (This result is not necessarily true for irreducible matrices.)

Proofs. Section 9.4.5.

   9.51. Bapat and Raghavan [1997: 66].

   9.52. Bapat and Raghavan [1997: 67].

9.5 LESLIE MATRIX

Definition 9.15. A $k \times k$ *Leslie matrix* for population growth in animal or human populations is a matrix $\mathbf{A}$ of the form

$$\mathbf{A} = \begin{pmatrix} f_1 & f_2 & f_3 & \cdots & f_{k-1} & f_k \\ p_1 & 0 & 0 & \cdots & 0 & 0 \\ 0 & p_2 & 0 & \cdots & 0 & 0 \\ . & . & . & \cdots & & . \\ 0 & 0 & 0 & \cdots & p_{k-1} & 0 \end{pmatrix},$$

where, for $i = 1, 2, \ldots, k$, f_i is the average number of daughters born to a single female during the time she is in age class i, and p_i is the proportion of females in the ith age class expected to survive and pass into the next age class. (Some authors start the sequences with f_0 and p_0.) These fertility and survival rates are said to be *age-specific*. Here each $f_i \geq 0$ and $0 < p_i \leq 1$, so that $\mathbf{A} \geq \mathbf{0}$. In some cases i may refer to a state (stage) rather than age class, and the model is then *stage-specific*.

The matrix $\mathbf{A}$, and those like it that describe population growth, are sometimes called *population projection* matrices. Typically, they will contain further non-negative elements such as down the diagonal.

9.53. Let $\mathbf{n}(t) = (n_1(t), n_2(t), \ldots, n_k(t))'$, where $n_i(t)$ is the number of females in the ith age class at time t (t a positive integer). Then

$$\mathbf{n}(t) = \mathbf{A}\mathbf{n}(t-1) = \mathbf{A}^t\mathbf{n}(0),$$

where $\mathbf{A}$ is a population projection matrix. The case when $\mathbf{A}$ is singular and we require $n(t-1)$ from $n(t)$ is considered by Campbell and Meyer [1979: 184–187].

9.54. A sufficient condition for the Leslie matrix $\mathbf{A}$ to be primitive is that two consecutive f_i's, say f_j and f_{j+1}, are positive.

9.55. Suppose $\mathbf{A}$ is primitive (i.e., $\mathbf{A}^p > 0$ for some positive integer p). In fact, most population projection matrices are primitive, and the only significant exceptions are age-classified matrices with a single reproductive age class (Caswell [2001: 81]).

(a) By (9.30g), there is a positive dominant eigenvalue ρ that is simple with $|\lambda| < \rho$ for every eigenvalue λ different from ρ.

(b) Setting $x_1 = 1$ and successfully solving $\mathbf{A}\mathbf{x} = \rho\mathbf{x}$ using the second through to the kth rows, a positive right eigenvector corresponding to ρ is

$$\mathbf{x} = (1, p_1\rho^{-1}, p_1 p_2 \rho^{-2}, \ldots, p_1 p_2 \cdots p_{k-1}\rho^{-(k-1)})'.$$

(c) Let $\mathbf{y}$ be the positive left eigenvector corresponding to ρ and scaled so that $\mathbf{x}'\mathbf{y} = 1$. Then, from (9.28),

$$\lim_{t\to\infty} \frac{\mathbf{A}^t}{\rho^t}\mathbf{n}(0) = \mathbf{x}\mathbf{y}'\mathbf{n}(0) = k\mathbf{x}, \text{ say.}$$

Thus for large t, $\mathbf{n}(t) = \mathbf{A}^t\mathbf{n}(0) \approx \rho^t k\mathbf{x}$, and $\mathbf{n}(t) \approx \rho\mathbf{n}(t-1)$.

(d) If $\mathbf{n}(t) = k\mathbf{x}$, then

$$\mathbf{n}(t+1) = \mathbf{A}\mathbf{n}(t) = k\mathbf{A}\mathbf{x} = k\rho\mathbf{x},$$

and a population with age distribution determined by $\mathbf{x}$ is said to have a *stable age distribution* as the age structure remains unchanged. According to (c), we see that as $t\to\infty$ the age distribution tends to the stable age distribution irrespective of the starting age distribution. Once the population reaches the stable age distribution, it increases, decreases, or remains constant in size depending on whether $\rho > 1$, $\rho < 1$, or $\rho = 1$. When $\rho = 1$, the population is said to be *stationary*. Also, $r = \ln\rho$ is called the *intrinsic rate of increase*.

9.56. (Diffusion Model) Suppose we have two identical patches of organisms coupled by diffusion. Suppose there are s stages and that the within-patch demography is described by the population projection matrix $\mathbf{A}$. Let $\mathbf{D} = \text{diag}(d_1, d_2, \ldots, d_s)$ be an $s \times s$ diffusion matrix, where d_i is the probability that an individual in stage i leaves its patch to go to the other patch. If $\mathbf{n}_i(t)$ is the stage abundance vector in patch i, then

$$\begin{pmatrix} \mathbf{n}_1(t+1) \\ \mathbf{n}_2(t+1) \end{pmatrix} = \begin{pmatrix} \mathbf{I}_s - \mathbf{D} & \mathbf{D} \\ \mathbf{D} & \mathbf{I}_s - \mathbf{D} \end{pmatrix} \begin{pmatrix} \mathbf{A} & 0 \\ 0 & \mathbf{A} \end{pmatrix} \begin{pmatrix} \mathbf{n}_1(t) \\ \mathbf{n}_2(t) \end{pmatrix}$$
$$= (\mathbf{K} \otimes \mathbf{D}\mathbf{A} + \mathbf{I}_s \otimes \mathbf{A}) \begin{pmatrix} \mathbf{n}_1(t) \\ \mathbf{n}_2(t) \end{pmatrix}, \quad \mathbf{K} = \begin{pmatrix} -1 & 1 \\ 1 & -1 \end{pmatrix}$$
$$= \mathbf{B} \begin{pmatrix} \mathbf{n}_1(t) \\ \mathbf{n}_2(t) \end{pmatrix},$$

where $\mathbf{B} \geq \mathbf{0}$ and "$\otimes$" is the Kronecker product. For a general modeling method for patches and stages, see Hunter and Caswell [2005].

An important application of the above theory is the life cycle graph described, for example, by Caswell [2001: chapter 4], where a matrix like $\mathbf{A}$ or $\mathbf{B}$ is constructed from the graph. A life cycle can also be described as an *absorbing finite-state Markov chain*, which involves a transition matrix (described below). Caswell [2006] discussed this demographic role of Markov chains in ecology.

Proofs. Section 9.5.

   9.54. Demetrius [1971].

   9.55. Caswell [2001: 84, section 4.5.2].

   9.56. Caswell [2001: 65–66].

9.6   STOCHASTIC MATRICES

9.6.1   Basic Properties

***Definition* 9.16.** A non-negative matrix with each of its row sums equal to 1 is called a *(row) stochastic matrix*. A common application is the *transition matrix* of a finite (discrete time) Markov chain in which the i, jth element of the matrix is the probability of going from state i to state j In what follows, $\mathbf{P}$ is an $n \times n$ stochastic matrix with $\mathbf{P1}_n = \mathbf{1}_n$. When the Markov chain is homogeneous, we are interested in powers $\mathbf{P}^k$ of $\mathbf{P}$. For example, if p_{i0} is the probability that the Markov chain is initially in state i, then $\mathbf{p}_{(0)} = (p_{10}, p_{20}, \ldots, p_{n0})'$ is the *initial probability distribution*; after k transitions, the corresponding probability distribution is $\mathbf{p}_{(k)}$, where $\mathbf{p}'_{(k)} = \mathbf{p}'_{(0)}\mathbf{P}^k$. If, as $k \to \infty$, $\mathbf{p}_{(k)}$ tends to a limit that does not depend on the initial probability distribution, we say that the process has the *ergodic property*. Ergodicity and the so-called *coefficient of ergodicity* play an important role in more general processes such as inhomogeneous Markov processes and products of inhomogeneous non-negative matrices (cf. Seneta [1981].) The matrix $\mathbf{I} - \mathbf{P}$ is called the *Markovian kernel* of the chain, and it has a useful group inverse as well as the usual weak inverse.

If $\mathbf{p}_{(0)}$ is such that $\mathbf{p}_{(k)} = \mathbf{p}_{(0)}$ for all k, we say that $\mathbf{p}^{(0)}$ is *stationary*, and a Markov chain with such an initial distribution is said to be stationary. We shall denote this stationary distribution by $\boldsymbol{\pi}$, where (setting $k = 1$) $\boldsymbol{\pi}'\mathbf{P} = \boldsymbol{\pi}'$ and $\boldsymbol{\pi}'\mathbf{1}_n = 1$.

9.57. If $\mathbf{P} = (p_{ij})$ is a stochastic matrix and $p = \min_i(p_{ii})$, then any eigenvalue λ_j of $\mathbf{P}$ satisfies $|\lambda_j - p| \leq 1 - p$. If p_{ii} and p_{jj} are the smallest main diagonal elements, then all the eigenvalues of $\mathbf{P}$ lie in the interior or on the boundary of the oval

$$|x - p_{ii}||x - p_{jj}| \leq (1 - p_{ii})(1 - p_{jj}).$$

9.58. A stochastic matrix $\mathbf{P}$ is irreducible and aperiodic if and only if $\mathbf{P}^k > 0$ for some positive integer k—that is, if and only if $\mathbf{P}$ is primitive.

9.59. If $\mathbf{P}$ is a stochastic matrix, then so is $\mathbf{P}^m$ for any positive integer m.

9.60. For any stochastic matrix $\mathbf{P}$,

$$\lim_{k \to \infty} \frac{1}{k}(\mathbf{I}_n + \mathbf{P} + \mathbf{P}^2 + \cdots + \mathbf{P}^{k-1}) = \mathbf{R},$$

where $\mathbf{R}$ is stochastic and $\mathbf{RP} = \mathbf{PR} = \mathbf{R} = \mathbf{R}^2$.

Proofs. Section 9.6.1.

     9.57. Quoted by Marcus and Minc [1964: 161].

     9.58. Bapat and Raghavan [1997: 49].

     9.59. $\mathbf{P}^m \mathbf{1}_n = \mathbf{P}^{m-1} \mathbf{1}_n = \cdots = \mathbf{1}_n$.

     9.60. Bapat and Raghavan [1997: 50].

9.6.2 Finite Homogeneous Markov Chain

There is a substantial literature on Markov chains, for example, Hunter [1983b], and more recently, Ching [2006], Hernández and Lasserre [2003], and Norris [1997], so that I shall consider just some basic results in this section.

9.61. Suppose $\mathbf{P}$, the $n \times n$ transition matrix of a finite Markov chain, is irreducible.

(a) Since $\mathbf{P1}_n = \mathbf{1}_n$, $\mathbf{P}$ has an eigenvalue equal to 1. However, since the row sums are all equal, it follows from (9.9b(iii)) that $\rho = 1$ with a positive right eigenvector of $\mathbf{1}_n$. If $\mathbf{q}$ is a positive left eigenvector (i.e., $\mathbf{q}'\mathbf{P} = \mathbf{q}'$), we can scale $\mathbf{q}$ such that $\mathbf{q}'\mathbf{1}_n = 1$; thus $\mathbf{q}$ represents a probability distribution.

(b) $\mathbf{q}'\mathbf{P}^k = \mathbf{q}'\mathbf{P}^{k-1} = \cdots = \mathbf{q}'$.

(c) The irreducible Markov chain has a unique stationary distribution $\boldsymbol{\pi}$, the solution of $\boldsymbol{\pi}'\mathbf{P} = \boldsymbol{\pi}'$ and $\boldsymbol{\pi}'\mathbf{1}_n = 1$. We can identify $\boldsymbol{\pi}$ with $\mathbf{q}$ of (a).

9.62. Suppose that the $n \times n$ transition matrix $\mathbf{P}$ is irreducible with stationary distribution $\boldsymbol{\pi}$. Then:

(a) If $\mathbf{t}$ and $\mathbf{u}$ are any $n \times 1$ vectors, then $(\mathbf{I}_n - \mathbf{P} + \mathbf{tu}')$ is nonsingular if and only if $\boldsymbol{\pi}'\mathbf{t} \neq 0$ and $\mathbf{u}'\mathbf{1}_n \neq 0$. If the latter conditions hold, then $(\mathbf{I}_n - \mathbf{P} + \mathbf{tu}')^{-1}$ is a weak inverse of $\mathbf{I}_n - \mathbf{P}$. Furthermore, any weak inverse can be expressed in the form

$$(\mathbf{I}_n - \mathbf{P} + \mathbf{tu}')^{-1} + \mathbf{1}_n \mathbf{f}' + \mathbf{g}\boldsymbol{\pi}',$$

where $\mathbf{f}$ and $\mathbf{g}$ are arbitrary vectors.

(b) $\mathbf{A}^+ = (\mathbf{I}_n - \mathbf{P} + \boldsymbol{\pi}\mathbf{1}_n')^{-1} + \dfrac{\mathbf{1}_n \boldsymbol{\pi}'}{n\boldsymbol{\pi}'\boldsymbol{\pi}}$.

In addition to the above weak and Moore–Penrose inverses, Hunter [1988] gives expressions for other types of generalized inverses. For further results see Hunter [1990, 1992].

9.63. Suppose $\mathbf{P}$ is a primitive stochastic matrix (i.e., irreducible and aperiodic). Using the above notation, we have the following special case of (9.28).

(a) $\lim_{k \to \infty} \mathbf{P}^k = \mathbf{1}_n \boldsymbol{\pi}'$ $(= \mathbf{Q}_0$, say), where $\boldsymbol{\pi}$ is the unique stationary distribution.

(b) If $\mathbf{p}$ is a probability distribution (i.e., $\mathbf{p}'\mathbf{1}_n - 1$), then
$$\lim_{k \to \infty} \mathbf{p}'\mathbf{P}^k = \mathbf{p}'\mathbf{1}_n \boldsymbol{\pi}' = \boldsymbol{\pi}'.$$

(c) $\mathbf{Q}_0$ is idempotent.

(d) $\mathbf{PQ}_0^m = \mathbf{Q}_0 \mathbf{P}^m = \mathbf{Q}_0$ for all integers $m \geq 1$.

(e) $\mathbf{Q}_0(\mathbf{P} - \mathbf{Q}_0) = \mathbf{0}$.

(f) Every nonzero eigenvalue of $\mathbf{P} - \mathbf{Q}_0$ is also an eigenvalue of $\mathbf{P}$.

9.64. Suppose that a general stochastic matrix $\mathbf{P}$ is expressed in the canonical form of (9.19), where $\mathbf{Q} \neq \mathbf{0}$. Here $\mathbf{Q}$ refers to the submatrix of $\mathbf{P}$ associated with transitions between the inessential states, and $\mathbf{P}$ is reducible.

(a) $\mathbf{Q}^k \to \mathbf{0}$ elementwise and geometrically fast as $k \to \infty$.

(b) $(\mathbf{I}_n - \mathbf{Q})^{-1}$ exists. In finite absorbing chains, this matrix is sometimes called the *fundamental matrix of absorbing chains*.

Definition 9.17. An $n \times n$ stochastic matrix $\mathbf{P}$ is said to be *regular* if its essential indices form a single class that is aperiodic. In this case $\mathbf{P}$ can be expressed in the canonical form (cf. 9.19)
$$\mathbf{P} = \begin{pmatrix} \mathbf{P}_1 & \mathbf{0} \\ \mathbf{R} & \mathbf{Q} \end{pmatrix},$$
where $\mathbf{P}_1$ is a stochastic irreducible aperiodic (primitive) matrix.

9.65. Suppose $\mathbf{P}$ is regular with canonical form described above. Let $\mathbf{q}_1$ be the unique stationary distribution of $\mathbf{P}_1$, and define $\mathbf{q}' = (\mathbf{q}_1, \mathbf{0}')$ to be an $1 \times n$ vector. Then, as $k \to \infty$,
$$\mathbf{P}^k \to \mathbf{1}_n \mathbf{q}'$$
elementwise, where $\mathbf{q}'$ is the unique stationary distribution corresponding to the matrix $\mathbf{P}$, the approach to the limit being geometrically fast. Thus the regularity of $\mathbf{P}$ is a sufficient condition for ergodicity; it is also a necessary condition.

9.66. If an n-state homogeneous Markov chain contains at least two essential classes, then any weighted linear combination of the stationary distribution vectors corresponding to each class, each appropriately augmented by zeros to give an $n \times 1$ vector, is a stationary distribution of the chain.

Proofs. Section 9.6.2.

9.61. Seneta [1981: 118–119].

9.63. Rao and Rao [1998: 483, with $\mathbf{Q}$ instead of $\mathbf{Q}_0$].

9.64. Seneta [1981: 120–123]

9.65. Seneta [1981: 127; 134, exercise 4.9].

9.66. Quoted by Seneta [1981: 134, exercise 4.12].

9.6.3   Countably Infinite Stochastic Matrix

In this section we consider a stochastic matrix with a countable (i.e., finite or denumerably infinite) index set $\{1, 2, \ldots\}$, with our focus on the infinite case. The matrix $\mathbf{P} = (p_{ij})$ will still represent a stochastic matrix, but with infinite row sums adding to unity. As matrix multiplication readily extends to infinite matrices, $\mathbf{P}^k = (p_{ij}^{(k)})$ is well-defined for $k = 1, 2, \ldots$, and it is also stochastic (Seneta [1981: chapter 5]). However, a more sensitive classification of indices is now required.

Definition 9.18. Let

$$l_{ij}^{(1)} = p_{ij} \quad \text{and} \quad l_{ij}^{(k+1)} = \sum_{r:r \neq i} l_{ir}^{(k)} p_{rj}, \quad k = 1, 2, \ldots,$$

with $l_{ij}^{(0)} = 0$, by definition, for all $i, j \in \{1, 2, \ldots\}$. Define, for each i and j,

$$L_{ij} = \sum_{k=0}^{\infty} l_{ij}^{(k)} \quad \text{and} \quad \mu_i = \sum_{k=1}^{\infty} k l_{ii}^{(k)} \leq \infty.$$

An index i (or state i) is said to be *recurrent* if $L_{ii} = 1$ and *transient* if $L_{ii} < 1$. A recurrent index i is said to be *positive-* or *null-recurrent* depending on whether $\mu_i < \infty$ or $\mu_i = \infty$, respectively. Here μ_i is called the *mean recurrence measure* of i. Note that in the Markov chain context, $l_{ij}^{(k)}$ is the probability of going from state i to state j in k steps (or in time k), without revisiting i in the meantime. Thus L_{ii} can be regarded as the probability of staying in or returning to state i for the first time. Also μ_i is the *mean recurrence time* of state i. Thus a state i is recurrent if, starting from state i, we will eventually return to state i with certainty. If state i is transient, then there is a positive probability that the system will never return to state i.

9.67. An inessential index is transient and a recurrent index is essential.

9.68. If i is a recurrent aperiodic index and j is any index such that $j \to i$ (cf. second part of Definition 9.5 in Section 9.4.1), then

$$\lim_{k \to \infty} p_{ij}^{(k)} = \mu_i^{-1} L_{ij}.$$

In particular

$$\lim_{k \to \infty} p_{ii}^{(k)} = \mu_i^{-1}.$$

Proofs. Section 9.6.3.

       9.67. Seneta [1981: 165–166].

       9.68. Seneta [1981: 171].

9.6.4   Infinite Irreducible Stochastic Matrix

The definition of irreducibility given by Definition 9.6 (second definition) applies to infinite matrices; that is, $\mathbf{A} \geq \mathbf{0}$ is irreducible if and only if every pair of indices i and j communicate.

9.69. The following hold for an infinite irreducible stochastic matrix.

(a) Every index has the same period.

(b) The indices are all transient, or all null-recurrent, or all positive-recurrent.

Definition 9.19. In the light of (9.69b) we say that an irreducible $\mathbf{P}$ is a *transient, or null-recurrent, or positive-recurrent matrix* depending on whether any one of its indices is transient, or null-recurrent, or positive-recurrent.

If $\mathbf{v}'\mathbf{P} = \mathbf{v}'$ and $\mathbf{v}$ is a nonzero non-negative (countably infinite) vector, we call $\mathbf{v}$ an *invariant measure*. Note that a multiple of such a measure is still a measure.

9.70. (General Ergodic Theorem)   We have the following series of limits.

(a) Let $\mathbf{P}$ be a primitive (i.e., irreducible and aperiodic) stochastic matrix. If $\mathbf{P}$ is transient or null-recurrent, then for any pair of indices i, j, we have $p_{ij}^{(k)} \to 0$ as $k \to \infty$. If $\mathbf{P}$ is positive-recurrent,

$$\lim_{k \to \infty} p_{ij}^{(k)} = \mu_j^{-1},$$

and the vector $\mathbf{x} = (\mu_i^{-1})$ is the unique stationary distribution (invariant vector) satisfying $\mathbf{x}'\mathbf{P} = \mathbf{x}'$ and $\sum_{i=1}^{\infty} x_i = 1$. The question of computing a finite dimensional approximation for $\mathbf{x}$ is discussed by Seneta [1981: section 7.2].

(b) If $\mathbf{P}$ is irreducible and periodic with period d, then

$$\lim_{k \to \infty} p_{ii}^{(dk)} = d/\mu_i.$$

9.71. If $\mathbf{P}$ is an irreducible transient or null-recurrent matrix, then there exists no invariant measure $\mathbf{v}'$ satisfying $\mathbf{v}'\mathbf{1} < \infty$.

Proofs. Section 9.6.4.

9.69. Seneta [1981: 172].

9.70. Seneta [1981: 177 for (a); 196, exercise 5.1, for (b)].

9.71. Seneta [1981: 178].

9.7   DOUBLY STOCHASTIC MATRICES

Definition 9.20. A square $n \times n$ matrix $\mathbf{A} = (a_{ij})$ is doubly stochastic if $\mathbf{A} \geq 0$ and all its column sums and row sums are 1. Some examples of doubly stochastic matrices are given by Marshall and Olkin [1979: 45–48]. For a reference to doubly stochastic matrices see Bapat and Raghavan [1997: chapter 2].

Definition 9.21. The *diagonal* of a matrix $\mathbf{A}$ associated with the permutation π is the set $\{a_{1\pi(1)}, a_{2\pi(2)}, \ldots, a_{n\pi(1)}\}$, and the corresponding *diagonal product* is $\prod_{i=1}^{n} a_{i\pi(i)}$. A diagonal is said to be *positive* if each element $a_{i\pi(i)}$ in the diagonal is positive.

Definition 9.22. The matrix $\mathbf{A}$ is said to have a *doubly stochastic pattern* if there exists a doubly stochastic matrix with the same pattern of zeros as $\mathbf{A}$.

9.72. A doubly stochastic matrix has a positive diagonal. An algorithm for finding such a diagonal is also available.

9.73. Let $\mathbf{A} = (a_{ij})$ be an $n \times n$ doubly stochastic matrix, and let $y_1 \geq y_2 \geq \cdots \geq y_n$. Then

$$\sum_{i=1}^{k} y_i \geq \sum_{i=1}^{k} \sum_{j=1}^{n} a_{ij} y_j, \quad k = 1, 2, \ldots, n.$$

9.74. The product of a finite number of doubly stochastic matrices is doubly stochastic.

9.75. If $n \times n$ $\mathbf{A}$ is doubly stochastic and nonsingular, then $\mathbf{A}^{-1}$ has row and column sums equal to 1, but it need not have non-negative elements.

9.76. If $\mathbf{A} \geq \mathbf{0}$ is $n \times n$ matrix with row totals and column totals not exceeding unity, then there exists a doubly stochastic $n \times n$ matrix $\mathbf{B}$ such that $\mathbf{B} \geq \mathbf{A}$.

9.77. (Scaling) If $\mathbf{A}$ is a non-negative $n \times n$ matrix with doubly stochastic pattern, then there exist diagonal matrices $\mathbf{D}_1$ and $\mathbf{D}_2$ with positive diagonal elements such that $\mathbf{C} = \mathbf{D}_1 \mathbf{A} \mathbf{D}_2$ is doubly stochastic.

9.78. If $\mathbf{A}$ is non-negative definite and doubly stochastic and has $a_{ii} \leq 1/(n-1)$ for each i, then the non-negative definite square root $\mathbf{A}^{1/2}$ (cf. 10.8) is doubly stochastic.

9.79. Every permutation matrix is a doubly stochastic matrix, because there is a single 1 in every row and column and the remaining elements are zero.

9.80. (Birkhoff–von Neumann) A matrix is doubly stochastic if and only if it is a convex combination of the permutation matrices.

9.81. The set of doubly stochastic matrices is the convex hull of all $n \times n$ permutation matrices (of which there are n), and the latter constitute the extreme points of this set.

9.82. Let $\mathbf{A} = (a_{ij})$ be a doubly stochastic $n \times n$ matrix.

(a) The permanent (cf. Section 4.5) of $\mathbf{A}$ is positive.

(b) $\mathrm{per}(\mathbf{A}) \geq \dfrac{n,}{n^n},$

   with equality if and only if $a_{ij} = n^{-1}$ for all i, j.

9.83. The matrix $(a_{ij}) = (n^{-1})$ is the unique irreducible idempotent $n \times n$ doubly stochastic matrix.

9.84. If $\mathbf{T} = (t_{ij})$ is a real orthogonal matrix, then $\mathbf{A} = (t_{ij}^2)$ is doubly stochastic.

9.85. If $\mathbf{A} \geq \mathbf{0}$ is $n \times n$ $(\mathbf{A} \neq \mathbf{0})$, then $\mathbf{A}$ has a doubly stochastic pattern if and only if every positive entry of $\mathbf{A}$ is contained in a positive diagonal.

***Definition* 9.23.** A matrix $\mathbf{A} = (a_{ij})$ is said to be *orthostochastic* if there exists an orthogonal matrix $\mathbf{T}$ such that $a_{ij} = t_{ij}^2$. If there exists a unitary matrix $\mathbf{U} = (u_{ij})$ such that $a_{ij} = |u_{ij}|^2$, then $\mathbf{A}$ is said to be *unitary-stochastic* (Marshall and Olkin [1979: 23]).

***Definition* 9.24.** A square matrix $\mathbf{A}$ is said to be *doubly substochastic* if $\mathbf{A} \geq 0$ and all row and column sums are at most 1.

9.86. The set of all $n \times n$ doubly substochastic matrices is a convex set.

9.87. From the definition we have that any square submatrix of a doubly substochastic matrix is doubly substochastic.

9.88. If $\mathbf{A} = (a_{ij})$ is doubly substochastic, then there exists a doubly stochastic matrix $\mathbf{B} = (b_{ij})$ such that $a_{ij} \leq b_{ij}$ for all i, j.

9.89. If $\mathbf{A} = (a_{ij})$ and $\mathbf{B} = (b_{ij})$ are doubly substochastic, then their Hadamard (Schur) product $\mathbf{A} \circ \mathbf{B} = (a_{ij}b_{ij})$ is doubly substochastic.

Proofs. Section 9.7.

9.72. Bapat and Raghavan [1998: 63–66].

9.73. Anderson [2003: 646].

9.74. Marshall and Olkin [1979: 20].

9.75. $\mathbf{A}\mathbf{1}_n = \mathbf{1}_n$ implies $\mathbf{1}_n = \mathbf{A}^{-1}\mathbf{1}_n$, and $\mathbf{A}'\mathbf{1}_n = \mathbf{1}_n$ implies $\mathbf{1}_n = \mathbf{A}^{-1\prime}\mathbf{1}_n$.

9.76. Bapat and Raghavan [1997: 75].

9.77. Bapat and Raghavan [1997: 87].

9.78. Marshall and Olkin [1979: 51].

9.80. Bapat and Raghavan [1997: 63], Rao and Rao [1998: 314–315], and Zhang [1999: 127].

9.81. Marshall and Olkin [1979: 19] and Rao and Rao [1998: 308–309].

9.82. Bapat and Raghavan [1997: 93] and Rao and Rao [1998: 314].

9.83. Marshall and Olkin [1979: 19].

9.84. Follows from the fact that the rows and columns of an orthogonal matrix each have unit length.

9.85. Bapat and Raghavan [1997: 68].

9.86. Follows immediately from the idea of a convex combination.

9.88. Horn and Johnson [1991: 165] and Marshall and Olkin [1979: 25].

9.89. Since $b_{ij} \leq 1$, $a_{ij}b_{ij} \leq a_{ij}$.

CHAPTER 10

POSITIVE DEFINITE AND NON-NEGATIVE DEFINITE MATRICES

Quadratic forms that are non-negative definite play an important role in statistical theory, particularly those related to chi-square distributions. They can also be used for establishing a wide variety of inequalities, such as those in Chapter 12.

10.1 INTRODUCTION

***Definition* 10.1.** Let $\mathbf{A}$ be an $n \times n$ Hermitian matrix, and let $\mathbf{x} \in \mathbb{C}^n$. Then $\mathbf{x}^* \mathbf{A} \mathbf{x}$ is said to be a *Hermitian non-negative definite (n.n.d.)* quadratic form if $\mathbf{x}^* \mathbf{A} \mathbf{x} \geq 0$ for all $\mathbf{x}$. If $\mathbf{x}^* \mathbf{A} \mathbf{x}$ is Hermitian n.n.d. we say that $\mathbf{A}$ is Hermitian n.n.d. and we write $\mathbf{A} \succeq \mathbf{0}$. (Some authors use the term positive semi-definite instead of n.n.d. We reserve the former for the following definition.)

If $\mathbf{A}$ is Hermitian and n.n.d., and there exists $\mathbf{x}$, $\mathbf{x} \neq \mathbf{0}$ such that $\mathbf{x}^* \mathbf{A} \mathbf{x} = 0$, we say that $\mathbf{A}$ is *Hermitian positive semidefinite* or positive indefinite. An alternative definition is that $\mathbf{A}$ is n.n.d. and $\det \mathbf{A} = 0$.

If $\mathbf{x}^* \mathbf{A} \mathbf{x} > 0$ for all $\mathbf{x} \neq \mathbf{0}$, then we say that $\mathbf{A}$ is *Hermitian positive definite (p.d.)* definite and write $\mathbf{A} \succ \mathbf{0}$.

Given $n \times n$ Hermitian matrices $\mathbf{A}$ and $\mathbf{B}$, we say that $\mathbf{A} \succeq \mathbf{B}$ if $\mathbf{A} - \mathbf{B} \succeq \mathbf{0}$. Similarly we say that $\mathbf{A} \succ \mathbf{B}$ if $\mathbf{A} - \mathbf{B} \succ \mathbf{0}$. This is referred to as the *(partial) Löwner ordering of matrices*. There are many applications of Löwner ordering in statistics such as estimability and efficiency of estimation.

In most applications, $\mathbf{A}$ is a real symmetric matrix, in which case we simply replace * by $'$, assume $\mathbf{x} \in \mathbb{R}^n$, and drop the term Hermitian in the above definitions and in what follows. Thus a positive definite matrix without the adjective Hermitian will always represent a real symmetric matrix. The same is true for a non-negative definite matrix.

10.1. The following matrices are all assumed to be Hermitian.

(a) $\mathbf{A}_1 \succeq \mathbf{B}_1$ and $\mathbf{A}_2 \succeq \mathbf{B}_2$ imply that $\mathbf{A}_1 + \mathbf{A}_2 \succeq \mathbf{B}_1 + \mathbf{B}_2$.

(b) If $\mathbf{A} \succeq \mathbf{B}$ and $\mathbf{B} \succeq \mathbf{C}$, then $\mathbf{A} \succeq \mathbf{C}$.

(c) $\mathbf{A}_1 \succeq \mathbf{B}_1$ and $\mathbf{A}_2 \succeq \mathbf{B}_2$ do not necessarily imply that $\mathbf{A}_1 \mathbf{A}_2 \succeq \mathbf{B}_1 \mathbf{B}_2$, even if $\mathbf{A}_1 \mathbf{A}_2$ and $\mathbf{B}_1 \mathbf{B}_2$ are Hermitian. Thus $\mathbf{A} \succeq \mathbf{B}$ does not neccessarily imply that $\mathbf{A}^2 \succeq \mathbf{B}^2$.

Proofs. Section 10.1.

10.1a. Consider the corresponding quadratics.

10.1b. $\mathbf{x}^*(\mathbf{A} - \mathbf{C})\mathbf{x} = \mathbf{x}^*(\mathbf{A} - \mathbf{B} + \mathbf{B} - \mathbf{C})\mathbf{x} = \mathbf{x}^*(\mathbf{A} - \mathbf{B})\mathbf{x} + \mathbf{x}^*(\mathbf{B} - \mathbf{C})\mathbf{x} \geq 0$.

10.2   NON-NEGATIVE DEFINITE MATRICES

10.2.1   Some General Properties

In this section we assume that $\mathbf{A}$ is a Hermitian $n \times n$ matrix, unless otherwise stated. The results hold for a real symmetric matrix if we replace $*$ by $'$.

10.2. $\mathbf{A} \succeq \mathbf{0}$ if and only if all its eigenvalues are real and non-negative .

10.3. If $\mathbf{A} \succeq \mathbf{0}$, then from (10.2), $\det \mathbf{A} = \prod_i \lambda_i \geq 0$.

10.4. If $\mathbf{A} \succeq \mathbf{0}$, then trace $\mathbf{A} = \sum_i \lambda_i \geq 0$.

10.5. Given $\mathbf{A} \succeq \mathbf{0}$, then any principal submatrix of $\mathbf{A}$, including $\mathbf{A}$ itself, is non-negative definite. In particular, the diagonal elements of $\mathbf{A}$ are non-negative.

10.6. $\mathbf{A} \succeq \mathbf{0}$ if and only if all principal minors (including $\mathbf{A}$ itself) of $\mathbf{A}$, and not just the leading ones, are non-negative. Note that $\mathbf{A} = \begin{pmatrix} 0 & 0 \\ 0 & -1 \end{pmatrix}$ has non-negative leading principal minors including $\mathbf{A}$ itself, but it is not non-negative definite.

10.7. (Fejer) $\mathbf{A} \succeq \mathbf{0}$ if and only if

$$\sum_{i=1}^{n} \sum_{j=1}^{n} a_{ij} b_{ij} \geq 0$$

for all $n \times n$ non-negative definite matrices $\mathbf{B} = (b_{ij})$.

10.8. If $\mathbf{A} \succeq \mathbf{0}$ and k is a positive integer, there exists a unique non-negative definite matrix $\mathbf{A}^{1/k} \succeq \mathbf{0}$ such that $(\mathbf{A}^{1/k})^k = \mathbf{A}$. In particular, if $\mathbf{A} = \mathbf{U}\mathbf{\Lambda}\mathbf{U}^*$, where $\mathbf{\Lambda} = \mathrm{diag}(\lambda_1, \ldots, \lambda_n)$ and the λ_i are the non-negative eigenvalues of $\mathbf{A}$, then $\mathbf{A}^{1/k} = \mathbf{U}\mathbf{\Lambda}^{1/k}\mathbf{U}^*$, where $\mathbf{\Lambda}^{1/k} = \mathrm{diag}(\lambda_1^{1/k}, \ldots, \lambda_n^{1/k})$. The case $k = 2$ arises frequently in statistics. We note the following.

(a) $\mathbf{A}$ and $\mathbf{A}^{1/k}$ commute.

(b) The eigenvalues of $\mathbf{A}^{1/k}$ are the kth roots of those of $\mathbf{A}$.

(c) rank $\mathbf{A} = \mathrm{rank}(\mathbf{A}^{1/k})$.

(d) If $\mathbf{A}$ is real, then so is $\mathbf{A}^{1/k}$.

(e) For $k = 2$ and $\mathbf{A}$ real, another way of deriving $\mathbf{A}^{1/2}$ is to obtain the Cholesky decomposition $\mathbf{A} = \mathbf{R}\mathbf{R}'$. Then, if $\mathbf{R} = \mathbf{P}\mathbf{\Sigma}\mathbf{Q}'$ is the singular value decomposition of $\mathbf{R}$, we have $\mathbf{A}^{1/2} = \mathbf{P}\mathbf{\Sigma}\mathbf{P}'$ as $(\mathbf{A}^{1/2})^2 = \mathbf{P}\mathbf{\Sigma}^2\mathbf{P}' = \mathbf{R}\mathbf{R}'$.

(f) If $\mathbf{A} \succ \mathbf{0}$, then $(\mathbf{A}^{-1})^{1/2} = (\mathbf{A}^{1/2})^{-1}$.

10.9. If $\mathbf{A} \succeq \mathbf{0}$, then the matrix (a_{ij}^k) for k a positive integer is non-negative definite.

10.10. If $\mathbf{A}$ is of rank r, then $\mathbf{A} \succeq \mathbf{0}$ if and only if $\mathbf{A} = \mathbf{R}\mathbf{R}^*$, where $\mathbf{R}$ is $n \times n$ of rank r. The result is also true if we replace $\mathbf{R}$ by an $n \times r$ matrix of rank r, as we have a full-rank factorization of $\mathbf{A}$.

10.11. $\mathbf{A} \succeq \mathbf{0}$ is of rank r if and only if there exists an $n \times r$ matrix $\mathbf{S}$ of rank r such that $\mathbf{S}^*\mathbf{A}\mathbf{S} = \mathbf{I}_r$.

10.12. Let $\mathbf{A} \succeq \mathbf{0}$.

(a) $\mathbf{C}\mathbf{A}\mathbf{C}^* \succeq \mathbf{0}$.

(b) If $\mathbf{C}\mathbf{A}\mathbf{C}^* = \mathbf{0}$, then $\mathbf{C}\mathbf{A} = \mathbf{0}$; in particular, $\mathbf{C}\mathbf{C}^* = \mathbf{0}$ implies that $\mathbf{C} = \mathbf{0}$.

(c) $\mathrm{rank}(\mathbf{C}^*\mathbf{A}\mathbf{C}) = \mathrm{rank}(\mathbf{A}\mathbf{C})$.

10.13. If $\mathbf{A} \succeq \mathbf{0}$ and $a_{ii} = 0$, then $a_{ij} = 0$ for all $j = 1, 2, \ldots, n$. Since $\mathbf{A}$ is Hermitian, $a_{ii} = 0$ if and only if the row and column containing a_{ii} consist entirely of zeros.

10.14. If $\mathbf{A} \succeq \mathbf{0}$, then $\mathbf{A}^k \succeq \mathbf{0}$ for k a positive integer.

10.15. If $\mathrm{trace}\,\mathbf{A} \geq \Re\,\mathrm{trace}(\mathbf{A}\mathbf{U})$ for all unitary matrices $\mathbf{U}$ (i.e., $\mathbf{U}^*\mathbf{U} = \mathbf{I}_n$), then $\mathbf{A}$ is non-negative definite. (Here $\Re$ denotes "real part of.")

10.16. Let $\mathbf{A}$ be any $m \times n$ real matrix, and let $\mathbf{V}$ be an $m \times m$ non-negative definite matrix. If $\mathbf{Z}$ is any matrix such that $\mathcal{C}(\mathbf{Z}) = \mathcal{N}(\mathbf{A}')$ (i.e., the columns of $\mathbf{Z}$ span the null space of $\mathbf{A}'$), then $\mathcal{C}(\mathbf{A}) \cap \mathcal{C}(\mathbf{V}\mathbf{Z}) = \mathbf{0}$ and

$$\mathcal{C}(\mathbf{A}, \mathbf{V}) = \mathcal{C}(\mathbf{A}, \mathbf{V}\mathbf{Z}) = \mathcal{C}(\mathbf{A}) \oplus \mathcal{C}(\mathbf{V}\mathbf{Z}).$$

We can express $\mathbf{Z}$ in the form $\mathbf{Z} = \mathbf{I} - (\mathbf{A}')^-\mathbf{A}'$.

10.17. Let $\mathbf{A} = (a_{ij}) \succeq \mathbf{0}$. If $f(z) = a_0 + a_1 z + a_2 z^2 + \ldots$ is an analytic function with non-negative coefficients and radius of convergence $R > 0$, then the matrix with (i, j)th elements $f(a_{ij})$ is n.n.d. if all $|a_{ij}| < R$.

10.18. We have the following results.

(a) Given the real matrix $\mathbf{A} \succeq \mathbf{0}$, then

$$\mathcal{C}(\mathbf{B}\mathbf{A}\mathbf{B}') = \mathcal{C}(\mathbf{B}\mathbf{A}) \quad \text{and} \quad \mathrm{rank}(\mathbf{B}\mathbf{A}\mathbf{B}') = \mathrm{rank}(\mathbf{B}\mathbf{A}) = \mathrm{rank}(\mathbf{A}\mathbf{B}').$$

(b) If $\begin{pmatrix} \mathbf{A} & \mathbf{B} \\ \mathbf{B}' & \mathbf{C} \end{pmatrix}$ is n.n.d., then $\mathcal{C}\mathbf{B}) \subseteq \mathcal{C}(\mathbf{A})$ and $\mathcal{C}(\mathbf{B}') \subseteq \mathcal{C}(\mathbf{C})$.

10.19. Let $\mathbf{A}$ be an $n \times n$ real symmetric idempotent matrix, and suppose that $\{\mathbf{B}_1, \mathbf{B}_2, \ldots, \mathbf{B}_k\}$ is a set of real $n \times n$ non-negative definite matrices such that

$$\mathbf{I}_n = \mathbf{A} + \sum_{i=1}^{k} \mathbf{B}_i.$$

Then $\mathbf{AB}_i = \mathbf{B}_i\mathbf{A} = \mathbf{0}$ for $i = 1, 2, \ldots, k$.

Proofs. Section 10.2.1.

10.2. Horn and Johnson [1985: 402] and Rao and Rao [1998: 181].

10.5. Set appropriate elements of $\mathbf{x}$ in $\mathbf{x}^*\mathbf{Ax}$ equal to zero.

10.6. Abadir and Magnus [2005: 223] and Zhang [1999: 160].

10.7. Horn and Johnson [1985: 459].

10.8. Horn and Johnson [1985: 405 for (a–(d)], Golub and Van Loan [1996: 149 for (e)], and Abadir and Magnus [2005: 221 for (f)],

10.9. Horn and Johnson [1985: 461].

10.10. Seber and Lee [2003: 460, real case]. We can also choose $\mathbf{R}$ such that $\mathbf{R}^*\mathbf{R} = \mathbf{\Lambda}$, where $\mathbf{\Lambda}$ is a diagonal matrix with diagonal elements the positive eigenvalues of $\mathbf{A}$. (cf. Abadir and Magnus [2005: 219]).

10.11. Seber and Lee [2003: 460, real case].

10.12. Abadir and Magnus [2005: 221, real case].

10.13. Zhang [1999: 161].

10.14. From (10.8) with $\mathbf{B} = \mathbf{A}^{1/2}$, we have $\mathbf{A}^k = (\mathbf{B}^2)^k = (\mathbf{B}^k)^2 = \mathbf{C}^2$, say, where $\mathbf{C}$ is symmetric.

10.15. Rao and Rao [1998: 343].

10.16. Harville [1997: 387].

10.17. Rao and Rao [1998: 214].

10.18. Sengupta and Jammalamadaka [2003: 45, $\mathbf{A}$ and $\mathbf{B}$ interchanged in (a)].

10.19. Graybill [1983: 398].

10.2.2   Gram Matrix

***Definition* 10.2.** Let $\{\mathbf{v}_1, \mathbf{v}_2, \ldots, \mathbf{v}_k\}$ be a set of n vectors in an inner product space $\mathcal{V}$ with inner product $\langle \cdot, \cdot \rangle$. Then the *Gram* matrix of the vectors $\mathbf{v}_i$ is the $k \times k$ matrix $\mathbf{G} = (g_{ij})$, where $g_{ij} = \langle \mathbf{v}_i, \mathbf{v}_j \rangle$.

10.20. Let $\mathbf{G}$ be the Gram matrix of the vectors $\{\mathbf{w}_1, \mathbf{w}_2, \ldots, \mathbf{w}_k\}$ in $\mathbb{C}^n$ with respect to the inner product $\langle \cdot, \cdot \rangle$, and let $\mathbf{W} = (\mathbf{w}_1, \mathbf{w}_2, \ldots, \mathbf{w}_k)$ be an $n \times k$ matrix.

(a) $\mathbf{G}$ is Hermitian non-negative definite.

(b) $\mathbf{G}$ is nonsingular if and only if the vectors $\mathbf{w}_1, \ldots, \mathbf{w}_k$ are linearly independent.

(c) There exists a Hermitian positive definite $n \times n$ matrix $\mathbf{A}$ such that

$$\mathbf{G} = \mathbf{W}^* \mathbf{A} \mathbf{W}.$$

(d) If r is the maximum number of linearly independent vectors in the set of vectors $\{\mathbf{w}_1, \ldots, \mathbf{w}_k\}$, then $\operatorname{rank} \mathbf{G} = \operatorname{rank} \mathbf{W} = r$.

(e) If $\langle \mathbf{x}, \mathbf{y} \rangle = \mathbf{x}^* \mathbf{y}$, then $\mathbf{A} = \mathbf{I}_n$ in (c).

Proofs. Section 10.2.2.

10.20. Horn and Johnson [1985: 407–408].

10.2.3   Doubly Non-negative Matrix

***Definition* 10.3.** A (real) non-negative definite matrix that is also non-negative (i.e., $\mathbf{A} \geq \mathbf{0}$ with non-negative elements) is referred to as *doubly non-negative* matrix. A square matrix $\mathbf{A}$ is said to be *completely positive* if there exists an $n \times k$ matrix $\mathbf{B}$ such that $\mathbf{B} \geq \mathbf{0}$ and $\mathbf{A} = \mathbf{B}\mathbf{B}'$. (The smallest value of k is called the *cp-rank* of $\mathbf{A}$ and its properties are considered by Berman and Shaked-Monderer [2003: chapter 3].)

Completely positive matrices arise in relation to graph theory, block designs, some maximum efficiency-robust tests, and in a Markovian model for DNA evolution (cf. Berman and Shaked-Monderer [2003: 68–70], who also give further references). The following results make full use of (10.10).

10.21. It follows immediately from the definition that $\mathbf{A}$ is completely positive if and only if it can be expressed in the form

$$\mathbf{A} = \sum_{i=1}^{k} \mathbf{b}_i \mathbf{b}_i', \quad \mathbf{b}_i \geq \mathbf{0}, \quad i = 1, \ldots, k,$$

where $\mathbf{b}_i$ is the ith column of $\mathbf{B}$.

10.22. A completely positive matrix is doubly non-negative. However, the converse is not necessarily true, except in some cases. For example, a rank 1 or rank 2 doubly non-negative matrix is completely positive.

10.23. We have the following results for completely positive matrices.

(a) The sum of completely positive matrices is completely positive.

(b) The Kronecker product of two completely positive matrices is completely positive.

(c) If $\mathbf{A}$ is a completely positive $n \times n$ matrix, and $\mathbf{C}$ is an $m \times n$ non-negative matrix, then $\mathbf{CAC}'$ is completely positive. Two special cases that are of interest when $m = n$ are when C is a permutation matrix, or a diagonal matrix with non-negative elements.

(d) If $\mathbf{A}$ is completely positive, then so is $\mathbf{A}^k$, where k is a positive integer.

(e) Let $\mathbf{A}$ and $\mathbf{B}$ be $n \times n$ completely positive matrices with columns $\mathbf{a}_i$ and $\mathbf{b}_i$, respectively. Then the Hadamard product

$$\mathbf{A} \circ \mathbf{B} = \sum_{i=1}^{n} \sum_{i=1}^{n} (\mathbf{a}_i \circ \mathbf{b}_j)(\mathbf{a}_i \circ \mathbf{b}_j)'$$

is also completely positive.

(f) The principal submatrices of a completely positive matrix are completely positive.

10.24. If $\mathbf{A}$ is a symmetric $n \times n$ *totally non-negative matrix* (i.e. every minor is positive), then $\mathbf{A}$ is completely positive. Furthermore, since $\mathbf{A} = \mathbf{BB}'$, we can choose $\mathbf{B}$ to be either a non-negative upper-triangular matrix or a non-negative lower-triangular matrix.

Proofs. Section 10.2.3.

10.22. Berman and Shaked-Monderer [2003: 64].

10.23a. This follows from (10.21).

10.23b. This follows from $(\mathbf{BB}') \otimes (\mathbf{CC}') = (\mathbf{B} \otimes \mathbf{C})(\mathbf{B} \otimes \mathbf{C})'$.

10.23c. Since $\mathbf{A} = \mathbf{BB}'$, the result follows from $\mathbf{CBB}'\mathbf{C}' = (\mathbf{CB})(\mathbf{CB})'$.

10.23d. The result is obvious when $k = 2l$ and follows from (c) when $k = 2l+1$.

10.23e. This follows from (10.21) and the fact that $\mathbf{cc}'$, where $\mathbf{c} = \mathbf{a}_i \circ \mathbf{b}_j$, is non-negative definite and the sum of non-negative definite matrices is non-negative.

10.23f. Berman and Shaked-Monderer [2003: 64–66].

10.24. Berman and Shaked-Monderer [2003: 126].

10.3 POSITIVE DEFINITE MATRICES

In this section we assume that $\mathbf{A}$ is a Hermitian $n \times n$ matrix, unless otherwise stated. Note that the eigenvalues of a Hermitian matrix are real.

10.25. There exists a real number a such that $\mathbf{I}_n + a\mathbf{A} \succ \mathbf{0}$.

10.26. (Kato) If $\mathbf{A}$ has no eigenvalue in the interval $[a, b]$, then $(\mathbf{A} - a\mathbf{I}_n)(\mathbf{A} - b\mathbf{I}_n) \succ \mathbf{0}$.

10.27. If $\mathbf{A} = (a_{ij}) \succ \mathbf{0}$, then so are $\mathbf{A}'$, $\overline{\mathbf{A}} = (\overline{a}_{ij})$, and $\mathbf{A}^{-1}$.

10.28. Given the inner product $\langle \mathbf{x}, \mathbf{y} \rangle = \mathbf{x}^*\mathbf{y}$, then $\mathbf{A} \succ \mathbf{0}$ if and only if $\mathbf{A}$ is the Gram matrix (cf. Section 10.2.2) of n linearly independent vectors.

10.29. $\mathbf{A} \succ \mathbf{0}$ if and only if all its eigenvalues are positive.

10.30. $\mathbf{A} \succ \mathbf{0}$ if and only if there exists a nonsingular matrix $\mathbf{R}$ such that $\mathbf{A} = \mathbf{R}\mathbf{R}^*$.

10.31. Let $\mathbf{A} \succ \mathbf{0}$, and let $\mathbf{C}$ be $p \times n$ of rank q $(q \le p)$. Then:

(a) $\mathbf{C}\mathbf{A}\mathbf{C}^* \succeq \mathbf{0}$.

(b) $\operatorname{rank}(\mathbf{C}\mathbf{A}\mathbf{C}^*) = \operatorname{rank}(\mathbf{C})$.

(c) $\mathbf{C}\mathbf{A}\mathbf{C}^* \succ \mathbf{0}$ if $q = p$.

(d) $\mathbf{B}^*\mathbf{A}\mathbf{B} = \mathbf{0}$ if and only if $\mathbf{B} = \mathbf{0}$.

10.32. If $\mathbf{A} \succ \mathbf{0}$ and k is a positive integer, then, arguing as in (10.8), there exists a unique $\mathbf{A}^{1/k} \succ \mathbf{0}$ such that $(\mathbf{A}^{1/k})^k = \mathbf{A}$. A particularly useful case is $k = 2$.

10.33. Consider the quadratic $\mathbf{x}^*\mathbf{A}\mathbf{x}$, where $\mathbf{A}$ is Hermitian.

(a) By relabeling the elements of $\mathbf{x}$, we see that if $\mathbf{A} \succ \mathbf{0}$, then so is any matrix obtained by interchanging any rows and the corresponding columns.

(b) By setting some of the x_i equal to zero, we see that the principal submatrices of $\mathbf{A}$ are all Hermitian p.d. In particular, the diagonal elements of $\mathbf{A}$ are all positive.

10.34. If $\mathbf{A} \succ \mathbf{0}$ then, since the diagonal elements are positive, we have:

(a) (Hadamard) $0 < \det \mathbf{A} \le a_{11}a_{22}\cdots a_{nn}$ with equality if and only if $\mathbf{A}$ is diagonal (see also (12.27)).

(b) $\operatorname{trace} \mathbf{A} > 0$.

(c) $|a_{ij}| < \sqrt{a_{ii}a_{jj}} \le \max\{a_{ii}, a_{jj}\}$, $i \ne j$.

10.35. $\mathbf{A} \succ \mathbf{0}$ if and only if all the leading principal minors are positive (including $\det \mathbf{A}$).

10.36. $\mathbf{A} \succ \mathbf{0}$ if and only if the principal minors in any nested sequence of n principal minors are positive.

10.37. If $\mathbf{A} \succ \mathbf{0}$, then from (10.27) we have $\mathbf{A}^{-1} = (a^{ij}) \succ \mathbf{0}$. Furthermore:

(a) $a^{ii} a_{ii} \geq 1$.

(b) If $\mathbf{A}$ is real and $a_{ij} < 0$ for all $i \neq j$, then $a^{ij} > 0$ for all i, j.

(c) Let

$$\mathbf{A} = \begin{pmatrix} \mathbf{A}_{11} & \mathbf{A}_{12} \\ \mathbf{A}_{21} & \mathbf{A}_{22} \end{pmatrix} \quad \text{and} \quad \mathbf{A}^{-1} = \mathbf{B} = \begin{pmatrix} \mathbf{B}_{11} & \mathbf{B}_{12} \\ \mathbf{B}_{21} & \mathbf{B}_{22} \end{pmatrix},$$

where $\mathbf{A}_{11}$ and $\mathbf{B}_{11}$ are $m \times m$ matrices. Then the ith diagonal element of $\mathbf{A}_{11}$ is greater than or equal to the ith diagonal element of $\mathbf{B}_{11}^{-1}$.

10.38. If $\mathbf{A} \succ \mathbf{0}$, then $\mathbf{A} + \mathbf{A}^{-1} \succeq 2\mathbf{I}_n$.

10.39. If $\mathbf{A} \succ \mathbf{0}$ is a real $n \times n$ matrix, then

$$\log(\det \mathbf{A}) \leq \operatorname{trace} \mathbf{A} - n,$$

with equality if and only if $\mathbf{A} = \mathbf{I}_n$.

10.40. If $\mathbf{A} \succ \mathbf{0}$ is a real $n \times n$ matrix, α a real scalar, and $\mathbf{a}$ a real $n \times 1$ vector, then

$$\alpha \mathbf{A} - \mathbf{a}\mathbf{a}' \succeq \mathbf{0} \quad \text{if and only if} \quad \mathbf{a}' \mathbf{A}^{-1} \mathbf{a} \leq \alpha.$$

10.41. If $\mathbf{A}$ is a real symmetric matrix, then there exists a scalar t such that $\mathbf{A} + t\mathbf{I}_n \succ \mathbf{0}$.

10.42. If $\mathbf{A}$ is any $m \times n$ matrix of rank r, then, from the corresponding quadratic form, $\mathbf{A}^*\mathbf{A}$ is Hermitian n.n.d. of rank r if $r < n$ and Hermitian p.d. if $r = n$.

10.43. (Otrowski–Taussky) If $\mathbf{A}$ is any $n \times n$ matrix such that $\mathbf{B} = \frac{1}{2}(\mathbf{A} + \mathbf{A}^*) \succ \mathbf{0}$, then

$$\det \mathbf{B} \leq |\det \mathbf{A}|,$$

with equality if and only if $\mathbf{A}$ is Hermitian.

10.44. If $\mathbf{A}$ is an $n \times n$ real symmetric matrix that is d.d. namely, strictly row or column diagonally dominant (cf. Section 8.4) and if $a_{ii} > 0$ for all $i = 1, 2, \ldots, n$, then it follows from (8.53b) and (10.35) that $\mathbf{A} \succ \mathbf{0}$.

***Definition* 10.4.** (Hilbert Matrix) The $n \times n$ matrix $\mathbf{H}(n) = (h_{ij})$, where $h_{ij} = 1/(i + j - 1)$, is called a *Hilbert matrix* of order n. It is well known that $\mathbf{H}(n)$ is highly ill-conditioned (e.g., Seber and Lee [2003: 166, 372]) and has a condition number of approximately $e^{3.5n}$ for large n. It arises in the fitting of polynomial regression models.

10.45. The Hilbert matrix $\mathbf{H}(n)$ is positive definite.

Proofs. Section 10.3.

10.25. Abadir and Magnus [2005: 218].

10.26. Abadir and Magnus [2005: 218, real case].

10.27. We have $a_{ji} = \bar{a}_{ij}$ so that $\sum \sum (a_{ij} + \bar{a}_{ij}) \bar{x}_i x_j$ is unchanged if we replace a_{ij} by a_{ji} or $\bar{a}_{ij}$. Also, if $\mathbf{x} = \mathbf{A}\mathbf{y}$, then $\mathbf{x}^* \mathbf{A}^{-1} \mathbf{x} = \mathbf{y}^* \mathbf{A}\mathbf{y}$.

10.28. Quoted by Berman and Shaked-Monderer [2003: 16] and proved by Horn and Johnson [1983: 407–408].

10.29. Horn and Johnson [1983: 402].

10.30. Horn and Johnson [1983: 406].

10.31. Abadir and Magnus [2005: 221, real case], Horn and Johnson [1983: 399, complex case], and Seber and Lee [2003: 461, real case].

10.34a. Abadir and Magnus [2005: 337] and Horn and Johnson [1983: 477].

10.34b. The eigenvalues are positive so that their sum (the trace) is positive.

10.34c. Harville [2001: 101, exercise 39].

10.35. Abadir and Magnus [2005: 223] and Horn and Johnson [1983: 404].

10.36. Permute rows and corresponding columns and note that $\mathbf{\Pi}'\mathbf{A}\mathbf{\Pi} \succ \mathbf{0}$ if and only if $\mathbf{A} \succ \mathbf{0}$ for the permutation matrix $\mathbf{\Pi}$; see Horn and Johnson [1985: 404].

10.37. Graybill [1983: 402–403, real case].

10.38. This follows from $\mathbf{U}^*(\mathbf{A} + \mathbf{A}^{-1})\mathbf{U} - 2\mathbf{I}_n = \mathbf{\Lambda} + \mathbf{\Lambda}^{-1} - 2\mathbf{I}_n \succeq \mathbf{0}$, since $\lambda_i + \lambda_i^{-1} - 2 = (\lambda_i^{1/2} - \lambda_i^{-1/2})^2 \geq 0$.

10.39. Abadir and Maganus [2005: 333].

10.40. Farebrother [1976].

10.41. Graybill [1983: 408–409].

10.43. Horn and Johnson [1985: 481].

10.45. This follows from the fact that if $\mathcal{V}$ is the space of continuous functions on $[0, 1]$, with inner product

$$\langle f, g \rangle = \int_0^1 f(x)g(x)\,dx,$$

then $\mathbf{H}(n)$ is the Gram matrix of $f_i(x) = x^{i-1}$, $i = 1, \ldots, n$ (Berman and Shaked-Monderer [2003: 16]).

10.4   PAIRS OF MATRICES

10.4.1   Non-negative or Positive Definite Difference

In this and subsequent sections I give a number of results for pairs of matrices. I have tried to be systematic with the consequence that some of the results overlap.

10.46. Suppose $\mathbf{A}$ and $\mathbf{B}$ are Hermitian $n \times n$ matrices.

(a) $\mathbf{A} \succeq \mathbf{B}$ if and only if $\mathbf{R}^*\mathbf{A}\mathbf{R} \succeq \mathbf{R}^*\mathbf{B}\mathbf{R}$ for nonsingular $\mathbf{R}$.

(b) Let $\mathbf{S}$ be any $n \times m$ matrix, then:

    (i) $\mathbf{A} \succ \mathbf{B}$ implies that $\mathbf{S}^* \mathbf{A} \mathbf{S} \succeq \mathbf{S}^* \mathbf{B} \mathbf{S}$.

    (ii) If $m \leq n$ and rank $\mathbf{S} = m$, then $\mathbf{A} \succ \mathbf{B}$ implies $\mathbf{S}^* \mathbf{A} \mathbf{S} \succ \mathbf{S}^* \mathbf{B} \mathbf{S}$.

10.47. Let $\mathbf{A}$ and $\mathbf{B}$ be $n \times n$ Hermitian matrices.

(a) If $\mathbf{A} \succeq \mathbf{B}$, then the following hold.

    (i) $\lambda_i(\mathbf{A}) \geq \lambda_i(\mathbf{B})$, where in each case the λ_i are ordered $\lambda_1 \geq \lambda_2 \geq \cdots \geq \lambda_n$.

    (ii) From (i) and (6.17b) we have trace $\mathbf{A} \geq$ trace $\mathbf{B}$.

    (iii) From (i) and (6.17a) we have $\|\mathbf{A}\|_F \geq \|\mathbf{B}\|_F$, where $\|\cdot\|_F$ is the Frobenius norm.

    (iv) If $\mathbf{A} \succ \mathbf{B}$, then the above inequalities are strict.

(b) If $\lambda_i(\mathbf{A}) \geq \lambda_i(\mathbf{B})$ for each i, then there exists a unitary matrix $\mathbf{U}$ such that

$$\mathbf{U}^* \mathbf{A} \mathbf{U} \succeq \mathbf{B}.$$

10.48. Let $\mathbf{A}$ and $\mathbf{B}$ be Hermitian non-negative $n \times n$ matrices. If $\mathbf{A} \succeq \mathbf{B}$, then the following hold.

(a) rank $\mathbf{A} \geq$ rank $\mathbf{B}$.

(b) det $\mathbf{A} \geq$ det $\mathbf{B}$.

(c) $\mathbf{A}^{1/2} \succeq \mathbf{B}^{1/2}$ (cf. 10.8).

(d) trace $\mathbf{A} \geq$ trace $\mathbf{B}$.

(e) It is not true in general that $\mathbf{A}^2 \succeq \mathbf{B}^2$.

(f) Suppose $\mathbf{A}$ and $\mathbf{B}$ commute, then $\mathbf{A}^k \succeq \mathbf{B}^k$ for $k = 2, 3, \ldots$.

10.49. Suppose $\mathbf{A}$ and $\mathbf{B}$ are Hermitian $n \times n$ matrices. If $\mathbf{B} \succ \mathbf{0}$ and $\mathbf{A} \succeq \mathbf{B}$, then:

(a) $\mathbf{A} = \mathbf{A} - \mathbf{B} + \mathbf{B} \succ \mathbf{0}$.

(b) If $\mathbf{A}_r$ is a principal submatrix of $\mathbf{A}$ of order r and $\mathbf{B}_r$ is the corresponding submatrix of $\mathbf{B}$, then $\mathbf{A}_r \succeq \mathbf{B}_r$.

10.50. Let $\mathbf{B} \succ \mathbf{0}$ be Hermitian and $\mathbf{A}$ be Hermitian n.n.d. (respectively p.d.).

(a) The eigenvalues of $\mathbf{A} \mathbf{B}^{-1}$, namely the roots of $\det(\mathbf{A} - \lambda \mathbf{B}) = 0$, are real and non-negative (respectively positive) because they are the same as those of $\mathbf{B}^{-1/2} \mathbf{A} \mathbf{B}^{-1/2}$, which is n.n.d. (respectively p.d.)

(b) $\mathbf{B} - \mathbf{A}$ is n.n.d. (respectively p.d.) if and only if the eigenvalues λ_i of $\mathbf{A} \mathbf{B}^{-1}$ all satisfy $\lambda_i \leq 1$ (respectively $\lambda_i < 1$).

10.51. Let $\mathbf{A}$ and $\mathbf{B}$ be $n \times n$ Hermitian p.d. matrices. Then:

(a) $\mathbf{A} \succeq \mathbf{B}$ if and only if $\mathbf{B}^{-1} \succeq \mathbf{A}^{-1}$.

(b) $\mathbf{A} \succ \mathbf{B}$ if and only if $\mathbf{B}^{-1} \succ \mathbf{A}^{-1}$.

(c) If $\mathbf{A} \succeq \mathbf{B}$, then $\lambda_i(\mathbf{A}) \geq \lambda_i(\mathbf{B}) > 0$ (cf. 10.47a(i)).

(d) If $\mathbf{A} \succeq \mathbf{B}$, then, from (c), trace $\mathbf{A} \geq$ trace $\mathbf{B}$ and (from 6.17c) det $\mathbf{A} \geq$ det $\mathbf{B}$. Equality occurs in each case if and only if $\mathbf{A} = \mathbf{B}$.

10.52. Let $\mathbf{A}$ and $\mathbf{B}$ be $n \times n$ real n.n.d. matrices.

(a) The following two statements are equivalent:

  (1) $\mathbf{A} \succeq \mathbf{B}$.
  (2) $\mathcal{C}(\mathbf{B}) \subseteq \mathcal{C}(\mathbf{A})$ and $\lambda_{\max}(\mathbf{B}\mathbf{A}^-) \leq 1$, where $\lambda_{\max}(\mathbf{B}\mathbf{A}^-)$ is independent of the choice of weak inverse $\mathbf{A}^-$. For example, we can choose $\mathbf{A}^+$, the Moore–Penrose inverse.

(b) If $\mathbf{A} \succeq \mathbf{B} \succeq \mathbf{0}$, then $\mathbf{B}^+ \succeq \mathbf{A}^+$ if and only if $\mathcal{C}(\mathbf{A}) = \mathcal{C}(\mathbf{B})$.

10.53. If $\mathbf{A}$ and $\mathbf{B}$ are real symmetric $n \times n$ nonsingular matrices and $\mathbf{A} \succ \mathbf{B}$, then $\mathbf{B}^{-1} \succ \mathbf{A}^{-1}$.

10.54. Given real $n \times n$ matrices $\mathbf{A} \succ \mathbf{0}$ and symmetric $\mathbf{B} = (b_{ij})$, then $\mathbf{A} - \mathbf{B} \succ \mathbf{0}$ provided that the $|b_{ij}|$ are all sufficiently small. In particular, $\mathbf{A} - t\mathbf{B} \succ \mathbf{0}$ for $|t|$ sufficiently small. Similarly, for sufficiently small positive t, $\mathbf{A} + t\mathbf{B} \succ \mathbf{0}$.

10.55. (Regression) Let $\mathbf{V} \succ \mathbf{0}$ be $n \times n$, and let $\mathbf{X}$ be an $n \times p$ matrix of rank p. Then:

(a) $\mathbf{V} \succeq \mathbf{X}(\mathbf{X}'\mathbf{V}^{-1}\mathbf{X})^{-1}\mathbf{X}'$.

(b) $\mathbf{X}'\mathbf{V}\mathbf{X} \succeq (\mathbf{X}'\mathbf{V}^{-1}\mathbf{X})^{-1}$ for any $\mathbf{X}$ such that $\mathbf{X}'\mathbf{X} = \mathbf{I}_p$.

Proofs. Section 10.4.1.

10.46. Horn and Johnson [1985: 470]. We have $\mathbf{x}^*\mathbf{R}^*\mathbf{A}\mathbf{R}\mathbf{x} = \mathbf{y}^*\mathbf{A}\mathbf{y}$ and $\mathbf{x} = \mathbf{0}$ if and only if $\mathbf{y} = \mathbf{0}$.

10.47a(i). Horn and Johnson [1985: 182, with $\mathbf{A}$ and $\mathbf{B}$ relabeled, $\mathbf{B}$ becoming $\mathbf{A} - \mathbf{B} \succeq \mathbf{0}$, and eigenvalues in the reverse order] and Zhang [1999: 227].

10.47b. Zhang [1999: 235].

10.48. Abadir and Magnus [2005: 332, for (c), (e), and (f)] and Zhang [1999: 169–170, for (a)–(d)].

10.49a. We have $\mathbf{x}^*(\mathbf{A} - \mathbf{B})\mathbf{x} + \mathbf{x}^*\mathbf{B}\mathbf{x} \geq \mathbf{x}^*\mathbf{B}\mathbf{x} > 0$.

10.49b. This follows by appropriately choosing $\mathbf{x}$ in $\mathbf{x}^*(\mathbf{A} - \mathbf{B})\mathbf{x}$.

10.50. Dhrymes [2000: 86–89, real case] and Horn and Johnson [1985: 471, with $\mathbf{A}$ and $\mathbf{B}$ interchanged].

10.51. Dhrymes [2000: 89, for (a)] and Horn and Johnson [1983: 471].

10.52. Liski and Puntanen [1989].

10.53. Graybill [1983: 409].

10.54. Graybill [1983: 409] and Seber [1977: 388].

10.55. Abadir and Magnus [2005: 342].

10.4.2   One or More Non-negative Definite Matrices

In this section we consider a number of inequalities for non-negative definite matrices. For further such inequalities, the reader should refer to Chapter 12, and to Chapter 6 for those relating to eigenvalues.

10.56. Suppose $\mathbf{A}$ and $\mathbf{B}$ are $n \times n$ Hermitian matrices with $\mathbf{B} \succeq \mathbf{0}$.

(a) $\lambda_i(\mathbf{A} + \mathbf{B}) \geq \lambda_i(\mathbf{A})$, $i = 1, 2, \ldots, n$, where $\lambda_1 \geq \lambda_2 \geq \ldots \geq \lambda_n$ are the (real) ordered eigenvalues of the particular matrix. If $\mathbf{B} \succ \mathbf{0}$, then the inequality is strict.

(b) From (a) we have $\text{trace}(\mathbf{A} + \mathbf{B}) \geq \text{trace } \mathbf{A}$.

(c) $\|\mathbf{A} + \mathbf{B}\|_F \geq \|\mathbf{A}\|_F$, where $\|\cdot\|_F$ is the Frobenius norm.

10.57. Let $\mathbf{A}$ and $\mathbf{B}$ be $n \times n$ Hermitian matrices.

(a) The eigenvalues of $\mathbf{AB}$ are real if either $\mathbf{A}$ or $\mathbf{B}$ is Hermitian non-negative definite.

(b) If $\mathbf{B} \succ \mathbf{0}$, then the roots of $\det(\mathbf{A} - \lambda\mathbf{B}) = 0$ are real.

10.58. Let $\mathbf{A} \succ \mathbf{0}$ and $\mathbf{B} \succeq \mathbf{0}$ be $n \times n$ Hermitian matrices. Then:

(a) $\mathbf{A} + \mathbf{B} \succ \mathbf{0}$.

(b) $\det(\mathbf{A} + \mathbf{B}) \geq \det \mathbf{A}$ with equality if and only if $\mathbf{B} = \mathbf{0}$.

(c) If $\mathbf{A} - \mathbf{B} \succ \mathbf{0}$, then $\det(\mathbf{A} - \mathbf{B}) < \det \mathbf{A}$.

10.59. Suppose $\mathbf{A} \succeq \mathbf{0}$ and $\mathbf{B} \succeq \mathbf{0}$ are Hermitian $n \times n$ matrices.

(a) The eigenvalues of $\mathbf{AB}$ are non-negative.

(b) $\text{trace}(\mathbf{AB}) \leq \text{trace } \mathbf{A} \text{ trace } \mathbf{B}$.

(c) $\det(\mathbf{A} + \mathbf{B}) \geq \det \mathbf{A} + \det \mathbf{B}$ with equality if and only if $\mathbf{A} + \mathbf{B}$ is singular or $\mathbf{A} = \mathbf{0}$ or $\mathbf{B} = \mathbf{0}$.

(d) $\frac{1}{4}(\mathbf{A}^{-1} + \mathbf{B}^{-1}) \succeq (\mathbf{A} + \mathbf{B})^{-1}$ if $\mathbf{A}$ and $\mathbf{B}$ are nonsingular, with equality if and only if $\mathbf{A} = \mathbf{B}$.

10.60. Given a real symmetric matrix $\mathbf{A} \succ \mathbf{0}$ and real skew-symmetric $\mathbf{B}$ (i.e., $\mathbf{B}' = -\mathbf{B}$), then $\det(\mathbf{A} + \mathbf{B}) \geq \det \mathbf{A}$ with equality if and only if $\mathbf{B} = \mathbf{0}$.

10.61. Given real symmetric $\mathbf{A} \succ \mathbf{0}$ and $\mathbf{B} \succ \mathbf{0}$, then

$$\alpha\mathbf{A}^{-1} + (1 - \alpha)\mathbf{B}^{-1} \succeq [\alpha\mathbf{A} + (1 - \alpha)\mathbf{B}]^{-1}$$

for all $0 \leq \alpha \leq 1$. A special case of historical interest is $\alpha = \frac{1}{2}$ (cf. (10.59d)).

10.62. If $\mathbf{A} \succ \mathbf{0}$ and $\mathbf{A} + \mathbf{B} \succ \mathbf{0}$ are real symmetric matrices, then

$$\det(\mathbf{A} + \mathbf{B})/(\det \mathbf{A}) \leq \exp[\text{trace}(\mathbf{A}^{-1}\mathbf{B})],$$

with equality if and only if $\mathbf{B} = \mathbf{0}$

10.63. (Haynsworth) If $\mathbf{A}$, $\mathbf{B}$, and $\mathbf{A} - \mathbf{B}$ are all real $n \times n$ p.d. matrices, then

$$\det(\mathbf{A} + \mathbf{B}) > \det \mathbf{A} + n \det \mathbf{B}.$$

10.64. (Hartfiel) If $\mathbf{A}$ and $\mathbf{B}$ are real $n \times n$ p.d. matrices, then

$$\det(\mathbf{A} + \mathbf{B}) \geq \det \mathbf{A} + \det \mathbf{B} + (2^n - 2)(\det \mathbf{A} \cdot \det \mathbf{B})^{1/2}.$$

10.65. (Olkin) If $\mathbf{A} \succ \mathbf{0}$ and $\mathbf{B}$ is symmetric with $\det(\mathbf{A} + \mathbf{B}) \neq 0$, then

$$\mathbf{A}^{-1} - (\mathbf{A} + \mathbf{B})^{-1} \succeq (\mathbf{A} + \mathbf{B})^{-1}\mathbf{B}(\mathbf{A} + \mathbf{B})^{-1}.$$

The inequality is strict if and only if $\mathbf{B}$ is nonsingular.

10.66. Let $\mathbf{A}$ and $\mathbf{B}$ be $n \times n$ real non-negative definite matrices. Then any two of the following conditions implies the third.

(1) rank $\mathbf{A} = $ rank $\mathbf{B}$.

(2) $\mathbf{A} \succeq \mathbf{B} \succeq \mathbf{0}$.

(3) $\mathbf{B}^+ \succeq \mathbf{A}^+ \succeq \mathbf{0}$.

10.67. Let $\mathbf{C}$ be any real symmetric matrix. There exist two unique matrices $\mathbf{A} \succeq \mathbf{0}$ and $\mathbf{B} \succeq \mathbf{0}$ such that $\mathbf{AB} = \mathbf{0}$ and

$$\mathbf{C} = \mathbf{A} - \mathbf{B}.$$

Proofs. Section 10.4.2.

10.56a. Horn and Johnson [1985: 182] and Magnus and Neudecker [1999: 208, real case].

10.56c. Follows from (10.47a(iii)) by relabelling $\mathbf{B} \to \mathbf{A}$, $\mathbf{A} - \mathbf{B} \to \mathbf{B}$ and $\mathbf{A} \to \mathbf{A} + \mathbf{B}$.

10.57. Graybill [1983: 404, real case].

10.58a. Use $\mathbf{x}^*(\mathbf{A} + \mathbf{B})\mathbf{x} \geq \mathbf{x}^*\mathbf{A}\mathbf{x}$.

10.58b. Follows from (10.56a) as $\lambda_i(\mathbf{A}) > 0$. Magnus and Neudecker [1999: 21, real case].

10.58c. We replace $\mathbf{A}$ by $\mathbf{A} - \mathbf{B}$ in (b) and (a).

10.59. Zhang [1999: 166, 168–169].

10.60. $\det(\mathbf{A} + \mathbf{B}) = \det \mathbf{A} \det(\mathbf{I}_n + \mathbf{A}^{-1/2}\mathbf{B}\mathbf{A}^{-1/2}) \geq \det \mathbf{A}$ by (5.24c), since $\mathbf{A}^{-1/2}\mathbf{B}\mathbf{A}^{-1/2}$ is skew-symmetric.

10.61. Marshall and Olkin [1979: 469–471 and Styan 1985: 41].

10.62. Abadir and Magnus [2005: 339].

10.63. Ouellette [1981: 216].

10.64. Ouellette [1981: 218].

10.65. Abadir and Magnus [2005: 340].

10.66. Oeullette [1981: 251] and Styan [1985: 47].

10.67. Graybill [1983: 339–401].

CHAPTER 11

SPECIAL PRODUCTS AND OPERATORS

In order to handle a number of complicated manipulations, which typically arise for example in multivariate statistical analysis, a number of special products and operators have been developed, along with rules for using them. Being able to treat a matrix like a stacked vector is one such example that arises when one is finding derivatives and Jacobians in later chapters.

11.1 KRONECKER PRODUCT

11.1.1 Two Matrices

We shall consider a number of operators on pairs of matrices that have the following product properties shared by the real numbers, $\mathbb{R}$.

(i) The product is associative, i.e., $a(bc) = (ab)c$ for all $a, b, c \in \mathbb{R}$.

(ii) The product is distributive with respect to addition, that is, $a(b+c) = ab+ac$ and $(a + b)c = ac + bc$ for all $a, b, c \in \mathbb{R}$.

(iii) There exist 0 and 1 such that for all $a \in \mathbb{R}$, $a(0) = 0$ and $a(1) = a$.

The following product has these properties.

A Matrix Handbook for Statisticians. By George A. F. Seber
Copyright © 2008 John Wiley & Sons, Inc.

Definition 11.1. If $\mathbf{A}$ is an $m \times n$ and $\mathbf{B}$ is $p \times q$, then the *Kronecker product* of $\mathbf{A}$ and $\mathbf{B}$ is defined by the $mp \times nq$ matrix

$$\mathbf{A} \otimes \mathbf{B} = \begin{pmatrix} a_{11}\mathbf{B} & a_{12}\mathbf{B} & \cdots & a_{1n}\mathbf{B} \\ a_{21}\mathbf{B} & a_{22}\mathbf{B} & \cdots & a_{2n}\mathbf{B} \\ \cdot & \cdot & \cdots & \cdot \\ a_{m1}\mathbf{B} & a_{m2}\mathbf{B} & \cdots & a_{mn}\mathbf{B} \end{pmatrix} = (a_{ij}\mathbf{B}).$$

The matrices $\mathbf{A}$ and $\mathbf{B}$ may be complex and we note that, in general, $\mathbf{A} \otimes \mathbf{B} \neq \mathbf{B} \otimes \mathbf{A}$. Also $\mathbf{A}$ and $\mathbf{B}$ can be replaced by vectors in the above definition.

The terms *direct product* and *tensor product* are also used in the literature. It should be noted that Graybill [1982: 216] defines the direct product to be $\mathbf{A} \times \mathbf{B}$, which is actually $\mathbf{B} \otimes \mathbf{A}$ in our notation. Although Kronecker's name is associated with the above product, Henderson et al. [1983] suggest that Zehfuss should perhaps have the honor (see also Horn and Johnson [1991: 254]). In addition to the following results, further properties are listed in this chapter under star product, vec and vech operators, vec-permutation matrix, Jacobians and matrix linear equations. Many of the proofs of the properties given below are straightforward, and details are given in Abadir and Magnus [2005: chapter 10], Brewer [1978], Harville [1997: chapter 16], Horn and Johnson [1991: section 4.2, complex case], Kollo and von Rosen [2005: chapter 1], Magnus and Neudecker [1999: chapter 2], and Schott [2005: chapter 8]. The product rule of (11.11a) is particularly useful.

Knonecker products have been used extensively in statistics—for example, in experimental design, analysis of variance modeling (e.g., Rogers [1984], Ryan [1996], and Schott [2005: 288-290]), and multivariate moment problems.

Definition 11.2. If $\mathbf{A}_i$ is $n_i \times n_i$ $(i = 1, 2, \ldots, r)$, then

$$\mathrm{diag}(\mathbf{A}_1, \ldots, \mathbf{A}_r) = \begin{pmatrix} \mathbf{A}_1 & \mathbf{0} & \cdots & \mathbf{0} \\ \mathbf{0} & \mathbf{A}_2 & \cdots & \mathbf{0} \\ \cdot & \cdot & \cdots & \cdot \\ \mathbf{0} & \mathbf{0} & \cdots & \mathbf{A}_r \end{pmatrix}$$

is said to be the *direct sum* of $\mathbf{A}_1, \ldots, \mathbf{A}_r$, and is sometimes written in the form $\mathrm{diag}(\mathbf{A}_1, \ldots, \mathbf{A}_r) = \mathbf{A}_1 \oplus \cdots \oplus \mathbf{A}_r$.

11.1. (General Properties)

(a) $c \otimes \mathbf{A} = c\mathbf{A} = \mathbf{A} \otimes c$.

(b) $\mathbf{x}' \otimes \mathbf{y} = \mathbf{y}\mathbf{x}' = \mathbf{y} \otimes \mathbf{x}'$.

(c) $a\mathbf{A} \otimes b\mathbf{B} = ab\mathbf{A} \otimes \mathbf{B}$.

(d) $\mathbf{I}_m \otimes \mathbf{I}_n = \mathbf{I}_{mn}$.

(e) $(\mathbf{A} \otimes \mathbf{B})' = \mathbf{A}' \otimes \mathbf{B}'$.

(f) $\overline{(\mathbf{A} \otimes \mathbf{B})} = \overline{\mathbf{A}} \otimes \overline{\mathbf{B}}$ and $(\mathbf{A} \otimes \mathbf{B})^* = \mathbf{A}^* \otimes \mathbf{B}^*$. Here $\overline{\mathbf{A}}$ is the complex conjugate of $\mathbf{A}$ and $\mathbf{A}^*$ is the conjugate tranpose.

(g) $(\mathbf{A} \otimes \mathbf{B})^- = \mathbf{A}^- \otimes \mathbf{B}^-$, where $\mathbf{A}^-$ and $\mathbf{B}^-$ are any weak inverses of $\mathbf{A}$ and $\mathbf{B}$, respectively.

(h) $(\mathbf{A} \otimes \mathbf{B})^+ = \mathbf{A}^+ \otimes \mathbf{B}^+$, where $\mathbf{A}^+$ and $\mathbf{B}^+$ are Moore–Penrose inverses.

(i) If $\mathbf{A}$ and $\mathbf{B}$ are nonsingular, then so is $\mathbf{A} \otimes \mathbf{B}$ and
$(\mathbf{A} \otimes \mathbf{B})^{-1} = \mathbf{A}^{-1} \otimes \mathbf{B}^{-1}$.

(j) $\mathbf{B} \otimes \mathbf{A} = \mathbf{H}_1 \mathbf{A} \otimes \mathbf{B} \mathbf{H}_2$, where $\mathbf{H}_1$ and $\mathbf{H}_2$ are permutation matrices that are independent of $\mathbf{A}$ and $\mathbf{B}$ except for their sizes.

(k) $\mathrm{rank}(\mathbf{A} \otimes \mathbf{B}) = \mathrm{rank}(\mathbf{A})\,\mathrm{rank}(\mathbf{B})$.

(l) If $\mathbf{A}$ is $m \times m$ and $\mathbf{B}$ is $p \times p$, then

    (i) $\det(\mathbf{A} \otimes \mathbf{B}) = (\det \mathbf{A})^p (\det \mathbf{B})^m$.

    (ii) $\mathrm{trace}(\mathbf{A} \otimes \mathbf{B}) = \mathrm{trace}(\mathbf{A})\,\mathrm{trace}(\mathbf{B})$.

(m) $\|\mathbf{A} \otimes \mathbf{B}\|_F = \|\mathbf{A}\|_F \|\mathbf{B}\|_F$, where $\|\cdot\|_F$ is the Frobenius norm.

(n) $\mathbf{I}_n \otimes \mathbf{A} = \mathrm{diag}(\mathbf{A}, \mathbf{A}, \dots, \mathbf{A})$, where there are n diagonal blocks.

(o) $(\mathbf{A} \otimes \mathbf{B})^k = \mathbf{A}^k \otimes \mathbf{B}^k$, for positive integer k.

11.2. (Direct Sum) $(\mathbf{A} \oplus \mathbf{B}) \otimes \mathbf{C} = (\mathbf{A} \otimes \mathbf{C}) \oplus (\mathbf{B} \otimes \mathbf{C})$. However, in general, $\mathbf{A} \otimes (\mathbf{B} \oplus \mathbf{C}) \neq (\mathbf{A} \otimes \mathbf{B}) \oplus (\mathbf{A} \otimes \mathbf{C})$.

11.3. (Partitioned Matrices)

(a) $(\mathbf{A}_1, \mathbf{A}_2) \otimes \mathbf{B} = (\mathbf{A}_1 \otimes \mathbf{B}, \mathbf{A}_2 \otimes \mathbf{B})$.

(b) $(\mathbf{A} \otimes \mathbf{x})\mathbf{B} = (\mathbf{A} \otimes \mathbf{x})(\mathbf{B} \otimes 1) = \mathbf{A}\mathbf{B} \otimes \mathbf{x}$.

(c) Suppose $\mathbf{A}$ is partitioned into submatrices, say

$$\mathbf{A} = \begin{pmatrix} \mathbf{A}_{11} & \cdots & \mathbf{A}_{1s} \\ \cdot & \cdots & \cdot \\ \mathbf{A}_{r1} & \cdots & \mathbf{A}_{rs} \end{pmatrix},$$

then

$$\mathbf{A} \otimes \mathbf{B} = \begin{pmatrix} \mathbf{A}_{11} \otimes \mathbf{B} & \cdots & \mathbf{A}_{1s} \otimes \mathbf{B} \\ \cdot & \cdots & \cdot \\ \mathbf{A}_{r1} \otimes \mathbf{B} & \cdots & \mathbf{A}_{rs} \otimes \mathbf{B} \end{pmatrix}.$$

(d) If $\mathbf{B} = (\mathbf{B}_1, \mathbf{B}_2, \dots, \mathbf{B}_r)$, then $\mathbf{a} \otimes \mathbf{B} = (\mathbf{a} \otimes \mathbf{B}_1, \dots, \mathbf{a} \otimes \mathbf{B}_r)$.

11.4. (Singular Value Decomposition) Let $\mathbf{A}$ be an $m \times n$ matrix of rank r_1 with a singular value decomposition $\mathbf{A} = \mathbf{V}_1 \mathbf{\Sigma}_1 \mathbf{W}_1^*$, and let $\mathbf{B}$ be a $p \times q$ matrix of rank r_2 with singular value decomposition $\mathbf{B} = \mathbf{V}_2 \mathbf{\Sigma}_2 \mathbf{W}_2^*$, where $\mathbf{V}_i$ and $\mathbf{W}_i$ ($i = 1, 2$) are unitary matrices. Let $\sigma_i(\mathbf{C})$ be the ith singular value of $\mathbf{C}$ for $\mathbf{C} = \mathbf{A}$ or $\mathbf{B}$. Then

$$\mathbf{A} \otimes \mathbf{B} = (\mathbf{V}_1 \otimes \mathbf{V}_2)(\mathbf{\Sigma}_1 \otimes \mathbf{\Sigma}_2)(\mathbf{W}_1 \otimes \mathbf{W}_2)^*,$$

where the nonzero singular values of $\mathbf{A} \otimes \mathbf{B}$ are the $r_1 r_2$ positive numbers $\{\sigma_i(\mathbf{A})\sigma_j(\mathbf{B})\}$ ($i = 1, \dots, r_1; j = 1, \dots, r_2$) (including multiplicities). Zero is a singular value of $\mathbf{A} \otimes \mathbf{B}$ with multiplicity $\min\{mp, nq\} - r_1 r_2$. In particular, the singular values of $\mathbf{A} \otimes \mathbf{B}$ are the same as those of $\mathbf{B} \otimes \mathbf{A}$, and $\mathrm{rank}(\mathbf{A} \otimes \mathbf{B}) = \mathrm{rank}(\mathbf{B} \otimes \mathbf{A}) = r_1 r_2$.

11.5. (Eigenvalues and Vectors) Let $\{\lambda_i\}$ and $\{\mathbf{x}_i\}$ be the eigenvalues and the corresponding right eigenvectors of the $m \times m$ matrix $\mathbf{A}$, and $\{\mu_j\}$ and let $\{\mathbf{y}_j\}$ be the eigenvalues and corresponding right eigenvectors for the $n \times n$ matrix $\mathbf{B}$.

(a) $\mathbf{A} \otimes \mathbf{B}$ has eigenvalues $\{\lambda_i \mu_j\}$ (including algebraic multiplicities), and $\{\mathbf{x}_i \otimes \mathbf{y}_j\}$ $(i = 1, 2, \ldots, m; j = 1, 2, \ldots, n)$ are right eigenvectors of $\mathbf{A} \otimes \mathbf{B}$ (but not necessarily all of them). Note that $\mathbf{B} \otimes \mathbf{A}$ also has eigenvalues $\{\lambda_i \mu_j\}$. It should be noted that not every eigenvector of $\mathbf{A} \otimes \mathbf{B}$ is of the form $\mathbf{x} \otimes \mathbf{y}$, where $\mathbf{x}$ is an eigenvector of $\mathbf{A}$ and $\mathbf{y}$ is an eigenvector of $\mathbf{B}$. Abadir and Magnus [2005: 279] give a counterexample.

(b) The so-called *Kronecker sum* $(\mathbf{A} \otimes \mathbf{I}_n + \mathbf{I}_m \otimes \mathbf{B})$ has eigenvalues $\{\lambda_i + \mu_j\}$ with corresponding right eigenvectors $\{\mathbf{x}_i \otimes \mathbf{y}_j\}$.

11.6. Let $\mathbf{A}$ be $m \times m$ and $\mathbf{B}$ be $n \times n$ matrices. We have the following results, some of which are also listed elsewhere under the appropriate matrix topic.

(a) If $\mathbf{A}$ and $\mathbf{B}$ are both diagonal matrices, then so is $\mathbf{A} \otimes \mathbf{B}$.

(b) If $\mathbf{A}$ and $\mathbf{B}$ are both upper (respectively lower) triangular matrices, then $\mathbf{A} \otimes \mathbf{B}$ is also upper (respectively lower) triangular.

(c) If $\mathbf{A}$ and $\mathbf{B}$ are non-negative definite (respectively positive definite), then so is $\mathbf{A} \otimes \mathbf{B}$.

(d) If $\mathbf{A}$ and $\mathbf{B}$ are both symmetric (respectively Hermitian), then so is $\mathbf{A} \otimes \mathbf{B}$.

(e) If $\mathbf{A}$ and $\mathbf{B}$ are both orthogonal (respectively unitary), then so is $\mathbf{A} \otimes \mathbf{B}$.

(f) If $\mathbf{A}$ and $\mathbf{B}$ are idempotent, then so is $\mathbf{C} = \mathbf{A} \otimes \mathbf{B}$. In fact

$$\mathbf{P_C} = \mathbf{P_A} \otimes \mathbf{P_B},$$

where $\mathbf{P_C}$ is the projection onto $\mathcal{C}(\mathbf{C})$.

11.7. If $\mathbf{A}$ and $\mathbf{B}$ are non-negative definite, then $\mathbf{A} \otimes \mathbf{A} \succeq \mathbf{B} \otimes \mathbf{B}$ if and only if $\mathbf{A} \succeq \mathbf{B}$, where $\mathbf{A} \succeq \mathbf{B}$ means that $\mathbf{A} - \mathbf{B}$ is non-negative definite.

11.8. If $\mathbf{A}$ and $\mathbf{B}$ are $n \times n$ non-negative definite matrices, then:

(a) $\operatorname{trace}(\mathbf{A} \otimes \mathbf{B}) \leq \frac{1}{4}(\operatorname{trace} \mathbf{A} + \operatorname{trace} \mathbf{B})^2$.

(b) $\operatorname{trace}(\mathbf{A} \otimes \mathbf{B}) \leq \frac{1}{2} \operatorname{trace}(\mathbf{A} \otimes \mathbf{A} + \mathbf{B} \otimes \mathbf{B})$.

Definition 11.3. The function f is analytic in an open set if it can be expressed as a power series, namely $f(z) = a_0 + a_1 z + a_2 z^2 + \cdots$.

11.9. If f is analytic and $f(\mathbf{A})$ exists, where $\mathbf{A}$ is $m \times m$, then:

(a) $f(\mathbf{I}_p \otimes \mathbf{A}) = \mathbf{I}_p \otimes f(\mathbf{A})$.

(b) $f(\mathbf{A} \otimes \mathbf{I}_p) = f(\mathbf{A}) \otimes \mathbf{I}_p$.

For example,

(i) $\exp(\mathbf{I}_p \otimes \mathbf{A}) = \mathbf{I}_p \otimes \exp(\mathbf{A})$.

(ii) $(\mathbf{A} \otimes \mathbf{I})^k = \mathbf{A}^k \otimes \mathbf{I}$, $k = 1, 2, \ldots$.

Proofs. Section 11.1.1.

11.1. For proofs see Abadir and Magnus [2005: section 101, Harville [1997: section 16.1], Rao and Rao [1998: chapter 6], and Schott [2005: section 8.2]. Some of the results follow using the product rule $(\mathbf{A} \otimes \mathbf{B})(\mathbf{C} \otimes \mathbf{D}) = \mathbf{AC} \otimes \mathbf{BD})$ from (11.11a). For example, to prove (g), $(\mathbf{A} \otimes \mathbf{B})(\mathbf{A}^- \otimes \mathbf{B}^-)(\mathbf{A} \otimes \mathbf{B}) = \mathbf{AA}^- \mathbf{A} \otimes \mathbf{BB}^- \mathbf{B} = \mathbf{A} \otimes \mathbf{B}$; (h) is similar. For (i), $(\mathbf{A}^{-1} \otimes \mathbf{B}^{-1})(\mathbf{A} \otimes \mathbf{B}) = \mathbf{A}^{-1} \mathbf{A} \otimes \mathbf{B}^{-1} \mathbf{B} = \mathbf{I}$; (k) and (l) are proved by Schott [2005: 286–2880]; and (m) is proved by Harville [2001: 143, exercise 9].

11.3c. Abadir and Magnus [2005: 278] and Harville [1997: 338–339].

11.3d. Harville [1997: 339] and Turkington [2002: 9].

11.4. Horn and Johnson [1991: 246].

11.5a. Horn and Johnson [1991: 245, m and n interchanged] and Rao and Rao [1998: 195]. For eigenvalues see Schott [2005: 286].

11.5b. Horn and Johnson [1991: 268–269, $\mathbf{A}$ and $\mathbf{B}$ interchanged].

11.6. The proofs follow by checking the appropriate property using (11.1) and applying the product rule (11.11a). For example, if $\mathbf{A}$ and $\mathbf{B}$ are orthogonal, then from (11.1e) we have $(\mathbf{A} \otimes \mathbf{B})'(\mathbf{A} \otimes \mathbf{B}) = \mathbf{A}'\mathbf{A} \otimes \mathbf{B}'\mathbf{B} = \mathbf{I}_{2n}$. For (c) use $\mathbf{A} = \mathbf{RR}^*$, and so on, and apply (11.1f) and (11.1k). Also, for (d), $(\mathbf{A} \otimes \mathbf{B})^* = \mathbf{A}^* \otimes \mathbf{B}^* = \mathbf{A} \otimes \mathbf{B}$. Harville [2001: 141, exercise 6] proves the second part of (f).

11.7. Abadir and Magnus [2005: 280].

11.8a. We use (11.1l(ii)), namely , $\text{trace}(\mathbf{A} \otimes \mathbf{B}) = \text{trace}\,\mathbf{A}\,\text{trace}\,\mathbf{B}$, and expand $(\text{trace}\,\mathbf{A} - \text{trace}\,\mathbf{B})^2 \geq 0$.

11.8b. Use the trace of a sum is the sum of the traces, and apply $\text{trace}(\mathbf{A} \otimes \mathbf{B}) = \text{trace}\,\mathbf{A}\,\text{trace}\,\mathbf{B}$.

11.9. Expand $f(\mathbf{B})$ as a matrix power series, apply the product rule to each term, as for example in (ii), and then use (11.10b). For (i) we use the power series given in Section 19.6.

11.1.2   More than Two Matrices

The following apply to any conformable matrices, provided the appropriate products and additions exist.

11.10. (Distributive Rules)

(a) Let $\mathbf{A}$ be $m \times n$, $\mathbf{B}$ be $p \times q$, and $\mathbf{C}$ be $r \times s$. Then

$$\mathbf{A} \otimes (\mathbf{B} \otimes \mathbf{C}) = (\mathbf{A} \otimes \mathbf{B}) \otimes \mathbf{C}.$$

We can therefore write each expression as $\mathbf{A} \otimes \mathbf{B} \otimes \mathbf{C}$.

(b) Let $\mathbf{A}$ and $\mathbf{B}$ be $m \times n$, and let $\mathbf{C}$ and $\mathbf{D}$ be $p \times q$, then

$$(\mathbf{A} + \mathbf{B}) \otimes (\mathbf{C} + \mathbf{D}) - \mathbf{A} \otimes \mathbf{C} + \mathbf{A} \otimes \mathbf{D} + \mathbf{B} \otimes \mathbf{C} \mid \mathbf{B} \otimes \mathbf{D}.$$

Special cases follow by setting $\mathbf{A} = \mathbf{0}$ or $\mathbf{C} = \mathbf{0}$.

(c) $\mathbf{A} \otimes \left(\sum_{i=1}^{r} \mathbf{B}_i \right) = \sum_{i=1}^{r} (\mathbf{A} \otimes \mathbf{B}_i)$ and $\left(\sum_{i=1}^{r} \mathbf{A}_i \right) \otimes \mathbf{B} = \sum_{i=1}^{r} (\mathbf{A}_i \otimes \mathbf{B})$.

(d) $\left(\sum_{i=1}^{r} \mathbf{A}_i \right) \otimes \left(\sum_{j=1}^{r} \mathbf{B}_j \right) = \sum_{i=1}^{r} \sum_{j=1}^{r} (\mathbf{A}_i \otimes \mathbf{B}_j)$.

11.11. (Mixed Product)

(a) (Product Rule) Let $\mathbf{A}$, $\mathbf{B}$, $\mathbf{C}$, and $\mathbf{D}$ be $m \times n$, $p \times q$, $n \times r$, and $q \times s$, respectively. Then

$$(\mathbf{A} \otimes \mathbf{B})(\mathbf{C} \otimes \mathbf{D}) = \mathbf{AC} \otimes \mathbf{BD}.$$

This leads to the following special cases.

  (i) From (11.15c), $(\mathbf{A} \otimes \mathbf{b}')(\mathbf{c} \otimes \mathbf{D}) = \mathbf{Acb}'\mathbf{D}$.

  (ii) If $\mathbf{A}$ is $m \times n$ and $\mathbf{B}$ is $p \times q$, then

$$\mathbf{A} \otimes \mathbf{B} = (\mathbf{A} \otimes \mathbf{I}_p)(\mathbf{I}_n \otimes \mathbf{B}).$$

(b) $(\mathbf{A}_1 \otimes \mathbf{B}_1)(\mathbf{A}_2 \otimes \mathbf{B}_2) \cdots (\mathbf{A}_k \otimes \mathbf{B}_k) = \mathbf{A}_1 \mathbf{A}_2 \cdots \mathbf{A}_k \otimes \mathbf{B}_1 \mathbf{B}_2 \cdots \mathbf{B}_k$.

11.12. Let $\mathbf{L}$ be a nonsingular $n \times n$ matrix $(n \geq 2)$, $\mathbf{A}$ and $\mathbf{B}$ be $m \times m$ matrices, and $\mathbf{a}$ and $\mathbf{b}$ be $n \times 1$ vectors. Then the $nm \times nm$ matrix

$$\mathbf{G} = \mathbf{L} \otimes \mathbf{B} + \mathbf{ab}' \otimes \mathbf{A}$$

has determinant

$$\det \mathbf{G} = (\det \mathbf{L})^m (\det \mathbf{B})^{n-1} \det \mathbf{C},$$

where

$$\mathbf{C} = \mathbf{B} + \alpha \mathbf{A} \quad \text{and} \quad \alpha = \mathbf{b}' \mathbf{L}^{-1} \mathbf{a}.$$

If $\mathbf{G}$ is nonsingular, then

$$\mathbf{G}^{-1} = \mathbf{L}^{-1} \otimes \mathbf{B}^{-1} - \mathbf{L}^{-1} \mathbf{ab}' \mathbf{L}^{-1} \otimes \mathbf{E},$$

where

$$\mathbf{E} = \mathbf{C}^{-1} \mathbf{AB}^{-1} = \mathbf{B}^{-1} \mathbf{AC}^{-1} \quad = \quad \mathbf{B}^{-1} \mathbf{AB}^{-1} \quad \text{if} \quad \alpha = 0,$$
$$= \quad \frac{1}{\alpha} (\mathbf{B}^{-1} - \mathbf{C}^{-1}) \quad \text{if} \quad \alpha \neq 0.$$

Definition 11.4. The *Kronecker power* of an $m \times n$ matrix $\mathbf{A}$ is defined as follows:

$$\mathbf{A}^{[2]} \quad = \quad \mathbf{A} \otimes \mathbf{A},$$
$$\mathbf{A}^{[k]} \quad = \quad \mathbf{A} \otimes \mathbf{A} \otimes \cdots \otimes \mathbf{A} = \mathbf{A} \otimes \mathbf{A}^{[k-1]} = \mathbf{A}^{[k-1]} \otimes \mathbf{A},$$

for $k = 2, 3, \ldots$.

11.13. $(\mathbf{AB})^{[k]} = \mathbf{A}^{[k]}\mathbf{B}^{[k]}$ for $k = 1, 2, \ldots$.

11.14. If $\mathbf{A}$ and $\mathbf{B}$ are non-negative definite, then $\mathbf{A} - \mathbf{B}$ is non-negative definite if and only if and $\mathbf{A}^{[2]} - \mathbf{B}^{[2]}$ is non-negative definite.

Proofs. Section 11.1.2.

    11.10. Prove directly from the definition of the Kronecker product (cf. Abadir and Magnus [2005: 275–276].

    11.11. Harville [1997: 337].

    11.12. Magnus [1982: 243, 271].

    11.13. Follows from the product rule (11.11a).

    11.14. Abadir and Magnus [2005: 280].

11.2 VEC OPERATOR

Definition **11.5.** Let $\mathbf{A} = (\mathbf{a}_1, \mathbf{a}_2, \ldots, \mathbf{a}_n)$ be an $m \times n$ matrix. Then vec $\mathbf{A}$ is a vector obtained by stacking the columns of $\mathbf{A}$, namely

$$\text{vec } \mathbf{A} = \begin{pmatrix} \mathbf{a}_1 \\ \mathbf{a}_2 \\ . \\ \mathbf{a}_n \end{pmatrix},$$

an $mn \times 1$ vector. Various other notations have been used for the above concept, and some history and references are given by Henderson and Searle [1981a]. Here vec $\mathbf{A}$ stands for "vector of columns of $\mathbf{A}$".

    Turkington [2002: 10] introduced the operator devec$\mathbf{A}$ that stacks the rows of $\mathbf{A}$ alongside each other so that $(\text{vec } \mathbf{A}')' = \text{devec}\mathbf{A}$.

    The following properties are proved by Henderson and Searle [1979: 67]), except where labeled otherwise. We assume that $\mathbf{A}_{m \times n}$ is $m \times n$, $\mathbf{B}_{n \times q}$ is $n \times q$, and $\mathbf{C}_{q \times r}$ is $q \times r$.

11.15. (Some General Properties)

  (a) vec $\mathbf{A} = (\mathbf{I}_n \otimes \mathbf{A})\text{vec } \mathbf{I}_n = (\mathbf{A}' \otimes \mathbf{I}_m)\text{vec } \mathbf{I}_m$.

  (b) vec $\mathbf{x} = \text{vec } \mathbf{x}' = \mathbf{x}$.

  (c) vec $(\mathbf{x}\mathbf{y}') = \mathbf{y} \otimes \mathbf{x}$.

  (d) From (c),

$$\text{vec } [(\mathbf{A}\mathbf{x})(\mathbf{y}'\mathbf{B})] = (\mathbf{B}'\mathbf{y}) \otimes (\mathbf{A}\mathbf{x}) = (\mathbf{B}' \otimes \mathbf{A})(\mathbf{y} \otimes \mathbf{x}) = (\mathbf{B}' \otimes \mathbf{A})\text{vec } (\mathbf{x}\mathbf{y}').$$

  (e) If $\mathbf{A}$ is nonsingular, we apply (11.16b) to vec $(\mathbf{A}^{-1}\mathbf{A}\mathbf{A}^{-1})$ to get

$$\text{vec } \mathbf{A}^{-1} = (\mathbf{A}^{-1'} \otimes \mathbf{A}^{-1})\text{vec } \mathbf{A}.$$

$$\text{(f)}\quad \text{vec}\,[\mathbf{A}(\mathbf{b}_1, \mathbf{b}_2, \ldots, \mathbf{b}_q)] = \begin{pmatrix} \mathbf{A}\mathbf{b}_1 \\ \mathbf{A}\mathbf{b}_2 \\ \cdot \\ \cdot \\ \mathbf{A}\mathbf{b}_q \end{pmatrix}.$$

11.16. (Products)

(a)

$$\begin{aligned}
\text{vec}\,(\mathbf{A}_{m\times n}\mathbf{B}_{n\times q}) &= (\mathbf{I}_q \otimes \mathbf{A})\text{vec}\,\mathbf{B} \\
&= (\mathbf{B}' \otimes \mathbf{A})\text{vec}\,\mathbf{I}_n \\
&= (\mathbf{B}' \otimes \mathbf{I}_m)\text{vec}\,\mathbf{A}.
\end{aligned}$$

(b) We highlight the following result as it is used extensively.

$$\text{vec}\,(\mathbf{A}_{m\times n}\mathbf{B}_{n\times q}\mathbf{C}_{q\times r}) = (\mathbf{C}' \otimes \mathbf{A})\text{vec}\,\mathbf{B}.$$

(c) Using (a), we have

$$\begin{aligned}
\text{vec}\,(\mathbf{A}_{m\times n}\mathbf{B}_{n\times q}\mathbf{C}_{q\times r}) &= (\mathbf{I}_r \otimes \mathbf{AB})\text{vec}\,\mathbf{C} \\
&= (\mathbf{C}'\mathbf{B}' \otimes \mathbf{I}_m)\text{vec}\,\mathbf{A}.
\end{aligned}$$

(d) Using the above results, we obtain

$$\begin{aligned}
\text{vec}\,(\mathbf{A}_{m\times n}\mathbf{B}_{n\times q}\mathbf{C}_{q\times r}\mathbf{D}_{r\times s}) &= (\mathbf{I} \otimes \mathbf{ABC})\text{vec}\,\mathbf{D} \\
&= (\mathbf{D}' \otimes \mathbf{AB})\text{vec}\,\mathbf{C} \\
&= (\mathbf{D}'\mathbf{C}' \otimes \mathbf{A})\text{vec}\,\mathbf{B} \\
&= (\mathbf{D}'\mathbf{C}'\mathbf{B}' \otimes \mathbf{I})\text{vec}\,\mathbf{A}.
\end{aligned}$$

(e) Using (a), we have

$$\begin{aligned}
\text{vec}\,[(\mathbf{A} + \mathbf{B})(\mathbf{C} + \mathbf{D})] &= [(\mathbf{I} \otimes \mathbf{A}) + (\mathbf{I} \otimes \mathbf{B})][\text{vec}\,\mathbf{C} + \text{vec}\,\mathbf{D}] \\
&= [(\mathbf{C}' \otimes \mathbf{I}) + (\mathbf{D}' \otimes \mathbf{I})][\text{vec}\,\mathbf{A} + \text{vec}\,\mathbf{B}].
\end{aligned}$$

Clearly, (a), (b), and (c) can be deduced from (d) by replacing appropriate matrices by identity matrices. However, (a)–(c) are listed for convenient reference.

11.17. (Trace)

(a) $\text{trace}(\mathbf{A}_{m\times n}\mathbf{B}_{n\times q}) = (\text{vec}\,\mathbf{A}')'\text{vec}\,\mathbf{B} = (\text{vec}\,\mathbf{B}')'\text{vec}\,\mathbf{A}.$

As noted by Henderson and Searle [1979: 67], the above can be expressed in an alternative form that is easier to remember, namely

$$\text{trace}(\mathbf{A}'\mathbf{B}) = (\text{vec}\,\mathbf{A})'\text{vec}\,\mathbf{B}.$$

This result can be used along with (11.16) to deduce the following.

(b)

$$\text{trace}(\mathbf{A}_{m \times n}\mathbf{B}_{n \times q}\mathbf{C}_{q \times s}) = (\text{vec } \mathbf{A}')'(\mathbf{I}_s \otimes \mathbf{B})\text{vec } \mathbf{C}$$
$$= (\text{vec } \mathbf{B}')'(\mathbf{I} \otimes \mathbf{C})\text{vec } \mathbf{A}$$
$$= (\text{vec } \mathbf{C}')'(\mathbf{I} \otimes \mathbf{A})\text{vec } \mathbf{B}$$
$$= (\text{vec } \mathbf{A}')'(\mathbf{C}' \otimes \mathbf{I})\text{vec } \mathbf{B}$$
$$= (\text{vec } \mathbf{B}')'(\mathbf{A}' \otimes \mathbf{I})\text{vec } \mathbf{C}$$
$$= (\text{vec } \mathbf{C}')'(\mathbf{B}' \otimes \mathbf{I})\text{vec } \mathbf{A}.$$

We can use such results as $\text{trace}(\mathbf{ABC}) = (\text{vec } \mathbf{A}')'\text{vec }(\mathbf{BC})$ and $\text{trace}(\mathbf{ABC}) = \text{trace}(\mathbf{BCA}) = \text{trace}(\mathbf{CAB})$.

(c)

$$\text{trace}(\mathbf{ABCD}) = (\text{vec } \mathbf{A}')'(\mathbf{D}' \otimes \mathbf{B})\text{vec } \mathbf{C}$$
$$= \text{trace}(\mathbf{D}(\mathbf{ABC})) = (\text{vec } \mathbf{D}')'(\mathbf{C}' \otimes \mathbf{A})\text{vec } \mathbf{B}$$
$$= \text{trace}(\mathbf{D}'(\mathbf{C}'\mathbf{B}'\mathbf{A}')) = (\text{vec } \mathbf{D})'(\mathbf{A} \otimes \mathbf{C}')\text{vec } \mathbf{B}'.$$

(d) From (c) and (11.16b) we have:

(i) $\text{trace}(\mathbf{AXBX}'\mathbf{C}) = \text{trace}(\mathbf{X}'\mathbf{CAXB}) = (\text{vec } \mathbf{X})'(\mathbf{B}' \otimes \mathbf{CA})\text{vec } \mathbf{X}.$

(ii) $\text{trace}(\mathbf{AX}'\mathbf{BXC}) = \text{trace}(\mathbf{X}'\mathbf{BXCA}) = (\text{vec } \mathbf{X})'(\mathbf{A}'\mathbf{C}' \otimes \mathbf{B})\text{vec } \mathbf{X}.$

The above can also be transposed to obtain further results.

Proofs. Section 11.2.

11.15a. Follows from (11.16a) with $\mathbf{A} = \mathbf{I}$ or $\mathbf{B} = \mathbf{I}$.

11.15d–e. Use (11.16b) with a suitable substitution.

11.16a. Abadir and Magnus [2005: 282] and Magnus and Neudecker [1999: 31].

11.16b. Harville [1997: 341].

11.16c–d. Dhrymes [2000: 118–120].

11.16e. Expand and use (a).

11.17b. Dhrymes [2000: 121].

11.17c. Abadir and Magnus [2006: 283], Harville [1997: 342], Magnus and Neudecker [1999: 31], and Schott [2003: 294].

11.17d. We use a result like $\text{trace}(\mathbf{X}'\mathbf{CAXB}) = (\text{vec } \mathbf{X})'\text{vec }(\mathbf{CAXB})$ (Henderson and Searle [1979: 67]).

11.3   VEC-PERMUTATION (COMMUTATION) MATRIX

We now introduce a permutation matrix that is particularly useful for dealing with matrices of random variables and their moments.

Definition 11.6. Let $\mathbf{A}$ be an $m \times n$ matrix. We define $\mathbf{I}_{(m,n)}$ as the the $mn \times mn$ permutation matrix such that vec $\mathbf{A} = \mathbf{I}_{(m,n)}$ vec $\mathbf{A}'$. Henderson and Searle [1979, 1981a], who give a useful historical background and a summary of its properties, call $\mathbf{I}_{(m,n)}$ the *vec-permutation matrix*. It is also called a *commutation matrix* by Abadir and Magnus [2005], Magnus and Neudecker [1999], and Schott [2005], who denote it by $\mathbf{K}_{nm}$ and, when $m = n$, $\mathbf{K}_n$; we shall mention both notations in our discussion. (Many of the results given in this section are also proved in Graybill [1983: section 9.3], though, as previously mentioned, he uses an alternative definition, namely $\mathbf{A} \times \mathbf{B}$ instead of $\mathbf{B} \otimes \mathbf{A}$.) The use of of the commutation matrix in statistics was discussed in Magnus and Neudecker [1979].

If $\mathbf{A}_{2 \times 3} = \begin{pmatrix} a_{11} & a_{12} & a_{13} \\ a_{21} & a_{22} & a_{23} \end{pmatrix}$, then

$$
\text{vec}\,\mathbf{A}_{2\times 3} = \begin{pmatrix} a_{11} \\ a_{21} \\ a_{12} \\ a_{22} \\ a_{13} \\ a_{23} \end{pmatrix} = \begin{pmatrix} 1 & 0 & 0 & 0 & 0 & 0 \\ 0 & 0 & 0 & 1 & 0 & 0 \\ 0 & 1 & 0 & 0 & 0 & 0 \\ 0 & 0 & 0 & 0 & 1 & 0 \\ 0 & 0 & 1 & 0 & 0 & 0 \\ 0 & 0 & 0 & 0 & 0 & 1 \end{pmatrix} \begin{pmatrix} a_{11} \\ a_{12} \\ a_{13} \\ a_{21} \\ a_{22} \\ a_{23} \end{pmatrix} = \mathbf{I}_{(2,3)} \text{vec}\,\mathbf{A}'.
$$

Thus $\mathbf{I}_{(m,n)}$ is a rearrangment of $\mathbf{I}_{mn}$ obtained by taking every nth row starting at the first, then every nth row starting at the second, and so on. Thus $\mathbf{I}_{(2,3)}$ consists of rows $1, 4, 2, 5, 3, 6$ of $\mathbf{I}_6$. As a permutation matrix, it has all the standard properties of a permutation matrix (cf. Section 8.2).

11.18. (Some Basic Properties)

(a) $\mathbf{I}_{(m,n)}$ $(= \mathbf{K}_{nm})$ is orthogonal, being a permutation matrix.

(b) $\mathbf{I}_{(m,n)}\mathbf{I}_{(n,m)} = \mathbf{I}_{mn}$ (i.e., $\mathbf{K}_{nm}\mathbf{K}_{mn} = \mathbf{I}_{mn}$)   so that

   (i) $\mathbf{I}'_{(m,n)} = \mathbf{I}^{-1}_{(m,n)} = \mathbf{I}_{(n,m)}$ (i.e., $\mathbf{K}'_{nm} = \mathbf{K}_{mn}$).

   (ii) vec $\mathbf{A}' = \mathbf{I}_{(n,m)}$ vec $\mathbf{A}$.

   (iii) $\mathbf{I}^2_{(n,n)} = \mathbf{I}_{(n,n)}$.

   (iv) $\mathbf{I}'_{(n,n)} = \mathbf{I}_{(n,n)}$.

(c) $\mathbf{I}_{(m,1)} = \mathbf{I}_{(1,m)} = \mathbf{I}_m$.

(d) If $\mathbf{E}_{ij}$ is the $m \times n$ matrix with 1 in the the i, jth position and zeros elsewhere, then

$$
\mathbf{I}_{(m,n)} = \sum_{i=1}^{m} \sum_{j=1}^{n} (\mathbf{E}'_{ij} \otimes \mathbf{E}_{ij}) \quad (= \mathbf{K}'_{mn}).
$$

In particular (Harville [1997: 345, transposed]),

$$\mathbf{I}_{(m,n)} = \begin{pmatrix} \mathbf{E}_{11} & \mathbf{E}_{21} & \cdots & \mathbf{E}_{m1} \\ \mathbf{F}_{12} & \mathbf{F}_{22} & \cdots & \mathbf{E}_{m2} \\ \cdot & \cdot & \cdots & \cdot \\ \mathbf{E}_{1n} & \mathbf{E}_{2n} & \cdots & \mathbf{E}_{mn} \end{pmatrix}.$$

As already noted,

$$\mathbf{K}_{mn} = \sum_{i=1}^{m}\sum_{j=1}^{n}(\mathbf{E}_{ij}\otimes\mathbf{E}_{ij}').$$

This result can be used to define $\mathbf{K}_{mn}$, and Schott [2005: 308] proves the equivalence of the two definitions.

(e) If $\mathbf{e}_{i,m}$ is the ith column of $\mathbf{I}_m$ and $\mathbf{e}_{j,n}$ is the jth column of $\mathbf{I}_n$, then

$$\begin{aligned} \mathbf{I}_{(m,n)} &= \sum_{i=1}^{m}(\mathbf{e}_{i,n}'\otimes\mathbf{I}_n\otimes\mathbf{e}_{i,n}) \\ &= \sum_{j=1}^{n}(\mathbf{e}_{j,m}\otimes\mathbf{I}_m\otimes\mathbf{e}_{j,n}'). \end{aligned}$$

(f) (i) $\det\mathbf{I}_{(m.n)} = (-1)^{\frac{1}{2}m(m-1)(n-1)}\det\mathbf{I}_{(m,n-1)} = (-1)^{\frac{1}{4}m(m-1)n(n-1)}$.

   (ii) $\det\mathbf{I}_{(n,n)} = (-1)^{\frac{1}{2}n(n-1)}$.

(g) (i) $\operatorname{trace}(\mathbf{I}_{(m,n)}) = 1 + g(m-1, n-1)$,

   where $g(a,b)$ is the greatest common divisor of a and b.

   (ii) $\operatorname{trace}\mathbf{I}_{(n,n)} = n$.

(h) $\mathbf{I}_{(n,n)}$ has eigenvalues ± 1 with respective multiplicities $\frac{1}{2}n(n\pm 1)$.

11.19. Let $\mathbf{A}_{m\times n}$ and $\mathbf{B}_{p\times q}$ be $m\times n$ and $p\times q$ matrices, and let $\mathbf{a}$ and $\mathbf{b}$ be $m\times 1$ and $p\times 1$ vectors.

(a) $\mathbf{I}_{(m,p)}(\mathbf{A}_{m\times n}\otimes\mathbf{B}_{p\times q}) = (\mathbf{B}_{p\times q}\otimes\mathbf{A}_{m\times n})\mathbf{I}_{(n,q)}$.

(b) $\mathbf{I}_{(m,p)}(\mathbf{A}_{m\times n}\otimes\mathbf{B}_{p\times q})\mathbf{I}_{(q,n)} = \mathbf{B}_{p\times q}\otimes\mathbf{A}_{m\times n}$.

(c) We have the following special cases of the above.

   (i) $\mathbf{I}_{(m,p)}(\mathbf{A}_{m\times n}\otimes\mathbf{b}_{p\times 1}) = \mathbf{b}_{p\times 1}\otimes\mathbf{A}_{m\times n}$.

   (ii) Multiplying (i) on the left by $\mathbf{I}_{(p,m)}$ and using (11.18b) gives us
   $\mathbf{I}_{(p,m)}(\mathbf{b}\otimes\mathbf{A}) = \mathbf{A}\otimes\mathbf{b}$.

   (iii) $(\mathbf{A}_{m\times n}\otimes\mathbf{b}_{1\times p}')\mathbf{I}_{(p,n)} = \mathbf{b}'\otimes\mathbf{A}$.

   (iv) $(\mathbf{b}'\otimes\mathbf{A})\mathbf{I}_{(n,p)} = \mathbf{A}\otimes\mathbf{b}'$.

   (v) $\mathbf{I}_{(m,p)}(\mathbf{a}_{m\times 1}\otimes\mathbf{b}_{p\times 1}) = \mathbf{b}_{p\times 1}\otimes\mathbf{a}_{m\times 1}$.

11.20. $\mathbf{A}_{m\times m}\otimes\mathbf{B}_{p\times p} = \mathbf{I}_{(p,m)}(\mathbf{I}_p\otimes\mathbf{A}_{m\times m})\mathbf{I}_{(m,p)}(\mathbf{I}_m\otimes\mathbf{B}_{p\times p})$.

11.21. For handling more than two matrices, we introduce $\mathbf{I}_{(ab,n)} = \mathbf{I}_{(m,n)}(= \mathbf{K}_{nm})$, where $m = ab$. Since $\mathbf{I}_{(mp,s)} = \mathbf{I}_{(pm,s)}$, we can interchange m and p in some of the following results.

(a) $\mathbf{I}_{(p,ms)}\mathbf{I}_{(m,ps)} = \mathbf{I}_{(mp,s)} = \mathbf{I}_{(m,ps)}\mathbf{I}_{(p,ms)}$.

(b) $\mathbf{I}_{(m,ps)}\mathbf{I}_{(p,sm)}\mathbf{I}_{(s,mp)} = \mathbf{I}_{(mp,s)}$.

(c) $\mathbf{I}_{(mp,s)} = (\mathbf{I}_{(p,s)} \otimes \mathbf{I}_m)(\mathbf{I}_p \otimes \mathbf{I}_{(m,s)}) = (\mathbf{I}_{(m,s)} \otimes \mathbf{I}_p)(\mathbf{I}_m \otimes \mathbf{I}_{(p,s)})$.

(d) Any two $\mathbf{I}$ matrices with the same set of three indices commute, for example, $\mathbf{I}_{(m,ps)}\mathbf{I}_{(p,sm)} = \mathbf{I}_{(p,sm)}\mathbf{I}_{(m,ps)}$.

11.22.

$$
\begin{aligned}
\mathbf{C}_{s\times t} \otimes \mathbf{A}_{m\times n} \otimes \mathbf{B}_{p\times q} &= \mathbf{I}_{(mp,s)}[(\mathbf{A}_{m\times n} \otimes \mathbf{B}_{p\times q}) \otimes \mathbf{C}_{s\times t}]\mathbf{I}_{(t,nq)} \\
&= \mathbf{I}_{(p,ms)}[\mathbf{B}_{p\times q} \otimes (\mathbf{C}_{s\times t} \otimes \mathbf{A}_{m\times n})]\mathbf{I}_{(nt,q)}.
\end{aligned}
$$

11.23. Using (11.16b), we obtain

$$
\begin{aligned}
(\mathbf{B}_{p\times q} \otimes \mathbf{A}_{m\times n})\mathrm{vec}\,\mathbf{X}_{n\times q} &= \mathrm{vec}\,(\mathbf{AXB'}) \\
&= \mathbf{I}_{(m,p)}\mathrm{vec}\,(\mathbf{BX'A'}) \\
&= \mathbf{I}_{(m,p)}(\mathbf{A} \otimes \mathbf{B})\mathrm{vec}\,\mathbf{X'} \\
&= \mathbf{I}_{(m,p)}(\mathbf{A} \otimes \mathbf{B})\mathbf{I}_{(q,n)}\mathrm{vec}\,\mathbf{X}.
\end{aligned}
$$

11.24. $\mathrm{vec}\,(\mathbf{A}_{m\times n} \otimes \mathbf{B}_{p\times q}) = (\mathbf{I}_n \otimes \mathbf{I}_{(m,q)} \otimes \mathbf{I}_p)(\mathrm{vec}\,\mathbf{A}_{m\times n} \otimes \mathrm{vec}\,\mathbf{B}_{p\times q})$.

11.25. (Products)

(a)

$$
\begin{aligned}
\mathbf{W}_{r\times s}\mathbf{Z}_{s\times t} &\otimes \mathbf{X}_{m\times n}\mathbf{Y}_{n\times p} = \\
&[\mathbf{I}_{rm} \otimes (\mathrm{vec}\,\mathbf{Y}')'][\mathbf{I}_r \otimes \mathrm{vec}\,\mathbf{X}'(\mathrm{vec}\,\mathbf{Z}')' \otimes \mathbf{I}_p][\mathrm{vec}\,\mathbf{W}' \otimes \mathbf{I}_{pt}].
\end{aligned}
$$

(b)

$$
\begin{aligned}
(\mathbf{W}_{r\times s}\mathbf{Z}_{s\times t} &\otimes \mathbf{X}_{m\times n}\mathbf{Y}_{n\times p})\mathbf{I}_{(p,t)} = \\
&[\mathbf{I}_{rm} \otimes (\mathrm{vec}\,\mathbf{Z}')'][\mathbf{I}_r \otimes (\mathbf{X} \otimes \mathbf{I}_s)\mathbf{I}_{(s,n)}(\mathbf{I}_s \otimes \mathbf{Y}) \otimes \mathbf{I}_t][\mathrm{vec}\,\mathbf{W}' \otimes \mathbf{I}_{pt}].
\end{aligned}
$$

(c)

$$
\begin{aligned}
\mathbf{I}_{(p,m)}(\mathbf{B}_{p\times q}\mathbf{C}_{q\times s} &\otimes \mathbf{A}_{m\times n}\mathbf{D}_{n\times t}) = (\mathbf{A} \otimes \mathbf{B})\mathbf{I}_{(q,n)}(\mathbf{C} \otimes \mathbf{D}) \\
&= (\mathbf{AD} \otimes \mathbf{BC})\mathbf{I}_{(s,t)} = (\mathbf{I}_m \otimes \mathbf{BC})\mathbf{I}_{(s,m)}(\mathbf{I}_s \otimes \mathbf{AD}) \\
&= (\mathbf{AD} \otimes \mathbf{I}_p)\mathbf{I}_{(p,t)}(\mathbf{BC} \otimes \mathbf{I}_t).
\end{aligned}
$$

(d) $\mathbf{I}_{(m,n)}(\mathbf{A}_{m\times p} \otimes \mathbf{b}_{n\times 1}\mathbf{c}'_{1\times q}) = \mathbf{b} \otimes \mathbf{A} \otimes \mathbf{c}'$ and $\mathbf{I}_{(n.m)}(\mathbf{bc}' \otimes \mathbf{A}) = \mathbf{c}' \otimes \mathbf{A} \otimes \mathbf{b}$.

11.26. (Trace) For any $m \times n$ matrices $\mathbf{A}$ and $\mathbf{B}$ we have

$$
\begin{aligned}
\mathrm{trace}[(\mathbf{A'} \otimes \mathbf{B})\mathbf{I}_{(n,m)}] &= \mathrm{trace}[\mathbf{I}_{(n,m)})(\mathbf{A'} \otimes \mathbf{B})] \\
&= \mathrm{trace}(\mathbf{A'B}).
\end{aligned}
$$

Proofs. Section 11.3.

11.18. For proofs see Magnus [1988: chapter 3] and Magnus and Neudecker [1979]. Also some proofs are given by Abadir and Magnus [2005: section 11.1], Harville [1997: section 16.3], Harville [2001: 149–153], Magnus and Neudecker [1999: 47], and Schott [2005: 306–307, 310].

11.19. Abadir and Magnus [2005: 301], Harville [1997: 347–348], and Schott [2005: 308].

11.20. Use $(\mathbf{A} \otimes \mathbf{I}_p)(\mathbf{I}_m \otimes \mathbf{B}) = \mathbf{A} \otimes \mathbf{B}$ and (11.19b).

11.21. Abadir and Magnus [2005: 306] and Henderson and Searle [1981a: 284–285].

11.22. Henderson and Searle [1981a: 284] and Magnus [1988: 44].

11.23. Henderson and Searle [1981a: 281] and Magnus [1988: 44].

11.24. Harville [1997: 349], Magnus [1988: 43], and Schott [2005: 309].

11.25a–b. Rogers [1980: 23].

11.25c–d. Abadir and Magnus [2005: 302, 304].

11.26. Abadir and Magnus [2005: 304].

11.4 GENERALIZED VEC-PERMUTATION MATRIX

***Definition* 11.7.** Let $\mathbf{I}_{(n)}$ be the matrix obtained from $\mathbf{I}_r$ by taking every nth row starting with the first, then every nth row starting with the second, and so on (cf. Tracey and Dwyer [1969] and Henderson and Searle [1981a]). Then $\mathbf{I}_{(n)}$ is called a *generalized vec-permutation* matrix. For example, if $r = 5$ and $n = 3$, $\mathbf{I}_{(3)} = (\mathbf{e}_1, \mathbf{e}_4, \mathbf{e}_2, \mathbf{e}_5, \mathbf{e}_3)$, where the $\mathbf{e}_i$ are the columns of $\mathbf{I}_r$.

We can apply the same procedure to any matrix $\mathbf{M}$ and obtain $\mathbf{M}_{(n)}$. In fact, $\mathbf{M}_{(n)} = \mathbf{I}_{(n)}\mathbf{M}$. We can also define $\mathbf{M}_{(m,n)} = \mathbf{I}_{(m,n)}\mathbf{M}$. When $r = mn$, $\mathbf{I}_{(m,n)} = \mathbf{I}_{(n)}$, and when $\mathbf{M}$ has mn rows, $\mathbf{M}_{(m,n)} = \mathbf{M}_{(n)}$.

11.27. In the following, $\mathbf{A}$ is $m \times n$, $\mathbf{B}$ is $p \times q$, $\mathbf{C}$ is $s \times t$, $\mathbf{a}$ is $m \times 1$, and $\mathbf{b}$ is $p \times 1$.

(a) $\operatorname{vec} \mathbf{A} = (\operatorname{vec} \mathbf{A}')_{(m,n)}$.

(b) $(\mathbf{A} \otimes \mathbf{B})_{(m,p)} = (\mathbf{B}' \otimes \mathbf{A}')_{(q,n)}$.

(c) $(\mathbf{a} \otimes \mathbf{B})_{(m,p)} = \mathbf{I}_{(m,n)}(\mathbf{a} \otimes \mathbf{B}) = \mathbf{B} \otimes \mathbf{a}$.

(d) $(\mathbf{A} \otimes \mathbf{b})_{(m,p)} = \mathbf{I}_{(m,p)}(\mathbf{A} \otimes \mathbf{b}) = \mathbf{b} \otimes \mathbf{A}$.

(e) $(\mathbf{a} \otimes \mathbf{b})_{(m,p)} = \mathbf{I}_{(m,p)}(\mathbf{a} \otimes \mathbf{b}) = \mathbf{b} \otimes \mathbf{a}$.

(f) $(\mathbf{a} \otimes \mathbf{b}' \otimes \mathbf{C})_{(m,s)} = (\mathbf{ab}' \otimes \mathbf{C})_{(m,s)} = \mathbf{b}' \otimes \mathbf{C} \otimes \mathbf{a}$.

11.28. $\mathbf{I}_{(m,n)} = (\mathbf{I}_m \otimes \mathbf{I}_n)_{(m,n)}$.

Proofs. Section 11.4.

11.27. Henderson and Searle [1981a: 283–284 and equation (49)]. Use (11.1b) for (f).

11.5   VECH OPERATOR

Definition 11.8. If $\mathbf{A}$ is an $n \times n$ matrix, then vech $\mathbf{A}$ (*vector-half*) is the $k = n(n+1)/2$ -dimensional vector obtained by stacking the columns of the lower triangle of $\mathbf{A}$, including the diagonal, one below the other; Magnus and Neudecker [1999] and Schott [2005] use the notation $\nu(\mathbf{A})$. For example, if

$$\mathbf{A} = \begin{pmatrix} a_{11} & a_{12} & a_{13} \\ a_{21} & a_{22} & a_{23} \\ a_{31} & a_{32} & a_{33} \end{pmatrix}$$

is symmetric, then

$$\text{vech } \mathbf{A} = \begin{pmatrix} a_{11} \\ a_{21} \\ a_{31} \\ a_{22} \\ a_{32} \\ a_{33} \end{pmatrix}.$$

This approach is useful for symmetric and lower-triangular matrices; for upper-triangular matrices we use vech $(\mathbf{A}')$.

11.5.1   Symmetric Matrix

A major application of the above definition is to symmetric matrices, so we now *assume* $\mathbf{A} = \mathbf{A}'$. For this case, vech $\mathbf{A}$ lists all the distinct elements of $\mathbf{A}$. As the elements of vec $\mathbf{A}$ are those of vech $\mathbf{A}$ with some repetitions, it follows that vec $\mathbf{A}$ and vech $\mathbf{A}$ are linear transformations of one another. This leads to the following definitions.

Definition 11.9. We have vech $\mathbf{A} = \mathbf{H}_n$ vec $\mathbf{A}$ and vec $\mathbf{A} = \mathbf{G}_n$ vech $\mathbf{A}$. The matrix $\mathbf{H}_n$ is $k \times n^2$, and Magnus and Neudecker [1999: 48–51] call the $n^2 \times k$ matrix $\mathbf{G}_n$ the *duplication matrix* $\mathbf{D}_n$ (see also Magnus [1988: chapter 4] and Schott [2005: section 8.7]). We shall also use the term duplication matrix. Examples of $\mathbf{G}_n$ and $\mathbf{H}_n$ for $n = 3$ ($k = 6$) together with $\mathbf{I}_{(3,3)}$, with which they have several relationships, are

$$\mathbf{D}_3 = \mathbf{G}_3(9 \times 6) = \left(\begin{array}{ccc|ccc} 1 & \cdot & \cdot & \cdot & \cdot & \cdot \\ \cdot & 1 & \cdot & \cdot & \cdot & \cdot \\ \cdot & \cdot & 1 & \cdot & \cdot & \cdot \\ \hline \cdot & 1 & \cdot & \cdot & \cdot & \cdot \\ \cdot & \cdot & \cdot & 1 & \cdot & \cdot \\ \cdot & \cdot & \cdot & \cdot & 1 & \cdot \\ \hline \cdot & \cdot & 1 & \cdot & \cdot & \cdot \\ \cdot & \cdot & \cdot & \cdot & 1 & \cdot \\ \cdot & \cdot & \cdot & \cdot & \cdot & 1 \end{array} \right),$$

$$
\mathbf{H}_3(6 \times 9) =
\left(
\begin{array}{ccc|ccc|ccc}
1 & \cdot & \cdot & \cdot & \cdot & \cdot & \cdot & \cdot & \cdot \\
\cdot & \alpha_1 & \cdot & 1-\alpha_1 & \cdot & \cdot & \cdot & \cdot & \cdot \\
\cdot & \cdot & \alpha_2 & \cdot & \cdot & 1-\alpha_2 & \cdot & \cdot & \cdot \\
\hline
\cdot & \cdot & \cdot & \cdot & 1 & \cdot & \cdot & \cdot & \cdot \\
\cdot & \cdot & \cdot & \cdot & \alpha_3 & \cdot & 1-\alpha_3 & \cdot \\
\hline
\cdot & \cdot & \cdot & \cdot & \cdot & \cdot & \cdot & \cdot & 1
\end{array}
\right),
$$

and

$$
\mathbf{I}_{(3,3)} =
\left(
\begin{array}{ccc|ccc|ccc}
1 & \cdot & \cdot & \cdot & \cdot & \cdot & \cdot & \cdot & \cdot \\
\cdot & \cdot & \cdot & 1 & \cdot & \cdot & \cdot & \cdot & \cdot \\
\cdot & \cdot & \cdot & \cdot & \cdot & 1 & \cdot & \cdot \\
\hline
\cdot & 1 & \cdot & \cdot & \cdot & \cdot & \cdot & \cdot & \cdot \\
\cdot & \cdot & \cdot & \cdot & 1 & \cdot & \cdot & \cdot & \cdot \\
\cdot & \cdot & \cdot & \cdot & \cdot & \cdot & 1 & \cdot \\
\hline
\cdot & \cdot & 1 & \cdot & \cdot & \cdot & \cdot & \cdot & \cdot \\
\cdot & \cdot & \cdot & \cdot & 1 & \cdot & \cdot & \cdot \\
\cdot & \cdot & \cdot & \cdot & \cdot & \cdot & \cdot & \cdot & 1
\end{array}
\right),
$$

where the dots represent zeros and the α_i's are arbitrary, except for $0 < \alpha_i < 1$ ($i = 1, 2, 3$).

The matrix $\mathbf{G}_n$ can be described as follows (e.g., Harville [1997: 352, with a correction]). For $i \geq j$, the $[(j-1)n+i]$th and $[(i-1)n+j]$th rows of $\mathbf{G}_n$ equal the $[(j-1)(n-j/2)+i]$th row of $\mathbf{I}_k$, that is, they equal the k-dimensional row vector whose $[(j-1)(n-j/2)+i]$th element is 1 and whose remaining elements are zero. For $j \geq i$ the $[(j-1)(n-j/2)+i]$th column is an n^2-dimensional column vector whose $[(j-1)n+i]$th and $[(i-1)n+j]$th elements are 1 and whose remaining elements are 0.

Another related matrix is $\mathbf{N}_n$, where $\mathbf{N}_n = \text{vec} \left(\frac{1}{2}(\mathbf{A} + \mathbf{A}') \right)$ transforms $\mathbf{A}$ into a symmetric matrix. This matrix is called the *symmetrizer*. As shown below, $\mathbf{N}_n$ turns out to be symmetric and idempotent, so I shall also denote it by $\mathbf{P}_n$ to remind us that it represents an orthogonal projection (see also Schott [2005: 312]).

11.29. (General Properties) For handy reference, we frequently have in the literature $\mathbf{G}_n \equiv \mathbf{D}_n$, $\mathbf{H}_n \equiv \mathbf{D}_n^+$, and $\mathbf{P}_n \equiv \mathbf{N}_n$; also $k = n(n+1)/2$.

(a) $\mathbf{H}_n$ is a left inverse of $\mathbf{G}_n$, i.e., $\mathbf{H}_n\mathbf{G}_n = \mathbf{I}_k$. Thus $\mathbf{H}_n$ is a weak inverse of $\mathbf{G}_n$ as $\mathbf{G}_n\mathbf{H}_n\mathbf{G}_n = \mathbf{G}_n$.

(b) Every row of $\mathbf{G}_n$ contains only one nonzero element, so that the columns of $\mathbf{G}_n$ are orthogonal.

(c) The $n^2 \times k$ matrix $\mathbf{G}_n$ is unique, of rank k.

(d) $\mathbf{I}_{(n,n)}\mathbf{G}_n = \mathbf{G}_n$ (i.e., $\mathbf{K}_{nn}\mathbf{D}_n = \mathbf{D}_n$).

(e) $\mathbf{G}_{n+1}'\mathbf{G}_{n+1} = \begin{pmatrix} 1 & 0 & 0 \\ 0 & 2\mathbf{I}_n & 0 \\ 0 & 0 & \mathbf{G}_n'\mathbf{G}_n \end{pmatrix}.$

(f) $(\mathbf{G}_{n+1}'\mathbf{G}_{n+1})^{-1} = \mathbf{G}_{n+1}^+\mathbf{G}_{n+1}^{+\,\prime} = \begin{pmatrix} 1 & 0 & 0 \\ 0 & \frac{1}{2}\mathbf{I}_n & 0 \\ 0 & 0 & (\mathbf{G}_n'\mathbf{G}_n)^{-1} \end{pmatrix}.$

(g) The $k \times n^2$ matrix $\mathbf{H}_n$ is not unique and has rank k.

(h) A useful form of $\mathbf{H}_n$ is $\mathbf{H}_n = \mathbf{G}_n^+ = (\mathbf{G}_n'\mathbf{G}_n)^{-1}\mathbf{G}_n'$, the Moore–Penrose inverse of $\mathbf{G}_n$ (Schott [2005: 313]). (This is the form taken by $\mathbf{H}_3(6 \times 9)$ above when all the α_i's are set equal to $\frac{1}{2}$.) Then:

   (i) $\mathbf{G}_n^+\mathbf{G}_n = \mathbf{I}_k$.

   (ii) $\mathbf{G}_n\mathbf{G}_n^+ = \frac{1}{2}(\mathbf{I}_{n^2} + \mathbf{I}_{(n,n)})(= \mathbf{P}_n)$.

   (iii) $\mathbf{G}_n^+\mathbf{I}_{(n,n)} = \mathbf{G}_n^+$.

(i) $\mathbf{P}_n = \mathbf{G}_n\mathbf{G}_n^+ = \mathbf{G}_n(\mathbf{G}_n'\mathbf{G}_n)^{-1}\mathbf{G}_n$. Then:

   (i) $\mathbf{P}_n\mathbf{G}_n = \mathbf{G}_n$ and $\mathbf{G}_n^+\mathbf{P}_n = \mathbf{G}_n^+$.

   (ii) $\mathbf{P}_n\mathrm{vec}\,\mathbf{A} = \mathrm{vec}\,[\frac{1}{2}(\mathbf{A} + \mathbf{A}')]$ for any $n \times n$ matrix $\mathbf{A}$.

   (iii) The symmetrizer $\mathbf{P}_n$ is symmetric and idempotent, that is, a projection matrix projecting orthogonally onto $\mathcal{C}(\mathbf{G}_n)$.

   (iv) $\mathrm{rank}\,\mathbf{P}_n = \mathrm{trace}\,\mathbf{P}_n = \frac{1}{2}n(n + 1)$.

   (v) $\mathbf{P}_n\mathbf{I}_{(n,n)} = \mathbf{P}_n = \mathbf{I}_{(n,n)}\mathbf{P}_n$ (i.e., $\mathbf{N}_n\mathbf{K}_{nn} = \mathbf{N}_n = \mathbf{K}_{nn}\mathbf{N}_n$).

   (vi) If $\mathbf{A}$ and $\mathbf{B}$ are $n \times n$, then $\mathbf{P}_n(\mathbf{A} \otimes \mathbf{B})\mathbf{P}_n = \mathbf{P}_n(\mathbf{B} \otimes \mathbf{A})\mathbf{P}_n$ and $\mathbf{P}_n(\mathbf{A} \otimes \mathbf{A})\mathbf{P}_n = \mathbf{P}_n(\mathbf{A} \otimes \mathbf{A}) = (\mathbf{A} \otimes \mathbf{A})\mathbf{P}_n$.

For further properties of $\mathbf{G}_n$, $\mathbf{G}_n'$, $\mathbf{G}_n^+$, $\mathbf{G}_n'\mathbf{G}_n$, and $\mathbf{G}_n\mathbf{G}_n'$, where $\mathbf{G}_n = \mathbf{D}_n$, see Abadir and Magnus [2005: section 11.3] and Magnus [1988: chapter 4].

11.30. Suppose $\mathbf{A}$ and $\mathbf{X}$ are both $n \times n$, and $\mathbf{X}$ is symmetric, then

$$
\begin{aligned}
\mathrm{vech}\,(\mathbf{AXA}') &= \mathbf{H}_n\mathrm{vec}\,(\mathbf{AXA}) \\
&= \mathbf{H}(\mathbf{A} \otimes \mathbf{A})\mathrm{vec}\,\mathbf{X} \\
&= \mathbf{H}_n(\mathbf{A} \otimes \mathbf{A})\mathbf{G}_n\mathrm{vech}\,\mathbf{X} \\
&= \mathbf{C}\mathrm{vech}\,\mathbf{X}, \quad \text{say.}
\end{aligned}
$$

Properties of $\mathbf{C}$ are given in (11.31c) below.

11.31. Suppose $\mathbf{A}$ is $n \times n$. Then:

(a) $\mathbf{G}_n\mathbf{G}_n^+(\mathbf{A} \otimes \mathbf{A})\mathbf{G}_n = (\mathbf{A} \otimes \mathbf{A})\mathbf{G}_n$.

(b) $\mathbf{G}_n\mathbf{G}_n^+(\mathbf{A} \otimes \mathbf{A})\mathbf{G}_n^{+\prime} = (\mathbf{A} \otimes \mathbf{A})\mathbf{G}_n^{+\prime}$.

(c) Let $\mathbf{C} = \mathbf{H}_n(\mathbf{A} \otimes \mathbf{A})\mathbf{G}_n$, a $k \times k$ matrix, where $k = n(n + 1)/2$. Then:

   (i) $\mathbf{C}$ is invariant with respect to the choice of $\mathbf{H}_n$, so we can choose $\mathbf{H} = \mathbf{G}^+$ (cf. 11.29h).

   (ii) $\mathbf{C}$ is nonsingular if and only if $\mathbf{A}$ is nonsingular. Then

$$
\mathbf{C}^{-1} = \mathbf{G}_n^+(\mathbf{A}^{-1} \otimes \mathbf{A}^{-1})\mathbf{G}_n.
$$

   (iii) If $\mathbf{A}$ is upper-triangular, lower-triangular, or diagonal, then $\mathbf{C}$ is respectively upper-triangular, lower-triangular, or diagonal with diagonal elements $a_{ii}a_{jj}$, $i = 1, 2, \ldots, n$; $j = 1, 2, \ldots, n$.

(iv) The eigenvalues of $\mathbf{C}$ are $\lambda_i\lambda_j$ $(1 \leq j \leq i \leq n)$, where λ_i $(i = 1, 2, \ldots, n)$ are the eigenvalues of $\mathbf{A}$.

(v) $\det \mathbf{C} = \det[\mathbf{G}_n^+(\mathbf{A} \otimes \mathbf{A})\mathbf{G}_n] = (\det \mathbf{A})^{n+1}$.

(vi) $\operatorname{trace} \mathbf{C} = \frac{1}{2}[(\operatorname{trace} \mathbf{A})^2 + \operatorname{trace}(\mathbf{A}^2)]$.

(vii) $\operatorname{rank} \mathbf{C} = \frac{1}{2}[(\operatorname{rank} \mathbf{A})^2 + \operatorname{rank} \mathbf{A}]$.

(viii) $\mathbf{C}^- = \mathbf{H}_n(\mathbf{A}^- \otimes \mathbf{A}^-)\mathbf{G}_n$ is a weak inverse of $\mathbf{C}$.

(d) If $\mathbf{A}$ is nonsingular, then:

(i) $[\mathbf{G}_n'(\mathbf{A} \otimes \mathbf{A})\mathbf{G}_n]^{-1} = \mathbf{G}_n^+(\mathbf{A}^{-1} \otimes \mathbf{A}^{-1})\mathbf{G}_n^{+\prime}$.

(ii) $\det[\mathbf{G}_n^+(\mathbf{A} \otimes \mathbf{A})\mathbf{G}_n^{+\prime}] = 2^{-n(n-1)/2}(\det \mathbf{A})^{n+1}$.

11.32. If $\mathbf{A}$ is any $n \times n$ matrix, then the following hold.

(a) $(\mathbf{A} \otimes \mathbf{A})\mathbf{G}_n = \mathbf{G}_n\mathbf{H}_n(\mathbf{A} \otimes \mathbf{A})\mathbf{G}_n$.

(b) $\mathbf{G}_n\mathbf{H}_n(\mathbf{A} \otimes \mathbf{A}) = (\mathbf{A} \otimes \mathbf{A})\mathbf{G}_n\mathbf{H}_n$.

(c) $\mathbf{H}_n(\mathbf{A} \otimes \mathbf{A}) = \mathbf{H}_n(\mathbf{A} \otimes \mathbf{A})\mathbf{G}_n\mathbf{H}_n$.

We can set $\mathbf{H}_n = \mathbf{G}_n^+$ in the above.

For some properties of $\mathbf{G}_n^+(\mathbf{I}_n \otimes \mathbf{A} + \mathbf{A} \otimes \mathbf{I}_n)\mathbf{G}_n$, $\mathbf{G}_n^+(\mathbf{A} \otimes \mathbf{B})\mathbf{G}_n$, and some related matrices (with $\mathbf{D}_n = \mathbf{G}_n$), including further relationships between $\mathbf{D}_{n+1}$ and $\mathbf{D}_n$, see Magnus [1988: 65–72].

Proofs. Section 11.5.1.

11.29a. $\operatorname{vech} \mathbf{A} = \mathbf{H}_n \operatorname{vec} \mathbf{A} = \mathbf{H}_n\mathbf{G}_n\operatorname{vech} \mathbf{A}$ for all symmetric $\mathbf{A}$. Henderson and Searle [1979: 69].

11.29b. Follows from the definition of $\mathbf{G}_n$.

11.29c. Schott [2005: 313].

11.29d. Henderson and Searle [1979: 69].

11.29e. Harville [1997: 355], Magnus [1988: 72], and Magnus and Neudecker [1999: 51].

11.29f. Magnus [1988: 72] and Magnus and Neudecker [1999: 51].

11.29g. Henderson and Searle [1979: 69].

11.29h. Abadir and Magnus [2006: 312–313], Harville [1997: 354–357], and Magnus [1988: 56].

11.29i. Abadir and Magnus [2005: 307], Magnus [1988: 48–49], and Schott [2005: 312]. For (v) see also Abadir and Magnus [2006: 308] and Magnus and Neudecker [1999: 50].

11.31a–b. Abadir and Magnus [2006: 315], Magnus [1988: chapter 3], and Magnus and Neudecker [1999: 49–50].

11.31c(i). Henderson and Searle [1979: 70].

11.31c(ii). Abadir and Magnus [2006: 315], Harville [1997: 358], Magnus [1988: chapter 3], and Magnus and Neudecker [1999: 49–50].

11.31c(iii). Magnus [1988: 63].

11.31c(iv). Magnus [1988: 64].

11.31c(v). Abadir and Magnus [2006: 316], Harville [1997: 362], Henderson and Searle [1979: 70], and Magnus [1988: 64-65].

11.31c(vi). Abadir and Magnus [2005: 316], Harville [1997: 358], and Magnus [1988: 64].

11.31c(vii)–(viii). Harville [1997: 358].

11.31d. Abadir and Magnus [2005: 317], Magnus [1988: 65], and Schott [2005: 315].

11.32. Harville [1997: 358] and Henderson and Searle [1979: 70].

11.5.2 Lower-Triangular Matrix

Definition 11.10. Let $\mathbf{A}$ be an $n \times n$ lower-triangular matrix. If $k = n(n+1)/2$, the $k \times n^2$ matrix $\mathbf{L}_n$ is called the *elimination matrix* if $\operatorname{vec} \mathbf{A} = \mathbf{L}'_n \operatorname{vech} \mathbf{A}$. The difference between $\mathbf{G}_n$ and $\mathbf{L}'_n$ is that $\operatorname{vec} \mathbf{A}$ now contains some zeros. Thus $\mathbf{L}'_n$ can be obtained from $\mathbf{G}_n$ ($= \mathbf{D}_n$) by replacing $n(n-1)/2$ rows of $\mathbf{G}_n$ by zeros; (d) below gives a clearer picture.

11.33. We have the following properties for $\mathbf{L}_n$.

(a) $\mathbf{L}_n$ has full row rank k.

(b) $\mathbf{L}\mathbf{L}'_n = \mathbf{I}_k$.

(c) $\mathbf{L}_n^{+} = \mathbf{L}'_n$.

(d) From (b) we have $\operatorname{vech} \mathbf{A} = \mathbf{L}_n \operatorname{vec} \mathbf{A}$, so that $\mathbf{L}_n$ eliminates the zeros from $\operatorname{vec} \mathbf{A}$.

(e) $\mathbf{L}_n \mathbf{G}_n = \mathbf{I}_k$, $k = n(n+1)/2$.

(f) $\mathbf{G}_n \mathbf{L}_n \mathbf{P}_n = \mathbf{P}_n$, where $\mathbf{P}_n = \frac{1}{2}(\mathbf{I}_{n^2} + \mathbf{I}_{(n,n)})$ (i.e., $\mathbf{D}_n \mathbf{L}_n \mathbf{N}_n = \mathbf{N}_n$).

(g) $\mathbf{G}_n^{+} = \mathbf{L}_n \mathbf{P}_n$.

Similar properties apply to the situation where $\mathbf{A}$ is *strictly lower-triangular*, that is lower-triangular but with zero diagonal elements (Schott 2005: 317-318]). Triangular matrices, and in fact any patterned matrix can be handled using a general kind of vcc operator (cf. Section 18.3.5).

Proofs. Section 11.5.2.

11.33. Magnus [1988: 77, 80] and Schott [2005: 316–317].

11.6 STAR OPERATOR

Definition **11.11.** Let $\mathbf{A} = (a_{ij})$ be $m \times n$ and $\mathbf{B}$ be $mp \times nq$. Then we define the $p \times q$ matrix (MacRae [1974])

$$\mathbf{A} * \mathbf{B} = \sum_{i=1}^{m} \sum_{j=1}^{n} a_{ij} \mathbf{B}_{ij},$$

where $\mathbf{B}_{ij}$ is the (i,j)th submatrix of $\mathbf{B}$ when $\mathbf{B}$ is partitioned into submatrices of size $p \times q$.
When $\mathbf{A}$ and $\mathbf{B}$ are the same size, $\mathbf{A} * \mathbf{B} = \text{trace}(\mathbf{A}'\mathbf{B})$.

11.34. If $\mathbf{C}$ is $r \times s$, then $(\mathbf{A} * \mathbf{B}) \otimes \mathbf{C} = \mathbf{A} * (\mathbf{B} \otimes \mathbf{C})$.

11.35. $\mathbf{A} * \mathbf{B} = \mathbf{B} * (\mathbf{A} \otimes \text{vec}\,\mathbf{I}_p (\text{vec}\,\mathbf{I}_q)')$.

11.36. If $\mathbf{x}$ is $p \times 1$ and $\mathbf{y}$ is $q \times 1$, then

$$
\begin{aligned}
\mathbf{x}'(\mathbf{A} * \mathbf{B})\mathbf{y} &= \mathbf{A} * (\mathbf{I}_m \otimes \mathbf{x}')\mathbf{B}(\mathbf{I}_n \otimes \mathbf{y}) \\
&= (\mathbf{I}_m \otimes \mathbf{x})\mathbf{A}(\mathbf{I}_n \otimes \mathbf{y}') * \mathbf{B}.
\end{aligned}
$$

11.37.

$$
\begin{aligned}
\mathbf{X}_{m \times n} \mathbf{Y}_{n \times p} \mathbf{Z}_{p \times q} &= \mathbf{Y} * \text{vec}\,\mathbf{X}(\text{vec}\,\mathbf{Z}')' \\
&= \mathbf{Y}' * (\mathbf{Z} \otimes \mathbf{I}_m)\mathbf{I}_{(m,q)}(\mathbf{X} \otimes \mathbf{I}_q).
\end{aligned}
$$

Proofs. Section 11.6.

    11.34–11.37. MacRae [1974].

11.7 HADAMARD PRODUCT

We now consider a particular product that arises in a wide variety of mathematical applications such as covariance matrices for independent zero mean random vectors and characteristic functions in probability theory (Horn and Johnson [1985: 301, 393–394]). Further mathematical applications are described by Horn and Johnson [1985: 455–457] and Horn and Johnson [1991: chapter 5]. The Hadamard product appears in several places in this book. In this section $\mathbf{A} \succeq \mathbf{B}$ means that $\mathbf{A} - \mathbf{B}$ is non-negative definite.

Definition **11.12.** If $\mathbf{A} = (a_{ij})$ and $\mathbf{B} = (b_{ij})$ are $m \times n$ real or complex matrices, then their *Hadamard product* (also referred to as the *Schur product*) is the $m \times n$ matrix $\mathbf{A} \circ \mathbf{B} = (a_{ij}b_{ij})$. The results below, where proofs are not referenced, follow from the definition by simply multiplying out the appropriate matrices.

11.38. Let $\mathbf{A}$ and $\mathbf{B}$ be $m \times n$ matrices, and let $\mathbf{e}_i$ be the ith column of $\mathbf{I}_m$.

  (a) Let $\boldsymbol{\Phi}_m = \sum_{i=1}^{m} \mathbf{e}_i(\mathbf{e}_i \otimes \mathbf{e}_i)' = \sum_{i=1}^{m} \mathbf{e}_i(\text{vec}\,\mathbf{E}_{ii})'$, where $\mathbf{E}_{ii} = \mathbf{e}_i\mathbf{e}_i'$. Then:

   (i) $\mathbf{A} \circ \mathbf{B} = \boldsymbol{\Phi}_m(\mathbf{A} \otimes \mathbf{B})\boldsymbol{\Phi}_n'$.

(ii) $\mathbf{\Psi}_m \mathbf{\Psi}'_m = \mathbf{I}_m$.

(iii) If $\mathbf{C} = (c_{ij})$ is $m \times m$, then

$$\mathbf{\Psi}_m \text{vec } \mathbf{C} - (c_{11}, c_{22}, \cdots, c_{mm})' = (\text{diag } \mathbf{C}) \mathbf{1}_m$$

(iv) $\mathbf{\Psi}_m \mathbf{I}_{(m,m)} = \mathbf{\Psi}_m$.

(b) $\mathbf{A} \circ \mathbf{B}$ is a submatrix of $\mathbf{A} \otimes \mathbf{B}$. In fact

$$\mathbf{A} \circ \mathbf{B} = (\mathbf{A} \otimes \mathbf{B})_{\alpha,\beta},$$

where (α, β) denotes the submatrix formed by the intersection of the rows of $\mathbf{A} \otimes \mathbf{B}$ in α with the columns in β, where $\alpha = \{1, m+2, 2m+3, \ldots, m^2\}$ and $\beta = \{1, n+2, 2n+3, \ldots, n^2\}$.

(c) If $m = n$, then $\mathbf{A} \circ \mathbf{B}$ is a principal submatrix of $\mathbf{A} \otimes \mathbf{B}$.

The above results can be used to prove results about $\mathbf{A} \circ \mathbf{B}$ using $\mathbf{A} \otimes \mathbf{B}$.

11.39. Let $\mathbf{A}$ and $\mathbf{B}$ be $m \times n$ matrices. Then the following hold.

(a) $\mathbf{A} \circ \mathbf{B} = \mathbf{B} \circ \mathbf{A}$.

(b) $(\mathbf{A} \circ \mathbf{B})' = \mathbf{A}' \circ \mathbf{B}'$.

(c) $\text{trace}(\mathbf{A}\mathbf{B}') = \mathbf{1}'_m (\mathbf{A} \circ \mathbf{B}) \mathbf{1}_n$.

(d) $\text{rank}(\mathbf{A} \circ \mathbf{B}) \leq \text{rank } \mathbf{A} \cdot \text{rank } \mathbf{B}$.

11.40. If all matrices are the same size, then

$$(\mathbf{A} + \mathbf{B}) \circ (\mathbf{C} + \mathbf{D}) = \mathbf{A} \circ \mathbf{C} + \mathbf{A} \circ \mathbf{D} + \mathbf{B} \circ \mathbf{C} + \mathbf{B} \odot \mathbf{D}.$$

11.41. If $\mathbf{A}$, $\mathbf{B}$, and $\mathbf{C}$ are all $m \times n$ matrices, then

$$\text{trace}[(\mathbf{A} \circ \mathbf{B})\mathbf{C}'] = \text{trace}[(\mathbf{A} \circ \mathbf{C})\mathbf{B}'].$$

11.42. (Multiplication by Diagonal Matrices)   Suppose $\mathbf{A}$ and $\mathbf{B}$ are $m \times n$, $\mathbf{D}$ is $m \times m$, and $\mathbf{E}$ is $n \times n$, where $\mathbf{D}$ and $\mathbf{E}$ are diagonal matrices, then

$$\mathbf{D}(\mathbf{A} \circ \mathbf{B})\mathbf{E} = (\mathbf{D}\mathbf{A}\mathbf{E}) \circ \mathbf{B} = (\mathbf{D}\mathbf{A}) \circ (\mathbf{B}\mathbf{E}) = (\mathbf{A}\mathbf{E}) \circ (\mathbf{D}\mathbf{B}) = \mathbf{A} \circ (\mathbf{D}\mathbf{B}\mathbf{E}).$$

11.43. If $\mathbf{A}$ is square, then $\mathbf{A} \circ \mathbf{1}\mathbf{1}' = \mathbf{A} = \mathbf{1}\mathbf{1}' \circ \mathbf{A}$.

11.44. (Quadratics)   Let $\mathbf{A}$ and $\mathbf{B}$ be $n \times n$ matrices, and suppose $\mathbf{y}, \mathbf{z} \in \mathbb{C}^n$. Then

$$\mathbf{y}^*(\mathbf{A} \circ \mathbf{B})\mathbf{z} = \text{trace}(\mathbf{D}^*_\mathbf{y} \mathbf{A} \mathbf{D}_\mathbf{z} \mathbf{B}'),$$

where $\mathbf{D}_\mathbf{y} = \text{diag}(\mathbf{y})$ and $\mathbf{D}_\mathbf{z} = \text{diag}(\mathbf{z})$.

11.45. If $\mathbf{A}$ and $\mathbf{B}$ are Hermitian matrices, then so is $\mathbf{A} \circ \mathbf{B}$.

11.46. Let $\mathbf{A}$ and $\mathbf{B}$ be Hermitian non-negative definite $n \times n$ matrices, that is, $\mathbf{A} \succeq \mathbf{0}$ and $\mathbf{B} \succeq \mathbf{0}$. Then:

(a) $\mathbf{A} \circ \mathbf{B} \succeq \mathbf{0}$. The same results apply to $\mathbf{A} \circ \mathbf{A} \circ \cdots \circ \mathbf{A}$ to any number of terms.

(b) $\det(\mathbf{A} \circ \mathbf{B}) + \det \mathbf{A} \cdot \det \mathbf{B} \geq b_{11}b_{22} \cdots b_{nn} \det \mathbf{A} + a_{11}a_{22} \cdots a_{nn} \det \mathbf{B}.$

(c) $\displaystyle\prod_{i=1}^{n} a_{ii}b_{ii} \geq \det(\mathbf{A} \circ \mathbf{B}) \geq b_{11}b_{22} \cdots b_{nn} \det \mathbf{A} \geq \det \mathbf{A} \det \mathbf{B}.$

(Note that $\mathbf{A}$ and $\mathbf{B}$ can be interchanged.) The left-hand side follows from (12.27).

(d) $\mathbf{A}^2 \circ \mathbf{B}^2 \succeq (\mathbf{A} \circ \mathbf{B})^2.$

(e) If $\mathbf{A}$ and $\mathbf{B}$ are positive definite, then:

  (i) $\mathbf{A} \circ \mathbf{B}$ is positive definite.
  (ii) $\mathbf{A}^{-1} \circ \mathbf{B}^{-1} \succeq (\mathbf{A} \circ \mathbf{B})^{-1}.$

(f) If $\mathbf{A} \succ \mathbf{0}$, and $\mathbf{B} \succeq \mathbf{0}$ with r nonzero diagonal entries, then $\operatorname{rank}(\mathbf{A} \circ \mathbf{B}) = r$.

(g) If $\mathbf{B} \succ \mathbf{0}$, and $\mathbf{A} \succeq \mathbf{0}$ with positive diagonal elements, then $\mathbf{A} \circ \mathbf{B} \succ \mathbf{0}$.

(h If $\mathbf{A} \succ \mathbf{0}$, then $\mathbf{A} \circ \mathbf{A}^{-1} \succeq \mathbf{I}_n \succeq (\mathbf{A}^{-1} \circ \mathbf{A})^{-1}$. Horn and Johnson [1991: section 5.4] discuss the properties of $\mathbf{A} \circ \mathbf{A}^{-1}$ and $\mathbf{A} \circ (\mathbf{A}^{-1})'$.

11.47. (Fejér's Theorem) Let $\mathbf{A}$ be any $n \times n$ matrix. Then $\mathbf{A}$ is Hermitian non-negative definite if and only if $\operatorname{trace}(\mathbf{A} \circ \mathbf{B}) \geq \mathbf{0}$ for all Hermitian non-negative definite $n \times n$ matrices $\mathbf{B}$.

11.48. (Eigenvalues) Let $\mathbf{A}$ and $\mathbf{B}$ be $n \times n$ Hermitian non-negative definite matrices.

(a) Let $b_{\max}$ and $b_{\min}$ be maximum and minimum entries of the diagonal elements of $\mathbf{B}$. Then, for all j,

$$b_{\min}\lambda_{\min}(\mathbf{A}) \leq \lambda_j(\mathbf{A} \circ \mathbf{B}) \leq b_{\max}\lambda_{\max}(\mathbf{A}).$$

(b) $\lambda_{\min}(\mathbf{A})\lambda_{\min}(\mathbf{B}) \leq \lambda_j(\mathbf{A} \circ \mathbf{B}) \leq \lambda_{\max}(\mathbf{A})\lambda_{\max}(\mathbf{B})$ for all j.

(c) Let $\mathbf{R} = (\rho_{ij})$ be any $n \times n$ correlation matrix.

  (i) Since $\rho_{ii} = 1$ for all i, it follows from (a) that
  $\lambda_{\min}(\mathbf{A}) \leq \lambda_j(\mathbf{A} \circ \mathbf{R}) \leq \lambda_{\max}(\mathbf{A}).$
  (ii) Setting $\mathbf{R} = \mathbf{I}_n$ we have $\lambda_{\min}(\mathbf{A}) \leq a_{jj} \leq \lambda_{\max}(\mathbf{A}).$

(d) $\lambda_{\min}(\mathbf{A} \circ \mathbf{B}) \geq \lambda_{\min}(\mathbf{A}\mathbf{B}).$

11.49. (Singular Values) Let $\mathbf{A}$ and $\mathbf{B}$ be $m \times n$ matrices, and let $\sigma_j(\mathbf{C})$ be the jth singular value of $\mathbf{C}$, where the singular values are listed in decreasing order of magnitude. Then

$$\sum_{j=1}^{i} \sigma_j(\mathbf{A} \circ \mathbf{B}) \leq \sum_{j=1}^{i} \sigma_j(\mathbf{A})\sigma_j(\mathbf{B}), \quad i = 1, 2, \ldots, n.$$

11.50. If $\mathbf{A}$ and $\mathbf{B}$ are real or complex $m \times n$ matrices, then

$$(\mathbf{A}\mathbf{A}^*) \circ (\mathbf{B}\mathbf{B}^*) \succeq (\mathbf{A} \circ \mathbf{B})(\mathbf{A}^* \circ \mathbf{B}^*).$$

11.51. Let $\mathbf{A}$ and $\mathbf{B}$ be $n \times n$ Hermitian positive definite matrices, and let $\mathbf{C}$ and $\mathbf{D}$ be any $m \times n$ real or complex matrices. Then

$$(\mathbf{CA}^{-1}\mathbf{C}^*) \circ (\mathbf{DB}^{-1}\mathbf{D}^*) \succeq (\mathbf{C} \circ \mathbf{D})(\mathbf{A} \circ \mathbf{B})^{-1}(\mathbf{C} \circ \mathbf{D})^*.$$

Proofs. Section 11.7.

11.38a. Magnus [1988: 110] and Schott [2005: 297, (i)] .

11.38b–c. Horn and Johnson [1991: 304].

11.39. Here (a) and (b) are obvious, (c) and (d) are given by Schott [2005: 297], and (d) is given by Horn and Johnson [1991: 307].

11.40. Follows directly from the definition of "∘".

11.41. Horn and Johnson [1991: 305–306].

11.42. Let $\mathbf{C} = \mathbf{A} \circ \mathbf{B}$. Then $(\mathbf{DCE})_{ij} = \sum_r \sum_s d_{ir}c_{rs}e_{sj}$, which can be expressed in the form $\sum_r \sum_s d_{ir}a_{rs}e_{sj} \cdot b_{rs} = [(\mathbf{DAE}) \circ \mathbf{B}]_{ij}$, and so on.

11.43. Follows from $a_{ij} \cdot 1 = a_{ij}$.

11.44. Horn and Johnson [1991: 306] and Schott [2005: 298].

11.46a. Horn and Johnson [1985: 458], Rao and Rao [1998: 204, 215], Schott [2005: 299, real case], and Zhang [1999: 192].

11.46b. Rao and Rao [1998: 212].

11.46c. Rao and Rao [1998: 210], Schott [2005: 302], and Zhang [1999: 200].

11.46d. Zhang [1999: 193].

11.46e(i). Abadir and Magnus [2005: 340] and Rao and Rao [1998: 204].

11.46e(ii). Horn and Johnson [1985: 475] and Zhang [1999: 193].

11.46f. Rao and Rao [1998: 213].

11.46g. Horn and Johnson [1991: 309] and Schott [2005: 300, real case].

11.46h. Horn and Johnson [1985: 475], Schott [2005: 304], and Zhang [1999: 193].

11.47. Rao and Rao [1998: 214].

11.48a. Rao and Rao [1998: 206] and Schott [2005: 303, real case].

11.48b. Horn and Johnson [1991: 312] and Rao and Rao [1998: 207].

11.48c. Rao and Rao [1998: 207].

11.48d. Bapat and Raghavan [1997: 142, real case] and Schott [2005: 305, real case].

11.49. Horn and Johnson [1991: 334].

11.50. Zhang [1999: 194].

11.51. Zhang [1999: 198].

11.8   RAO–KHATRI PRODUCT

***Definition* 11.13.** Let $\mathbf{A} = (\mathbf{a}_1, \mathbf{a}_2, \ldots, \mathbf{a}_n)$ be a $p \times n$ and $\mathbf{B} = (\mathbf{b}_1, \mathbf{b}_2, \ldots, \mathbf{b}_n)$ be $m \times n$ matrices. Then the *Rao–Khatri product*, denoted by $\mathbf{A} \odot \mathbf{B}$, of $\mathbf{A}$ and $\mathbf{B}$ is the $mp \times n$ partitioned matrix

$$\mathbf{A} \odot \mathbf{B} = (\mathbf{a}_1 \otimes \mathbf{b}_1, \mathbf{a}_2 \otimes \mathbf{b}_2, \ldots, \mathbf{a}_n \otimes \mathbf{b}_n).$$

11.52. Let $\mathbf{A}_{p \times n}$, $\mathbf{B}_{m \times n}$, $\mathbf{C}_{m \times p}$, and $\mathbf{D}_{n \times m}$ be four matrices. Then

$$(\mathbf{C} \otimes \mathbf{D})_{mn \times mp}(\mathbf{A} \odot \mathbf{B})_{mp \times n} = (\mathbf{CA})_{m \times n} \odot (\mathbf{DB})_{n \times n}.$$

11.53. Let $\mathbf{A}$ and $\mathbf{B}$ be non-negative definite $n \times n$ matrices of ranks r and s, respectively. Let $\mathbf{A} = \mathbf{R}'\mathbf{R}$, where $\mathbf{R}$ is $r \times n$, and let $\mathbf{B} = \mathbf{S}'\mathbf{S}$, where $\mathbf{S}$ is $s \times n$ [cf. (10.10)]. Then

$$\mathbf{A} \circ \mathbf{B} = (\mathbf{R} \odot \mathbf{S})'(\mathbf{R} \odot \mathbf{S}).$$

Proofs. Section 11.8.

11.52–11.53. Rao and Rao [1998: 216].

CHAPTER 12

INEQUALITIES

Inequalities are used extensively in statistics and, because they relate to almost every chapter in this book, they are difficult to categorize. Those concerned with general inner products and norms are considered in Sections 2.2.1 and 4.6. Those involved with ranks are discussed in Chapter 3, while those for eigenvalues appear in Chapter 6. Some inequalities for non-negative definite matrices appear in Chapter 9, and those relating to majorization appear in Chapter 23. There are a large number of inequalities involving probability and random variables and a selection of these appear in Chapters 22 and 23. Optimization in Chapter 24 generates further inequalities. So what is in this chapter? I have collected here some of the more traditional inequalities such as Cauchy–Schwarz, Kantorovich, Hölder, Minkowski, and so on, and their extensions. At the end I have listed a few identities that can be useful in setting up inequalities.

12.1 CAUCHY–SCHWARZ INEQUALITIES

The inequalities given below are fairly basic ones. However, for further extensions and refinements, including those for complex numbers, the reader is referred to Dragomir [2004: chapters 1–3].

12.1.1 Real Vector Inequalities and Extensions

12.1. Let $\mathbf{x} = (x_i)$ and $\mathbf{y} = (y_i)$ be real n-dimensional vectors. In addition to the basic inequality in (a) below, we can obtain various extensions from (2.17) by using a different vector space and a different inner product.

(a) (Cauchy–Schwarz) $(\mathbf{x}'\mathbf{y})^2 \leq (\mathbf{x}'\mathbf{x})(\mathbf{y}'\mathbf{y})$,

with equality if and only if $\mathbf{x} \propto \mathbf{y}$. Many different proofs of this result are available. For example, we can use Lagrange's identity of (12.44a). Alternatively, we also have

$$\mathbf{x}'\mathbf{x} - (\mathbf{x}'\mathbf{y})(\mathbf{y}'\mathbf{y})^{-1}(\mathbf{y}'\mathbf{x}) = \mathbf{x}'(\mathbf{I}_n - \mathbf{P_y})\mathbf{x} \geq 0,$$

since the projection matrix $\mathbf{I}_n - \mathbf{P_y} = \mathbf{P}_{\mathcal{C}(\mathbf{y})^{\perp}}$ that projects orthogonally onto $\mathcal{C}(\mathbf{y})^{\perp}$ is non-negative definite (cf. 2.49f).

(b) Let $\mathbf{A}$ be non-negative definite and, using (10.10), let $\mathbf{A} = \mathbf{B}'\mathbf{B}$.

  (i) $(\mathbf{x}'\mathbf{A}\mathbf{y})^2 \leq (\mathbf{x}'\mathbf{A}\mathbf{x})(\mathbf{y}'\mathbf{A}\mathbf{y})$, with equality if and only if $\mathbf{B}\mathbf{x} \propto \mathbf{B}\mathbf{y}$. A sufficient condition for equality is $\mathbf{x} \propto \mathbf{y}$. Furthermore, if $\mathbf{A}$ is positive definite, then

$$\sup_{\mathbf{x}:\mathbf{x}\neq\mathbf{0}} \frac{(\mathbf{x}'\mathbf{A}\mathbf{y})^2}{\mathbf{x}'\mathbf{A}\mathbf{x}} = \mathbf{y}'\mathbf{A}\mathbf{y}.$$

  (ii) From (i) we can deduce $|a_{ij}| \leq \max_i |a_{ii}|$.

(c) If $\mathbf{A}$ is non-negative definite and $\mathbf{y} \in \mathcal{C}(\mathbf{A})$, then for any weak inverse $\mathbf{A}^-$,

$$(\mathbf{x}'\mathbf{y})^2 \leq (\mathbf{x}'\mathbf{A}\mathbf{x})(\mathbf{y}'\mathbf{A}^-\mathbf{y}),$$

with equality if and only if $\mathbf{y} \propto \mathbf{A}\mathbf{x}$.

(d) If $\mathbf{A}$ is positive definite, then

$$(\mathbf{x}'\mathbf{y})^2 \leq (\mathbf{x}'\mathbf{A}\mathbf{x})(\mathbf{y}'\mathbf{A}^{-1}\mathbf{y}),$$

with equality if and only if $\mathbf{x} \propto \mathbf{A}^{-1}\mathbf{y}$ or, equivalently, $\mathbf{y} \propto \mathbf{A}\mathbf{x}$.

(e) If $\mathbf{A}$ is positive definite, then from (d) we have

$$(\mathbf{x}'\mathbf{x})^2 \leq (\mathbf{x}'\mathbf{A}\mathbf{x})(\mathbf{x}'\mathbf{A}^{-1}\mathbf{x}),$$

with equality when $\mathbf{x} \propto \mathbf{A}\mathbf{x}$, that is, when $\mathbf{x}$ is an eigenvector of $\mathbf{A}$.

(f) Let $\alpha_i \geq 0$ $(i = 1, 2, \ldots, n)$ such that $\sum_i \alpha_i = 1$. Let $z_1, z_2, \ldots, z_n$ be real numbers.

  (i) Setting $x_i = \sqrt{\alpha_i}$ and $y_i = \sqrt{\alpha_i}z_i$ in (a) leads to

$$\left(\sum_i \alpha_i z_i\right)^2 \leq \sum_i \alpha_i z_i^2,$$

with equality if and only if $z_1 = z_2 = \cdots = z_n$.

(ii) We can set $\alpha_i = 1/n$ to get

$$\left(\sum_{i=1}^{n} z_i\right)^2 < n \sum_{i=1}^{n} z_i^2,$$

with equality if and only if the $z_i s$ are all equal.

(iii) If all the eigenvalues λ_i of the $n \times n$ matrix $\mathbf{A}$ are real, then, from (i) with $\alpha_i = n^{-1}$ and $z_i = \lambda_i$, we have

$$\left(\frac{1}{n}\operatorname{trace}\mathbf{A}\right)^2 \le \frac{1}{n}\operatorname{trace}(\mathbf{A}^2),$$

with equality if and only if the eigenvalues are all equal.

(iv) If $\mathbf{A}$ is symmetric and nonzero with rank r, then, from (ii) with $n = r$, we have $(\sum_{i=1}^{r} \lambda_i)^2 \le r \sum_{i=1}^{r} \lambda_i^2$, where the λ_i are the nonzero eigenvalues of $\mathbf{A}$. Hence

$$\operatorname{rank}\mathbf{A} \ge \frac{(\operatorname{trace}\mathbf{A})^2}{\operatorname{trace}(\mathbf{A}^2)}.$$

Equality occurs if and only if $\mathbf{A}$ is proportional to a symmetric idempotent matrix.

(g) If $p_i \ge 0$ for all i, then, replacing x_i by $\sqrt{p_i}x_i$ and y_i by $\sqrt{p_i}y_i$ in (a), we have

$$(\sum_{i=1}^{n} p_i x_i y_i)^2 \le (\sum_{i=1}^{n} p_i x_i^2)(\sum_{i=1}^{n} p_i y_i^2).$$

(h) Suppose $x_i > 0$ for all i and $\boldsymbol{\mu} = (\mu_i)$ is arbitrary. Then, replacing x_i by $\sqrt{x_i}/\mu_i$ and y_i by $1/\sqrt{x_i}$ in (a)(i) and rearranging, we get

$$(\sum_{i=1}^{n} x_i^{-1})^{-1} \le (\sum_{i=1}^{n} x_i \mu_i^{-2})(\sum_{i=1}^{n} \mu_i^{-1})^{-2}.$$

If $\mathrm{E}(x_i) = \mu_i$, then taking expected values shows that expected value of a harmonic mean does not exceed the harmonic mean of their expected values.

(i) (Constrained Version) Let $\mathbf{A}$ be an $n \times n$ matrix, and let $\mathbf{y} \in \mathcal{C}(\mathbf{A})$. If $\mathbf{P_A}$ represents the orthogonal projection onto $\mathcal{C}(\mathbf{A})$, so that we have $\mathbf{P_A} = \mathbf{A}(\mathbf{A}'\mathbf{A})^{-}\mathbf{A}$ (cf. 2.49f), then

$$(\mathbf{x}'\mathbf{y})^2 \le (\mathbf{x}'\mathbf{P_A}\mathbf{x})(\mathbf{y}'\mathbf{y}).$$

Equality occurs when $\mathbf{y} \propto \mathbf{P_A}\mathbf{x}$.

(j) When $\mathbf{A}$ is positive definite and $\mathbf{x}'\mathbf{y} = 0$, then

$$(\mathbf{x}'\mathbf{A}\mathbf{y})^2 \le \left(\frac{\lambda_1 - \lambda_n}{\lambda_1 + \lambda_n}\right) \mathbf{x}'\mathbf{A}\mathbf{x} \cdot \mathbf{y}'\mathbf{A}\mathbf{y},$$

where λ_1 and λ_n are the maximum and minimum (positive) eigenvalues of $\mathbf{A}$.

12.2. (Some Lower Bounds) The results above give us upper bounds for $(\mathbf{x}'\mathbf{x})^2$ and $(\mathbf{x}'\mathbf{y})^2$. We now consider some lower bounds. Further details and extensions are given by Dragomir [2004: chapters 4 and 5].

(a) (Kantorovich) Let $\mathbf{A}$ be an $n \times n$ real positive definite (p.d.) matrix with maximum and minimum eigenvalues of λ_1 and λ_n, respectively. Let $\mathbf{x}$ and $\mathbf{y}$ be any nonzero vectors in $\mathbb{R}^n$.

 (i) $(\mathbf{x}'\mathbf{x})^2 \geq \dfrac{4\lambda_1\lambda_n}{(\lambda_1 + \lambda_n)^2}(\mathbf{x}'\mathbf{A}\mathbf{x})(\mathbf{x}'\mathbf{A}^{-1}\mathbf{x}).$

 There is a unit vector $\mathbf{x}$ for which there is an equality. The result also holds for a Hermitian p.d. matrix with $'$ replaced by $*$. For a generalization see Pronzato et al. [2005].

 (ii) If $\mathbf{A} = \operatorname{diag}(\mathbf{a})$ $(\mathbf{a} > \mathbf{0})$, then the ordered eigenvalues are the same as the ordered a_i for a diagonal matrix. Let $a_{max} = \max_i\{a_i\}$, and so on. Then, from (i), we obtain

$$(\mathbf{x}'\mathbf{x})^2 \geq \frac{4a_{max}a_{min}}{(a_{max} + a_{min})^2}(\sum_{i=1}^{n} a_i x_i^2)(\sum_{i=1}^{n} a_i^{-1} x_i^2).$$

 (iii) If $x_i = 1$ for all i in (ii), we have

$$n^2 \geq \frac{4a_{max}a_{min}}{(a_{max} + a_{min})^2}(\sum_{i=1}^{n} a_i)(\sum_{i=1}^{n} a_i^{-1}).$$

(b) (Polya–Szegö) Let $\mathbf{x}$ and $\mathbf{y}$ have positive elements. Then

$$(\mathbf{x}'\mathbf{y})^2 \geq \frac{4x_{min}x_{max}y_{min}y_{max}}{(x_{max}y_{max} + x_{min}y_{min})^2}(\mathbf{x}'\mathbf{x})(\mathbf{y}'\mathbf{y}).$$

(c) (Greub and Rheinboldt) Let $\mathbf{A}$ and $\mathbf{B}$ be $n \times n$ positive definite commuting matrices $(\mathbf{AB} = \mathbf{BA})$ with eigenvalues $\lambda_1 \geq \cdots \geq \lambda_n > 0$ and $\mu_1 \geq \cdots \geq \mu_n > 0$, respectively. Then $\mathbf{AB}$ is symmetric and

$$(\mathbf{x}'\mathbf{AB}\mathbf{x})^2 \geq \frac{4\lambda_1\lambda_n\mu_1\mu_n}{(\lambda_1\mu_1 + \lambda_n\mu_n)^2}(\mathbf{x}'\mathbf{A}^2\mathbf{x})(\mathbf{x}'\mathbf{B}^2\mathbf{x}).$$

(d) If $\mathbf{A}$ is an $n \times n$ nonsingular matrix with maximum and minimum singular values of σ_1 and σ_n, respectively, then

$$\frac{(\mathbf{x}'\mathbf{A}\mathbf{y})(\mathbf{y}'\mathbf{A}^{-1}\mathbf{x})}{(\mathbf{x}'\mathbf{x})(\mathbf{y}'\mathbf{y})} \leq \frac{(\sigma_1 + \sigma_n)^2}{4\sigma_1\sigma_n}.$$

 Rao [2005: 67] uses the above result to define antisingular values and vectors.

(e) Suppose $\mathbf{x} = (x_i) > \mathbf{0}$, $\mathbf{y} = (y_i) > \mathbf{0}$, and $\mathbf{w} = (w_i) \geq \mathbf{0}$. Let

$$m = \min_i \left\{\frac{x_i}{y_i}\right\} \quad \text{and} \quad M = \max_i \left\{\frac{x_i}{y_i}\right\}.$$

 Then $\quad \dfrac{\sum_{i=1}^{n} w_i x_i^2 \sum_{i=1}^{n} w_i y_i^2}{(\sum_{i=1}^{n} w_i x_i y_i)^2} \leq \dfrac{(m+M)^2}{4mM}.$

Proofs. Section 12.1.1.

12.1a. Abadir and Magnus [2005: 7].

12.1b. Abadir and Magnus [2005: 323].

12.1c. Neudecker and Liu [1994: 351]. Use $\mathbf{y}'\mathbf{A}\mathbf{y} = \mathbf{y}'\mathbf{A}\mathbf{A}^-\mathbf{A}\mathbf{y}$ in (b)(i) and set $\mathbf{z} = \mathbf{A}\mathbf{y}$.

12.1d. Replace $\mathbf{x}$ by $\mathbf{A}^{1/2}\mathbf{x}$ and $\mathbf{y}$ by $\mathbf{A}^{-1/2}\mathbf{y}$ in (a).

12.1f(iv). Abadir and Magnus [2005: 324].

12.1h. Rao and Rao [1998: 461].

12.1i. Use (b) with $\mathbf{A}$ replaced by $\mathbf{P_A}$. If $\mathbf{y} \in \mathcal{C}(\mathbf{A})$, then $\mathbf{y} = \mathbf{P_A}\mathbf{y}$.

12.1j. Drury et al. [2002: 97].

12.2a. Abadir and Magnus [2005: 331], Horn and Johnson [1985: 444], Rao and Rao [1998: 462], and Zhang [1999: 204].

12.2b. Dragomir [2004: 93] and quoted by Rao and Rao [1998: 456].

12.2c. Greub and Rheinbolt [1959] and quoted by Rao and Rao [1998: 456].

12.2d. Strang [1960] and quoted by Rao and Rao [1998: 465].

12.2e. Dragomir [2004: 91].

12.1.2   Complex Vector Inequalities

Many of the above inequalities can be generalized to the complex case. By the same token, the following results for complex vectors will hold for their real counterparts.

12.3. Let $\mathbf{x}$ and $\mathbf{y}$ be two complex vectors in $\mathbb{C}^n$, and let $\mathbf{A}$ be a Hermitian non-negative definite $n \times n$ matrix.

(a) There are two versions of the Cauchy–Schwarz inequality.

(i) Using the inner product $\langle \mathbf{x}, \mathbf{y} \rangle = \mathbf{x}^*\mathbf{y}$, we have from (2.17) that

$$\mathbf{x}^*\mathbf{x} - (\mathbf{x}^*\mathbf{y})(\mathbf{y}^*\mathbf{y})^{-1}(\mathbf{y}^*\mathbf{x}) \geq 0.$$

Equality occurs when $\mathbf{x} \propto \mathbf{y}$.

(ii) Since $|\bar{a}b| = |a||b|$, we have from (5.1b),

$$\left| \sum_{i=1}^{n} x_i y_i \right|^2 \leq \left(\sum_{i=1}^{n} |x_i y_i| \right)^2 \leq \sum_{i=1}^{n} |x_i|^2 \sum_{i=1}^{n} |y_i|^2.$$

Equality occurs when $\mathbf{x} = c\bar{\mathbf{y}}$ for any complex scalar c.

(b) $|\mathbf{x}^*\mathbf{y}|^2 \leq (\mathbf{x}^*\mathbf{x})(\mathbf{y}^*\mathbf{y})$. Equality occurs when $\mathbf{x} \propto \mathbf{y}$.

(c) $|\mathbf{x}^*\mathbf{A}\mathbf{y}|^2 \leq (\mathbf{x}^*\mathbf{A}\mathbf{x})(\mathbf{y}^*\mathbf{A}\mathbf{y})$, with equality when $\mathbf{x} \propto \mathbf{y}$.

(d) Let $\mathbf{A}$ be Hermitian positive definite.

(i) $|\mathbf{x}^*\mathbf{y}|^2 \leq (\mathbf{x}^*\mathbf{A}\mathbf{x})(\mathbf{y}^*\mathbf{A}^{-1}\mathbf{y})$. Equality occurs when $\mathbf{y} \propto \mathbf{A}\mathbf{x}$.

(ii) $(\mathbf{x}^*\mathbf{x})^2 \leq (\mathbf{x}^*\mathbf{A}\mathbf{x})(\mathbf{x}^*\mathbf{A}^{-1}\mathbf{x})$, which implies $(\mathbf{x}^*\mathbf{A}\mathbf{x})^{-1} \leq \mathbf{x}^*\mathbf{A}^{-1}\mathbf{x}$ when $\mathbf{x}^*\mathbf{x} = 1$. Equality occurs when $\mathbf{x} \propto \mathbf{A}\mathbf{x}$, that is, when $\mathbf{x}$ is an eigenvector.

(e) (Wielandt) If $\mathbf{A}$ is Hermitian positive definite and $\mathbf{x}^*\mathbf{y} = 0$, then

$$|\mathbf{x}^*\mathbf{A}\mathbf{y}|^2 \leq \left(\frac{\lambda_1 - \lambda_n}{\lambda_1 + \lambda_n}\right)^2 (\mathbf{x}^*\mathbf{A}\mathbf{x})(\mathbf{y}^*\mathbf{A}\mathbf{y}).$$

Equality occurs when $\mathbf{x} = (\mathbf{x}_1 + \mathbf{x}_n)/\sqrt{2}$ and $\mathbf{y} = (\mathbf{x}_1 - \mathbf{x}_n)/\sqrt{2}$, where $\mathbf{x}_1$ and $\mathbf{x}_n$ are the eigenvectors corresponding to λ_1 and λ_n, respectively, the maximum and minimum eigenvalues of $\mathbf{A}$. Rao [2005: 63] applies the above result to sphericity tests in multivariate analysis. Along with references, he also gives a matrix generalization of the above result (Rao [2005: 62]).

Note that $|\cdot|$ represents the modulus.

12.4. Let $\mathbf{A}$ and $\mathbf{C}$ be Hermitian positive definite $n \times n$ and $m \times m$ matrices, respectively, and let $\mathbf{B}$ be $n \times m$. The following statements are equivalent:

(1) $(\mathbf{x}^*\mathbf{A}\mathbf{x})(\mathbf{y}^*\mathbf{C}\mathbf{y}) \geq |\mathbf{x}^*\mathbf{B}\mathbf{y}|^2$ for all $\mathbf{x} \in \mathbb{C}^n$ and all $\mathbf{y} \in \mathbb{C}^m$.

(2) $\mathbf{x}^*\mathbf{A}\mathbf{x} + \mathbf{y}^*\mathbf{C}\mathbf{y} \geq 2|\mathbf{x}^*\mathbf{B}\mathbf{y}|$ for all $\mathbf{x} \in \mathbb{C}^n$ and all $\mathbf{y} \in \mathbb{C}^m$.

(3) $\rho(\mathbf{B}^*\mathbf{A}^{-1}\mathbf{B}\mathbf{C}^{-1}) \leq 1$, where $\rho(\cdot)$ is the spectral radius.

(4) $\begin{pmatrix} \mathbf{A} & \mathbf{B} \\ \mathbf{B}^* & \mathbf{C} \end{pmatrix} \succeq \mathbf{0}$ (i.e., non-negative definite).

Proofs. Section 12.1.2.

12.3. For (a) see Dragomir [2004: 2–3]; for (a)–(d) see Zhang [1999: 203] (and quoted by Rao and Rao [1998: 455]); and for (c), Horn and Johnson [1985] and Rao [2005: 61, real case].

12.4. Horn and Johnson [1985: 473].

12.1.3   Real Matrix Inequalities

In this section we give a number of matrix inequalities that might be regarded as extensions of the Cauchy–Schwarz inequality for vectors.

12.5. Let $\mathbf{A}$ and $\mathbf{B}$ be any real $m \times n$ matrices.

(a) $(\operatorname{trace} \mathbf{A}'\mathbf{B})^2 \leq (\operatorname{trace} \mathbf{A}'\mathbf{A})(\operatorname{trace} \mathbf{B}'\mathbf{B})$,   with equality if and only if one of the matrices is a multiple of the other.

This inequality can also be expressed in the form $|\langle \mathbf{A}, \mathbf{B} \rangle| \leq \|\mathbf{A}\|_F \|\mathbf{B}\|_F$ (cf. (2.20) and Harville [1997: 62]), where $\| \cdot \|_F$ is the Frobenius norm. For some generalizations see Rao and Rao [1998: 494–495].

(b) $\text{trace}[(\mathbf{A}'\mathbf{B})^2] \leq \text{trace}[(\mathbf{A}'\mathbf{A})(\mathbf{B}'\mathbf{B})]$, with equality if and only if $\mathbf{AB}'$ is symmetric. Furthermore, since $\text{trace}(\mathbf{A}'\mathbf{B}) = \text{trace}(\mathbf{B}'\mathbf{A}) = \text{trace}(\mathbf{AB}')$, we have

$$\text{trace}[(\mathbf{A}'\mathbf{B})^2] \leq \text{trace}[(\mathbf{AA}')(\mathbf{BB}')],$$

with equality if and only if $\mathbf{A}'\mathbf{B}$ is symmetric.

Setting $m = n$ and $\mathbf{A} = \mathbf{I}_n$, we have

$$\text{trace}(\mathbf{B}^2) \leq \text{trace}(\mathbf{B}'\mathbf{B}),$$

with equality if and only if $\mathbf{B}$ is symmetric.

(c) $(\det \mathbf{A}'\mathbf{B})^2 \leq (\det \mathbf{A}'\mathbf{A})(\det \mathbf{B}'\mathbf{B})$, with equality if and only if $\mathbf{A}'\mathbf{A}$ or $\mathbf{B}'\mathbf{B}$ are singular, or if $\mathbf{B} = \mathbf{AR}$ for some nonsingular $\mathbf{R}$.

(d) From (2.15a) we have $\|\mathbf{A} + \mathbf{B}\|_F \leq \|\mathbf{A}\|_F + \|\mathbf{B}\|_F$, where $\| \cdot \|_F$ is the Frobenius norm.

(e) $\mathbf{A}'[\mathbf{I}_m - \mathbf{B}(\mathbf{B}'\mathbf{B})^-\mathbf{B}']\mathbf{A} = \mathbf{A}'(\mathbf{I}_m - \mathbf{P_B})\mathbf{A}$ is non-negative definite since $(\mathbf{I}_m - \mathbf{P_B})$ is non-negative definite (cf. 2.49f). Hence, from (10.48b),

$$\det(\mathbf{A}'\mathbf{A}) \geq \det[\mathbf{A}'\mathbf{B}(\mathbf{B}'\mathbf{B})^-\mathbf{B}'\mathbf{A}].$$

12.6. (Measures of Relative Efficiency in Regression) Consider the linear regression model of Section 20.7, namely $\mathbf{y} = \mathbf{X}\boldsymbol{\beta} + \boldsymbol{\epsilon}$, where $\mathbf{X}$ is $n \times p$ of rank p, $\text{var}(\boldsymbol{\epsilon}) = \sigma^2 \mathbf{V}$, and $\mathbf{V}$ is positive definite. We define the eigenvalues of $\mathbf{V}$ to be $\lambda_i = \lambda_i(\mathbf{V})$ and we impose the usual order $\lambda_1 \geq \lambda_2 \geq \cdots \geq \lambda_n > 0$. Then the variance matrix of the generalized (weighted) least squares estimate of $\boldsymbol{\beta}$ is $(\mathbf{X}'\mathbf{V}^{-1}\mathbf{X})^{-1}$ and that of the ordinary least squares estimator is $(\mathbf{X}'\mathbf{X})^{-1}(\mathbf{X}'\mathbf{V}\mathbf{X})(\mathbf{X}'\mathbf{X})^{-1}$. Measures of the relative efficiency of the ordinary least squares estimate with respect to the generalized least squares estimate have been based on the roots of

$$\det[(\mathbf{X}'\mathbf{X})^{-1}(\mathbf{X}'\mathbf{V}\mathbf{X})(\mathbf{X}'\mathbf{X})^{-1} - \theta(\mathbf{X}'\mathbf{V}^{-1}\mathbf{X})^{-1}] = 0.$$

Four such measures E_i ($i = 1, 2, 3, 4$) taken from Rao and Rao [1998: 464] are given below.

(a) $E_1 = \dfrac{\det(\mathbf{X}'\mathbf{V}\mathbf{X}) \det(\mathbf{X}'\mathbf{V}^{-1}\mathbf{X})}{[\det(\mathbf{X}'\mathbf{X})]^2} = \prod_{i=1}^{p} \theta_i$. If $s = \min\{p, n - p\}$, then

$$1 \leq E_1 \leq \prod_{i=1}^{s} \frac{(\lambda_i + \lambda_{n-i+1})^2}{4\lambda_i \lambda_{n-i+1}}.$$

(b) $E_2 = \sum_{i=1}^{p} \theta_i$. If $s = \min\{p, n - p\}$, $t = 0$ if $s = p$, and $t = 2p - n$ if $s = n - p$, then

$$p \leq \sum_{i=1}^{p} \theta_i \leq \sum_{i=1}^{s} \frac{(\lambda_i + \lambda_{n-i+1})^2}{4\lambda_i \lambda_{n-i+1}} + t.$$

(c) $E_3 = \text{trace}[(\mathbf{X}'\mathbf{X})^{-1}(\mathbf{X}'\mathbf{V}\mathbf{X})(\mathbf{X}'\mathbf{X})^{-1} - (\mathbf{X}'\mathbf{V}^{-1}\mathbf{X})^{-1}]$.

When $\mathbf{X}'\mathbf{X} = \mathbf{I}_n$,

$$0 \le E_3 \le \sum_{i=1}^{s}(\sqrt{\lambda_i} - \sqrt{\lambda_{n-i+1}})^2,$$

where $s = \min\{p, n - p\}$.

(d) $E_4 = \text{trace}[\mathbf{P}\mathbf{V}^2\mathbf{P} - (\mathbf{P}\mathbf{V}\mathbf{P})(\mathbf{P}\mathbf{V}\mathbf{P})]$, where $\mathbf{P} = \mathbf{X}(\mathbf{X}'\mathbf{X})^{-1}\mathbf{X}'$ represents the orthogonal projection onto $\mathcal{C}(\mathbf{X})$. Then

$$0 \le E_4 \le \frac{1}{4}\sum_{i=1}^{s}(\lambda_i - \lambda_{n-i+1})^2.$$

12.7. (Matrix Kantorovich Inequality) Let $\mathbf{A}$ be a positive definite $n \times n$ matrix, and let $\mathbf{U}$ be an $n \times p$ matrix such that $\mathbf{U}'\mathbf{U} = \mathbf{I}_p$. If $\lambda_1 = \lambda_{\max}(\mathbf{A})$ and $\lambda_n = \lambda_{\min}(\mathbf{A})$, then

$$\mathbf{U}'\mathbf{A}\mathbf{U} \preceq \frac{(\lambda_1 + \lambda_n)^2}{4\lambda_1\lambda_n}(\mathbf{U}'\mathbf{A}^{-1}\mathbf{U})^{-1}.$$

Interchanging $\mathbf{A}$ and $\mathbf{A}^{-1}$ so that $\lambda_1^{-1} = \lambda_{\min}(\mathbf{A}^{-1})$ and $\lambda_n^{-1} = \lambda_{\max}(\mathbf{A}^{-1})$, we have

$$\mathbf{U}'\mathbf{A}^{-1}\mathbf{U} \preceq \frac{(\lambda_1 + \lambda_n)^2}{4\lambda_1\lambda_n}(\mathbf{U}'\mathbf{\Lambda}\mathbf{U})^{-1}.$$

Also

$$(\mathbf{U}'\mathbf{A}\mathbf{U})^{-1} \preceq \mathbf{U}'\mathbf{A}^{-1}\mathbf{U}.$$

(Note that $\mathbf{B} \preceq \mathbf{C}$ means that $\mathbf{C} - \mathbf{B}$ is non-negative definite.) For further extensions see Baksalary and Puntanen [1991] and Drury et al. [2002].

12.8. (Further Matrix Kantorovich-Type Inequalities)

(a) Let $\mathbf{A}$, $\mathbf{B}$, and $\mathbf{C}$ be $n \times n$ positive definite matrices, and let $\mathbf{X}$ be an $n \times k$ matrix of rank k. Then:

(i)

$$\sup_{\mathbf{X}} \frac{\det(\mathbf{X}'\mathbf{B}^{-1}\mathbf{A}\mathbf{B}^{-1}\mathbf{X})\det(\mathbf{X}'\mathbf{A}^{-1}\mathbf{X})}{[\det(\mathbf{X}'\mathbf{B}^{-1}\mathbf{X})]^2} = \sum_{i=1}^{m} \frac{(\mu_i + \mu_{n-i+1})^2}{4\mu_i\mu_{n-i+1}},$$

where $m = \min\{k, n - k\}$ and $\mu_1 \ge \cdots \ge \mu_n > 0$ are the roots of $\det(\mathbf{B} - \mu\mathbf{A}) = 0$, that is, the eigenvalues of $\mathbf{B}\mathbf{A}^{-1}$ (and of $\mathbf{A}^{-1}\mathbf{B}$).

(ii)

$$\sup_{\mathbf{X}} \frac{\det(\mathbf{X}'\mathbf{B}^2\mathbf{X})\det(\mathbf{X}'\mathbf{C}^2\mathbf{X})}{[\det(\mathbf{X}'\mathbf{B}\mathbf{C}\mathbf{X})]^2} = \sum_{i=1}^{m} \frac{(\mu_i + \mu_{n-i+1})^2}{4\mu_i\mu_{n-i+1}},$$

where $m = \min\{k, n - k\}$, $\mathbf{B}\mathbf{C} = \mathbf{C}\mathbf{B}$, and the μ_i are the eigenvalues of $\mathbf{B}\mathbf{C}^{-1}$.

(b) Let $\mathbf{B}$ be an $n \times n$ non-negative definite matrix of rank b, and let $\mathbf{A}$ be $n \times r$ of rank a ($a \leq \min\{b, r\}$) such that $\mathcal{C}(\mathbf{A}) \subseteq \mathcal{C}(\mathbf{B})$. Furthermore, let $\lambda_1 \geq \cdots \geq \lambda_b > 0$ be the eigenvalues of $\mathbf{B}$. Then, if $\mathbf{A}^+$ and $\mathbf{B}^+$ are Moore–Penrose inverses, we have:

(i) $\mathbf{A}^+\mathbf{B}^+(\mathbf{A}^+)' \preceq \dfrac{(\lambda_1 + \lambda_b)^2}{4\lambda_1\lambda_b}(\mathbf{A}'\mathbf{B}\mathbf{A})^+$,

with equality if and only if $\mathbf{A} = \mathbf{0}$, or $\mathbf{A}'\mathbf{B}\mathbf{A} = \frac{1}{2}(\lambda_1 + \lambda_b)\mathbf{A}'\mathbf{A}$ and $\mathbf{A}'\mathbf{B}^+\mathbf{A} = \frac{\lambda_1+\lambda_b}{2\lambda_1\lambda_b}\mathbf{A}'\mathbf{A}$.

(ii) $\mathbf{A}^+\mathbf{B}(\mathbf{A}^+)' - (\mathbf{A}'\mathbf{B}^+\mathbf{A})^+ \preceq (\sqrt{\lambda_1} - \sqrt{\lambda_b})^2(\mathbf{A}'\mathbf{A})^+$,

with equality if and only if $\mathbf{A} = \mathbf{0}$, or $\lambda_1 = \lambda_b$, or $\mathbf{A}'\mathbf{B}\mathbf{A} = (\lambda_1 + \lambda_b - \sqrt{\lambda_1\lambda_b})\mathbf{A}'\mathbf{A}$ and $\mathbf{A}'\mathbf{B}^+\mathbf{A} = (\lambda_1\lambda_b)^{-1/2}\mathbf{A}'\mathbf{A}$.

The above, along with two further results, are quoted by Rao and Rao [1998: 496]. They also give a Kantorovich-type inequality for complex matrices. See also Liu [2002a] and Liu and Neudecker [1996].

Proofs. Section 12.1.3.

12.5a–b. Abadir and Magnus [2005: 325] and Magnus and Neudecker [1999: 201–202].

12.5c. Abadir and Magnus [2005: 330] and Magnus and Neudecker [1999: 201].

12.6a. Bloomfield and Watson [1975] and Knott [1975].

12.6b. Khatri and Rao [1981, 1982].

12.6c. Rao [1985]; see also Drury et al. [2002: section 3] for further details and related work.

12.6d. Bloomfield and Watson [1975].

12.7. Marshall and Olkin [1990] and Zhang [1999: 204].

12.8a(i). Lin [1984].

12.8a(ii). Khatri and Rao [1981]. This result follows from (i) by replacing $\mathbf{A}^{-1}$ by $\mathbf{C}^2$ and $\mathbf{B}^{-1}$ by $\mathbf{BC}$. Here $\mathbf{BC}$ is symmetric and positive definite when $\mathbf{BC} = \mathbf{CB}$ as the eigenvalues of $\mathbf{B}^{1/2}\mathbf{CB}^{1/2}$ are positive.

12.1.4 Complex Matrix Inequalities

12.9. Let $\mathbf{X}$ and $\mathbf{Y}$ be $n \times p$ and $n \times q$ complex matrices, respectively. Then generalizing (12.5e), we have

$$\mathbf{X}^*\mathbf{X} - \mathbf{X}^*\mathbf{Y}(\mathbf{Y}^*\mathbf{Y})^-\mathbf{Y}^*\mathbf{X} = \mathbf{X}^*(\mathbf{I}_n - \mathbf{P}_\mathbf{Y})\mathbf{X} \succeq \mathbf{0},$$

i.e., non-negative definite as the orthogonal projector $\mathbf{I}_n - \mathbf{P}_\mathbf{Y}$ is Hermitian non-negative definite (cf. 2.49f). Equality occurs if and only if $\mathcal{C}(\mathbf{X}) \subseteq \mathcal{C}(\mathbf{Y})$. A generalization of this result follows.

12.10. Let $\mathbf{A}$ be an $n \times n$ Hermitian non-negative definite matrix with $\mathbf{P_A} = \mathbf{A}(\mathbf{A}^*\mathbf{A})^-\mathbf{A}^*$, and let $\mathbf{U}$ be an $n \times p$ matrix. Then

$$\mathbf{U}^*\mathbf{A}^+\mathbf{U} \succeq \mathbf{U}^*\mathbf{P_A}\mathbf{U}(\mathbf{U}^*\mathbf{A}\mathbf{U})^+\mathbf{U}^*\mathbf{P_A}\mathbf{U},$$

with equality if and only if $\mathcal{C}(\mathbf{A}\mathbf{U}) = \mathcal{C}(\mathbf{P_A}\mathbf{U})$.

12.11. Let $\mathbf{A}$ be an $n \times n$ Hermitian positive definite matrix, and let $\mathbf{X}$ be $n \times p$ and $\mathbf{Y}$ be $n \times q$ satisfying $\mathbf{X}^*\mathbf{Y} = \mathbf{0}$. Then

$$(\mathbf{X}^*\mathbf{A}\mathbf{Y})(\mathbf{Y}^*\mathbf{A}\mathbf{Y})^-(\mathbf{Y}^*\mathbf{A}\mathbf{X}) \preceq \left(\frac{\lambda_1 - \lambda_n}{\lambda_1 + \lambda_n}\right)^2 \mathbf{X}^*\mathbf{A}\mathbf{X},$$

where $(\mathbf{Y}^*\mathbf{A}\mathbf{Y})^-$ is any weak inverse, with equality when

$$(\mathbf{u}_1 + \mathbf{u}_n) \propto (\lambda_1^{-1}\mathbf{P}\mathbf{u}_1 + \lambda_n^{-1}\mathbf{P}\mathbf{u}_n).$$

Here $\mathbf{P} = \mathbf{X}(\mathbf{X}^*\mathbf{X})^-\mathbf{X}^*$ is the orthogonal projector onto $\mathcal{C}(\mathbf{X})$ and λ_1 and λ_n are the largest and smallest eigenvalues, respectively, of $\mathbf{A}$ with corresponding eigenvectors $\mathbf{u}_1$ and $\mathbf{u}_n$.

12.12. Let $\mathbf{A}$ and $\mathbf{B}$ be $n \times n$ real or complex matrices.

(a) $|\operatorname{trace}(\mathbf{A}\mathbf{B})|^2 \leq \operatorname{trace}(\mathbf{A}^*\mathbf{A})\operatorname{trace}(\mathbf{B}^*\mathbf{B})$.

(b) If $\mathbf{A}$ and $\mathbf{B}$ are Hermitian, then

$$\operatorname{trace}[(\mathbf{A}\mathbf{B})^2] \leq \operatorname{trace}(\mathbf{A}^2\mathbf{B}^2),$$

with equality if and only if $\mathbf{A}\mathbf{B} = \mathbf{B}\mathbf{A}$. For a generalization see (12.33d).

12.13. (Unitarily Invariant Norm) Let $\|\cdot\|_{ui}$ be any unitarily invariant norm defined on the vector space of $m \times n$ complex matrices (Section 4.6.3), and let $\mathbf{A}$ and $\mathbf{B}$ be $m \times n$ matrices.

(a) If $|\mathbf{A}|$ represents the general square root of $\mathbf{A}$ (i.e., $|\mathbf{A}| = (\mathbf{A}^*\mathbf{A})^{1/2}$), then:

   (i) $\| |\mathbf{A}^*\mathbf{B}|^p \|_{ui}^2 \leq \|(\mathbf{A}^*\mathbf{A})^p\|_{ui}\|(\mathbf{B}^*\mathbf{B})^p\|_{ui}$ for all $p > 0$.

   (ii) Setting $p = \frac{1}{2}$ in (i) and using $\| |\mathbf{A}| \|_{ui} = \|\mathbf{A}\|_{ui}$ (cf. 4.86), we have

$$\| |\mathbf{A}^*\mathbf{B}|^{\frac{1}{2}} \|_{ui}^2 \leq \|\mathbf{A}\|_{ui}\|\mathbf{B}\|_{ui}.$$

   (iii) If $p = 1$ in (i), we have a Cauchy–Schwarz type of inequality

$$\|\mathbf{A}^*\mathbf{B}\|_{ui}^2 \leq \|\mathbf{A}^*\mathbf{A}\|_{ui}\|\mathbf{B}^*\mathbf{B}\|_{ui}.$$

(b) (Hadamard Product) $\|\mathbf{A} \circ \mathbf{B}\|_{ui}^2 \leq \|\mathbf{A}\mathbf{A}^*\|_{ui}\|\mathbf{B}^*\mathbf{B}\|_{ui}$.

Proofs. Section 12.1.4.

12.10. Baksalary and Puntanen [1991: 104], who also give some special cases and variations on the result.

12.11. Wang and Ip [2000] (see also Drury [2002]).

12.12. Zhang [1999: 25, 213].

12.13a. Horn and Johnson [1985: 212, exercises 6 and 7, hint for proof only].

12.13b. Horn and Johnson [1991: 212, exercise 8, hint for proof only].

12.2 HÖLDER'S INEQUALITY AND EXTENSIONS

Let $\mathbf{a}, \mathbf{b}, \ldots, \mathbf{g}$ be m real $n \times 1$ vectors of non-negative elements, and let $\alpha_i > 0$ $(i = 1, 2, \ldots, n)$ such that $\sum_{i=1}^{m} \alpha_i = 1$.

12.14.

$$\prod_{i=1}^{n} a_i^{\alpha_i} + \prod_{i=1}^{n} b_i^{\alpha_i} + \cdots + \prod_{i=1}^{n} g_i^{\alpha_i} \leq \prod_{i=1}^{n} (a_i + b_i + \cdots + g_i)^{\alpha_i}.$$

Equality occurs if and only if either every pair of vectors $\mathbf{a}$, $\mathbf{b}$, and so on, are proportional, or there is a k such that $a_k = b_k = \cdots = g_k = 0$. If $\mathbf{A} = (\mathbf{a}, \mathbf{b}, \ldots, \mathbf{g})$, then the conditions for equality are either rank $\mathbf{A} = 1$ or $\mathbf{A}$ contains a row of zeros.

12.15. Interchanging the rows and columns of $\mathbf{A}$ in the previous result leads to the following.

(a) $\displaystyle\sum_{i=1}^{n} a_i^{\alpha_1} b_i^{\alpha_2} \cdots g_i^{\alpha_m} \leq \Big(\sum_{i=1}^{n} a_i\Big)^{\alpha_1} \Big(\sum_{i=1}^{n} b_i\Big)^{\alpha_2} \cdots \Big(\sum_{i=1}^{n} g_i\Big)^{\alpha_m},$

with equality if and only if rank $\mathbf{A} = 1$ or $\mathbf{A}$ contains a column of zeros.

(b) Putting $m = 2$ in (a) leads to

$$\sum_{i=1}^{n} a_i^{\alpha} b_i^{1-\alpha} \leq \Big(\sum_{i=1}^{n} a_i\Big)^{\alpha} \Big(\sum_{i=1}^{n} b_i\Big)^{1-\alpha} \qquad (0 < \alpha < 1),$$

with equality if and only if $\mathbf{a} = k\mathbf{b}$.

(c) (Hölder's Inequality) Replacing a_i by $a_i^{1/\alpha}$ and b_i by $b_i^{1/(1-\alpha)}$ in (b) leads to

$$\sum_{i=1}^{n} a_i b_i \ \leq \ \Big(\sum_{i=1}^{n} a_i^{1/\alpha}\Big)^{\alpha} \Big(\sum_{i=1}^{n} b_i^{1/(1-\alpha)}\Big)^{1-\alpha}$$
$$= \ \Big(\sum_{i=1}^{n} a^r\Big)^{1/r} \Big(\sum_{i=1}^{n} b^s\Big)^{1/s},$$

where $r \,(= 1/\alpha) > 1$ and $r^{-1} + s^{-1} = 1$. Equality occurs if and only if $a_i^r = k b_i^s$ for $i = 1, 2, \ldots, n$, or either $\mathbf{a}$ or $\mathbf{b}$ is $\mathbf{0}$. The inequality in (c) is reversed if $r \neq 0$, $r < 1$ (and $s < 0$). We can deduce the previous results from (c).

(d) If $\mathbf{a}$ and $\mathbf{b}$ are vectors of complex numbers, then replacing a_i by $|a_i|$, and so on, in (c), we have for $r > 1$ and $r^{-1} + s^{-1} = 1$,

$$\sum_{i=1}^{n} |a_i b_i| \leq \Big(\sum_{i=1}^{n} |a_i|^r\Big)^{1/r} \Big(\sum_{i=1}^{n} |b_i|^s\Big)^{1/s}.$$

Equality occurs if and only if

$$|a_i|^r = k|b_i|^s$$

for $i = 1, 2, \ldots, n$, and $\arg(a_i b_i)$ is independent of i.

12.16. (Matrix Analogues) Let $\mathbf{A}$ and $\mathbf{B}$ any two $n \times n$ non-negative definite matrices, and let $0 < \alpha < 1$.

(a) (Magnus) $\operatorname{trace}(\mathbf{A}^\alpha \mathbf{B}^{1-\alpha}) \leq (\operatorname{trace} \mathbf{A})^\alpha (\operatorname{trace} \mathbf{B})^{1-\alpha}$

   with equality if and only if $\mathbf{B} = k\mathbf{A}$ for some $k > 0$.

(b) $\operatorname{trace}(\mathbf{A}^\alpha \mathbf{B}^{1-\alpha}) \leq \operatorname{trace}[\alpha \mathbf{A} + (1 - \alpha)\mathbf{B}]$

   with equality if and only if $\mathbf{A} = \mathbf{B}$.

(c) $(\det \mathbf{A})^\alpha (\det \mathbf{B})^{1-\alpha} \leq \det(\alpha \mathbf{A} + (1 - \alpha)\mathbf{B})$

   with equality if and only if $\mathbf{A} = \mathbf{B}$ or $\det(\alpha \mathbf{A} + (1 - \alpha)\mathbf{B}) = 0$. The result is obviously true if either $\mathbf{A}$ or $\mathbf{B}$ is singular, so it is more applicable to positive definite matrices. In this case it follows that $\phi(\mathbf{A}) = \log \det \mathbf{A}$ is concave on the space of positive definite matrices.

(d) Let $\mathbf{A}_i$ be positive definite and $\alpha_i > 0$ for $(i = 1, 2, \ldots, k)$, where $\sum_i \alpha_i = 1$. Then

$$(\det \mathbf{A}_1)^{\alpha_1} (\det \mathbf{A}_2)^{\alpha_2} \cdots (\det \mathbf{A}_k)^{\alpha_k} \leq \det(\alpha_1 \mathbf{A}_1 + \alpha_2 \mathbf{A}_2 + \cdots + \alpha_k \mathbf{A}_k),$$

   with equality if and only if the $\mathbf{A}_i$ are all equal.

Proofs. Section 12.2.

12.14. Hardy et al. [1952: section 2.7] and Magnus and Neudecker [1999: 220–221].

12.15c. For a direct proof see, for example, Marcus and Minc [1964: 108], Rao and Bhimasankaram [2000: 254], and Rao and Rao [1998: 457].

12.16a. Magnus and Neudecker [1999: 221].

12.16b. Magnus and Neudecker [1999: 222].

12.16c. Abadir and Magnus [2005: 334] and Magnus and Neudecker [1999: 222].

12.16d. Abadir and Magnus [2005: 334–335].

12.3   MINKOWSKI'S INEQUALITY AND EXTENSIONS

12.17. Let $\mathbf{X}$ be an $m \times n$ real matrix whose elements are non-negative and not all zero. If $p > 1$, then

$$[\sum_{i=1}^{m} (\sum_{j=1}^{n} x_{ij})^p]^{1/p} < \sum_{j=1}^{n} (\sum_{i=1}^{m} x_{ij}^p)^{1/p},$$

with equality if and only if rank $\mathbf{X} = 1$. The inequality reverses if $p < 1$ ($p \neq 0$). If $p < 0$, then the x_{ij} are assumed to be all positive. A number of special cases follow below.

(a) Putting $n = 2$, $a_i = x_{i1}$, and $b_i = x_{i2}$, we have

$$[\sum_{i=1}^{m}(a_i + b_i)^p]^{1/p} \leq (\sum_{i=1}^{m}a_i^p)^{1/p} + (\sum_{i=1}^{m}b_i^p)^{1/p} \qquad (p > 1),$$

with equality if and only if $a_i = kb_i$ for $i = 1, 2, \ldots, n$.

(b) Putting $m = 2$, $c_j = x_{1j}$, and $d_j = x_{2j}$, we have

$$[(\sum_{j=1}^{n}c_j)^p + (\sum_{j=1}^{n}d_j)^p]^{1/p} \leq \sum_{j=1}^{n}(c_j^p + d_j^p)^{1/p} \qquad (p > 1),$$

with equality if and only if $c_j = kd_j$ for $j = 1, 2, \ldots, n$.

(c) If $\alpha_i \geq 0$ $(i = 1, 2, \ldots, m)$ such that $\sum_{i=1}^{m}\alpha_i = 1$, then replacing x_{ij} by $\alpha_i^{1/p}x_{ij}$ leads to

$$[\sum_{i=1}^{m}\alpha_i(\sum_{j=1}^{n}x_{ij})^p)]^{1/p} \leq \sum_{j=1}^{n}(\sum_{i=1}^{m}\alpha_i x_{ij}^p)^{1/p} \qquad (p > 1),$$

with equality if and only if rank $\mathbf{X} = 1$. The inequality reverses for $p < 1$ $(p \neq 0)$.

12.18. (Matrix Analogues) Let $\mathbf{A}$ and $\mathbf{B}$ any two $n \times n$ Hermitian non-negative definite matrices.

(a) (Magnus) $[\operatorname{trace}(\mathbf{A} + \mathbf{B})^p]^{1/p} \leq (\operatorname{trace}\mathbf{A}^p)^{1/p} + (\operatorname{trace}\mathbf{B}^p)^{1/p}$ $(p > 1)$, with equality if and only if $\mathbf{A} = k\mathbf{B}$ for some $k > 0$.

(b) $[\det(\mathbf{A} + \mathbf{B})]^{1/n} \geq (\det\mathbf{A})^{1/n} + (\det\mathbf{B})^{1/n}$, with equality if and only if $\det(\mathbf{A} + \mathbf{B}) = 0$ or $\mathbf{A} = k\mathbf{B}$ for some $k > 0$.

(c) $[\det(\alpha\mathbf{A} + (1 - \alpha)\mathbf{B})]^{1/n} \geq \alpha(\det\mathbf{A})^{1/n} + (1 - \alpha)(\det\mathbf{B})^{1/n}$, $0 \leq \alpha \leq 1$.

Proofs. Section 12.3.

12.17. Hardy et al. [1952: 30] and Marcus and Minc [1964: 109, $p > 1$]. See also Rao and Bhimasankaram [2000: 254] for (c).

12.18a. Magnus and Neudecker [1999: 224].

12.18b. Abadir and Magnus [2005: 329] and Magnus and Neudecker [1999: 227].

12.18c. Marcus and Minc [1964: 115].

12.4 WEIGHTED MEANS

Let $x_1, x_2, \ldots, x_n$ be non-negative real numbers, and let $\alpha_i > 0$ $(i = 1, 2, \ldots, n)$ be such that $\sum_{i=1}^{n} \alpha_i = 1$. Define

$$M_p(\mathbf{x}) = \begin{cases} \prod_i x_i^{\alpha_i}, & p = 0, \\ (\sum_i \alpha_i x_i^p)^{1/p}, & p \neq 0. \end{cases}$$

If $p < 0$, we assume that the x_is are all positive. An important special case is $\alpha_i = 1/n$ for all i. For further details see Bullen [2003], Hardy et al. [1952: chapter II], and Magnus and Neudecker [1999: 227–231].

12.19. For every $\lambda > 0$, $M_p(\lambda \mathbf{x}) = \lambda M_p(\mathbf{x})$.

12.20. Equality occurs in each of the following two inequalities if and only if the x_i's are all equal.

(a) $M_0(\mathbf{x}) \le M_1(\mathbf{x})$, so that $\prod_i x_i^{\alpha_i} \le \sum_i \alpha_i x_i$.

 Setting each $\alpha_i = n^{-1}$, we see that the geometric mean is less than or equal to the arithmetic mean. Note the special case $x^\alpha y^{1-\alpha} \le \alpha x + (1 - \alpha)y$.

(b) $M_p(\mathbf{x}) \le M_q(\mathbf{x})$ for $p < q$.

 Setting each $\alpha_i = n^{-1}$, $p = -1$, and $q = 0$, we see that the harmonic mean is less than or equal to the geometric mean.

(c) (Matrix Version) If $\mathbf{A}_i$ $(i = 1, 2, \ldots, n)$ are positive definite pairwise commuting matrices (i.e., $\mathbf{A}_i \mathbf{A}_j = \mathbf{A}_j \mathbf{A}_i$ for all $i, j, i \neq j$), then

$$\frac{1}{n} \sum_{i=1}^{n} \mathbf{A}_i \succeq (\mathbf{A}_1 \cdots \mathbf{A}_n)^{1/n} \succeq n \left(\sum_{i=1}^{n} \mathbf{A}^{-1} \right)^{-1}.$$

 Equality occurs if and only if the $\mathbf{A}_i$ are all equal. (Here $\mathbf{A} \succeq \mathbf{B}$ means that $\mathbf{A} - \mathbf{B}$ is non-negative definite.)

12.21. (Limits)

(a) $\lim_{p \to 0} M_p(\mathbf{x}) = M_0(\mathbf{x})$.

(b) Let $x_{\min}$ be the smallest x_i and $x_{\max}$ the largest. Then

$$\lim_{p \to \infty} M_p(\mathbf{x}) = x_{\max}, \quad \lim_{p \to -\infty} M_p(\mathbf{x}) = x_{\min}, \quad \text{and} \quad x_{\min} \le M_r(\mathbf{x}) \le x_{\max}.$$

12.22. $M_p(\mathbf{x})$ is a concave function of $\mathbf{x}$ for $p \le 1$ and a convex function for $p \ge 1$. In particular,

$$\begin{aligned} M_p(\mathbf{x}) + M_p(\mathbf{y}) &\le M_p(\mathbf{x} + \mathbf{y}) & (p < 1) \\ M_p(\mathbf{x}) + M_p(\mathbf{y}) &\ge M_p(\mathbf{x} + \mathbf{y}) & (p > 1), \end{aligned}$$

with equality if and only if $\mathbf{x}$ and $\mathbf{y}$ are linearly dependent.

 Also $p \log M_p(\mathbf{x})$ is a convex function of p.

Proofs. Section 12.4.

12.19. Magnus and Neudecker [1999: 228].

12.20a. Magnus and Neudecker [1999: 202].

12.20b. Magnus and Neudecker [1999: 230]. These inequalities can also be deduced from likelihood ratio test inequalities (Stefanski [1996]).

12.20c. Rao and Rao [1998: 499].

12.21. Magnus and Neudecker [1999: 228–229].

12.22. Magnus and Neudecker 1999: 230–231].

12.5 QUASILINEARIZATION (REPRESENTATION) THEOREMS

The representation of a nonlinear function as an envelope of linear functions is called *quasilinearization* or *representation*. The method is useful in proving a number of inequalities.

12.23. Let $p > 1$, $q = p/(p - 1)$, and $a_i \geq 0$ for $i = 1, 2, \ldots, n$. Then

$$\sum_{i=1}^{n} a_i x_i \leq (\sum_{i=1}^{n} a_i^p)^{1/p}$$

for every set of non-negative $x_1, x_2, \ldots, x_n$ satisfying $\sum_i x_i^q = 1$. Equality occurs if and only if all the a_i are zero or

$$x_i^q = \frac{a_i^p}{\sum_{j=1}^{n} a_j^p} \qquad \text{for } i = 1, 2, \ldots, n.$$

Hence

$$\max_{\mathcal{R}} \sum_{i=1}^{n} a_i x_i = (\sum_{i=1}^{n} a_i^p)^{1/p},$$

where $\mathcal{R}$ is the region defined by $\sum_i x_i^q = 1$, $x_i \geq 0$ $(i = 1, 2, \ldots, n)$.

12.24. (Matrix Versions) Let $\mathbf{A}$ be a non-negative definite $n \times n$ matrix.

(a) If $p > 1$ and $q = p/(p - 1)$, then

$$\text{trace}(\mathbf{AX}) \leq (\text{trace } \mathbf{A}^p)^{1/p}$$

for every non-negative definite $n \times n$ matrix $\mathbf{X}$ satisfying $\text{trace}(\mathbf{X}^q) = 1$. Equality occurs if and only if $\mathbf{X}^q = \mathbf{A}^p/(\text{trace } \mathbf{A}^p)$. Hence

$$\max_{\mathcal{R}} \text{trace}(\mathbf{AX}) = (\text{trace } \mathbf{A}^p)^{1/p},$$

where $\mathcal{R}$ is the region of all non-negative definite matrices $\mathbf{X}$ of the same size satisfying $\text{trace } \mathbf{X}^q = 1$.

(b) If $\mathbf{A}$ is also positive definite, then for every positive definite $n \times n$ matrix $\mathbf{X}$ satisfying $\det \mathbf{X} = 1$ we have

$$n^{-1} \operatorname{trace}(\mathbf{AX}) \geq (\det \mathbf{A})^{1/n},$$

with equality if and only if $\mathbf{X} = (\det \mathbf{A})^{1/n} \mathbf{A}^{-1}$.

If $\mathbf{X} = \mathbf{I}_n$, then $n^{-1} \operatorname{trace}(\mathbf{A}) \geq (\det \mathbf{A})^{1/n}$ with equality if and only if $\mathbf{A} = k\mathbf{I}_n$ for some $k \geq 0$.

Therefore, given $\mathbf{A}$ positive definite, we have

$$\min_{\mathcal{R}} n^{-1} \operatorname{trace}(\mathbf{AX}) = (\det \mathbf{A})^{1/n},$$

where the minimization is over the space of all positive definite matrices $\mathbf{X}$ such that $\det \mathbf{X} = 1$.

(c) If $\mathbf{A}$ is a positive definite $n \times n$ matrix and $\mathbf{B}$ is any $m \times n$ matrix of rank m, then

$$\operatorname{trace}(\mathbf{X'AX}) \geq \operatorname{trace}[(\mathbf{BA}^{-1}\mathbf{B'})^{-1}]$$

for every $n \times m$ matrix $\mathbf{X}$ satisfying $\mathbf{BX} = \mathbf{I}_m$ with equality if and only if $\mathbf{X} = \mathbf{A}^{-1}\mathbf{B'}(\mathbf{BA}^{-1}\mathbf{B'})^{-1}$.

(d) Let $\mathbf{A}$ be an $n \times n$ symmetric matrix with (not necessarily distinct) eigenvalues $\lambda_1 \geq \lambda_2 \geq \cdots \geq \lambda_n$. Then, for any $n \times k$ matrix $\mathbf{X}$ such that $\mathbf{X'X} = \mathbf{I}_k$ $(k \leq n)$, we obtain

$$\operatorname{trace}(\mathbf{X'AX}) \leq \sum_{i=1}^{k} \lambda_i,$$

with equality when the columns of $\mathbf{X}$ are orthonormal right eigenvectors corresponding to $\lambda_1, \dots, \lambda_k$, respectively.

Proofs. Section 12.5.

12.23. Magnus and Neudecker [1999: 218].

12.24a. Magnus and Neudecker [1999: 219].

12.24b. Abadir and Magnus [2005: 328] and Magnus and Neudecker [1999: 225].

12.24c. Quoted by Magnus and Neudecker [1999: 237, exercise 10].

12.24d. Harville [1997: 556].

12.6 SOME GEOMETRICAL PROPERTIES

12.25. (Ellipsoids) Let $\mathbf{a}$, $\mathbf{y}$, and $\boldsymbol{\theta}$ be n-dimensional real vectors, and let $\mathbf{L}$ be a positive definite $n \times n$ matrix. Then, for $r > 0$, $\boldsymbol{\theta}$ satisfies

$$\mathbf{a'y} - r(\mathbf{a'La})^{1/2} \leq \mathbf{a'\theta} \leq \mathbf{a'y} + r(\mathbf{a'La})^{1/2}$$

for all $\mathbf{a}$ if and only if $(\mathbf{y}-\boldsymbol{\theta})'\mathbf{L}^{-1}(\mathbf{y}-\boldsymbol{\theta}) \leq r^2$. Geometrically, this result states that a point $\mathbf{y}$ lies in an ellipsoid with center $\boldsymbol{\theta}$ if and only if it lies between every pair of parallel tangent planes. This result was originally proved geometrically by Scheffé [1953]. When $\mathbf{L} = \mathbf{I}_n$, the ellipsoid becomes a sphere, and Hsu [1996: 231–233] gives a simple proof of this case.

12.26. (Rectangles) Let $\mathbf{a}$, $\mathbf{c}\,(\mathbf{c} \geq \mathbf{0})$, and $\mathbf{z}$ be n-dimensional real vectors, then

$$\max_{1\leq i\leq n} |z_i| \leq c_i \quad \text{if and only if} \quad |\mathbf{a}'\mathbf{z}| \leq \sum_{i=1}^{n} c_i|a_i| \text{ for all } \mathbf{a}.$$

This result is useful for the construction of simultaneous confidence intervals (Hsu [1996: 233]).

Proofs. Section 12.6.

12.25. Seber and Lee [2003: 123].

12.26. Miller [1981: 74].

12.7 MISCELLANEOUS INEQUALITIES

12.7.1 Determinants

12.27. (Hadamard) Let $\mathbf{A} = (a_{ij})$ be a non-negative definite $n \times n$ Hermitian matrix. Then

$$\det \mathbf{A} \leq a_{11}a_{22}\cdots a_{nn},$$

with equality if and only if some $a_{ii} = 0$ or $\mathbf{A}$ is diagonal.

12.28. (Hadamard) If $\mathbf{A} = (a_{ij})$ is any $n \times n$ complex matrix, then

$$|\det \mathbf{A}| \leq \prod_{i=1}^{n}(\sum_{j=1}^{n}|a_{ij}|^2)^{1/2} \quad \text{and}$$

$$|\det \mathbf{A}| \leq \prod_{j=1}^{n}(\sum_{i=1}^{n}|a_{ij}|^2)^{1/2},$$

with equality if and only if $\mathbf{A}\mathbf{A}^*$ is diagonal or $\mathbf{A}$ has a zero row; alternatively, if $\mathbf{A}^*\mathbf{A}$ is diagonal or $\mathbf{A}$ has a zero column.

12.29. Let $\mathbf{A}$ and $\mathbf{B}$ be Hermitian non-negative definite $n \times n$ matrices. Then:

(a) $\det(\mathbf{A}+\mathbf{B}) \geq \det \mathbf{A}+\det \mathbf{B}$, with equality if and only if $n = 1$ or $\det(\mathbf{A}+\mathbf{B}) = 0$.

(b) If $\mathbf{A} - \mathbf{B}$ is non-negative definite, then $\det \mathbf{A} \geq \det \mathbf{B}$ with equality if and only if $\mathbf{A}$ and $\mathbf{B}$ are nonsingular (i.e., positive definite) and $\mathbf{A} = \mathbf{B}$, or if $\mathbf{A}$ and $\mathbf{B}$ are both singular.

12.30. If $\mathbf{A}$ and $\mathbf{B}$ are $n \times n$ real or complex matrices, then

$$\det(\mathbf{I}_n + \mathbf{A}\mathbf{A}^*)\det(\mathbf{I}_n + \mathbf{B}^*\mathbf{B}) \geq |\det(\mathbf{A} + \mathbf{B})|^2 + |(\det(\mathbf{I}_n - \mathbf{A}\mathbf{B}^*)|^2,$$

with equality if and only if $n = 1$, or $\mathbf{A} + \mathbf{B} = \mathbf{0}$, or $\mathbf{A}\mathbf{B}^* = \mathbf{I}_n$.

12.31. If $\mathbf{X}$ is $m \times n$ and $\mathbf{Y}$ is $n \times p$, both real or complex matrices, then from (12.9) we have that $\mathbf{X}^*\mathbf{X} - \mathbf{X}^*\mathbf{Y}(\mathbf{Y}^*\mathbf{Y})^-\mathbf{Y}^*\mathbf{X}$ is non-negative definite. Hence, by (12.29b),

$$\det(\mathbf{X}^*\mathbf{X}) \geq \det(\mathbf{X}^*\mathbf{Y}(\mathbf{Y}^*\mathbf{Y})^-\mathbf{Y}^*\mathbf{X}),$$

with equality when $\mathcal{C}(\mathbf{X}) \subseteq \mathcal{C}(\mathbf{Y})$.

Proofs. Section 12.7.1.

12.27. Horn and Johnson [1985: 477] and Zhang [1999: 176].

12.28. This follows from the previous inequality (12.27) applied to $\mathbf{A}\mathbf{A}^*$, and so on. See also Basilevsky [1983: 100], Horn and Johnson [1985: 477-478], and Magnus and Neudecker [1999: 214, real case].

12.29. Abadir and Magnus [2005: 326, real case].

12.30. Zhang [1999: 184–185].

12.7.2   Trace

12.32. If $\mathbf{A} = (a_{ij})$ is a non-negative definite matrix, then

$$\operatorname{trace}(\mathbf{A}^p) \geq \sum_{i=1}^{n} a_{ii}^p \qquad (p > 1),$$

$$\operatorname{trace}(\mathbf{A}^p) \leq \sum_{i=1}^{n} a_{ii}^p \qquad (0 < p < 1),$$

with equality if and only if $\mathbf{A}$ is diagonal.

12.33. Let $\mathbf{A}$ and $\mathbf{B}$ be $n \times n$ non-negative definite matrices.

(a) $0 \leq \operatorname{trace}(\mathbf{A}\mathbf{B}) \leq (\operatorname{trace}\mathbf{A})(\operatorname{trace}\mathbf{B})$.

(b) $\sqrt{\operatorname{trace}(\mathbf{A}\mathbf{B})} \leq \frac{1}{2}(\operatorname{trace}\mathbf{A} + \operatorname{trace}\mathbf{B})$, with equality if $\mathbf{A} = \mathbf{0}$ and $\operatorname{trace}\mathbf{B} = 0$, or if $\mathbf{B} = \mathbf{0}$ and $\operatorname{trace}\mathbf{A} = 0$, but also if $\mathbf{A} = \mathbf{B} = \mathbf{a}\mathbf{a}'$ for some $\mathbf{a} \neq \mathbf{0}$.

(c) (Araki–Lieb–Thirring)

$$\operatorname{trace}[(\mathbf{B}^{1/2}\mathbf{A}\mathbf{B}^{1/2})^{st}] \leq \operatorname{trace}[(\mathbf{B}^{t/2}\mathbf{A}^t\mathbf{B}^{t/2})^s],$$

where s and t are positive real numbers with $t \geq 1$.

(d) (Lieb–Thirring) Let m and k be positive integers with $m \geq k$. Then

$$\operatorname{trace}[(\mathbf{A}^k\mathbf{B}^k)^m] \leq [\operatorname{trace}(\mathbf{A}^m\mathbf{B}^m)]^k.$$

In particular,
$$\text{trace}[(\mathbf{AB})^m] \leq \text{trace}(\mathbf{A}^m\mathbf{B}^m).$$

Proofs. Section 12.7.2.

    12.32. Magnus and Neudecker [1999: 217].

    12.33a—b. Abadir and Magnus [2005: 329–330].

    12.33c. Quoted by Bhatia [1997: 258].

    12.33d. Quoted by Bhatia [1997: 279].

12.7.3   Quadratics

12.34. (Bergstrom) If $\mathbf{A}$ and $\mathbf{B}$ are both positive definite, then
$$\mathbf{x}'(\mathbf{A} + \mathbf{B})^{-1}\mathbf{x} \leq \frac{(\mathbf{x}'\mathbf{A}^{-1}\mathbf{x})(\mathbf{x}'\mathbf{B}^{-1}\mathbf{x})}{\mathbf{x}'(\mathbf{A}^{-1} + \mathbf{B}^{-1})\mathbf{x}}.$$

12.35. Let $\mathbf{A} \geq \mathbf{0}$ (i.e., has non-negative elements) be an $n \times n$ matrix and let $\mathbf{x} \geq \mathbf{0}$ be an $n \times 1$ vector. Then, for any positive integer k,
$$(\mathbf{x}'\mathbf{A}^k\mathbf{x})(\mathbf{x}'\mathbf{x})^{k-1} \geq (\mathbf{x}'\mathbf{A}\mathbf{x})^k,$$

with equality if and only if $\mathbf{x}$ is an eigenvector of $\mathbf{A}$.

Proofs. Section 12.7.3.

    12.34. Abadir and Magnus [2005: 323].

    12.35. Mulholland and Smith [1959].

12.7.4   Sums and Products

12.36. (Triangle Inequality) For all $a_i, b_i, \ldots, g_i$ $(i = 1, 2, \ldots, n)$,
$$\left(\sum_{i=1}^{n}(a_i + b_i + \cdots + g_i)^2\right)^{1/2} \leq (\sum_{i=1}^{n}a_i^2)^{1/2} + (\sum_{i=1}^{n}b_i^2)^{1/2} + \cdots + (\sum_{i=1}^{n}g_i^2)^{1/2}.$$

12.37. For all non-negative $a_i, b_i, \ldots, g_i$ $(i = 1, 2, \ldots, n)$,
$$\sum_{i=1}^{n}(a_i + b_i + \cdots + g_i)^r \geq \sum_{i=1}^{n}a_i^r + \sum_{i=1}^{n}b_i^r + \cdots + \sum_{i=1}^{n}g_i^r, \qquad r > 1,$$
$$\sum_{i=1}^{n}(a_i + b_i + \cdots + g_i)^r \leq \sum_{i=1}^{n}a_i^r + \sum_{i=1}^{n}b_i^r + \cdots + \sum_{i=1}^{n}g_i^r, \qquad 0 < r < 1,$$

with equality if and only if all the numbers but one of each set $a_j, b_j, \ldots, g_j$ $(j = 1, 2, \ldots, n)$ are zero.

12.38. (Ordered Numbers) Let $a_1 \geq a_2 \geq \cdots \geq a_n \geq 0$ and $b_1 \geq b_2 \geq \ldots \geq b_n \geq 0$. If

$$\prod_{i=1}^{k} a_i \leq \prod_{i=1}^{k} b_i, \quad k = 1, 2, \ldots, n,$$

then

$$\sum_{i=1}^{k} a_i \leq \sum_{i=1}^{k} b_i, \quad k = 1, 2, \ldots, n.$$

12.39. (Information Inequalities) Let $\mathbf{a} = (a_1, a_2, \ldots, a_n)'$ and $\mathbf{b} = (b_1, b_2, \ldots, b_n)'$ be two vectors.

(a) Supose $\mathbf{a} > \mathbf{0}$ and $\mathbf{b} > \mathbf{0}$ (i.e., have positive elements) such that $\sum_i a_i \geq \sum_i b_i$. Then

$$\sum_{i=1}^{n} a_i \log \frac{b_i}{a_i} \leq 0,$$

with equality being attained if and only if $a_i = b_i$ for all i. Also, if $a_i < 1$ and $b_i \leq 1$ for all i, then

$$2 \sum_{i=1}^{n} a_i \log \frac{a_i}{b_i} \leq \sum_{i=1}^{n} a_i (a_i - b_i)^2.$$

(b) Suppose $\mathbf{a} \geq \mathbf{0}$ and $\mathbf{b} \geq \mathbf{0}$ (i.e., have non-negative elements) such that $\sum_i a_i = \sum_i b_i > 0$, then

$$\prod_{i=1}^{n} a_i^{a_i} \geq \prod_{i=1}^{n} b_i^{a_i},$$

with equality if and only if $\mathbf{a} = \mathbf{b}$.

12.40. (Jensen) Let $x_i \geq 0$ $(i = 1, 2, \ldots, n)$, then

$$\left(\sum_{i=1}^{n} x_i^r \right)^{1/r} \geq \left(\sum_{i=1}^{n} x_i^s \right)^{1/s}, \quad 0 < r < s,$$

with equality if and only if all the x_i are zero except one. Also

$$\lim_{r \to \infty} \left(\sum_{i=1}^{n} x_i^r \right)^{1/r} = \max_i x_i.$$

12.41. If $\lambda_1 \geq \lambda_2 \geq \cdots \geq \lambda_n > 0$, then

$$\max_{i,j} \left[\frac{(\lambda_i + \lambda_j)^2}{4 \lambda_i \lambda_j} \right] = \frac{(\lambda_1 + \lambda_n)^2}{4 \lambda_1 \lambda_n}.$$

12.42. If $x_i \geq 0$ for $i = 1, 2, \ldots, n$, then

$$\prod_{i=1}^{n} (1 + x_i) \geq \left(1 + (x_1 x_2 \cdots x_n)^{1/n} \right)^n,$$

with equality if and only if $x_1 = x_2 = \cdots = x_n$.

12.43. Suppose $x_1 \geq x_2 \geq \cdots \geq x_n > 0$ and y_i/x_i is decreasing in i. Let $\alpha_i \geq 0$ for $i = 1, 2, \ldots, n$ such that $\sum_{i=1}^{n} \alpha_i = 1$, and define

$$g(r) = \begin{cases} (\sum_{i=1}^{n} \alpha_i x_i^r / \sum_{i=1}^{n} \alpha_i y_i^r)^{1/r}, & \text{if} \quad r \neq 0, \\ \prod_{i=1}^{n} x_i^{\alpha_i} / \prod_{i=1}^{n} y_i^{\alpha_i}, & \text{if} \quad r = 0. \end{cases}$$

Then $g(r)$ increases as r increases.

Proofs. Section 12.7.4.

12.36. Follows from (12.17).

12.37. Hardy et al. [1952: 32].

12.38. Horn and Johnson [1991: 174].

12.39a. Rao and Rao [1998: 458].

12.39b. Bapat and Raghavan [1997: 81].

12.40. Hardy et al. [1952: 28].

12.41. Quoted by Rao and Rao [1998: 466].

12.42. Marshall and Olkin [1979: 72].

12.43. Marshall and Olkin [1979: 131].

12.8   SOME IDENTITIES

12.44. Let $\mathbf{a} = (a_i)$ and $\mathbf{b} = (b_i)$.

(a) (Lagrange Identity)

    (i) (Real Vectors) $(\mathbf{a}'\mathbf{a})(\mathbf{b}'\mathbf{b}) - (\mathbf{a}'\mathbf{b})^2 = \frac{1}{2} \sum_i \sum_j (a_i b_j - a_j b_i)^2$.

    (ii) (Complex Vectors)
$\sum_i |a_i|^2 \sum_i |b_i|^2 - |\sum_i a_i b_i|^2 = \frac{1}{2} \sum_i \sum_j |\bar{a}_i b_j - \bar{a}_j b_i|^2$.

(b) (Abel's Identity)

$$\mathbf{a}'\mathbf{b} = \sum_{i=1}^{n-1} \left((a_i - a_{i+1}) \sum_{j=1}^{i} b_j \right) + a_n \sum_{j=1}^{n} b_j.$$

12.45. If $\mathbf{a}$, $\mathbf{b}$, and $\mathbf{c}$ are $n \times 1$ vectors then

$$\sum_i a_i \sum_j a_j b_j c_j - \sum_i a_i b_i \sum_j a_j c_j = \frac{1}{2} \sum_i \sum_j a_i a_j (b_i - b_j)(c_i - c_j).$$

12.46. If $\mathbf{A}$ is symmetric and nonsingular, we have from (24.26a)

$$\mathbf{x}'\mathbf{A}\mathbf{x} - 2\mathbf{b}'\mathbf{x} = (\mathbf{x} - \mathbf{A}^{-1}\mathbf{b})'\mathbf{A}(\mathbf{x} - \mathbf{A}^{-1}\mathbf{b}) - \mathbf{b}'\mathbf{A}^{-1}\mathbf{b}.$$

12.47. Suppose that $\mathbf{A}$ and $\mathbf{B}$ are $n \times n$ symmetric matrices, and $\mathbf{A} + \mathbf{B}$ is non-singular. Let $\mathbf{a}$, $\mathbf{b}$, and $\mathbf{x}$ be $n \times 1$ vectors. Then

$$(\mathbf{x} - \mathbf{a})'\mathbf{A}(\mathbf{x} - \mathbf{a}) + (\mathbf{x} - \mathbf{b})'\mathbf{B}(\mathbf{x} - \mathbf{b})$$
$$= (\mathbf{x} - \mathbf{c})'(\mathbf{A} + \mathbf{B})(\mathbf{x} - \mathbf{c}) + (\mathbf{a} - \mathbf{b})'\mathbf{A}(\mathbf{A} + \mathbf{B})^{-1}\mathbf{B}(\mathbf{a} - \mathbf{b}),$$

where $\mathbf{c} = (\mathbf{A} + \mathbf{B})^{-1}(\mathbf{Aa} + \mathbf{Bb})$.

12.48. Suppose $\mathbf{A}$ and $\mathbf{B}$ positive definite matrices, and let $\mathbf{a}$, $\mathbf{b}$, and $\mathbf{x}$ be $n \times 1$ vectors. Define

$$\mathbf{C}^{-1} = \mathbf{A}^{-1} + \mathbf{B}^{-1} \quad \text{and} \quad \mathbf{D} = \mathbf{A} + \mathbf{B}.$$

Then

$$(\mathbf{x} - \mathbf{a})'\mathbf{A}^{-1}(\mathbf{x} - \mathbf{a}) + (\mathbf{x} - \mathbf{b})'\mathbf{B}^{-1}(\mathbf{x} - \mathbf{b}) = (\mathbf{x} - \mathbf{c})'\mathbf{C}^{-1}(\mathbf{x} - \mathbf{c}) + (\mathbf{a} - \mathbf{b})'\mathbf{D}^{-1}(\mathbf{a} - \mathbf{b}),$$

where $\mathbf{c} = \mathbf{C}(\mathbf{A}^{-1}\mathbf{a} + \mathbf{B}^{-1}\mathbf{b})$.

Proofs. Section 12.8.

12.44a. Dragomir [2004: 3].

12.44b. Rao and Rao [1998: 385].

12.46. Use $\mathbf{x} = \mathbf{x} - \mathbf{A}^{-1}\mathbf{b} + \mathbf{A}^{-1}\mathbf{b}$.

12.47. Multiply out and use (15.4c).

12.48. Abadir and Magnus [2005: 217]. Follows from (12.47) by replacing $\mathbf{A}$ by $\mathbf{A}^{-1}$ and $\mathbf{B}$ by $\mathbf{B}^{-1}$ and using $\mathbf{A}^{-1}(\mathbf{A}^{-1} + \mathbf{B}^{-1})^{-1}\mathbf{B}^{-1} = (\mathbf{A} + \mathbf{B})^{-1}$ (cf. 15.4c).

CHAPTER 13

LINEAR EQUATIONS

In this chapter we investigate the solution of various linear equations with a vector or matrix of unknown variables. Nonlinear matrix equations are not considered in this book except in (13.24) and (13.25), and the reader is referred to Horn and Johnson [1991: Section 6.4] for some background on this topic.

13.1 UNKNOWN VECTOR

13.1.1 Consistency

Definition 13.1. In this section we consider the problem of solving the equation $\mathbf{A}_{m \times n} \mathbf{x}_{n \times 1} = \mathbf{b}_{m \times 1}$ for $\mathbf{x}$ when rank $\mathbf{A} = r$ ($r \leq \min(m, n)$) and $\mathbf{b} \neq \mathbf{0}$. The equation is said to be *consistent* if there exists at least one solution. Otherwise, the equation is said to be *inconsistent*. Clearly we must have $\mathbf{b} \in \mathcal{C}(\mathbf{A})$ for consistency. Note that this section is a special case of Section 13.2, which considers the equation $\mathbf{AXB} = \mathbf{C}$.

13.1. Using the above notation, the following are equivalent.

(a) The equation $\mathbf{Ax} = \mathbf{b}$ is consistent.

(b) rank$(\mathbf{A}, \mathbf{b}) = $ rank $\mathbf{A}$.

(c) $\mathbf{AA}^{-}\mathbf{b} = \mathbf{b}$, where $\mathbf{A}^{-}$ is any weak inverse of $\mathbf{A}$.

A Matrix Handbook for Statisticians. By George A. F. Seber
Copyright © 2008 John Wiley & Sons, Inc.

13.2. From (16.33) we can find nonsingular $\mathbf{P}$ and $\mathbf{Q}$ such that

$$\mathbf{PAQ} = \begin{pmatrix} \mathbf{I}_r & \mathbf{0} \\ \mathbf{0} & \mathbf{0} \end{pmatrix}.$$

Then the equation $\mathbf{Ax} = \mathbf{b}$ is consistent if and only if the last $m - r$ elements of $\mathbf{Pb}$ are zero.

13.3. The equation $\mathbf{Ax} = \mathbf{b}$ has a unique solution if and only if $\mathbf{A}$ has full column rank (i.e., $n = r$). When $\mathbf{A}$ has full column rank, it has a left inverse $\mathbf{L}$ such that $\mathbf{LA} = \mathbf{I}_n$. Then $\tilde{\mathbf{x}} = \mathbf{Lb}$ is the solution. In particular, we can choose $\mathbf{L} = (\mathbf{A'A})^{-1}\mathbf{A'}$.

Proofs. Section 13.1.1.

13.1. Graybill [1983: 151–152], Schott [2005: 222], and Searle [1982: 232].

13.2. Searle [1982: 232].

13.3. Schott [2005: 227].

13.1.2   Solutions

13.4. All possible solutions of the consistent equation $\mathbf{Ax} = \mathbf{b}$ can be generated from

$$\tilde{\mathbf{x}} = \mathbf{A}^-\mathbf{b} + (\mathbf{I}_n - \mathbf{A}^-\mathbf{A})\mathbf{z}$$

for any specific weak inverse $\mathbf{A}^-$ by using all possible values of the arbitrary $n \times 1$ vector $\mathbf{z}$ (including $\mathbf{z} = \mathbf{0}$). Thus every solution of $\mathbf{Ax} = \mathbf{b}$ can be expressed in the above form for some $\mathbf{z}$.

13.5. All possible solutions of the consistent equation $\mathbf{Ax} = \mathbf{b}$ can be generated from $\tilde{\mathbf{x}} = \mathbf{A}^-\mathbf{b}$ by using all possible weak inverses $\mathbf{A}^-$ of $\mathbf{A}$.

13.6. If $\tilde{\mathbf{x}}_1, \tilde{\mathbf{x}}_2, \ldots, \tilde{\mathbf{x}}_t$ are any t solutions of the consistent equation $\mathbf{Ax} = \mathbf{b}$, then $\sum_{i=1}^t a_i\tilde{\mathbf{x}}_i$ is a solution if and only if $\sum_{i=1}^t a_i = 1$

13.7. If $\mathbf{A}$ is $m \times n$ of rank r, the consistent equation $\mathbf{Ax} = \mathbf{b}$ has exactly $n - r + 1$ linearly independent solutions.

(a) One possible set of such solutions is $\mathbf{A}^-\mathbf{b}$ along with the set

$$\tilde{\mathbf{x}}_i = \mathbf{A}^-\mathbf{b} + (\mathbf{I}_n - \mathbf{A}^-\mathbf{A})\mathbf{z}_i, \quad i = 1, 2, \ldots, n - r,$$

where the $\mathbf{z}_i$ are arbitrary, but chosen so that the $(\mathbf{I}_n - \mathbf{A}^-\mathbf{A})\mathbf{z}_i$ are all linearly independent.

(b) Every solution can be expressed as a linear combination of the linearly independent solutions.

13.8. The value of $\mathbf{a'}\tilde{\mathbf{x}}$ is the same for all solutions $\tilde{\mathbf{x}}$ to $\mathbf{Ax} = \mathbf{b}$ if and only if $\mathbf{a'} = \mathbf{a'}\mathbf{A}^-\mathbf{A}$. There are only r linearly independent vectors $\mathbf{a}_i$ satisfying $\mathbf{a}_i' = \mathbf{a}_i'\mathbf{A}^-\mathbf{A}$.

13.9. (Methods of Solution for Consistent Equations) These methods generally involve some factorization of $\mathbf{A}$.

(a) (Singular Value Decomposition) Suppose $\mathbf{A}$ is $m \times n$ with singular value decomposition $\mathbf{A} = \mathbf{P\Sigma Q}'$, where $\mathbf{P}$ and $\mathbf{Q}$ are orthogonal $m \times m$ and $n \times n$ matrices, respectively, with columns $\mathbf{p}_i$ and $\mathbf{q}_i$, and $\mathbf{\Sigma}$ is an $m \times n$ diagonal matrix with positive or zero diagonal elements σ_i, the singular values of $\mathbf{A}$. Then $\mathbf{Ax} = \mathbf{b}$ implies that $\mathbf{\Sigma Q}'\mathbf{x} = \mathbf{P}'\mathbf{b}$, or $\mathbf{\Sigma y} = \mathbf{c}$. This simplified form can be used to determine the nature of the solutions of the original equations (Schott [2005: 242]).

If $\mathbf{A}$ is nonsingular and $n \times n$, and $\mathbf{P} = (\mathbf{p}_1, \mathbf{p}_2, \ldots, \mathbf{p}_n)$, then

$$\mathbf{x} = \mathbf{A}^{-1}\mathbf{b} = \mathbf{Q\Sigma}^{-1}\mathbf{P}'\mathbf{b} = \sum_{i=1}^{n} \frac{\mathbf{p}_i'\mathbf{b}}{\sigma_i}\mathbf{q}_i,$$

so that if σ_n, the smallest singular value of $\mathbf{A}$, is small, a small change in $\mathbf{A}$ or $\mathbf{b}$ can induce a relatively large change in $\mathbf{x}$ (Golub and Van Loan [1996: 80]).

(b) (LU Factorization) We can use the factorization $\mathbf{A} = \tilde{\mathbf{L}}\mathbf{U}$, where $\tilde{\mathbf{L}}$ is a lower-triangular matrix with unit diagonal elements and $\mathbf{U}$ is an upper-triangular matrix (cf. Section 16.4). Since $\mathbf{Ax} = \tilde{\mathbf{L}}\mathbf{Ux} = \tilde{\mathbf{L}}\mathbf{y} = \mathbf{b}$, we simply solve $\tilde{\mathbf{L}}\mathbf{y} = \mathbf{b}$ for $\mathbf{y}$ and $\mathbf{Ux} = \mathbf{y}$ for $\mathbf{x}$. The process used for carrying out the calculations is called *Gaussian elimination* with the related ideas of *pivoting* and *sweeping*. It can also be applied to $m \times n$ matrices (Golub and Van Loan [1996: chapter 3] and Rao and Bhimasankaram [2000: section 5.6]). The method can be used for solving normal equations that arise in least squares estimation for linear regression (Seber and Lee [2003: section 11.2]).

Proofs. Section 13.1.2.

13.4–13.5. Schott [2005: 225] and Searle [1982: 238].

13.6. Searle [1982: 238].

13.7. Schott [2005: 228] and Searle [1982: 240–241].

13.8. Searle [1982: 242–244].

13.1.3 Homogeneous Equations

13.10. We consider solutions of $\mathbf{Ax} = \mathbf{0}$, where $\mathbf{A}$ is $m \times n$ of rank r.

(a) The solutions form the null space $\mathcal{N}(\mathbf{A})$ of $\mathbf{A}$ of dimension $n - r$. Any orthonormal basis for $\mathcal{N}(\mathbf{A})$ will give a set of $n - r$ orthogonal solutions.

(b) A nonzero solution exists if and only if $\det \mathbf{A} = 0$.

(c) All the solutions to $\mathbf{Ax} = \mathbf{0}$ are of the form $\mathbf{x}_0 = (\mathbf{I}_n - \mathbf{A}^-\mathbf{A})\mathbf{z}$ for arbitrary $\mathbf{z}$ and any weak inverse $\mathbf{A}^-$ of $\mathbf{A}$. For $\mathbf{z}_i \neq \mathbf{0}$, there exist $q - r$ linearly independent such solutions $(\mathbf{I}_n - \mathbf{A}^-\mathbf{A})\mathbf{z}_i$.

Proofs. Section 13.1.3.

13.10. Searle [1982: section 9.7].

13.1.4   Restricted Equations

13.11. Given $\mathbf{A}$ is $m \times n$, we wish to solve the consistent equation $\mathbf{Ax} = \mathbf{b}$ with the restriction that $\mathbf{x} \in \mathcal{V}$, a vector subspace of $\mathbb{R}^n$. Here $\mathcal{V}$ could represent the column space or null space of a matrix.

(a) If $\mathbf{P}_\mathcal{V}$ is the orthogonal projection onto $\mathcal{V}$, then $\mathbf{I}_n - \mathbf{P}_\mathcal{V}$ is the orthogonal projection onto $\mathcal{V}^\perp$. We are now interested in the solution of

$$\begin{pmatrix} \mathbf{Ax} \\ (\mathbf{I} - \mathbf{P}_\mathcal{V})\mathbf{x} \end{pmatrix} = \begin{pmatrix} \mathbf{b} \\ \mathbf{0} \end{pmatrix}.$$

(b) The restricted equation is consistent if and only if the equation $\mathbf{AP}_\mathcal{V}\mathbf{z} = \mathbf{b}$ is consistent. If this is the case and $\mathbf{z}_0$ is a solution of the latter equation, then $\mathbf{x}_0$ is a solution of the restricted equation if and only if $\mathbf{x}_0 = \mathbf{P}_\mathcal{V}\mathbf{z}_0$.

(c) If the restricted equations are consistent, then a general solution is

$$\{\mathbf{x}_0 : \mathbf{x}_0 = \mathbf{P}_\mathcal{V}(\mathbf{AP}_\mathcal{V})^-\mathbf{b} + \mathbf{P}_\mathcal{V}[\mathbf{I} - (\mathbf{AP}_\mathcal{V})^-\mathbf{AP}_\mathcal{V}]\mathbf{y},$$

where $\mathbf{y}$ is an arbitrary $n \times 1$ vector and $(\mathbf{AP}_\mathcal{V})^-$ is any weak inverse of $\mathbf{AP}_\mathcal{V}$.

Proofs. Section 13.1.4.

13.11. Ben-Israel and Greville [2003: 88–89].

13.2   UNKNOWN MATRIX

We are interested in solving the equation $\mathbf{A}_{m \times n}\mathbf{X}_{n \times p}\mathbf{B}_{p \times q} = \mathbf{C}_{m \times q}$. When the appropriate matrices are square, special cases follow by setting $\mathbf{A} = \mathbf{I}$ or $\mathbf{B} = \mathbf{I}$, and using $\mathbf{I}^- = \mathbf{I}$ in the result below. We note that if $\mathbf{x} = \text{vec}\,\mathbf{X}$ and $\mathbf{c} = \text{vec}\,\mathbf{C}$, then, by 11.16b),

$$(\mathbf{B}' \otimes \mathbf{A})\mathbf{x} = \text{vec}\,(\mathbf{AXB}) = \text{vec}\,\mathbf{C} = \mathbf{c},$$

which reduces the problem to the case considered in the previous section. More generally, consider the system

$$\sum_{i=1}^r \mathbf{A}_i\mathbf{X}\mathbf{B}_i + \sum_{j=1}^s \mathbf{L}_j\mathbf{X}'\mathbf{M}_j = \mathbf{C},$$

where the $\mathbf{A}_i$ are $m \times n$, the $\mathbf{B}_i$ are $p \times q$, the $\mathbf{L}_j$ are $m \times p$, and the $\mathbf{M}_j$ are $n \times q$. This can be reexpressed in the form (cf. 11.18b(ii))

$$\left\{\sum_{i=1}^r (\mathbf{B}_i' \otimes \mathbf{A}_i) + \left[\sum_{j=1}^s (\mathbf{M}_j' \otimes \mathbf{L}_j)\right] \mathbf{I}_{(p,n)}\right\}\mathbf{x} = \mathbf{c}.$$

13.2.1   Consistency

13.12. The equation $\mathbf{AXB} = \mathbf{C}$ is said to be *consistent* if it has at least one solution for $\mathbf{X}$.

(a) A necessary and sufficient condition for $\mathbf{AXB} = \mathbf{C}$ to be consistent is that $\mathbf{AA^-CB^-B} = \mathbf{C}$ for any particular pair of weak inverses $\mathbf{A^-}$ and $\mathbf{B^-}$.

(b) $\mathbf{AXB} = \mathbf{C}$ is consistent if and only if $\mathcal{C}(\mathbf{C}) \subset \mathcal{C}(\mathbf{A})$ and $\mathcal{C}(\mathbf{C'}) \subset \mathcal{C}(\mathbf{B'})$.

(c) If the equation $\mathbf{AXB} = \mathbf{C}$ is consistent, then the following are general solutions for $\mathbf{X}$ with $\mathbf{X}_0 = \mathbf{A^-CB^-}$.

(i) $\mathbf{X}_0 + \mathbf{W} - \mathbf{A^-AWBB^-}$ for conformable arbitrary $\mathbf{W}$.

(ii) $\mathbf{X}_0 + (\mathbf{I} - \mathbf{A^-A})\mathbf{U} + \mathbf{V}(\mathbf{I} - \mathbf{BB^-})$ for conformable arbitrary $\mathbf{U}$ and $\mathbf{V}$. This result can also be expressed in the form $\mathbf{A^-CB^-} + \mathbf{Z}_0$, where $\mathbf{Z}_0$ is a solution of $\mathbf{AZB} = \mathbf{0}$.

(iii) $\mathbf{X}_0 + \mathbf{A^-AR}(\mathbf{I} - \mathbf{BB^-}) + (\mathbf{I} - \mathbf{A^-A})\mathbf{SBB^-} + (\mathbf{I} - \mathbf{A^-A})\mathbf{T}(\mathbf{I} - \mathbf{BB^-})$, for conformable arbitrary $\mathbf{R}$, $\mathbf{S}$, and $\mathbf{T}$.

(d) A number of special cases follow from the above results—for example, the general solution of $\mathbf{AX} = \mathbf{0}$ is $\tilde{\mathbf{X}} = (\mathbf{I} - \mathbf{A^-A})\mathbf{U}$, where $\mathbf{U}$ is arbitrary.

Proofs. Section 13.2.1.

13.12a–b. Harville [1997: 125–126].

13.12c. Harville [1997: section 11.12]. In each case we simply check that the solution satisfies $\mathbf{AXB} = \mathbf{C}$ using (a) for $\mathbf{X}_0$. For the second part of (ii), we simply show that $\mathbf{AZ}_0\mathbf{B} = \mathbf{0}$ using $\mathbf{AA^-A} = \mathbf{A}$, and so on.

13.2.2   Some Special Cases

13.13. Setting $\mathbf{B} = \mathbf{I}$ in (13.12) above, we see that the following conditions are equivalent.

(1) The equations $\mathbf{AX} = \mathbf{C}$ are consistent (i.e., have a solution).

(2) $\mathcal{C}(\mathbf{C}) \subset \mathcal{C}(\mathbf{A})$.

(3) $\mathbf{AA^-C} = \mathbf{C}$ for any particular weak inverse $\mathbf{A^-}$ (cf. 13.12a).

(4) $\mathbf{k'C} = \mathbf{0}$ for every row vector $\mathbf{k'}$ such that $\mathbf{k'A} = \mathbf{0}$. Harville [1997: 73] calls the equations *compatible* if they have this property.

The equations are also consistent if the rows of $\mathbf{A}$ are linearly independent.

13.14. Let $\mathbf{A}$ be an $n \times n$ matrix, which is possibly complex, of rank $n - 1$. Let $\mathbf{u}$ and $\mathbf{v}$ be any eigenvectors of $\mathbf{A}$ associated with the eigenvalue zero (not necessarily simple) such that $\mathbf{Au} = \mathbf{0}$ and $\mathbf{v^*A} = \mathbf{0'}$. Then the general solution of $\mathbf{AX} = \mathbf{0}$ is $\mathbf{X} = \mathbf{uz'}$, where $\mathbf{z}$ is arbitrary. Similarly, the general solution of $\mathbf{XA} = \mathbf{0}$ is $\mathbf{X} = \mathbf{wv^*}$, where $\mathbf{w}$ is arbitrary. Finally, the general solution of the equations $\mathbf{AX} = \mathbf{0}$ and $\mathbf{XA} = \mathbf{0}$ is $\mathbf{X} = c\mathbf{uv^*}$, where c is an arbitrary constant.

13.15. Let $\mathbf{X}$ be an unknown $n \times p$ matrix. For any $m \times n$ matrix $\mathbf{A}$ and any $m \times p$ matrix $\mathbf{C}$, the equations $\mathbf{A}'\mathbf{A}\mathbf{X} = \mathbf{A}'\mathbf{C}$ are consistent since from (2.35) $\mathcal{C}(\mathbf{A}') = \mathcal{C}(\mathbf{A}'\mathbf{A})$. These equations arise in multivariate least squares estimation.

13.16. If the following matrices are conformable and $\mathcal{C}(\mathbf{C}) \subset \mathcal{C}(\mathbf{L}')$, then the equations

$$\begin{pmatrix} \mathbf{A}'\mathbf{A} & \mathbf{L} \\ \mathbf{L}' & 0 \end{pmatrix} \begin{pmatrix} \mathbf{X} \\ \mathbf{Y} \end{pmatrix} = \begin{pmatrix} \mathbf{A}'\mathbf{B} \\ \mathbf{C} \end{pmatrix}$$

are consistent for the unknowns $\mathbf{X}$ and $\mathbf{Y}$. This result is used for restricted least squares theory.

13.17. Suppose $\mathbf{X}$ is an unknown $m \times n$ matrix.

(a) If "$\otimes$" is the Kronecker product, then, using (11.16a), the equation $\mathbf{A}_{m \times m}\mathbf{X} + \mathbf{X}\mathbf{B}_{n \times n} = \mathbf{C}_{m \times n}$ can be expressed in the form

$$(\mathbf{I}_n \otimes \mathbf{A} + \mathbf{B}' \otimes \mathbf{I}_m)\mathrm{vec}\,\mathbf{X} = \mathrm{vec}\,\mathbf{C},$$

or $\mathbf{F}\mathbf{x} = \mathbf{c}$, say, where $\mathbf{F}$ is called the *Kronecker sum*. Some properties of $\mathbf{F}$ are given by Horn and Johnson [1991: section 4.4]. The equation has a unique solution if and only if $\mathbf{A}$ and $-\mathbf{B}$ have no eigenvalues in common.

(b) We also have from (a),

$$(\mathbf{S}\mathbf{A}\mathbf{S}^{-1})\mathbf{S}\mathbf{X}\mathbf{T} + \mathbf{S}\mathbf{X}\mathbf{T}(\mathbf{T}^{-1}\mathbf{B}\mathbf{T}) = \mathbf{S}\mathbf{C}\mathbf{T},$$

which may be rewritten as $\mathbf{A}_1\mathbf{X}_1 + \mathbf{X}_1\mathbf{B}_1 = \mathbf{C}_1$. With suitable similarity transformations, the transformed equation may be easier to handle; the original solution is then readily recovered (Horn and Johnson [1991: 256]).

(c) A related equation is *Lyapunov's equation*

$$\mathbf{X}\mathbf{A} + \mathbf{A}^*\mathbf{X} = \mathbf{H},$$

where $\mathbf{A}$, $\mathbf{X}$, and $\mathbf{H}$ are all $n \times n$, and $\mathbf{H}$ is Hermitian. This equation arises in the study of matrix stability and is discussed in detail by Horn and Johnson [1991: chapter 4]. The equation

$$\mathbf{X}\mathbf{A} + \mathbf{A}^*\mathbf{X} = \mathbf{C}$$

has a unique solution for any $n \times n$ matrix $\mathbf{C}$ if and only if λ and $-\bar{\lambda}$ are not both eigenvalues of $\mathbf{A}$.

13.18. If $\mathbf{X}$ and $\mathbf{A}$ are $n \times n$, and the eigenvalues of $\mathbf{A}$ are λ_i, then the equation

$$\mathbf{A}\mathbf{X} - \mathbf{X}\mathbf{A} = a\mathbf{X}$$

has a nontrivial solution if and only if $a = \lambda_i - \lambda_j$.

13.19. Suppose $\mathbf{A}$ is $m \times m$, $\mathbf{X}$ is $m \times n$, and $\mathbf{B}$ is $n \times n$. If $\mathbf{A}$ and $\mathbf{B}$ have no eigenvalues in common, then $\mathbf{A}\mathbf{X} - \mathbf{X}\mathbf{B} = \mathbf{0}$ has a unique solution $\mathbf{X} = \mathbf{0}$. A nonzero solution exists if there are eigenvalues in common.

13.20. $\mathbf{AX} + \mathbf{YB} = \mathbf{C}$ if and only if $(\mathbf{I} \otimes \mathbf{A})\text{vec}\,\mathbf{X} + (\mathbf{B}' \otimes \mathbf{I})\text{vec}\,\mathbf{Y} = \text{vec}\,\mathbf{C}$, where "$\otimes$" is the Kronecker product.

13.21. The equation $\mathbf{AX} - \mathbf{YB} = \mathbf{C}$ has a solution for $\mathbf{X}$ and $\mathbf{Y}$ if and only if

$$\text{rank} \begin{pmatrix} \mathbf{A} & \mathbf{C} \\ \mathbf{0} & \mathbf{B} \end{pmatrix} = \text{rank} \begin{pmatrix} \mathbf{A} & \mathbf{0} \\ \mathbf{0} & \mathbf{B} \end{pmatrix}.$$

13.22. The matrix equation

$$\begin{pmatrix} \mathbf{A} & \mathbf{B} \\ \mathbf{B}' & \mathbf{0} \end{pmatrix} \begin{pmatrix} \mathbf{X}_1' \\ \mathbf{X}_2' \end{pmatrix} = \begin{pmatrix} \mathbf{G}_1' \\ \mathbf{G}_2' \end{pmatrix},$$

in $\mathbf{X}_1$ and $\mathbf{X}_2$, where $\mathbf{A}$, $\mathbf{B}$, $\mathbf{G}_1$, and $\mathbf{G}_2$ are given matrices of appropriate orders and $\mathbf{A}$ is non-negative definite, has a solution if and only if

$$\mathcal{C}(\mathbf{G}_1') \subset \mathcal{C}(\mathbf{A}, \mathbf{B}) \qquad \text{and} \qquad \mathcal{C}(\mathbf{G}_2') \subset \mathcal{C}(\mathbf{B}'),$$

in which case the general solution is

$$\begin{aligned} \mathbf{X}_1 &= \mathbf{G}_1(\mathbf{N}^+ - \mathbf{N}^+\mathbf{BC}^+\mathbf{B}'\mathbf{N}^+) + \mathbf{G}_2\mathbf{C}^+\mathbf{B}'\mathbf{N}^+ + \mathbf{Q}_1(\mathbf{I} - \mathbf{NN}^+) \text{ and} \\ \mathbf{X}_2 &= \mathbf{G}_1\mathbf{N}^+\mathbf{BC}^+ + \mathbf{G}_2(\mathbf{I} - \mathbf{C}^+) + \mathbf{Q}_2(\mathbf{I} - \mathbf{B}^+\mathbf{B}), \end{aligned}$$

where $\mathbf{N} = \mathbf{A} + \mathbf{BB}'$, $\mathbf{C} = \mathbf{B}'\mathbf{N}^+\mathbf{B}$, and $\mathbf{Q}_1$ and $\mathbf{Q}_2$ are arbitrary matrices of appropriate orders. (Note that $\mathbf{N}$, $\mathbf{N}^+$, $\mathbf{NN}^+$, $\mathbf{B}^+\mathbf{B}$, $\mathbf{C}$, and $\mathbf{C}^+$ are all symmetric.) Special cases are:

(a) If $\mathcal{C}(\mathbf{B}) \subset \mathcal{C}(\mathbf{A})$, then we can take $\mathbf{N} = \mathbf{A}$.

(b) If $\mathbf{G}_1 = \mathbf{0}$, then the original equations have a solution if and only if $\mathcal{C}(\mathbf{G}_2') \subseteq \mathcal{C}(\mathbf{B}')$, in which case the general solution for $\mathbf{X}_1$ is

$$\mathbf{X}_1 = \mathbf{G}_2(\mathbf{B}'\mathbf{N}^+\mathbf{B})^+\mathbf{B}'\mathbf{N}^+ + \mathbf{Q}(\mathbf{I} - \mathbf{NN}^+),$$

where $\mathbf{N} = \mathbf{A} + \mathbf{BB}'$ and $\mathbf{Q}$ is arbitrary of appropriate order. If, in addition, $\mathcal{C}(\mathbf{B}) \subseteq \mathcal{C}(\mathbf{A})$, then the general solution can be written as

$$\mathbf{X}_1 = \mathbf{G}_2(\mathbf{B}'\mathbf{A}^+\mathbf{B})^+\mathbf{B}'\mathbf{A}^+ + \mathbf{Q}(\mathbf{I} - \mathbf{AA}^+).$$

13.23. The equations $\mathbf{AX} = \mathbf{C}$ and $\mathbf{XB} = \mathbf{D}$ have a common solution if and only if each equation separately has a solution and $\mathbf{AD} = \mathbf{CB}$, in which case, the general expression for a common solution is

$$\begin{aligned} \mathbf{X} &= \mathbf{A}^-\mathbf{C} + \mathbf{DB}^- - \mathbf{A}^-\mathbf{ADB}^- + (\mathbf{I} - \mathbf{A}^-\mathbf{A})\mathbf{Z}(\mathbf{I} - \mathbf{BB}^-) \\ &= \mathbf{X}_0 + (\mathbf{I} - \mathbf{A}^-\mathbf{A})\mathbf{Z}(\mathbf{I} - \mathbf{BB}^-), \end{aligned}$$

where $\mathbf{X}_0$ is a common solution and $\mathbf{Z}$ is arbitrary.

13.24. If $\mathbf{B}$ is $m \times n$ and $\mathbf{X}$ is $n \times m$, then the general solution $\mathbf{X}$ of $\mathbf{XBX} = \mathbf{X}$ is

$$\mathbf{X} = \mathbf{C}(\mathbf{DBC})_{12}^-\mathbf{D},$$

where $(\cdot)_{12}^-$ is the reflexive inverse, and $n \times p$ $\mathbf{C}$ and $q \times m$ $\mathbf{D}$ are arbitrary matrices. The solution has the same rank as $\mathbf{DBC}$.

13.25. If $\mathbf{B}$ is $m \times n$ and $\mathbf{X}$ is $n \times m$, then the general solution of $\mathbf{XBX} = \mathbf{0}$ is $\mathbf{X} = \mathbf{YC}$, where p as well as the $p \times m$ matrix $\mathbf{C}$ are arbitrary, and $\mathbf{Y}$ is an arbitrary solution of $\mathbf{CBY} = \mathbf{0}$. If $\mathbf{X}$ also has to satisfy $\mathbf{WBX} = \mathbf{0}$, then $\mathbf{Y}$ is now an arbitrary solution of

$$\begin{pmatrix} \mathbf{C} \\ \mathbf{W} \end{pmatrix} \mathbf{BY} = \mathbf{0}.$$

13.26. The equations $\mathbf{A}_1 \mathbf{XB}_1 = \mathbf{C}_1$ and $\mathbf{A}_2 \mathbf{XB}_2 = \mathbf{C}_2$ have a common solution if and only if each equation is consistent and

$$\min_{\mathbf{A}_2 \mathbf{XB}_2 = \mathbf{C}_2} \operatorname{rank}(\mathbf{C}_1 - \mathbf{A}_1 \mathbf{XB}_1) = 0,$$

which is equivalent to

$$\operatorname{rank} \begin{pmatrix} \mathbf{C}_1 & \mathbf{0} & \mathbf{A}_1 \\ \mathbf{0} & -\mathbf{C}_2 & \mathbf{A}_2 \\ \mathbf{B}_1 & \mathbf{B}_2 & \mathbf{0} \end{pmatrix} = \operatorname{rank} \begin{pmatrix} \mathbf{A}_1 \\ \mathbf{A}_2 \end{pmatrix} + \operatorname{rank}(\mathbf{B}_1, \mathbf{B}_2).$$

A proof and further details relating to this problem are given by Tian [2002: 197]

13.27. (Two Unknowns) We wish to consider the solution of the matrix equation $\mathbf{AXB} + \mathbf{CYD} = \mathbf{M}$ for $\mathbf{X}$ and $\mathbf{Y}$. Since $\operatorname{vec}(\mathbf{AXB}) = (\mathbf{B}' \otimes \mathbf{A})\operatorname{vec}\mathbf{X}$, we can rewrite the matrix equation in the form

$$(\mathbf{B}' \otimes \mathbf{A}, \mathbf{D}' \otimes \mathbf{C}) \begin{pmatrix} \operatorname{vec}\mathbf{X} \\ \operatorname{vec}\mathbf{Y} \end{pmatrix} = \operatorname{vec}\mathbf{M},$$

which is solvable if and only if (cf. 13.1c)

$$(\mathbf{B}' \otimes \mathbf{A}, \mathbf{D}' \otimes \mathbf{C})(\mathbf{B}' \otimes \mathbf{A}, \mathbf{D}' \otimes \mathbf{C})^- \operatorname{vec}\mathbf{M} = \operatorname{vec}\mathbf{M}.$$

In this case, from (13.4), the general solution is

$$\begin{pmatrix} \operatorname{vec}\mathbf{X} \\ \operatorname{vec}\mathbf{Y} \end{pmatrix} = (\mathbf{B}' \otimes \mathbf{A}, \mathbf{D}' \otimes \mathbf{C})^- \operatorname{vec}\mathbf{M} + [\mathbf{I} - (\mathbf{B}' \otimes \mathbf{A}, \mathbf{D}' \otimes \mathbf{C})^- (\mathbf{B}' \otimes \mathbf{A}, \mathbf{D}' \otimes \mathbf{C})]\mathbf{v},$$

where $\mathbf{v}$ is an arbitrary vector. Using his extremal ranks method, Tian [2006b] gives necessary and sufficient rank conditions for solutions $\mathbf{X}$ and $\mathbf{Y}$ to exist and also provides methods for finding solutions.

Proofs. Section 13.2.2.

13.13. Harville [1997: 73].

13.14. Magnus and Neudecker [1988: 44].

13.16. Harville [1997: 75–76].

13.17a. Graham [1981: 38–39] and Horn and Johnson [1991: 270].

13.17c. Horn and Johnson [1991: 270].

13.18. Graham [1981: 40].

13.19. Zhang [1999: 139] and (b) quoted by Horn and Johnson [1991: 270].

13.20. Horn and Johnson [1991: 255].

13.21. Horn and Johnson [1991: 281–283].

13.22. Magnus and Neudecker [1999: 60–62].

13.23. Ben-Israel and Greville [2003: 54] and Rao and Mitra [1971: 25].

13.24–13.25. Rao and Mitra [1971: 56-57]. They also give solutions to $\mathbf{XBXB} = \mathbf{XB}$, $\mathbf{BXBX} = \mathbf{BX}$, $\mathbf{BXBXB} = \mathbf{BXB}$, and $\mathbf{XBXBX} = \mathbf{XBX}$.

CHAPTER 14

PARTITIONED MATRICES

Partitioned matrices arise frequently in statistics, especially in proofs. For some partitions and their relationship with ranks, the reader should consult Section 3.6. This chapter is closely linked to the next chapter on patterned matrices.

14.1 SCHUR COMPLEMENT

Definition **14.1.** Let

$$A = \begin{pmatrix} E & F \\ G & H \end{pmatrix},$$

where $\mathbf{A}$ is possibly rectangular. If $\mathbf{E}$ is square and nonsingular, then

$$S = H - GE^{-1}F = (A/E)$$

is called the *Schur complement* of $\mathbf{E}$ in $\mathbf{A}$. If $\mathbf{H}$ is nonsingular (instead of, or in addition to, $\mathbf{E}$), then

$$T = E - FH^{-1}G = (A/H)$$

is the Schur complement of $\mathbf{H}$ in $\mathbf{A}$.

Schur complements occur in various places in this book, sometimes using a different notation. Because of the wide applicability of Schur complements, we have collected some of the results together here in one place using the present notation, which is the one used in three key references, namely Ouellette [1981], Puntanen and Styan, [2005b], and Styan [1985]. These writers show how the Schur complement can be used to prove a number of matrix results that are typically proved by other methods. They also show how Schur complements arise naturally in statistics, especially in multivariate analysis and in linear models.

14.1. (Determinants) If $\mathbf{A}$ is nonsingular, we have that (see also 14.17)

$$\det \mathbf{A} = \begin{cases} \det \mathbf{E} \cdot \det(\mathbf{A}/\mathbf{E}), & \text{if } \mathbf{E} \text{ is nonsingular,} \\ \det \mathbf{H} \cdot \det(\mathbf{A}/\mathbf{H}), & \text{if } \mathbf{H} \text{ is nonsingular.} \end{cases}$$

Therefore if $\mathbf{A}$ and $\mathbf{E}$ are nonsingular, then so is $\mathbf{A}/\mathbf{E}$. The same applies to $\mathbf{A}$ and $\mathbf{H}$.

14.2. (Ranks) From (3.40a(vi) and (3.40(vii)) we have:

(a) If $\mathbf{E}$ is nonsingular, rank $\mathbf{A}$ = rank $\mathbf{E}$ + rank$(\mathbf{A}/\mathbf{E})$.

(b) If $\mathbf{H}$ is nonsingular, rank $\mathbf{A}$ = rank $\mathbf{H}$ + rank$(\mathbf{A}/\mathbf{H})$.

14.3. (Inverses) If $\mathbf{A}$, $\mathbf{E}$, and $\mathbf{H}$ are all nonsingular, then:

(a) $(\mathbf{A}/\mathbf{H})^{-1} = \mathbf{E}^{-1} + \mathbf{E}^{-1}\mathbf{F}(\mathbf{A}/\mathbf{E})^{-1}\mathbf{G}\mathbf{E}^{-1}$.

(b) $(\mathbf{A}/\mathbf{E})^{-1} = \mathbf{H}^{-1} + \mathbf{H}^{-1}\mathbf{G}(\mathbf{A}/\mathbf{H})^{-1}\mathbf{F}\mathbf{H}^{-1}$.

14.4. (Inertia) We recall that the inertia $\mathcal{I}n(\mathbf{A})$ of a symmetric matrix $\mathbf{A}$ is given by the triple (r_+, r_-, r_0), where r_+ is the number of positive eigenvalues, r_- is the number of negative eigenvalues, and r_0 is the number of zero eigenvalues. Then, if $\mathbf{A}$ is symmetric and $\mathbf{E}$ is nonsingular,

$$\mathcal{I}n(\mathbf{A}) = \mathcal{I}n(\mathbf{E}) + \mathcal{I}n(\mathbf{A}/\mathbf{E}).$$

Ouellette [1981: 207–210] extends the above result to the case when $(\mathbf{A}/\mathbf{E})$ is also partitioned.

14.5. (Non-negative Definite Matrices) Suppose $\mathbf{A}$ is symmetric and $\mathbf{E}$ is positive definite.

(a) $\mathbf{A}$ is non-negative definite if and only if $(\mathbf{A}/\mathbf{E})$ is non-negative definite.

(b) $\mathbf{A}$ is positive definite if and only if $(\mathbf{A}/\mathbf{E})$ is positive definite.

14.6. (Subpartition) Suppose that

$$\mathbf{A} = \left(\begin{array}{ccc} \mathbf{E} & \vdots & \mathbf{F} \\ \cdots & \cdots & \cdots \\ \mathbf{G} & \vdots & \mathbf{H} \end{array} \right) = \left(\begin{array}{cccc} \mathbf{K} & \mathbf{L} & \vdots & \mathbf{F}_1 \\ \mathbf{M} & \mathbf{N} & \vdots & \mathbf{F}_2 \\ \cdots & \cdots & \cdots & \cdots \\ \mathbf{G}_1 & \mathbf{G}_2 & \vdots & \mathbf{H} \end{array} \right),$$

where $\mathbf{E}$ and $\mathbf{K}$ are nonsingular. Then $(\mathbf{E}/\mathbf{K})$ is a nonsingular leading principal submatrix of $(\mathbf{A}/\mathbf{K})$, and

$$(\mathbf{A}/\mathbf{E}) = ((\mathbf{A}/\mathbf{K})/(\mathbf{E}/\mathbf{K})).$$

14.7. (Sum)   Let

$$A = \begin{pmatrix} E & F \\ F' & H \end{pmatrix} \quad \text{and} \quad B = \begin{pmatrix} K & L \\ L' & N \end{pmatrix}$$

be symmetric $(m + n) \times (m + n)$ matrices, where E and K are $m \times m$. Suppose that A and B are non-negative definite (n.n.d.) and E and K are positive definite.

(a) $F'E^{-1}F + L'K^{-1}L - (F + L)'(E + K)^{-1}(F + L)$ is n.n.d. with the same rank as $F - EK^{-1}L$.

(b) $((A + B)/(E + K)) - (A/E) - (B/K)$ is n.n.d.

(c) $\det[(A + B)/(E + K)] = \dfrac{\det(A + B)}{\det(E + K)} \geq \dfrac{\det A}{\det E} + \dfrac{\det B}{\det K}$.

***Definition* 14.2.** (Generalized Schur Complement)   Referring to Definition 14.1, if E is rectangular, or square and singular, then we replace E^{-1} by any weak inverse E^- and call (A/E) the *generalized Schur complement* of E in A. We have a similar definition for (A/H).

We shall use the following notation below:

$$S = (A/E) = H - GE^-F \quad \text{and} \quad T = (A/H) = E = FH^-G.$$

14.8. (General Properties of the Generalized Schur Complement)

(a) If A and E are both square and either $\mathcal{C}(F) \subseteq \mathcal{C}(E)$ or $\mathcal{C}(G') \subseteq \mathcal{C}(E')$, then S is invariant for all weak inverses E^- and

$$\det A = \det E \cdot \det S.$$

(b) If A and H are both square and either $\mathcal{C}(G) \subseteq \mathcal{C}(H)$ or $\mathcal{C}(F') \subseteq \mathcal{C}(H')$, then T is invariant for all weak inverses H^- and

$$\det A = \det H \cdot \det T.$$

(c) If A is non-negative definite and E and H are both square (i.e., $G = F'$), then $\mathcal{C}(F) \subseteq \mathcal{C}(E)$ and $\mathcal{C}(F') \subseteq \mathcal{C}(H)$; also (a) and (b) hold.

14.9. Suppose A is non-negative definite and E and H are both square, then:

(a) S and T are invariant with respect to the weak inverses E^- and H^-.

(b) (Rank)

    (i) rank A = rank E + rank(A/E) .

    (ii) rank A = rank H + rank(A/H).

(c) (Inertia)

    (i) $\mathcal{In}(A) = \mathcal{In}(E) + \mathcal{In}(A/E)$.

    (ii) $\mathcal{In}(A) = \mathcal{In}(H) + \mathcal{In}(A/H)$.

(d) If $\nu(\mathbf{A})$ refers to the nullity of $\mathbf{A}$, then:

    (i) $\nu(\mathbf{A}) = \nu(\mathbf{E}) + \nu[(\mathbf{A}/\mathbf{E})]$.

    (ii) $\nu(\mathbf{A}) = \nu(\mathbf{H}) + \nu[(\mathbf{A}/\mathbf{H})]$.

Proofs. Section 14.1.

    14.1. Ouellette [1981: 195, 209].

    14.2. Ouellette [1981: 199].

    14.3. Abadir and Magnus [2005: 107]. See also (15.3c).

    14.4. Ouellette [1981: 207–210]

    14.5. Abadir and Magnus [2005: 228–229] and Ouellette [1981: 208].

    14.6. Ouellette [1981: 210].

    14.7. Ouellette [1981: 211–212].

    14.8a–b. Ouellette [1981: 224–225].

    14.8c. Follows from (a) and (b) and (14.26g).

    14.9a–d. Puntanen and Styan [2005b: section 6.0.4]; for (a) see Ouellette [1981: 242] and Styan [1985: 45]; for (b) see Styan [1985: 45] and (4.40a(iii)); for (c) see Ouellette [1981: 238, theorem 4.7]; and (d) follows from (b) and the fact that the rank plus the nullity of a matrix is equal to the number of columns.

14.2 INVERSES

The notation used so far for Schur complements is sometimes not so helpful for the more general results in this section, as it is not easy to see the patterns. I now introduce a subscript notation as well, as both are used in the literature. Some of the above results will appear again under a different guise. The results on inverses in this section are established by simply checking that $\mathbf{A}\mathbf{A}^{-1} = \mathbf{I}$. The other results are verified by multiplying out the matrices concerned and using (14.11).

14.10. Let

$$\mathbf{A} = \begin{pmatrix} \mathbf{A}_{11} & \mathbf{A}_{12} \\ \mathbf{A}_{21} & \mathbf{A}_{22} \end{pmatrix},$$

where $\mathbf{A}$, $\mathbf{A}_{11}$, and $\mathbf{A}_{22}$ are all real or complex matrices that are not necessarily square.

(a) If $\mathbf{A}_{11}$ is nonsingular and $\mathbf{A}_{22\cdot 1} = \mathbf{A}_{22} - \mathbf{A}_{21}\mathbf{A}_{11}^{-1}\mathbf{A}_{12}\ (= \mathbf{A}/\mathbf{A}_{11})$, then

    (i) $\mathbf{A} = \begin{pmatrix} \mathbf{I} & \mathbf{0} \\ \mathbf{A}_{21}\mathbf{A}_{11}^{-1} & \mathbf{I} \end{pmatrix} \begin{pmatrix} \mathbf{A}_{11} & \mathbf{0} \\ \mathbf{0} & \mathbf{A}_{22\cdot 1} \end{pmatrix} \begin{pmatrix} \mathbf{I} & \mathbf{A}_{11}^{-1}\mathbf{A}_{12} \\ \mathbf{0} & \mathbf{I} \end{pmatrix}.$

       This is sometimes called the *Aitken block-diagonalization formula*. When

$\mathbf{A}$ is non-negative definite, the above result still holds with $\mathbf{A}_{21} = \mathbf{A}'_{12}$ and $\mathbf{A}_{11}^{-1}$ replaced by $\mathbf{A}_{11}^-$ throughout.

(ii) $\left(\begin{array}{cc} \mathbf{A}_{11}^{-1} & \mathbf{0} \\ -\mathbf{A}_{21}\mathbf{A}_{11}^{-1} & \mathbf{I} \end{array} \right) \mathbf{A} = \left(\begin{array}{cc} \mathbf{I} & \mathbf{A}_{11}^{-1}\mathbf{A}_{12} \\ \mathbf{0} & \mathbf{A}_{22\cdot1} \end{array} \right).$

(iii) If $\mathbf{A}^{-1}$ exists, then

$$\mathbf{A}^{-1} = \left(\begin{array}{cc} \mathbf{I} & -\mathbf{A}_{11}^{-1}\mathbf{A}_{12} \\ \mathbf{0} & \mathbf{I} \end{array} \right) \left(\begin{array}{cc} \mathbf{A}_{11}^{-1} & \mathbf{0} \\ -\mathbf{A}_{22\cdot1}^{-1}\mathbf{A}_{21}\mathbf{A}_{11}^{-1} & \mathbf{A}_{22\cdot1}^{-1} \end{array} \right).$$

(iv) If $\mathbf{A}^{-1}$ exists, then

$$\mathbf{A}^{-1} = \left(\begin{array}{cc} \mathbf{A}_{11}^{-1} & \mathbf{0} \\ \mathbf{0} & \mathbf{0} \end{array} \right) + \left(\begin{array}{c} -\mathbf{A}_{11}^{-1}\mathbf{A}_{12} \\ \mathbf{I} \end{array} \right) \mathbf{A}_{22\cdot1}^{-1}(-\mathbf{A}_{21}\mathbf{A}_{11}^{-1}, \mathbf{I}).$$

(v) If $\mathbf{A}$ and $\mathbf{A}_{11}$ have rank r and $\mathbf{A}_{11}$ is $r \times r$, then $\mathbf{A}_{22\cdot1} = \mathbf{0}$.

(b) If $\mathbf{A}_{22}$ is nonsingular and $\mathbf{A}_{11\cdot2} = \mathbf{A}_{11} - \mathbf{A}_{12}\mathbf{A}_{22}^{-1}\mathbf{A}_{21} (= (\mathbf{A}/\mathbf{A}_{22}))$, then

(i) $\mathbf{A} = \left(\begin{array}{cc} \mathbf{I} & \mathbf{A}_{12}\mathbf{A}_{22}^{-1} \\ \mathbf{0} & \mathbf{I} \end{array} \right) \left(\begin{array}{cc} \mathbf{A}_{11\cdot2} & \mathbf{0} \\ \mathbf{0} & \mathbf{A}_{22} \end{array} \right) \left(\begin{array}{cc} \mathbf{I} & \mathbf{0} \\ \mathbf{A}_{22}^{-1}\mathbf{A}_{21} & \mathbf{I} \end{array} \right).$

When $\mathbf{A}$ is non-negative definite, the above result still holds with $\mathbf{A}_{21} = \mathbf{A}'_{12}$ and $\mathbf{A}_{22}^{-1}$ replaced by $\mathbf{A}_{22}^-$ throughout.

(ii) $\mathbf{A} \left(\begin{array}{cc} \mathbf{I} & \mathbf{0} \\ -\mathbf{A}_{22}^{-1}\mathbf{A}_{21} & \mathbf{A}_{22}^{-1} \end{array} \right) = \left(\begin{array}{cc} \mathbf{A}_{11\cdot2} & \mathbf{A}_{12}\mathbf{A}_{22}^{-1} \\ \mathbf{0} & \mathbf{I} \end{array} \right).$

(iii) If $\mathbf{A}^{-1}$ exists, then

$$\mathbf{A}^{-1} = \left(\begin{array}{cc} \mathbf{I} & \mathbf{0} \\ -\mathbf{A}_{22}^{-1}\mathbf{A}_{21} & \mathbf{I} \end{array} \right) \left(\begin{array}{cc} \mathbf{A}_{11\cdot2}^{-1} & -\mathbf{A}_{11\cdot2}^{-1}\mathbf{A}_{12}\mathbf{A}_{22}^{-1} \\ \mathbf{0} & \mathbf{A}_{22}^{-1} \end{array} \right).$$

(iv) If $\mathbf{A}^{-1}$ exists, then

$$\mathbf{A}^{-1} = \left(\begin{array}{cc} \mathbf{0} & \mathbf{0} \\ \mathbf{0} & \mathbf{A}_{22}^{-1} \end{array} \right) \left(\begin{array}{c} \mathbf{I} \\ -\mathbf{A}_{22}^{-1}\mathbf{A}_{21} \end{array} \right) \mathbf{A}_{11\cdot2}^{-1}(\mathbf{I}, -\mathbf{A}_{12}\mathbf{A}_{22}^{-1}).$$

(v) If $\mathbf{A}$ and $\mathbf{A}_{22}$ have rank r and $\mathbf{A}_{22}$ is $r \times r$, then $\mathbf{A}_{11\cdot2} = \mathbf{0}$.

14.11. Suppose $\mathbf{A}$ is partitioned as above and is nonsingular.

(a) If $\mathbf{A}_{11}$ is nonsingular and $\mathbf{A}_{22\cdot1} = \mathbf{A}_{22} - \mathbf{A}_{21}\mathbf{A}_{11}^{-1}\mathbf{A}_{12}$, then

$$\mathbf{A}^{-1} = \left(\begin{array}{cc} \mathbf{A}_{11}^{-1} + \mathbf{A}_{11}^{-1}\mathbf{A}_{12}\mathbf{A}_{22\cdot1}^{-1}\mathbf{A}_{21}\mathbf{A}_{11}^{-1} & -\mathbf{A}_{11}^{-1}\mathbf{A}_{12}\mathbf{A}_{22\cdot1}^{-1} \\ -\mathbf{A}_{22\cdot1}^{-1}\mathbf{A}_{21}\mathbf{A}_{11}^{-1} & \mathbf{A}_{22\cdot1}^{-1} \end{array} \right).$$

(b) If $\mathbf{A}_{22}$ is nonsingular and $\mathbf{A}_{11\cdot2} = \mathbf{A}_{11} - \mathbf{A}_{12}\mathbf{A}_{22}^{-1}\mathbf{A}_{21}$, then

$$\mathbf{A}^{-1} = \left(\begin{array}{cc} \mathbf{A}_{11\cdot2}^{-1} & -\mathbf{A}_{11\cdot2}^{-1}\mathbf{A}_{12}\mathbf{A}_{22}^{-1} \\ -\mathbf{A}_{22}^{-1}\mathbf{A}_{21}\mathbf{A}_{11\cdot2}^{-1} & \mathbf{A}_{22}^{-1} + \mathbf{A}_{22}^{-1}\mathbf{A}_{21}\mathbf{A}_{11\cdot2}^{-1}\mathbf{A}_{12}\mathbf{A}_{22}^{-1} \end{array} \right).$$

(c) If $\mathbf{A}_{11}$ and $\mathbf{A}_{22}$ are both nonsingular, then we have the following.

  (i) $\mathbf{A}_{22\cdot1}^{-1} = \mathbf{A}_{22}^{-1} + \mathbf{A}_{22}^{-1}\mathbf{A}_{21}\mathbf{A}_{11\cdot2}^{-1}\mathbf{A}_{12}\mathbf{A}_{22}^{-1}$.

  (ii) Interchanging 1 and 2 above,

$$\mathbf{A}_{11\cdot2}^{-1} = \mathbf{A}_{11}^{-1} + \mathbf{A}_{11}^{-1}\mathbf{A}_{12}\mathbf{A}_{22\cdot1}^{-1}\mathbf{A}_{21}\mathbf{A}_{11}^{-1}.$$

When $\mathbf{A}_{11}$ and $\mathbf{A}_{22}$ are both nonsingular, the two representations of $\mathbf{A}^{-1}$ given by (a) and (b) above are identical, by the uniqueness of the inverse, even though the off-diagonal blocks may not look equal. Thus, for example, it can be shown that

  (iii) $\mathbf{A}_{22\cdot1}^{-1}\mathbf{A}_{21}\mathbf{A}_{11}^{-1} = \mathbf{A}_{22}^{-1}\mathbf{A}_{21}\mathbf{A}_{11\cdot2}^{-1}$.

For this reason the reader will find various versions of $\mathbf{A}^{-1}$ in the literature. (e.g., compare Graybill [1983: 184] and Muirhead [1982: 580] with Anderson [2003: 638] and Zhang [1999: 184, where $\mathbf{A}$ is positive definite). When $\mathbf{A}$ is symmetric or Hermitian we have $\mathbf{A}_{21} = \mathbf{A}_{12}^*$.

Some special cases follow.

14.12. If $\mathbf{A}_{11}$ and $\mathbf{A}_{22}$ are nonsingular, then the following inverses below exist (by 14.18) below, and

$$\begin{pmatrix} \mathbf{A}_{11} & \mathbf{0} \\ \mathbf{A}_{21} & \mathbf{A}_{22} \end{pmatrix}^{-1} = \begin{pmatrix} \mathbf{A}_{11}^{-1} & \mathbf{0} \\ -\mathbf{A}_{22}^{-1}\mathbf{A}_{21}\mathbf{A}_{11}^{-1} & \mathbf{A}_{22}^{-1} \end{pmatrix}.$$

Similarly,

$$\begin{pmatrix} \mathbf{A}_{11} & \mathbf{A}_{12} \\ \mathbf{0} & \mathbf{A}_{22} \end{pmatrix}^{-1} = \begin{pmatrix} \mathbf{A}_{11}^{-1} & -\mathbf{A}_{11}^{-1}\mathbf{A}_{12}\mathbf{A}_{22}^{-1} \\ \mathbf{0} & \mathbf{A}_{22}^{-1} \end{pmatrix}.$$

We get special cases if we set $\mathbf{A}_{11}$ and/or $\mathbf{A}_{22}$ equal to identity matrices.

Nonsingular block-triangular matrices with more than two blocks can be inverted by applying the above method iteratively (cf. Harville [1997: 94]).

14.13. Suppose $\mathbf{A}$ and $\mathbf{D}$ are nonsingular.

  (a) If $\alpha = d - \mathbf{c}'\mathbf{A}^{-1}\mathbf{b} \neq 0$, we have from (14.10a(iv))

$$\begin{pmatrix} \mathbf{A} & \mathbf{b} \\ \mathbf{c}' & d \end{pmatrix}^{-1} = \begin{pmatrix} \mathbf{A}^{-1} & \mathbf{0} \\ \mathbf{0}' & 0 \end{pmatrix} + \frac{1}{\alpha}\begin{pmatrix} \mathbf{A}^{-1}\mathbf{b} \\ -1 \end{pmatrix}(\mathbf{c}'\mathbf{A}^{-1}, -1).$$

  (b) If $\beta = a - \mathbf{b}'\mathbf{D}^{-1}\mathbf{c} \neq 0$, we have from (14.10b(iv))

$$\begin{pmatrix} a & \mathbf{b}' \\ \mathbf{c} & \mathbf{D} \end{pmatrix}^{-1} = \begin{pmatrix} 0 & \mathbf{0}' \\ \mathbf{0} & \mathbf{D}^{-1} \end{pmatrix} + \frac{1}{\beta}\begin{pmatrix} -1 \\ \mathbf{D}^{-1}\mathbf{c} \end{pmatrix}(-1, \mathbf{b}'\mathbf{D}^{-1}).$$

14.14. Let $(\mathbf{A}, \mathbf{B})$ be an $n \times (k+m)$ matrix of full column rank, where $\mathbf{A}$ is $n \times k$. Define

$$\mathbf{Z} = (\mathbf{A}, \mathbf{B})'(\mathbf{A}, \mathbf{B}) = \begin{pmatrix} \mathbf{A}'\mathbf{A} & \mathbf{A}'\mathbf{B} \\ \mathbf{B}'\mathbf{A} & \mathbf{B}'\mathbf{B} \end{pmatrix}.$$

Let $\mathbf{M_C} = \mathbf{I}_n - \mathbf{C}(\mathbf{C'C})^{-1}\mathbf{C}$ for $\mathbf{C} = \mathbf{A}, \mathbf{B}$, and define $\mathbf{E} = \mathbf{B'M_A B}$ and $\mathbf{F} = \mathbf{A'M_B A}$. Then, from (14.11a,b),

$$
\mathbf{Z}^{-1} = \begin{pmatrix} (\mathbf{A'A})^{-1} + (\mathbf{A'A})^{-1}\mathbf{A'BE}^{-1}\mathbf{B'A}(\mathbf{A'A})^{-1} & -(\mathbf{A'A})^{-1}\mathbf{A'BE}^{-1} \\ -\mathbf{E}^{-1}\mathbf{B'A}(\mathbf{A'A})^{-1} & \mathbf{E}^{-1} \end{pmatrix}
$$

$$
= \begin{pmatrix} \mathbf{F}^{-1} & -\mathbf{F}^{-1}\mathbf{A'B}(\mathbf{B'B})^{-1} \\ -(\mathbf{B'B})^{-1}\mathbf{B'AF}^{-1} & (\mathbf{B'B})^{-1} + (\mathbf{B'B})^{-1}\mathbf{B'AF}^{-1}\mathbf{A'B}(\mathbf{B'B})^{-1} \end{pmatrix}
$$

14.15. Given conformable matrices and the existence of the appropriate inverses, we have

$$
\begin{pmatrix} \mathbf{A} & \mathbf{B} & \mathbf{C} \\ \mathbf{B'} & \mathbf{D} & \mathbf{0} \\ \mathbf{C'} & \mathbf{0} & \mathbf{E} \end{pmatrix}^{-1}
$$

$$
= \begin{pmatrix} \mathbf{Q}^{-1} & -\mathbf{Q}^{-1}\mathbf{BD}^{-1} & -\mathbf{Q}^{-1}\mathbf{CE}^{-1} \\ -\mathbf{D}^{-1}\mathbf{B'Q}^{-1} & \mathbf{D}^{-1} + \mathbf{D}^{-1}\mathbf{B'Q}^{-1}\mathbf{BD}^{-1} & \mathbf{D}^{-1}\mathbf{B'Q}^{-1}\mathbf{CE}^{-1} \\ -\mathbf{E}^{-1}\mathbf{C'Q}^{-1} & \mathbf{E}^{-1}\mathbf{C'Q}^{-1}\mathbf{BD}^{-1} & \mathbf{E}^{-1} + \mathbf{E}^{-1}\mathbf{C'Q}^{-1}\mathbf{CE}^{-1} \end{pmatrix},
$$

where $\mathbf{Q} = \mathbf{A} - \mathbf{BD}^{-1}\mathbf{B'} - \mathbf{CE}^{-1}\mathbf{C'}$.

14.16. (Powers) Suppose $\mathbf{A}$ is $m \times m$ and $\mathbf{D}$ is $n \times n$.

(a)

$$
\begin{pmatrix} \mathbf{A} & \mathbf{B} \\ \mathbf{0} & \mathbf{D} \end{pmatrix}^k = \begin{pmatrix} \mathbf{A}^k & \mathbf{Q}_k \\ \mathbf{0} & \mathbf{D}^k \end{pmatrix}, \quad k = 1, 2, \ldots,
$$

where $\mathbf{Q}_k = \sum_{i=1}^k \mathbf{A}^{k-i}\mathbf{BD}^{i-1}$.

(b) If, in (a), $\mathbf{D} = \mathbf{I}_n$ and $\mathbf{I}_m - \mathbf{A}$ is nonsingular, then

$$
\mathbf{Q}_k = (\mathbf{I}_m - \mathbf{A})^{-1}(\mathbf{I}_m - \mathbf{A}^k)\mathbf{B}.
$$

(c) If $\mathbf{A}$ and $\mathbf{B}$ are nonsingular,

$$
\begin{pmatrix} \mathbf{A} & \mathbf{B} \\ \mathbf{0} & \mathbf{D} \end{pmatrix}^{-k} = \begin{pmatrix} \mathbf{A}^{-k} & \mathbf{R}_k \\ \mathbf{0} & \mathbf{D}^{-k} \end{pmatrix},
$$

where $\mathbf{R}_k = -\sum_{i=1}^k \mathbf{A}^{-(k-i+1)}\mathbf{BD}^{-i}$.

Proofs. Section 14.2.

14.10a(v) and b(v). Graybill [1983: 126–127].

14.13. Abadir and Magnus [2005: 105].

14.14. Abadir and Magnus [2005: 107].

14.15. Magnus and Neudecker [1999: 12].

14.16. Abadir and Magnus [2005: 109].

14.3   DETERMINANTS

14.17. Suppose $\mathbf{A}$ is partitioned as in (14.10).

(a) If $\mathbf{A}_{11}$ is nonsingular,

$$\det \mathbf{A} = \det(\mathbf{A}_{11}) \det(\mathbf{A}_{22\cdot1}).$$

If, in addition, $\mathbf{A}$ is nonsingular, then so is $\mathbf{A}_{22\cdot1}$, the Schur complement of $\mathbf{A}_{11}$.

(b) If $\mathbf{A}_{22}$ is nonsingular,

$$\det \mathbf{A} = \det(\mathbf{A}_{22}) \det(\mathbf{A}_{11\cdot2}).$$

If, in addition, $\mathbf{A}$ is nonsingular, then so is $\mathbf{A}_{11\cdot2}$, the Schur complement of $\mathbf{A}_{22}$.

(c) If $\mathbf{A}_{11}^-$ and $\mathbf{A}_{22}^-$ are any weak inverses of $\mathbf{A}_{11}$ and $\mathbf{A}_{22}$, then:

(i) If $\mathcal{C}(\mathbf{A}_{21}) \subseteq \mathcal{C}(\mathbf{A}_{22})$ or $\mathcal{C}(\mathbf{A}_{12}') \subseteq \mathcal{C}(\mathbf{A}_{22}')$, we have

$$\det \mathbf{A} = (\det \mathbf{A}_{22}) \det(\mathbf{A}_{11} - \mathbf{A}_{12}\mathbf{A}_{22}^-\mathbf{A}_{21}).$$

(ii) If $\mathcal{C}(\mathbf{A}_{12}) \subseteq \mathcal{C}(\mathbf{A}_{11})$ or $\mathcal{C}(\mathbf{A}_{21}') \subseteq \mathcal{C}(\mathbf{A}_{11}')$, we have

$$\det \mathbf{A} = (\det \mathbf{A}_{11}) \det(\mathbf{A}_{22} - \mathbf{A}_{21}\mathbf{A}_{11}^-\mathbf{A}_{12}).$$

14.18. The following two results are often useful.

(a) If $\mathbf{A}$ and $\mathbf{B}$ are $m \times m$ and $n \times n$, respectively, then, for conformable matrices,

$$\det \begin{pmatrix} \mathbf{A} & \mathbf{0} \\ \mathbf{E} & \mathbf{B} \end{pmatrix} = \det \begin{pmatrix} \mathbf{A} & \mathbf{F} \\ \mathbf{0} & \mathbf{B} \end{pmatrix} = \det \mathbf{A} \cdot \det \mathbf{B}.$$

We can set $\mathbf{A}$ or $\mathbf{B}$ equal to the identity matrix.

Note that the two matrices on the left are nonsingular if and only if both $\mathbf{A}$ and $\mathbf{B}$ are nonsingular.

(b) Using a similar notation to (a),

$$\det \begin{pmatrix} \mathbf{0} & \mathbf{F} \\ \mathbf{E} & \mathbf{B} \end{pmatrix} = \det \begin{pmatrix} \mathbf{B} & \mathbf{E} \\ \mathbf{F} & \mathbf{0} \end{pmatrix} = (-1)^{mn} \det \mathbf{E} \cdot \det \mathbf{F}.$$

14.19. If $\mathbf{B}$ and $\mathbf{C}$ are $n \times n$ matrices, then

$$\det \begin{pmatrix} \mathbf{0} & \mathbf{B} \\ -\mathbf{I}_n & \mathbf{C} \end{pmatrix} = \det \mathbf{B}.$$

14.20. If $\mathbf{C} = (\mathbf{A}, \mathbf{B})$ is square, then from $\det(\mathbf{CC}') = \det(\mathbf{C}'\mathbf{C}) = \det(\mathbf{C})^2$ we have

$$\det(\mathbf{AA}' + \mathbf{BB}') = \det \begin{pmatrix} \mathbf{A}'\mathbf{A} & \mathbf{A}'\mathbf{B} \\ \mathbf{B}'\mathbf{A} & \mathbf{B}'\mathbf{B} \end{pmatrix}.$$

14.21. Let $\mathbf{A}$ and $\mathbf{D}$ be square matrices. Then:

(a) $\det \begin{pmatrix} \mathbf{A} & \mathbf{b} \\ \mathbf{c}' & d \end{pmatrix} = d \det \mathbf{A} - \mathbf{c}'(\mathrm{adj}\mathbf{A})\mathbf{b}$,

   or $\det \mathbf{A}(d - \mathbf{c}'\mathbf{A}^{-1}\mathbf{b})$ if $\mathbf{A}$ is nonsingular, where $\mathrm{adj}\mathbf{A}$ is the adjoint matrix of $\mathbf{A}$.

(b) $\det \begin{pmatrix} a & \mathbf{b}' \\ \mathbf{c} & \mathbf{D} \end{pmatrix} = \det \mathbf{D}(a - \mathbf{b}'\mathbf{D}^{-1}\mathbf{c})$ if $\mathbf{D}$ is nonsingular.

(c) From (a) we have $\det \begin{pmatrix} \mathbf{A} & \mathbf{u} \\ \mathbf{u}' & -1 \end{pmatrix} = -\det(\mathbf{A} + \mathbf{uu}')$.

14.22. (Adjoint) Let

$$\mathbf{A} = \begin{pmatrix} \mathbf{E} & \mathbf{F} \\ \mathbf{G} & \mathbf{H} \end{pmatrix},$$

be an $n \times n$ matrix such that $\mathbf{E}$ is $m \times m$. If

$$\mathrm{adj}\mathbf{A} = \begin{pmatrix} \mathbf{E}_1 & \mathbf{F}_1 \\ \mathbf{G}_1 & \mathbf{H}_1 \end{pmatrix},$$

where $\mathbf{E}_1$ is $m \times m$, then:

(a) $\det \mathbf{H}_1 = (\det \mathbf{A})^{n-m-1} \det \mathbf{E}$ for $m = 0, 1, 2, \ldots, n - 1$.

(b) $\det \mathbf{E}_1 = (\det \mathbf{A})^m \det \mathbf{H}$, for $m = 0, 1, 2, \ldots, n - 1$.

14.23. If $\mathbf{AC} = \mathbf{CA}$, then

$$\det \begin{pmatrix} \mathbf{A} & \mathbf{B} \\ \mathbf{C} & \mathbf{D} \end{pmatrix} = \det(\mathbf{AD} - \mathbf{CB}).$$

If we set $\mathbf{A} = \mathbf{I}$, then the above is true.

14.24. If $\mathbf{A}$ and $\mathbf{B}$ are $n \times n$ matrices, then

$$\det \begin{pmatrix} \mathbf{A} & \mathbf{B} \\ \mathbf{B} & \mathbf{A} \end{pmatrix} = \det(\mathbf{A} + \mathbf{B}) \cdot \det(\mathbf{A} - \mathbf{B}).$$

14.25. The determinant of the matrix inversed in (14.15) is

$$\det \mathbf{D} \cdot \det \mathbf{E} \cdot \det(\mathbf{A} - \mathbf{BD}^{-1}\mathbf{B}' - \mathbf{CE}^{-1}\mathbf{C}').$$

Proofs. Section 14.3.

14.17a. We take determinants in (14.10a(i)) and use the fact that the determinant of a triangular matrix is the product of its diagonal elements.

14.17b. Similar to (a), but using (14.10b(i)).

14.17c. Schott [2005: 263]; see also (14.8a,b).

14.18a. Harville [1997: 185], Rao and Bhimasankaram [2000: 234], and Searle [1982: 97].

14.18b. Harville [1987: 187].

14.19. This follows from (14.18b) and the fact that $n^2 + n = n(n+1)$ is even. See also Searle [1982: 98].

14.21. Abadir and Magnus [2005: 113].

14.22. Ouellette [1981: 205–206].

14.23. Abadir and Magnus [2005: 116].

14.24. Abadir and Magnus [2005: 117].

14.25. Abadir and Magnus [2005: 118].

14.4   POSITIVE AND NON-NEGATIVE DEFINITE MATRICES

Schur complements arise in this section using a different notation, and the results should be compared with those in Section 14.1. Note that $\mathbf{A} \succeq \mathbf{B}$ means that $\mathbf{A} - \mathbf{B}$ is non-negative definite.

14.26. Let

$$\mathbf{A} = \left(\begin{array}{cc} \mathbf{A}_{11} & \mathbf{A}_{12} \\ \mathbf{A}_{21} & \mathbf{A}_{22} \end{array} \right)$$

be a real symmetric matrix (i.e., $\mathbf{A}_{12} = \mathbf{A}'_{21}$, with $\mathbf{A}_{11}$ and $\mathbf{A}_{22}$ square matrices).

(a) $\mathbf{A} \succ \mathbf{0}$ (i.e., is positive definite or p.d.) if and only if $\mathbf{A}_{11}$ and $\mathbf{A}_{22} - \mathbf{A}_{21}\mathbf{A}_{11}^{-1}\mathbf{A}_{12} (= (\mathbf{A}/\mathbf{A}_{11}))$ are p.d.

(b) $\mathbf{A} \succ \mathbf{0}$ if and only if $\mathbf{A}_{22}$ and $\mathbf{A}_{11} - \mathbf{A}_{12}\mathbf{A}_{22}^{-1}\mathbf{A}_{21}$ are p.d.

(c) If $\mathbf{A} \succ \mathbf{0}$, then

$$\mathbf{A}_{22} \succeq \mathbf{A}_{22} - \mathbf{A}_{21}\mathbf{A}_{11}^{-1}\mathbf{A}_{12}.$$

(d) If $\mathbf{A} \succ \mathbf{0}$ and $\mathbf{A}^{11}$ is the leading principal submatrix of $\mathbf{A}^{-1}$ with the same size as $\mathbf{A}_{11}$, then

$$\mathbf{A}^{11} - \mathbf{A}_{11}^{-1} \succeq \mathbf{0}.$$

(e) (Fischer Inequality) If $\mathbf{A} \succ \mathbf{0}$, then

$$\det \mathbf{A} \leq \det \mathbf{A}_{11} \cdot \det \mathbf{A}_{22},$$

with equality if and only if both sides vanish or $\mathbf{A}_{12} = \mathbf{0}$.

(f) If $\mathbf{A} \succeq \mathbf{0}$ and the blocks $\mathbf{A}_{11}$, $\mathbf{A}_{12}$, and $\mathbf{A}_{22}$ are square matrices of the same size, then

$$|\det \mathbf{A}_{12}|^2 \leq \det \mathbf{A}_{11} \det \mathbf{A}_{22}.$$

(g) If $\mathbf{A} \succeq \mathbf{0}$, then $\mathcal{C}(\mathbf{A}_{12}) \subseteq \mathcal{C}(\mathbf{A}_{11})$ and $\mathcal{C}(\mathbf{A}_{21}) \subseteq \mathcal{C}(\mathbf{A}_{22})$.

The above results will also hold if $\mathbf{A}$ is Hermitian.

14.27. Let the real symmetric matrix $\mathbf{A}$ be partitioned as in (14.10) above, where $\mathbf{A}_{11} \succ \mathbf{0}$. Then for any square matrix $\mathbf{A}_{22} \succeq \mathbf{0}$,

$$\mathbf{A} \succeq \mathbf{0} \quad \text{if and only if} \quad \mathbf{A}_{22} \succeq \mathbf{A}_{21} \mathbf{A}_{11}^{-1} \mathbf{A}_{12}.$$

14.28. Let

$$\mathbf{A} = \left(\begin{array}{cc} \mathbf{A}_{11} & \mathbf{A}_{12} \\ \mathbf{A}_{12}' & \mathbf{0} \end{array} \right),$$

where $\mathbf{A}_{11}$ is a non-negative definite $m \times m$ matrix and $\mathbf{A}_{12}$ is $m \times n$. The symmetric $(m+n) \times (m+n)$ matrix $\mathbf{A}$ is sometimes referred to as a *borderd Gramian matrix*.

(a) $\mathbf{A}$ is nonsingular if and only if rank $\mathbf{A}_{12} = n$ and $\mathbf{A}_{11} + \mathbf{A}_{12}\mathbf{A}_{12}'$ is positive definite.

(b) If $\mathbf{A}$ is nonsingular, and setting $\mathbf{A}_{21} = \mathbf{A}_{12}'$, then

$$\mathbf{A}^{-1} = \left(\begin{array}{cc} \mathbf{B}_{11}^{-1} - \mathbf{B}_{11}^{-1}\mathbf{A}_{12}\mathbf{B}_{22}^{-1}\mathbf{A}_{21}\mathbf{B}_{11}^{-1} & \mathbf{B}_{11}^{-1}\mathbf{A}_{12}\mathbf{B}_{22}^{-1} \\ \mathbf{B}_{22}^{-1}\mathbf{A}_{21}\mathbf{B}_{11}^{-1} & \mathbf{I}_n - \mathbf{B}_{22}^{-1} \end{array} \right),$$

where $\mathbf{B}_{11} = \mathbf{A}_{11} + \mathbf{A}_{12}\mathbf{A}_{21}$ and $\mathbf{B}_{22} = \mathbf{A}_{21}\mathbf{B}_{11}^{-1}\mathbf{A}_{12}$.

(c) If $\mathbf{B}_{11}$ above is nonsingular, then

$$\det \mathbf{A} = (-1)^n \det \mathbf{B}_{11} \cdot \det \mathbf{B}_{22}.$$

(d) If $\mathbf{A}$ is nonsingular, then

$$\det \mathbf{A} = (-1)^n \det \mathbf{A}_{11} \cdot \det(\mathbf{A}_{21}\mathbf{A}_{11}^{-1}\mathbf{A}_{12}).$$

14.29. Let $\mathbf{A}$ be positive definite and let $\mathbf{B} = \left(\begin{array}{cc} \mathbf{A} & \mathbf{b} \\ \mathbf{b}' & c \end{array} \right)$, where $\mathbf{A}$, $\mathbf{b}$, and c are real.

(a) $\det \mathbf{B} = \det \mathbf{A}(c - \mathbf{b}'\mathbf{A}^{-1}\mathbf{b}) \leq c \det \mathbf{A}$, with equality if and only if $\mathbf{b} = \mathbf{0}$.

(b) $\mathbf{B}$ is positive definite if and only if $\det \mathbf{B} > 0$.

(c) If $c = \mathbf{b}'\mathbf{A}^{-1}\mathbf{b}$, then $\mathbf{B}$ is non-negative definite.

(d) $\mathbf{x}'\mathbf{A}\mathbf{x} - 2\mathbf{b}'\mathbf{x} \geq -\mathbf{b}'\mathbf{A}^{-1}\mathbf{b}$.

14.30. Let $\mathbf{A}$ and $\mathbf{B}$ be real $n \times n$ matrices, and let

$$\mathbf{A} = \left(\begin{array}{cc} \mathbf{A}_1 & \mathbf{a}_1 \\ \mathbf{a}_1' & a \end{array} \right) \quad \text{and} \quad \mathbf{B} = \left(\begin{array}{cc} \mathbf{B}_1 & \mathbf{b}_1 \\ \mathbf{b}_1' & b \end{array} \right),$$

where $\mathbf{A}_1$ and $\mathbf{B}_1$ are positive definite. Then

$$\mathbf{a}_1'\mathbf{A}_1^{-1}\mathbf{a}_1 + \mathbf{b}_1'\mathbf{B}_1^{-1}\mathbf{b}_1 - (\mathbf{a}_1 + \mathbf{b}_1)'(\mathbf{A}_1 + \mathbf{B}_1)^{-1}(\mathbf{a}_1 + \mathbf{b}_1)$$
$$= (\mathbf{A}_1^{-1}\mathbf{a}_1 - \mathbf{B}_1^{-1}\mathbf{b}_1)'(\mathbf{A}_1^{-1} + \mathbf{B}_1^{-1})^{-1}(\mathbf{A}_1^{-1}\mathbf{a}_1 - \mathbf{B}_1^{-1}\mathbf{b}_1).$$

Anderson [2003: 419] gives an application to testing that several multivariate normal populations are identical.

14.31. Let $\mathbf{A}$ and $\mathbf{B}$ be $n \times n$ positive definite matrices. There exists a unique matrix $\mathbf{C}$ such that

$$
\begin{aligned}
c_{ij} &= a_{ij}, \quad (i,j) \in \{1, 2. \ldots, t\} \\
c^{ij} &= b^{ij}, \quad (i,j) \notin \{1, 2, \ldots, t\}
\end{aligned}
$$

where $\mathbf{C}^{-1} = c^{ij}$ and $\mathbf{B}^{-1} = b^{ij}$. This result has an application to graphical models for determining patterns of independence.

Proofs. Section 14.4.

14.26a. Horn and Johnson [1985: 472, complex case] and Zhang [1999: 175, complex case].

14.26b. Same as (a) with the subscripts 1 and 2 interchanged.

14.26c. Horn and Johnson [1985: 474, in proof of theorem 7.7.8] and Zhang [1999: 175, complex case].

14.26d. Follows from (14.11a); see also Zhang [1999: 175, complex case].

14.26e. Horn and Johnson [1985: 478] and Zhang [1999: 175, complex case].

14.26f. Abadir and Magnus [2005: 228, 341].

14.26g. Sengupta and Jammalamadaka [2003: 45]; see also (14.8c).

14.27. Zhang [1999: 178, complex case].

14.28. Abadir and Magnus [2005: 230–231].

14.29. Magnus and Neudecker [1988: 23–24].

14.30. Anderson [2003: 419].

14.31. Anderson [2003: 614, 616].

14.5   EIGENVALUES

In this section we assume that the $n \times n$ matrix $\mathbf{A}$ is partitioned as in (14.10) with $\mathbf{A}_{ii}$ being $n_i \times n_i$, for $i = 1, 2$ ($n_1 + n_2 = n$). We also continue with the notation $\lambda_1(\mathbf{A}) \geq \cdots \geq \lambda_n(\mathbf{A})$ for ordering the eigenvalues when they are real, which is the case for a symmetric matrix.

14.32. Suppose $\mathbf{A}$ is non-negative definite. If h and i are integers between 1 and n inclusive, then:

(a) $\lambda_{h+i-1}(\mathbf{A}) \leq \lambda_h(\mathbf{A}_{11}) + \lambda_i(\mathbf{A}_{22})$, if $h + i \leq n + 1$,

(b) $\lambda_{h+i-n}(\mathbf{A}) \geq \lambda_h(\mathbf{A}_{11}) + \lambda_i(\mathbf{A}_{22})$, if $h + i \geq n + 1$,

where $\lambda_h(\mathbf{A}_{11}) = 0$ if $h > n_1$ and $\lambda_i(\mathbf{A}_{22}) = 0$ if $i > n_2$.

14.33. Suppose $\mathbf{A}$ is non-negative definite and $i_1, i_2, \ldots, i_k$ are distinct integers beween 1 and n, inclusive. Then for $k = 1, 2, \ldots, n$,

$$\sum_{j=1}^{k}[\lambda_{i_j}(\mathbf{A}_{11}) + \lambda_{n-k+j}(\mathbf{A}_{22})] \;\leq\; \sum_{j=1}^{k}\lambda_{i_j}(\mathbf{A})$$

$$\leq\; \sum_{j=1}^{k}[\lambda_{i_j}(\mathbf{A}_{11}) + \lambda_j(\mathbf{A}_{22})],$$

where $\lambda_j(\mathbf{A}_{11}) = 0$ if $j > n_1$ and $\lambda_j(\mathbf{A}_{22}) = 0$ if $j > n_2$.

14.34. Suppose $\mathbf{A}$ is symmetric and $\lambda_{n_1}(\mathbf{A}_{11}) > \lambda_1(\mathbf{A}_{22})$.

(a) For $j = 1, 2, \ldots, n_1$,

$$0 \leq \lambda_j(\mathbf{A}) - \lambda_j(\mathbf{A}_{11}) \leq \frac{\lambda_1(\mathbf{A}_{12}\mathbf{A}'_{12})}{\lambda_j(\mathbf{A}_{11}) - \lambda_1(\mathbf{A}_{22})},$$

and for $j = 1, 2, \ldots, n_2$,

$$0 \;\leq\; \lambda_{n_2-j+1}(\mathbf{A}_{22}) - \lambda_{n-j+1}(\mathbf{A}) \leq \frac{\lambda_1(\mathbf{A}_{12}\mathbf{A}'_{12})}{\lambda_{n_1}(\mathbf{A}_{11}) - \lambda_{n_2-j+1}(\mathbf{A}_{22})}.$$

Tighter bounds are given by Dümbgen [1995]. The above bounds are useful in obtaining the asymptotic distribution of the eigenvalues of a random symmetric matrix (Eaton and Tyler [1991]).

(b) For $k = 1, 2, \ldots, n_1$,

$$0 \leq \sum_{j=1}^{k}[\lambda_j(\mathbf{A}) - \lambda_j(\mathbf{A}_{11})] \leq \frac{\sum_{j=1}^{k}\lambda_j(\mathbf{A}_{12}\mathbf{A}'_{12})}{\lambda_k(\mathbf{A}_{11}) - \lambda_1(\mathbf{A}_{22})}.$$

14.35. Suppose $\mathbf{A}$ is positive definite, and let $\mathbf{B}_1 = \mathbf{A}_{11} - \mathbf{A}_{12}\mathbf{A}_{22}^{-1}\mathbf{A}_{21}$, $\mathbf{B}_2 = \mathbf{A}_{22} - \mathbf{A}_{21}\mathbf{A}_{11}^{-1}\mathbf{A}_{12}$, and $\mathbf{C} = -\mathbf{B}_1^{-1}\mathbf{A}_{12}\mathbf{A}_{22}^{-1}$, where $\mathbf{A}_{12} = \mathbf{A}'_{21}$. Then if $\lambda_1(\mathbf{B}_1) < \lambda_{n_2}(\mathbf{B}_2)$,

$$0 \leq \sum_{j=1}^{k}[\lambda_{n_1-j+1}(\mathbf{B}_1) - \lambda_{n-j+1}(\mathbf{A})] \leq \frac{\lambda_{n_1-k+1}^2(\mathbf{B}_1)}{\lambda_{n_1-k+1}^{-1}(\mathbf{B}_1) - \lambda_{n_2}^{-1}(\mathbf{B}_2)} \sum_{j=1}^{k}\lambda_j(\mathbf{C}\mathbf{C}'),$$

for $k = 1, 2, \ldots, n_1$.

Proofs. Section 14.5.

14.32–14.34. Schott [2005: 271–273].

14.35. Schott [2005: 275–276].

14.6 GENERALIZED INVERSES

14.6.1 Weak Inverses

14.36. Let $\mathbf{A} = (\mathbf{A}_{11}, \mathbf{A}_{12})$, where $\mathbf{A}_{11}$ is nonsingular. Then

$$\mathbf{A}^- = \begin{pmatrix} \mathbf{A}_{11}^{-1} - \mathbf{A}_{11}^{-1}\mathbf{A}_{12}\mathbf{Y} \\ \mathbf{Y} \end{pmatrix}$$

is a weak inverse of $\mathbf{A}$ for arbitrary $\mathbf{Y}$.

14.37. Let $\mathbf{A} = \begin{pmatrix} \mathbf{A}_{11} \\ \mathbf{A}_{21} \end{pmatrix}$, where $\mathbf{A}_{11}$ is nonsingular. Then

$$\mathbf{A}^- = (\mathbf{A}_{11}^{-1} - \mathbf{X}\mathbf{A}_{21}\mathbf{A}_{11}^{-1}, \mathbf{X})$$

is a weak inverse of $\mathbf{A}$ for arbitrary $\mathbf{X}$.

14.38. Let $\mathbf{A}$ be $m \times n$.

(a) $\begin{pmatrix} \mathbf{A}^- \\ \mathbf{B}^- \end{pmatrix}$ is a weak inverse of $(\mathbf{A}, \mathbf{B})$ if and only if $\mathbf{A}\mathbf{A}^-\mathbf{B} = \mathbf{0}$ and $\mathbf{B}\mathbf{B}^-\mathbf{A} = \mathbf{0}$.

(b) $(\mathbf{A}^-, \mathbf{C}^-)$ is a weak inverse of $\begin{pmatrix} \mathbf{A} \\ \mathbf{C} \end{pmatrix}$ if and only if $\mathbf{C}\mathbf{A}^-\mathbf{A} = \mathbf{0}$ and $\mathbf{A}\mathbf{C}^-\mathbf{C} = \mathbf{0}$.

For conditions on the ranks for the above weak inverses to hold, see Tian [2005b].

14.39. Let

$$\mathbf{A} = \begin{pmatrix} \mathbf{A}_{1(p \times n)} \\ \mathbf{A}_{2(q \times n)} \end{pmatrix} \quad \text{and} \quad \mathbf{G} = (\mathbf{G}_{1(n \times p)}, \mathbf{G}_{2(n \times q)}),$$

with $p + q = m$. Then $\mathcal{C}(\mathbf{A}_1') \cap \mathcal{C}(\mathbf{A}_2') = \mathbf{0}$ and $\mathbf{G}$ is a weak inverse of $\mathbf{A}$ if and only if

$$\mathbf{A}_1\mathbf{G}_1\mathbf{A}_1 = \mathbf{A}_1, \ \mathbf{A}_2\mathbf{G}_1\mathbf{A}_1 = \mathbf{0}, \ \mathbf{A}_2\mathbf{G}_2\mathbf{A}_2 = \mathbf{A}_2, \text{ and } \mathbf{A}_1\mathbf{G}_2\mathbf{A}_2 = \mathbf{0}.$$

If rank $\mathbf{A}_1 = p$, the first two equations above become $\mathbf{A}_1\mathbf{G}_1 = \mathbf{I}_p$ and $\mathbf{A}_2\mathbf{G}_1 = \mathbf{0}$.

14.40. Let

$$\mathbf{A} = (\mathbf{A}_{1(m \times p)}, \mathbf{A}_{2(m \times q)}) \quad \text{and} \quad \mathbf{G} = \begin{pmatrix} \mathbf{G}_{1(p \times m)} \\ \mathbf{G}_{2(q \times m)} \end{pmatrix},$$

with $p + q = n$. Then $\mathcal{C}(\mathbf{A}_1) \cap \mathcal{C}(\mathbf{A}_2) = \mathbf{0}$ and $\mathbf{G}$ is a weak inverse of $\mathbf{A}$ if and only if

$$\mathbf{A}_1\mathbf{G}_1\mathbf{A}_1 = \mathbf{A}_1, \ \mathbf{A}_1\mathbf{G}_1\mathbf{A}_2 = \mathbf{0}, \ \mathbf{A}_2\mathbf{G}_2\mathbf{A}_2 = \mathbf{A}_2, \text{ and } \mathbf{A}_2\mathbf{G}_2\mathbf{A}_1 = \mathbf{0}.$$

If rank $\mathbf{A}_1 = p$, the first two equations above become $\mathbf{G}_1\mathbf{A}_1 = \mathbf{I}_p$ and $\mathbf{G}_1\mathbf{A}_2 = \mathbf{0}$.

14.41. Let $\mathbf{A}$ be partitioned in the form of (14.10).

(a) If $\mathcal{C}(\mathbf{A}_{12}) \subseteq \mathcal{C}(\mathbf{A}_{11})$, $\mathcal{C}(\mathbf{A}_{21}') \subseteq \mathcal{C}(\mathbf{A}_{11}')$, $\mathbf{A}_{11}^-$ is a particular weak inverse of $\mathbf{A}_{11}$, and $\mathbf{A}_{22 \cdot 1} = \mathbf{A}_{22} - \mathbf{A}_{21}\mathbf{A}_{11}^-\mathbf{A}_{12}$, we have

$$\mathbf{A}^- = \begin{pmatrix} \mathbf{A}_{11}^- + \mathbf{A}_{11}^-\mathbf{A}_{12}\mathbf{A}_{22 \cdot 1}^-\mathbf{A}_{21}\mathbf{A}_{11}^- & -\mathbf{A}_{11}^-\mathbf{A}_{12}\mathbf{A}_{22 \cdot 1}^- \\ -\mathbf{A}_{22 \cdot 1}^-\mathbf{A}_{21}\mathbf{A}_{11}^- & \mathbf{A}_{22 \cdot 1}^- \end{pmatrix}.$$

(b) If $\mathcal{C}(\mathbf{A}_{21}) \subseteq \mathcal{C}(\mathbf{A}_{22})$, $\mathcal{C}(\mathbf{A}'_{12}) \subseteq \mathcal{C}(\mathbf{A}'_{22})$, $\mathbf{A}_{22}^-$ is a particular weak inverse of $\mathbf{A}_{22}$, and $\mathbf{A}_{11\cdot 2} = \mathbf{A}_{11} - \mathbf{A}_{12}\mathbf{A}_{22}^-\mathbf{A}_{21}$, we have

$$\mathbf{A}^- = \begin{pmatrix} \mathbf{A}_{11\cdot 2}^- & -\mathbf{A}_{11\cdot 2}^-\mathbf{A}_{12}\mathbf{A}_{22}^- \\ -\mathbf{A}_{22}^-\mathbf{A}_{21}\mathbf{A}_{11\cdot 2}^- & \mathbf{A}_{22}^- + \mathbf{A}_{22}^-\mathbf{A}_{21}\mathbf{A}_{11\cdot 2}^-\mathbf{A}_{12}\mathbf{A}_{22}^- \end{pmatrix}.$$

Necessary and sufficient conditions are given in (14.44) below using a different notation. Some rank conditions for the above to hold are given by Tian and Takane [2005].

14.42. Let $\mathbf{A}$ be an $n \times n$ non-negative definite matrix partitioned as in (14.10), where $\mathbf{A}_{11}$ is $p \times p$. Suppose that the $n \times n$ matrix $\mathbf{G}$ is a weak inverse of $\mathbf{A}$ and is partitioned in exactly the same way as $\mathbf{A}$. If each of the first p rows of $\mathbf{A}$ is nonzero and is not a linear combination of the remaining rows of $\mathbf{A}$, then

$$\mathbf{G}_{11} = (\mathbf{A}_{11} - \mathbf{A}_{12}\mathbf{A}_{22}^-\mathbf{A}_{21})^{-1}$$

for any weak inverse $\mathbf{A}_{22}^-$ of $\mathbf{A}_{22}$. Also $\mathbf{G}_{11}$ is unique.

14.43. Let

$$\mathbf{A} = \begin{pmatrix} \mathbf{V} & \mathbf{X} \\ \mathbf{X}' & \mathbf{0} \end{pmatrix} \quad \text{and} \quad \mathbf{G} = \begin{pmatrix} \mathbf{G}_{11} & \mathbf{G}_{12} \\ \mathbf{G}_{21} & \mathbf{G}_{22} \end{pmatrix},$$

where $\mathbf{V}$ is an $n \times n$ non-negative definite matrix, $\mathbf{X}$ is $n \times p$, and $\mathbf{G}_{11}$ is $n \times n$.

(a) If $\mathbf{G}$ is a weak inverse of $\mathbf{A}$, we have the following.

(i) $\mathbf{G}_{12}$ is weak inverse of of $\mathbf{X}'$.

(ii) $\mathbf{G}_{21}$ is weak inverse of of $\mathbf{X}$.

(iii) $\mathbf{VG}_{12}\mathbf{X}' = \mathbf{XG}_{21}\mathbf{V} = -\mathbf{XG}_{22}\mathbf{X}'$.

(iv) $\mathbf{VG}_{11}\mathbf{X} = \mathbf{0}$, $\mathbf{X}'\mathbf{G}_{11}\mathbf{V} = \mathbf{0}$, and $\mathbf{X}'\mathbf{G}_{11}\mathbf{X} = \mathbf{0}$.

(v) $\mathbf{V} = \mathbf{VG}_{11}\mathbf{V} - \mathbf{XG}_{22}\mathbf{X}'$.

(vi) $\mathbf{VG}_{11}\mathbf{V}$, $\mathbf{VG}_{12}\mathbf{X}'$, $\mathbf{XG}_{21}\mathbf{V}$, and $\mathbf{XG}_{22}\mathbf{X}'$ are symmetric and invariant to the choice of the weak inverse $\mathbf{G}$.

(b) If $\mathbf{U}$ is any $p \times p$ matrix such that $\mathcal{C}(\mathbf{X}) \subseteq \mathcal{C}(\mathbf{V} + \mathbf{XUX}')$, and $\mathbf{W}$ is any weak inverse of $\mathbf{V} + \mathbf{XUX}'$, then

$$\begin{pmatrix} \mathbf{W} - \mathbf{WX}(\mathbf{X}'\mathbf{WX})^-\mathbf{X}'\mathbf{W} & \mathbf{WX}(\mathbf{X}'\mathbf{WX})^- \\ (\mathbf{X}'\mathbf{WX})^-\mathbf{X}'\mathbf{W} & -(\mathbf{X}'\mathbf{WX})^- + \mathbf{U} \end{pmatrix}$$

is a weak inverse of $\mathbf{A}$.

14.44. Let

$$\mathbf{A} = \begin{pmatrix} \mathbf{E} & \mathbf{F} \\ \mathbf{G} & \mathbf{H} \end{pmatrix}.$$

(a) Let $\tilde{\mathbf{E}}$ be a particular weak inverse of $\mathbf{E}$ and $\mathbf{S} = \mathbf{H} - \mathbf{G}\tilde{\mathbf{E}}\mathbf{F}$, the generalized Schur complement. Then

$$\mathbf{B} = \begin{pmatrix} \tilde{\mathbf{E}} + \tilde{\mathbf{E}}\mathbf{FS}^-\mathbf{G}\tilde{\mathbf{E}} & -\tilde{\mathbf{E}}\mathbf{FS}^- \\ -\mathbf{S}^-\mathbf{G}\tilde{\mathbf{E}} & \mathbf{S}^- \end{pmatrix},$$

is a weak inverse of $\mathbf{A}$ for a particular weak inverse $\mathbf{S}^-$ if and only if rank is additive on the Schur complement (i.e., rank $\mathbf{A}$ = rank $\mathbf{E}$ + rank $\mathbf{S}$), and then $\mathbf{B}$ is a weak inverse of $\mathbf{A}$ for any weak inverse $\mathbf{S}^-$. Sufficient conditions are $\mathcal{C}(\mathbf{F}) \subseteq \mathcal{C}(\mathbf{E})$ and $\mathcal{C}(\mathbf{G}') \subseteq \mathcal{C}(\mathbf{E}')$, as in (14.41a).

(b) Let $\tilde{\mathbf{H}}$ be a particular weak inverse of $\mathbf{H}$ and $\mathbf{T} = \mathbf{E} - \mathbf{F}\tilde{\mathbf{H}}\mathbf{G}$, the generalized Schur complement. Then

$$\mathbf{C} = \left(\begin{array}{cc} \mathbf{T}^- & -\mathbf{T}^-\mathbf{F}\tilde{\mathbf{H}} \\ -\tilde{\mathbf{H}}\mathbf{G}\mathbf{T}^- & \tilde{\mathbf{H}} + \tilde{\mathbf{H}}\mathbf{G}\mathbf{T}^-\mathbf{F}\tilde{\mathbf{H}} \end{array} \right),$$

is a weak inverse of $\mathbf{A}$ for a particular weak inverse $\mathbf{T}^-$ if and only if rank is additive on the Schur complement (i.e., rank $\mathbf{A}$ = rank $\mathbf{H}$ + rank $\mathbf{T}$), and then $\mathbf{C}$ is a weak inverse of $\mathbf{A}$ for any weak inverse $\mathbf{T}^-$. Sufficient conditions are $\mathcal{C}(\mathbf{G}) \subseteq \mathcal{C}(\mathbf{H})$ and $\mathcal{C}(\mathbf{F}') \subseteq \mathcal{C}(\mathbf{H}')$, as in (14.41b).

We can obtain (b) from (a) by simply interchanging $\mathbf{E}$ and $\mathbf{H}$, $\mathbf{F}$ and $\mathbf{G}$, and $\mathbf{S}$ and $\mathbf{T}$.

Proofs. Section 14.6.1.

14.36–14.37. Harville [1997: 111].

14.38. Harville [1997: 119].

14.39. Rao and Rao [1998: 270, 272].

14.40. Rao and Rao [1998: 271, 273].

14.41. Schott [2005: 267–268].

14.42. Rao and Rao [1998: 275].

14.43. Harville [1997: 473–476].

14.44. Harville [2001: 41, exercise 8] and Marsaglia and Styan [1974b: 438–439].

14.6.2 Moore–Penrose Inverses

We consider just a few special cases below. For further results relating to partitioned matrices see Baksalary and Styan [2002] and Groß [2000]. The Moore–Penrose inverse of $(\mathbf{A}, \mathbf{B})$ is considered in detail by Campbell and Meyer [1979: 58–59] and Schott [2005: 192-195]. A number of general rank conditions for Moore–Penrose inverses to exist are given by Tian [2004].

14.45. Let $\mathbf{A}$ and $\mathbf{B}$ be defined as in (14.44) above.

(a) $\mathbf{B} = \mathbf{A}^+$ if and only if $\tilde{\mathbf{E}} = \mathbf{E}^+$, $\mathbf{S}^- = \mathbf{S}^+$,

$$\mathrm{rank}\left(\begin{array}{c} \mathbf{E} \\ \mathbf{G} \end{array} \right) = \mathrm{rank}(\mathbf{E}, \mathbf{F}) = \mathrm{rank}\,\mathbf{E} \quad \text{and} \quad \mathrm{rank}\left(\begin{array}{c} \mathbf{F} \\ \mathbf{H} \end{array} \right) = \mathrm{rank}(\mathbf{G}, \mathbf{H}) = \mathrm{rank}\,\mathbf{S}.$$

(b) Since $\mathbf{A}^+$ is unique, we get the same result if we do the interchanges described at the end of (14.44). This leads to the following result.

If $\mathbf{S} = \mathbf{H} - \mathbf{G}\mathbf{E}^+\mathbf{F}$ and $\mathbf{T} = \mathbf{E} - \mathbf{F}\mathbf{H}^+\mathbf{G}$, then

$$\mathbf{A}^+ = \left(\begin{array}{cc} \mathbf{T}^+ & -\mathbf{E}^+\mathbf{F}\mathbf{S}^+ \\ -\mathbf{H}^+\mathbf{G}\mathbf{T}^+ & \mathbf{S}^+ \end{array} \right),$$

if and only if

$$\mathrm{rank}\left(\begin{array}{c} \mathbf{E} \\ \mathbf{G} \end{array} \right) = \mathrm{rank}(\mathbf{E}, \mathbf{F}) = \mathrm{rank}\,\mathbf{E} = \mathrm{rank}\,\mathbf{T}$$

and

$$\mathrm{rank}\left(\begin{array}{c} \mathbf{F} \\ \mathbf{H} \end{array} \right) = \mathrm{rank}(\mathbf{G}, \mathbf{H}) = \mathrm{rank}\,\mathbf{H} = \mathrm{rank}\,\mathbf{S}.$$

14.46. If $\mathbf{A} = \left(\begin{smallmatrix} \mathbf{B} \\ \mathbf{C} \end{smallmatrix} \right)$ and $\mathbf{B}\mathbf{C}' = \mathbf{0}$, then

(a) $\mathbf{A}^+ = (\mathbf{B}^+, \mathbf{C}^+)$

(b) $\mathbf{A}^+\mathbf{A} = \mathbf{B}^+\mathbf{B} + \mathbf{C}^+\mathbf{C}$.

(c)

$$\mathbf{A}\mathbf{A}^+ = \left(\begin{array}{cc} \mathbf{B}\mathbf{B}^+ & \mathbf{0} \\ \mathbf{0} & \mathbf{C}\mathbf{C}^+ \end{array} \right).$$

14.47. Suppose $\mathbf{A}$ is an $m \times n$ matrx of rank r, where $r < \min\{m, n\}$, and $\mathbf{A}$ is partitioned as

$$\mathbf{A} = \left(\begin{array}{cc} \mathbf{A}_{11} & \mathbf{A}_{12} \\ \mathbf{A}_{21} & \mathbf{A}_{22} \end{array} \right),$$

where $\mathbf{A}_{11}$ is $r \times r$ of rank r. Then

$$\mathbf{A}^+ = \left(\begin{array}{cc} \mathbf{A}_{11}'\mathbf{B}\mathbf{A}_{11}' & \mathbf{A}_{11}'\mathbf{B}\mathbf{A}_{21}' \\ \mathbf{A}_{12}'\mathbf{B}\mathbf{A}_{11}' & \mathbf{A}_{12}'\mathbf{B}\mathbf{A}_{21}' \end{array} \right),$$

where

$$\mathbf{B} = (\mathbf{A}_{11}\mathbf{A}_{11}' + \mathbf{A}_{12}\mathbf{A}_{12}')^{-1}\mathbf{A}_{11}(\mathbf{A}_{11}'\mathbf{A}_{11} + \mathbf{A}_{21}'\mathbf{A}_{21})^{-1}.$$

Proofs. Section 14.6.2.

14.45. Ouellettte [1981: 233–234].

14.46a. Quoted by Dhrymes [2000: 104].

14.46b–c. Quoted by Graybill [1983: 115; his $\mathbf{A}^-$ is our $\mathbf{A}^+$].

14.47. Graybill [1983: 127].

14.7 MISCELLANEOUS PARTITIONS

We close this chapter with a few partitions that may provide some ideas in algebraic manipulations.

14.48. $\mathbf{A} + \mathbf{B} = (\mathbf{A}, \mathbf{B}) \begin{pmatrix} \mathbf{I} \\ \mathbf{I} \end{pmatrix} = (\mathbf{I}, \mathbf{I}) \begin{pmatrix} \mathbf{A} \\ \mathbf{B} \end{pmatrix}.$

14.49. $\begin{pmatrix} \mathbf{I} & -\mathbf{A} \\ \mathbf{0} & \mathbf{I} \end{pmatrix} \begin{pmatrix} \mathbf{0} & \mathbf{AB} \\ \mathbf{BC} & \mathbf{B} \end{pmatrix} \begin{pmatrix} \mathbf{I} & \mathbf{0} \\ -\mathbf{C} & \mathbf{I} \end{pmatrix} = \begin{pmatrix} -\mathbf{ABC} & \mathbf{0} \\ \mathbf{0} & \mathbf{B} \end{pmatrix}.$

We can set $\mathbf{B} = \mathbf{I}$.

14.50. $\begin{pmatrix} \mathbf{I} & \mathbf{A} \\ \mathbf{0} & \mathbf{I} \end{pmatrix} \begin{pmatrix} \mathbf{A} & \mathbf{0} \\ -\mathbf{I} & \mathbf{B} \end{pmatrix} = \begin{pmatrix} \mathbf{0} & \mathbf{AB} \\ -\mathbf{I} & \mathbf{B} \end{pmatrix}.$

CHAPTER 15

PATTERNED MATRICES

15.1 INVERSES

Matrices that have a particular pattern occur frequently in statistics. Such matrices are typically used as intermediary steps in proofs and in perturbation techniques, when one is interested in the effect of making a small structural change to a matrix. Patterned matrices also occur in experimental designs and in certain variance matrices of random vectors. A related chapter is Chapter 14.

15.1. (Some Identities) There are a number of identities that are useful and which can be used to prove the results in this section. It is assumed that all inverses exist.

(a)　(i) $\mathbf{VA}^{-1}(\mathbf{A} - \mathbf{UD}^{-1}\mathbf{V}) = (\mathbf{D} - \mathbf{VA}^{-1}\mathbf{U})\mathbf{D}^{-1}\mathbf{V}$,
　　　or taking the inverse of both sides,

　　(ii) $\mathbf{D}^{-1}\mathbf{V}(\mathbf{A} - \mathbf{UD}^{-1}\mathbf{V})^{-1} = (\mathbf{D} - \mathbf{VA}^{-1}\mathbf{U})\mathbf{VA}^{-1}$.

(b) Setting $\mathbf{A} = \mathbf{I}$, $\mathbf{D} = -\mathbf{I}$, and interchanging $\mathbf{U}$ and $\mathbf{V}$ in (a)(ii), we have that

$$\mathbf{U}(\mathbf{I} + \mathbf{VU})^{-1} = (\mathbf{I} + \mathbf{UV})^{-1}\mathbf{U}.$$

(c) If $\mathbf{I} + \mathbf{U}$ is nonsingular,

$$(\mathbf{I} + \mathbf{U})^{-1} = \mathbf{I} - (\mathbf{I} + \mathbf{U})^{-1}\mathbf{U} = \mathbf{I} - \mathbf{U}(\mathbf{I} + \mathbf{U})^{-1}.$$

(d) $\mathbf{U}'\mathbf{A}^{-1}\mathbf{U}(\mathbf{I} + \mathbf{U}'\mathbf{A}^{-1}\mathbf{U})^{-1} = \mathbf{I} - (\mathbf{I} + \mathbf{U}'\mathbf{A}^{-1}\mathbf{U})^{-1}$.

A Matrix Handbook for Statisticians. By George A. F. Seber
Copyright © 2008 John Wiley & Sons, Inc.

(e) If $\mathbf{A}$ and $\mathbf{B}$ are $n \times n$ complex matrices, then

$$\mathbf{I}_n + \mathbf{A}\mathbf{A}^* = (\mathbf{A} + \mathbf{B})(\mathbf{I}_n + \mathbf{B}^*\mathbf{B})^{-1}(\mathbf{A} + \mathbf{B})^*$$
$$+ (\mathbf{I}_n - \mathbf{A}\mathbf{B}^*)(\mathbf{I}_n + \mathbf{B}\mathbf{B}^*)^{-1}(\mathbf{I}_n - \mathbf{A}\mathbf{B}^*)^*.$$

Note that the right-hand side does not depend on of $\mathbf{B}$.

(f) $(\alpha\mathbf{I}_n - \mathbf{A})^{-1} - (\beta\mathbf{I}_n - \mathbf{A})^{-1} = (\beta - \alpha)(\beta\mathbf{I}_n - \mathbf{A})^{-1}(\alpha\mathbf{I}_n - \mathbf{A})^{-1}.$

15.2. If $\mathbf{A}$ is nonsingular and the other matrices are conformable square or rectangular matrices (e.g., $\mathbf{A}$ is $n \times n$, $\mathbf{U}$ is $n \times p$, $\mathbf{B}$ is $p \times q$, and $\mathbf{V}$ is $q \times n$), then we have the following inverses from Henderson and Searle [1981b: 57–58].

(a)

$$
\begin{aligned}
(\mathbf{A} + \mathbf{U}\mathbf{B}\mathbf{V})^{-1} &= \mathbf{A}^{-1} - (\mathbf{I} + \mathbf{A}^{-1}\mathbf{U}\mathbf{B}\mathbf{V})^{-1}\mathbf{A}^{-1}\mathbf{U}\mathbf{B}\mathbf{V}\mathbf{A}^{-1} \\
&= \mathbf{A}^{-1} - \mathbf{A}^{-1}(\mathbf{I} + \mathbf{U}\mathbf{B}\mathbf{V}\mathbf{A}^{-1})^{-1}\mathbf{U}\mathbf{B}\mathbf{V}\mathbf{A}^{-1} \\
&= \mathbf{A}^{-1} - \mathbf{A}^{-1}\mathbf{U}(\mathbf{I} + \mathbf{B}\mathbf{V}\mathbf{A}^{-1}\mathbf{U})^{-1}\mathbf{B}\mathbf{V}\mathbf{A}^{-1} \\
&= \mathbf{A}^{-1} - \mathbf{A}^{-1}\mathbf{U}\mathbf{B}(\mathbf{I} + \mathbf{V}\mathbf{A}^{-1}\mathbf{U}\mathbf{B})^{-1}\mathbf{V}\mathbf{A}^{-1} \\
&= \mathbf{A}^{-1} - \mathbf{A}^{-1}\mathbf{U}\mathbf{B}\mathbf{V}(\mathbf{I} + \mathbf{A}^{-1}\mathbf{U}\mathbf{B}\mathbf{V})^{-1}\mathbf{A}^{-1} \\
&= \mathbf{A}^{-1} - \mathbf{A}^{-1}\mathbf{U}\mathbf{B}\mathbf{V}\mathbf{A}^{-1}(\mathbf{I} + \mathbf{U}\mathbf{B}\mathbf{V}\mathbf{A}^{-1})^{-1}
\end{aligned}
$$

All results follow from the first by repeatedly applying (15.1b).

If the left-hand side exists, then the inverses on the right-hand side exist. This is because each inverse on the right-hand side is the inverse of the sum of $\mathbf{I}$ and a cyclic permutation of $\mathbf{A}^{-1}\mathbf{U}\mathbf{B}\mathbf{V}$, and it exists because its determinant is nonzero. For example,

$$\det(\mathbf{I} + \mathbf{A}^{-1}\mathbf{U}\mathbf{B}\mathbf{V}) = \det(\mathbf{A}^{-1})\det(\mathbf{A} + \mathbf{U}\mathbf{B}\mathbf{V}) \neq 0.$$

We can then obtain the other determinants using

$$\det(\mathbf{I} + \mathbf{C}\mathbf{D}) = \det(\mathbf{I} + \mathbf{D}\mathbf{C})$$

from (4.33) and (15.10b)

(b) A number of special cases are readily available by setting $\mathbf{B} = \mathbf{I}$ and/or $\mathbf{V} = \mathbf{I}$, and replacing matrices by vectors. For example, Steerneman and van Perlo-ten Kleij [2005] consider a matrix of the form $\mathbf{V} = \mathbf{A} - \mathbf{X}\mathbf{Y}^*$, where $\mathbf{V}$ is a nonsingular complex matrix and $\mathbf{X}$ and $\mathbf{Y}$ are $n \times p$ complex matrices. They consider various special cases, and give eigenvalues and eigenvectors of the real matrix $\mathbf{D} - \mathbf{x}\mathbf{y}'$, where $\mathbf{D}$ is diagonal matrix (cf. 15.6).

We can also set $\mathbf{V} = \mathbf{U}'$ in (a), in which case we get $(\mathbf{A} + \mathbf{U}\mathbf{B}\mathbf{U}')^{-1}$ that arises, for example, as a dispersion matrix for many mixed models in the analysis of variance. The following are special cases of (a).

(i) (Sherman–Morrison)

$$(\mathbf{A} + b\mathbf{u}\mathbf{v}')^{-1} = \mathbf{A}^{-1} - b\mathbf{A}^{-1}\mathbf{u}\mathbf{v}'\mathbf{A}^{-1}/(1 + b\mathbf{v}'\mathbf{A}^{-1}\mathbf{u}).$$

This is used as an "updating" formula discussed further in (15.11). The situation when $\mathbf{A}$ or the modified matrix is singular is investigated by Baksalary and Baksalary [2004c] (see also Section 15.5.2).

(ii) $\mathbf{x}'(\mathbf{A} + \mathbf{x}\mathbf{x}')^{-1}\mathbf{x} = \dfrac{\mathbf{x}'\mathbf{A}^{-1}\mathbf{x}}{1 + \mathbf{x}'\mathbf{A}^{-1}\mathbf{x}}.$

(iii) $(\mathbf{A} + \mathbf{U}\mathbf{B}\mathbf{U}')^{-1} = \mathbf{A}^{-1} - \mathbf{A}^{-1}\mathbf{U}\mathbf{B}(\mathbf{I} + \mathbf{U}'\mathbf{A}^{-1}\mathbf{U}\mathbf{B})^{-1}\mathbf{U}'\mathbf{A}^{-1}.$

For a good historical discussion and further results see Henderson and Searle [1981b]. They also give some statistical applications of these identities such as inverting the variance matrix for a multinomial vector, inverting a matrix with the pattern of an intraclass correlation matrix, and obtaining the generalized least squares estimates for a variance component model.

15.3. Another set of results can be derived by assuming that $\mathbf{B}$ is also nonsingular. From the fourth equation of (15.2a) we have:

(a)   (i) $(\mathbf{A} + \mathbf{U}\mathbf{B}\mathbf{V})^{-1} = \mathbf{A}^{-1} - \mathbf{A}^{-1}\mathbf{U}\mathbf{B}(\mathbf{B} + \mathbf{B}\mathbf{V}\mathbf{A}^{-1}\mathbf{U}\mathbf{B})^{-1}\mathbf{B}\mathbf{V}\mathbf{A}^{-1}.$

    (ii) $(\mathbf{A} + \mathbf{U}\mathbf{B}\mathbf{V})^{-1} = \mathbf{A}^{-1} - \mathbf{A}^{-1}\mathbf{U}(\mathbf{B}^{-1} + \mathbf{V}\mathbf{A}^{-1}\mathbf{U})^{-1}\mathbf{V}\mathbf{A}^{-1}.$

    (iii) Setting $\mathbf{B} = \mathbf{I}$, we have the so-called *Sherman–Morrison–Woodbury formula*

$$(\mathbf{A} + \mathbf{U}\mathbf{V})^{-1} = \mathbf{A}^{-1} - \mathbf{A}^{-1}\mathbf{U}(\mathbf{I} + \mathbf{V}\mathbf{A}^{-1}\mathbf{U})^{-1}\mathbf{V}\mathbf{A}^{-1}.$$

This result also holds with $\mathbf{A}$ Hermitian.

    (iv) Setting $\mathbf{A}$ and $\mathbf{B}$ equal to identity matrices in (i) or (ii), we have

$$(\mathbf{I} + \mathbf{U}\mathbf{V})^{-1} = \mathbf{I} - \mathbf{U}(\mathbf{I} + \mathbf{V}\mathbf{U})^{-1}\mathbf{V}.$$

(b) Setting $\mathbf{V} = \mathbf{U}'$ in (a)(ii) gives us

  (i) $(\mathbf{A} + \mathbf{U}\mathbf{B}\mathbf{U}')^{-1} = \mathbf{A}^{-1} - \mathbf{A}^{-1}\mathbf{U}(\mathbf{B}^{-1} + \mathbf{U}'\mathbf{A}^{-1}\mathbf{U})^{-1}\mathbf{U}'\mathbf{A}^{-1}.$

  (ii) If $\mathbf{C} = (\mathbf{U}'\mathbf{A}^{-1}\mathbf{U})^{-1}$ exists, then using (15.4b) with $\mathbf{B}$ and $\mathbf{C}$ instead of $\mathbf{A}$ and $\mathbf{B}$, we have

$$
\begin{aligned}
(\mathbf{A} + \mathbf{U}\mathbf{B}\mathbf{U}')^{-1} &= \mathbf{A}^{-1} - \mathbf{A}^{-1}\mathbf{U}(\mathbf{B}^{-1} + \mathbf{C}^{-1})^{-1}\mathbf{U}'\mathbf{A}^{-1} \\
&= \mathbf{A}^{-1} - \mathbf{A}^{-1}\mathbf{U}\mathbf{B}(\mathbf{B} + \mathbf{C})^{-1}\mathbf{C}\mathbf{U}'\mathbf{A}^{-1} \\
&= \mathbf{A}^{-1} - \mathbf{A}^{-1}\mathbf{U}(\mathbf{B} + \mathbf{C} - \mathbf{C})(\mathbf{B} + \mathbf{C})^{-1}\mathbf{C}\mathbf{U}'\mathbf{A}^{-1} \\
&= \mathbf{A}^{-1} - \mathbf{A}^{-1}\mathbf{U}\mathbf{C}\mathbf{U}'\mathbf{A}^{-1} + \mathbf{A}^{-1}\mathbf{U}\mathbf{C}(\mathbf{B} + \mathbf{C})^{-1} \\
&\quad \times \mathbf{C}\mathbf{U}'\mathbf{A}^{-1},
\end{aligned}
$$

  (iii) In particular,

$$(\mathbf{A} + \mathbf{B})^{-1} = \mathbf{A}^{-1} - \mathbf{A}^{-1}(\mathbf{A}^{-1} + \mathbf{B}^{-1})^{-1}\mathbf{A}^{-1}.$$

We can also interchange $\mathbf{A}$ and $\mathbf{B}$ and can replace $\mathbf{A}$ by $\mathbf{A}^{-1}$ and $\mathbf{B}$ by $\mathbf{B}^{-1}$.

We note that in (b) (and (a)) we can replace $\mathbf{B}$ by $-\mathbf{B}$.

(c) Setting $\mathbf{B} = -\mathbf{D}^{-1}$ in (a)(ii) leads to a number of results like the following:

  (i) $(\mathbf{A} - \mathbf{U}\mathbf{D}^{-1}\mathbf{V})^{-1} = \mathbf{A}^{-1} + \mathbf{A}^{-1}\mathbf{U}(\mathbf{D} - \mathbf{V}\mathbf{A}^{-1}\mathbf{U})^{-1}\mathbf{V}\mathbf{A}^{-1}$.

  Note that the left-hand side of the above is the inverse of a Schur complement. As a special case we have

  (ii) $(\mathbf{I} - \mathbf{U}\mathbf{V})^{-1} = \mathbf{I} + \mathbf{U}(\mathbf{I} - \mathbf{V}\mathbf{U})^{-1}\mathbf{V}$, as in (a)(iv) with a sign change.

15.4. Gentle [1998: 62] notes that in linear regression we often need inverses of various sums of matrices and gives the following additional identities for nonsingular $\mathbf{A}$ and $\mathbf{B}$.

(a) $(\mathbf{A} + \mathbf{B}\mathbf{B}')^{-1}\mathbf{B} = \mathbf{A}^{-1}\mathbf{B}(\mathbf{I} + \mathbf{B}'\mathbf{A}^{-1}\mathbf{B})^{-1}$.

(b) $(\mathbf{A}^{-1} + \mathbf{B}^{-1})^{-1} = \mathbf{A}(\mathbf{A} + \mathbf{B})^{-1}\mathbf{B}$.

(c) $\mathbf{A}(\mathbf{A} + \mathbf{B})^{-1}\mathbf{B} = \mathbf{B}(\mathbf{A} + \mathbf{B})^{-1}\mathbf{A}$.

(d) $\mathbf{A}^{-1} + \mathbf{B}^{-1} = \mathbf{A}^{-1}(\mathbf{A} + \mathbf{B})\mathbf{B}^{-1}$

We can also add, for nonsingular $\mathbf{A} + \mathbf{B}$,

(e) $\mathbf{A} - \mathbf{A}(\mathbf{A} + \mathbf{B})^{-1}\mathbf{A} = \mathbf{B} - \mathbf{B}(\mathbf{A} + \mathbf{B})^{-1}\mathbf{B}$.

15.5. (Non-negative Definite Matrices)

(a) If $\mathbf{A}$ is positive definite, then $\mathbf{A} - \mathbf{b}\mathbf{b}'$ is positive definite if and only if $\mathbf{b}'\mathbf{A}^{-1}\mathbf{b} < 1$.

(b) If $\mathbf{A}$ is non-negative definite (n.n.d.), then $\mathbf{A} - \mathbf{b}\mathbf{b}'$ is n.n.d. if and only if $\mathbf{b} \in \mathcal{C}(\mathbf{A})$ and $\mathbf{b}'\mathbf{A}^{-}\mathbf{b} \le 1$.

15.6. If $\mathbf{A} = \mathrm{diag}(a_1, a_2, \ldots, a_n)$ is a nonsingular diagonal matrix and $\mathbf{C} = \mathbf{A} + \alpha\mathbf{u}\mathbf{v}'$, then we have the following.

(a) $\mathbf{C}^{-1} = \mathbf{A}^{-1} + \alpha\beta^{-1}\mathbf{f}\mathbf{g}'$, where $\beta = -(1 + \alpha\sum_{i=1}^{n}(u_i v_i / a_i)$ $(\neq 0)$, $f_i = u_i / a_i$, and $g_i = v_i / a_i$.

(b) $\det \mathbf{C} = \left(1 + \alpha\sum_{i=1}^{n}\dfrac{u_i v_i}{a_i}\right)\prod_{i=1}^{n} a_i$.

(c) The characteristic equation of $\mathbf{C}$ is given by

$$\det(\mathbf{C} - \lambda\mathbf{I}_n) = (1 + \alpha\sum_{i=1}^{n}\frac{u_i v_i}{a_i - \lambda})\prod_{i=1}^{n}(a_i - \lambda) = 0.$$

  For further details relating to the eigenvalues and eigenvectors of $\mathbf{C}$ see Steerneman and van Perlo-ten Kleij [2005] and the references therein.

(d) If $a_1 = a_2 = \cdots = a_n = a$, then $\mathbf{C}$ has $n - 1$ eigenvalues equal to a and one eigenvalue equal to $a + \alpha\sum_{i=1}^{n} u_i v_i$.

15.7. Let $p_i > 0$ for $i = 1, 2, \ldots, k-1$, where $\sum_{i=1}^{k-1} p_i < 1$, and let $p_k = 1 - \sum_{i=1}^{k-1} p_i$. Then the variance matrix for a nonsingular $(k-1)$-dimensional multinomial random variable is

$$\mathbf{V} = n\{\operatorname{diag}(p_1, p_2, \ldots, p_{k-1}) - \mathbf{p}\mathbf{p}'\},$$

where $\mathbf{p}' = (p_1, p_2, \ldots, p_{k-1})$. From (15.6a) with $n = k - 1$ and $\alpha = -1$, we have

$$\mathbf{V}^{-1} = n^{-1} \begin{pmatrix} (p_1^{-1} + p_k^{-1}) & p_k^{-1} & p_k^{-1} & \cdots & p_k^{-1} \\ p_k^{-1} & (p_2^{-1} + p_k^{-1}) & p_k^{-1} & \cdots & p_k^{-1} \\ p_k^{-1} & p_k^{-1} & (p_3^{-1} + p_k^{-1}) & \cdots & p_k^{-1} \\ p_k^{-1} & p_k^{-1} & p_k^{-1} & \cdots & p_k^{-1} \\ \vdots & \vdots & \vdots & \vdots & \vdots \\ p_k^{-1} & p_k^{-1} & p_k^{-1} & \cdots & (p_{k-1}^{-1} + p_k^{-1}) \end{pmatrix}.$$

Proofs. Section 15.1.

15.1a. Henderson and Searle [1981b: 56].

15.1b. Henderson and Searle [1981b: 57].

15.1c. Use the identity $\mathbf{I} = \mathbf{I} + \mathbf{U} - \mathbf{U}$, multiply on the left by $(\mathbf{I} + \mathbf{U})^{-1}$, and then multiply on the right.

15.1d. We take the inverse term on the right-hand side over to the left.

15.1e. Zhang [1999: 185].

15.1f. Multiply $(\beta \mathbf{I}_n - \mathbf{A}) - (\alpha \mathbf{I}_n - \mathbf{A}) = (\beta - \alpha)\mathbf{I}_n$ on the left by $(\beta \mathbf{I}_n - \mathbf{A})^{-1}$ and on the right by $(\alpha \mathbf{I}_n - \mathbf{A})^{-1}$.

15.3a. Harville [1997: 424–425].

15.4a. We take the inverses of both sides.

15.4b. Take inverses of both sides.

15.4c. Interchange $\mathbf{A}$ and $\mathbf{B}$ in (b).

15.4d. Simply multiply out.

15.4e. This follows from $(\mathbf{A} + \mathbf{B})(\mathbf{A} + \mathbf{B})^{-1}\mathbf{C} = \mathbf{C}$ for $\mathbf{C} = \mathbf{A}, \mathbf{B}$ and from (c).

15.5. Abadir and Magnus [2005: 227] and Rao and Bhimasankaram [2000: 345, see solution to exercise 15].

15.6. Graybill [1983: 189, 203, 206].

15.7. Graybill [1983: 189].

15.2   DETERMINANTS

15.8. If $\mathbf{A}$ has rank 1, then from (3.4b) and (4.33),

$$\det(\mathbf{I}_n + x\mathbf{A}) = 1 + x\,\text{trace}(\mathbf{A}).$$

15.9. For general $\mathbf{A}$, $\det(\mathbf{I} + x\mathbf{A}) = 1 + x\,\text{trace}\,\mathbf{A} + O(x^2)$.

15.10. Suppose $\mathbf{C}$ is $n \times m$, $\mathbf{D}$ is $m \times n$, $\mathbf{u}$ is $n \times 1$, $\mathbf{w}$ is $n \times 1$, and $\mathbf{v}$ is $m \times 1$, then we have the following results.

(a) $\det(\mathbf{I}_n + \mathbf{CD}) = \det \begin{pmatrix} \mathbf{I}_n & \mathbf{C} \\ -\mathbf{D} & \mathbf{I}_m \end{pmatrix}$.

(b) We have from (4.33), $\det(\mathbf{I}_n \pm \mathbf{CD}) = \det(\mathbf{I}_m \pm \mathbf{DC})$.

(c) Setting $\mathbf{C} = \mathbf{u}'$ and $\mathbf{D} = \mathbf{v}'\mathbf{A}$, we have from (b) and (15.8)

$$\det(\mathbf{I}_n \pm \mathbf{uv}'\mathbf{A}) = \det(1 \pm \mathbf{v}'\mathbf{Au}) = 1 \pm \text{trace}(\mathbf{v}'\mathbf{Au}) = 1 \pm \mathbf{v}'\mathbf{Au}.$$

(d) If $\mathbf{A}$ is $n \times n$, $\mathbf{B}$ is $m \times m$, and $\mathbf{A}$ and $\mathbf{B}$ are nonsingular, then

$$\begin{aligned} \det(\mathbf{A} + \mathbf{UBV}) &= \det(\mathbf{A})\det(\mathbf{I}_m + \mathbf{VA}^{-1}\mathbf{UB}) \\ &= \det(\mathbf{A})\det(\mathbf{B}^{-1} + \mathbf{VA}^{-1}\mathbf{U})\det(\mathbf{B}). \end{aligned}$$

We have the following special cases.

(i) $\det(\mathbf{A} \pm \mathbf{uu}') = \det(\mathbf{A})(1 \pm \mathbf{u}'\mathbf{A}^{-1}\mathbf{u})$.

(ii) $\mathbf{u}'\mathbf{A}^{-1}\mathbf{u} = 1 - (\det(\mathbf{A} - \mathbf{uu}')/\det(\mathbf{A}))$.

(iii) $\det(\mathbf{A} + \alpha\mathbf{uw}') = \det(\mathbf{A})(1 + \alpha\mathbf{w}'\mathbf{A}^{-1}\mathbf{u}) = \det(\mathbf{A}) + \alpha\mathbf{w}'(\text{adj}\mathbf{A})\mathbf{u}$, where $\text{adj}\mathbf{A}$ is the adjoint matrix of $\mathbf{A}$.

Proofs. Section 15.2.

15.8. Abadir and Magnus [172–173].

15.9. Anderson [2003: 646].

15.10a–b. Muirhead [1982: 578].

15.10d. Harville [1997: 416].

15.3   PERTURBATIONS

Definition **15.1.** Suppose we have a matrix $\mathbf{A}$ involved in a system of equations and we wish to know what happens to the system if we change $\mathbf{A}$ to $\mathbf{A} + \delta\mathbf{A}$. If the matrix $\delta\mathbf{A}$ is of rank one (cf. (3.4b)), then it is called a *rank one* perturbation. Other kinds of perturbations may consist of adding or subtracting an observation to see what effect this has on any inference or diagnostics. Clearly, such perturbations have many uses in statistics, and although the theory underlying these is given above and elsewhere, it is helpful to collect some of the results used generally and in linear regression together here. For a historical overview and some computational aspects see Hager [1989].

15.11. (General) Let $\mathbf{A}$ be an $n \times n$ nonsingular matrix. We consider the effect on the inverse of $\mathbf{A}$ of three modifications using (15.2b(i)).

(a) (Add to an Element) If we add h to a_{ij}, then $\mathbf{A}$ becomes $\mathbf{A} + h\mathbf{E}_{ij} = \mathbf{A} + h\mathbf{e}_i\mathbf{e}_j'$, where $\mathbf{e}_i$ is the ith colum of $\mathbf{I}_n$, and

$$(\mathbf{A} + h\mathbf{e}_i\mathbf{e}_j')^{-1} = \mathbf{A}^{-1} - \frac{h\mathbf{A}^{-1}\mathbf{e}_i\mathbf{e}_j'\mathbf{A}^{-1}}{1 + h\mathbf{e}_j'\mathbf{A}^{-1}\mathbf{e}_i}.$$

(b) (Add to a Column) If $\mathbf{f}$ is added to the jth column of $\mathbf{A}$, then

$$(\mathbf{A} + \mathbf{f}\mathbf{e}_j')^{-1} = \mathbf{A}^{-1} - \frac{\mathbf{A}^{-1}\mathbf{f}\mathbf{e}_j'\mathbf{A}^{-1}}{1 + \mathbf{e}_j'\mathbf{A}^{-1}\mathbf{f}}.$$

(c) (Add to a Row) If row $\mathbf{g}'$ is added to the ith row of $\mathbf{A}$, then

$$(\mathbf{A} + \mathbf{e}_i\mathbf{g}')^{-1} = \mathbf{A}^{-1} + \frac{\mathbf{A}^{-1}\mathbf{e}_i\mathbf{g}'\mathbf{A}^{-1}}{1 + \mathbf{g}'\mathbf{A}^{-1}\mathbf{e}_i}.$$

(d) (Diagonal Increment) If the inverses exist, we have from (15.3a(ii)),

$$(\mathbf{A} + k\mathbf{I}_n)^{-1} = \mathbf{A}^{-1} - \mathbf{A}^{-1}(k^{-1}\mathbf{I}_n + \mathbf{A}^{-1})^{-1}\mathbf{A}.$$

It is assumed that all the above denominators are nonzero.

15.12. (Sample Mean and Variance Matrix) Suppose we have a set of d-dimensional observations $\mathbf{x}_1, \mathbf{x}_2, \ldots, \mathbf{x}_n$, and we define

$$W_n = \sum_{i=1}^{n} w_i, \quad \overline{\mathbf{x}}_n = \sum_{i=1}^{n} w_i\mathbf{x}_i/W_n, \quad \text{and} \quad \mathbf{S}_n = \sum_{i=1}^{n} w_i(\mathbf{x}_i - \overline{\mathbf{x}}_n)(\mathbf{x}_i - \overline{\mathbf{x}}_n)'.$$

We want to know what happens to these quantities when we add an observation $\mathbf{x}_{n+1}$ or subtract $\mathbf{x}_n$. Setting $\mathbf{d}_{n+1} = \mathbf{x}_{n+1} - \overline{\mathbf{x}}_n$ and $\mathbf{f}_n = \mathbf{x}_n - \overline{\mathbf{x}}_n$, we have the following.

(a) (Add an Observation)

(i) $\overline{\mathbf{x}}_{n+1} = \overline{\mathbf{x}}_n + \frac{w_{n+1}}{W_{n+1}}\mathbf{d}_{n+1}.$

(ii) $\mathbf{S}_{n+1} = \mathbf{S}_n + w_{n+1}(1 - \frac{w_{n+1}}{W_{n+1}})\mathbf{d}_{n+1}\mathbf{d}_{n+1}'.$

(b) (Subtract an Observation)

(i) $\overline{\mathbf{x}}_{n-1} = \overline{\mathbf{x}}_n - \frac{w_n}{W_{n-1}}\mathbf{f}_n.$

(ii) $\mathbf{S}_{n-1} = \mathbf{S}_n - w_n(1 - \frac{w_n}{W_{n-1}}\mathbf{f}_n\mathbf{f}_n').$

(c) (Equal Weights) With equal weights we have $w_i = 1/n$, $W_n = 1$, and so on. Let $\overline{\mathbf{x}}_k = \sum_{i=1}^{k} \mathbf{x}_i/k$ and $\mathbf{S}_k = \sum_{i=1}^{k}(\mathbf{x}_i - \overline{\mathbf{x}}_k)(\mathbf{x}_i - \overline{\mathbf{x}}_k)'/k$ for $k = n-1, n, n+1$. Then:

(i) $\overline{\mathbf{x}}_{n+1} = \overline{\mathbf{x}}_n + \frac{1}{n+1}\mathbf{d}_{n+1}$ and $\overline{\mathbf{x}}_{n-1} = \overline{\mathbf{x}}_n - \frac{1}{n-1}\mathbf{f}_n$.

(ii) $\frac{n+1}{n}\mathbf{S}_{n+1} = \mathbf{S}_n + \frac{1}{n+1}\mathbf{d}_{n+1}\mathbf{d}'_{n\,|\,1}$ and $\frac{n-1}{n}\mathbf{S}_{n-1} = \mathbf{S}_n - \frac{1}{n-1}\mathbf{f}_n\mathbf{f}'_n$.

15.13. (Regression) Let $\mathbf{X} = (\mathbf{x}_1, \mathbf{x}_2, \ldots, \mathbf{x}_n)' = (\mathbf{x}^{(1)}, \mathbf{x}^{(2)}, \ldots \mathbf{x}^{(p)})$ be an $n \times p$ matrix of rank p. We are interested what in happens to $(\mathbf{X}'\mathbf{X})^{-1}$ and related quantities when the rows and columns of $\mathbf{X}$ are modified.

(a) (Add or Delete a Row) Suppose that ith row $\mathbf{x}'_i$ is deleted giving us $\mathbf{X}(i)$ instead of $\mathbf{X}$, then $\mathbf{X}(i)'\mathbf{X}(i) = \mathbf{X}'\mathbf{X} - \mathbf{x}_i\mathbf{x}'_i$. Let $h_{ii} = \mathbf{x}'_i(\mathbf{X}'\mathbf{X})^{-1}\mathbf{x}_i$.

  (i) $(\mathbf{X}'\mathbf{X} - \mathbf{x}_i\mathbf{x}'_i)^{-1} = (\mathbf{X}'\mathbf{X})^{-1} + \dfrac{(\mathbf{X}'\mathbf{X})^{-1}\mathbf{x}_i\mathbf{x}'_i(\mathbf{X}'\mathbf{X})^{-1}}{1 - h_{ii}}$.

  (ii) $\det(\mathbf{X}'\mathbf{X} - \mathbf{x}_i\mathbf{x}'_i)) = \det(\mathbf{X}'\mathbf{X})(1 - h_{ii})$ (from 15.10d(iii)).

  (iii) If an extra row $\mathbf{x}'$ is added to $\mathbf{X}$, then one simply replaces $\mathbf{x}_i$ by $\mathbf{x}$ and changes all the signs in (i) and (ii) above.

  (iv) Let $\widehat{\boldsymbol{\beta}} = (\mathbf{X}'\mathbf{X})^{-1}\mathbf{X}'\mathbf{y}$ and $\widehat{\boldsymbol{\beta}}(i) = (\mathbf{X}(i)'\mathbf{X}(i))^{-1}\mathbf{X}(i)'\mathbf{y}(i)$, where $\mathbf{y}(i)$ is $\mathbf{y}$ without its ith element y_i. Here $\widehat{\boldsymbol{\beta}}$ and $\widehat{\boldsymbol{\beta}}(i)$ are the respective least squares estimates of $\boldsymbol{\beta}$ under a regression model with full rank design matrix $\mathbf{X}$, and under the same model but with the ith case deleted, the so-called *leaving-one-out model*. Then

  $$\widehat{\boldsymbol{\beta}}(i) = \widehat{\boldsymbol{\beta}} - \frac{(\mathbf{X}'\mathbf{X})^{-1}\mathbf{x}_i(y_i - \mathbf{x}'_i\widehat{\boldsymbol{\beta}})}{1 - h_{ii}}.$$

  This result forms the basis of a number of regression diagnostics (e.g., Seber and Lee [2003: section 10.6]).

(b) (Substitute One Row for Another) If we replace row $\mathbf{x}'_-$ by row $\mathbf{x}'_+$, we can combine (a)(ii) and (iii) to get

$$\begin{aligned}\det(\mathbf{X}'\mathbf{X} + \mathbf{x}_+\mathbf{x}'_+ - \mathbf{x}_-\mathbf{x}'_-) = {} & \det(\mathbf{X}'\mathbf{X})\big[(1 + \mathbf{x}'_+(\mathbf{X}'\mathbf{X})^{-1}\mathbf{x}_+ \\ & - \mathbf{x}'_-(\mathbf{X}'\mathbf{X})^{-1}\mathbf{x}_-(1 + \mathbf{x}'_+(\mathbf{X}'\mathbf{X})^{-1}\mathbf{x}_+) \\ & + (\mathbf{x}'_+(\mathbf{X}'\mathbf{X})^{-1}\mathbf{x}_-)^2\big],\end{aligned}$$

a result given by Gentle [1998: 171]. He indicates how this result is used in a stepwise method for maximizing $\det(\mathbf{X}'\mathbf{X})$, a problem that arises, for example, in optimal design theory (D-optimality, cf. Section 24.5).

(c) (Add or Delete a Column)

  (i) If an extra column $\mathbf{x}$ is added to $\mathbf{X}$ giving $\mathbf{X}_1 = (\mathbf{X}, \mathbf{x})$, then by (14.11),

  $$\begin{aligned}(\mathbf{X}'_1\mathbf{X}_1)^{-1} &= \begin{pmatrix} \mathbf{X}'\mathbf{X} & \mathbf{X}'\mathbf{x} \\ \mathbf{x}'\mathbf{X} & \mathbf{x}'\mathbf{x} \end{pmatrix}^{-1} \\ &= \begin{pmatrix} (\mathbf{X}'\mathbf{X})^{-1} + v\mathbf{u}\mathbf{u}', & -v\mathbf{u} \\ -v\mathbf{u}', & v \end{pmatrix},\end{aligned}$$

  with $\mathbf{u} = (\mathbf{X}'\mathbf{X})^{-1}\mathbf{X}'\mathbf{x}$, $v = [\mathbf{x}'(\mathbf{I}_n - \mathbf{P})\mathbf{x}]^{-1}$, and $\mathbf{P} = \mathbf{X}(\mathbf{X}'\mathbf{X})^{-1}\mathbf{X}'$.

(ii) Suppose the last column $\mathbf{x} = \mathbf{x}^{(p)}$ is deleted from $\mathbf{X}$ giving us $\mathbf{X}^{(p)}$ so that $\mathbf{X} = (\mathbf{X}^{(p)}, \mathbf{x})$. Then

$$\mathbf{X}'\mathbf{X} = \begin{pmatrix} \mathbf{X}^{(p)\prime}\mathbf{X}^{(p)} & \mathbf{X}^{(p)\prime}\mathbf{x} \\ \mathbf{x}'\mathbf{X}^{(p)} & \mathbf{x}'\mathbf{x} \end{pmatrix},$$

and we can use (i) with $\mathbf{X}$ now playing the role of $\mathbf{X}_1$ to pick out from $(\mathbf{X}'\mathbf{X})^{-1}$ the values of $\mathbf{u}$ and v and obtain $(\mathbf{X}^{(p)\prime}\mathbf{X}^{(p)})^{-1}$ by subtraction.

(d) (Diagonal Increments) The expression $(\mathbf{X}'\mathbf{X} + k\mathbf{I}_p)^{-1}$ occurs in the context of *ridge regression* and *Bayes regression* estimators, and can be expressed in terms of $(\mathbf{X}'\mathbf{X})^{-1}$ using (15.11d) above.

The above expressions do not indicate how they are actually computed, as one avoids finding the inverse of a matrix directly. Computational details are given by Seber and Lee [2003: section 11.6; they involve using the *sweep operator* (Seber [1977: 351] or Seber and Lee [2003: 335]) and modifiying the QR decomposition. One can also use a weighted least squares approach (Escobar and Moser [1993]). The above theory applies to the linear model $(\mathbf{y}, \mathbf{X}\boldsymbol{\beta}, \sigma^2\mathbf{I}_n)$ (cf. Section 20.7), where $\mathbf{X}$ has full column rank. For updates relating to the more general model $(\mathbf{y}, \mathbf{X}\boldsymbol{\beta}, \sigma^2\mathbf{V})$, with $\mathbf{X}$ being less than full rank and $\mathbf{V}$ being possibly singular, the reader is referred to Sengupta and Jammalamadaka [2003: chapter 9], who also include changes produced in various other statistics.

15.14. (Interchanges in Design Models) Let $\mathbf{A} = \mathbf{X}'\mathbf{P}\mathbf{X}$, where $\mathbf{X}$ is $n \times p$, $\mathbf{P}$ is an $n \times n$ symmetric idempotent matrix, and $\mathbf{A}\mathbf{1}_n = \mathbf{0}$. Suppose we interchange two rows of $\mathbf{X}$ so that $\mathbf{A}$ becomes $\mathbf{A}_2$. We can assume, without any loss of generality, that it is the first two rows. Let $\mathbf{X}$ and $\mathbf{P}$ be partitioned as follows:

$$\mathbf{X} = \begin{pmatrix} \mathbf{x}'_1 \\ \mathbf{x}'_2 \\ \mathbf{X}_3 \end{pmatrix} \quad \text{and} \quad \mathbf{P} = \begin{pmatrix} c_1 & c_2 & \mathbf{d}'_1 \\ c_2 & c_3 & \mathbf{d}'_2 \\ \mathbf{d}_1 & \mathbf{d}_2 & \mathbf{D}_3 \end{pmatrix}.$$

Then we find that

$$\mathbf{A} - \mathbf{A}_2 = (c_1 - c_3)(\mathbf{x}_1\mathbf{x}'_1 - \mathbf{x}_2\mathbf{x}'_2) + \mathbf{B} + \mathbf{B}',$$

where

$$\mathbf{B} = (\mathbf{x}_1 - \mathbf{x}_2)(\mathbf{d}_1 - \mathbf{d}_2)'\mathbf{X}_3.$$

Suppose that the spectral decomposition of the symmetric matrix $\mathbf{A} - \mathbf{A}_2$ is $\mathbf{T}\boldsymbol{\Lambda}\mathbf{T}'$, where $\boldsymbol{\Lambda}$ is a diagonal matrix whose diagonal elements are the eigenvalues of $\mathbf{A} - \mathbf{A}_2$, with corresponding eigenvectors given by the columns of the orthogonal matrix $\mathbf{T}$. Then, from (15.25),

$$\mathbf{A}_2^+ = (\mathbf{A} - \mathbf{T}\boldsymbol{\Lambda}\mathbf{T})^+ = \mathbf{A}^+ + \mathbf{A}^+\mathbf{T}\boldsymbol{\Psi}^+\mathbf{T}'\mathbf{A}^+,$$

where $\boldsymbol{\Psi} = \boldsymbol{\Lambda}^{-1} - \mathbf{T}'\mathbf{A}^+\mathbf{T}$, provided $\mathcal{C}(\mathbf{T}\boldsymbol{\Lambda}\mathbf{T}') \subset \mathcal{C}(\mathbf{A})$. For further computational details and applications with regard to experimental designs with blocking structure, see John [2001]. This method of interchanging two rows is particularly useful in searching for the most efficient designs. It has been also applied to so-called α-designs, where one is involved with block circulants and Hermitian matrices (Williams and John [2000]).

15.15. (Perturbed Identity Matrix)   Let $\mathbf{T}(\theta) = \mathbf{I}_n - \theta \mathbf{e}_i \mathbf{e}_j'$ be an $n \times n$ matrix, where θ is a real scalar and $\mathbf{e}_r$ is the rth column of $\mathbf{I}_n$.

(a) If $i \neq j$, then $\mathbf{T}^{-1}(\theta) = \mathbf{T}(-\theta)$.

(b) Let $\mathbf{A} = (a_{ij})$ be an $n \times n$ upper-triangular matrix. When $i < j$,

$$\mathbf{T}^{-1}(\theta)\mathbf{A}\mathbf{T}(\theta) = \mathbf{A} + \theta(\mathbf{e}_i \mathbf{e}_j' \mathbf{A} - \mathbf{A}\mathbf{e}_i \mathbf{e}_j') = \mathbf{A} + \theta\mathbf{P},$$

where

$$\mathbf{P} = \begin{pmatrix} \mathbf{0} & \mathbf{Q} \\ \mathbf{0} & \mathbf{0} \end{pmatrix}, \quad \mathbf{Q} = \begin{pmatrix} -a_{1i} & 0 & \cdots & 0 \\ -a_{2i} & 0 & \cdots & 0 \\ \cdot & & \cdots & \cdot \\ -a_{i-1,i} & 0 & \cdots & 0 \\ a_{jj} - a_{ii} & a_{j,j+1} & \cdots & a_{jn} \end{pmatrix},$$

and the the submatrices in $\mathbf{P}$ have i and $n-i$ rows, and $j-1$ and $n-j+1$ columns, respectively.

(c) If $a_{ii} \neq a_{jj}$ and $\theta = a_{ij}/(a_{ii} - a_{jj})$, then $[\mathbf{T}^{-1}(\theta)\mathbf{A}\mathbf{T}(\theta)]_{ij} = 0$ and the only elements in $\mathbf{A}$ to be disturbed are in row i to the right of a_{ij} and in column j above a_{ij}.

15.16. The effect of a perturbation on a finite irreducible discrete time Markov chain is examined by Hunter [2005] with reference to mean first passage times and the stationary distribution. He also gives references to the literature on the subject. In a random environment there could be small random perturbations to a transition matrix and an example of this is considered by Hoppensteadt et al. [1996].

Proofs. Section 15.3.

15.12. Clarke [1971], Seber [1984: 15], and Trenkler and Puntanen [2005: 145].

15.13a. Seber and Lee [2003: 268].

15.15. Abadir and Magnus [2006: 184–185].

15.4   MATRICES WITH REPEATED ELEMENTS AND BLOCKS

15.17. If $\mathbf{A} = \begin{pmatrix} a\mathbf{I}_n & b\mathbf{I}_n \\ c\mathbf{I}_n & d\mathbf{I}_n \end{pmatrix}$, then $\det \mathbf{A} = (ad - bc)^n$.

15.18. Let $\mathbf{J}_{m,n}$ be an $m \times n$ matrix of ones, i.e., $\mathbf{J}_{m,n} = \mathbf{1}_m \mathbf{1}_n'$, and define $\mathbf{J}_m$ to be $\mathbf{J}_{m,m}$. We now consider a number of results that use these matrices.

(a) If $\mathbf{A} = a\mathbf{I}_n + b\mathbf{J}_n$ ($b \neq 0$), that is, we have $a + b$ on the diagonal and b everywhere else, then:

(i) $\det \mathbf{A} = a^{n-1}(a + nb)$.

(ii) $\mathbf{A}^{-1} = \dfrac{1}{a}\left(\mathbf{I}_n - \dfrac{b}{a + nb}\mathbf{J}_n\right)$, $(a \neq 0, a \neq -nb)$.

(iii) $\det(\lambda\mathbf{I}_n - \mathbf{A}) = (\lambda - a - nb)(\lambda - a)^{n-1}$, so that the eigenvalue $a + nb$ has algebraic multiplicity 1 and a has multiplicity $n - 1$. The eigenvector $n^{-1/2}\mathbf{1}_n$ corresponds to the eigenvalue $\lambda = a + nb$. A set of eigenvectors of $\mathbf{A}$ are the rows of the Helmert matrix (8.10).

(iv) Sometimes we have c on the diagonal and b everywhere else. In this case we set $a = c - b$ in the above. For example, a common case is the correlation matrix that arises, for example, in a one-way random effects model, namely

$$\mathbf{R} = (1 - \rho)\mathbf{I}_n + \rho\mathbf{J}_n.$$

This has eigenvalues $(1 - \rho)$ and $1 + \rho(n - 1)$, so that $\mathbf{R}$ is positive definite if the eigenvalues are positive, that is, when

$$-\frac{1}{n-1} < \rho < 1.$$

(b) If

$$\mathbf{A} = \begin{pmatrix} m\mathbf{I}_m & \mathbf{J}_{m,m-1} & \mathbf{J}_{m,m-1} \\ \mathbf{J}'_{m,m-1} & m\mathbf{I}_{m-1} & \mathbf{J}_{m,m-1} \\ \mathbf{J}'_{m,m-1} & \mathbf{J}_{m-1,m} & m\mathbf{I}_{m-1} \end{pmatrix},$$

then

$$m\mathbf{A}^{-1} = \begin{pmatrix} \mathbf{I}_m + \frac{2(m-1)}{m}\mathbf{J}_m & -\mathbf{J}_{m,m-1} & -\mathbf{J}_{m,m-1} \\ -\mathbf{J}'_{m,m-1} & \mathbf{I}_{m-1} + \mathbf{J}_{m-1} & 0 \\ -\mathbf{J}'_{m,m-1} & 0 & \mathbf{I}_{m-1} + \mathbf{J}_{m-1} \end{pmatrix}.$$

This kind of pattern arises in Latin square designs.

(c) Consider the $(m + n) \times (m + n)$ matrix

$$\mathbf{A} = \begin{pmatrix} a_1\mathbf{I}_m & a_2\mathbf{J}_{m,n} \\ a_2\mathbf{J}'_{m,n} & a_3\mathbf{I}_n \end{pmatrix},$$

where $a_1 \neq 0$ and $a_3 \neq 0$.

(i) If $d = a_1 a_3 - mna_2^2 \neq 0$,

$$\mathbf{A}^{-1} = \begin{pmatrix} a_1^{-1}\mathbf{I}_m + b_1\mathbf{J}_m & b_2\mathbf{J}_{m,n} \\ b_2\mathbf{J}'_{m,n} & a_3^{-1}\mathbf{I}_n + b_3\mathbf{J}_n \end{pmatrix},$$

where $b_1 = na_2^2/(a_1 d_1)$, $b_2 = -a_2/d$ and $b_3 = ma_2^2/(a_3 d)$.

(ii) $\det \mathbf{A} = a_1^{m-1} a_3^{n-1} d$.

(iii) The matrix $\mathbf{A}$ above occurs in the form $\mathbf{X}'\mathbf{X}$, where $\mathbf{X}$ is the so-called design matrix, in a 2-way ANOVA with equal numbers of observations per cell by setting $a_1 = n$, $a_3 = m$, and $a_2 = 1$. This matrix is singular as $d = 0$, and it is in fact non-negative definite. A generalized inverse is

$$\mathbf{A}^- = \begin{pmatrix} \mathbf{I}_m/n & 0 \\ 0 & \mathbf{C}_n/m \end{pmatrix},$$

where $\mathbf{C}_n$ is the centering matrix $(\mathbf{I}_n - \mathbf{1}_n\mathbf{1}'_n/n)$.

(iv) If we set $a_1 = a_3 = 1$, and $a_2 = \rho$, we get the so-called *intraclass correlation matrix*. The eigenvalues of $\mathbf{A}$ are then 1 with algebraic multiplicity $m + n - 2$ and $1 \pm \rho\sqrt{mn}$, each with multiplicity 1. $\mathbf{A}$ is then positive definite if and only if

$$-(mn)^{-1/2} < \rho < (mn)^{-1/2}.$$

(d) Let

$$\mathbf{A} = \begin{pmatrix} a_1 & a_2\mathbf{1}'_m & a_3\mathbf{1}'_m & \cdots & a_n\mathbf{1}'_n \\ a_2\mathbf{1}_m & b_2\mathbf{I}_m + c_2\mathbf{J}_m & b_3\mathbf{I}_m + c_3\mathbf{J}_m & \cdots & b_n\mathbf{I}_m + c_n\mathbf{J}_m \\ a_3\mathbf{1}_m & b_3\mathbf{I}_m + c_3\mathbf{J}_m & d_3\mathbf{I}_m + e_3\mathbf{J}_m & \cdots & d_n\mathbf{I}_m + e_n\mathbf{J}_m \\ \vdots & \vdots & \vdots & \vdots & \vdots \\ a_n\mathbf{1}_m & b_n\mathbf{I}_m + c_n\mathbf{J}_m & d_n\mathbf{I}_m + e_n\mathbf{J}_m & \cdots & y_n\mathbf{I}_m + z_n\mathbf{J}_m \end{pmatrix}.$$

If the inverse of $\mathbf{A}$ exists, then it has the same pattern as $\mathbf{A}$. For example,

$$\begin{bmatrix} a & b\mathbf{1}'_{n-1} \\ b\mathbf{1}_{n-1} & (c-d)\mathbf{I}_{n-1} + d\mathbf{J}_{n-1} \end{bmatrix}^{-1} = \begin{bmatrix} e & f\mathbf{1}'_{n-1} \\ f\mathbf{1}_{n-1} & (g-h)\mathbf{I}_{n-1} + h\mathbf{J}_{n-1} \end{bmatrix},$$

where $e = \frac{1}{a}[1 + \lambda b^2(n-1)]$, $f = -\lambda b$, $g = \frac{1}{c-d}[1 - \lambda(ad - b^2)]$, $h = \frac{1}{c-d}\lambda(b^2 - ad)$, and $\lambda = \{a(c-d) + (n-1)(ad - b^2)\}^{-1}$. This example arises in Latin square models and response surfaces; Graybill [1983: 195–196] gives a numerical example.

15.19. Let $\mathbf{A}$ and $\mathbf{B}$ be $m \times m$ matrices, and let

$$\mathbf{C} = \begin{pmatrix} \mathbf{A} + \mathbf{B} & \mathbf{B} & \cdots & \mathbf{B} & \mathbf{B} \\ \mathbf{B} & \mathbf{A} + \mathbf{B} & \cdots & \mathbf{B} & \mathbf{B} \\ \cdot & \cdot & \cdots & \cdot & \cdot \\ \mathbf{B} & \mathbf{B} & \cdots & \mathbf{B} & \mathbf{A} + \mathbf{B} \end{pmatrix},$$

where $\mathbf{C}$ has n diagonal blocks.

(a) (i) $\det \mathbf{C} = (\det \mathbf{A})^{n-1} \det(\mathbf{A} + n\mathbf{B})$.

(ii) $\mathbf{C}$ has eigenvalues λ_i $(i = 1, 2, \ldots, m)$ each of algebraic multiplicity 1 and eigenvalues μ_i $(i = 1, 2, \ldots, m)$ each of multiplicity $n - 1$, where the λ_i are the eigenvalues of $\mathbf{A} + n\mathbf{B}$ and the μ_i are the eigenvalues of $\mathbf{A}$.

(b) Consider the special case $\mathbf{A} + \mathbf{B} = \mathbf{I}_m$ and $\mathbf{B} = \mathbf{J}_m$.

(i) Using (15.18a(i)),

$$\begin{aligned} \det \mathbf{C} &= \det(\mathbf{A} + \mathbf{B} - \mathbf{B})^{n-1} \det[\mathbf{A} + \mathbf{B} + (n-1)\mathbf{B}] \\ &= (1 - m)^{n-1}[1 + m(n-1)]. \end{aligned}$$

(ii) $\mathbf{C}$ is nonsingular if and only if $m > 1$.

(iii) If $\mathbf{C}^{-1}$ exists, it has the same block structure as $\mathbf{C}$ with $\mathbf{I}_m + a\mathbf{J}_m$ in each of the diagonal block positions and $b\mathbf{J}_m$ in all the off-diagonal block positions, where

$$a = \frac{m(m-1)(n-1)}{(n-1)m + 1} \quad \text{and} \quad b = \frac{-(m-1)}{(n-1)m + 1}(m-1).$$

15.20. Let

$$\mathbf{A} = \begin{pmatrix} a_1b_1 & 0 & 0 & \cdots & 0 \\ a_2b_1 & a_2b_2 & 0 & \cdots & 0 \\ . & . & & \cdots & . \\ a_nb_1 & a_nb_2 & a_nb_3 & \cdots & a_nb_n \end{pmatrix},$$

where all the a_i and b_i are nonzero. Then $\det \mathbf{A} = \prod_{i=1}^n (a_ib_i)$ and

$$\mathbf{A}^{-1} = \begin{pmatrix} (a_1b_1)^{-1} & 0 & 0 & \cdots & 0 & 0 \\ -(a_1b_2)^{-1} & (a_2b_2)^{-1} & 0 & \cdots & 0 & 0 \\ 0 & -(a_2b_3)^{-1} & (a_3b_3)^{-1} & \cdots & 0 & 0 \\ . & . & . & \cdots & . \\ 0 & 0 & 0 & \cdots & (a_{n-1}b_{n-1})^{-1} & 0 \\ 0 & 0 & 0 & \cdots & -(a_{n-1}b_n)^{-1} & (a_nb_n)^{-1} \end{pmatrix}.$$

For other patterned matrices that are either tridiagonal or have a tridiagonal inverse see Section 8.11.

15.21. Let $\mathbf{A} = (a_{ij})$ be any $n \times n$ matrix. If $\mathbf{B} = (a_{ij} - a_{i.}a_{.j}/a_{..})$, where $a_{i.} = \sum_j a_{ij}$, $a_{.j} = \sum_i a_{ij}$, and $a_{..} = \sum_i \sum_j a_{ij}$, then

$$\mathbf{B} = \mathbf{A} - \mathbf{A}\mathbf{1}_n(\mathbf{1}'_n\mathbf{A}\mathbf{1}_n)^{-1}\mathbf{1}'_n\mathbf{A}.$$

Proofs. Section 15.4.

15.17. Graybill [1983: 185].

15.18a. Graybill [1983: 191, 204, and a special case of 206, with $a-b$ replaced by a].

15.18b. Roy and Sarhan [1956: 230].

15.18c(i)–(ii). Graybill [1983: 193, 205] and Roy and Sarhan [1956].

15.18c(iii). Ouellette [1981: 284].

15.18c(iv). Ouellette [1981: 285].

15.18d. Roy and Sarhan [1956: 230].

15.19a(i). Graybill [1983: 231, with $\mathbf{A} - \mathbf{B}$ replaced by $\mathbf{A}$].

15.19a(ii). Simply replace $\mathbf{A}$ by $\mathbf{A} - \lambda\mathbf{I}_m$ in (a).

15.19b. Graybill [1983: 231].

15.20. Graybill [1983:186] and Roy and Sarhan [1956].

15.21. Given the vector $n \times 1$ vector $\mathbf{x} = (x_i)$, we have $\sum_i x_i = \mathbf{1}'_n\mathbf{x}$. We use this for the rows and columns of $\mathbf{A}$.

15.5 GENERALIZED INVERSES

15.5.1 Weak Inverses

15.22. Suppose that $\mathcal{C}(\mathbf{UBV}) \subset \mathcal{C}(\mathbf{A})$ (or equivalently $\mathbf{AA^-UBV} = \mathbf{UBV}$) and $\mathcal{C}[(\mathbf{UBV})'] \subset \mathcal{C}(\mathbf{A}')$ (or equivalently $\mathbf{UBVA^-A} = \mathbf{UBV}$), then we have the following weak inverses of $(\mathbf{A} + \mathbf{UBV})$.

$$
\begin{aligned}
\mathbf{G}_1 &= \mathbf{A}^- - \mathbf{A}^-(\mathbf{A}^- + \mathbf{A}^-\mathbf{UBVA}^-)^-\mathbf{A}^-\mathbf{UBVA}^-, \\
\mathbf{G}_2 &= \mathbf{A}^- - \mathbf{A}^-\mathbf{U}(\mathbf{U} + \mathbf{UBVA}^-\mathbf{U})^-\mathbf{UBVA}^-, \\
\mathbf{G}_3 &= \mathbf{A}^- - \mathbf{A}^-\mathbf{UB}(\mathbf{B} + \mathbf{BVA}^-\mathbf{UB})^-\mathbf{BVA}^-, \\
\mathbf{G}_4 &= \mathbf{A}^- - \mathbf{A}^-\mathbf{UBV}(\mathbf{V} + \mathbf{VA}^-\mathbf{UBV})^-\mathbf{VA}^-, \\
\mathbf{G}_5 &= \mathbf{A}^- - \mathbf{A}^-\mathbf{UBVA}^-(\mathbf{A}^- + \mathbf{A}^-\mathbf{UBVA}^-)^-\mathbf{A}^-.
\end{aligned}
$$

The above sufficient conditions are satisfied if $\mathbf{A}$ is nonsingular.

15.23. Let $\mathbf{X}$ be an $n \times p$ matrix of rank r, and let $\mathbf{H}$ be a $q \times p$ matrix of rank $p - r$ such that $\mathcal{C}(\mathbf{X}') \cap \mathcal{C}(\mathbf{H}') = \mathbf{0}$.

(a) $\begin{pmatrix} \mathbf{X} \\ \mathbf{H} \end{pmatrix}$ has rank p so that $\mathbf{A} = \mathbf{X}'\mathbf{X} + \mathbf{H}'\mathbf{H}$ is nonsingular.

(b) $\mathbf{A}^{-1}$ is a weak inverse of $\mathbf{X}'\mathbf{X}$.

(c) $\begin{pmatrix} \mathbf{X}'\mathbf{X} & \mathbf{H}' \\ \mathbf{H} & \mathbf{0} \end{pmatrix}$ is nonsingular if $q = p - r$, and its inverse is then a weak inverse of

the matrix $\mathbf{X}'\mathbf{X}$.

The above results arise in studying identifiability constraints in analysis of variance models.

15.24. Let $\mathbf{A}$ be $m \times n$, and let $\mathbf{x}$ and $\mathbf{y}$ be $m \times 1$ and $n \times 1$ vectors, respectively. If either $\mathbf{x} \in \mathcal{C}(\mathbf{A})$ or $\mathbf{y} \in \mathcal{C}(\mathbf{A}')$, then, for any weak inverse $\mathbf{A}^-$ of $\mathbf{A}$,

$$
(\mathbf{A} + \mathbf{xy}')^- = \mathbf{A}^- - \frac{\mathbf{A}^-\mathbf{xy}'\mathbf{A}^-}{1 + \mathbf{y}'\mathbf{A}^-\mathbf{x}},
$$

provided $1 + \mathbf{y}'\mathbf{A}^-\mathbf{x} \neq 0$.

Proofs. Section 15.5.1.

15.22. Quoted by Henderson and Searle [1981b: 58]; see also Harville [1997: 426–428] for some proofs.

15.23. Seber [1977: 74, 77].

15.24. Quoted in Rao and Rao [1998: 281]. Setting $\mathbf{C} = \mathbf{A} + \mathbf{xy}'$, we can show, after some algebra, that $\mathbf{CC^-C} = \mathbf{C}$. We make use of the fact that $\mathbf{AA}^-$ projects onto $\mathcal{C}(\mathbf{A})$. In particular, if $\mathbf{x} = \mathbf{Ay}$, then $\mathbf{AA^-x} = \mathbf{AA^-Ay} = \mathbf{Ay} = \mathbf{x}$.

15.5.2   Moore–Penrose Inverses

15.25. If $\mathbf{B}$ is nonsingular,

$$(\mathbf{A} + \mathbf{U}\mathbf{B}\mathbf{U}')^+ = \mathbf{A}^+ - \mathbf{A}^+\mathbf{U}(\mathbf{B}^{-1} + \mathbf{U}'\mathbf{A}^+\mathbf{U})^+\mathbf{U}'\mathbf{A}^+,$$

if and only if $\mathcal{C}(\mathbf{U}\mathbf{B}\mathbf{U}') \subseteq \mathcal{C}(\mathbf{A})$, or equivalently $\mathbf{A}^+\mathbf{A}\mathbf{U}\mathbf{B}\mathbf{U}' = \mathbf{U}\mathbf{B}\mathbf{U}'$. The result also holds if $\mathbf{A}$ is Hermitian (Williams and John [2000: 697]).

15.26. Let $\mathbf{A}$ be an $n \times n$ nonsingular matrix, and let $\mathbf{c}$ and $\mathbf{d}$ be $n \times 1$ vectors. Then $\mathbf{A} + \mathbf{c}\mathbf{d}'$ is singular if and only if $1 + \mathbf{d}'\mathbf{A}^{-1}\mathbf{c} = 0$ and, if this is the case, then

$$(\mathbf{A} + \mathbf{c}\mathbf{d}')^+ = (\mathbf{I}_m - \mathbf{y}\mathbf{y}^+)\mathbf{A}^{-1}(\mathbf{I}_n - \mathbf{x}\mathbf{x}^+),$$

where $\mathbf{x} = \mathbf{A}^{-1}\mathbf{d}$, $\mathbf{y} = \mathbf{A}^{-1}\mathbf{c}$, and $\mathbf{x}^+ = (\mathbf{x}'\mathbf{x})^{-1}\mathbf{x}'$ etc.

15.27. Let $\mathbf{A}$ be an $n \times n$ symmetric matrix, and suppose that $\mathbf{c}$ and $\mathbf{d}$ are $n \times 1$ vectors in $\mathcal{C}(\mathbf{A})$. If $1 + \mathbf{d}'\mathbf{A}^+\mathbf{c} \neq 0$, then

$$(\mathbf{A} + \mathbf{c}\mathbf{d}')^+ = \mathbf{A}^+ - \frac{\mathbf{A}^+\mathbf{c}\mathbf{d}'\mathbf{A}^+}{1 + \mathbf{d}'\mathbf{A}^+\mathbf{c}}.$$

15.28. Let $\mathbf{A}$ be an $m \times n$ complex matrix, $\mathbf{c} \in \mathbb{C}^m$, $\mathbf{d} \in \mathbb{C}^n$, and $\beta = 1 + \mathbf{d}^*\mathbf{A}^+\mathbf{c}$. Define $\mathbf{k} = \mathbf{A}^+\mathbf{c}$, $\mathbf{h}' = \mathbf{d}^*\mathbf{A}^+$, $\mathbf{u} = (\mathbf{I}_m - \mathbf{A}\mathbf{A}^+)\mathbf{c}$, and $\mathbf{v}' = \mathbf{d}^*(\mathbf{I}_n - \mathbf{A}^+\mathbf{A})$. Then:

(a) $\operatorname{rank}(\mathbf{A} + \mathbf{c}\mathbf{d}^*) = \operatorname{rank}\begin{pmatrix} \mathbf{A} & \mathbf{u} \\ \mathbf{d}^* & -\beta \end{pmatrix} - 1.$

(b) $(\mathbf{A} + \mathbf{c}\mathbf{d}^*)^+ = \mathbf{A}^+ - \mathbf{k}\mathbf{u}^+ - (\mathbf{h}\mathbf{v}^+)' + \beta\mathbf{v}^{+\prime}\mathbf{u}^+.$

Note that $\mathbf{x}^+ = \mathbf{x}^*/(\mathbf{x}^*\mathbf{x})$.

15.29. If $\mathbf{A}$ is block diagonal, then $\mathbf{A}^+$ is also block diagonal. For example,

$$\mathbf{A} = \begin{pmatrix} \mathbf{A}_1 & \mathbf{0} & \mathbf{0} \\ \mathbf{0} & \mathbf{A}_2 & \mathbf{0} \\ \mathbf{0} & \mathbf{0} & \mathbf{0} \end{pmatrix} \quad \text{if and only if} \quad \mathbf{A}^+ = \begin{pmatrix} \mathbf{A}_1^+ & \mathbf{0} & \mathbf{0} \\ \mathbf{0} & \mathbf{A}_2^+ & \mathbf{0} \\ \mathbf{0} & \mathbf{0} & \mathbf{0} \end{pmatrix}.$$

15.30. (Multinomial Distribution) Consider the variance matrix $\boldsymbol{\Sigma} = (\sigma_{ij})$ of an n-dimensional (singular) multinomial distribution. Here $\sigma_{ii} = np_i(1 - p_i)$ and $\sigma_{ij} = -np_ip_j$ $(i \neq j)$, where $0 < p_i < 1$ for all i and $p_1 + p_2 + \cdots + p_n = 1$. If $\mathbf{D}_\mathbf{p} = \operatorname{diag}(p_1, p_2, \ldots, p_n)$, then $\boldsymbol{\Sigma} = n(\mathbf{D}_\mathbf{p} - \mathbf{p}\mathbf{p}')$ is singular and

$$\boldsymbol{\Sigma}^+ = n^{-1}(\mathbf{I}_n - n^{-1}\mathbf{1}_n\mathbf{1}_n')\mathbf{D}_\mathbf{p}^{-1}(\mathbf{I}_n - n^{-1}\mathbf{1}_n\mathbf{1}_n').$$

15.31. Let

$$\mathbf{B} = \begin{pmatrix} \mathbf{V} & \mathbf{C}^* \\ \mathbf{C} & \mathbf{0} \end{pmatrix},$$

where $\mathbf{V}$ is $n \times n$ Hermitian non-negative definite and $\mathbf{C}$ is $r \times n$. Then

$$\mathbf{B}^+ = \begin{pmatrix} \mathbf{0} & \mathbf{C}^+ \\ \mathbf{C}^{+*} & -\mathbf{C}^{+*}\mathbf{V}\mathbf{C}^+ \end{pmatrix} + \begin{pmatrix} \mathbf{I}_n \\ -\mathbf{C}^{+*}\mathbf{V} \end{pmatrix} \mathbf{Q}(\mathbf{I}_n, -\mathbf{V}^+\mathbf{C}),$$

where $\mathbf{E} = \mathbf{I}_n - \mathbf{C}^+\mathbf{C}$ and $\mathbf{Q} = (\mathbf{E}\mathbf{V}\mathbf{E})^+$.

Proofs. Section 15.5.2.

15.25. John [2001: 1175].

15.26. Schott [2005: 197–198].

15.27. Quoted by Schott [2005: 217, exercise 5.32]. Can be proved in a manner similar to that of (15.24).

15.28. Campbell and Meyer [1979: 47–48]. They also list several special cases in which one or more of $\mathbf{u}$, $\mathbf{v}$, and β are zero. They also give Moore–Penrose inverses of $\begin{pmatrix} \mathbf{A} & \mathbf{c} \\ \mathbf{d}^* & \alpha \end{pmatrix}$.

15.30. Follows from (15.26) above, along with $\mathbf{D_p}^{-1}\mathbf{p} = \mathbf{1}_n$.

15.31. Campbell and Meyer [1979: 64].

CHAPTER 16

FACTORIZATION OF MATRICES

The factorization of a matrix $\mathbf{A}$ can be expressed two ways; either as a reduction $\mathbf{XAY} = \mathbf{C}$ or as a factorization $\mathbf{A} = \mathbf{URV}$. In many cases these are equivalent because of the presence of nonsingular matrices—for example, $\mathbf{A} = \mathbf{X}^{-1}\mathbf{C}\mathbf{Y}^{-1}$ if $\mathbf{X}$ and $\mathbf{Y}$ are nonsingular. Authors tend to have different preferences for which form they use. Useful summaries of some of the factorizations are given by Abadir and Magnus [2005: 158], Horn and Johnson [1985: 157], and Rao and Rao [1998: 190–193].

16.1 SIMILARITY REDUCTIONS

As eigenvalues are used in this section, we remind the reader of Definition 6.1. In what follows, we assume that an $n \times n$ matrix has eigenvalues $\lambda_1, \lambda_2, \ldots, \lambda_n$ with $|\lambda_1| \geq |\lambda_2| \geq \cdots \geq |\lambda_n| \geq 0$ and has distinct eigenvalues $\mu_1, \mu_2, \ldots, \mu_s$, similarly ordered, with algebraic and geometric multiplicities $m(\mu_j)$ and $g(\mu_j)$, respectively.

Definition 16.1. Let $\mathbf{A}$ and $\mathbf{B}$ be $n \times n$ matrices over $\mathcal{F}$. We say that $\mathbf{A}$ is *similar* to $\mathbf{B}$ if there exists a nonsingular matrix $\mathbf{K}$ over $\mathcal{F}$ such that $\mathbf{K}^{-1}\mathbf{AK} = \mathbf{B}$.

16.1. Let $\mathbf{A}$ be an $n \times n$ real or complex matrix.

(a) $\mathbf{A}$ is similar to its transpose.

(b) $\mathbf{A}^*\mathbf{A}$ is similar to $\mathbf{AA}^*$.

A Matrix Handbook for Statisticians. By George A. F. Seber
Copyright © 2008 John Wiley & Sons, Inc.

(c) $\overline{\mathbf{A}}\mathbf{A}$ is similar to $\mathbf{A}\overline{\mathbf{A}}$.

(d) $\mathbf{A}$ is similar to a symmetric matrix.

(e) $\mathbf{A}$ is similar to a complex triangular matrix (either upper or lower) whose diagonal elements are the eigenvalues of $\mathbf{A}$.

16.2. Let $\mathbf{A}$ and $\mathbf{B}$ be real $n \times n$ matrices. If $\mathbf{R}$ is a complex nonsingular matrix such that $\mathbf{R}^{-1}\mathbf{A}\mathbf{R} = \mathbf{B}$, then there exists a real nonsingular matrix $\mathbf{S}$ such that $\mathbf{S}^{-1}\mathbf{A}\mathbf{S} = \mathbf{B}$.

16.3. Let $\mathbf{A}$ be an upper-triangular matrix with distinct diagonal elements $\mathrm{diag}(\mathbf{A})$. Then there exists a unit upper-triangular matrix $\mathbf{R}$ (i.e., with ones on the diagonal) such that $\mathbf{R}^{-1}\mathbf{A}\mathbf{R} = \mathrm{diag}(\mathbf{A})$.

Definition 16.2. Let $\mathbf{J}_m(\lambda)$ be an $m \times m$ matrix of the form

$$\mathbf{J}_m(\lambda) = \begin{pmatrix} \lambda & 1 & 0 & \cdots & 0 & 0 \\ 0 & \lambda & 1 & \cdots & 0 & 0 \\ . & . & . & \cdots & . & . \\ 0 & 0 & 0 & \cdots & \lambda & 1 \\ 0 & 0 & 0 & \cdots & 0 & \lambda \end{pmatrix},$$

where $\mathbf{J}_1(\lambda) = \lambda$. Then $\mathbf{J}_m(\lambda)$ is said to be a *Jordan block matrix*. We find it convenient to include the case $m = 1$.

16.4. Every Jordan block $\mathbf{J}_m(\lambda)$ $(m > 1)$ is not diagonalizable because it has only one linearly independent eigenvector $\mathbf{x} = (x_1, 0, \ldots, 0)'$, where x_1 is arbitrary (cf. Definition 16.3 above (16.10)). This follows from the fact the diagonal elements of the upper-triangular matrix $\mathbf{J}_m(\lambda)$ are its eigenvalues, so that it has one eigenvalue λ repeated m times, and $\mathbf{x}$ satisfies $\mathbf{J}_m(\lambda)\mathbf{x} = \lambda\mathbf{x}$.

16.5. Every Jordan block is permutation similar to its transpose since $\mathbf{J}_m(\lambda)' = \mathbf{\Pi}\mathbf{J}_m(\lambda)\mathbf{\Pi}$, where $\mathbf{\Pi} = (\mathbf{e}_m, \mathbf{e}_{m-1}, \ldots, \mathbf{e}_1)$ is the backward identity permutation matrix, where $(\mathbf{e}_1, \mathbf{e}_2, \ldots, \mathbf{e}_m) = \mathbf{I}_m$.

16.6. Let $\mathbf{x} = (x_i)$ be an $m \times 1$ vector. The Jordan block $\mathbf{J} = \mathbf{J}_m(0)$ has the following properties.

(a) $\mathbf{J}\mathbf{x} = (x_2, x_3, \ldots, x_m, 0)'$, representing a *forward shift*.

(b) $\mathbf{J}'\mathbf{x} = (0, x_1, x_2, \ldots, x_{m-1})'$, representing a *backward shift*.

(c) $(\mathbf{I}_m - \mathbf{J}')\mathbf{x} = (x_1, x_2 - x_1, x_3 - x_2, \ldots, x_m - x_{m-1})'$, representing a *difference operator*.

(d) $(\mathbf{I}_m - \mathbf{J}')^{-1}\mathbf{x} = (x_1, x_1 + x_2, x_1 + x_2 + x_3, \ldots, x_1 + x_2 + \cdots + x_m)'$, which can be called a *partial sum operator*.

16.7. (Jordan Canonical Form) If $\mathbf{A}$ is a real or complex $n \times n$ matrix, then there exists a nonsingular matrix $\mathbf{R}$ such that

$$\mathbf{R}^{-1}\mathbf{A}\mathbf{R} = \begin{pmatrix} \mathbf{J}_{n_1}(\lambda_1) & \mathbf{0} & \cdots & \mathbf{0} \\ \mathbf{0} & \mathbf{J}_{n_2}(\lambda_2) & \cdots & \mathbf{0} \\ . & . & \cdots & . \\ \mathbf{0} & \mathbf{0} & \cdots & \mathbf{J}_{n_k}(\lambda_k) \end{pmatrix} = \mathbf{J}_O,$$

where $\sum_{i=1}^{k} n_i = n$ and the λ_i are the (not necessarily distinct) eigenvalues of $\mathbf{A}$; that is, $\mathbf{A}$ is similar to $\mathbf{J}_O$. The matrix $\mathbf{J}_O$ is said to be in *Jordan canonical form*, which is unique apart from the order of the blocks. One application of the Jordan canonical form is in the analysis of a system of ordinary differential equations with constant coefficients (Horn and Johnson [1985: 132–133]). The topic of *Jordan chains* is considered by Abadir and Magnus [2005: section 7.6]

If $\mu_1, \mu_2, \ldots, \mu_s$ are the distinct λ_i, then we have the following.

(a) The number k of Jordan blocks (including multiple occurrences of the same blocks) is the number of linearly independent eigenvectors of $\mathbf{J}_O$.

(b) The matrix $\mathbf{J}_O$ is diagonalizable (cf. Definition 16.3 above (16.10)) if and only if $k = n$.

(c) The number of Jordan blocks correponding to μ_j is the geometric multiplicity $g(\mu_j)$. The sum of the orders (sizes) of all the Jordan blocks corresponding to μ_j is the algebraic multiplicity $m(\mu_j)$.

(d) $\mathbf{J}_O$ is not completely determined in general by a knowledge of the eigenvalues and their algebraic and geometric multiplicities. We must also know the sizes of the Jordan blocks corresponding to each λ_i.

(e) The minimal polynomial of $\mathbf{J}_O$ (and therefore of $\mathbf{A}$, as similar matrices have the same minimal polynomial, cf. (6.12)) is

$$f(\mu) = \prod_{j=1}^{s} (\mu - \mu_j)^{r_j},$$

where r_j is the order of the largest Jordan block of $\mathbf{J}_O$ corresponding to μ_j.

(f) The sizes of the Jordan blocks corresponding to a given μ_j are determined by a knowledge of the ranks of certain powers.

(g) If $\mathbf{A}$ is a real matrix with only real eigenvalues, then the similarity matrix $\mathbf{R}$ can be taken to be real.

(h) It is convenient to standardize the order of the Jordan blocks as follows. For each μ_j we have $g(\mu_j)$ blocks that we order in decreasing size, and we order these s groups of blocks according to our convention $|\mu_1| > \cdots > |\mu_s|$; $\mathbf{J}_O$ is then unique.

16.8. If $\mathbf{A}$ and $\mathbf{B}$ are $n \times n$ similar matrices, then they have the same Jordan canonical form.

16.9. Let $\mathbf{A}$ be an $n \times n$ upper-triangular matrix with zeros on the main diagonal (sometimes called a *strictly upper-triangular matrix*).

(a) There exists a nonsingular $n \times n$ matrix $\mathbf{S}$ and integers $n_1, n_2, \ldots, n_m$ with $n_1 \geq n_2 \geq \cdots \geq n_m$ and $n_1 + n_2 + \cdots + n_m = n$ such that

$$\mathbf{S}^{-1}\mathbf{A}\mathbf{S} = \begin{pmatrix} \mathbf{J}_{n_1}(0) & \mathbf{0} & \cdots & \mathbf{0} \\ \mathbf{0} & \mathbf{J}_{n_2}(0) & \cdots & \mathbf{0} \\ . & . & \cdots & . \\ \mathbf{0} & \mathbf{0} & \cdots & \mathbf{J}_{n_m}(0) \end{pmatrix}.$$

(b) If **A** is nilpotent with nilpotency index k, then $m = \dim \mathcal{N}(\mathbf{A})$, the size of the largest block is $k \times k$, and each block is nilpotent.

Definition 16.3. If **A** is similar to a diagonal matrix, then **A** is said to be *diagonalizable*. Other terms used are *diagonable, simple, semi-simple* or *nondefective*. Note that **A** can be real or complex.

16.10. **A** is diagonalizable if and only if one, and hence all, of the following equivalent conditions are satisfied (cf. Definition 6.1):

(1) $m(\mu_j) = g(\mu_j)$ for each j; that is, the eigenvalues of **A** are all regular.

(2) $g(\mu_1) + g(\mu_2) + \cdots + g(\mu_s) = n$, that is, the sum of the eigenspaces of **A** is $\mathbb{C}^n$.

(3) $\operatorname{rank}(\mathbf{A} - \mu_j \mathbf{I}_n) = n - m(\mu_j)$ for $j = 1, 2, \ldots, s$. The equivalence with (1) follows from (6.1b)

(4) **A** possesses n linearly independent right (respectively left) eigenvectors.

16.11. If **A** has n distinct eigenvalues, then they are simple and therefore regular, so that by (16.10(1)) above **A** is diagonalizable.

16.12. A matrix **A** is diagonalizable if and only if $h(\mathbf{A}) = \mathbf{0}$, where

$$h(x) = (x - \mu_1)(x - \mu_2) \cdots (x - \mu_s),$$

and the μ_j are the distinct eigenvalues of **A**. If $h(\mathbf{A}) = \mathbf{0}$, then $h(\lambda) = q(\lambda)$, the minimal polynomial.

16.13. It follows from (16.12) above that an idempotent matrix **A** is diagonalizable since $h(\mathbf{A}) = \mathbf{0}$, where

$$h(x) = x(x - 1),$$

and the eigenvalues of **A** are $\lambda = 0, 1$.

16.14. (Product) Suppose **A** and **B** are $n \times n$ Hermitian matrices with **A** positive-definite. Then **AB** has real eigenvalues and is diagonalizable. Also, **AB** has the same number of positive, negative, and zero eigenvalues as **B**. Furthermore, any diagonalizable matrix with real eigenvalues is the product of a positive definite Hermitian matrix and a Hermitian matrix.

16.15. (Approximation) Let $\mathbf{A} = (a_{ij})$ be an $n \times n$ matrix. For every $\epsilon > 0$, there exists a matrix $\mathbf{A}(\epsilon) = (a_{ij}(\epsilon))$ with distinct eigenvalues (and is therefore diagonalizable) such that

$$\sum_{i=1}^{n} \sum_{j=1}^{n} |a_{ij} - a_{ij}(\epsilon)|^2 < \epsilon.$$

16.16. Let **A** be $n \times n$ and **B** be $m \times m$ matrices, and let

$$\mathbf{C} = \begin{pmatrix} \mathbf{A} & \mathbf{0} \\ \mathbf{0} & \mathbf{B} \end{pmatrix}.$$

Then **C** is diagonalizable if and only if **A** and **B** are both diagonalizable.

16.17. (Spectral Decomposition) Suppose $\mathbf{A}$ is diagonalizable of rank r.

(a) There exist linearly independent right eigenvectors $\mathbf{x}_1, \mathbf{x}_2, \ldots, \mathbf{x}_n$ and linearly independent left eigenvectors $\mathbf{y}_1', \mathbf{y}_2', \ldots, \mathbf{y}_n'$ such that $\mathbf{y}_i'\mathbf{x}_j = \delta_{ij}$, where $\delta_{ij} = 1$ when $i = j$ and 0 otherwise. Also

$$
\begin{aligned}
\mathbf{A} &= \sum_{i=1}^{n} \lambda_i \mathbf{x}_i \mathbf{y}_i' \\
&= \sum_{i=1}^{n} \lambda_i \mathbf{F}_i \\
&= \sum_{i=1}^{r} \lambda_i \mathbf{F}_i
\end{aligned}
$$

for nonzero λ_i, where the rank one $\mathbf{F}_i$ are not unique unless all the eigenvalues are distinct. Here the $\mathbf{F}_i$ are idempotent, mutually orthogonal, and $\sum_{i=1}^{n} \mathbf{F}_i = \mathbf{I}_n$. If $\mathbf{R} = (\mathbf{x}_1, \mathbf{x}_2, \ldots, \mathbf{x}_n)$ and $\mathbf{S} = (\mathbf{y}_1, \mathbf{y}_2, \ldots, \mathbf{y}_n)'$, then $\mathbf{A} = \mathbf{R}\boldsymbol{\Lambda}\mathbf{S}$, where $\mathbf{SR} = \mathbf{I}_n$ implies that $\mathbf{S} = \mathbf{R}^{-1}$. Note that as the rank is unchanged when multiplying by a nonsingular matrix, rank $\mathbf{A} = $ rank $\boldsymbol{\Lambda}$, and rank $\mathbf{A}$ is the number of nonzero eigenvalues of $\mathbf{A}$.

(b) (Unique Decomposition)

(i) We can also write

$$
\mathbf{A} = \sum_{j=1}^{s} \mu_j \mathbf{E}_j,
$$

where $\mathbf{E}_j$ represents the sum of the $\mathbf{F}_i$ corresponding to the same eigenvalue and the μ_j are the distinct eigenvalues (including zero). The $\mathbf{E}_j$, called the *spectral set*, are unique, idempotent, and mutually orthogonal (i.e., $\mathbf{E}_j \mathbf{E}_k = \delta_{jk} \mathbf{E}_j$) and satisfy $\sum_{j=1}^{s} \mathbf{E}_j = \mathbf{I}_n$.

(ii) Also, for $k = 1, 2, \ldots$,

$$
\mathbf{A}^k = \sum_{j=1}^{s} \mu_j^k \mathbf{E}_j.
$$

(c) If $\mathbf{A}$ is nonsingular, then

$$
\mathbf{A}^{-1} = \sum_{j=1}^{s} \mu_j^{-1} \mathbf{E}_j.
$$

If $\mathbf{A}$ is singular, then

$$
\mathbf{A}^{+} = \sum_{j} \mu_j^{-1} \mathbf{F}_j,
$$

where the summation is over the nonzero eigenvalues, and $\mathbf{A}^{+}$ is the Moore–Penrose inverse of $\mathbf{A}$. This is proved directly from the definition of $\mathbf{A}^{+}$.

For a spectral decomposition of an arbitrary matrix see Rao and Mitra [1971: 38].

16.18. Suppose $\mathbf{A}$ is an $n \times n$ diagonalizable matrix with distinct eigenvalues λ_i. Then the Vandermonde matrix (cf. Section 8.12.1)

$$
\mathbf{B} = \begin{pmatrix}
1 & 1 & \cdots & 1 \\
\lambda_1 & \lambda_2 & \cdots & \lambda_n \\
\lambda_1^2 & \lambda_2^2 & \cdots & \lambda_n^2 \\
. & . & \cdots & . \\
\lambda_1^{n-1} & \lambda_2^{n-1} & \cdots & \lambda_n^{n-1}
\end{pmatrix}
$$

is nonsingular with inverse $\mathbf{B}^{-1} = (\beta_{ij})$, say. From the previous result (16.17b), with $s = n$ and $\mathbf{E}_j = \mathbf{F}_j$, we have

$$
\begin{pmatrix}
\mathbf{I}_n \\
\mathbf{A} \\
\vdots \\
\mathbf{A}^{n-1}
\end{pmatrix} = (\mathbf{B} \otimes \mathbf{I}_n) \begin{pmatrix}
\mathbf{F}_1 \\
\mathbf{F}_2 \\
\vdots \\
\mathbf{F}_n
\end{pmatrix},
$$

where "$\otimes$" is the Knonecker product. Now $\mathbf{B} \otimes \mathbf{I}_n$ can be expressed symbolically as $(b_{ij}\mathbf{I}_n)$ so that using $(\mathbf{B} \otimes \mathbf{I}_n)^{-1} = \mathbf{B}^{-1} \otimes \mathbf{I}_n$ and defining $\mathbf{A}^0 = \mathbf{I}_n$ we have

$$
\mathbf{F}_i = \sum_{j=1}^{n} \beta_{ij} \mathbf{A}^{j-1}.
$$

Substituting in $\mathbf{A}^k$ from (16.17b(ii)) gives us

$$
\mathbf{A}^k = \sum_{i=1}^{n} \sum_{j=1}^{n} \lambda_i^k \beta_{ij} \mathbf{A}^{j-1}, \quad k = 1, 2, \ldots.
$$

16.19. If $\mathbf{A}$ is real symmetric (respectively Hermitian) $n \times n$ matrix with distinct eigenvalues μ_j $(j = 1, 2, \ldots, s)$, then $m(\mu_j) = g(\mu_j)$ for $j = 1, 2, \ldots, s$. Hence, by (16.10(1)) above, all real symmetric (respectively Hermitian) matrices are diagonalizable.

Proofs. Section 16.1.

16.1a. Horn and Johnson [1985: 134–135], Meyer [2000a: 596], and Zhang [1999: 83].

16.1b–c. Zhang [1999: 83].

16.1d. Quoted by Rao and Rao [1998: 192].

16.1e. Rao and Bhimasankaram [2000: 288–289].

16.2. Zhang [1999: 152].

16.3. Abadir and Magnus [2006: 186].

16.5. Horn and Johnson [1985: 134].

16.6. Abadir and Magnus [2005: 193].

16.7. For proofs, references, and comments see Abadir and Magnus [2005: section 7.5], Horn and Johnson [1985: section 3.1], Meyer [2000a: sections 7.7 and 7.8], and Rao and Bhimasankaram [2000: section 8.6].

16.8. The result follows from the uniqueness of the Jordan canonical form.

16.9a. Abadir and Magnus [2005: 195–196] and Horn and Johnson [1985: 123].

16.9b. Meyer[2000a: 579].

16.10. Rao and Bhimasankaram [2000: 296–297].

16.11. Horn and Johnson [1985: 48].

16.12. Horn and Johnson [1985: 145] and Rao and Bhimasankaram [2000: 296–297].

16.13. Rao and Bhimasankaram [2000: 297].

16.14. Horn and Johnson [1985: 465].

16.15. Horn and Johnson [1985: 89].

16.16. Horn and Johnson [1985: 49].

16.17a. Harville [1997: section 21.5].

16.17b. Rao and Bhimasankaram [2000: 299–300].

16.2 REDUCTION BY ELEMENTARY TRANSFORMATIONS

16.2.1 Types of Transformation

Definition 16.4. An *elementary row transformation* of an $m \times n$ matrix $\mathbf{A}$ over $\mathcal{F}$ is one of the following operations:

(1) Multiply row i by a scalar c in $\mathcal{F}$. This achieved by left-multiplying $\mathbf{A}$ by the identity matrix $\mathbf{I}_m$ with its ith diagonal element replaced by c. The latter has determinant c.

(2) Add row j to row i. This is achieved by left-multiplying by the matrix $\mathbf{I}_m + \mathbf{E}_{ij}$, where $\mathbf{E}_{ij}$ has 1 in the (i, j)th position and zeros elsewhere. This transformation has determinant 1.

(3) Interchange the ith and jth rows. This is achieved by left-multiplying by the permutation matrix $\mathbf{\Pi}_{ij}$, where $\mathbf{\Pi}_{ij}$ is $\mathbf{I}_m$ with its ith and jth rows interchanged. (Technically the third transformation can be carried out using a sequence of the previous two transformations, but that route is less convenient.)

These operations can also be extended to submatrices of partitioned matrices (cf. Zhang 1999: 30]).

16.20. Note the following:

(a) $\mathbf{\Pi}_{ij}$ is symmetric and orthogonal so that $\mathbf{\Pi}_{ij}^{-1} = \mathbf{\Pi}_{ij}$.

(b) $\mathbf{E}_{ij}^2 = \mathbf{0}$ and $(\mathbf{I}_m + c\mathbf{E}_{ij})^{-1} = \mathbf{I}_m - c\mathbf{E}_{ij}$.

***Definition* 16.5.** An *elementary (row) transformation matrix* $\mathbf{M}$ is defined to be one of the above three types of matrices, referred to as types (1), (2), and (3). For further details see, for example, Abadir and Magnus [2005: section 6.1]

Elementary column transformations can be carried out by right-multiplying $\mathbf{A}$ by an elementary transformation matrix (but using $\mathbf{E}_{ji} = \mathbf{E}_{ij}'$ instead of $\mathbf{E}_{ij}$).

16.21. The inverse of an elementary transformation matrix is also an elementary transformation matrix. Also, as such matrices are all nonsingular, a product of such matrices is nonsingular. Therefore multiplying $\mathbf{A}$ by such a matrix does not change the rank of $\mathbf{A}$ (cf. 3.14a).

16.2.2   Equivalence Relation

***Definition* 16.6.** Let $\mathbf{A}$ and $\mathbf{B}$ be $m \times n$ real or complex matrices. If $\mathbf{B}$ is obtained from $\mathbf{A}$ by elementary row or column transformations matrices, then $\mathbf{A}$ is said to be equivalent to $\mathbf{B}$, and we write $\mathbf{A} \sim \mathbf{B}$.

16.22. Any one of the following statements implies the other two.

(1) $\mathbf{A} \sim \mathbf{B}$.

(2) $\mathbf{B} = \mathbf{RAS}$ for some non-singular matrices $\mathbf{R}$ and $\mathbf{S}$.

(3) $\operatorname{rank} \mathbf{A} = \operatorname{rank} \mathbf{B}$ (cf. 3.14a).

***Definition* 16.7.** From (16.22(2)) above we see that: (i) $\mathbf{A} \sim \mathbf{A}$ (reflexive), (ii) if $\mathbf{A} \sim \mathbf{B}$, then $\mathbf{B} \sim \mathbf{A}$ (symmetric), and (iii) if $\mathbf{A} \sim \mathbf{B}$ and $\mathbf{B} \sim \mathbf{C}$, then $\mathbf{A} \sim \mathbf{C}$ (transitive). Any relation that satisfies these three conditions is called an *SRT relation*. Thus the equivalence relation "$\sim$" is an SRT relation.

Other SRT relations for square matrices are summarised as follows:

(1) If $\mathbf{B} = \mathbf{R}^{-1}\mathbf{AR}$ for nonsingular $\mathbf{R}$, then $\mathbf{B}$ is said to be *similar* to $\mathbf{A}$. This is discussed in Section 16.1 above.

(2) If $\mathbf{B} = \mathbf{R}'\mathbf{AR}$ for nonsingular $\mathbf{R}$, then $\mathbf{B}$ is said to be *congruent* to $\mathbf{A}$. Its main application is for real matrices. If $\mathbf{A}$ and $\mathbf{B}$ are complex matrices such that $\mathbf{B} = \mathbf{R}^*\mathbf{AR}$, then $\mathbf{B}$ is said to be *Hermitian congruent* to $\mathbf{A}$.

(3) If $\mathbf{B} = \mathbf{U}^*\mathbf{AU}$, where $\mathbf{U}$ is unitary, then $\mathbf{B}$ is said to be *unitarily similar* to $\mathbf{A}$. If, for real matrices, $\mathbf{B} = \mathbf{T}'\mathbf{AT}$, where $\mathbf{T}$ is orthogonal, we say that $\mathbf{B}$ is *orthogonally similar* to $\mathbf{A}$.

16.2.3   Echelon Form

***Definition* 16.8.** Using elementary row transformations, a real or complex $m \times n$ matrix $\mathbf{A}$ can be reduced to a matrix $\mathbf{B}$ with the following properties:

(1) If a row contains at least one nonzero entry, then the first nonzero entry is 1.

(2) The zero rows, if any, come last.

(3) In any two consecutive nonzero rows, the leading 1 in the lower row occurs further to the right than the leading 1 in the upper row.

A matrix in the above form is said to be in *(row) echelon form*. For example,

$$\mathbf{B}_1 = \begin{pmatrix} 0 & 1 & * & * & * & * \\ 0 & 0 & 0 & 1 & * & * \\ 0 & 0 & 0 & 0 & 1 & * \\ 0 & 0 & 0 & 0 & 0 & 0 \end{pmatrix},$$

where the elements denoted * are arbitrary. If we now subtract multiples of the second and third rows from the first, we obtain

$$\mathbf{B}_2 = \begin{pmatrix} 0 & 1 & * & 0 & 0 & * \\ 0 & 0 & 0 & 1 & 0 & * \\ 0 & 0 & 0 & 0 & 1 & * \\ 0 & 0 & 0 & 0 & 0 & 0 \end{pmatrix}.$$

This matrix has the additional property:

(4) Each column that contains a leading 1 has zeros elsewhere.

A matrix with the above four properties is said to be in *reduced (row) echelon form* We shall omit the word "row" in using the above definitions. Rao and Bhimasankaram [2000: 167–170] give a number of algorithms for carrying out various reductions. It should be noted that the terminology relating to echelon forms is not consistent in the literature.

We see that the first three rows of $\mathbf{B}_2$ give a row basis for the original matrix $\mathbf{A}$, and the three columns each containing 1 form a column basis for $\mathbf{A}$.

16.23. Any matrix $\mathbf{A}$ can be reduced to a unique matrix in reduced echelon form by elementary row transformations.

16.24. The rank of a matrix in reduced echelon form is the number of nonzero rows. This is the same as the rank of the original matrix.

16.25. If $\mathbf{A}$ is a nonsingular matrix of order n, then its reduced echelon form is $\mathbf{I}_n$. Hence there exist elementary transformation matrices $\mathbf{M}_k$, $k = 1, 2, \ldots, K$, such that $\mathbf{M}_K \mathbf{M}_{K-1} \cdots \mathbf{M}_1 \mathbf{A} = \mathbf{I}_n$, i.e., $\mathbf{MA} = \mathbf{I}_n$, where $\mathbf{M}$ is nonsingular. Also, taking $\mathbf{A}$ over to the right-hand side, $\mathbf{M}_K \mathbf{M}_{K-1} \cdots \mathbf{M}_1 \mathbf{I}_n = \mathbf{A}^{-1}$. Thus any sequence of elementary row transformations that transforms $\mathbf{A}$ to $\mathbf{I}_n$ transforms $\mathbf{I}_n$ to $\mathbf{A}^{-1}$.

16.26. For any two $n \times p$ matrices $\mathbf{A}$ and $\mathbf{B}$, the following statements are equivalent.

(1) $\mathcal{C}(\mathbf{A}') = \mathcal{C}(\mathbf{B}')$.

(2) The reduced echelon forms of $\mathbf{A}$ and $\mathbf{B}$ are the same.

(3) $\mathbf{B}$ can be obtained from $\mathbf{A}$ by a finite sequence of elementary row operations.

(4) $\mathbf{B} = \mathbf{KA}$ for some nonsingular matrix $\mathbf{K}$.

Proofs. Section 16.2.3.

16.23. Rao and Bhimasankaram [2000: 172].

16.26. Rao and Bhimasankaram [2000: 171–172].

16.2.4   Hermite Form

Definition 16.9. A square matrix $\mathbf{H}$ is said to be in *(upper) Hermite form* if (a) it is upper-triangular, (b) its principal diagonal elements are all zeros or ones, (c) when a diagonal element is zero the entire row is zero, and (d) when a diagonal element is one, the rest of the elements in the column are all zeros. For example, a Hermite form for a 5×5 matrix $\mathbf{A}$ could take the form

$$
\mathbf{H_A} =
\begin{pmatrix}
0 & 0 & 0 & 0 & 0 \\
0 & 1 & 0 & * & 0 \\
0 & 0 & 1 & * & 0 \\
0 & 0 & 0 & 0 & 0 \\
0 & 0 & 0 & 0 & 1
\end{pmatrix},
$$

where the starred elements are arbitrary. If $\mathbf{H}$ comes from $\mathbf{A}$ we shall write $\mathbf{H_A}$.

There is a close relationship between the reduced echelon form and the Hermite form of a matrix. For example, the reduced echelon form corresponding to $\mathbf{H_A}$ would be

$$
\mathbf{B} =
\begin{pmatrix}
0 & 1 & 0 & * & 0 \\
0 & 0 & 1 & * & 0 \\
0 & 0 & 0 & 0 & 1 \\
0 & 0 & 0 & 0 & 0 \\
0 & 0 & 0 & 0 & 0
\end{pmatrix}.
$$

We see that $\mathbf{H_A}$ can be obtained from $\mathbf{B}$ by simply interchanging rows, i.e., by carrying out elementary row transformations. This is the case in general so that many of the results for reduced echelon forms apply to Hermite forms, as we shall see later.

16.27. $\mathbf{H_A^2} = \mathbf{H_A}$.

16.28. If $\mathbf{A}$ is a square matrix over $\mathcal{F}$, there exists a nonsingular matrix $\mathbf{K}$ such that $\mathbf{KA} = \mathbf{H_A}$. The matrix $\mathbf{K}$ is a product of elementary row transformation matrices.

16.29. Two real $n \times n$ matrices $\mathbf{A}$ and $\mathbf{B}$ have the same Hermite form if and only if $\mathcal{C}(\mathbf{A}') = \mathcal{C}(\mathbf{B}')$. The following are consequences of this result.

(a) $\mathbf{A'A}$, $\mathbf{A^{-}A}$ and $\mathbf{A}$ have the same Hermite form.

(b) If $\mathbf{B}$ is nonsingular, then $\mathbf{BA}$ and $\mathbf{A}$ have the same Hermite form.

16.30. Let $\mathbf{A}$ be $n \times n$. Since $\mathbf{H_A^2} = \mathbf{H_A}$ we have the following.

(a) $\mathbf{AH_A} = \mathbf{A}$.

(b) The identity matrix $\mathbf{I}_n$ is the only $n \times n$ matrix in Hermite form that is nonsingular. Thus if $\mathbf{A}$ is nonsingular, then $\mathbf{H_A} = \mathbf{I}_n$.

16.31. (Rank)

(a) rank $\mathbf{H_A}$ = rank $\mathbf{A}$.

(b) The rank of a matrix in Hermite form is the number of non-null rows in it, or the number of diagonal elements equal to one. Thus reducing a matrix to echelon form is a method of finding its rank. (For an algorithm see Rao and Bhimasankaram [2000: 181–182].)

(c) If the $i_1, i_2, \ldots, i_k$ diagonal elements of $\mathbf{H_A}$ are each equal to one, and the remaining diagonal elements of $\mathbf{H_A}$ are equal to zero, then the $i_1, i_2, \ldots, i_k$ columns of $\mathbf{A}$ are linearly independent.

16.32. (Idempotency)

(a) $\mathbf{A}$ is idempotent if and only if $\mathbf{H_A}$ is a weak inverse of $\mathbf{A}$.

(b) $\mathbf{A}$ is idempotent if and only if $\mathbf{H_A A} = \mathbf{H_A}$.

***Definition* 16.10.** An $n \times n$ matrix $\mathbf{H}$ is said to be in *(upper) Hermite canonical form* if takes the form

$$\mathbf{H} = \begin{pmatrix} \mathbf{I}_r & \mathbf{C} \\ \mathbf{0} & \mathbf{0} \end{pmatrix}.$$

By looking at the example given in Definition 16.9, we see that a Hermite form can be transformed into a Hermite canonical form by carrying out suitable row and column interchanges. This process can be carried further to transform $\mathbf{C}$ into the zero matrix, as we see in the next result.

16.33. (Reduction to Diagonal Form) Let $\mathbf{A}$ be an $m \times n$ matrix of rank r defined over $\mathcal{F}$.

(a) There exist nonsingular matrices $\mathbf{F}$ and $\mathbf{G}$ of sizes $m \times m$ and $n \times n$, respectively, such that

$$\mathbf{FAG} = \begin{pmatrix} \mathbf{I}_r & \mathbf{0} \\ \mathbf{0} & \mathbf{0} \end{pmatrix},$$

so that $\mathbf{A}$ is equivalent to a diagonal matrix. Thus

$$\begin{aligned} \mathbf{A} &= \mathbf{F}^{-1} \begin{pmatrix} \mathbf{I}_r & \mathbf{0} \\ \mathbf{0} & \mathbf{0} \end{pmatrix} \mathbf{G}^{-1} \\ &= \mathbf{R} \begin{pmatrix} \mathbf{I}_r & \mathbf{0} \\ \mathbf{0} & \mathbf{0} \end{pmatrix} \mathbf{S}, \end{aligned}$$

say. (Some bordering matrices are absent if $\mathbf{A}$ has full row or column rank.) The matrices $\mathbf{F}$ and $\mathbf{G}$ and their respective inverses $\mathbf{R}$ and $\mathbf{S}$ are all products of elementary transformation matrices.

(b) (Full Rank Factorization) From (a) we have

$$\mathbf{A} = (\mathbf{R}_1, \mathbf{R}_2) \begin{pmatrix} \mathbf{I}_r & \mathbf{0} \\ \mathbf{0} & \mathbf{0} \end{pmatrix} \begin{pmatrix} \mathbf{S}_1 \\ \mathbf{S}_2 \end{pmatrix} = \mathbf{R}_1 \mathbf{S}_1,$$

where $\mathbf{R}_1$ is $m \times r$ of rank r and $\mathbf{S}_1$ is an $r \times n$ of rank r.

(c) (Singular Value Decomposition) If $\mathbf{A}$ is real (respectively complex), we can choose $\mathbf{R}$ with orthogonal columns and $\mathbf{S}$ with orthogonal rows. If we then incorporate the lengths of the columns of $\mathbf{R}$ and the rows of $\mathbf{S}$ into $\mathbf{I}_r$, we get what is effectively the singular value decomposition of $\mathbf{A}$, namely

$$A = P \begin{pmatrix} D_r & 0 \\ 0 & 0 \end{pmatrix} Q,$$

where $\mathbf{P}$ and $\mathbf{Q}$ are orthogonal (respectively unitary) matrices and $\mathbf{D}_r$ is a diagonal matrix with positive elements. This decomposition is discussed in more detail in Section 16.3. We note that

$$A = (P_1, P_2) \begin{pmatrix} D_r & 0 \\ 0 & 0 \end{pmatrix} \begin{pmatrix} Q_1 \\ Q_2 \end{pmatrix} = (P_1 D_r^{1/2})(D_r^{1/2} Q_1) = P_2 Q_2.$$

Thus $\mathbf{A}$ can be expressed in the form $\mathbf{P}_2 \mathbf{Q}_2$, where $\mathbf{P}_2$ has orthogonal columns and $\mathbf{Q}_2$ has orthogonal rows. We can choose $\mathbf{P}_2$ (respectively $\mathbf{Q}_2$) to have orthonormal columns (respectively rows).

Proofs. Section 16.2.4.

16.27. Quoted by Graybill [1983: 131].

16.28. Graybill [1983: 130] and Rao [1973: 18].

16.29–16.31. Graybill [1983: 138–140].

16.32. Graybill [1983: 140–141].

16.33a. Marsaglia and Styan [1974a: 280, theorem 10] and Rao [1973: 19].

16.33b. Marsaglia and Styan [1974a: 271, theorem 1] and Rao [1973: 19].

16.3   SINGULAR VALUE DECOMPOSITION (SVD)

The singular value decomposition is regarded by many as one of the most useful factorizations for real or complex matrices. For example, the SVD has many applications in statistics such as SAS (single-spectrum analysis) in times series (cf. Golyandina et al. [2001: chapter 4]), matrix approximation in dimension reduction techniques, least squares approximation of a square matrix by a scalar multiple of an orthogonal or unitary matrix (Horn and Johnson [1985: 429], and procrustes analysis (Gower and Dijksterhuis [2004] and Seber [1984: section 5.6]). It is also a useful computational tool for calculating various quantities. In what follows, we interpret the transpose as conjugate transpose when dealing with complex matrices.

Definition 16.11. Any $m \times n$ real (respectively complex) matrix of rank r ($r \le p = \min\{m, n\}$) can be expressed in the form

$$\mathbf{A}_{m \times n} = \mathbf{P}_{m \times m} \boldsymbol{\Sigma}_{m \times n} \mathbf{Q}'_{n \times n},$$

where $\mathbf{P}$ is an $m \times m$ orthogonal (respectively unitary) matrix, $\mathbf{Q}$ is an $n \times n$ orthogonal (respectively unitary) matrix, and $\boldsymbol{\Sigma} = (\sigma_{ij})$ is an $m \times n$ matrix with

$$\sigma_{11} \ge \sigma_{22} \cdots \ge \sigma_{rr} > 0 = \sigma_{r+1,r+1} = \cdots = \sigma_{pp}, \text{ and } \sigma_{ij} = 0 \text{ for all } i, j, i \ne j.$$

This factorization of $\mathbf{A}$ is called the *singular value decomposition*. The σ_{ii}, abbreviated to $\sigma_i = \sigma_i(\mathbf{A})$ $(i = 1, 2, \ldots, p)$, are called the *singular values* of $\mathbf{A}$, which are defined to be the positive square roots of the ranked eigenvalues of $\mathbf{AA}'$. These eigenvalues are non-negative as $\mathbf{AA}'$ is non-negative definite (by 10.10).

The columns of $\mathbf{p}_i$ of $\mathbf{P}$ are the orthonormalized right eigenvectors associated with $\mathbf{AA}'$, and the columns $\mathbf{q}_i$ of $\mathbf{Q}$ are the orthonormalized right eigenvectors associated with $\mathbf{A}'\mathbf{A}$. The first r columns in each case correspond to the nonzero σ_i. Note that $\mathbf{Aq}_i = \sigma_i\mathbf{p}_i$ and $\mathbf{A}'\mathbf{p}_i = \sigma_i\mathbf{q}_i$ $(i = 1, 2, \ldots, r)$. The vectors $\mathbf{p}_i$ and $\mathbf{q}_i$ are also called the left and right singular vectors, respectively, associated with σ_i.

We note that $\mathbf{AA}'$ and $\mathbf{A}'\mathbf{A}$ have p common eigenvalues (cf. 6.54c), including some zeros when $r < p$. Any remaining eigenvalues of $\mathbf{AA}'$ (if $m > n$) or $\mathbf{A}'\mathbf{A}$ (if $m < n$) are zero.

Existence proofs are given by Horn and Johnson [1985: 411], Rao and Rao [1998: 172, complex case], Schott [2005: 140–141, real case], Searle [1982: 316, real case], and Seber and Lee [2003: 471, real case]. For some computational details see Gentle [1998: section 4.4], Golub and Van Loan [1996], and Stewart [1998, 2001].

In practice, several versions of the SVD are given in the literature, which we give below.

(1) Let $\mathbf{\Delta}_p = \text{diag}(\sigma_1, \sigma_2, \cdots, \sigma_p)$. If $p = n \leq m$, then

$$\mathbf{A} = \mathbf{P}\begin{pmatrix} \mathbf{\Delta}_n \\ \mathbf{0} \end{pmatrix}\mathbf{Q}' = \mathbf{P}_n\mathbf{\Delta}_n\mathbf{Q}',$$

where the $m \times n$ matrix $\mathbf{P}_n$ consists of the first n columns of $\mathbf{P}$, and $\mathbf{\Delta}_n$ and $\mathbf{Q}$ are both $n \times n$. The zero matrix is omitted if $m = n$. If $p = m < n$, then

$$\mathbf{A} = \mathbf{P}(\mathbf{\Delta}_m, \mathbf{0})\mathbf{Q}' = \mathbf{P}\mathbf{\Delta}_m\mathbf{Q}'_m,$$

where $\mathbf{Q}_m$ consists of the first m columns of $\mathbf{Q}$. We note that $\mathbf{P}'_n\mathbf{P}_n = \mathbf{I}_n$ and $\mathbf{Q}'_m\mathbf{Q}_m = \mathbf{I}_m$. These two versions are often referred to as the *thin singular value decompositions*.

Is the decomposition unique? If $m \geq n$ (i.e., $p = n$), then $\mathbf{\Sigma}$ will be unique as the eigenvalues of $\mathbf{A}'\mathbf{A}$ are unique. However, the eigenvectors making up $\mathbf{P}_n$ and $\mathbf{Q}$ will not be unique unless the eigenvalues are distinct and an appropriate sign convention is adopted for eigenvectors.

(2) If $\mathbf{P} = (p_{ij})$ and $\mathbf{Q} = (q_{ij})$, with respective columns $\mathbf{p}_i$ and $\mathbf{q}_i$, then

$$\mathbf{A} = \mathbf{P}\begin{pmatrix} \mathbf{\Delta}_r & \mathbf{0} \\ \mathbf{0} & \mathbf{0} \end{pmatrix}\mathbf{Q}' = \mathbf{P}_r\mathbf{\Delta}_r\mathbf{Q}'_r = \sum_{k=1}^{r}\sigma_k\mathbf{p}_k\mathbf{q}'_k$$

and $a_{ij} = \sum_{k=1}^{r}\sigma_k p_{ik}q_{jk}$.

If $\mathbf{A}$ is complex and $\mathbf{P}$ and $\mathbf{Q}$ are unitary matrices, then $\mathbf{A} = \sum_{k=1}^{r}\sigma_k\mathbf{p}_k\mathbf{q}_k^*$ and $a_{ij} = \sum_{k=1}^{r}\sigma_k p_{ik}\bar{q}_{jk}$.

Note that $\mathbf{AA}'\mathbf{P}_r = \mathbf{P}_r\mathbf{\Delta}_r^2$ and $\mathbf{A}'\mathbf{AQ}_r = \mathbf{Q}_r\mathbf{\Delta}_r^2$. The correct procedure is to find $\mathbf{P}_r$ and $\mathbf{\Delta}_r^2$ from $\mathbf{AA}'\mathbf{P}_r = \mathbf{P}_r\mathbf{\Delta}_r^2$ and then define $\mathbf{Q}_r = \mathbf{A}'\mathbf{P}_r\mathbf{\Delta}_r^{-1}$. Alternatively, we can obtain $\mathbf{Q}_r$ and $\mathbf{\Delta}_r^2$ from $\mathbf{A}'\mathbf{AQ}_r = \mathbf{Q}_r\mathbf{\Delta}_r^2$ and then define $\mathbf{P}_r = \mathbf{AQ}_r\mathbf{\Delta}_r^{-1}$ (Abadir and Magnus [2005: 226]).

16.34. Let $\mathbf{A}$ be an $m \times n$ matrix. From the above we have the following useful information.

(a) The number of nonzero singular values is the rank of $\mathbf{A}$. This provides a useful computational method for finding the rank of a matrix.

(b) The r columns of $\mathbf{P}_r$ and $\mathbf{Q}_r$ are orthonormal bases for $\mathcal{C}(\mathbf{A})$ and $\mathcal{C}(\mathbf{A}')$, respectively, while the remaining columns of $\mathbf{P}$ and $\mathbf{Q}$ span $\mathcal{N}(\mathbf{A}')$ and $\mathcal{N}(\mathbf{A})$, respectively.

(c) $\mathbf{P_A} = \mathbf{P}_r \mathbf{P}_r'$, the orthogonal projection onto $\mathcal{C}(\mathbf{A})$.

(d) $\mathbf{A}$ and $(\mathbf{A}^*\mathbf{A})^{1/2}$ have the same singular values.

(e) Two full-rank factorizations of $\mathbf{A}$ (cf. 3.5) are $(\mathbf{P}_r \mathbf{\Delta}_r)(\mathbf{Q}_r')$ and $(\mathbf{P}_r)(\mathbf{\Delta}_r \mathbf{Q}_r')$.

Proofs. Section 16.3.

16.34a–b. Schott [2005: 140–141].

16.34c. Sengupta and Jammalamadaka [2003: 43].

16.34d. Follows from the fact that $(\mathbf{A}^*\mathbf{A})^{1/2}((\mathbf{A}^*\mathbf{A})^{1/2})^* = \mathbf{A}^*\mathbf{A}$ is Hermitian with eigenvalues $\sigma_i^2(\mathbf{A})$.

16.4   TRIANGULAR FACTORIZATIONS

16.35. (*LU* and *LDU* factorizations) Under certain conditions, a real or complex $m \times n$ matrix can be expressed in the form $\mathbf{A} = \mathbf{L}_1 \mathbf{U}_1$, where $\mathbf{L}_1$ is lower-triangular and $\mathbf{U}_1$ is upper-triangular. If $m < n$, then $\mathbf{L}_1$ is $m \times m$, while if $m > n$, $\mathbf{L}_1$ is $m \times n$.

(a) A sufficient condition for such a factorization to exist is that for $k = 1, 2, \ldots, p$ ($p = \min\{m, n\}$), each $k \times k$ leading principal submatrix $\mathbf{A}_k$ of $\mathbf{A}$ is nonsingular.

  (i) The usual factorisation is to have either the diagonal elements of $\mathbf{L}_1$ all ones (and written as $\hat{\mathbf{L}}$), or the diagonal elements of $\mathbf{U}_1$ all ones (and written as $\tilde{\mathbf{U}}$) so that
$$\mathbf{A} = \hat{\mathbf{L}}\mathbf{U} = \mathbf{L}\tilde{\mathbf{U}}.$$

  (ii) If we put the diagonal elements from both matrices into a single diagonal matrix $\mathbf{D}$, then
$$\mathbf{A} = \hat{\mathbf{L}}\mathbf{D}\tilde{\mathbf{U}}.$$

(b) If $m < n$, which is often the case, and $\mathbf{A}_k$ is nonsingular for $k = 1, 2, \ldots, m$, then $\mathbf{A} = \mathbf{L}\tilde{\mathbf{U}}$, where $\mathbf{L}$ is $m \times m$ and nonsingular, $\tilde{\mathbf{U}}$ is an $m \times n$ matrix such that the first m columns form an upper-triangular matrix with unit diagonal elements, and $\mathbf{L}$ and $\tilde{\mathbf{U}}$ are unique.

(c) A typical application of the above theory is the solution of linear equations, for example $\mathbf{Bx} = \mathbf{b}$. If we set $\mathbf{A} = (\mathbf{B}, \mathbf{I}_m, \mathbf{b})$ and then factorize $\mathbf{A}$, we can obtain $\mathbf{B}^{-1}$ as a bonus (Rao and Bhimasankaram [2000: 213]). If $\mathbf{B}$ is square and nonsingular and $\mathbf{B} = \mathbf{LU}$, then we can solve $\mathbf{Ly} = \mathbf{b}$ for $\mathbf{y}$ using *forward substitution* and solve $\mathbf{Ux} = \mathbf{y}$ for $\mathbf{x}$ by *back substitution*. The process is usually refered to as *Gaussian elimination*.

(d) The matrix $\mathbf{M}_k = \mathbf{I}_n - \boldsymbol{\tau}\mathbf{e}'_k$, where $\boldsymbol{\tau} \in \mathbb{R}^n$ and $\mathbf{e}_k$ has 1 for its kth element and zeros elsewhere, is a *Gauss transformation* if the first k elements of $\boldsymbol{\tau}$ are zero. If this is the case, and $\tau_i = x_i/x_k$ $(x_k \neq 0)$ for $i = k + 1, \ldots, n$, then

$$\mathbf{M}_k\mathbf{x} = (x_1, x_2, \ldots, x_k, 0, \ldots, 0)'.$$

16.36. (Square Matrix LU Factorizations) Let $\mathbf{A}$ be a real or complex $n \times n$ matrix and let $\mathbf{A}_k$ be its leading $k \times k$ principal submatrix.

(a) Suppose $\mathbf{A}$ has rank r. If $\mathbf{A}_k$ is nonsingular for $k = 1, 2, \ldots, r$, then $\mathbf{A}$ may be factored as $\mathbf{A} = \mathbf{LU}$, where $\mathbf{L}$ and $\mathbf{U}$ are $n \times n$. Furthermore, the factorization may be chosen so that either $\mathbf{L}$ or $\mathbf{U}$ is nonsingular. Both $\mathbf{L}$ and $\mathbf{U}$ can be chosen to be nonsingular if $\mathbf{A}$ is nonsingular (i.e., $r = n$).

(b) There exist $n \times n$ permutation matrices $\boldsymbol{\Pi}_1$ and $\boldsymbol{\Pi}_2$ such that $\mathbf{A} = \boldsymbol{\Pi}_1\mathbf{LU}\boldsymbol{\Pi}_2$. If $\mathbf{A}$ is nonsingular, it may be written as

$$\mathbf{A} = \boldsymbol{\Pi}_1\mathbf{LU}.$$

(c) Suppose $\mathbf{A}_k$ is nonsingular for $k = 1, 2, \ldots, n - 1$.

   (i) $\mathbf{A}$ can be expressed in the form $\mathbf{A} = \tilde{\mathbf{L}}\mathbf{U} = \mathbf{L}\tilde{\mathbf{U}}$, where all the triangular matrices are unique.

   (ii) Also, $\mathbf{A}$ can also be expressed in the form $\mathbf{A} = \tilde{\mathbf{L}}\mathbf{D}\tilde{\mathbf{U}}$, where $\mathbf{D}$ is diagonal and all the matrices are unique. (Note that it is possible for $\mathbf{A}$ be singular.)

   (iii) If $\mathbf{A}$ is a real symmetric matrix, then we can also write $\mathbf{A} = \mathbf{U}'\mathbf{D}_1\mathbf{U}$, where $\mathbf{U}$ is real and the diagonal matrix $\mathbf{D}_1$ has elements ± 1.

(d) Suppose $\mathbf{A}$ is nonsingular and $\mathbf{A} = \mathbf{L}\tilde{\mathbf{U}}$. If $\mathbf{L} = (l_{ij})$, then, since $\det \mathbf{A} = \det \mathbf{L} \cdot \det \tilde{\mathbf{U}}$, we have $\det \mathbf{A} = \prod_{i=1}^n l_{ii} \neq 0$, $\det \mathbf{A}_k = \prod_{i=1}^k l_{ii} \neq 0$ for $k = 1, 2, \ldots, n - 1$, and $\mathbf{L}$ and $\tilde{\mathbf{U}}$ are unique, by (c).

(e) If $\mathbf{A}$ is Hermitian with an LDU factorization, then we can express it in the form $\mathbf{A} = \tilde{\mathbf{U}}^*\mathbf{D}_2\tilde{\mathbf{U}}$, where $\mathbf{D}_2$ is a diagonal matrix.

16.37. (Schur Decomposition Theorems) We now consider a series of powerful theorems that can be used to provide shorter proofs for a wide range of other results (e.g., Abadir and Magnus [2005: section 7.4]).

Let $\lambda_1, \lambda_2, \ldots, \lambda_n$ be the eigenvalues of the $n \times n$ matrix $\mathbf{A}$ in a prescribed order.

(a) If $\mathbf{A}$ is a real or complex matrix, there exists a unitary matrix $\mathbf{Q}$ such that $\mathbf{Q}^*\mathbf{AQ} = \mathbf{T}$ is upper-triangular with diagonal elements the eigenvalues of $\mathbf{A}$ in the same order. Neither $\mathbf{Q}$ nor $\mathbf{T}$ are unique.

(b) If $\mathbf{A}$ is real with real eigenvalues, then $\mathbf{Q}$ can be chosen to be real and orthogonal. The upper-triangular matrix is also real.

(c) If $\mathbf{A}$ is real with k real eigenvalues $\lambda_1, \lambda_2, \ldots, \lambda_k$ and complex eigenvalues $x_j + iy_j$ for $j > k$, there exists a real orthogonal $\mathbf{T}$ such that $\mathbf{T}'\mathbf{A}\mathbf{T} = \mathbf{R}$, where $\mathbf{R}$ resembles an upper-triangular matrix, but with diagonal blocks of the form $\mathbf{R}_{11}, \mathbf{R}_{22}, \ldots, \mathbf{R}_{tt}$. Here $\mathbf{R}_{jj} = \lambda_j$, for $j = 1, 2, \ldots, k$; and for $j > k$

$$\mathbf{R}_{jj} = \begin{pmatrix} x_j & b_j \\ -c_j & x_j \end{pmatrix}, \quad \sqrt{b_j c_j} = y_j,$$

where $b_j \geq c_j$ and $b_j c_j > 0$. The elements below these blocks are zero so that $\mathbf{R}$ is of upper Hessenberg form. Golub and Van Loan [1996: 341] refer to such a matrix as *quasi-triangular* and show how to compute it using QR iterations (cf. Section 16.5). For an application to probability theory see Edelman [1997].

16.38. (Cholesky Decomposition for Non-negative Definite Matrices) If $\mathbf{A}$ is an $n \times n$ non-negative definite matrix, there exist $n \times n$ upper-triangular matrices $\mathbf{U}$ and $\mathbf{U}_1$ with non-negative diagonal elements such that

$$\mathbf{A} = \mathbf{U}'\mathbf{U} = \mathbf{U}_1\mathbf{U}_1'.$$

If $\mathbf{A}$ is positive definite, the matrix $\mathbf{U}$ is unique if its diagonal elements are all positive (or all negative); the same applies to $\mathbf{U}_1$. Some writers prefer to use lower-triangular matrices $\mathbf{L} = \mathbf{U}'$ or $\mathbf{L}_1 = \mathbf{U}_1'$. The result also holds for $\mathbf{A}$ Hermitian non-negative definite, that is, there exists an upper-triangular matrix $\mathbf{U}$ such that $\mathbf{A} = \mathbf{U}^*\mathbf{U}$. If $\mathbf{A}$ is positive definite, then $\mathbf{U}$ is unique if its diagonal elements are positive (Rao and Rao [1998: 173]). For some computational aspects when $\mathbf{A}$ is non-negative definite see Smith [2001].

16.39. A scaled version of the above when $\mathbf{A}$ is positive definite is also used. If $\tilde{\mathbf{D}} = \operatorname{diag}(u_{11}, u_{22}, \ldots, u_{nn})$, then $\tilde{\mathbf{D}}$ has an inverse. Let

$$\tilde{\mathbf{U}} = \tilde{\mathbf{D}}^{-1}\mathbf{U} = \begin{pmatrix} 1 & \tilde{u}_{12} & \tilde{u}_{13} & \cdots & \tilde{u}_{1n} \\ 0 & 1 & \tilde{u}_{23} & \cdots & \tilde{u}_{2n} \\ . & . & . & \cdots & . \\ 0 & 0 & . & \cdots & 1 \end{pmatrix},$$

so that

$$\mathbf{A} = \mathbf{U}'\mathbf{U} = \tilde{\mathbf{U}}'\tilde{\mathbf{D}}^2\tilde{\mathbf{U}} = \tilde{\mathbf{U}}'\mathbf{D}\tilde{\mathbf{U}} = \tilde{\mathbf{L}}\mathbf{D}\tilde{\mathbf{L}}',$$

where $\mathbf{D}$ is a diagonal matrix with positive diagonal elements.

16.40. (Algorithm for the Cholesky Decomposition) If $\mathbf{A}$ is a positive definite $n \times n$ matrix, and the diagonal elements of $\mathbf{U}$ are all positive, we have the following steps.

Step 1: Set

$$u_{11} = a_{11}^{1/2},$$
$$u_{1j} = \frac{a_{1j}}{u_{11}} \quad (j = 2, 3, \ldots, n).$$

Step 2: For $i = 2, 3, \ldots, p-1$ set

$$u_{ij} = 0 \quad (j = 1, 2, \ldots, i-1),$$

$$u_{ii} = \left(a_{ii} - \sum_{k=1}^{i-1} u_{ki}^2\right)^{1/2},$$

$$u_{ij} = \frac{a_{ij} - \sum_{k=1}^{i-1} u_{ki}u_{kj}}{u_{ii}} \quad (j = i+1, \ldots, n).$$

Step 3: Set

$$u_{nn} = \left(a_{nn} - \sum_{k=1}^{n-1} u_{ki}^2\right)^{1/2}.$$

The decomposition $\mathbf{A} = \tilde{\mathbf{U}}'\mathbf{D}\tilde{\mathbf{U}}$ can be used to avoid the computation of square roots. This modification is called the *Banachiewicz factorization* or the *root-free Cholesky decomposition* (Gentle [1998: 93–94]).

16.41. (Matrix Inverse) If $\mathbf{A} = \mathbf{U}'\mathbf{U}$ is a positive definite $n \times n$ matrix, we have $\mathbf{A}^{-1} = \mathbf{U}^{-1}(\mathbf{U}')^{-1} = \mathbf{T}\mathbf{T}'$, where $\mathbf{T}$ is upper-triangular. From $\mathbf{UT} = \mathbf{I}_n$ we find that $\mathbf{T}$ is given by

$$
\begin{aligned}
t_{ii} &= u_{ii}^{-1} \quad (i = 1, 2, \ldots, n), \\
t_{ij} &= 0 \quad (i > j), \\
t_{ij} &= -\frac{\sum_{k=i+1}^{j} u_{ik}t_{kj}}{u_{ii}} \quad (j = i+1, \ldots, n).
\end{aligned}
$$

Then

$$
\begin{aligned}
(\mathbf{A}^{-1})_{rs} &= \sum_{k=s}^{n} t_{rk}t'_{ks} \\
&= \sum_{k=s}^{n} t_{rk}t_{sk} \quad (s = r, r+1, \ldots, n),
\end{aligned}
$$

which is the product of the rth and sth rows of $\mathbf{T}$.

Proofs. Section 16.4.

16.35a. Golub and Van Loan [1996: 102].

16.35b. Rao and Bhimasankaram [2000: 211–212].

16.35c. Golub and Van Loan [1996: 88–103].

16.35d. Golub and Van Loan [1996: 94–95].

16.36a. Horn and Johnson [1985: 160].

16.36b. Horn and Johnson [1985: 163].

16.36c(i). Graybill [1983: 207] and Rao and Bhimasankaram [2000: 216].

16.36c(iii). Graybill [1983: 210].

16.36d. Golub and Van Loan [1996: 97].

16.37a. Abadir and Magnus [2005: 187], Horn and Johnson [1985: 79], Rao and Rao [1998: 174–175], Schott [2005: 157], and Zhang [1999: 64–65].

16.37b. Muirhead [1982: 587] and Schott [2005: 158].

16.37c. Rao and Rao [1998: 189–190].

16.38. For an inductive proof for the positive-definite case see Schott [2005: 147] and Seber [1977: 388].

16.40. Seber and Lee [2003: 336].

16.41. Seber [1977: 305–306].

16.5   ORTHOGONAL–TRIANGULAR REDUCTIONS

The so-called QR decomposition plays an important role in the analysis of regression models, particularly in statistical computing packages. In fact, many of the regression theorems can actually be derived via the QR decomposition (e.g., Ansley [1985], Eubank and Webster [1985], Mandel [1982], and Nelder [1985]).

Definition 16.12. Any $n \times p$ real matrix $\mathbf{A}$ of rank r can be expressed in the form $\mathbf{A} = \mathbf{Q}\mathbf{R}$, where $\mathbf{Q}$ is an $n \times n$ orthogonal matrix and $\mathbf{R}$ is an $n \times p$ upper-triangular matrix. This is called the *QR decomposition*. If $n \geq p$, then

$$\mathbf{Q}\mathbf{R} = (\mathbf{Q}_p, \mathbf{Q}_{n-p}) \begin{pmatrix} \mathbf{R}_1 \\ \mathbf{0} \end{pmatrix}$$
$$= \mathbf{Q}_p \mathbf{R}_1,$$

where $\mathbf{Q}_p$ consists of the first p columns of $\mathbf{Q}$, and $\mathbf{R}_1$ is a $p \times p$ upper-triangular matrix. Harville [1997: 66–68], Horn and Johnson [1985: 112-113], and Seber and Lee [2003: 338] give algorithmic proofs, while Seber [1977: 388] gives an inductive proof. Some authors refer to $\mathbf{A} = \mathbf{Q}_p \mathbf{R}_1$ as the QR decomposition.

If $n \leq p$, we replace $\mathbf{R}$ by $(\mathbf{R}_2, \mathbf{S})$, where $\mathbf{R}_2$ is an $n \times n$ upper-triangular matrix. The above results and those below are also true for complex $\mathbf{A}$ if $\mathbf{Q}$ is now unitary and we replace $'$ by $*$ (cf. Rao and Rao [1998: 168]).

Note that $\mathbf{Q}'\mathbf{A} = \mathbf{R}$, and the reduction of $\mathbf{A}$ can be carried out using a variety of algorithms. For example, the orthogonal matrix $\mathbf{Q}'$ could consist of a product of Householder reflections or Givens rotations, or one could use the Gram–Schmidt algorithm. For further details of the real case see Seber and Lee [2003: chapter 11].

16.42. We use the above notation in what follows, and we assume $n \geq p$.

(a) Suppose $r = p$.

   (i) Since $\mathbf{R}_1$ has full rank p, $\mathbf{A}'\mathbf{A} = \mathbf{R}_1'\mathbf{Q}_p'\mathbf{Q}_p\mathbf{R}_1 = \mathbf{R}_1'\mathbf{R}_1$ is positive definite, and $\mathbf{R}_1'\mathbf{R}_1$ is the Cholesky decomposition of $\mathbf{A}'\mathbf{A}$. Therefore, if the diagonal elements of $\mathbf{R}_1$ are all positive (or all negative), then $\mathbf{R}_1$ is unique and $\mathbf{Q}_p = \mathbf{A}\mathbf{R}_1^{-1}$ is unique. Hence the decomposition $\mathbf{A} = \mathbf{Q}_p\mathbf{R}_1$ is also unique. However, the matrix $\mathbf{Q}_{n-p}$ is not unique because any permutation of its columns will still give $\mathbf{A} = \mathbf{Q}\mathbf{R}$.

(ii) The Moore–Penrose inverse of $\mathbf{A}$ is

$$\mathbf{A}^+ = (\mathbf{R}_1^{-1}, \mathbf{0})\mathbf{Q}' = \mathbf{R}_1^{-1}\mathbf{Q}_p'.$$

(iii) If $\mathbf{A}$ is $n \times n$ and nonsingular, then

$$\det \mathbf{A} = \det \mathbf{Q} \det \mathbf{R}_1 = \prod_{i=1}^{n} r_{ii},$$

where $\mathbf{R}_1 = (r_{ij})$. This is a useful method of finding a determinant. One application in statistics is in optimal experimental designs. For example, the D-optimal criterion chooses the design matrix $\mathbf{X}$ such that $\det(\mathbf{X}'\mathbf{X})$ is maximized.

(b) Suppose $r < p$.

(i) We first note that $\mathbf{A}'\mathbf{A} = \mathbf{R}_1'\mathbf{R}_1$ as above, but now $\mathbf{A}'\mathbf{A}$ is non-negative definite. However, $\mathbf{R}_1'\mathbf{R}_1$ is still the Cholesky decomposition of $\mathbf{A}'\mathbf{A}$ and $\mathbf{R}_1$ is unique if the diagonal entries are non-negative. An inductive proof for the case $n = p$ is given by Graybill [1983: 210].

(ii) We can permutate the columns of $\mathbf{A}$ by postmultiplying by a permutation matrix $\mathbf{\Pi}$ so that the first r columns of the permutated matrix $\tilde{\mathbf{A}} = \mathbf{A}\mathbf{\Pi}$ are linearly independent. Then $\tilde{\mathbf{A}} = \mathbf{Q}\tilde{\mathbf{R}}$, where

$$\tilde{\mathbf{R}} = \left(\begin{array}{cc} \mathbf{R}_{11} & \mathbf{R}_{12} \\ \mathbf{0} & \mathbf{0} \end{array} \right),$$

and $\mathbf{R}_{11}$ is an $r \times r$ nonsingular upper-triangular matrix. Thus $\tilde{\mathbf{R}}$ is upper-triangular, but with its bottom $n - r$ rows all zeros. Since $\mathbf{\Pi}^{-1} = \mathbf{\Pi}'$ we have

$$\begin{aligned} \mathbf{A} &= \mathbf{Q}\tilde{\mathbf{R}}\mathbf{\Pi}' \\ &= \mathbf{Q}\left(\begin{array}{cc} \mathbf{R}_{11} & \mathbf{R}_{12} \\ \mathbf{0} & \mathbf{0} \end{array} \right)\mathbf{\Pi}' \\ &= \mathbf{Q}_r(\mathbf{R}_{11}, \mathbf{R}_{12})\mathbf{\Pi}', \end{aligned}$$

where $\mathbf{Q}_r$ consists of the first r columns of $\mathbf{Q}$. As $\mathbf{\Pi}$ is not unique, $\mathbf{Q}_r$ will not be unique.

(iii) A weak inverse of $\mathbf{A}$ is given by

$$\mathbf{A}^- = \mathbf{\Pi}\left(\begin{array}{cc} \mathbf{R}_{11}^{-1} & \mathbf{0} \\ \mathbf{0} & \mathbf{0} \end{array} \right)\mathbf{Q}',$$

as $\mathbf{A}\mathbf{A}^-\mathbf{A} = \mathbf{A}$.

(iv) Additional orthogonal transformations can be applied to $\mathbf{A} = \mathbf{Q}\tilde{\mathbf{R}}\mathbf{\Pi}'$ to get

$$\mathbf{A} = \mathbf{Q}\left(\begin{array}{cc} \mathbf{R}_0 & \mathbf{0} \\ \mathbf{0} & \mathbf{0} \end{array} \right)\mathbf{P}',$$

where $\mathbf{P}$ is orthogonal and $\mathbf{R}_0$ is $r \times r$ and nonsingular. This is a convenient method of finding r.

(v) From (iv),

$$\mathbf{A}^+ = \mathbf{P} \begin{pmatrix} \mathbf{R}_0^{-1} & \mathbf{0} \\ \mathbf{0} & \mathbf{0} \end{pmatrix} \mathbf{Q}'.$$

For further computational details see Gentle [1998: 95–102] and Golub and Van Loan [1996: section 5.2].

There is also a symmetric QR iterative process that is a useful computational tool (Golub and Van loan [1996: section 8.2]).

16.43. (Tridiagonal Matrix) If $\mathbf{T} = \mathbf{QR}$ is a QR decomposition of a symmetric tridiagonal matrix $\mathbf{T}$, all matrices being $n \times n$, then $\mathbf{Q}$ has lower bandwidth 1, $\mathbf{R}$ has upper bandwidth 2, and

$$\mathbf{T}_1 = \mathbf{RQ} = (\mathbf{Q}'\mathbf{Q})\mathbf{RQ} = \mathbf{Q}'\mathbf{TQ}$$

is also symmetric and tridiagonal. (The upper bandwidth is the number of nonzero diagonals above the main diagonal, and the lower bandwidth is the number of nonzero diagonals below the main diagonal; all other elements except possibly those in the main diagonal are zero.) The factorization can be computed by applying a sequence of $n - 1$ Givens rotations.

Proofs. Section 16.5.

16.42a(ii). Bates [1983].

16.42a(iii). Gentle [1998: 115].

16.42b(i)–(iii). Gentle [1998: 96].

16.42b(iv). Gentle [1998: 115].

16.43. Golub and Van Loan [1996: 417].

16.6   FURTHER DIAGONAL OR TRIDIAGONAL REDUCTIONS

16.44. (Spectral Decomposition Theorem) Let $\mathbf{A}$ be any $n \times n$ real symmetric (respectively Hermitian) matrix. Then there exists an orthogonal (respectively unitary) matrix $\mathbf{Q} = (\mathbf{q}_1, \mathbf{q}_2, \ldots, \mathbf{q}_n)$ such that

$$\mathbf{Q}'\mathbf{AQ} = \mathrm{diag}(\lambda_1, \lambda_1, \ldots, \lambda_n) = \mathbf{\Lambda}, \text{ say,}$$

where $\lambda_1 \geq \lambda_2 \geq \cdots \geq \lambda_n$ are the ordered eigenvalues of $\mathbf{A}$ (which we know are real). When $\mathbf{A}$ is Hermitian, $\mathbf{Q}'$ is replaced by $\mathbf{Q}^*$. With the above ordering, $\mathbf{\Lambda}$ is unique and $\mathbf{Q}$ is unique up to a postfactor of

$$\mathbf{S} = \begin{pmatrix} \mathbf{S}_1 & \mathbf{0} & \cdots & \mathbf{0} \\ \mathbf{0} & \mathbf{S}_2 & \cdots & \mathbf{0} \\ . & . & \cdots & . \\ \mathbf{0} & \mathbf{0} & \cdots & \mathbf{S}_k \end{pmatrix}, \quad \mathbf{S}_i \in \mathbf{Q}(m_i),$$

where k is the number of different eigenvalues of $\mathbf{A}$; $m_1, m_2, \ldots, m_k$ are the algebraic multiplicities, that is $\lambda_1 = \lambda_2 = \cdots = \lambda_{m_1} > \lambda_{m_1+1} = \cdots = \lambda_{m_1+m_2}$, and so on;

and $\mathbf{Q}(m_i)$ stands for the set of all $m_i \times m_i$ orthogonal (respectively unitary) matrices.

If all the eigenvalues are distinct, each $m_i = 1$ and $\mathbf{S}$ reduces to a diagonal matrix with diagonal elements equal to ± 1. In this case the columns $\mathbf{q}_i$ of $\mathbf{Q}$ are unique except for their signs. If we stipulate, for example, that the element of $\mathbf{q}_i$ with the largest magnitude is positive, then $\mathbf{S} = \mathbf{I}_n$ and $\mathbf{Q}$ is unique. We note that:

(a) $\mathbf{A} = \mathbf{Q}\mathbf{\Lambda}\mathbf{Q}' = \sum_{i=1}^{n} \lambda_i \mathbf{q}_i \mathbf{q}_i' = \sum_{i=1}^{n} \lambda_i \mathbf{F}_i$, where the $\mathbf{F}_i$ are symmetric, idempotent, and satisfy $\mathbf{F}_i \mathbf{F}_j = \mathbf{0}$ for all $i, j, j \neq i$.

(b) If $\mathbf{x} = \mathbf{Q}\mathbf{y}$, then $\mathbf{x}'\mathbf{A}\mathbf{x} = \mathbf{y}'\mathbf{Q}'\mathbf{A}\mathbf{Q}\mathbf{y} = \lambda_1 y_1^2 + \cdots + \lambda_n y_n^2$. An algorithm for carrrying out this reduction by completing the square rather than finding eigenvalues and eigenvectors (known as Lagrange's reduction) is described by Rao and Bhimasankaram [2000: 333].

(c) If $\mathbf{A}$ has rank r, $\mathbf{\Lambda}_r$ contains the r nonzero eigenvalues, and $\mathbf{Q}_r$ contains the corresponding right eigenvectors of $\mathbf{A}$, then $\mathbf{A} = \mathbf{Q}_r \mathbf{\Lambda}_r \mathbf{Q}_r'$, where $\mathbf{Q}_r' \mathbf{Q}_r = \mathbf{I}_r$.

16.45. (Tridiagonal Reduction) Suppose $\mathbf{A}$ is a real symmetric $n \times n$ matrix.

(a) There exists an orthogonal matrix $\mathbf{Q}$ such that

$$\mathbf{Q}'\mathbf{A}\mathbf{Q} = \mathbf{B},$$

where $\mathbf{B}$ is tridiagonal. This is a very useful reduction used in numerical analysis because it provides an intermediate step for speeding up a diagonalization process. If $\mathbf{Q} = (\mathbf{q}_1, \mathbf{q}_2, \ldots, \mathbf{q}_n)$, then the $\mathbf{q}_i$ are called *Lanczos vectors* (cf. Golub and Van Loan [1996: 473])

(b) If $\mathbf{q}_1$ is defined in (a), then

$$\mathbf{Q}'(\mathbf{q}_1, \mathbf{A}\mathbf{q}_1, \ldots, \mathbf{A}^{n-1}\mathbf{q}_1) = \mathbf{R},$$

where $\mathbf{R}$ is upper-triangular. The matrix in brackets is a Krylov matrix. If $\mathbf{R}$ is nonsingular, then $\mathbf{B}$ of (a) has no zero subdiagonal elements.

16.46. (Normal Matrix)

(a) (Diagonal Reduction) An $n \times n$ complex matrix is normal (i.e., $\mathbf{A}\mathbf{A}^* = \mathbf{A}^*\mathbf{A}$) if and only if there exists a unitary matrix $\mathbf{Q}$ such that $\mathbf{Q}^*\mathbf{A}\mathbf{Q} = \text{diag}(\lambda_1, \lambda_2, \ldots, \lambda_n)$, where the λ_i are the eigenvalues of $\mathbf{A}$.

(b) (Tridiagonal Reduction)

   (i) Let $\mathbf{A}$ be a real $n \times n$ matrix. Then $\mathbf{A}$ is normal (i.e., $\mathbf{A}\mathbf{A}' = \mathbf{A}'\mathbf{A}$) if and only if there is a real orthogonal matrix $\mathbf{Q}$ such that

$$\mathbf{Q}'\mathbf{A}\mathbf{Q} = \text{diag}(\mathbf{A}_1, \mathbf{A}_2, \ldots, \mathbf{A}_k) = \mathbf{D}_1, \quad 1 \leq k \leq n,$$

   where tridiagonal $\mathbf{D}_1$ is a real block-diagonal matrix, and $\mathbf{A}_j$ is either a real 1×1 matrix or a real 2×2 matrix of the form

$$\mathbf{A}_j = \begin{pmatrix} \alpha_j & \beta_j \\ -\beta_j & \alpha_j \end{pmatrix}.$$

(ii) If $\mathbf{A}$ is a real skew-symmetric matrix (i.e., $\mathbf{A}' = -\mathbf{A}$), then $\mathbf{A}$ is normal. It then follows that $\mathbf{A}$ is skew-symmetric if and only if there exists a real orthogonal matrix $\mathbf{Q}$ such that

$$\mathbf{Q}'\mathbf{A}\mathbf{Q} = \mathrm{diag}(0, 0, \ldots, 0, \mathbf{A}_1, \mathbf{A}_2, \ldots, \mathbf{A}_t) = \mathbf{D}_2,$$

where $\mathbf{D}_2$ is a real block diagonal matrix with each $\mathbf{A}_j$ having the form

$$\mathbf{A}_j = \begin{pmatrix} 0 & \beta_j \\ -\beta_j & 0 \end{pmatrix}.$$

(iii) If $\mathbf{A}$ is an orthogonal matrix, then it is normal. It follows that $\mathbf{A}$ is orthogonal if and only if there exists a real orthogonal matrix $\mathbf{Q}$ such that

$$\mathbf{Q}'\mathbf{A}\mathbf{Q} = \mathrm{diag}(\lambda_i, \lambda_2, \ldots, \lambda_r, \mathbf{A}_1, \mathbf{A}_2, \ldots, \mathbf{A}_s) = \mathbf{D}_3,$$

where $\mathbf{D}_3$ is a real block diagonal matrix with each $\lambda_j = \pm 1$ and each matrix $\mathbf{A}_j$ having the form

$$\mathbf{A}_j = \begin{pmatrix} \cos\theta_j & \sin\theta_j \\ -\sin\theta_j & \cos\theta_j \end{pmatrix}.$$

16.47. (Hermitian matrix) If $\mathbf{A}$ is an $n \times n$ Hermitian matrix of rank r, then there exists a nonsingular matrix $\mathbf{S}$ such that $\mathbf{S}^*\mathbf{A}\mathbf{S} = \mathbf{D}$, where

$$\mathbf{D} = \mathrm{diag}(1, 1, \ldots, 1, -1, -1, \ldots, -1, 0, 0, \ldots, 0).$$

The number of $+1$'s and $-1's$ are the same as the number of positive and negative eigenvalues of $\mathbf{A}$ (say r_+ and r_-, respectively), and the number of zeros is $r_0 = n - r$. The result obviously holds for a real symmetric matrix and real $\mathbf{S}$ (e.g., Anderson [2003: 640]). Clearly the signature, defined below, is unique.

Definition 16.13. Refering to (16.47) above, if $\mathbf{A}$ is a Hermitian matrix, the triple (r_+, r_-, r_0) is called the *inertia* of $\mathbf{A}$, while $r_+ - r_-$ is called the *signature* of $\mathbf{A}$.

Proofs. Section 16.6.

16.44. Harville [1997: 534–539].

16.45a. Golub and Van Loan [1996: 414].

16.45b. Golub and Van Loan [1996: 416].

16.46a. Rao and Bhimasankaram [2000: 313] and Rao and Rao [1998: 175, 190].

16.46b(i). Horn and Johnson [1985: 105].

16.46b(ii). Horn and Johnson [1985: 107–108].

16.46b(iii). Horn and Johnson [1985: 108].

16.47. Horn and Johnson [1985: 221–222].

16.7 CONGRUENCE

16.48. (Sylvester's Law of Inertia) Let $\mathbf{A}$ and $\mathbf{B}$ be $n \times n$ Hermitian matrices. There exists an $n \times n$ nonsingular matrix $\mathbf{S}$ such that $\mathbf{A} = \mathbf{SBS}^*$ if and only if $\mathbf{A}$ and $\mathbf{B}$ have the same inertia (cf. 16.47).

16.49. (Ostrowski) Let $\mathbf{A}$ be Hermitian and $\mathbf{S}$ nonsingular, both $n \times n$ matrices. Then, for each $i = 1, 2, \ldots, n$, there exists a positive real number θ_i such that $\lambda_{\max}(\mathbf{SS}^*) \geq \theta_i \geq \lambda_{\min}(\mathbf{SS}^*)$ and

$$\lambda_i(\mathbf{SAS}^*) = \theta_i \lambda_i(\mathbf{A}).$$

16.50. Let $\mathbf{A}$ and $\mathbf{B}$ be $n \times n$ real or complex symmetric matrices. There exists a nonsingular $\mathbf{S}$ such that $\mathbf{A} = \mathbf{SBS}'$ if and only if $\mathbf{A}$ and $\mathbf{B}$ have the same rank.

Proofs. Section 16.7.

  16.48–16.50. Horn and Johnson [1985: 223, 224, 225].

16.8 SIMULTANEOUS REDUCTIONS

16.51. Let $\mathbf{A}$ and $\mathbf{B}$ be $n \times n$ real symmetric matrices.

(a) (i) There exists a real orthogonal matrix $\mathbf{Q}$ such that $\mathbf{Q}'\mathbf{AQ}$ and $\mathbf{Q}'\mathbf{BQ}$ are both diagonal if and only if $\mathbf{AB} = \mathbf{BA}$ (that is $\mathbf{AB}$ is symmetric).

  (ii) The previous result holds for more than two matrices. A set of real symmetric matrices are simultaneously diagonalizable by the same orthogonal matrix $\mathbf{Q}$ if and only if they commute pairwise.

  (iii) The above result also holds for Hermitian matrices and unitary $\mathbf{Q}$.

(b) If a real linear combination of $\mathbf{A}$ and $\mathbf{B}$ is positive definite, then there exists a nonsingular matrix $\mathbf{R}$ such that $\mathbf{R}'\mathbf{AR}$ and $\mathbf{R}'\mathbf{BR}$ are diagonal.

(c) If $\mathbf{A}$ is also positive definite, there exists a nonsingular $\mathbf{S}$ such that $\mathbf{S}'\mathbf{AS} = \mathbf{I}_n$ and $\mathbf{S}'\mathbf{BS} = \operatorname{diag}(\lambda_1, \lambda_2, \ldots, \lambda_n)$, where the λ_i are the roots of $|\lambda\mathbf{A} - \mathbf{B}| = 0$, i.e., are the eigenvalues of $\mathbf{A}^{-1}\mathbf{B}$ (or $\mathbf{BA}^{-1}$ or $\mathbf{A}^{-1/2}\mathbf{BA}^{-1/2}$). The λ_i are real.

(d) If $\mathbf{A}$ and $\mathbf{B}$ are both non-negative definite, there exists a nonsingular matrix $\mathbf{R}$ such that $\mathbf{R}'\mathbf{AR}$ and $\mathbf{R}'\mathbf{BR}$ are both diagonal.

16.52. Let $\mathbf{A}$ and $\mathbf{B}$ be $n \times n$ complex matrices.

(a) If $\mathbf{A}$ and $\mathbf{B}$ are both symmetric, there exists a unitary $\mathbf{U}$ such that $\mathbf{UAU}'$ and $\mathbf{UBU}'$ are both diagonal if and only if $\mathbf{A}\bar{\mathbf{B}}$ is normal; that is, $\mathbf{A}\bar{\mathbf{B}}\mathbf{B}\bar{\mathbf{A}} = \bar{\mathbf{B}}\mathbf{A}\bar{\mathbf{A}}\mathbf{B}$.

(b) If $\mathbf{A}$ is Hermitian and $\mathbf{B}$ is symmetric, there exists a unitary $\mathbf{U}$ such that $\mathbf{UAU}^*$ and $\mathbf{UBU}'$ are both diagonal if and only if $\mathbf{AB}$ is symmetric; that is $\mathbf{AB} = \mathbf{B}\bar{\mathbf{A}}$.

(c) If $\mathbf{A}$ is Hermitian positive definite and $\mathbf{B}$ is symmetric, then there exists a nonsingular matrix $\mathbf{S}$ such that $\mathbf{S}^*\mathbf{AS}$ and $\mathbf{S}'\mathbf{BS}$ are both diagonal.

(d) Let $\mathbf{A}$ be a Hermitian matrix, $\mathbf{B}$ be a Hermitian non-negative definite matrix with rank $r \leq n$, and $\mathbf{N}$ be an $n \times n - r$ matrix of rank $n - r$ such that $\mathbf{N}^*\mathbf{B} = \mathbf{0}$. Then:

  (i) There exists an $n \times r$ matrix $\mathbf{L}$ such that $\mathbf{L}^*\mathbf{BL} = \mathbf{I}_r$ and $\mathbf{L}^*\mathbf{AL} = \boldsymbol{\Delta}$, where $\boldsymbol{\Delta}$ is an $r \times r$ diagonal matrix.

  (ii) A necessary and sufficient condition that there exists a nonsingular matrix $\mathbf{R}$ such that $\mathbf{R}^*\mathbf{AR}$ and $\mathbf{R}^*\mathbf{BR}$ are both diagonal is that

$$\mathrm{rank}(\mathbf{N}^*\mathbf{A}) = \mathrm{rank}(\mathbf{N}^*\mathbf{AN}).$$

  (iii) A necessary and sufficient condition that there exists a nonsingular matrix $\mathbf{R}$ such that $\mathbf{R}^*\mathbf{BR}$ and $\mathbf{R}^{-1}\mathbf{A}(\mathbf{R}^{-1})^*$ are both diagonal is

$$\mathrm{rank}(\mathbf{BA}) = \mathrm{rank}(\mathbf{BAB}).$$

  (iv) If, in addition, $\mathbf{A}$ is Hermitian non-negative definite, there exists a nonsingular matrix $\mathbf{R}$ such that $\mathbf{R}^*\mathbf{AR}$ and $\mathbf{R}^*\mathbf{BR}$ are both diagonal.

  (v) If, in addition, $\mathbf{A}$ is Hermitian non-negative definite, then there exists a nonsingular matrix $\mathbf{R}$ such that $\mathbf{R}^*\mathbf{BR}$ and $\mathbf{R}^{-1}\mathbf{A}(\mathbf{R}^{-1})^*$ are both diagonal.

For other results like (a)-(c) see the table of Horn and Johnson [1985: 229].

16.53. (Simultaneous Upper-Triangular Reductions) Let $\mathbf{A}$ and $\mathbf{B}$ be $n \times n$ complex matrices.

(a) There exist unitary matrices $\mathbf{P}$ and $\mathbf{Q}$ such that $\mathbf{P}^*\mathbf{AQ} = \mathbf{T}$ and $\mathbf{P}^*\mathbf{BQ} = \mathbf{S}$ are upper-triangular. If the diagonal elements s_{ii} of $\mathbf{S}$ are all nonzero, then $\lambda_i(\mathbf{AB}^{-1}) = t_{ii}/s_{ii}$ for $i = 1, 2, \ldots, n$.

(b) If $\mathbf{A}$ and $\mathbf{B}$ are real, there exist real orthogonal matrices $\mathbf{P}$ and $\mathbf{Q}$ such that $\mathbf{P}^*\mathbf{AQ}$ is upper quasi-triangular (upper Hessenberg) and $\mathbf{P}'\mathbf{BQ}$ is upper-triangular.

(c) If $\mathbf{AB} = \mathbf{BA}$, then there exists a unitary matrix $\mathbf{U}$ such that $\mathbf{U}^*\mathbf{AU}$ and $\mathbf{U}^*\mathbf{BU}$ are both upper-triangular. This result holds for any family of commuting matrices (Horn and Johnson [1985: 81]).

16.54. (Simultaneous Singular Value Decompositions)

(a) (Two Matrices) Let $\mathbf{A}$ and $\mathbf{B}$ be $m \times n$ matrices. There exist unitary matrices $\mathbf{P}_{m \times m}$ and $\mathbf{Q}_{n \times n}$ such that $\mathbf{A} = \mathbf{P}\boldsymbol{\Sigma}_1\mathbf{Q}^*$ and $\mathbf{B} = \mathbf{P}\boldsymbol{\Sigma}_2\mathbf{Q}^*$, where $\boldsymbol{\Sigma}_i$ $(i = 1, 2)$ are $m \times n$ diagonal matrices, if and only if if $\mathbf{AB}^*$ and $\mathbf{B}^*\mathbf{A}$ are both normal.

(b) (More Than Two Matrices) Given $m \times n$ matrices $\mathbf{A}_i$ $(i = 1, 2, \ldots, k)$, there exist unitary matrices $\mathbf{P}$ and $\mathbf{Q}$ such that $\mathbf{A} = \mathbf{P}\boldsymbol{\Sigma}_i\mathbf{Q}^*$ for all i, where the $\boldsymbol{\Sigma}_i$ are all diagonal, if and only if each $\mathbf{A}_i^*\mathbf{A}_j$ $(i \neq j)$ is normal and all the pairs of $\mathbf{A}_i\mathbf{A}_j^*$ $(i \neq j)$ commute.

16.55. (Diagonalizable Matrices)

(a) (Two Matrices) Two diagonalizable $n \times n$ matrices are simultaneously diagonalizable; that is, there is a single nonsingular matrix $\mathbf{R}$ such that $\mathbf{R}^{-1}\mathbf{AR}$ and $\mathbf{R}^{-1}\mathbf{BR}$ are diagonal, if and only if $\mathbf{A}$ and $\mathbf{B}$ commute (i.e., $\mathbf{AB} = \mathbf{BA}$). Commuting matrices play a major role in simultaneous factorizations as we have seen in (16.53c) and (16.54) above. For details see Horn and Johnson [1985: chapter 2].

(b) Let $\mathcal{S}$ be an arbitrary (finite or infinite) set of $n \times n$ matrices in which every pair commutes. Then:

(i) There is a vector $\mathbf{x} \in \mathbb{C}^n$ that is an eigenvector of every $\mathbf{A} \in \mathcal{S}$.

(ii) The members of $\mathcal{S}$ can be simultaneously diagonalized.

Proofs. Section 16.8.

16.51a(i). Abadir and Magnus [2005: 180] and Searle [1982: 312–313].

16.51a(ii). Rao and Bhimasankaram [2000: 355–356] and Schott [2005: 163–165].

16.48a(iii). Horn and Johnson [1985: 228; they also give other equivalent conditions for the simultaneous diagonalization of two Hermitian matrices] and Rao and Rao [1998: 185-186].

16.51b. Horn and Johnson [1985: 465, complex case with a real linear combination and $\mathbf{R}^*$ instead of $\mathbf{R}'$] and Schott [2005: 161–162, real case].

16.51c. Abadir and Magnus [2005: 225] and Searle [1982: 313]. This result also holds for Hermitian matrices (cf. Horn and Johnson [1985: 250–251] and Rao and Rao [1998: 185–186]).

16.51d. Schott [2005: 162] and Searle [1982: 313–314].

16.52a–b. Horn and Johnson [1985: 235].

16.52c. Horn and Johnson [1985: 466].

16.52d. For proofs of (d) and further results, see Rao and Mitra [1971: chapter 6].

16.53a. Golub and Van Loan [1996: 377].

16.53b. Stewart [1972].

16.53c. Zhang [1999: 61] and Meyer [2000a: 522, exercise 7.2.15].

16.54a. Horn and Johnson [1985: 426, exercise 26].

16.54b. Quoted by Rao and Rao [1998: 192].

16.55a. Horn and Johnson [1985: 50] and Meyer [2000a: 522, exercise 7.2.16].

16.55b. Horn and Johnson [1985: 51–52].

16.9   POLAR DECOMPOSITION

16.56. Let $\mathbf{A}$ be an $m \times n$ complex matrix of rank r ($r \leq \min\{m, n\}$).

(a) Suppose $m \leq n$. Then, using the thin complex version of the singular value decomposition (cf. Section 16.3) we have the *polar decomposition*

$$\mathbf{A} = \mathbf{P}\boldsymbol{\Delta}_m\mathbf{Q}_m^* = (\mathbf{P}\boldsymbol{\Delta}_m\mathbf{P}^*)(\mathbf{P}\mathbf{Q}_m^*) = \mathbf{BW},$$

where $\mathbf{B} = (\mathbf{AA}^*)^{1/2}$ is an $m \times m$ unique Hermitian non-negative definitive matrix of rank r, and $\mathbf{W}$ is $m \times n$ with orthonormal rows (that is, $\mathbf{WW}^* = \mathbf{I}_m$). If rank $\mathbf{A} = m$, then the matrix $\mathbf{W}$ is unique and $\mathbf{B}$ is Hermitian positive definite. If $\mathbf{A}$ is real, then both $\mathbf{B}$ and $\mathbf{W}$ can be taken as real.

(b) If $m = n$, then $\mathbf{W}$ is unitary. Furthermore, if $\mathbf{A}$ is nonsingular, then $\mathbf{W}$ is uniquely determined as $\mathbf{B}^{-1}\mathbf{A}$.

(c) Let $m \geq n$. By applying (a) to $\mathbf{A}^*$ we can write $\mathbf{A} = \mathbf{VC}$, where the $m \times n$ matrix $\mathbf{V}$ has orthonormal columns and $\mathbf{C}$ is an $n \times n$ unique non-negative definite Hermitian matrix of rank r. If $\mathbf{A}$ nonsingular, then $\mathbf{V} = \mathbf{W}$.

(d) $\mathbf{B} = \mathbf{C}$ if and only if $\mathbf{A}$ is normal.

16.57. Suppose that the $n \times n$ matrix $\mathbf{A}$ has a polar decomposition $\mathbf{A} = \mathbf{BW}$. Then it follows from (16.56d) above that $\mathbf{A}$ is normal if and only if $\mathbf{BW} = \mathbf{WB}$.

Proofs. Section 16.9.

16.56. Horn and Johnson [1985: 412–414].

16.57. Abadir and Magnus [2005: 226, real case] and Horn and Johnson [1985: 414].

16.10   MISCELLANEOUS FACTORIZATIONS

16.58. (Takagi Factorization) Let $\mathbf{A} = (a_{ij})$ be a real or complex symmetric $n \times n$ matrix. Then $\mathbf{A}$ can be expressed in the form $\mathbf{A} = \mathbf{QDQ}'$ (note $\mathbf{Q}'$ and not $\mathbf{Q}^*$), where $\mathbf{Q}$ is an $n \times n$ unitary matrix and $\mathbf{D}$ is a real non-negative diagonal matrix. The columns of $\mathbf{Q}$ are an orthogonal set of right eigenvectors of $\mathbf{A}\bar{\mathbf{A}}$, and the corresponding diagonal elements of $\mathbf{D}$ are the non-negative square roots of the corresponding eigenvalues of $\mathbf{A}\bar{\mathbf{A}}$.

16.59. Any square matrix $\mathbf{A}$ can be factorized as $\mathbf{A} = \mathbf{SQDQ}'\mathbf{S}^{-1}$, where $\mathbf{S}$ is nonsingular, $\mathbf{Q}$ is unitary, and $\mathbf{D}$ is diagonal with non-negative main diagonal entries; all matrices are $n \times n$.

16.60. For any square matrix $\mathbf{A}$, there exists a unitary $\mathbf{Q}$ and upper-triangular matrix $\mathbf{V}$ such that $\mathbf{A} = \mathbf{QVQ}'$, where all matrices are $n \times n$, if and only if the eigenvalues of $\mathbf{A}\bar{\mathbf{A}}$ are real and non-negative. When this condition is true, the main diagonal elements of $\mathbf{V}$ may be chosen to be non-negative.

16.61. If **H** is Hermitian, there exists a unitary marix **Q** such that $\mathbf{Q}^*\mathbf{AQ}$ is tridiagonal (and also Hermitian).

16.62. (Upper Hessenberg Reduction)   For any square matrix **A** there exists a unitary matrix **Q** such that $\mathbf{QAQ}^*$ is upper Hessenberg.

Proofs. Section 16.10.

> 16.58. Horn and Johnson [1985: 157, 204] and quoted by Rao and Rao [1998: 192].

> 16.59. Horn and Johnson [1985: 157, 210].

> 16.60. Quoted by Rao and Rao [1998: 192].

> 16.61. Quoted by Rao and Rao [1998: 190]. The real case (Jacobi's reduction) is discussed by Meyer [2000a: 353].

> 16.62. Quoted by Rao and Rao [1998: 190]. It is also described by Meyer [2000a: 351, real case].

CHAPTER 17

DIFFERENTIATION

Methods of differentiation and differentials involving scalars, vectors, and matrices are used extensively in statistics. Applications include maximum likelihood and least squares estimation, large sample theory, statistical computing, and Jacobians, the subject of the next chapter. Turkington [2002], for example, applies first and second order differentiation to find maximum likelihood estimates and variance estimates for linear regression models, autoregressive time series, seemingly unrelated regression equations, and linear simultaneous equations models. Magnus and Neudecker [1999] do a similar thing with multivariate models, errors-in-variables models, nonlinear regression, and simultaneous equation models.

Differentiation is also used in sensitivity analysis and perturbation methods, which endeavor to determine the perturbation in a system when there are small changes in the parameters. It is also used in the derivation of elasticities (a term from economics), where one determines the proportional perturbation when there is a proportional change in a parameter. Some examples are model fittting (e.g., Seber and Wild [1989: 121 126, 668]), ecological population dynamics (Caswell [2001, 2007]), and multivariate elliptical linear regression models (Liu [2002b]). The chapter closes with a few results on difference equations.

17.1 INTRODUCTION

I have endeavored to categorize the methods of differentation for easy reference, though some results, especially relating to a function of a function, fit into more than

A Matrix Handbook for Statisticians. By George A. F. Seber
Copyright © 2008 John Wiley & Sons, Inc.

one category. There is also some overlap of topics as one can consider differentiation either with respect to a vector or matrix, or with respect to an element of a vector or a matrix. A helpful survey of the subject including an historical overview is given by Nel [1980]. He also considers differentiation with respect to patterned matrices (cf. Section 18.3.5).

17.2   SCALAR DIFFERENTIATION

For some analytical background to the subject in a statistical context, the reader is referred to Abadir and Magnus [2005: chapter 13], Magnus and Neudecker [1999], and Schott [2005: chapter 9].

17.2.1   Differentiation with Respect to t

Definition **17.1.** We first define the derivative of a matrix or vector with respect to a scalar. If $\mathbf{A}(t) = (a_{ij}(t))$, then $\partial \mathbf{A}(t)/\partial t$ is defined to be $(\partial a_{ij}(t)/\partial t)$; that is, the derivative of $\mathbf{A}(t)$ is obtained by differentiating each element of $\mathbf{A}$. The same is true for a vector $\mathbf{a}(t) = (a_i(t))$.

Unless specified (e.g., $\mathbf{A}$ is symmetric), we assume that the elements of all the matrices differentiated are functionally independent (i.e., unconstrained). Also, the following apply when we have a vector $\mathbf{t} = (t_i, t_2, \dots, t_r)'$ and ∂t is replaced by ∂t_i.

17.1. We have from the definition:

(a) $\dfrac{\partial \{\mathbf{A X}(t)\mathbf{B}\}}{\partial t} = \mathbf{A} \dfrac{\partial \mathbf{X}(t)}{\partial t} \mathbf{B}.$

(b) $\dfrac{\partial \mathrm{vec}\, \mathbf{X}(t)}{\partial t} = \mathrm{vec}\, \dfrac{\partial \mathbf{X}(t)}{\partial t}.$

17.2. (Products) Noting that $\otimes$ is the Kronecker product, the following result is used extensively in the next section.

(a) $\dfrac{\partial \mathbf{A}(t)\mathbf{B}(t)\mathbf{C}(t)}{\partial t} = \dfrac{\partial \mathbf{A}}{\partial t}\mathbf{B}\mathbf{C} + \mathbf{A}\dfrac{\partial \mathbf{B}}{\partial t}\mathbf{C} + \mathbf{A}\mathbf{B}\dfrac{\partial \mathbf{C}}{\partial t}.$

(b) $\dfrac{\partial (\mathbf{A}(t) \otimes \mathbf{B}(t))}{\partial t} = \dfrac{\partial \mathbf{A}}{\partial t} \otimes \mathbf{B} + \mathbf{A} \otimes \dfrac{\partial \mathbf{B}}{\partial t}.$

17.3. (Inverse)

(a) Differentiating $\mathbf{A}\mathbf{A}^{-1} = \mathbf{I}$ for nonsingular $\mathbf{A}(t)$, we get

$$\frac{\partial \mathbf{A}^{-1}(t)}{\partial t} = -\mathbf{A}^{-1}\frac{\partial \mathbf{A}(t)}{\partial t}\mathbf{A}^{-1}.$$

(b) If $\mathbf{R}$ does not depend on t, then differentiating $[\mathbf{R}'\mathbf{A}(t)\mathbf{R}]^{-1}\mathbf{R}'\mathbf{A}(t)\mathbf{R} = \mathbf{I}$ gives us

$$\frac{\partial [\mathbf{R}'\mathbf{A}(t)\mathbf{R}]^{-1}}{\partial t} = [\mathbf{R}'\mathbf{A}(t)\mathbf{R}]^{-1}\mathbf{R}'\frac{\partial \mathbf{A}(t)}{\partial t}\mathbf{R}[\mathbf{R}'\mathbf{A}(t)\mathbf{R}]^{-1}.$$

(c) If $\mathbf{A}$ is symmetric and $\mathbf{B}(t) = \mathbf{R}[\mathbf{R}'\mathbf{A}(t)\mathbf{R}]^{-1}\mathbf{R}'$, where $\mathbf{R}$ does not depend on t, then, using (b), we obtain

$$\frac{\partial \mathbf{B}}{\partial t} = -\mathbf{B}\frac{\partial \mathbf{A}}{\partial t}\mathbf{B}.$$

(d) If $\mathbf{A}^{-}(t)$ is a weak inverse of $\mathbf{A}(t)$, then differentiating $\mathbf{A}\mathbf{A}^{-}\mathbf{A} = \mathbf{A}$ gives us

$$\mathbf{A}\frac{\partial \mathbf{A}^{-}(t)}{\partial t}\mathbf{A} = -\mathbf{A}\mathbf{A}^{-}\frac{\partial \mathbf{A}}{\partial t}\mathbf{A}^{-}\mathbf{A}.$$

Further details are given in (17.8) below.

(e) $\mathbf{A}\dfrac{\partial[(\mathbf{A}'\mathbf{A})^{-}]}{\partial t}\mathbf{A}' = -\mathbf{A}(\mathbf{A}'\mathbf{A})^{-}\dfrac{\partial(\mathbf{A}\mathbf{A}')}{\partial t}(\mathbf{A}'\mathbf{A})^{-}\mathbf{A}'.$

17.4. (Determinants)   If $\mathbf{A}(t)$ is nonsingular, then

$$\frac{\partial \log(\det \mathbf{A})}{\partial t} = \operatorname{trace}\left[\mathbf{A}^{-1}\frac{\partial \mathbf{A}}{\partial t}\right].$$

The result is also true if $\mathbf{A}$ is symmetric. A further result follows by noting that

$$\frac{\partial \det \mathbf{A}}{\partial t} = \det \mathbf{A}\frac{\partial \log(\det \mathbf{A})}{\partial t}.$$

17.5. (Trace) $\dfrac{\partial[\operatorname{trace}(\mathbf{A}(t))]}{\partial t} = \operatorname{trace}\left[\dfrac{\partial \mathbf{A}(t)}{\partial t}\right].$

17.6. (Exponential) $\dfrac{\partial e^{\mathbf{A}t}}{\partial t} = \mathbf{A}e^{\mathbf{A}t}.$

Proofs. Section 17.2.1.

17.2. Graham [1981: 38].

17.3d–e. Searle [1982: 335].

17.4. Searle [1982: 337–338].

17.6. Abadir and Magnus [2005: 368].

17.2.2   Differentiation with Respect to a Vector Element

We now consider the special case when t is an element of an $n \times 1$ vector $\mathbf{x} = (x_i)$. The results in this section still apply if $\mathbf{F}$ is a matrix function of a matrix $\mathbf{X} = (x_{ij})$, and we replace x_i by x_{ij}.

17.7. Let $\mathbf{F}$ be a square matrix function of a vector $\mathbf{x} = (x_1, x_2, \ldots, x_n)'$, then from (17.4) and (17.3a) we have the following.

(a) $\dfrac{\partial \operatorname{trace}[\mathbf{F}(\mathbf{x})]}{\partial x_i} = \operatorname{trace}\left(\dfrac{\partial \mathbf{F}}{\partial x_i}\right).$

(b) Suppose $\mathbf{F}$ is nonsingular.

(i) $\dfrac{\partial \det[\mathbf{F}(\mathbf{x})]}{\partial x_i} = (\det \mathbf{F}) \operatorname{trace}\left(\mathbf{F}^{-1}\dfrac{\partial \mathbf{F}}{\partial x_i}\right).$

(ii) $\dfrac{\partial \log \det[\mathbf{F}(\mathbf{x})]}{\partial x_i} = \dfrac{1}{\det \mathbf{F}}\dfrac{\partial \det \mathbf{F}}{\partial x_i} = \operatorname{trace}\left(\mathbf{F}^{-1}\dfrac{\partial \mathbf{F}}{\partial x_i}\right).$

(iii) $\dfrac{\partial \mathbf{F}^{-1}}{\partial x_i} = -\mathbf{F}^{-1}\dfrac{\partial \mathbf{F}}{\partial x_i}\mathbf{F}^{-1}.$

(iv) $\dfrac{\partial \log \det(\mathbf{A}\mathbf{F}^{-1}\mathbf{B})}{\partial x_i} = -\operatorname{trace}\left[\mathbf{F}^{-1}\mathbf{B}(\mathbf{A}\mathbf{F}^{-1}\mathbf{B})^{-1}\mathbf{A}\mathbf{F}^{-1}\dfrac{\partial \mathbf{F}}{\partial x_i}\right].$

(v) If $\operatorname{adj}\mathbf{F}$ is the adjoint matrix of $\mathbf{F}$, then $\operatorname{adj}\mathbf{F} = (\det \mathbf{F})\mathbf{F}^{-1}$ and

$$\frac{\partial \operatorname{adj}\mathbf{F}}{\partial x_i} = \frac{\partial \det \mathbf{F}}{\partial x_i}\mathbf{F}^{-1} + \det \mathbf{F}\frac{\partial \mathbf{F}^{-1}}{\partial x_i}.$$

(c) (Kronecker Product) Let $\mathbf{F}$ and $\mathbf{G}$ be $p \times q$ and $r \times s$ matrix functions of $\mathbf{x}$.

(i) $\dfrac{\partial (\mathbf{F} \otimes \mathbf{G})}{\partial x_i} = \mathbf{F} \otimes \dfrac{\partial \mathbf{G}}{\partial x_i} + \dfrac{\partial \mathbf{F}}{\partial x_i} \otimes \mathbf{G}.$

We can replace $\mathbf{F}$ by $\operatorname{vec}\mathbf{F}$ and $\mathbf{G}$ by $\operatorname{vec}\mathbf{G}$ in the above equation.

(ii) $\dfrac{\partial \operatorname{vec}(\mathbf{F} \otimes \mathbf{G})}{\partial x_i} = (\mathbf{I}_q \otimes \mathbf{K}_{sp} \otimes \mathbf{I}_r)\dfrac{\partial (\operatorname{vec}\mathbf{F} \otimes \operatorname{vec}\mathbf{G})}{\partial x_i},$

where $\mathbf{K}_{sp}\,(=\mathbf{I}_{(p,s)})$ is the commutation matrix.

17.8. Let $\mathbf{F}$ be a $p \times q$ matrix function of $\mathbf{x}$. If $\mathbf{F}^-$ is a weak inverse of $\mathbf{F}$, then, under certain analytical conditions including continuous differentiability and constant rank in some neighborhood, we have

$$\mathbf{F}\frac{\partial \mathbf{F}^-}{\partial x_i}\mathbf{F} = -\mathbf{F}\mathbf{F}^-\frac{\partial \mathbf{F}}{\partial x_i}\mathbf{F}^-\mathbf{F}.$$

In particular, there exists a weak inverse $\mathbf{G}$ of $\mathbf{F}$ such that $\dfrac{\partial \mathbf{G}}{\partial x_i} = -\mathbf{G}\dfrac{\partial \mathbf{F}}{\partial x_i}\mathbf{G}.$

17.9. Let $\mathbf{F}$ be a $p \times q$ matrix function of $\mathbf{x}$. If $\mathbf{F}^+$ is the Moore–Penrose inverse of $\mathbf{F}$, then, under certain analytical conditions including continuous differentiability and constant rank in some neighborhood, we obtain

$$\frac{\partial \mathbf{F}^+}{\partial x_i} =$$

$$-\mathbf{F}^+\frac{\partial \mathbf{F}}{\partial x_i}\mathbf{F}^+ + \mathbf{F}^+(\mathbf{F}^+)'\left(\frac{\partial \mathbf{F}}{\partial x_i}\right)'(\mathbf{I}_p - \mathbf{F}\mathbf{F}^+) + (\mathbf{I}_q - \mathbf{F}^+\mathbf{F})\left(\frac{\partial \mathbf{F}}{\partial x_i}\right)'(\mathbf{F}^+)'\mathbf{F}^+.$$

17.10. (Eigenvalue and Eigenvector) Let $\mathbf{F}$ be a symmetric matrix function of an $n \times 1$ vector $\mathbf{x}$. Let λ be a simple eigenvalue of $\mathbf{F}$ (i.e., one with an algebraic multiplicity of 1) and corresponding right eigenvector $\mathbf{u}$ of unit length. Then, given $\mathbf{F}\mathbf{u} = \lambda\mathbf{u}$ and $\mathbf{H}^+$, the Moore–Penrose inverse of $\mathbf{H} = \mathbf{F} - \lambda\mathbf{I}$, we obtain

$$\frac{\partial \lambda}{\partial x_i} = \mathbf{u}'\frac{\partial \mathbf{F}}{\partial x_i}\mathbf{u} \quad \text{and} \quad \frac{\partial \mathbf{u}}{\partial x_i} = -\mathbf{H}^+\frac{\partial \mathbf{F}}{\partial x_i}\mathbf{u}.$$

17.11. Consider the idempotent matrix $\mathbf{P} = \mathbf{X}(\mathbf{X}'\mathbf{W}\mathbf{X})^{-}\mathbf{X}'\mathbf{W}$, where $\mathbf{X}$ is $n \times p$ and $\mathbf{W}$ is an $n \times n$ positive definite matrix such that the elements of $\mathbf{W}$ and/or $\mathbf{X}$ are functions of a vector $\mathbf{z}$. Then, under certain analytical conditions including continuous differentiability and constant rank in some neighborhood, we obtain

$$\frac{\partial \mathbf{P}}{\partial z_i} = (\mathbf{I}_n - \mathbf{P})\frac{\partial \mathbf{X}}{\partial z_i}(\mathbf{X}'\mathbf{W}\mathbf{X})^{-}\mathbf{X}'\mathbf{W} + \mathbf{X}(\mathbf{X}'\mathbf{W}\mathbf{X})^{-}(\frac{\partial \mathbf{X}}{\partial z_i})'\mathbf{W}(\mathbf{I}_n - \mathbf{P})$$
$$+ \mathbf{X}(\mathbf{X}'\mathbf{W}\mathbf{X})^{-}\mathbf{X}'\frac{\partial \mathbf{W}}{\partial z_i}(\mathbf{I}_n - \mathbf{P}).$$

17.12. Suppose $\mathbf{X} = \mathbf{X}(\phi)$, where $\mathbf{X}$ is $n \times p$ of rank p and is a function of $\phi = (\phi_1, \phi_2, \ldots, \phi_k)'$. Then,

$$\frac{\partial \det(\mathbf{X}'\mathbf{X})}{\partial \phi_i} = \det(\mathbf{X}'\mathbf{X}) \operatorname{trace}\left(\mathbf{X}^{+}\frac{\partial \mathbf{X}}{\partial \phi_i}\right),$$

where $\mathbf{X}^{+}$ is the Moore–Penrose inverse of $\mathbf{X}$. This theory arises in nonlinear modeling.

Proofs. Section 17.2.2.

17.7. Harville [1997: 305, 307–308] and Harville [2001: 158, exercise 32].

17.8. Harville [1997: 312].

17.9. Harville [1997: 511].

17.10. Harville [1997: section 21.15 for proofs and analytical background].

17.11. Harville [1997: 315]. Derivatives are also given for $\mathbf{WP}$ and $\mathbf{W} - \mathbf{WP}$.

17.12. Bates and Watts [1987] and Bates and Watts [1988: chapter 4] give further details. For a summary see Seber and Wild [1989: 543–544].

17.2.3   Differentiation with Respect to a Matrix Element

***Definition* 17.2.** We define the matrix $\mathbf{E}_{ij}$ to be an $m \times n$ matrix with 1 in the i,jth position and zeros elsewhere. Thus $\mathbf{E}_{ij} = \mathbf{e}_{i,m}\mathbf{e}'_{n,j}$, where $\mathbf{e}_{i,m}$ is the ith column of $\mathbf{I}_m$ and $\mathbf{e}_{j,n}$ is the jth column of $\mathbf{I}_n$.

In what follows, we consider the special case of $t = x_{ij}$, an element of the real $m \times n$ matrix $\mathbf{X}$, and include differentiation with respect to a vector element. Results in this section can be derived using the properties given in the previous section along with (17.13) below. We assume that the elements of $\mathbf{X}$ are functionally independent (i.e., are "unconstrained"), unless stated otherwise. When $m = n$, then $\mathbf{E}_{ij} = \mathbf{E}'_{ji}$.

17.13. (Basic Result)

(a) It is straightforward to show that

$$\frac{\partial \mathbf{X}}{\partial x_{ij}} = \begin{cases} \mathbf{E}_{ij}, & \mathbf{X} \text{ unconstrained,} \\ \mathbf{E}_{ij} + \mathbf{E}'_{ij} - \delta_{ij}\mathbf{E}_{ii}, & \mathbf{X} \text{ symmetric,} \end{cases}$$

where $\delta_{ij} = 1$ when $i = j$ and $\delta_{ij} = 0$ when $i \neq j$.

(b) $\dfrac{\partial \mathbf{X}'}{\partial x_{ij}} = \left(\dfrac{\partial \mathbf{X}}{\partial x_{ij}} \right)'.$

(c) To convert a result given below about an unconstrained $\mathbf{X}$ into one for symmetric $\mathbf{X}$, we simply replace $\mathbf{E}_{ij}$ by $\mathbf{E}_{ij} + \mathbf{E}'_{ij} - \delta_{ij}\mathbf{E}_{ii}$.

17.14. (Products) We assume that the following matrices are conformable and $\mathbf{X}$ is unconstrained. The results follow directly from (17.1) and (17.2). Further results can be obtained by setting $\mathbf{A}$ and/or $\mathbf{B}$ equal to the identity matrix.

(a) $\dfrac{\partial (\mathbf{AXB})}{\partial x_{ij}} = \mathbf{A}\dfrac{\partial \mathbf{X}}{\partial x_{ij}}\mathbf{B} = \mathbf{A}\mathbf{E}_{ij}\mathbf{B}.$

(b) $\dfrac{\partial (\mathbf{AX'B})}{\partial x_{ij}} = \mathbf{A}\mathbf{E}'_{ij}\mathbf{B}.$

(c) $\dfrac{\partial (\mathbf{X'AXB})}{\partial x_{ij}} = \dfrac{\partial \mathbf{X}'}{\partial x_{ij}}\mathbf{AXB} + \mathbf{X}'\dfrac{\partial (\mathbf{AXB})}{\partial x_{ij}} = \mathbf{E}'_{ij}\mathbf{AXB} + \mathbf{X}'\mathbf{AE}_{ij}\mathbf{B}.$

(d) $\dfrac{\partial (\mathbf{XAX'B})}{\partial x_{ij}} = \mathbf{E}_{ij}\mathbf{AX'B} + \mathbf{XAE}'_{ij}\mathbf{B}.$

(e) $\dfrac{\partial (\mathbf{AXX'B})}{\partial x_{ij}} = \mathbf{A}\dfrac{\partial \mathbf{XX}'}{\partial x_{ij}}\mathbf{B} = \mathbf{A}\mathbf{E}_{ij}\mathbf{X'B} + \mathbf{AXE}'_{ij}\mathbf{B}.$

(f) $\dfrac{\partial (\mathbf{XAXB})}{\partial x_{ij}} = \mathbf{E}_{ij}\mathbf{AXB} + \mathbf{XAE}_{ij}\mathbf{B}.$

(g) $\dfrac{\partial (\mathbf{X'AX'B})}{\partial x_{ij}} = \mathbf{E}'_{ij}\mathbf{AX'B} + \mathbf{X}'\mathbf{AE}'_{ij}\mathbf{B}.$

(h) $\dfrac{\partial \mathbf{XX'X}}{\partial x_{ij}} = \mathbf{E}_{ij}\mathbf{X'X} + \mathbf{XE}'_{ij}\mathbf{X} + \mathbf{XX'E}_{ij}.$

17.15. (Inverses)

(a) If $\mathbf{BXC}$ is nonsingular, we differentiate $(\mathbf{BXC})(\mathbf{BXC})^{-1} = \mathbf{I}$ to get (cf. 17.3a)

$$\dfrac{\partial \{\mathbf{A}(\mathbf{BXC})^{-1}\mathbf{D}\}}{\partial x_{ij}} = -\mathbf{A}(\mathbf{BXC})^{-1}\mathbf{BE}_{ij}\mathbf{C}(\mathbf{BXC})^{-1}\mathbf{D}.$$

(b) Suppose $\mathbf{X}$ is $m \times m$ and nonsingular. From (a) we have

$$\dfrac{\partial \mathbf{X}^{-1}}{\partial x_{ij}} = -\mathbf{X}^{-1}\mathbf{E}_{ij}\mathbf{X}^{-1} = -\mathbf{X}^{-1}\mathbf{e}_{i,m}\mathbf{e}'_{j,m}\mathbf{X}^{-1} = -\mathbf{y}_i\mathbf{z}'_j,$$

where $\mathbf{y}_i$ is the ith column and $\mathbf{z}'_j$ is the jth row of $\mathbf{X}^{-1}$.

If $\mathbf{X}$ is symmetric, then using (17.13c) we have

$$\dfrac{\partial \mathbf{X}^{-1}}{\partial x_{ij}} = \begin{cases} -\mathbf{y}_i\mathbf{y}'_i, & \text{if } i = j, \\ -\mathbf{y}_i\mathbf{y}'_j - \mathbf{y}_j\mathbf{y}'_i, & \text{if } i > j, \end{cases}$$

where $\mathbf{y}_i$ is the ith column of $\mathbf{X}^{-1}$.

17.16. (Determinants) Suppose $\mathbf{X}$ is square and ξ_{ij} is the cofactor of x_{ij}. Then

$$\frac{\partial \det \mathbf{X}}{\partial x_{ij}} = \begin{cases} \xi_{ij}, & \mathbf{X} \text{ unconstrained,} \\ (2 - \delta_{ij})\xi_{ij}, & \mathbf{X} = \mathbf{X}', \end{cases}$$

where $\delta_{ij} = 1$ when $i = j$ and 0 otherwise.

17.17. (Powers) Let $\mathbf{X}$ be nonsingular, and let k be a positive integer.

(a) We can prove by induction that

$$\frac{\partial \mathbf{X}^k}{\partial x_{ij}} = \sum_{r=0}^{k-1} \mathbf{X}^r \mathbf{E}_{ij} \mathbf{X}^{k-r-1}$$

for $k = 1, 2, \ldots$, where $\mathbf{X}^0 = \mathbf{I}_n$.

(b) Differentiating $\mathbf{X}^k \mathbf{X}^{-k} = \mathbf{I}$ gives us

$$\frac{\partial \mathbf{X}^{-k}}{\partial x_{ij}} = -\mathbf{X}^{-k} \frac{\partial \mathbf{X}^k}{\partial x_{ij}} \mathbf{X}^{-k}.$$

17.18. (Some Matrix Functions)   Let $\mathbf{Y}$ be a nonsingular matrix function of $\mathbf{X}$, where $\mathbf{X}$ is unconstrained.

(a) $\dfrac{\partial \det \mathbf{Y}}{\partial x_{ij}} = \operatorname{trace}\left[(\operatorname{adj}\mathbf{Y})\left(\dfrac{\partial \mathbf{Y}}{\partial x_{ij}} \right) \right] = (\det \mathbf{Y}) \operatorname{trace}\left[\mathbf{Y}^{-1} \dfrac{\partial \mathbf{Y}}{\partial x_{ij}} \right].$

When $\mathbf{X}$ is symmetric we can use (17.13c) in the following applications.

(i) If $\mathbf{Y} = \mathbf{AXB}$, then (from 17.14a)

$$\begin{aligned} \frac{\partial \det(\mathbf{AXB})}{\partial x_{ij}} &= \det(\mathbf{AXB}) \operatorname{trace}[(\mathbf{AXB})^{-1} \mathbf{AE}_{ij}\mathbf{B}] \\ &= \det(\mathbf{AXB})\{[\mathbf{B}(\mathbf{AXB})^{-1}\mathbf{A}]'\}_{ij}. \end{aligned}$$

(ii) If $\mathbf{Y} = \mathbf{X}'\mathbf{AX}$, then (from 17.14c)

$$\frac{\partial \det(\mathbf{X}'\mathbf{AX})}{\partial x_{ij}} = \det(\mathbf{X}'\mathbf{AX}) \operatorname{trace}\{(\mathbf{X}'\mathbf{AX})^{-1}[\mathbf{E}'_{ij}\mathbf{AX} + \mathbf{X}'\mathbf{AE}_{ij}]\}.$$

(b) $\dfrac{\partial \operatorname{vec} \mathbf{Y}}{\partial x_{ij}} = \operatorname{vec} \dfrac{\partial \mathbf{Y}}{\partial x_{ij}}.$

(c) $\dfrac{\partial \operatorname{trace} \mathbf{Y}}{\partial x_{ij}} = \operatorname{trace}\left(\dfrac{\partial \mathbf{Y}}{\partial x_{ij}} \right).$

(d) $\dfrac{\partial \mathbf{Y}^{-1}}{\partial x_{ij}} = -\mathbf{Y}^{-1} \dfrac{\partial \mathbf{Y}}{\partial x_{ij}} \mathbf{Y}^{-1}.$

We can get a corresponding result for $(\mathbf{Y}^{-1})'$ by simply replacing $\mathbf{Y}$ by $\mathbf{Y}'$, as $(\mathbf{Y}^{-1})' = (\mathbf{Y}')^{-1}$.

17.19. (Eigenvalue and Eigenvector)   Let $\mathbf{X}$ be a symmetric $n \times n$ matrix with simple eigenvalue λ (i.e., has an algebraic multiplicity of 1), and corresponding eigenvector $\mathbf{u} = (u_i)$ of unit length. Then

$$\frac{\partial \lambda}{\partial x_{ij}} = 2\mathbf{u}\mathbf{u}' - \text{diag}(u_1^2, u_2^2, \ldots, u_n^2).$$

Also, if $\mathbf{g}_j$ is the jth column of $\mathbf{H}^+$, the Moore–Penrose inverse of $\mathbf{H} = \mathbf{X} - \lambda\mathbf{I}_n$, we have

$$\frac{\partial \mathbf{u}}{\partial x_{ij}} = \begin{cases} -u_i\mathbf{g}_i, & \text{if} \quad j = i, \\ -(u_j\mathbf{g}_i + u_i\mathbf{g}_j), & \text{if} \quad j < i. \end{cases}$$

Proofs. Section 17.2.3.

   17.14. Graham [1981: 60–64, 69].

   17.15b. Harville [2001: 130, exercise 21].

   17.16. Searle [1982: 336].

   17.17. Graham [1981: 67–68].

   17.18. Harville [1997: section 15.8].

   17.19. Harville [1997: 567].

17.3   VECTOR DIFFERENTIATION: SCALAR FUNCTION

17.3.1   Basic Results

Definition 17.3. If f is a function of $\mathbf{x}$, we denote the vector of partial derivatives $(\partial f/\partial x_i)$ by the column vector $\partial f/\partial\mathbf{x}$, that is, $\dfrac{\partial f}{\partial\mathbf{x}} = \left(\dfrac{\partial f}{\partial x_i}\right)$. We also define the row vector $\partial f/\partial\mathbf{x}' = (\partial f/\partial\mathbf{x})'$. Some authors (e.g., Dhrymes [2000]) reverse the notation.

17.20. (Basic Results) Let $\mathbf{x}$ and $\mathbf{a}$ be $n \times 1$ vectors, and let $\mathbf{A}$ an $n \times n$ matrix.

(a) $\dfrac{\partial\mathbf{x}'\mathbf{a}}{\partial\mathbf{x}} = \mathbf{a} = \dfrac{\partial\mathbf{a}'\mathbf{x}}{\partial\mathbf{x}}$.

(b) $\dfrac{\partial\mathbf{x}'\mathbf{A}\mathbf{x}}{\partial\mathbf{x}} = (\mathbf{A} + \mathbf{A}')\mathbf{x}$, or $2\mathbf{A}\mathbf{x}$ if $\mathbf{A}$ is symmetric.

17.21. (Chain Rule) If z is a differentiable scalar function of $\mathbf{y}$, and $\mathbf{y}$ is a differentiable function of $\mathbf{x}$, then

$$\frac{\partial z}{\partial x_i} = \sum_j \frac{\partial z}{\partial y_j}\frac{\partial y_j}{\partial x_i},$$

which can be expressed in the form of the row vector

$$\frac{\partial z}{\partial\mathbf{x}'} = \frac{\partial z}{\partial\mathbf{y}'} \cdot \frac{\partial\mathbf{y}}{\partial\mathbf{x}'}.$$

In terms of column vectors,

$$\frac{\partial z}{\partial \mathbf{x}} = \left(\frac{\partial \mathbf{y}}{\partial \mathbf{x}'} \right)' \frac{\partial z}{\partial \mathbf{y}}.$$

The function z might include functions such as the trace, the determinant, and quadratic expressions.

Proofs. Section 17.3.1.

17.20. Abadir and Magnus [2005: 356, transposed].

17.3.2   x = vec X

In applying the following results using the chain rule above, it can be more convenient to work with $\partial (\text{vec } \mathbf{X})'$ instead of $\partial \text{ vec } \mathbf{X}$. The right-hand side is then transposed, as indicated in (17.22) below. Some authors use the reverse notation (e.g., Dhrymes [2000]). Note that the following derivatives are all column vectors.

17.22. If $f(\mathbf{X})$ is a scalar function of the matrix $\mathbf{X}$, then

$$\frac{\partial f(\mathbf{X})}{\partial \text{ vec } \mathbf{X}} = \text{vec} \left(\frac{\partial f(\mathbf{X})}{\partial \mathbf{X}} \right) = \left(\frac{\partial f(\mathbf{X})}{\partial (\text{vec } \mathbf{X})'} \right)'.$$

17.23. (Trace)

(a) $\dfrac{\partial \text{ trace}(\mathbf{AXB})}{\partial \text{ vec } \mathbf{X}} = \text{vec} \, (\mathbf{A}'\mathbf{B}').$

We can obtain this result directly by noting that

$$\text{trace}(\mathbf{AXB}) = \text{trace}(\mathbf{BAX}) = \text{vec} \, (\mathbf{A}'\mathbf{B}')'\text{vec } \mathbf{X},$$

and using (17.20a). We can set $\mathbf{A}$ or $\mathbf{B}$ equal to $\mathbf{I}$.

(b) $\dfrac{\partial \text{ trace}(\mathbf{X}'\mathbf{AXB})}{\partial \text{ vec } \mathbf{X}} = [(\mathbf{B}' \otimes \mathbf{A}) + (\mathbf{B} \otimes \mathbf{A}')]\text{vec } \mathbf{X}.$

Provided that the appropriate matrices are square, other results follow from $\text{trace}(\mathbf{CD}) = \text{trace}(\mathbf{DC})$ and $\text{trace } \mathbf{C} = \text{trace } \mathbf{C}'$. For example,

$$\text{trace}(\mathbf{X}'\mathbf{AXB}) = \text{trace}(\mathbf{AXBX}') = \text{trace}(\mathbf{XBX}'\mathbf{A}) = \text{trace}(\mathbf{BX}'\mathbf{AX}).$$

17.24. (Determinants and Log Determinants)   The following matrices $\mathbf{X}$ and $\mathbf{Y}$ are nonsingular, and we use the result $\text{vec} \, (\mathbf{AXB}) = (\mathbf{B}' \otimes \mathbf{A})\text{vec } \mathbf{X}$.

(a) $\dfrac{\partial \det \mathbf{X}}{\partial \text{ vec } \mathbf{X}} = \text{vec} \, [(\text{adj}\mathbf{X})'] = (\det \mathbf{X})\text{vec} \, (\mathbf{X}^{-1'}).$

(b) If $\mathbf{Y} = \mathbf{X}'\mathbf{AX}$, then

$$\frac{\partial \det \mathbf{Y}}{\partial \text{vec } \mathbf{X}} = \det \mathbf{Y}[(\mathbf{Y}^{-1'} \otimes \mathbf{A}) + (\mathbf{Y}^{-1} \otimes \mathbf{A}')]\text{vec } \mathbf{X}.$$

When $\mathbf{A}$ is symmetric, then $\mathbf{Y}$ is also symmetric and

$$\frac{\partial \det \mathbf{Y}}{\partial \operatorname{vec} \mathbf{X}} = 2 \det \mathbf{Y}[(\mathbf{Y}^{-1} \otimes \mathbf{A}) \operatorname{vec} \mathbf{X}.$$

(c) If $\mathbf{Y} = \mathbf{XBX}'$, then

$$\frac{\partial \det \mathbf{Y}}{\partial \operatorname{vec} \mathbf{X}} = \det \mathbf{Y}[\mathbf{B} \otimes (\mathbf{Y}^{-1})' + (\mathbf{B}' \otimes \mathbf{Y}^{-1})] \operatorname{vec} \mathbf{X}.$$

When $\mathbf{B}$ is symmetric, then $\mathbf{Y}$ is symmetric and

$$\frac{\partial \det \mathbf{Y}}{\partial \operatorname{vec} \mathbf{X}} = 2 \det \mathbf{Y}[\mathbf{B} \otimes (\mathbf{Y}^{-1})] \operatorname{vec} \mathbf{X}.$$

(d) If $\mathbf{Y}$ is one of the above functions, then

$$\frac{\partial (\log \det \mathbf{Y})}{\partial \operatorname{vec} \mathbf{X}} = \frac{1}{\det \mathbf{Y}} \frac{\partial \det \mathbf{Y}}{\partial \operatorname{vec} \mathbf{X}}.$$

Proofs. Section 17.3.2.

17.23. Dhrymes [2000: 156–157, transposed] and Rogers [1980: 54].

17.24a. Rao and Rao [1998: 229] and Schott [2005: 360].

17.24b. Abadir and Magnus [2005: 372–373, transposed].

17.24a–d. Turkington [2002: chapter 4],

17.3.3   Function of a Function

17.25. Suppose $y = \mathbf{w}'\mathbf{Az}$, where $\mathbf{A}$ is $m \times n$ and $\mathbf{w}$, $\mathbf{z}$, and $\mathbf{A}$ are all functions of $\mathbf{x}$. We wish to find the row vector $\partial y/\partial \mathbf{x}'$. We first note from (11.16b) and (11.15c) that

$$y = \operatorname{vec} y = (\mathbf{z}' \otimes \mathbf{w}') \operatorname{vec} \mathbf{A} = (\mathbf{z} \otimes \mathbf{w})' \operatorname{vec} \mathbf{A} = [\operatorname{vec}(\mathbf{wz}')]' \operatorname{vec} \mathbf{A}.$$

Then using $\mathbf{w}'\mathbf{Az} = \mathbf{z}'\mathbf{A}'\mathbf{w}$, we get

$$\frac{\partial y}{\partial \mathbf{x}'} = \mathbf{z}'\mathbf{A}'\frac{\partial \mathbf{w}}{\partial \mathbf{x}'} + [\operatorname{vec}(\mathbf{wz}')]'\frac{\partial \operatorname{vec} \mathbf{A}}{\partial \mathbf{x}'} + \mathbf{w}'\mathbf{A}\frac{\partial \mathbf{z}}{\partial \mathbf{x}'}.$$

17.26. Let $y = \operatorname{trace}[\mathbf{F}(\mathbf{Z})]$, where $\mathbf{F}$ is a square matrix function of $\mathbf{Z}$ and $\mathbf{Z}$ is a function of $\mathbf{x}$. Then, by the chain rule (17.21), we obtain

$$\frac{\partial y}{\partial \mathbf{x}'} = \frac{\partial y}{\partial (\operatorname{vec} \mathbf{Z})'} \cdot \frac{\partial \operatorname{vec} \mathbf{Z}}{\partial \mathbf{x}'},$$

where $\partial y/\partial(\operatorname{vec} \mathbf{Z})'$ can be obtained from (17.23) and transposing. We give three examples from Dhrymes [2000: section 5.4].

(a) $\dfrac{\partial \operatorname{trace}(\mathbf{AZB})}{\partial \mathbf{x}'} = \operatorname{vec}(\mathbf{A}'\mathbf{B}')'\dfrac{\partial \operatorname{vec} \mathbf{Z}}{\partial \mathbf{x}'}.$

(b) $\dfrac{\partial \operatorname{trace}(\mathbf{A}\mathbf{Z}'\mathbf{B}\mathbf{Z})}{\partial \mathbf{x}'} = (\operatorname{vec}\mathbf{Z})'(\mathbf{A}' \otimes \mathbf{B} + \mathbf{A} \otimes \mathbf{B}')\dfrac{\partial \operatorname{vec}\mathbf{Z}}{\partial \mathbf{x}'}.$

(c) $\dfrac{\partial \det \mathbf{Z}}{\partial \mathbf{x}'} = \operatorname{vec}\left[(\operatorname{adj}\mathbf{Z})'\right]'\dfrac{\partial \operatorname{vec}\mathbf{Z}}{\partial \mathbf{x}'}.$

Proofs. Section 17.3.3.

17.26. Dhrymes [2000: section 5.4].

17.26a. The result follows from (17.23a).

17.26b. We use (17.23b) transposed with $\mathbf{A}$ and $\mathbf{B}$ interchanged.

17.26c. We use (17.24a).

17.4 VECTOR DIFFERENTIATION: VECTOR FUNCTION

Definition 17.4. Let $\mathbf{x}$ and $\mathbf{y}$ be $n \times 1$ vectors. We define

$$\frac{\partial \mathbf{y}}{\partial \mathbf{x}'} = \left(\frac{\partial y_i}{\partial x_j}\right).$$

I find this notation easy to remember because $\mathbf{y}$, being a column vector, means that i refers to the row number, and $\mathbf{x}'$, being a row vector, means that j refers to the column number. This notation is used, for example, by Magnus and Nuedecker [19] and Harville [1997]. However, other notations are used in the literature. For example, Dhrymes [200] calls the above expression $\partial \mathbf{y}/\partial \mathbf{x}$, while Graham [1981], Searle [1982], and Turkington [2002] define $\partial \mathbf{y}/\partial \mathbf{x} = (\partial y_j/\partial x_i)$, the transpose of our definition. However, such a definition does not adapt so well to the chain rule below in (17.29) and in the derivation of Jacobians, which are discussed in the next chapter.

If $\mathbf{Y} = \mathbf{F}(\mathbf{X})$ is a matrix function of $\mathbf{X}$, we shall also be interested in the derivative $\partial \operatorname{vec}\mathbf{Y}/(\partial \operatorname{vec}\mathbf{X})'$. Rao and Rao [1998: Section 6.5] denoted the latter expression by $^*\partial \mathbf{Y}/\partial \mathbf{X}$ and list a number of results. The Kronecker product "$\otimes$" is very useful in this regard, along with (17.60). Many of the results are proved using the method of differentials (Section 17.8).

17.27. Since the Kronecker product $\mathbf{x} \otimes \mathbf{a}$ is a vector, we have the following.

(a) $\dfrac{\partial(\mathbf{x} \otimes \mathbf{a})}{\partial \mathbf{x}'} = \mathbf{I}_n \otimes \mathbf{a}.$

(b) $\dfrac{\partial(\mathbf{a} \otimes \mathbf{x})}{\partial \mathbf{x}'} = \mathbf{a} \otimes \mathbf{I}_n.$

17.28. If $\mathbf{y} = \mathbf{A}\mathbf{x}$, then

$$\partial \mathbf{y}/\partial \mathbf{x}' = \mathbf{A}.$$

Similarly, if $\operatorname{vec}\mathbf{Y} = \mathbf{B}\operatorname{vec}\mathbf{X}$, then

$$\frac{\partial \operatorname{vec}\mathbf{Y}}{\partial(\operatorname{vec}\mathbf{X})'} = \mathbf{B}.$$

17.29. (Chain Rule) If z is a differentiable vector function of $\mathbf{y}$ and $\mathbf{y}$ is a differentiable function of $\mathbf{x}$, then, arguing as in (17.21),

$$\frac{\partial \mathbf{z}}{\partial \mathbf{x}'} = \frac{\partial \mathbf{z}}{\partial \mathbf{y}'} \cdot \frac{\partial \mathbf{y}}{\partial \mathbf{x}'}.$$

This result also holds if z is a scalar (cf. 17.21), and then $\partial z / \partial \mathbf{x}'$ is a row vector.

17.30. (Matrices with Functionally Independent Elements) In what follows we can obtain special cases by putting some of the matrices equal to the identity matrix. Also $\mathbf{I}_{(m,n)} (= \mathbf{K}_{mn})$ is the vec-permutation (commutation) matrix.

(a) $\dfrac{\partial \operatorname{vec} \mathbf{X}}{\partial (\operatorname{vec} \mathbf{X})'} = \mathbf{I}.$

(b) If $\mathbf{Y} = \mathbf{AXB}$, then from (11.16b), $\operatorname{vec} \mathbf{Y} = (\mathbf{B}' \otimes \mathbf{A}) \operatorname{vec} \mathbf{X}$ and

$$\frac{\partial \operatorname{vec} \mathbf{Y}}{\partial (\operatorname{vec} \mathbf{X})'} = \mathbf{B}' \otimes \mathbf{A}.$$

(c) If $\mathbf{Y} = \mathbf{AX}'\mathbf{B}$ and $\mathbf{X}$ is $m \times n$, then $\operatorname{vec} \mathbf{Y} = (\mathbf{B}' \otimes \mathbf{A}) \operatorname{vec} \mathbf{X}'$, $\operatorname{vec} \mathbf{X}' = \mathbf{I}_{(n,m)} \operatorname{vec} \mathbf{X}$ and, from (a) and (11.16b),

$$\frac{\partial \operatorname{vec} \mathbf{Y}}{\partial (\operatorname{vec} \mathbf{X})'} = (\mathbf{B}' \otimes \mathbf{A}) \mathbf{I}_{(n,m)}.$$

(d) If $\mathbf{X}$ is nonsingular and $\mathbf{Y} = \mathbf{AX}^{-1}\mathbf{B}$, then

$$\frac{\partial \operatorname{vec} \mathbf{Y}}{\partial (\operatorname{vec} \mathbf{X})'} = -(\mathbf{X}^{-1}\mathbf{B})' \otimes (\mathbf{AX}^{-1}).$$

We can set $\mathbf{A} = \mathbf{B} = \mathbf{I}$.

(e) If $\mathbf{X}$ is nonsingular and $\mathbf{Y} = \mathbf{X}^k$, where k is a positive integer, then

$$\frac{\partial \operatorname{vec} \mathbf{Y}}{\partial (\operatorname{vec} \mathbf{X})'} = \sum_{i=1}^{k} ((\mathbf{X}')^{k-i} \otimes \mathbf{X}^{i-1}).$$

(f) If $\mathbf{X}$ is $m \times n$ and $\mathbf{Y} = \mathbf{X}'\mathbf{AX}$, then

$$\frac{\partial \operatorname{vec} \mathbf{Y}}{\partial (\operatorname{vec} \mathbf{X})'} = (\mathbf{X}'\mathbf{A}' \otimes \mathbf{I}_n) \mathbf{I}_{(n,m)} + (\mathbf{I}_n \otimes \mathbf{X}'\mathbf{A}).$$

If $\mathbf{A}$ is symmetric, we get $(\mathbf{I}_{n^2} + \mathbf{I}_{(n,n)})(\mathbf{I}_n \otimes \mathbf{X}'\mathbf{A})$.

(g) If $\mathbf{X}$ is $m \times n$ and $\mathbf{Y} = \mathbf{XBX}'$, then

$$\frac{\partial \operatorname{vec} \mathbf{Y}}{\partial (\operatorname{vec} \mathbf{X})'} = (\mathbf{XB}' \otimes \mathbf{I}_m) + (\mathbf{I}_m \otimes \mathbf{XB}) \mathbf{I}_{(n,m)}.$$

If $\mathbf{B}$ is symmetric, we get $(\mathbf{I}_{m^2} + \mathbf{I}_{(m,m)})(\mathbf{XB} \otimes \mathbf{I}_m)$.

(h) If $\mathbf{X}$ is $m \times n$, $\mathbf{U}$ is a $p \times q$ matrix function of $\mathbf{X}$, and $\mathbf{V}$ is a $q \times r$ matrix function of $\mathbf{X}$, then

$$\frac{\partial \text{vec}\,(\mathbf{UV})}{\partial(\text{vec}\,\mathbf{X})'} = (\mathbf{V} \otimes \mathbf{I}_p)'\frac{\partial \text{vec}\,\mathbf{U}}{\partial(\text{vec}\,\mathbf{X})'} + (\mathbf{I}_r \otimes \mathbf{U})\frac{\partial \text{vec}\,\mathbf{V}}{\partial(\text{vec}\,\mathbf{X})'}.$$

17.31. Let $\mathbf{F}(\mathbf{X}) = \mathbf{Z}(\mathbf{Y}(\mathbf{X}))$, then by the chain rule (17.29),

$$\frac{\partial \text{vec}\,\mathbf{F}}{\partial(\text{vec}\,\mathbf{X})'} = \frac{\partial \text{vec}\,\mathbf{Z}(\mathbf{V})}{\partial(\text{vec}\,\mathbf{V})'}\Bigg|_{\mathbf{V}=\mathbf{Y}(\mathbf{X})}\frac{\partial \text{vec}\,\mathbf{Y}(\mathbf{X})}{\partial(\text{vec}\,\mathbf{X})'}.$$

17.32. (Symmetric Matrices) Let $\mathbf{X}$ be an $n \times n$ symmetric matrix.

(a) If $\mathbf{Y} = \mathbf{AXA}'$, then $\text{vech}\,\mathbf{Y} = \mathbf{H}_n(\mathbf{A} \otimes \mathbf{A})\mathbf{G}_n\,\text{vech}\,\mathbf{X}$ (cf. 11.30), where $\mathbf{H}_n$ can be replaced by $\mathbf{G}_n^+ (= \mathbf{D}_n^+)$, and

$$\frac{\partial\,\text{vech}\,\mathbf{Y}}{\partial(\text{vech}\,\mathbf{X})'} = \mathbf{H}_n(\mathbf{A} \otimes \mathbf{A})\mathbf{G}_n.$$

Here $\mathbf{G}_n$ is the duplication matrix.

(b) If $\mathbf{Y} = \mathbf{X}^{-1}$, then $\mathbf{Y} = \mathbf{X}^{-1}\mathbf{X}\mathbf{X}^{-1}$, and from (a),

$$\frac{\partial\,\text{vech}\,\mathbf{Y}}{\partial(\text{vech}\,\mathbf{X})'} = -\mathbf{H}_n(\mathbf{X}^{-1} \otimes \mathbf{X}^{-1})\mathbf{G}_n.$$

(c) If $\mathbf{Y} = \mathbf{X}^k$, where $k = 2, 3, \ldots$, then

$$\frac{\partial\,\text{vech}\,\mathbf{Y}}{\partial(\text{vech}\,\mathbf{X})'} = \mathbf{H}_n\sum_{i=1}^{k}(\mathbf{X}^{k-i} \otimes \mathbf{X}^{i-1})\mathbf{G}_n.$$

(d) If $\mathbf{Y} = \mathbf{X}^+$, then

$$\begin{aligned}\frac{\partial\,\text{vech}\,\mathbf{Y}}{\partial\text{vech}\,\mathbf{X})'} = &\ \mathbf{G}_n^+\{[\mathbf{X}^+\mathbf{X}^+ \otimes (\mathbf{I}_n - \mathbf{X}^+\mathbf{X}) \\ &+ (\mathbf{I}_n - \mathbf{XX}^+) \otimes \mathbf{X}^+\mathbf{X}^+]\mathbf{I}_{n,n} - (\mathbf{X}^+ \otimes \mathbf{X}^+)\}\mathbf{G}_n,\end{aligned}$$

where $\mathbf{I}_{(n,n)}$ is the vec-permutation (commutation) matrix.

17.33. Let $\mathbf{F}$ be a $p \times q$ matrix whose elements are a function of $\mathbf{x} = (x_1, x_2, \ldots, x_n)'$. (Here $\mathbf{x}$ can be $\text{vec}\,\mathbf{X}$.) The following results mirror (17.30) and (17.32):

(a) $\dfrac{\partial\,\text{vec}\,(\mathbf{AFB})}{\partial\mathbf{x}'} = (\mathbf{B}' \otimes \mathbf{A})\dfrac{\partial\,\text{vec}\,\mathbf{F}}{\partial\mathbf{x}'}.$

(b) If $\mathbf{F}$ is nonsingular, $\dfrac{\partial\,\text{vec}\,\mathbf{F}^{-1}}{\partial\mathbf{x}'} = -(\mathbf{F}^{-1\prime} \otimes \mathbf{F}^{-1})\dfrac{\partial\,\text{vec}\,\mathbf{F}}{\partial\mathbf{x}'}.$

(c) If $\mathbf{F} = \mathbf{X}$ and $\mathbf{x} = \text{vec}\,\mathbf{X}$, then $\dfrac{\partial\,\text{vec}\,\mathbf{F}}{\partial\mathbf{x}'} = \mathbf{I}_{pq}.$

(d) If $\mathbf{F}$ is $n \times n$, then

$$\frac{\partial\,\text{vec}\,(\mathbf{F}^k)}{\partial\mathbf{x}'} = \sum_{i=1}^{k}[(\mathbf{F}^{i-1})' \otimes \mathbf{F}^{k-i}]\frac{\partial\,\text{vec}\,\mathbf{F}}{\partial\mathbf{x}'} \quad (\mathbf{F}^0 = \mathbf{I}_n).$$

(e) If $\mathbf{F}$ is symmetric and $n \times n$, then:

(i) $\dfrac{\partial \operatorname{vech} (\mathbf{AFA}')}{\partial \mathbf{x}'} = \mathbf{H}_n (\mathbf{A} \otimes \mathbf{A}) \mathbf{G}_n \dfrac{\partial \operatorname{vech} \mathbf{F}}{\partial \mathbf{x}'}.$

(ii) $\dfrac{\partial \operatorname{vech} \mathbf{F}^{-1}}{\partial \mathbf{x}'} = -\mathbf{H}_n (\mathbf{F}^{-1} \otimes \mathbf{F}^{-1}) \mathbf{G}_n \dfrac{\partial \operatorname{vech} \mathbf{F}}{\partial \mathbf{x}'}.$

17.34. Let $\mathbf{F}$ and $\mathbf{G}$ be $m \times n$ and $p \times q$ matrices, respectively, which are functions of $\mathbf{x}$.

(a) (Kronecker Product) From (17.7c) we have:

(i) $\dfrac{\partial (\operatorname{vec} \mathbf{F} \otimes \operatorname{vec} \mathbf{G})}{\partial \mathbf{x}'} = \left(\operatorname{vec} \mathbf{F} \otimes \dfrac{\partial \operatorname{vec} \mathbf{G}}{\partial \mathbf{x}'} \right) + \left(\dfrac{\partial \operatorname{vec} \mathbf{F}}{\partial \mathbf{x}'} \otimes \operatorname{vec} \mathbf{G} \right).$

(ii) $\dfrac{\partial \operatorname{vec} (\mathbf{F} \otimes \mathbf{G})}{\partial \mathbf{x}'} = (\mathbf{I}_n \otimes \mathbf{I}_{(m,q)} \otimes \mathbf{I}_p) \dfrac{\partial (\operatorname{vec} \mathbf{F} \otimes \operatorname{vec} \mathbf{G})}{\partial \mathbf{x}'}.$

(b) (Hadamard Product) If $\mathbf{F}$ and $\mathbf{G}$ are both $m \times n$ matrix functions of $\mathbf{x}$, then

$$\dfrac{\partial \operatorname{vec} (\mathbf{F} \circ \mathbf{G})}{\partial \mathbf{x}'} = \mathbf{D}(\mathbf{F}) \dfrac{\partial \operatorname{vec} \mathbf{G}}{\partial \mathbf{x}'} + \mathbf{D}(\mathbf{G}) \dfrac{\partial \operatorname{vec} \mathbf{F}}{\partial \mathbf{x}'},$$

where "$\circ$" represents the Hadamard product, and for any $m \times n$ matrix $\mathbf{A}$,

$$\mathbf{D}(\mathbf{A}) = \operatorname{diag}(a_{11}, a_{12}, \ldots, a_{1n}, a_{21}, a_{22}, \ldots, a_{2n}, \ldots, a_{m1}, a_{m2}, \ldots, a_{mn}).$$

17.35. Suppose $\mathbf{y} = \mathbf{Az}$, where $\mathbf{A}$ is $m \times n$ and $\mathbf{A}$ and $\mathbf{z}$ are functions of $\mathbf{x}$. We want to find $\partial \mathbf{y}/\partial \mathbf{x}'$. Since $\mathbf{y} = \operatorname{vec} \mathbf{y} = \operatorname{vec} (\mathbf{Az}) = (\mathbf{z}' \otimes \mathbf{I}_m) \operatorname{vec} \mathbf{A}$, we have

$$
\begin{aligned}
\dfrac{\partial \mathbf{y}}{\partial \mathbf{x}'} &= \left. \dfrac{\partial \operatorname{vec} (\mathbf{Az})}{\partial \mathbf{x}'} \right]_{\mathbf{z} \text{ constant}} + \left. \dfrac{\partial (\mathbf{Az})}{\partial \mathbf{x}'} \right]_{\mathbf{A} \text{ constant}} \\
&= (\mathbf{z}' \otimes \mathbf{I}_m) \dfrac{\partial \operatorname{vec} \mathbf{A}}{\partial \mathbf{x}'} + \mathbf{A} \dfrac{\partial \mathbf{z}}{\partial \mathbf{x}'}.
\end{aligned}
$$

Proofs. Section 17.4.

17.30b. Abadir and Magnus [2005: 362], Harville [1997: 366], and Henderson and Searle [1979: 73].

17.30d. Abadir and Magnus [2005: 366] and Turkington [2002: 73, transposed].

17.30e. Abadir and Magnus [2005: 362-363] and Henderson and Searle [1979: 73].

17.30f–g. Abadir and Magnus [2005: 366] and Turkington [2002: 74, transposed].

17.30h. Rao and Rao [1998: 234, with typo corrected].

17.32a. Harville [1997: 366] and Henderson and Searle [1979: 74].

17.32b. Harville [1997: 368].

17.32c. Henderson and Searle [1979: 74].

17.32d. Schott [2005: 364].

17.33. Harville [1997: section 16.6]; for (d) see Harville [2001: 157, exercise 31].

17.34a(ii). Harville [2001: 158, exercise 32].

17.34b. Quoted by Rao and Rao [1998: 235].

17.5   MATRIX DIFFERENTIATION: SCALAR FUNCTION

17.5.1   General Results

Definition 17.5. Let $y = f(\mathbf{X})$ be a scalar function of the elements x_{ij} of the $m \times n$ matrix $\mathbf{X}$. Then the derivative of y with respect to $\mathbf{X}$, written $\partial y / \partial \mathbf{X}$, is the matrix with (i, j)th element $\partial y / \partial x_{ij}$, that is,

$$\frac{\partial f(\mathbf{X})}{\partial \mathbf{X}} = \left(\frac{\partial f(\mathbf{X})}{\partial x_{ij}} \right).$$

If $\mathbf{X}$ is a vector $\mathbf{x}$, then we write $\partial y / \partial \mathbf{x}$, a column vector with ith element $\partial y / \partial x_i$. Thus

$$\frac{\partial f(\mathbf{x})}{\partial \mathbf{x}} = \left(\frac{\partial f(\mathbf{x})}{\partial x_i} \right).$$

It is assumed that $\mathbf{X}$ and $\mathbf{x}$ have functionally dependent elements, unless stated to the contrary (e.g., $\mathbf{X}$ is symmetric). Note that

$$\frac{\partial f(\mathbf{X})}{\partial \mathbf{X}} = \left(\frac{\partial f(\mathbf{X})}{\partial \mathbf{X}'} \right)'.$$

A special case is when $y_{rs} = F_{rs}(\mathbf{X})$, where y_{rs} is the (r, s)th element of $\mathbf{Y} = \mathbf{F}(\mathbf{X})$.

We remind the reader that $\mathrm{diag}(\mathbf{A})$ is the diagonal matrix whose diagonal elements are the same as the diagonal elements of $\mathbf{A}$. Such matrices feature frequently below. Many of the results in this section can be derived using the method of differentials, as demonstrated in (17.57).

17.36. (Chain Rules)

(a) If $y = f(\mathbf{X})$ and $z = g(y)$, then $\dfrac{\partial z}{\partial x_{ij}} = \dfrac{\partial z}{\partial y} \cdot \dfrac{\partial y}{\partial x_{ij}}$, which leads to

$$\frac{\partial z}{\partial \mathbf{X}} = \frac{\partial z}{\partial y} \cdot \frac{\partial y}{\partial \mathbf{X}}.$$

(b) If $\mathbf{Y} = \mathbf{F}(\mathbf{X})$ and $z = g(\mathbf{Y})$, then

$$\frac{\partial z}{\partial x_{ij}} = \sum_r \sum_s \frac{\partial z}{\partial y_{rs}} \frac{\partial y_{rs}}{\partial x_{ij}} = \mathrm{trace} \left(\frac{\partial z}{\partial \mathbf{Y}} \cdot \frac{\partial \mathbf{Y}'}{\partial x_{ij}} \right).$$

We can also write the above equation in the form

$$\frac{\partial g(\mathbf{Y})}{\partial \mathbf{X}} = \sum_r \sum_s \frac{\partial g}{\partial y_{rs}} \frac{\partial y_{rs}}{\partial \mathbf{X}}.$$

Nel [1980: 150–151] used this equation to derive some of the results below.

17.37. (Symmetric $\mathbf{X}$) If $y = f(\mathbf{X})$, where $\mathbf{X}$ is symmetric, then

$$\frac{\partial f}{\partial \mathbf{X}} = \left\{ \frac{\partial f(\mathbf{Y})}{\partial \mathbf{Y}} + \frac{\partial f(\mathbf{Y})}{\partial \mathbf{Y}'} - \operatorname{diag}\left(\frac{\partial f(\mathbf{Y})}{\partial \mathbf{Y}} \right) \right\}_{\mathbf{Y}=\mathbf{X}}.$$

In working out the derivative $\partial f(\mathbf{Y})/\partial \mathbf{Y}$, we pretend that the function $f(\cdot)$ is defined on the class of matrices $\mathbf{Y}$ with all independent components, and then the derivative is formed. Rao and Rao [1998: 231] give some helpful examples of the method.

Proofs. Section 17.5.1.

17.37. Rao and Rao [1998: 230–231].

17.5.2 f = trace

We now give various matrix derivatives for the trace of matrix products. Variations of the following can be obtained by using the results $\operatorname{trace} \mathbf{C} = \operatorname{trace} \mathbf{C}'$, $\operatorname{trace}(\mathbf{DE}) = \operatorname{trace}(\mathbf{ED})$, $\operatorname{trace}(\mathbf{AXB}) = \operatorname{trace}(\mathbf{BAX})$, and $\mathbf{a}'\mathbf{Xb} = \operatorname{trace}(\mathbf{a}'\mathbf{Xb})$ for square $\mathbf{C}$, $\mathbf{DE}$, and $\mathbf{AXB}$. In what follows, we assume $\mathbf{X}$ to be $m \times n$ and unconstrained, unless otherwise stated. If $\mathbf{X}$ is symmetric, we assume it to be $n \times n$. We can also set $\mathbf{A} = \mathbf{I}_n$ and/or $\mathbf{B} = \mathbf{I}_n$ to get special cases. The following can be readily derived from the basic simple result

$$\frac{\partial \operatorname{trace} \mathbf{Y}}{\partial x_{ij}} = \operatorname{trace}\left(\frac{\partial \mathbf{Y}}{\partial x_{ij}} \right)$$

and then using the results of (17.13c) and (17.14). We also use the fact that if $\mathbf{W} = (w_{ij})$, then $\operatorname{trace}(\mathbf{E}'_{ij}\mathbf{W}) = \operatorname{trace}(\mathbf{WE}'_{ij}) = w_{ij}$ and $\operatorname{trace}(\mathbf{E}_{ij}\mathbf{W}) = \operatorname{trace}(\mathbf{WE}_{ij}) = w_{ji}$, where $\mathbf{E}_{ij}$ has 1 in the i, jth position and zeros eslewhere.

17.38. If $y = \operatorname{trace}[(\mathbf{U}(\mathbf{X})\mathbf{V}(\mathbf{X})]$, where $\mathbf{U}$ and $\mathbf{V}$ are matrix functions of $\mathbf{X}$, then

$$\frac{\partial y}{\partial \mathbf{X}} = \frac{\partial \operatorname{trace}[\mathbf{U}(\mathbf{X})\mathbf{V}(\mathbf{Y})]}{\partial \mathbf{X}}\Big|_{\mathbf{Y}=\mathbf{X}} + \frac{\partial \operatorname{trace}[\mathbf{U}(\mathbf{Y})\mathbf{V}(\mathbf{X})]}{\partial \mathbf{X}}\Big|_{\mathbf{Y}=\mathbf{X}}.$$

17.39. Using (17.14a) and (17.13), we obtain

$$\frac{\partial \operatorname{trace}(\mathbf{AXB})}{\partial \mathbf{X}} = \begin{cases} \mathbf{C}', & \mathbf{X} \text{ unconstrained }, \\ \mathbf{C} + \mathbf{C}' - \operatorname{diag} \mathbf{C}, & \mathbf{X} \text{ symmetric}; \mathbf{A}, \mathbf{B} \text{ square,} \end{cases}$$

where $\mathbf{C} = \mathbf{BA}$.

To obtain further results we use $\operatorname{trace}(\mathbf{AX}'\mathbf{B}) = \operatorname{trace}(\mathbf{B}'\mathbf{XA}')$, and also set $\mathbf{A}$ or $\mathbf{B}$ equal to $\mathbf{I}$.

17.40. If $\mathbf{A}$ is $m \times m$, $\mathbf{B}$ is $n \times n$, and $\mathbf{X}$ is unconstrained, we have from (17.14c) that
$$\frac{\partial \operatorname{trace}(\mathbf{X}'\mathbf{A}\mathbf{X}\mathbf{B})}{\partial \mathbf{X}} = \frac{\partial \operatorname{trace}(\mathbf{X}\mathbf{B}\mathbf{X}'\mathbf{A})}{\partial \mathbf{X}} = \mathbf{A}\mathbf{X}\mathbf{B} + \mathbf{A}'\mathbf{X}\mathbf{B}'.$$
An important special case is when $\mathbf{B} = \mathbf{I}_n$ and $\mathbf{A}$ is symmetric. Then
$$\frac{\partial \operatorname{trace}(\mathbf{X}'\mathbf{A}\mathbf{X})}{\partial \mathbf{X}} = 2\mathbf{A}\mathbf{X}.$$

17.41. Using (17.14f) and (17.13), we obtain
$$\frac{\partial \operatorname{trace}(\mathbf{X}\mathbf{A}\mathbf{X}\mathbf{B})}{\partial \mathbf{X}} = \begin{cases} \mathbf{H}', & \mathbf{X} \text{ unconstrained}, \\ \mathbf{H} + \mathbf{H}' - \operatorname{diag}\mathbf{H}, & \mathbf{X} \text{ symmetric}, \end{cases}$$
where $\mathbf{H} = \mathbf{B}\mathbf{X}\mathbf{A} + \mathbf{A}\mathbf{X}\mathbf{B}$. We can get the special case of $\operatorname{trace}[(\mathbf{A}\mathbf{X})^2]$ by noting that $\operatorname{trace}[(\mathbf{A}\mathbf{X})^2] = \operatorname{trace}(\mathbf{A}\mathbf{X}\mathbf{A}\mathbf{X}) = \operatorname{trace}(\mathbf{X}\mathbf{A}\mathbf{X}\mathbf{A})$. Also, we can set $\mathbf{B} = \mathbf{I}$.

17.42. If $\mathbf{X}$ is nonsingular and unconstrained, we have from (17.15a):

(a) $\dfrac{\partial \operatorname{trace}[\mathbf{A}(\mathbf{B}\mathbf{X}\mathbf{C})^{-1}\mathbf{D}]}{\partial \mathbf{X}} = -[\mathbf{C}(\mathbf{B}\mathbf{X}\mathbf{C})^{-1}\mathbf{D}\mathbf{A}(\mathbf{B}\mathbf{X}\mathbf{C})^{-1}\mathbf{B}]'.$

(b) From (a) we have $\dfrac{\partial \operatorname{trace}(\mathbf{A}\mathbf{X}^{-1}\mathbf{B})}{\partial \mathbf{X}} = -(\mathbf{X}^{-1}\mathbf{B}\mathbf{A}\mathbf{X}^{-1})'.$

(c) A useful special case is $\dfrac{\partial \operatorname{trace}\mathbf{X}^{-1}}{\partial \mathbf{X}} = -(\mathbf{X}^{-2})'.$

(d) When $\mathbf{X}$ is symmetric, we have from (17.13) that
$$\frac{\partial \operatorname{trace}\mathbf{X}^{-1}}{\partial \mathbf{X}} = -2(\mathbf{X}^{-2}) + \operatorname{diag}(\mathbf{X}^{-2}).$$

17.43. Using (17.17a) and (17.13), we have for $k = 2, 3, \ldots$
$$\frac{\partial \operatorname{trace}\mathbf{X}^k}{\partial \mathbf{X}} = \begin{cases} k(\mathbf{X}^{k-1})', & \mathbf{X} \text{ unconstrained}, \\ 2k\mathbf{X}^{k-1} - k\operatorname{diag}(\mathbf{X}^{k-1}), & \mathbf{X} \text{ symmetric}. \end{cases}$$

17.44. Suppose $\mathbf{X}$ is unconstrained.

(a) $\dfrac{\partial \operatorname{trace}e^{\mathbf{X}}}{\partial \mathbf{X}} = (e^{\mathbf{X}})'.$

(b) $\dfrac{\partial e^{\operatorname{trace}(\mathbf{X}^2)}}{\partial \mathbf{X}} = 2e^{\operatorname{trace}(\mathbf{X}^2)}\mathbf{X}'.$

Proofs. Section 17.5.2.

17.38. Rao and Rao [1998: 232].

17.39. Harville [2001: 116, exercise 8].

17.40. Graham [1981: 77–78].

17.44a. Abadir and Magnus [2005: 368, exercise 13.29].

17.44b. The derivative is $e^{\operatorname{trace}(\mathbf{X}^2)}\partial \operatorname{trace}(\mathbf{X}^2)/\partial \mathbf{X}$, and then use (17.14f) with $\mathbf{A} = \mathbf{B} = \mathbf{I}$.

17.5.3   f = determinant

In this section we assume that all the determinants are nonzero and that $\mathbf{X}$ is unconstrained, unless otherwise stated. Most of the following results for $\mathbf{X}$ unconstrained are derived in Dwyer [1967]. The constrained case follows from the unconstrained case using (17.13) above.

17.45. From (17.18a(i)),

$$\frac{\partial \det \mathbf{X}}{\partial \mathbf{X}} = \begin{cases} (\mathrm{adj}\mathbf{X})' = (\det \mathbf{X})(\mathbf{X}^{-1})', & \mathbf{X} \text{ unconstrained}, \\ \det(\mathbf{X})[2\mathbf{X}^{-1} - \mathrm{diag}(\mathbf{X}^{-1})], & \mathbf{X} \text{ symmetric}. \end{cases}$$

17.46. From (17.45),

$$\frac{\partial \log(\det \mathbf{X})}{\partial \mathbf{X}} = (\det \mathbf{X})^{-1} \frac{\partial \det \mathbf{X}}{\partial \mathbf{X}} = \begin{cases} (\mathbf{X}^{-1})', & \mathbf{X} \text{ unconstrained}, \\ 2\mathbf{X}^{-1} - \mathrm{diag}(\mathbf{X}^{-1}), & \mathbf{X} \text{ symmetric}. \end{cases}$$

17.47.

$$\frac{\partial \det(\mathbf{AXB})}{\partial \mathbf{X}} = \begin{cases} \det(\mathbf{AXB})\,\mathbf{C}', & \mathbf{X} \text{ unconstrained}, \\ \det(\mathbf{AXB})[\mathbf{C} + \mathbf{C}' - \mathrm{diag}\,\mathbf{C}], & \mathbf{X} \text{ symmetric}, \end{cases}$$

where $\mathbf{C} = \mathbf{B}(\mathbf{AXB})^{-1}\mathbf{A}$.

17.48. If k is a positive integer and $\mathbf{X}$ is unconstrained,

$$\frac{\partial (\det \mathbf{X})^k}{\partial \mathbf{X}} = k(\det \mathbf{X})^{k-1} \frac{\partial \det \mathbf{X}}{\partial \mathbf{X}}.$$

In particular,

$$\frac{\partial (\det \mathbf{X})^2}{\partial \mathbf{X}} = 2(\det \mathbf{X})^2 (\mathbf{X}^{-1})'.$$

17.49. Assuming $\mathbf{X}'\mathbf{AX}$ is nonsingular,

(a) $$\frac{\partial \det(\mathbf{X}'\mathbf{AX})}{\partial \mathbf{X}} = \det(\mathbf{X}'\mathbf{AX})\{\mathbf{AX}(\mathbf{X}'\mathbf{AX})^{-1} + \mathbf{A}'\mathbf{X}[(\mathbf{X}'\mathbf{AX})^{-1}]'\}.$$

This result is linked to (17.24b).

(b) Setting $\mathbf{A} = \mathbf{I}$ we get

$$\begin{aligned} \frac{\partial \det(\mathbf{X}'\mathbf{X})}{\partial \mathbf{X}} &= 2[\det(\mathbf{X}'\mathbf{X})]\mathbf{X}(\mathbf{X}'\mathbf{X})^{-1} \\ &= 2\det(\mathbf{X}'\mathbf{X})\,\mathbf{X}^+, \end{aligned}$$

where $\mathbf{X}^+$ is the Moore–Penrose inverse of $\mathbf{X}$ (cf. 17.57e). Bates [1983] gave computational details.

(c) Replacing $\mathbf{X}$ by $\mathbf{X}'$, we get

$$\frac{\partial \det(\mathbf{XX}')}{\partial \mathbf{X}} = 2[\det(\mathbf{XX}')](\mathbf{XX}')^{-1}\mathbf{X}.$$

17.50. Let $\mathbf{F}(\mathbf{X})$ be a square nonsingular matrix function of $\mathbf{X}$, and let $\mathbf{G}(\mathbf{X}) = \mathbf{C}[\mathbf{F}(\mathbf{X})]^{-1}\mathbf{A}$. Then

$$\frac{\partial \det[(\mathbf{F}(\mathbf{X})]}{\partial \mathbf{X}} = \begin{cases} \det[\mathbf{F}(\mathbf{X})](\mathbf{GXB} + \mathbf{G'XB'}), & \text{if} \quad \mathbf{F}(\mathbf{X}) = \mathbf{AXBX'C}, \\ \det[\mathbf{F}(\mathbf{X})](\mathbf{BXG} + \mathbf{B'XG'}), & \text{if} \quad \mathbf{F}(\mathbf{X}) = \mathbf{AX'BXC}, \\ \det[\mathbf{F}(\mathbf{X})](\mathbf{GXB} + \mathbf{BXG})', & \text{if} \quad \mathbf{F}(\mathbf{X}) = \mathbf{AXBXC}. \end{cases}$$

17.51. If $\mathbf{X}$ is nonsingular,

$$\frac{\partial \det(\mathbf{AX^{-1}B})}{\partial \mathbf{X}} = \begin{cases} -\det(\mathbf{AX^{-1}B})\mathbf{C'}, & \mathbf{X} \text{ unconstrained}, \\ -\det(\mathbf{AX^{-1}B})[\mathbf{C} + \mathbf{C'} - \operatorname{diag}\mathbf{C}], & \mathbf{X} \text{ symmetric}, \end{cases}$$

where $\mathbf{C} = [\mathbf{X^{-1}B(AX^{-1}B)^{-1}AX^{-1}}]$.

17.52. If $\mathbf{F}$ is a nonsingular matrix function of $\mathbf{X}$ with $\det \mathbf{F} > 0$, then

$$\frac{\partial \log \det \mathbf{F}}{\partial \mathbf{X}} = (\det \mathbf{F})^{-1} \frac{\partial \det \mathbf{F}}{\partial \mathbf{X}}.$$

This can be applied to all the previous results.

Proofs. Section 17.5.3.

17.45. Mathai [1997: 9] and Searle [1982: 337]; see also (17.57b).

17.46. Henderson and Searle [1979: 76] and Searle [1982: 33]; see also (17.57d).

17.47. Rogers [1980: 52]; see also (17.57d).

17.48. Graham [1981: 75–76] and Magnus and Neudecker [1999: 179, $k = 2$].

17.49a. This result also follows from (17.50).

17.49b. Magnus and Neudecker [1999: 179] and Rogers [1980: 52].

17.49c. Magnus and Neudecker [1999: 179].

17.50. Quoted by Magnus and Neudecker [1999: 180].

17.51. Rogers [1980: 52].

17.5.4 $\mathbf{f} = y_{rs}$

17.53. In what follows we assume that $\mathbf{X}$ and $\mathbf{E}_{ij}$ (with 1 in the (i,j)th position and zeros elsewhere) are both $m \times n$ matrices.

(a) $\dfrac{\partial (\mathbf{AXB})_{rs}}{\partial \mathbf{X}} = \mathbf{A'E}_{rs}\mathbf{B'}$.

(b) $\dfrac{\partial (\mathbf{AX'B})_{rs}}{\partial \mathbf{X}} = \mathbf{BE}'_{rs}\mathbf{A}$.

(c) $\dfrac{\partial (\mathbf{AX^{-1}B})_{rs}}{\partial \mathbf{X}} = -(\mathbf{X^{-1}})'\mathbf{A'E}_{rs}\mathbf{B'}(\mathbf{X^{-1}})'$.

(d) $\dfrac{\partial (\mathbf{X}'\mathbf{A}\mathbf{X})_{rs}}{\partial \mathbf{X}} = \mathbf{A}\mathbf{X}\mathbf{E}'_{rs} + \mathbf{A}'\mathbf{X}\mathbf{E}_{rs}.$

(e) $\dfrac{\partial (\mathbf{X}\mathbf{A}\mathbf{X})_{rs}}{\partial \mathbf{X}} = \mathbf{E}_{rs}\mathbf{X}'\mathbf{A}' + \mathbf{A}'\mathbf{X}'\mathbf{E}_{rs}.$

(f) $\dfrac{\partial (\mathbf{X}'\mathbf{A}\mathbf{X}')_{rs}}{\partial \mathbf{X}} = \mathbf{A}\mathbf{X}'\mathbf{E}'_{rs} + \mathbf{E}'_{rs}\mathbf{X}'\mathbf{A}'.$

(g) If k is a positive integer, $\dfrac{\partial (\mathbf{X}^k)_{rs}}{\partial \mathbf{X}} = \displaystyle\sum_{j=0}^{k-1} (\mathbf{X}')^j \mathbf{E}_{rs}(\mathbf{X}')^{n-j-1}$, where $\mathbf{X}^0 = \mathbf{I}.$

Proofs. Section 17.5.4.

17.53. Graham [1981: 60–68].

17.5.5   f = eigenvalue

17.54. If λ is a nonrepeated (simple) eigenvalue of the square matrix $\mathbf{X}$ with left eigenvector $\mathbf{v}$ and right eigenvector $\mathbf{u}$, then

(a)

$$\frac{\partial \lambda}{\partial \mathbf{X}} - \mathbf{v}(\mathbf{v}'\mathbf{u})^{-1}\mathbf{u}'.$$

(b) If λ_i is a simple eigenvalue, $\mathbf{X}$ is symmetric, and $\mathbf{u}_0$ is the normalized right eigenvector (i.e., $\mathbf{u}'_0\mathbf{u}_0 = 1$), which is also the left eigenvalue, then

$$\frac{\partial \lambda}{\partial \mathbf{X}} = \mathbf{u}_0\mathbf{u}'_0.$$

Proofs. Section 17.5.5.

17.54a. Lancaster [1964] and Nel [1980: 141].

17.54b. Lancaster [1964] and Magnus and Neudecker [1999: 180].

17.6   TRANSFORMATION RULES

We now give some transformation rules that enable us to use the results from one type of differentiation to obtain results for other types.

17.55. Let $\mathbf{X}$ be an $m \times n$ matrix, and let $\mathbf{Y}$ be a function of $\mathbf{X}$. The following equivalent expressions (adapted from Graham's [1981: 65, 74] two "transformation principles") apply for all conformable $\mathbf{A}_t$, $\mathbf{B}_t$, $\mathbf{C}_v$, and $\mathbf{D}_v$, including functions of $\mathbf{X}$, and are simply different ways of writing $\partial y_{rs}/\partial x_{ij}$. If we obtain an expression like (1) or (2) below, for example, in the process of differentiation, then we can immediately obtain (3), which may be more difficult to get directly.

(1) $\quad \dfrac{\partial y_{rs}}{\partial \mathbf{X}} = \sum_t \mathbf{A}'_t \mathbf{E}_{rs} \mathbf{B}'_t + \sum_v \mathbf{C}_v \mathbf{E}'_{rs} \mathbf{D}_v.$

(2) $\quad \dfrac{\partial \mathbf{Y}}{\partial x_{ij}} - \sum_t \mathbf{A}_t \mathbf{E}_{ij} \mathbf{B}_t \mid \sum_v \mathbf{D}_v \mathbf{E}'_{ij} \mathbf{C}_v.$

(3) $\quad \dfrac{\partial \operatorname{vec} \mathbf{Y}}{\partial (\operatorname{vec} \mathbf{X})'} = \sum_t \mathbf{B}'_t \otimes \mathbf{A}_t + \sum_v \mathbf{I}_{(m,n)} (\mathbf{D}_v \otimes \mathbf{C}'_v).$

It should be noted that $\mathbf{E}_{rs}$ and $\mathbf{E}_{ij}$ may be of different sizes. We also recall (cf. 11.19a) that if $\mathbf{C}$ and $\mathbf{D}$ are both $m \times n$, then

$$\mathbf{I}_{(m,n)}(\mathbf{D} \otimes \mathbf{C}') = (\mathbf{C}' \otimes \mathbf{D}) \mathbf{I}_{(n,m)}.$$

17.7   MATRIX DIFFERENTIATION: MATRIX FUNCTION

***Definition* 17.6.** Let $\mathbf{Y} = \mathbf{F}(\mathbf{X})$, where $\mathbf{Y}$ is $p \times q$ and $\mathbf{X}$ is $m \times n$. Then the derivative of $\mathbf{Y}$ with respect to $\mathbf{X}$ can be defined in different ways. One method is to use the $mp \times nq$ matrix (MacRae [1974] and Rogers [1980])

$$\frac{\mathrm{d}\mathbf{Y}}{\mathrm{d}\mathbf{X}} = \begin{pmatrix} \dfrac{\partial y_{11}}{\partial \mathbf{X}} & \dfrac{\partial y_{12}}{\partial \mathbf{X}} & \cdots & \dfrac{\partial y_{1q}}{\partial \mathbf{X}} \\ \dfrac{\partial y_{21}}{\partial \mathbf{X}} & \dfrac{\partial y_{22}}{\partial \mathbf{X}} & \cdots & \dfrac{\partial y_{2q}}{\partial \mathbf{X}} \\ \cdot & \cdot & \cdots & \cdot \\ \dfrac{\partial y_{p1}}{\partial \mathbf{X}} & \dfrac{\partial y_{p2}}{\partial \mathbf{X}} & \cdots & \dfrac{\partial y_{pq}}{\partial \mathbf{X}} \end{pmatrix} = \mathbf{Y} \otimes \frac{\partial}{\partial \mathbf{X}},$$

where the multiplication of a matrix element by a derivative operator corresponds to the operation of differentiation. Some authors—for example, Vetter [1970] — have used the reverse order $(\partial / \partial \mathbf{X}) \otimes \mathbf{Y}$ in the above definition. Rogers defines the latter to be

$$\begin{pmatrix} \dfrac{\partial \mathbf{Y}}{\partial x_{11}} & \dfrac{\partial \mathbf{Y}}{\partial x_{12}} & \cdots & \dfrac{\partial \mathbf{Y}}{\partial x_{1n}} \\ \dfrac{\partial \mathbf{Y}}{\partial x_{21}} & \dfrac{\partial \mathbf{Y}}{\partial x_{22}} & \cdots & \dfrac{\partial \mathbf{Y}}{\partial x_{2n}} \\ \cdot & \cdot & \cdots & \cdot \\ \dfrac{\partial \mathbf{Y}}{\partial x_{m1}} & \dfrac{\partial \mathbf{Y}}{\partial x_{m2}} & \cdots & \dfrac{\partial \mathbf{Y}}{\partial x_{mn}} \end{pmatrix}.$$

This is the definition for $\mathrm{d}\mathbf{Y}/\mathrm{d}\mathbf{X}$ used by Graham [1981: chapter 6].

The above definitions can also be used when $\mathbf{X}$ or $\mathbf{Y}$ are vectors. Magnus and Neudecker [1999: chapter 9] and Rao and Rao [1998: 233] discuss the relative merits of the above definitions and recommend a third alternative definition of a matrix derivative, namely $\partial \operatorname{vec} \mathbf{Y}/\partial (\operatorname{vec} \mathbf{X})'$ as the only appropriate definition. This ties in nicely with the use of Jacobians; such derivatives and Jacobians are discussed in the next chapter. Kollo and von Rosen [2005: 127] define $\mathrm{d}\mathbf{Y}/\mathrm{d}\mathbf{X} = \partial (\operatorname{vec} \mathbf{Y})'/\partial \operatorname{vec} \mathbf{X}$, the transpose of the former definition. Their notation has the advantage in that it is consistent with the case $\partial \phi(\mathbf{X})/\partial \operatorname{vec} \mathbf{X}$, where ϕ is a scalar function. For those interested in results relating to the two previous displayed definitions, the reader is referred to Graham [1981], MacRae [1974], Neudecker [2003], Rogers [1980], and the references therein.

17.8 MATRIX DIFFERENTIALS

We mentioned some transformation rules for finding matrix derivatives in (17.55) above. There is, however, another powerful method for finding derivatives based on matrix differentials using another transformation rule given in (17.60) below. They can be used to derive some of the expressions given above, as indicated in the next chapter, Section 18.2. A good reference for this method is Abadir and Magnus [2005: chapter 13]).

Definition 17.7. If $y = f(\mathbf{x})$ is a scalar function of $\mathbf{x} = (x_1, x_2, \ldots, x_n)'$, then the differential dy is defined to be

$$dy = \sum_{i=1}^{n} \frac{\partial f}{\partial x_i} dx_i = \sum_{i=1}^{n} \frac{\partial y}{\partial x_i} dx_i.$$

If $\mathbf{X} = (x_{ij})$ is an $m \times n$ matrix, then we define the differential $d\mathbf{X}$ to be the matrix of differentials dx_{ij}, that is, $d\mathbf{X} = (dx_{ij})$. In the case of a vector $\mathbf{x} = (x_i)$, we have $d\mathbf{x} = (dx_i)$, so that we can therefore express $d\mathbf{X}$ as a vector using $\text{vec}\, d\mathbf{X}$ ($=$ $\text{dvec}\, \mathbf{X}$). For some analytical details see Abadir and Magnus [2005: chapter 13], Magnus and Neudecker [1999], and Schott [2005: sections 9.2, 9.3]. In what follows, $\mathbf{X}$ can be replaced by $\mathbf{F}(\mathbf{X})$, a matrix function of $\mathbf{X}$, when obtaining differentials so that $d\mathbf{X}$ is replaced by $d\mathbf{F}$.

17.56. (Basic Properties) Let $\mathbf{X}$ be an $m \times n$ matrix.

(a) If $\mathbf{A}$ is a matrix of constants, then $d(\mathbf{AX}) = \mathbf{A}d\mathbf{X}$.

(b) $d(\mathbf{X} \pm \mathbf{Y}) = d\mathbf{X} \pm d\mathbf{Y}$.

(c) $d(\mathbf{XY}) = (d\mathbf{X})\mathbf{Y} + \mathbf{X}d\mathbf{Y}$.

(d) $d(\mathbf{X}') = (d\mathbf{X})'$.

(e) $\text{dvec}\, \mathbf{X} = \text{vec}\, d\mathbf{X}$.

(f) $\text{dvec}\, \mathbf{X}' = \text{vec}\, (d\mathbf{X}') = \mathbf{I}_{(n,m)}\text{vec}\, (d\mathbf{X})$ (cf. Definition 11.6 above (11.18)).

(g) If $\mathbf{X}$ is an $n \times n$ matrix, we obtain

$$d(\text{trace}\, \mathbf{X}) = \text{trace}(d\mathbf{X}) = \text{trace}(\mathbf{I}_n d\mathbf{X}) = \text{vec}\, (\mathbf{I}_n)'d(\text{vec}\, \mathbf{X}),$$

from (11.17a).

(h) (Kronecker product) $d(\mathbf{X} \otimes \mathbf{Y}) = (d\mathbf{X}) \otimes \mathbf{Y} + \mathbf{X} \otimes d\mathbf{Y}$.

(i) (Hadamard product) $d(\mathbf{X} \circ \mathbf{Y}) = (d\mathbf{X}) \circ \mathbf{Y} + \mathbf{X} \circ d\mathbf{Y}$.

(j) $d(\det \mathbf{X}) = (\det \mathbf{X})\, \text{trace}(\mathbf{X}^{-1}d\mathbf{X})$.

(k) $d\mathbf{X}^{-1} = -\mathbf{X}^{-1}(d\mathbf{X})\mathbf{X}^{-1}$.

17.57. (A Scalar Transformation Rule) If $y = f(\mathbf{X})$ is a scalar function of $\mathbf{X}$, then $dy = \text{trace}(\mathbf{A}'d\mathbf{X})$ if and only if $\dfrac{\partial f}{\partial \mathbf{X}} = \mathbf{A}$. Furthermore, from (11.17a), we have

$$dy = \text{trace}(\mathbf{A}'d\mathbf{X}) = (\text{vec}\, \mathbf{A})'\text{dvec}\, \mathbf{X} \quad \text{if and only if} \quad \frac{\partial f}{\partial(\text{vec}\, \mathbf{X})} = \text{vec}\, \mathbf{A}.$$

Here $\mathbf{A}$ may be a function of $\mathbf{X}$. Examples follow with $\mathbf{X}$ unconstrained (Abadir and Magnus [2005: 357]).

(a) If $y = \text{trace}(\mathbf{X}'\mathbf{X})$, then $\mathrm{d}y = 2\,\text{trace}(\mathbf{X}'\mathrm{d}\mathbf{X})$ (by (17.56c) and (17.56g)), from which we get $\dfrac{\partial y}{\partial \mathbf{X}} = 2\mathbf{X}$.

(b) If $y = \det \mathbf{X}$, where $\mathbf{X}$ is nonsingular, then $\mathrm{d}y = (\det \mathbf{X})\,\text{trace}(\mathbf{X}^{-1}\mathrm{d}\mathbf{X})$ (by 17.56j) and $\dfrac{\partial \mathbf{y}}{\partial \mathbf{x}} = (\det \mathbf{X})\mathbf{X}^{-1\prime}$. We also have

$$\mathrm{d}\log(\det \mathbf{X}) = (\det \mathbf{X})^{-1}\mathrm{d}(\det \mathbf{X}).$$

(c) If $y = \text{trace}(\mathbf{X}\mathbf{A}\mathbf{X}'\mathbf{B})$, then

$$
\begin{aligned}
\mathrm{d}y &= \text{trace}[\mathrm{d}(\mathbf{X}\mathbf{A}\mathbf{X}'\mathbf{B})] = \text{trace}[(\mathrm{d}\mathbf{X})\mathbf{A}\mathbf{X}'\mathbf{B}] + \text{trace}[\mathbf{X}\mathbf{A}(\mathrm{d}\mathbf{X})'\mathbf{B}] \\
&= \text{trace}[(\mathbf{A}\mathbf{X}'\mathbf{B} + \mathbf{A}'\mathbf{X}'\mathbf{B}')\mathrm{d}\mathbf{X}]
\end{aligned}
$$

and

$$\frac{\partial y}{\partial \mathbf{X}} = (\mathbf{A}\mathbf{X}'\mathbf{B} + \mathbf{A}'\mathbf{X}'\mathbf{B}')'.$$

(d) If $y = \det(\mathbf{A}\mathbf{X}\mathbf{B})$, where $\mathbf{Y} = \mathbf{A}\mathbf{X}\mathbf{B}$ is nonsingular, then from (b)

$$
\begin{aligned}
\mathrm{d}(\det \mathbf{Y}) &= \det \mathbf{Y}\,\text{trace}(\mathbf{Y}^{-1}\mathrm{d}\mathbf{Y}) \\
&= \det \mathbf{Y}\,\text{trace}[\mathbf{Y}^{-1}\mathbf{A}(\mathrm{d}\mathbf{X})\mathbf{B}] \\
&= \det \mathbf{Y}\,\text{trace}[\mathbf{B}(\mathbf{A}\mathbf{X}\mathbf{B})^{-1}\mathbf{A}\mathrm{d}\mathbf{X}] \\
&= \det \mathbf{Y}\,\text{trace}(\mathbf{C}\mathrm{d}\mathbf{X}), \quad \text{say,}
\end{aligned}
$$

and $\dfrac{\partial y}{\partial \mathbf{X}} = (\det \mathbf{Y})\mathbf{C}'$, where $\mathbf{C} = \mathbf{B}(\mathbf{A}\mathbf{X}\mathbf{B})^{-1}\mathbf{A}$.

(e) If $y = \det(\mathbf{X}'\mathbf{X})$, where $\mathbf{Y} = \mathbf{X}'\mathbf{X}$ is nonsingular, then from (a) we obtain

$$
\begin{aligned}
\mathrm{d}(\det \mathbf{Y}) &= \det \mathbf{Y}\,\text{trace}(\mathbf{Y}^{-1}\mathrm{d}\mathbf{Y}) \\
&= \det \mathbf{Y}\,\text{trace}[\mathbf{Y}^{-1}\mathrm{d}(\mathbf{X}'\mathbf{X})] \\
&= 2\det \mathbf{Y}\,\text{trace}[\mathbf{Y}^{-1}\mathbf{X}'\mathrm{d}\mathbf{X}]
\end{aligned}
$$

and $\dfrac{\partial y}{\partial \mathbf{X}} = 2(\det \mathbf{Y})(\mathbf{X}\mathbf{Y}^{-1})$.

17.58. (A Vector Transformation Rule) If the vector $\mathbf{y}$ is a differentiable function of the vector $\mathbf{x}$, then we have $\mathrm{d}\mathbf{y} = \mathbf{A}\mathrm{d}\mathbf{x}$ if and only if

$$\frac{\partial \mathbf{y}}{\partial \mathbf{x}'} = \left(\frac{\partial y_i}{\partial x_j}\right) = \mathbf{A}.$$

Here $\mathbf{A}$ can be a function of $\mathbf{x}$, and we can substitute $\mathbf{x} = \text{vec}\,\mathbf{X}$, and so on, as in the next result. For example, if $\mathbf{y} = \mathbf{A}\mathbf{x}$, where $\mathbf{A}$ is a function of $\mathbf{x}$, then since $(\mathrm{d}\mathbf{A})\mathbf{x} = \text{vec}\,[(\mathrm{d}\mathbf{A})\mathbf{x}]$, we have from (11.16a, third equation)

$$\mathrm{d}\mathbf{y} = (\mathrm{d}\mathbf{A})\mathbf{x} + \mathbf{A}\mathrm{d}\mathbf{x} = (\mathbf{x}' \otimes \mathbf{I})\mathrm{d}\text{vec}\,\mathbf{A} + \mathbf{A}\mathrm{d}\mathbf{x}$$

and

$$\frac{\partial \mathbf{y}}{\partial \mathbf{x}'} = (\mathbf{x}' \otimes \mathbf{I})\frac{\partial \operatorname{vec} \mathbf{A}}{\partial \mathbf{x}'} + \mathbf{A}.$$

17.59. (A Matrix Transformation Rule) Let $\mathbf{Y}$ be a differentiable function of $\mathbf{X}_{m \times n}$. Then we have the following:

(a) $\operatorname{d} \operatorname{vec} \mathbf{Y} = \operatorname{vec}(\operatorname{d}\mathbf{Y}) = \mathbf{B} \operatorname{vec}(\operatorname{d}\mathbf{X}) = \mathbf{B} \operatorname{vec} \operatorname{d}\mathbf{X}$ if and only if

$$\frac{\partial \operatorname{vec} \mathbf{Y}}{\partial (\operatorname{vec} \mathbf{X})'} = \mathbf{B}.$$

(b) $\operatorname{vec}(\operatorname{d}\mathbf{Y}) = \mathbf{B} \operatorname{vec}(\operatorname{d}\mathbf{X}') = \mathbf{B}\mathbf{I}_{(n,m)} \operatorname{vec}(\operatorname{d}\mathbf{X})$ (by 17.56f) if and only if

$$\frac{\partial \operatorname{vec} \mathbf{Y}}{\partial (\operatorname{vec} \mathbf{X})'} = \mathbf{B}\mathbf{I}_{(n,m)}.$$

In the above, $\mathbf{B}$ may be a function of $\mathbf{X}$, but not of $\operatorname{d}\mathbf{X}$.

17.60. (Equivalent Representations) Let $\mathbf{X}$ be $m \times n$. If "$\otimes$" is the Kronecker product, then the following three statements are equivalent.

(1) $\operatorname{d}\mathbf{Y} = \mathbf{A}(\operatorname{d}\mathbf{X})\mathbf{B} + \mathbf{C}(\operatorname{d}\mathbf{X}')\mathbf{D}$.

(2) $\operatorname{vec}(\operatorname{d}\mathbf{Y}) = (\mathbf{B}' \otimes \mathbf{A})\operatorname{vec}(\operatorname{d}\mathbf{X}) + (\mathbf{D}' \otimes \mathbf{C})\operatorname{vec}(\operatorname{d}\mathbf{X}')$.

(3) $\dfrac{\partial \operatorname{vec} \mathbf{Y}}{\partial (\operatorname{vec} \mathbf{X})'} = \mathbf{B}' \otimes \mathbf{A} + (\mathbf{D}' \otimes \mathbf{C})\mathbf{I}_{(n,m)}$.

Here $\mathbf{I}_{(n,m)}$ is the vec-permutation (commutation) matrix, and the matrices $\mathbf{A}$, $\mathbf{B}$, $\mathbf{C}$, and $\mathbf{D}$ may all be functions of $\mathbf{X}$. Examples follow for $\mathbf{X}$ unconstrained.

(a) Let $\mathbf{R}$, $\mathbf{S}$, and $\mathbf{T}$ be matrices of constants. If $\mathbf{Y} = \mathbf{R}\mathbf{X}'\mathbf{S}\mathbf{X}\mathbf{T}$, where $\mathbf{X}$ is $m \times n$, then from (17.56a) and (17.56c) above we obtain

$$\operatorname{d}\mathbf{Y} = \mathbf{R}\mathbf{X}'\mathbf{S}(\operatorname{d}\mathbf{X})\mathbf{T} + \mathbf{R}\operatorname{d}(\mathbf{X}')\mathbf{S}\mathbf{X}\mathbf{T},$$

and by (17.60(3)) we obtain

$$\frac{\partial \operatorname{vec} \mathbf{Y}}{\partial (\operatorname{vec} \mathbf{X})'} = \mathbf{T}' \otimes (\mathbf{R}\mathbf{X}'\mathbf{S}) + [(\mathbf{S}\mathbf{X}\mathbf{T})' \otimes \mathbf{R}]\mathbf{I}_{(n,m)}.$$

(b) If $\mathbf{X}$ and $\mathbf{C}$ are $m \times n$, $\mathbf{B}$ is an $m \times m$ symmetric matrix, and $\mathbf{Y} = (\mathbf{X} - \mathbf{C})'\mathbf{B}(\mathbf{X} - \mathbf{C})$, then from (a) with $\mathbf{D} = \mathbf{B}(\mathbf{X} - \mathbf{C})$ we have

$$\operatorname{d}\mathbf{Y} = (\operatorname{d}\mathbf{X})'\mathbf{D} + \mathbf{D}'\operatorname{d}\mathbf{X},$$

so that

$$
\begin{aligned}
\frac{\partial \operatorname{vec} \mathbf{Y}}{\partial (\operatorname{vec} \mathbf{X}')} &= \mathbf{I}_n \otimes \mathbf{D}' + (\mathbf{D}' \otimes \mathbf{I}_n)\mathbf{I}_{(n,m)} \\
&= \mathbf{I}_{(n,n)}(\mathbf{I}_n \otimes \mathbf{D}') + \mathbf{I}_{n^2}(\mathbf{I}_n \otimes \mathbf{D}') \\
&= 2\mathbf{P}_n(\mathbf{I}_n \otimes \mathbf{D}').
\end{aligned}
$$

Here $\mathbf{P}_n (= \mathbf{N}_n)$ is the symmetrizer matrix in Definition 11.9 (see also (11.29h(ii))). The case $\mathbf{Y} = \mathbf{X}'\mathbf{X}$ was given by Abadir and Magnus [2005: 363].

(c) If $\mathbf{Y} = \mathbf{X}^{-1}$, where $\mathbf{X}$ is nonsingular, then from (17.56k) we have

$$d\mathbf{Y} = -\mathbf{X}^{-1}(d\mathbf{X})\mathbf{X}^{-1}.$$

(i) $\dfrac{\partial \operatorname{vec} \mathbf{X}^{-1}}{\partial (\operatorname{vec} \mathbf{X})'} = -(\mathbf{X}^{-1\prime} \otimes \mathbf{X}^{-1}).$

(ii) $\operatorname{trace}[\mathbf{X}\, d(\mathbf{X}^{-1})] = -\operatorname{trace}[\mathbf{X}\mathbf{X}^{-1}(d\mathbf{X})\mathbf{X}^{-1}] = -\operatorname{trace}(\mathbf{X}^{-1}d\mathbf{X}).$

(d) If $\mathbf{T}$ is orthogonal and $\det(\mathbf{T} + \mathbf{I}) \neq 0$, then there exists a one-to-one relation between $\mathbf{T}$ and the skew symmetric matrix $\mathbf{S}$, namely, $\mathbf{S} = 2(\mathbf{T} + \mathbf{I})^{-1} - \mathbf{I}$, where $\mathbf{T} = 2(\mathbf{S} + \mathbf{I})^{-1} - \mathbf{I}$ (cf. 5.19). Then from (b),

$$d\mathbf{T} = -\frac{1}{2}(\mathbf{T} + \mathbf{I})(d\mathbf{S})(\mathbf{T} + \mathbf{I}).$$

17.61. (Moore–Penrose Inverse) If $\mathbf{X}$ is $m \times n$ with Moore–Penrose $\mathbf{X}^+$, then, provided that rank $\mathbf{X}$ is constant (over a suitable set), we obtain

$$d\mathbf{X}^+ = (\mathbf{I}_n - \mathbf{X}^+\mathbf{X})(d\mathbf{X}')\mathbf{X}^{+\prime}\mathbf{X}^+ + \mathbf{X}^+\mathbf{X}^{+\prime}(d\mathbf{X}')(\mathbf{I}_m - \mathbf{X}\mathbf{X}^+) - \mathbf{X}^+(d\mathbf{X})\mathbf{X}^+.$$

Hence, using rule (3) in (17.60), we have

$$\frac{\partial \operatorname{vec} \mathbf{X}^+}{\partial (\operatorname{vec} \mathbf{X})'} = \{\mathbf{X}^{+\prime}\mathbf{X}^+ \otimes (\mathbf{I}_n - \mathbf{X}^+\mathbf{X}) + (\mathbf{I}_m - \mathbf{X}\mathbf{X}^+) \otimes \mathbf{X}^+\mathbf{X}^{+\prime}\}\mathbf{I}_{(n,m)} - (\mathbf{X}^{+\prime} \otimes \mathbf{X}^+).$$

17.62. (Idempotent Matrix) Let $\mathbf{X} = (\mathbf{x}_1, \mathbf{x}_2, \ldots, \mathbf{x}_p)$ be $n \times p$ of rank p, and define $\mathbf{M} = \mathbf{I}_n - \mathbf{X}(\mathbf{X}'\mathbf{X})^{-1}\mathbf{X}'$. Then:

(a) $d\mathbf{M} = -\mathbf{M}(d\mathbf{X})(\mathbf{X}'\mathbf{X})^{-1}\mathbf{X}' - \mathbf{X}(\mathbf{X}'\mathbf{X})^{-1}(d\mathbf{X})'\mathbf{M}.$

(b) $\dfrac{\partial \operatorname{vec} \mathbf{M}}{(\partial \operatorname{vec} \mathbf{X})'} = -(\mathbf{I}_{n^2} + \mathbf{I}_{(n,n)})[\mathbf{X}(\mathbf{X}'\mathbf{X})^{-1} \otimes \mathbf{M}].$

(c) From $d\mathbf{X} = (d\mathbf{x}_j)\mathbf{e}_j'$, we obtain

$$d\mathbf{M} = -\mathbf{M}(d\mathbf{x}_j)\mathbf{e}_j'(\mathbf{X}'\mathbf{X})^{-1}\mathbf{X}' - \mathbf{X}(\mathbf{X}'\mathbf{X})^{-1}\mathbf{e}_j(d\mathbf{x}_j)'\mathbf{M},$$

where $\mathbf{e}_j$ is the jth column of $\mathbf{I}_p$.

(d) $\dfrac{\partial \operatorname{vec} \mathbf{M}}{\partial \mathbf{x}_j'} = -(\mathbf{I}_{n^2} + \mathbf{I}_{(n,n)})[\mathbf{X}(\mathbf{X}'\mathbf{X})^{-1}\mathbf{e}_j \otimes \mathbf{M}].$

17.63. (Eigenvalue and Eigenvector) Let $\mathbf{X}$ be a symmetric $n \times n$ matrix with distinct eigenvalue λ_i and corresponding normalized eigenvector $\boldsymbol{\gamma}_i$ (with unit length). Then:

(a) $d\lambda_i = \boldsymbol{\gamma}_i'(d\mathbf{X})\boldsymbol{\gamma}_i$ and $d\lambda_i = d\operatorname{vec}\lambda_i = (\boldsymbol{\gamma}_i' \otimes \boldsymbol{\gamma}_i')d\operatorname{vec}\mathbf{X}$. Since $\operatorname{vec}\mathbf{X} = \mathbf{G}_n \operatorname{vech}\mathbf{X}$ we have

$$\frac{\partial \lambda_i}{\partial \operatorname{vech}\mathbf{X}} = (\boldsymbol{\gamma}_i' \otimes \boldsymbol{\gamma}_i')\mathbf{G}_n.$$

(b) $d\gamma_i = -(\mathbf{X} - \lambda_i\mathbf{I}_n)^+(d\mathbf{X})\gamma_i$ and $\dfrac{\partial\lambda_i}{\partial\,\text{vech}\,\mathbf{X}} = -\{\gamma_i' \otimes (\mathbf{X} - \lambda_i\mathbf{I}_n)^+\}\mathbf{G}_n,$

   where $(\mathbf{X} - \lambda_i\mathbf{I}_n)^+$ is the Moore–Penrose inverse of $(\mathbf{X} - \lambda_i\mathbf{I}_n)$.

17.64. (Sensitivity Analysis in Regression) From Section 20.7.1, the ordinary least squares (OLS) estimate of a full-rank regression model is $\widehat{\boldsymbol{\beta}} = (\mathbf{X}'\mathbf{X})^{-1}\mathbf{X}'\mathbf{y}$, where $\mathbf{X} = (\mathbf{x}_1, \mathbf{x}_2, \ldots, \mathbf{x}_p)$ is $n \times p$ of rank p, and the residual is $\mathbf{r} = (\mathbf{I}_n - \mathbf{X}(\mathbf{X}'\mathbf{X})^{-1}\mathbf{X}')\mathbf{y} = \mathbf{M}\mathbf{y}$. Then:

(a) $\dfrac{\partial\widehat{\boldsymbol{\beta}}}{\partial(\text{vec}\,\mathbf{X})'} = (\mathbf{X}'\mathbf{X})^{-1} \otimes \mathbf{r}' - [\widehat{\boldsymbol{\beta}}'(\otimes(\mathbf{X}'\mathbf{X})^{-1}\mathbf{X}'].$

(b) $\dfrac{\partial\widehat{\boldsymbol{\beta}}}{\partial\mathbf{x}_j'} = -(\mathbf{X}'\mathbf{X})^{-1}(\widehat{\beta}_j\mathbf{X}' - \mathbf{e}_j\mathbf{r}').$

(c) $\dfrac{\partial\mathbf{r}}{\partial(\text{vec}\,\mathbf{X})'} = -\widehat{\boldsymbol{\beta}}' \otimes \mathbf{M} - \mathbf{X}(\mathbf{X}'\mathbf{X})^{-1} \otimes \mathbf{r}'.$

(d) $\dfrac{\partial\mathbf{r}}{\partial\mathbf{x}_j'} = -(\widehat{\beta}_j\mathbf{M} + \mathbf{X}(\mathbf{X}'\mathbf{X})^{-1}\mathbf{e}_j\mathbf{r}').$

Proofs. Section 17.8.

   17.56. Abadir and Magnus [2005: 355, 362, 369] and Schott [2005: 356].

   17.57b. Mathai [1997: 71]

   17.57c. Abadir and Magnus [2005: 359].

   17.60d. Deemer and Olkin [1951: 364–365].

   17.61. Magnus and Neudecker [1999: 154] and Schott [2005: 361].

   17.62. Abadir and Magnus [2005: 365–366].

   17.63. Magnus and Neudecker [1999: 159–160, differentials only; they also give the complex case, and some second differentials] and Schott [2005: 369].

   17.64. Abadir and Magnus [2006: 375–376].

17.9   PERTURBATION USING DIFFERENTIALS

An important problem is that of finding a Taylor expansion for a function of $\mathbf{X}+d\mathbf{X}$, when the elements of $d\mathbf{X}$ are small. We begin by writing $d\mathbf{X} = \epsilon\mathbf{Y}$, where ϵ is small, so that $\mathbf{X} + \epsilon\mathbf{Y}$ represents a small perturbation of $\mathbf{X}$. If $\mathbf{f}$ is a vector function of $\mathbf{X}$, then a Taylor expansion would take the form

$$\mathbf{f}(\mathbf{X} + \epsilon\mathbf{Y}) = \mathbf{f}(\mathbf{X}) + \sum_{i=1}^{\infty} \epsilon^i\mathbf{g}_i(\mathbf{X}, \mathbf{Y}),$$

where $\mathbf{g}_i(\mathbf{X}, \mathbf{Y})$ represents some vector function of $\mathbf{X}$ and $\mathbf{Y}$. Similarly, if we have a matrix function $\mathbf{F}$, then the expansion would take the form

$$\mathbf{F}(\mathbf{X} + \epsilon\mathbf{Y}) = \mathbf{F}(\mathbf{X}) + \sum_{i=1}^{\infty} \epsilon^i \mathbf{G}_i(\mathbf{X}, \mathbf{Y}),$$

where $\mathbf{G}_i$ is now a matrix function. Schott [2005: section 9.6] demonstrated the method with several examples, and some of the results of these are given below. He also demonstrated how the method can be used for finding differentials and, ultimately, Jacobians.

17.65. Suppose that $\mathbf{X}$ is nonsingular and $\mathbf{F}(\mathbf{X}) = \mathbf{X}^{-1}$. Then

$$\begin{aligned}(\mathbf{X} + d\mathbf{X})^{-1} &= \mathbf{X}^{-1} - \mathbf{X}^{-1}(d\mathbf{X})\mathbf{X}^{-1} + \mathbf{X}^{-1}(d\mathbf{X})\mathbf{X}^{-1}(d\mathbf{X})\mathbf{X}^{-1} \\ &\quad - \mathbf{X}^{-1}(d\mathbf{X})\mathbf{X}^{-1}(d\mathbf{X})\mathbf{X}^{-1}(d\mathbf{X})\mathbf{X}^{-1} + \cdots.\end{aligned}$$

17.66. Let $\mathbf{X}$ be a real symmetric $n \times n$ matrix with spectral decomposition $\mathbf{X} = \mathbf{Q}\mathbf{\Lambda}\mathbf{Q}'$, where $\mathbf{\Lambda} = \text{diag}(\lambda_1, \lambda_2, \ldots, \lambda_n)$ and $\lambda_i = \lambda_i(\mathbf{X})$ is a distinct eigenvalue of $\mathbf{X}$ corresponding to the eigenvector $\mathbf{q}_i$, the ith column of $\mathbf{Q}$. Let $\lambda_i(\mathbf{X} + d\mathbf{X})$ and γ_i be the eigenvalue and corresponding eigenvector of $\mathbf{X} + d\mathbf{X}$. If $d\mathbf{X} = \mathbf{Q}'\mathbf{W}\mathbf{Q}$, where $\mathbf{W}$ is "small" and symmetric, then we have the following:

(a) $\lambda_i(\mathbf{X} + d\mathbf{X}) = \lambda_i + \mathbf{q}_i'\mathbf{W}\mathbf{q}_i + \cdots.$

(b) $\gamma_i(\mathbf{X} + d\mathbf{X}) = \mathbf{q}_i - (\mathbf{Z} - \lambda_i\mathbf{I}_n)^+\mathbf{W}\mathbf{q}_i + \cdots.$

Proofs. Section 17.9.

17.65–17.66. Schott [2005: 369].

17.10 MATRIX LINEAR DIFFERENTIAL EQUATIONS

17.67. If $\mathbf{x} = \mathbf{x}(t)$ is an $n \times 1$ vector with elements that are functions of t and $\mathbf{A}$ is an $n \times n$ constant matrix, then

$$\frac{\partial\mathbf{x}(t)}{\partial t} = \mathbf{A}\mathbf{x}(t), \qquad \mathbf{x}(0) = \mathbf{x}_0$$

has a formal solution $\mathbf{x} = e^{\mathbf{A}t}\mathbf{x}_0$. If $\mathbf{A}$ is not a diagonal matrix, then the system of equations is said to be *coupled*. This coupling, which links $\partial x_i(t)/\partial t$ to the other components of $\mathbf{x}(t)$, makes the solution harder to actually find. If $\mathbf{A}$ can be transformed to a diagonal or near diagonal form, then the problem may be easier to solve. For example, if $\mathbf{A} = \mathbf{S}\mathbf{J}_O\mathbf{S}^{-1}$, where $\mathbf{J}_O$ is the Jordan canonical form of $\mathbf{A}$, then the differential equation becomes

$$\frac{\partial\mathbf{y}(t)}{\partial t} = \mathbf{J}_O\mathbf{y}(t), \qquad \mathbf{y}(0) = \mathbf{y}_0,$$

where $\mathbf{x}(t) = \mathbf{S}\mathbf{y}(t)$ and $\mathbf{y}_0 = \mathbf{S}^{-1}\mathbf{x}_0$. However, if $\mathbf{A}$ is diagonalizable, then $\mathbf{J}_O = \text{diag}(\lambda_1, \ldots, \lambda_n)$, where the λ_i are the eigenvalues of $\mathbf{A}$. The transformed equations are now uncoupled and have solutions

$$y_i(t) = y_i(0)e^{\lambda_i t}, \quad i = 1, 2, \ldots, n.$$

For further details see Horn and Johnson [1985: 133-134].

17.68. If $\mathbf{x} = \mathbf{x}(t)$, then $\dfrac{\partial \mathbf{x}(t)}{\partial t} = \mathbf{A}\mathbf{x}(t) + \mathbf{b}(t)$ and $\mathbf{x}(t_0) = \mathbf{x}_0$ has solution

$$\mathbf{x}(t) = e^{\mathbf{A}(t-t_0)}\mathbf{x}_0 + e^{\mathbf{A}t}\int_{t_0}^{t} e^{-\mathbf{A}s}\mathbf{b}(s)ds.$$

Further details are given by Seber and Wild [1989: section 8.3]. For solutions of the more general case

$$\mathbf{A}\frac{\partial \mathbf{x}(t)}{\partial t} + \mathbf{B}\mathbf{x}(t) = \mathbf{b}(t),$$

where $\mathbf{A}$ may be nonsingular, see Campbell and Meyer [1979: section 9.2].

17.69. If $\mathbf{X} = \mathbf{X}(t)$ is an $m \times n$ matrix with elements that are functions of t, then

$$\frac{\partial \mathbf{X}(t)}{\partial t} = \mathbf{A}\mathbf{X} + \mathbf{X}\mathbf{B}, \qquad \mathbf{X}(0) = \mathbf{X}_0$$

has solution $\mathbf{X} = e^{\mathbf{A}t}\mathbf{X}_0\,e^{\mathbf{B}t}$.

Proofs. Section 17.10.

    17.68. Bellman [1970: 173] and Gantmacher [1959: 116–124, 153–154].

    17.69. Graham [1981: 41] and Horn and Johnson [1991: 503–511].

17.11 SECOND-ORDER DERIVATIVES

Second-order derivatives are often required for determining the stationary values of a function. Below we give some techniques for finding Hessians.

Definition 17.8. Let $f(\mathbf{X})$ be a scalar function of the $m \times n$ matrix $\mathbf{X}$ that is twice differentiable inside the domain of f. Then the *Hessian* of f is defined to be

$$\nabla^2 f(\mathbf{X}) = \frac{\partial^2 f(\mathbf{X})}{\partial \text{vec}\,\mathbf{X}\,\partial(\text{vec}\,\mathbf{X})'}$$

$$= \frac{\partial}{\partial \text{vec}\,\mathbf{X}}\left(\frac{\partial f(\mathbf{X})}{\partial(\text{vec}\,\mathbf{X})'}\right).$$

If $\mathbf{X}$ is a vector, say $\mathbf{x}$, then

$$\nabla^2 f(\mathbf{x}) = \frac{\partial^2 f}{\partial \mathbf{x}\partial \mathbf{x}'} = \left(\frac{\partial^2 f}{\partial x_i \partial x_j}\right).$$

If $\mathbf{Y}$ is a matrix function of $\mathbf{X}$, we also define the *second differential* $\mathrm{d}^2\mathbf{Y} = \mathrm{d}(\mathrm{d}\mathbf{Y})$; and in deriving this in applications, we note that $\mathrm{d}^2\mathbf{X} = \mathbf{0}$. For some analytical details see Abadir and Magnus [2005: chapter 13], Harville [1997: section 15.1], Magnus and Neudecker [1999: chapter 6], and Nel [1980]. A number of examples are given by Nel [1980: section 7.2].

17.70. The Hessian as defined above is symmetric.

17.71. (Identification Rules)

(a) $d^2 f(\mathbf{X}) = (\text{vec } d\mathbf{X})' \mathbf{B} (\text{vec } d\mathbf{X})$ if and only if $\nabla^2 f(\mathbf{X}) = \frac{1}{2}(\mathbf{B} + \mathbf{B}')$, where $\mathbf{B}$ may depend on $\mathbf{X}$ but not on $d\mathbf{X}$.

For example, if $f(\mathbf{X}) = \text{trace}(\mathbf{AXBX}'\mathbf{C})$, where $\mathbf{A}$, $\mathbf{B}$, and $\mathbf{C}$ are square matrices (not necessarily of the same size) of constants, then taking differentials twice and setting $d^2\mathbf{X} = \mathbf{0}$ and $d(d\mathbf{X})' = \mathbf{0}$ we have, interchanging "d" and "trace", and noting that $\text{trace } \mathbf{F} = \text{trace } \mathbf{F}'$,

$$df(\mathbf{X}) = \text{trace}[\mathbf{A}(d\mathbf{X})\mathbf{BX}'\mathbf{C} + \mathbf{AXB}(d\mathbf{X})'\mathbf{C}],$$

and

$$
\begin{aligned}
d^2 f(\mathbf{X}) &= 2\,\text{trace}[\mathbf{A}(d\mathbf{X})\mathbf{B}(d\mathbf{X})'\mathbf{C}] \\
&= 2\,\text{trace}[(d\mathbf{X})'\mathbf{CA}(d\mathbf{X})\mathbf{B}] \\
&= 2(\text{vec } d\mathbf{X})'(\mathbf{B}' \otimes \mathbf{CA})(\text{vec } d\mathbf{X}),
\end{aligned}
$$

since $\text{trace}(\mathbf{D}'\mathbf{E}) = (\text{vec }\mathbf{D})'\text{vec }\mathbf{E}$ and $\text{vec}(\mathbf{DEF}) = (\mathbf{F}' \otimes \mathbf{D})\text{vec }\mathbf{E}$. Here "$\otimes$" is the Kronecker product. We thus have from (a) the following rule:

(b) $d^2 f(\mathbf{X}) = \text{trace}[\mathbf{A}(d\mathbf{X})\mathbf{B}(d\mathbf{X})'\mathbf{C}]$ if and only if

$$\nabla^2 f(\mathbf{X}) = \frac{1}{2}(\mathbf{B}' \otimes \mathbf{CA} + \mathbf{B} \otimes \mathbf{A}'\mathbf{C}').$$

Similarly, by using $\text{trace}(\mathbf{FG}) = \text{trace}(\mathbf{GF})$, we see that

$d^2 f(\mathbf{X}) = \text{trace}[\mathbf{B}(d\mathbf{X}')\mathbf{C}(d\mathbf{X})]$ if and only if

$$\nabla^2 f(\mathbf{X}) = \frac{1}{2}(\mathbf{B}' \otimes \mathbf{C} + \mathbf{B} \otimes \mathbf{C}').$$

For example, if $f(\mathbf{X}) = \text{trace}(\mathbf{X}'\mathbf{AX})$, then $d^2 f(\mathbf{X}) = 2\,\text{trace}[(d\mathbf{X}')\mathbf{A}d\mathbf{X}]$ and

$$\nabla^2 f(\mathbf{X}) = \mathbf{I} \otimes (\mathbf{A} + \mathbf{A}').$$

(c) $d^2 f(\mathbf{X}) = \text{trace}[\mathbf{B}(d\mathbf{X})\mathbf{C}(d\mathbf{X})]$ if and only if

$$\nabla^2 f(\mathbf{X}) = \frac{1}{2}\mathbf{I}_{(m,n)}(\mathbf{B}' \otimes \mathbf{C} + \mathbf{C}' \otimes \mathbf{B}),$$

where $\mathbf{X}$ is $m \times n$ and $\mathbf{I}_{(m,n)}$ is the commutation (vec-permutation) matrix. We have the following examples for $n \times n$ $\mathbf{X}$.

(i) If $f(\mathbf{X}) = \text{trace}(\mathbf{X}^{-1})$, then $d^2 f(\mathbf{X}) = 2\,\text{trace}[\mathbf{X}^{-2}(d\mathbf{X})\mathbf{X}^{-1}d\mathbf{X}]$ and

$$\nabla^2 f(\mathbf{X}) = \mathbf{I}_{(n,n)}(\mathbf{X}'^{-2} \otimes \mathbf{X}^{-1} + \mathbf{X}'^{-1} \otimes \mathbf{X}^{-2}).$$

(ii) If $f(\mathbf{X}) = \det \mathbf{X}$, then

$$\nabla^2 f(\mathbf{X}) = -\det \mathbf{X}[\mathbf{I}_{(n,n)}(\mathbf{X}'^{-1} \otimes \mathbf{X}^{-1}) - (\text{vec } \mathbf{X}'^{-1})(\text{vec } \mathbf{X}'^{-1})'].$$

(d) We also have the following special case for vectors.

$$\mathrm{d}^2 f(\mathbf{x}) - (\mathrm{d}\mathbf{x})'\mathbf{A}(\mathrm{d}\mathbf{x}) \quad \text{if and only if} \quad \nabla^2 f(\mathbf{x}) = \tfrac{1}{2}(\mathbf{A} + \mathbf{A}').$$

For example, if $f(\mathbf{x}) = \mathbf{x}'\mathbf{B}\mathbf{x}$, where $\mathbf{B}$ is a symmetric constant matrix, then $\mathrm{d}f = 2\mathbf{x}'\mathbf{B}\mathrm{d}\mathbf{x}$ and $\mathrm{d}^2 f = 2(\mathrm{d}\mathbf{x})'\mathbf{B}\mathrm{d}\mathbf{x}$. Here $\mathbf{A}$ can be a function of $\mathbf{x}$.

17.72. Suppose $\mathbf{X}$ has L-structure (e.g., is symmetric or triangular) so that $\mathrm{vec}\,\mathbf{X} = \mathbf{\Delta}\phi(\mathbf{X})$ (cf. Section 18.3.5 for notation), then

$$\frac{\partial^2 f(\mathbf{X})}{\partial\phi(\mathbf{X})\partial(\phi(\mathbf{X}))'} = \mathbf{\Delta}'\nabla^2 f(\mathbf{X})\mathbf{\Delta}.$$

If $\mathbf{X}$ is symmetric and $n \times n$, then $\phi(\mathbf{X}) = \mathrm{vech}\,\mathbf{X}$ and $\mathbf{\Delta} = \mathbf{G}_n$, the duplication matrix.

We demonstrate the above theory with the example $f(\mathbf{X}) = \mathrm{trace}(\mathbf{X}^{-1})$. From (17.71c(i)) we obtain

$$\nabla^2 f(\mathbf{X}) = \mathbf{I}_{(n,n)}(\mathbf{X}'^{-1} \otimes \mathbf{X}^{-2} + \mathbf{X}'^{-2} \otimes \mathbf{X}^{-1}).$$

If $\mathbf{X}$ is symmetric, then

$$\frac{\partial^2 f(\mathbf{X})}{\partial\phi(\mathbf{X})\partial(\phi(\mathbf{X}))'} = \mathbf{\Delta}'\nabla^2 f(\mathbf{X})\mathbf{\Delta} = 2\mathbf{\Delta}'(\mathbf{X}^{-1} \otimes \mathbf{X}^{-2})\mathbf{\Delta},$$

since $\mathbf{I}_{(n,n)}\mathbf{\Delta} = \mathbf{I}_{(n,n)}\mathbf{G}_n = \mathbf{G}_n = \mathbf{\Delta}$, by (11.29d).

17.73. Let $\mathbf{F}(\mathbf{t})$, with r, sth elements $f_{rs}(\mathbf{t})$, be a nonsingular matrix function of $\mathbf{t}$.

(a)

$$\frac{\partial^2 \det \mathbf{F}}{\partial t_i \partial t_j} =$$

$$(\det \mathbf{F})\left[\mathrm{trace}\left(\mathbf{F}^{-1}\frac{\partial^2 \mathbf{F}}{\partial t_i \partial t_j}\right) + \mathrm{trace}\left(\mathbf{F}^{-1}\frac{\partial \mathbf{F}}{\partial t_i}\right)\mathrm{trace}\left(\mathbf{F}^{-1}\frac{\partial \mathbf{F}}{\partial t_j}\right)\right.$$

$$\left. - \mathrm{trace}\left(\mathbf{F}^{-1}\frac{\partial \mathbf{F}}{\partial t_i}\mathbf{F}^{-1}\frac{\partial \mathbf{F}}{\partial t_j}\right)\right].$$

(b)

$$\frac{\partial^2 \log(\det \mathbf{F})}{\partial t_i \partial t_j} = \mathrm{trace}\left(\mathbf{F}^{-1}(\frac{\partial^2 \mathbf{F}}{\partial t_i \partial t_j})\right) - \mathrm{trace}\left(\mathbf{F}^{-1}\frac{\partial \mathbf{F}}{\partial t_i}\mathbf{F}^{-1}\frac{\partial \mathbf{F}}{\partial t_j}\right).$$

(c)

$$\frac{\partial^2 \mathbf{F}^{-1}}{\partial t_i \partial t_j} =$$

$$-\mathbf{F}^{-1}\frac{\partial^2 \mathbf{F}}{\partial t_i \partial t_j}\mathbf{F}^{-1} + \mathbf{F}^{-1}\frac{\partial \mathbf{F}}{\partial t_i}\mathbf{F}^{-1}\frac{\partial \mathbf{F}}{\partial t_j}\mathbf{F}^{-1} + \mathbf{F}^{-1}\frac{\partial \mathbf{F}}{\partial t_j}\mathbf{F}^{-1}\frac{\partial \mathbf{F}}{\partial t_i}\mathbf{F}^{-1}.$$

17.74. Let $d(\boldsymbol{\theta}) = \det[\mathbf{X}(\boldsymbol{\theta})'\mathbf{X}(\boldsymbol{\theta})]$, where $\mathbf{X}$ is $n \times p$ of rank p for $\boldsymbol{\theta} \in \Omega$, and let $k(\boldsymbol{\theta}) = \frac{1}{2}\log d(\boldsymbol{\theta})$. Then from (17.12),

$$k_r = \frac{\partial k(\boldsymbol{\theta})}{\partial \theta_r} = \operatorname{trace}(\mathbf{X}^+ \mathbf{X}_r),$$

where $\mathbf{X}_r = \partial \mathbf{X}/\partial \theta_r$. Also

$$
\begin{aligned}
k_{rs} &= \frac{\partial^2 k(\boldsymbol{\theta})}{\partial \theta_r \partial \theta_s} \\
&= -\operatorname{trace}(\mathbf{X}^+ \mathbf{X}_r \mathbf{X}^+ \mathbf{X}_s) + \operatorname{trace}[\mathbf{X}^+(\mathbf{X}^+)'\mathbf{X}_r'(\mathbf{I}_n - \mathbf{X}\mathbf{X}^+)\mathbf{X}_s] \\
&\quad + \operatorname{trace}(\mathbf{X}^+ \mathbf{X}_{rs}),
\end{aligned}
$$

where $\mathbf{X}^+$ is the Moore–Penrose of $\mathbf{X}$. The Hessian $\mathbf{H}(\boldsymbol{\theta}) = (h_{rs})$ of $d(\boldsymbol{\theta})$ is given by

$$h_{rs} = \frac{\partial^2 d(\boldsymbol{\theta})}{\partial \theta_r \partial \theta_s} = 2d(\boldsymbol{\theta})(k_{rs} + 2k_r k_s).$$

Proofs. Section 17.11.

17.70. Magnus and Neudecker [1999: 105–106].

17.71a. Abadir and Magnus [2005: 353] and Magnus and Neudecker [1999: 190].

17.71b. Magnus and Neudecker [1999: 192–193].

17.71c. Abadir and Magnus [2005: 380-381] and Magnus and Neudecker [1999: 192].

17.71d. Abadir and Magnus [2005: 353]

17.72. Magnus [1988: 155].

17.73. Harville [1997: section 15.9].

17.74. Bates and Watts [1985] and Seber and Wild [1989: 543].

17.12   VECTOR DIFFERENCE EQUATIONS

Definition **17.9.** The *vector difference equation*

$$\mathbf{A}_0 \mathbf{y}_t + \mathbf{A}_1 \mathbf{y}_{t-1} + \ldots + \mathbf{A}_r \mathbf{y}_{t-r} = \mathbf{g}(t),$$

with $\mathbf{A}_0$ nonsingular and all vectors n-dimensional functions of t, is called an *rth-order vector difference equation with constant coefficients*. Since $\mathbf{A}_0$ is nonsingular, we can set $\mathbf{A}_0 = \mathbf{I}_d$ without loss of generality. Difference equations arise in discrete time stochastic processes and in iterative procedures that converge. The case when $\mathbf{A}_0$ is singular can be handled using the Drazin inverse of $\mathbf{A}_0$ (Campbell and Meyer [1979: 181–184]).

17.75. The above difference equation can be reduced to a first-order equation as follows. Let $\mathbf{z}_t = (\mathbf{y}_t', \mathbf{y}_{t-1}', \ldots, \mathbf{y}_{t-r+1}')'$ and

$$\mathbf{B} = \begin{pmatrix} -\mathbf{A}_1 & -\mathbf{A}_2 & \ldots & -\mathbf{A}_{r-1} & -\mathbf{A}_r \\ \mathbf{I}_d & \mathbf{0} & \ldots & \mathbf{0} & \mathbf{0} \\ \mathbf{0} & \mathbf{0} & \ldots & \mathbf{I}_d & \mathbf{0} \end{pmatrix}.$$

Then $\mathbf{z}_t = \mathbf{B}\mathbf{z}_{t-1} + \mathbf{e}_1 \otimes \mathbf{g}(t)$, where $\mathbf{e}_1$ is an r-dimensional vector $(1, 0, \ldots, 0)'$ and "$\otimes$" is the Kronecker product. The solution of this was studied by Dhrymes [2000: 175–178]. He applies it to the general linear structural econometric model.

17.76. If $\mathbf{x}_t = \mathbf{A}\mathbf{x}_{t-1} + \mathbf{d}$, where $\mathbf{A}$ and $\mathbf{x}_0$ are known, then provided that $\mathbf{A}^t \to \mathbf{0}$ as $t \to \infty$ and $(\mathbf{I} - \mathbf{A})^{-1}$ exists, we have $\mathbf{x}_t = \mathbf{A}^t \mathbf{x}_0 + (\mathbf{I} - \mathbf{A}^t)(\mathbf{I} - \mathbf{A})^{-1}\mathbf{d}$ and $\mathbf{x}_t \to (\mathbf{I} - \mathbf{A})^{-1}\mathbf{d}$ as $t \to \infty$.

Definition 17.10. (Linear Stationary Iterations) Let $\mathbf{A}\mathbf{x} = \mathbf{b}$, with $\mathbf{A}$ $n \times n$ and expressible in the form $\mathbf{A} = \mathbf{M} - \mathbf{N}$, where $\mathbf{M}^{-1}$ exists. Let $\mathbf{H} = \mathbf{M}^{-1}\mathbf{N}$, the *iteration matrix*, and let $\mathbf{d} = \mathbf{M}^{-1}\mathbf{b}$. Given an initial $n \times 1$ vector $\mathbf{x}_{(0)}$, then a *linear stationary iteration* is

$$\mathbf{x}_{(k)} = \mathbf{H}\mathbf{x}_{(k-1)} + \mathbf{d}, \quad k = 1, 2, 3, \ldots.$$

17.77. Given the notation above, if $\rho(\mathbf{H}) < 1$, where ρ is the spectral radius, then $\mathbf{A}$ is nonsingular and

$$\lim_{k \to \infty} \mathbf{x}_{(k)} = \mathbf{x} = \mathbf{A}^{-1}\mathbf{b}$$

for every initial vector $\mathbf{x}_{(0)}$. For details and methods see Meyer [2000a: 620–626].

Proofs. Section 17.12.

17.76. Searle [1982: 289].

17.77. Meyer [2000a: 621].

CHAPTER 18

JACOBIANS

Jacobians play a fundamental role in statistical distribution theory. Formulae for Jacobians and their proofs are given in many places, especially in the appendices of statistics books. In the case of complicated transformations, it is not always clear what the sign of the Jacobian is as it will depend on how the variables are ordered. Fortunately, the ordering only affects the sign of the Jacobian, which usually does not matter as in applications we are mainly interested in the absolute value of the Jacobian.

18.1 INTRODUCTION

Before listing a number of results, we look at the meaning of a Jacobian and give a number of different techniques for finding Jacobians.

Definition 18.1. Suppose $\mathbf{f} : \mathbf{x} \to \mathbf{y} = \mathbf{f}(\mathbf{x})$, where $\mathbf{x}$ and $\mathbf{y}$ belong to $\mathbb{R}^n$, is a one-to-one (bijective) differentiable function, i.e., it has an inverse function $g = f^{-1}$. Then $\partial \mathbf{x}/\partial \mathbf{y}' = (\partial x_i/\partial y_j)$ is called the *Jacobian matrix* of the transformation $\mathbf{x} \to \mathbf{y}$, and its determinant

$$ J_{\mathbf{x} \to \mathbf{y}} = \det \left(\frac{\partial \mathbf{x}}{\partial \mathbf{y}'} \right) $$

is called the *Jacobian* of the transformation. (Some authors call the absolute value $|J_{\mathbf{x} \to \mathbf{y}}|$ the Jacobian.) For further comments on this definition see Section 17.7.

A Matrix Handbook for Statisticians. By George A. F. Seber
Copyright © 2008 John Wiley & Sons, Inc.

In the above definition we want to differentiate $\mathbf{x}$ with respect to $\mathbf{y}$, so it is more natural to use $\mathbf{x} = \mathbf{g}(\mathbf{y})$, as does Muirhead [1982: chapter 2], for example. However, as most of the references use $\mathbf{y} = \mathbf{f}(\mathbf{x})$, I have decided to stay with the latter in this chapter. As it can be a source of possible confusion, Daniel L. Solomon gives the following mnemonic rule (Searle [1982: 339]) to help get the order of the variables right: If "o" represents the old coordinates ($\mathbf{x}$) and "n" represents the new cordinates ($\mathbf{y}$), then $J_o \rightarrow n$ is correct, but $J_n \rightarrow o$ is not (spells *no*).

If we interchange two elements of $\mathbf{y}$, we change the sign of $J_{\mathbf{y}\rightarrow\mathbf{x}}$. Since, in practice, we are generally more interested in the absolute value of the Jacobian, $|J_{\mathbf{y}\rightarrow\mathbf{x}}|$, it does not matter in what order we list the elements of $\mathbf{x}$ and $\mathbf{y}$. Several authors—for example, Mathai [1997], whom I will refer to frequently in this chapter—get around this problem by stating that the sign of a particular Jacobian should be ignored. I shall tend to use absolute values throughout.

As already mentioned, if we want to differentiate $\mathbf{x}$ with respect to $\mathbf{y}$, we can endeavor to express the transformation in the form $\mathbf{x} = \mathbf{g}(\mathbf{y})$. However, it may be easier to find $J_{\mathbf{y}} \rightarrow \mathbf{x}$ first as $J_{\mathbf{x}} \rightarrow \mathbf{y} = J_{\mathbf{y}}^{-1} \rightarrow \mathbf{x}$. To see this, we have $\mathbf{x} = \mathbf{g}(\mathbf{f}(\mathbf{x}))$ and

$$
\mathbf{I}_n = \left(\frac{\partial x_i}{\partial x_j} \right) = \left(\sum_r \frac{\partial g_i}{\partial y_r} \cdot \frac{\partial y_r}{\partial x_j} \right) = \frac{\partial \mathbf{x}}{\partial \mathbf{y}'} \cdot \frac{\partial \mathbf{y}}{\partial \mathbf{x}'}.
$$

Then

$$
1 = \det \left\{ \frac{\partial \mathbf{x}}{\partial \mathbf{y}'} \cdot \frac{\partial \mathbf{y}}{\partial \mathbf{x}'} \right\} = \det \left\{ \frac{\partial \mathbf{x}}{\partial \mathbf{y}'} \right\} \cdot \det \left\{ \frac{\partial \mathbf{y}}{\partial \mathbf{x}'} \right\} = J_{\mathbf{x}} \rightarrow \mathbf{y} \, J_{\mathbf{y}} \rightarrow \mathbf{x}.
$$

We note that $J_{\mathbf{y}} \rightarrow \mathbf{x}$ will be expressed in terms of $\mathbf{x}$, which in applications usually has to be replaced by its function of $\mathbf{y}$. For example, two important statistical applications of Jacobians are (i) change of variables in integration, that is,

$$
\int \cdots \int h(x_1, x_2, \ldots, x_n) \mathrm{d}x_1 \mathrm{d}x_2 \cdots \mathrm{d}x_n = \int \cdots \int h(\mathbf{g}(\mathbf{y})) |J_{\mathbf{x}} \rightarrow \mathbf{y}| \mathrm{d}y_1 \mathrm{d}y_2 \cdots \mathrm{d}y_n,
$$

and (ii) probability density functions for functions of random variables, namely

$$
f_{\mathbf{y}}(\mathbf{y}) = f_{\mathbf{x}}(\mathbf{g}(\mathbf{y})) |J_{\mathbf{x}} \rightarrow \mathbf{y}|.
$$

If $\mathbf{x}$ and $\mathbf{y}$ are replaced by matrices with $\mathbf{Y} = \mathbf{F}(\mathbf{X})$, we define

$$
J_{\mathbf{X}\rightarrow\mathbf{Y}} = J_{\text{vec }\mathbf{X}\rightarrow\text{vec }\mathbf{Y}} = \det \left(\frac{\partial \text{ vec } \mathbf{X}}{\partial (\text{vec } \mathbf{Y})'} \right),
$$

and, if $\mathbf{X}$ and $\mathbf{Y}$ are symmetric or lower-triangular, we define

$$
J_{\mathbf{X}\rightarrow\mathbf{Y}} = J_{\text{vech }\mathbf{X}\rightarrow\text{vech }\mathbf{Y}} = \det \left(\frac{\partial \text{ vech } \mathbf{X}}{\partial (\text{vech } \mathbf{Y})'} \right).
$$

For upper-triangular matrices we can use vech $(\mathbf{X}')$.

In order to evaluate the above Jacobians, various properties of the Kronecker product and of the vec and vech operators are required from Chapter 11. In this chapter we shall concentrate on finding $J_{\mathbf{Y}\rightarrow\mathbf{X}}$ or $|J_{\mathbf{Y}\rightarrow\mathbf{X}}|$, which can then be inverted. Unless otherwise stated, all matrices and scalars are real. We now give some techniques for finding Jacobians; some of these techniques are demonstrated by Olkin [2002].

18.2 METHOD OF DIFFERENTIALS

Differentials were introduced in Section 17.8 along with some rules that provide a powerful method for finding Jacobians.

18.1. A key result from (17.28) for vectors is as follows. If $d\mathbf{y} = \mathbf{A}d\mathbf{x}$, where $\mathbf{A}$ may be a function of $\mathbf{x}$, then

$$\frac{\partial \mathbf{y}}{\partial \mathbf{x}'} = \mathbf{A} \quad \text{and} \quad J_{\mathbf{y} \to \mathbf{x}} = \det \mathbf{A}.$$

In the case of matrices, if $d \operatorname{vec} \mathbf{Y} = \mathbf{B} d \operatorname{vec} \mathbf{X}$, then

$$\frac{\partial \operatorname{vec} \mathbf{Y}}{\partial (\operatorname{vec} \mathbf{X})'} = \mathbf{B} \quad \text{and} \quad J_{\mathbf{Y} \to \mathbf{X}} = \det \mathbf{B},$$

where $\mathbf{B}$ may be a function of $\mathbf{X}$. Also $d\operatorname{vec} \mathbf{X} = \operatorname{vec}(d\mathbf{X})$.

18.2. We recall the following equivalent statements from (17.60), where $\mathbf{X}$ is $m \times n$.

(1) $d\mathbf{Y} = \mathbf{A}(d\mathbf{X})\mathbf{B} + \mathbf{C}(d\mathbf{X}')\mathbf{D}.$

(2) $\operatorname{vec}(d\mathbf{Y}) = (\mathbf{B}' \otimes \mathbf{A})\operatorname{vec}(d\mathbf{X}) + (\mathbf{D}' \otimes \mathbf{C})\operatorname{vec}(d\mathbf{X}')$, or, equivalently,

$$d \operatorname{vec} \mathbf{Y} = \{(\mathbf{B}' \otimes \mathbf{A}) + (\mathbf{D}' \otimes \mathbf{C})\mathbf{I}_{(n,m)}\}d \operatorname{vec} \mathbf{X}.$$

(3) $\dfrac{\partial \operatorname{vec} \mathbf{Y}}{\partial (\operatorname{vec} \mathbf{X})'} = \mathbf{B}' \otimes \mathbf{A} + (\mathbf{D}' \otimes \mathbf{C})\mathbf{I}_{(n,m)}.$

When the above hold, we see from (2) and (3) that

$$J_{\mathbf{Y} \to \mathbf{X}} = J_{\operatorname{vec} \mathbf{Y} \to \operatorname{vec} \mathbf{X}} = J_{d \operatorname{vec} \mathbf{Y} \to d \operatorname{vec} \mathbf{X}}.$$

We demonstrate how these results can be used by working through two examples.

Example 1 Let $\mathbf{Y}_{m \times n} = \mathbf{A}_{m \times m}\mathbf{X}_{m \times n}\mathbf{B}_{n \times n}$, where $\mathbf{A}$ and $\mathbf{B}$ are nonsingular. Then $d\mathbf{Y} = \mathbf{A}(d\mathbf{X})\mathbf{B}$ and $\operatorname{vec}(d\mathbf{Y}) = (\mathbf{B}' \otimes \mathbf{A})\operatorname{vec}(d\mathbf{X})$. Using (17.1$l$(ii)), we get

$$J_{\mathbf{Y} \to \mathbf{X}} = \det(\mathbf{B}' \otimes \mathbf{A}) = (\det \mathbf{B})^m (\det \mathbf{A})^n.$$

Thus, $J_{\mathbf{X} \to \mathbf{Y}} = (\det \mathbf{B})^{-m}(\det \mathbf{A})^{-n}.$

Example 2 Let $\mathbf{Y} = \mathbf{X}^{-1}$, where $\mathbf{X}$ is a nonsingular $n \times n$ matrix. Since $\mathbf{Y}\mathbf{X} = \mathbf{I}_n$, we have $\mathbf{0} = d(\mathbf{Y}\mathbf{X}) = \mathbf{Y}d\mathbf{X} + (d\mathbf{Y})\mathbf{X}$, or $d\mathbf{Y} = -\mathbf{X}^{-1}(d\mathbf{X})\mathbf{X}^{-1}$. Thus, from Example 1 with $\mathbf{A} = -\mathbf{X}^{-1}$ and $\mathbf{B} = \mathbf{X}^{-1}$, we have

$$J_{\mathbf{Y} \to \mathbf{X}} = (-1)^{n^2} (\det \mathbf{X})^{-2n},$$

and $J_{\mathbf{X} \to \mathbf{Y}} = (-1)^{n^2}(\det \mathbf{Y})^{-2n}.$

18.3 FURTHER TECHNIQUES

In addition to the method of differentials, there are a number of other useful related techniques that we now describe.

18.3.1   Chain Rule

One useful technique makes use of the chain rule (17.21). This rule leads to the result that if we have the transformations $\mathbf{x} \to \mathbf{y}$ and $\mathbf{y} \to \mathbf{z}$, then $J_{\mathbf{X} \to \mathbf{z}} = J_{\mathbf{X} \to \mathbf{y}} J_{\mathbf{y} \to \mathbf{z}}$, i.e., Jacobians multiply. If the Jacobian of $\mathbf{x} \to \mathbf{z}$ is hard to derive, it may be possible to find an *intermediary* variable $\mathbf{y}$ such that $J_{\mathbf{X} \to \mathbf{y}}$ and $J_{\mathbf{y} \to \mathbf{z}}$ are easy to find. This method was used frequently by Deemer and Olkin [1951] and Olkin [1953], and several examples of it appear later.

18.3.2   Exterior (Wedge) Product of Differentials

This elegant technique is described in detail by Muirhead [1982: chapter 2] and used extensively by Mathai [1997] . We assume that $\mathbf{y}$ and $\mathbf{x}$ are $n \times 1$ vectors, and we introduce a skew-symmetric product "$\wedge$" of differentials called the *exterior* or *wedge* product, satisfying (i) $dy_i \wedge dy_j = -dy_j \wedge dy_i$ and (ii) $dy_i \wedge dy_i = 0$ (a consequence of (i)). To evaluate $\det(\partial\mathbf{y}/\partial\mathbf{x}')$, we begin with

$$dy_i = \sum_{j=1}^{n} \frac{\partial y_i}{\partial x_j} dx_j,$$

and multipy these together using the above two properties of the exterior product to get

$$dy_1 \wedge dy_2 \wedge \cdots \wedge dy_n = \det\left(\frac{\partial\mathbf{y}}{\partial\mathbf{x}'}\right) dx_1 \wedge dx_2 \wedge \cdots \wedge dx_n.$$

This result is readily demonstrated for $n = 2$. We have

$$
\begin{aligned}
dy_1 \wedge dy_2 &= \left(\frac{\partial y_1}{\partial x_1} dx_1 + \frac{\partial y_1}{\partial x_2} dx_2\right) \wedge \left(\frac{\partial y_2}{\partial x_1} dx_1 + \frac{\partial y_2}{\partial x_2} dx_2\right) \\
&= \frac{\partial y_1}{\partial x_1} \cdot \frac{\partial y_2}{\partial x_1} dx_1 \wedge dx_1 + \frac{\partial y_1}{\partial x_1} \cdot \frac{\partial y_2}{\partial x_2} dx_1 \wedge dx_2 \\
&\quad + \frac{\partial y_1}{\partial x_2} \cdot \frac{\partial y_2}{\partial x_1} dx_2 \wedge dx_1 + \frac{\partial y_1}{\partial x_2} \cdot \frac{\partial y_2}{\partial x_2} dx_2 \wedge dx_2 \\
&= 0 + \left(\frac{\partial y_1}{\partial x_1} \cdot \frac{\partial y_2}{\partial x_2} - \frac{\partial y_1}{\partial x_2} \cdot \frac{\partial y_2}{\partial x_1}\right) dx_1 \wedge dx_2 + 0 \\
&= \det\begin{pmatrix} \dfrac{\partial y_1}{\partial x_1} & \dfrac{\partial y_1}{\partial x_2} \\[2mm] \dfrac{\partial y_2}{\partial x_1} & \dfrac{\partial y_2}{\partial x_2} \end{pmatrix} dx_1 \wedge dx_2 \\
&= \det\left(\frac{\partial\mathbf{y}}{\partial\mathbf{x}'}\right) dx_1 \wedge dx_2.
\end{aligned}
$$

We shall define the wedge product for a vector $\mathbf{x}$ as

$$d_w\mathbf{x} = \wedge_{i=1}^{n} dx_i.$$

This approach extends to matrices by setting $\mathbf{y} = \text{vec}\,\mathbf{Y}$ and defining $d_w\mathbf{Y} = \wedge_{i,j \in \mathcal{D}} dy_{ij}$, where $\mathcal{D}$ denotes the ordered set of distinct elements of $\mathbf{Y}$, ordered according to $\mathbf{y}$. I have deliberately made this notation different from that of Muirhead

[1982] and Mathai [1997] to avoid confusion, because they use brackets like $(\mathrm{d}\mathbf{X})$ with opposite meanings.

If $\mathbf{Y}$ is symmetric or lower-triangular, we can use $\mathbf{y} = \operatorname{vech}\mathbf{Y}$ for the distinct elements of $\mathbf{Y}$ and define $\mathrm{d}_w\mathbf{Y} = \mathrm{d}_w\mathbf{y} = \mathrm{d}_w\operatorname{vech}\mathbf{Y}$. In the case of a skew-symmetric matrix, the diagonal elements are ignored, as they are zero. However, as already noted, the order in which the distinct elements of these matrices are listed is not important in applications.

18.3. It follows from the above that if $\operatorname{vec}\mathbf{Y} = \mathbf{y}$ and $\operatorname{vec}\mathbf{X} = \mathbf{x}$, then

$$\mathrm{d}_w\mathbf{Y} = \det\left(\frac{\partial\mathbf{y}}{\partial\mathbf{x}'}\right)\mathrm{d}_w\mathbf{X}.$$

Since, as already mentioned, the order of the variables can be arbitrary, we have

$$\mathrm{d}_w\mathbf{Y} = (\det\mathbf{C})\mathrm{d}_w\mathbf{X} \quad\Longrightarrow\quad |J_{\mathbf{Y}\to\mathbf{x}}| = |\det\mathbf{C}|.$$

Example 3 Let $(\mathbf{y}_1, \mathbf{y}_2, \ldots, \mathbf{y}_n) = \mathbf{Y} = \mathbf{AX} = (\mathbf{Ax}_1, \mathbf{Ax}_2, \ldots, \mathbf{Ax}_n)$. Then, from (18.1), $\mathrm{d}_w\mathbf{y}_i = (\det\mathbf{A})\mathrm{d}_w\mathbf{x}_i$, so that $\mathrm{d}_w\mathbf{Y} = \wedge_{i=1}^n\mathrm{d}_w\mathbf{y}_i = (\det\mathbf{A})^n\mathrm{d}_w\mathbf{X}$, and $|J_{\mathbf{Y}\leftarrow\mathbf{x}}| = |\det\mathbf{A}|^n$. Alternatively, $\operatorname{vec}\mathbf{Y} = \operatorname{diag}(\mathbf{A}, \mathbf{A}, \ldots, \mathbf{A})\operatorname{vec}\mathbf{X}$ and, from (18.1), $|\det\mathbf{C}| = |\det\{\operatorname{diag}(\mathbf{A}, \mathbf{A}, \ldots, \mathbf{A})\}| = |\det\mathbf{A}|^n$, as before.

Example 4 Let $\mathbf{Y} = \mathbf{BX}$, where the matrices are all $n\times n$ lower-triangular matrices so that $(\mathbf{y}_1, \mathbf{y}_2, \ldots, \mathbf{y}_n) = (\mathbf{Bx}_1, \mathbf{Bx}_2, \ldots, \mathbf{Bx}_n)$. For $r = 1, 2, \ldots, n$, let $\mathbf{y}_{(r)} = (y_{r,r}, y_{r+1,r}, \ldots, y_{n,r})'$ be $\mathbf{y}_r$ without its leading zeros; $\mathbf{x}_{(r)}$ is similarly defined. Note that $\mathbf{y}_{(n)} = y_{n,n} = b_{n,n}x_{n,n} = b_{n,n}\mathbf{x}_{(n)}$. Then

$$a_n = \det\left(\frac{\partial y_{n,n}}{\partial x_{n,n}}\right) = b_{n,n}$$

$$a_{n-1} = \det\left(\frac{\partial(y_{n-1,n-1}, y_{n-1,n})}{\partial(x_{n-1,n-1}, x_{n-1,n})}\right)$$

$$= \det\begin{pmatrix} b_{n-1,n-1} & b_{n-1,n} \\ 0 & b_{n,n} \end{pmatrix} = b_{n,n}b_{n-1,n-1},$$

and so on. Hence

$$\mathrm{d}_w\mathbf{Y} = \wedge_{r=1}^n\mathrm{d}_w\mathbf{y}_{(r)} = \prod_{r=1}^n a_i\mathrm{d}_w\mathbf{y}_{(r)} = \prod_{r=1}^n b_{r,r}^r\mathrm{d}_w\mathbf{X}$$

and $|J_{\mathbf{Y}\to\mathbf{x}}| = |\prod_{r=1}^n b_{r,r}^r|$.

18.3.3 Induced Functional Equations

Olkin and Sampson [1972] described a method whereby one sets up an equation satisfied by the Jacobian and then solves the equation.

Example 5 Suppose $\mathbf{Y} = \mathbf{AXA}'$, where $\mathbf{A}$ is nonsingular and $\mathbf{Y}$ is symmetric. To find the Jacobian of this transformation, let $\mathbf{Z} = \mathbf{BYB}'$, where $\mathbf{B}$ is nonsingular and $\mathbf{Z}$ is symmetric, so that $\mathbf{Z} = \mathbf{ABX}(\mathbf{AB})'$. Then

$$|J_{\mathbf{Z}\to\mathbf{x}}| = |J_{\mathbf{Z}\to\mathbf{Y}}| \cdot |J_{\mathbf{Y}\to\mathbf{x}}|.$$

As the transformation is linear in $\mathbf{X}$, $|J_{\mathbf{Y} \to \mathbf{X}}|$ is a positive function of $\mathbf{A}$ above, say $h(\mathbf{A})$. Then, by the above equation,

$$h(\mathbf{AB}) = h(\mathbf{A})h(\mathbf{B}),$$

which, for this example, has solution $h(\mathbf{A}) = |\det \mathbf{A}|^c$, for some c. Setting $\mathbf{A} = \operatorname{diag}(a, 1, 1, \ldots, 1)$ and finding $J_{\mathbf{Y} \to \mathbf{X}}$ for this simple case, leads to $c = n + 1$ (Olkin and Sampson [1972: 263]).

Olkin and Sampson [1972] derive solutions of the equation $h(\mathbf{CD}) = h(\mathbf{C})h(\mathbf{D})$ for diagonal, triangular, orthogonal, and symmetric matrices. Their paper can be referred to for details (see also Mathai [1997: 40–44] for a summary of the main results).

18.3.4   Jacobians Involving Transposes

Consider the transformation $\mathbf{Y} = \mathbf{X}'$, where $\mathbf{X}$ is an $m \times n$ unconstrained matrix. Then, by (11.18b(i)) and (11.18f(i)),

$$\operatorname{vec} \mathbf{Y} = \operatorname{vec} \mathbf{X}' = \mathbf{I}_{(n,m)} \operatorname{vec} \mathbf{X},$$

$J_{\mathbf{Y} \to \mathbf{X}} = \det \mathbf{I}_{(n,m)} = (-1)^{\frac{1}{4}m(m-1)n(n-1)}$, and $|J_{\mathbf{Y} \to \mathbf{X}}| = 1$.

Example 6 Consider the transformation $\mathbf{Y}_{m \times n} = \mathbf{A}_{m \times n}\mathbf{X}'_{m \times n}\mathbf{B}_{n \times n}$, where $\mathbf{A}$ and $\mathbf{B}$ are nonsingular. Setting $\mathbf{W} = \mathbf{X}'$ and $\mathbf{Y} = \mathbf{AWB}$, we have, from Example 1 above and the chain rule (17.21),

$$J_{\mathbf{Y} \to \mathbf{X}} = J_{\mathbf{Y} \to \mathbf{W}} \, J_{\mathbf{W} \to \mathbf{X}} = (\det \mathbf{B})^m (\det \mathbf{A})^n (-1)^{\frac{1}{4}m(m-1)n(n-1)}.$$

In practice, we are more interested in absolute values so that we do not need to distinguish between $\mathbf{X}$ and $\mathbf{X}'$ in linear transformations like the above, as $|J_{\mathbf{W} \to \mathbf{X}}| = |J_{\mathbf{X}' \to \mathbf{X}}| = 1$ and $|J_{\mathbf{Y} \to \mathbf{X}}| = |J_{\mathbf{Y} \to \mathbf{W}}|$.

Example 7 Suppose we know the Jacobian for the transformation $\mathbf{Y} = \mathbf{AX}$, and we want to find the Jacobian for $\mathbf{Y} = \mathbf{XA}$. Taking transposes, we obtain $\mathbf{Y}' = \mathbf{A}'\mathbf{X}'$ or $\mathbf{W} = \mathbf{A}'\mathbf{Z}$. Then

$$
\begin{aligned}
J_{\mathbf{Y} \to \mathbf{X}} &= J_{\mathbf{Y} \to \mathbf{Y}'} \, J_{\mathbf{Y}' \to \mathbf{X}'} \, J_{\mathbf{X}' \to \mathbf{X}} \\
&= J_{\mathbf{Y} \to \mathbf{Y}'} \, J_{\mathbf{W} \to \mathbf{Z}} \, J_{\mathbf{X}' \to \mathbf{X}}
\end{aligned}
$$

and, by Example 6, $|J_{\mathbf{Y} \to \mathbf{X}}| = |J_{\mathbf{W} \to \mathbf{Z}}|$. Hence the absolute value of the Jacobian for $\mathbf{Y} = \mathbf{XA}$ can be obtained from the one for $\mathbf{Y} = \mathbf{AX}$ by simply replacing $\mathbf{A}$ by $\mathbf{A}'$. Example 7 is, of course, a special case of Example 6, but the method is instructive.

18.3.5   Patterned Matrices and L-Structures

Sometimes the matrices involved are "patterned" or structured in some way such as symmetric or triangular. We are therefore interested in this case where $\operatorname{vec} \mathbf{X}$ (where $\mathbf{X}$ is $m \times n$) will be in a linear subspace $\mathcal{D}_s$ of $\mathbb{R}^{mn}$. For example, if $\mathbf{X}$ is $n \times n$ and lower-triangular, then $\operatorname{vec} \mathbf{X}$ will contain zeros in a certain pattern.

Magnus [1988] proposed a method based on linear structures, a linear structure being the set of all real matrices of a specified order, say $m \times n$, that satisfy a set of linear restrictions. He gives the following definition.

Definition 18.2. Let $\mathcal{D}_s$ be an s-dimensional subspace of $\mathbb{R}^{mn}$ and let $\boldsymbol{\delta}_1, \boldsymbol{\delta}_2, \ldots,$ $\boldsymbol{\delta}_s$ be a set of basis vectors for $\mathcal{D}_s$. Then the $mn \times s$ matrix

$$\boldsymbol{\Delta}_{mn \times s} = (\boldsymbol{\delta}_1, \boldsymbol{\delta}_2, \ldots, \boldsymbol{\delta}_s)$$

is called a basis matrix for $\mathcal{D}_s$, and the collection of real $m \times n$ matrices

$$L(\boldsymbol{\Delta}_{mn \times s}) = \{\mathbf{X} : \mathbf{X} \in \mathbb{R}^{m \times n}, \operatorname{vec} \mathbf{X} \in \mathcal{D}_s\}$$

is called a *linear structure (L-structure)*, s is called the *dimension* of the L-structure, and $m \times n$ is called the *order* of the L-structure. Here $\boldsymbol{\Delta}$ is not unique, but ,once defined, there exists a unique $s \times 1$ vector $\boldsymbol{\phi}(\mathbf{X})$ such that $\operatorname{vec} \mathbf{X} = \boldsymbol{\Delta} \boldsymbol{\phi}(\mathbf{X})$ or $\boldsymbol{\phi}(\mathbf{X}) = \boldsymbol{\Delta}^+ \operatorname{vec} \mathbf{X}$, where $\boldsymbol{\Delta}^+$ is the Moore–Penrose inverse. Typically, $\boldsymbol{\phi}(\mathbf{X})$ is the vector containing the "free" elements of $\mathbf{X}$ in an appropriate order so that $\boldsymbol{\Delta}$ is then unique.

For example, if $\mathbf{X}$ is a real symmetric $n \times n$ matrix, then among its n^2 elements x_{ij} there exist $\frac{1}{2}n(n-1)$ linear relationships of the form $x_{ij} = x_{ji}$ $(i < j)$. Here $s = n^2 - \frac{1}{2}n(n-1) = \frac{1}{2}n(n+1)$. In particular, for $n = 2$, $\mathcal{D}_3 \subset \mathbb{R}^4$ and a basis matrix for $\mathcal{D}_3$ is the 4×3 matrix

$$\boldsymbol{\Delta}_{4 \times 3} = \begin{pmatrix} 1 & 0 & 0 \\ 0 & 1 & 0 \\ 0 & 1 & 0 \\ 0 & 0 & 1 \end{pmatrix}.$$

Thus if $(a, b, c)' \in \mathbb{R}^3$, then

$$\boldsymbol{\Delta}_{4 \times 3} \begin{pmatrix} a \\ b \\ c \end{pmatrix} = \begin{pmatrix} a \\ b \\ b \\ c \end{pmatrix} = \operatorname{vec} \begin{pmatrix} a & b \\ b & c \end{pmatrix} = \operatorname{vec} \mathbf{X}.$$

We note that

$$\boldsymbol{\phi}(\mathbf{X}) = (a, b, c)' = \operatorname{vech} \mathbf{X}, \quad \text{and} \quad \boldsymbol{\Delta}_{4 \times 3} = \mathbf{G}_2,$$

the so-called duplication matrix. A general theory for finding such basis matrices is given by Kollo and von Rosen [2005: section 1.3.6]. We now give a key result.

18.4. Suppose $\mathbf{Y} = \mathbf{F}(\mathbf{X})$ is a one-to-one function representing a relationship between s variables x_{ij} and s variables y_{ij}, where $\mathbf{Y} \in L(\boldsymbol{\Delta}_2)$ for every $\mathbf{X} \in L(\boldsymbol{\Delta}_1)$ with the dimensions of $L(\boldsymbol{\Delta}_1)$ and $L(\boldsymbol{\Delta}_2)$ both equal to s. Then, from Magnus [1988: 34],

$$J_{\mathbf{Y} \to \mathbf{X}} = \det \left[\boldsymbol{\Delta}_2^+ \frac{\partial \operatorname{vec} \mathbf{Y}}{(\partial \operatorname{vec} \mathbf{X})'} \boldsymbol{\Delta}_1 \right],$$

where $\frac{\partial \operatorname{vec} \mathbf{Y}}{(\partial \operatorname{vec} \mathbf{X})'}$ is calculated ignoring the *a priori* knowledge about the L-structures.

Henderson and Searle [1979: 74–76] discussed the same idea, but from a slightly different perspective. They define $\text{vecp}_{\mathbf{z}}(\mathbf{Z})$ as the vector of the distinct elements with pattern $\text{p}_{\mathbf{z}}$, where $\mathbf{Z} = \mathbf{X}$ or $\mathbf{Y}$ (they use $\mathbf{X}_1$ and $\mathbf{X}_2$). Then

$$\text{vecp}_{\mathbf{z}}(\mathbf{Z}) = \mathbf{P}_{\mathbf{z}}\text{vec}\,\mathbf{Z} \text{ and vec}\,\mathbf{Z} = \mathbf{Q}_{\mathbf{z}}\text{vecp}_{\mathbf{z}}(\mathbf{Z}) = \mathbf{Q}_{\mathbf{z}}\mathbf{P}_{\mathbf{z}}\text{vec}\,\mathbf{Z},$$

where $\mathbf{P}_{\mathbf{z}}$ and $\mathbf{Q}_{\mathbf{z}}$ correspond to $\mathbf{H}$ and $\mathbf{G}$ of Section 11.5.1. In particular, $\mathbf{P}_{\mathbf{z}}\mathbf{Q}_{\mathbf{z}} = \mathbf{I}$, $\mathbf{Q}_{\mathbf{z}}$ has full column rank, and $\mathbf{P}_{\mathbf{z}} = (\mathbf{Q}'_{\mathbf{z}}\mathbf{Q}_{\mathbf{z}})^{-1}\mathbf{Q}'_{\mathbf{z}}$ is one possible choice for $\mathbf{P}_{\mathbf{z}}$. Finally,

$$J_{\mathbf{X}\to\mathbf{Y}} = \det\left[\mathbf{P}_{\mathbf{x}}\frac{\partial\,\text{vec}\,\mathbf{X}}{\partial(\text{vec}\,\mathbf{Y})'}\mathbf{Q}_{\mathbf{y}}\right].$$

Derivatives for patterned matrices are also discussed by Nel [1980: section 6]. Kollo and von Rosen [2005: 135–149] develop derivatives for structured matrices, but use a derivative notation $\mathbf{Y}\otimes\partial/\partial\mathbf{X}$.

We are now going to systematically list Jacobians. If the order of the variables is not well defined, $|J|$ will be quoted instead of J.

18.4   VECTOR TRANSFORMATIONS

The following transformations between $n\times 1$ vectors are one-to-one.

18.5. If $\mathbf{y} = \mathbf{A}\mathbf{x}$, where $\mathbf{A}$ is nonsingular, then from (17.28) we have

$$|J_{\mathbf{y}\leftarrow\mathbf{x}}| = |\det\mathbf{A}|.$$

When $\mathbf{A} = a\mathbf{I}_n$, $|J_{\mathbf{y}\leftarrow\mathbf{x}}| = |a|^n$.

18.6. (Symmetric Functions) Let $y_i = x_{(i\times)}$, $i = 1, 2, \ldots, n$, where $x_{(i\times)}$ is the so-called *(elementary) symmetric function* representing the sum of all the products of x_j taken i at a time. Thus $y_1 = x_1 + \cdots + x_n$, $y_2 = x_1x_2 + x_1x_3 + \cdots + x_{n-1}x_n$, and $y_n = x_1x_2\cdots x_n$. Then for each $x_j > 0$,

$$|J_{\mathbf{y}\leftarrow\mathbf{x}}| = \prod_{i=1}^{n-1}\prod_{j=i+1}^{n}|x_i - x_j|.$$

18.7. If

$$
\begin{aligned}
y_1 &= x_1 + x_2 + \cdots + x_n,\\
y_2 &= x_1^2 + x_2^2 + \cdots + x_n^2,\\
&\vdots \qquad \vdots\\
y_{n-1} &= x_1^{n-1} + x_2^{n-1} + \cdots + x_n^{n-1},\\
y_n &= x_1x_2\cdots x_n,
\end{aligned}
$$

then, for each $x_j > 0$,

$$|J_{\mathbf{y}\to\mathbf{x}}| = (n-1), \prod_{i=1}^{n-1}\prod_{j=i+1}^{n}|x_i - x_j|.$$

18.8. (Polar Coordinates)

(a) Consider the transformation

$$
\begin{aligned}
x_1 &= r\sin\theta_1\sin\theta_2\cdots\sin\theta_{n-2}\sin\theta_{n-1},\\
x_2 &= r\sin\theta_1\sin\theta_2\cdots\sin\theta_{n-2}\cos\theta_{n-1},\\
x_3 &= r\sin\theta_1\sin\theta_2\cdots\cos\theta_{n-2},\\
&\ \ \vdots \qquad \vdots\\
x_{n-1} &= r\sin\theta_1\cos\theta_2,\\
x_n &= r\cos\theta_1,
\end{aligned}
$$

where $r > 0$, $0 < \theta_i \le \pi$ $(i = 1, 2, \ldots, n-2)$, and $0 < \theta_{n-1} \le 2\pi$. Then, if $\boldsymbol{\theta} = (\theta_1, \theta_2, \ldots, \theta_n)'$, we have

$$
|J_{\mathbf{x}\,\to\,r,\boldsymbol{\theta}}| = r^{n-1}|(\sin\theta_1)^{n-2}(\sin\theta_2)^{n-3}\cdots\sin\theta_{n-2}|
$$

(b) If we reverse the order of the x_i and replace θ_i by $\frac{\pi}{2} - \theta_i$ in (a), we get

$$
\begin{aligned}
x_1 &= r\sin\theta_1,\\
x_j &= r\cos\theta_1\cos\theta_2\cdots\cos\theta_{j-1}\sin\theta_j, \quad j = 2, 3, \ldots, n-1,\\
x_n &= r\cos\theta_1\cos\theta_2\cdots\cos\theta_{n-1},
\end{aligned}
$$

where $r > 0$, $-\frac{\pi}{2} < \theta_i \le \frac{\pi}{2}$ $(i = 1, \ldots, n-2)$, and $-\pi < \theta_{n-1} \le \pi$. Then

$$
|J_{\mathbf{x}\,\to\,r,\boldsymbol{\theta}}| = r^{n-1}|(\cos\theta_1)^{n-2}(\cos\theta_2)^{n-3}\cdots\cos\theta_{n-2}|.
$$

Proofs. Section 18.4.

18.6. Mathai [1997: 43]

18.7. Mathai [1997: 45].

18.8a. Mathai [1997: 45] and Muirhead [1982: 55].

18.8b. Mathai [1997: 45].

18.5 JACOBIANS FOR COMPLEX VECTORS AND MATRICES

We demonstrate the meaning of a Jacobian for complex variables using a simple example taken from Mathai [1997: 175–176]. Let $\mathbf{y} = \mathbf{y}_1 + i\mathbf{y}_2$ and $\mathbf{x} = \mathbf{x}_1 + i\mathbf{x}_2$, where the $\mathbf{x}_i$ and $\mathbf{y}_i$ are all real $n \times n$ vectors. Consider the transformation $\mathbf{y} = \boldsymbol{\Lambda}\mathbf{x}$ where $\mathbf{A}$ is real. Then $\mathbf{y}_i = \mathbf{A}\mathbf{x}_i$ for $i = 1, 2$, and

$$
\begin{pmatrix} \mathbf{y}_1 \\ \mathbf{y}_2 \end{pmatrix} = \begin{pmatrix} \mathbf{A} & \mathbf{0} \\ \mathbf{0} & \mathbf{A} \end{pmatrix} \begin{pmatrix} \mathbf{x}_1 \\ \mathbf{x}_2 \end{pmatrix} = \mathbf{B}\mathbf{x}.
$$

We define the Jacobian of the transformation to be $J_{\mathbf{y}_1,\mathbf{y}_2\to\mathbf{x}_1,\mathbf{x}_2}$. From (18.5) this is $\det\mathbf{B} = (\det\mathbf{A})^2$ $(= |\det\mathbf{A}|^2$, say$)$.

If $\mathbf{A}$ is complex and $\mathbf{A} = \mathbf{A}_1 + i\mathbf{A}_2$, then

$$
\begin{aligned}
\mathbf{y} &= \mathbf{y}_1 + i\mathbf{y}_2 = (\mathbf{\Lambda}_1 + i\mathbf{A}_2)(\mathbf{x}_1 + i\mathbf{x}_2) \\
&= (\mathbf{A}_1\mathbf{x}_1 - \mathbf{A}_2\mathbf{x}_2) + i(\mathbf{A}_1\mathbf{x}_2 + \mathbf{A}_2\mathbf{x}_1),
\end{aligned}
$$

and we have $\mathbf{y}_1 = \mathbf{A}_1\mathbf{x}_1 - \mathbf{A}_2\mathbf{x}_2$ and $\mathbf{y}_2 = \mathbf{A}_1\mathbf{x}_2 + \mathbf{A}_2\mathbf{x}_1$. Then $\partial\mathbf{y}_1/\partial\mathbf{x}_1' = \mathbf{A}_1$, $\partial\mathbf{y}_1/\partial\mathbf{x}_2' = -\mathbf{A}_2$, $\partial\mathbf{y}_2/\partial\mathbf{x}_1' = \mathbf{A}_2$, and $\partial\mathbf{y}_2/\partial\mathbf{x}_2' = \mathbf{A}_1$. Hence, from Section 5.1.2,

$$
|\mathbf{J}_{\mathbf{y}_1,\mathbf{y}_2 \to \mathbf{x}_1,\mathbf{x}_2}| = \left| \det \begin{pmatrix} \mathbf{A}_1 & -\mathbf{A}_2 \\ \mathbf{A}_2 & \mathbf{A}_1 \end{pmatrix} \right| = (|\det \mathbf{A}|)^2 = |\det(\mathbf{A}\mathbf{A}^*)|.
$$

Thus the above equation is true for both the real and complex cases.

When vectors are replaced by matrices, the expression $\mathbf{J}_{\mathbf{Y}_1,\mathbf{Y}_2 \to \mathbf{X}_1,\mathbf{X}_2}$ denotes the Jacobian of the transformation, where $\mathbf{Y}_1$ and $\mathbf{Y}_2$ are written as functions of $\mathbf{X}_1$ and $\mathbf{X}_2$, or where $\mathbf{Y} = \mathbf{Y}_1 + i\mathbf{Y}_2$ is a function of $\mathbf{X} = \mathbf{X}_1 + i\mathbf{X}_2$, the elements of $\mathbf{X}$ being functionally independent. As we have seen from the above example, we can typically go from the real to the complex case by squaring absolute values of determinants or by replacing $|\det \mathbf{A}|$ by $|\det \mathbf{A}\mathbf{A}^*| = |\det \mathbf{A}|^2$. We shall also see below that a term like $|x_{ii}|$ for a real diagonal element x_{ii} remains the same for a complex element except that $|\cdot|$ now refers to the modulus of a complex number.

18.6   MATRICES WITH FUNCTIONALLY INDEPENDENT ELEMENTS

18.9. If $\mathbf{Y}_{m\times n} = a\mathbf{X}_{m\times n}$, then $J_{\mathbf{Y}\to\mathbf{X}} = a^{mn}$ and $|J_{\mathbf{Y}\to\mathbf{X}}| = |a|^{mn}$. For complex matrices the latter Jacobian becomes $|a|^{2mn}$.

18.10. If $\mathbf{Y}_{m\times n} = \mathbf{A}_{m\times m}\mathbf{X}_{m\times n}\mathbf{B}_{n\times n}$, where $\mathbf{A}$ and $\mathbf{B}$ are nonsingular then, from Example 1 above in Section 18.2,

$$
|J_{\mathbf{Y}\to\mathbf{X}}| = |\det \mathbf{B}|^m \cdot |\det \mathbf{A}|^n.
$$

The transformation is clearly one-to-one. In particular, if $\mathbf{y} = \mathbf{A}\mathbf{x}$, then

$$
|J_{\mathbf{y}\to\mathbf{x}}| = \left| \frac{\partial\mathbf{y}}{\partial\mathbf{x}'} \right| = |\det \mathbf{A}|.
$$

Other cases follow by setting $\mathbf{A}$ or $\mathbf{B}$ equal to the identity matrix.

If the matrices are complex, we find that we simply replace $\mathbf{A}$ by $\mathbf{A}\mathbf{A}^*$ and $\mathbf{B}$ by $\mathbf{B}\mathbf{B}^*$ in the above expressions.

18.11. Let $\mathbf{Y} = \mathbf{A}\mathbf{X}\mathbf{A}' \pm \mathbf{B}\mathbf{X}\mathbf{B}'$, where all the matrices are $n\times n$ and $\mathbf{A}\otimes\mathbf{A}\pm\mathbf{B}\otimes\mathbf{B}$ are nonsingular. Then $\mathrm{vec}\,\mathbf{Y} = (\mathbf{A}\otimes\mathbf{A}\pm\mathbf{B}\otimes\mathbf{B})\mathrm{vec}\,\mathbf{X}$, so that the transformation is one-to-one, and

$$
|J_{\mathbf{Y}\to\mathbf{X}}| = \prod_{i=1}^{n}\prod_{j=1}^{n} |\alpha_i\alpha_j \pm \beta_i\beta_j|,
$$

where the α_i and β_j are the respective eigenvalues of $\mathbf{A}$ and $\mathbf{B}$.

18.12. Let $\mathbf{Y} = \mathbf{X}^{-1}$, where $\mathbf{X}$ is $n\times n$ and nonsingular.

(a) From Example 2 in Section 18.2 above, $|J_{\mathbf{Y}\to\mathbf{X}}| = (|\det\mathbf{X}|)^{-2n}$.

(b) When $\mathbf{X}$ is complex, we replace $\mathbf{X}$ by $\mathbf{XX}^*$.

18.13. Let $\mathbf{Y} = (\det \mathbf{X})\mathbf{X}^{-1}$, where $\mathbf{X}$ is $n \times n$, and $\det \mathbf{X} > 0$ to ensure the transformation is one-to-one.

(a) (i) If $\mathbf{X}$ is real, $|J_{\mathbf{Y} \to \mathbf{X}}| = (n-1)(|\det \mathbf{X}|)^{n(n-2)}$.

(ii) If $\mathbf{X}$ is complex, then

$$|J_{\mathbf{Y}_1, \mathbf{Y}_2 \to \mathbf{X}_1, \mathbf{X}_2}| = (n-1)^2 |\det \mathbf{X}|^{2n(n-2)}.$$

(b) (i) If $\mathbf{X}$ is real and $\mathbf{Z} = \mathbf{Y}^{-1} = \mathbf{X}/\det \mathbf{X}$, then

$$|J_{\mathbf{Z} \to \mathbf{X}}| = (n-1)|\det \mathbf{X}|^{-n^2}.$$

This follows from $J_{\mathbf{Z} \to \mathbf{X}} = J_{\mathbf{Z} \to \mathbf{Y}} J_{\mathbf{Y} \to \mathbf{X}}$ and (18.12).

(ii) When $\mathbf{X}$ is complex, $|J_{\mathbf{Z}_1, \mathbf{Z}_2 \to \mathbf{X}_1, \mathbf{X}_2}| = (n-1)^2 |\det \mathbf{X}|^{-2n^2}$.

18.14. If $\mathbf{Y} = \mathbf{AX}^{-1}\mathbf{B}$ and all matrices are $n \times n$ and nonsingular, then the transformation is one-to-one and

$$|J_{\mathbf{Y} \to \mathbf{X}}| = |(\det \mathbf{B})^n (\det \mathbf{X})^{-2n} (\det \mathbf{A})^n|.$$

This can be proved from (18.12) using $\mathbf{Y} = \mathbf{AZB}$, $\mathbf{Z} = \mathbf{X}^{-1}$, and the chain rule.

18.15. Let $\mathbf{Y} = \mathbf{X}^k$, where $\mathbf{X}$ is $n \times n$ and nonsingular, and k is a positive integer.

(a) If $\mathbf{X}$ has nonzero, not necessarily distinct, real eigenvalues $\lambda_1, \lambda_2, \ldots, \lambda_n$, then

(i) $|J_{\mathbf{Y} \to \mathbf{X}}| = |\prod_{i=1}^{n} \prod_{j=1}^{n} \sum_{r=1}^{k} \lambda_i^{k-r} \lambda_j^{r-1}|$.

We note that the transformation is generally not one-to-one.

(ii) If the eigenvalues are distinct, an alternative expression is given by

$$
\begin{aligned}
|J_{\mathbf{Y} \to \mathbf{X}}| &= \left| k^n (\det \mathbf{X})^{k-1} \prod_{i=1}^{n} \prod_{j=i+1}^{n} \left(\frac{\lambda_i^k - \lambda_j^k}{\lambda_i - \lambda_j} \right)^2 \right| \\
&= \left| k^n (\det \mathbf{X})^{k-1} \prod_{i=1}^{n} \prod_{j \neq i}^{n} \left(\frac{\lambda_i^k - \lambda_j^k}{\lambda_i - \lambda_j} \right) \right|,
\end{aligned}
$$

which, by noting that $\det \mathbf{X} = \prod_i \lambda_i$, is readily shown to be the same as the expression in (i).

(b) Suppose $k = 2$, that is, $\mathbf{Y} = \mathbf{X}^2$.

(i) From (a),

$$|J_{\mathbf{Y} \to \mathbf{X}}| = \prod_{i=1}^{n} \prod_{j=1}^{n} |\lambda_i + \lambda_j|.$$

(ii) When $\mathbf{X}$ is complex and the eigenvalues are distinct, then

$$|J_{\mathbf{Y}_1, \mathbf{Y}_2 \to \mathbf{X}_1, \mathbf{X}_2}| = \prod_{i=1}^{n} \prod_{j=1}^{n} |\lambda_i + \lambda_j|^2.$$

Proofs. Section 18.6.

18.9. Deemer and Olkin [1951: 347] and Mathai [1997: 177–178, complex case].

18.10. Abadir and Magnus [2005: 373], Henderson and Searle [1979: 72], and Muirhead [1982: 58 [1997: 177]. The complex case is given by Mathai [1997: 177].

18.11. Mathai [1997: 75–77].

18.12a. Abadir and Magnus [2005: 373] and Mathai [1997: 54].

18.12b. Mathai [1997: 190].

18.13a(i). Abadir and Magnus [2005: 373] and Mathai [1997: 72].

18.13a(ii). Mathai [1997: 205].

18.14. Henderson and Searle [1979: 73] and Mathai [1997: 60–61].

18.15a(i). Henderson and Searle [1979: 73].

18.15a(ii). Mathai [1997: 98].

18.15b(i). Henderson and Searle [1979: 73].

18.15b(ii) Mathai [1997: 209].

18.7   SYMMETRIC AND HERMITIAN MATRICES

Let $\mathbf{X}$ and $\mathbf{Y}$ be $n \times n$ real symmetric matrices, unless otherwise stated. We note that if $\mathbf{X} = \mathbf{X}_1 + i\mathbf{X}_2$ is Hermitian, then $\mathbf{X}_1$ is real symmetric and $\mathbf{X}_2$ is real skew-symmetric.

18.16. The following transformations are one-to-one.

(a) If $\mathbf{Y} = a\mathbf{X}$, then $|J_{\mathbf{Y} \to \mathbf{X}}| = |a^{n(n+1)/2}|$.

(b) If $\mathbf{Y} = \mathbf{A}\mathbf{X}\mathbf{A}'$, where $\mathbf{A}$ is nonsingular, then $[|J_{\mathbf{Y} \to \mathbf{X}}| = (|\det \mathbf{A})|^{n+1}$.

(c) If $\mathbf{Y} = \mathbf{A}\mathbf{X}^{-1}\mathbf{A}'$, where $\mathbf{A}$ and $\mathbf{X}$ are nonsingular, then

$$|J_{\mathbf{Y} \to \mathbf{X}}| = |(\det \mathbf{A})^{n+1}(\det \mathbf{X})^{-(n+1)}|.$$

(d) If $\mathbf{X} = \mathbf{X}_1 + i\mathbf{X}_2$ is Hermitian, then $\mathbf{Y} = \mathbf{Y}_1 + i\mathbf{Y}_2 = \mathbf{A}\mathbf{X}\mathbf{A}^*$ is Hermitian and $|J_{\mathbf{Y}_1, \mathbf{Y}_2 \to \mathbf{X}_1, \mathbf{X}_2}| = |\det \mathbf{A}|^{2n}$.

18.17. Let $\mathbf{A}$ and $\mathbf{B}$ be real nonsingular matrices, and assume that $\mathbf{A} \otimes \mathbf{A} \pm \mathbf{B} \otimes \mathbf{B}$ are nonsingular.

(a) The transformation $\mathbf{Y} = \mathbf{A}\mathbf{X}\mathbf{A}' \pm \mathbf{B}\mathbf{X}\mathbf{B}'$ is one-to-one since, from (11.30),

$$\text{vech } \mathbf{Y} = [\mathbf{H}(\mathbf{A} \otimes \mathbf{A} \pm \mathbf{B} \otimes \mathbf{B})\mathbf{G}]\text{vech } \mathbf{X}.$$

The matrix in square brackets is nonsingular because, from (11.29c) and (11.29g), $\mathbf{H}$ has full row rank and $\mathbf{G}$ full column rank.

(i) If λ_i $(i = 1, 2, \ldots, n)$ are the eigenvalues of $\mathbf{AB}^{-1}$, then

$$|J_{\mathbf{X} \to \mathbf{Y}}| = |(\det \mathbf{B})^{n+1} \prod_{i=1}^{n} \prod_{j=i}^{n} (1 \pm \lambda_i \lambda_j)|.$$

(ii) Alternatively, if α_i and β_i $(i = 1, 2, \ldots, n)$ are the eigenvalues of $\mathbf{A}$ and $\mathbf{B}$, respectively, then

$$|J_{\mathbf{X} \to \mathbf{Y}}| = |\prod_{i=1}^{n} \prod_{j=i}^{n} (\alpha_i \alpha_j \pm \beta_i \beta_j)|.$$

When $\mathbf{B} = \mathbf{0}$, it can be shown that the above result reduces to (18.16b).

(iii) If $\mathbf{A}$ and $\mathbf{B}$ are lower-triangular with respective diagonal elements a_{ii} and b_{ii}, then

$$|J_{\mathbf{Y} \to \mathbf{X}}| = |\prod_{i=1}^{n} \prod_{j=i}^{n} (a_{ii} a_{jj} \pm b_{ii} b_{jj})|.$$

This is the same as (ii) as the diagonal elements of a triangular matrix are its eigenvalues.

(b) If $\mathbf{Y} = \mathbf{AXB'} + \mathbf{BXA'}$, then

$$|J_{\mathbf{X} \to \mathbf{Y}}| = |(\det \mathbf{B})^{n+1} \prod_{i=1}^{n} \prod_{j=i}^{n} (\lambda_i + \lambda_j)|,$$

where λ_i $(i = 1, 2, \ldots, n)$ are the eigenvalues of $\mathbf{AB}^{-1}$. We need $\mathbf{B} \otimes \mathbf{A} + \mathbf{A} \otimes \mathbf{B}$ to be nonsingular for the transformation to be one-to-one.

18.18. Suppose $\mathbf{Y} = \mathbf{X}^{-1}$, where $\mathbf{X}$ is nonsingular and symmetric.

(a) From (18.16c) with $\mathbf{A} = \mathbf{I}$, $|J_{\mathbf{Y} \to \mathbf{X}}| = |\det \mathbf{X}|^{-(n+1)}$.

(b) If $\mathbf{X} = \mathbf{X}_1 + i\mathbf{X}_2$ is Hermitian, then $\mathbf{Y} = \mathbf{Y}_1 + i\mathbf{Y}_2$ is also Hermitian and

$$|J_{\mathbf{Y}_1, \mathbf{Y}_2 \to \mathbf{X}_1, \mathbf{X}_2}| = |\det(\mathbf{XX}^*)|^{-n} = |\det \mathbf{X}|^{-2n}.$$

18.19. Let $\mathbf{Y} = (\det \mathbf{X})\mathbf{X}^{-1}$, where $\mathbf{X}$ is positive definite. (The latter condition is sufficient for the transformation to be one-to-one.)

(a) We have:

(i) $\det \mathbf{Y} = (\det \mathbf{X})^{n-1}$ and

$$|J_{\mathbf{Y} \to \mathbf{X}}| = (n-1)(\det \mathbf{X})^{(n+1)(n-2)/2}.$$

(ii) If $\mathbf{X} = \mathbf{X}_1 + i\mathbf{X}_2$ is Hermitian and positive definite, and $\mathbf{Y} = \mathbf{Y}_1 + i\mathbf{Y}_2$, then

$$|J_{\mathbf{Y}_1, \mathbf{Y}_2 \to \mathbf{X}_1, \mathbf{X}_2}| = (n-1)|\det \mathbf{X}|^{n(n-2)}.$$

(b) Suppose $\mathbf{Z} = \mathbf{Y}^{-1} = \mathbf{X}/\det \mathbf{X}$. Then

(i) $|J_{\mathbf{Z} \to \mathbf{X}}| = (n-1)|\det \mathbf{X}|^{-n(n+1)/2}$.

(ii) If $\mathbf{X} = \mathbf{X}_1 + i\mathbf{X}_2$ is Hermitian and positive definite, and $\mathbf{Z} = \mathbf{Z}_1 + i\mathbf{Z}_2$, then

$$|J_{\mathbf{Z}_1, \mathbf{Z}_2 \to \mathbf{X}_1, \mathbf{X}_2}| = (n-1)|\det \mathbf{X}|^{-n^2}.$$

18.20. Let $\mathbf{Y} = \mathbf{X}^k$, $k = 2, 3, \ldots$, and let λ_i ($i = 1, 2, \ldots, n$) be the eigenvalues of $\mathbf{X}$.

(a)   (i) $|J_{\mathbf{Y} \to \mathbf{X}}| = k^n |(\det \mathbf{X})^{k-1}| \prod_{i=1}^{n} \prod_{j=i+1}^{n} |\mu_{ij}|$,

where

$$\mu_{ij} = \begin{cases} (\lambda_i^k - \lambda_j^k)/(\lambda_i - \lambda_j), & \text{if } \lambda_i \neq \lambda_j, \\ k\lambda_i^{k-1}, & \text{if } \lambda_i = \lambda_j. \end{cases}$$

(ii) If the eigenvalues are distinct, then $k^n(\det \mathbf{X})^{k-1} = \prod_{i=1}^{n} k\lambda_i^{k-1}$ and

$$\begin{aligned} |J_{\mathbf{Y} \to \mathbf{X}}| &= |\prod_{i=1}^{n} \prod_{j=i+1}^{n} \left\{ \frac{\lambda_i^k - \lambda_j^k}{\lambda_i - \lambda_j} \right\} \prod_{i=1}^{n} k\lambda_i^{k-1}| \\ &= |\prod_{i=1}^{n} \prod_{j=i}^{n} (\lambda_i^{k-1} + \lambda_i^{k-2}\lambda_j + \ldots + \lambda_j^{k-1})| \\ &= |\prod_{i=1}^{n} \prod_{j=i}^{n} \sum_{r=1}^{k} \lambda_i^{k-r}\lambda_j^{r-1}|. \end{aligned}$$

(b) When $k = 2$ we have:

(i)

$$|J_{\mathbf{Y} \to \mathbf{X}}| = \prod_{i=1}^{n} \prod_{j=i}^{n} |\lambda_i + \lambda_j|.$$

The transformation $\mathbf{Y} = \mathbf{X}^2$ is generally not one-to-one.

(ii) When $\mathbf{X} = \mathbf{X}_1 + i\mathbf{X}_2$ is Hermitian, and $\mathbf{Y} = \mathbf{Y}_1 + i\mathbf{Y}_2$, then

$$|J_{\mathbf{Y}_1, \mathbf{Y}_2 \to \mathbf{X}_1, \mathbf{X}_2}| = 2^n |\det \mathbf{X}| \prod_{i=1}^{n} \prod_{j=i+1}^{n} |\lambda_i + \lambda_j|^2.$$

18.21. If $\mathbf{Y} = \mathbf{X} \mathbf{A} \mathbf{X}$, where $\mathbf{A}$ is symmetric, and λ_i ($i = 1, 2, \ldots, n$) are the eigenvalues of $\mathbf{X} \mathbf{A}$, then, since $\det(\mathbf{X} \mathbf{A}) = \prod_i \lambda_i$, we have

$$|J_{\mathbf{Y} \to \mathbf{X}}| = 2^n |\det(\mathbf{A}) \det(\mathbf{X}) \prod_{i=1}^{n} \prod_{j=i+1}^{n} (\lambda_i + \lambda_j)| = |\prod_{i=1}^{n} \prod_{j=i}^{n} (\lambda_i + \lambda_j)|.$$

If $\mathbf{A}$ and $\mathbf{X}$ are positive definite and the λ_i are such that $\lambda_i > \cdots \lambda_n > 0$, then the transformation is one-to-one.

Proofs. Section 18.7.

18.16a. Mathai [1997: 32].

18.16b. Abadir and Magnus [2005: 373], Magnus [1988: 128], and Mathai [1997: 32].

18.16c. Mathai [1997: 60].

18.17a(i). Magnus [1988: 128].

18.17a(ii). Mathai [1997: 75–77].

18.17a(iii). Magnus [1988: 128].

18.17b. Magnus [1988: 128].

18.18b. Mathai [1997: 190].

18.19a(i). Deemer and Olkin [1951: 357, theorem 4.4; they also give the Jacobian of $\mathbf{Y}^{-1}$ in corollary 4.4], Magnus [1988: 128], and Mathai [1997: 74].

18.19a(ii). Mathai [1997: 206].

18.19b(i). Mathai [1997: 75].

18.19b(ii). Mathai [1997: 206].

18.20a(i). Magnus [1988: 128].

18.20a(ii). Henderson and Searle [1979: 79] and Mathai [1997: 98].

18.20b(i). Mathai [1997: 66, 69].

18.20b(ii). Mathai [1997: 209].

18.21. Magnus [1988: 128] and Mathai [1997: 70].

18.8   SKEW-SYMMETRIC AND SKEW-HERMITIAN MATRICES

Let $\mathbf{X}$ and $\mathbf{Y}$ be $n \times n$ matrices with $\mathbf{X}$ real skew-symmetric, that is, $\mathbf{X}' = -\mathbf{X}$. Then, for the following transformations, $\mathbf{Y}$ is also skew-symmetric. Note that if $\mathbf{X} = \mathbf{X}_1 + i\mathbf{X}_2$ is skew-Hermitian, then $\mathbf{X}_1$ is real skew-symmetric and $\mathbf{X}_2$ is real symmetric.

18.22. If $\mathbf{Y} = a\mathbf{X}$, then $|J_{\mathbf{Y} \to \mathbf{X}}| = |a|^{n(n-1)/2}$.

18.23. Let $\mathbf{Y} = \mathbf{AXA}'$. Then the following hold.

(a) $|J_{\mathbf{Y} \to \mathbf{X}}| = (|\det \mathbf{A}|)^{n-1}$.

(b) If $\mathbf{X} = \mathbf{X}_1 + i\mathbf{X}_2$ is skew-Hermitian and $\mathbf{Y} = \mathbf{Y}_1 + i\mathbf{Y}_2 = \mathbf{AXA}^*$, then $\mathbf{Y}$ is skew-Hermitian and

$$|J_{\mathbf{Y} \to \mathbf{X}}| = |\det(\mathbf{AA}^*)|^n = |\det \mathbf{A}|^{2n},$$

which is the same as for the Hermitian case (cf. 18.16d).

18.24. Let $\mathbf{A}$ and $\mathbf{B}$ be nonsingular, and let λ_i $(i = 1, 2, \ldots, n)$ be the eigenvalues of $\mathbf{AB}^{-1}$.

(a) If $\mathbf{Y} = \mathbf{AXA}' \pm \mathbf{BXB}'$, then $|J_{\mathbf{X} \to \mathbf{Y}}| = |(\det \mathbf{B})^{n-1} \prod_{i=1}^{n} \prod_{j=i+1}^{n} (1 \pm \lambda_i \lambda_j)|$. If $\mathbf{A}$ and $\mathbf{B}$ are lower-triangular, then

$$|J_{\mathbf{Y} \to \mathbf{X}}| = |\prod_{i=1}^{n} \prod_{j=i+1}^{n} (a_{ii} a_{jj} \pm b_{ii} b_{jj})|.$$

The above transformations are one-to-one if $(\mathbf{A} \otimes \mathbf{A} \pm \mathbf{B} \otimes \mathbf{B})$ are nonsingular.

(b) If $\mathbf{Y} = \mathbf{AXB}' + \mathbf{BXA}'$, then

$$|J_{\mathbf{X} \to \mathbf{Y}}| = |(\det \mathbf{B})^{n-1} \prod_{i=1}^{n} \prod_{j=i+1}^{n} (\lambda_i + \lambda_j)|.$$

The above transformation is nonsingular if $(\mathbf{B} \otimes \mathbf{A} + \mathbf{A} \otimes \mathbf{B})$ is nonsingular.

18.25. Let $\mathbf{Y} = \mathbf{AX}^{-1}\mathbf{A}'$, where $\mathbf{A}$ is nonsingular.

(a) $|J_{\mathbf{Y} \to \mathbf{X}}| = |\det \mathbf{A}|^{n-1} |\det \mathbf{X}|^{-(n-1)}$.

(b) If $\mathbf{Y} = \mathbf{X}^{-1}$, we can set $\mathbf{A} = \mathbf{I}$ in (a).

  (i) $|J_{\mathbf{Y} \to \mathbf{X}}| = |\det \mathbf{X}|^{-(n-1)}$.

  (ii) If $\mathbf{X} = \mathbf{X} - 1 + i\mathbf{X}_2$ is skew-Hermitian and $\mathbf{Y} = \mathbf{Y}_1 + i\mathbf{Y}_2$, then $|J_{\mathbf{Y}_1, \mathbf{Y}_2 \to \mathbf{X}_1, \mathbf{X}_2}| = |\det(\mathbf{XX}^*)|^{-n} = |\det \mathbf{X}|^{-2n}$, the same as for the Hermitian case of (18.18b).

18.26. If $\mathbf{Y} = (\det \mathbf{X})\mathbf{X}^{-1}$, where $\det \mathbf{X} \neq 0$, then

$$|J_{\mathbf{Y} \to \mathbf{X}}| = (n-1)|\det \mathbf{X}|^{(n-1)(n-2)/2}.$$

18.27. Let $\mathbf{Y} = \mathbf{X}^k$, $k = 3, 5, \ldots$, and let λ_i $(i = 1, 2, \ldots, n)$ be the eigenvalues of $\mathbf{X}$. Then we have

$$|J_{\mathbf{Y} \to \mathbf{X}}| = \prod_{i=1}^{n} \prod_{j=i+1}^{n} \mu_{ij},$$

where

$$\mu_{ij} = \begin{cases} (\lambda_i^k + \lambda_j^k)/(\lambda_i + \lambda_j), & \text{if } \lambda_i \neq -\lambda_j, \\ k\lambda_i^{k-1}, & \text{if } \lambda_i = -\lambda_j. \end{cases}$$

If $\mathbf{Y} = \mathbf{X}^2$, then $\mathbf{Y}$ is symmetric and the transformation is not one-to-one.

18.28. If $\mathbf{Y} = \mathbf{XAX}$, where $\mathbf{A}$ is skew symmetric, then

$$|J_{\mathbf{Y} \to \mathbf{X}}| = |\prod_{i=1}^{n} \prod_{j=i+1}^{n} (\lambda_i + \lambda_j)|,$$

where λ_i $(i = 1, 2, \ldots, n)$ are the eigenvalues of $\mathbf{XA}$.

Proofs. Section 18.8.

18.22. Mathai [1997: 36].

18.23a. Deemer and Olkin [1951: 349], Magnus [1988: 135], Mathai [1997: 36], and Olkin and Sampson [1972: 263].

18.23b. Mathai [1997: 185].

18.24. Magnus [1988: 135].

18.25a. Mathai [1997: 60].

18.25b(ii). Mathai [1997: 190].

18.26. Magnus [1988: 136].

18.27. Magnus [1988: 136].

18.28. Magnus [1988: 135].

18.9 TRIANGULAR MATRICES

Any matrix with a "tilde"—for example, $\tilde{\mathbf{X}}$—will denote a $n \times n$ nonsingular lower-triangular matrix. Results for upper-triangular matrices can be obtained by taking transposes. In what follows we assume that the elements in the lower triangle of $\tilde{\mathbf{X}}$ are unconstrained (functionally independent). Also, the product of lower-triangular matrices is lower-triangular, and the inverse of a lower-triangular matrix is also lower-triangular.

18.9.1 Linear Transformations

18.29. Let $\tilde{\mathbf{Y}} = \tilde{\mathbf{P}}\tilde{\mathbf{X}}\tilde{\mathbf{Q}}$, where $\tilde{\mathbf{P}}$ and $\tilde{\mathbf{Q}}$ are lower-triangular and nonsingular.

(a) If the matrices are all real,

$$|J_{\tilde{\mathbf{Y}} \to \tilde{\mathbf{X}}}| = |\prod_{i=1}^{n} p_{ii}^{i} q_{ii}^{n-i+1}|.$$

We get special cases by setting $\tilde{\mathbf{P}} = \mathbf{I}_n$ or $\tilde{\mathbf{Q}} = \mathbf{I}_n$.

(b) If the matrices are all complex (i.e., $\tilde{\mathbf{Y}} = \tilde{\mathbf{Y}}_1 + i\tilde{\mathbf{Y}}_2$, etc.), then we have the following results.

(i) If $\tilde{\mathbf{Y}} = \tilde{\mathbf{P}}\tilde{\mathbf{X}}$, then $|J_{\tilde{\mathbf{Y}}_1, \tilde{\mathbf{Y}}_2 \to \tilde{\mathbf{X}}_1, \tilde{\mathbf{X}}_2}| = \prod_{i=1}^{n} |p_{ii}|^{2i}$
(or $\prod_{i=1}^{n} |p_{ii}|^{2i-1}$ if the p_{ii}'s and x_{ii}'s are real).

(ii) If $\tilde{\mathbf{Y}} = \tilde{\mathbf{X}}\tilde{\mathbf{Q}}$, then $|J_{\tilde{\mathbf{Y}}_1, \tilde{\mathbf{Y}}_2 \to \tilde{\mathbf{X}}_1, \tilde{\mathbf{X}}_2}| = \prod_{i=1}^{n} |q_{ii}|^{2(n-i+1)}$
(or $\prod_{i=1}^{n} |q_{ii}|^{2(n-i)+1}$ if the q_{ii}'s and x_{ii}'s are real).

(c) Given real matrices, if $\tilde{\mathbf{X}}$, and therefore $\tilde{\mathbf{Y}}$, has fixed diagonal elements, then

$$|J_{\tilde{\mathbf{Y}} \to \tilde{\mathbf{X}}}| = |\prod_{i=1}^{n} p_{ii}^{i-1} q_{ii}^{n-i}|.$$

(d) Given real matrices, if $\tilde{\mathbf{Y}} = a\tilde{\mathbf{X}}$, then

$$|J_{\tilde{\mathbf{Y}} \to \tilde{\mathbf{X}}}| = |a|^{n(n+1)/2}.$$

When the matrices and a are complex, we get $|a|^{n(n+1)}$.

18.30. (Upper-Triangular) If $\tilde{\mathbf{Z}} = \tilde{\mathbf{P}}'\tilde{\mathbf{X}}'\tilde{\mathbf{Q}}'$, where $\tilde{\mathbf{P}}$ and $\tilde{\mathbf{Q}}$ are nonsingular, then $\tilde{\mathbf{Z}}$ is upper-triangular.

(a) For real matrices,

$$|J_{\tilde{\mathbf{Z}} \to \tilde{\mathbf{X}}}| = |\prod_{i=1}^{n} q_{ii}^{i} p_{ii}^{n-i+1}|.$$

By interchanging $\tilde{\mathbf{P}}$ and $\tilde{\mathbf{Q}}$, taking the transpose, and noting that $|J_{\tilde{\mathbf{Z}} \to \tilde{\mathbf{X}}}| = |J_{\tilde{\mathbf{Z}} \to \tilde{\mathbf{X}}'}|$, we see that the above result is equivalent to (18.29a), but for upper-triangular matrices.

(b) The results for complex matrices are similar to those given in (18.29a) by transposing, and interchanging (i) and (ii).

18.31. Let $\tilde{\mathbf{Y}} = \tilde{\mathbf{P}}\tilde{\mathbf{X}}\tilde{\mathbf{Q}} + \tilde{\mathbf{R}}\tilde{\mathbf{X}}\tilde{\mathbf{S}}$, where the matrices are all real.

(a) $|J_{\tilde{\mathbf{Y}} \to \tilde{\mathbf{X}}}| = |\prod_{i=1}^{n} \prod_{j=1}^{i} (p_{ii}q_{jj} + r_{ii}s_{jj})|.$

The transformation is one-to-one if $\tilde{\mathbf{Q}}' \otimes \tilde{\mathbf{P}} + \tilde{\mathbf{S}}' \otimes \tilde{\mathbf{R}}$ is nonsingular. Also, if $\tilde{\mathbf{R}} = \mathbf{0}$,

$$\prod_{i=1}^{n} \prod_{j=1}^{i} (p_{ii}q_{jj}) = p_{11}(p_{22}q_{11}p_{22}q_{22})(p_{33}q_{11}p_{33}q_{22}p_{33}q_{33}) \cdots = \prod_{i=1}^{n} p_{ii}^{i} q_{ii}^{n-i+1},$$

as in (18.29a).

(b) If $\tilde{\mathbf{X}}$ has fixed diagonal elements, then

$$|J_{\tilde{\mathbf{Y}} \to \tilde{\mathbf{X}}}| = |\prod_{i=1}^{n} \prod_{j=1}^{i-1} (p_{ii}q_{jj} + r_{ii}s_{jj})|.$$

Proofs. Section 18.9.1.

18.29a. Magnus [1988: 131], Mathai [1997: 29], and Olkin and Sampson [1972: 264].

18.29b. Mathai [1997:179–180].

18.29c. Magnus [1988: 137].

18.29d. Mathai [1997: 179].

18.30a. Mathai [1997: 29].

18.30b. Mathai [1997: 180–181].

18.31a. Magnus [1988: 132].

18.31b. Magnus [1988: 137].

18.9.2   Nonlinear Transformations of X

All matrices are real, unless otherwise stated.

18.32. Let $\tilde{\mathbf{Y}} = \tilde{\mathbf{X}}\tilde{\mathbf{P}}\tilde{\mathbf{X}}$.

(a) $|J_{\tilde{\mathbf{Y}}\to\tilde{\mathbf{X}}}| = 2^n\,|(\det\tilde{\mathbf{P}})(\det\tilde{\mathbf{X}})\prod_{i=1}^{n}\prod_{j=i+1}^{n}(p_{ii}x_{ii} + p_{jj}x_{jj})|.$

(b) If $\tilde{\mathbf{X}}$ has fixed diagonal elements, then

$$|J_{\tilde{\mathbf{Y}}\to\tilde{\mathbf{X}}}| = |\prod_{i=1}^{n}\prod_{j=i+1}^{n}(p_{ii}x_{ii} + p_{jj}x_{jj})|.$$

18.33. Let $\tilde{\mathbf{Y}} = \tilde{\mathbf{X}}^{-1}$.

(a) We have

$$|J_{\tilde{\mathbf{Y}}\to\tilde{\mathbf{X}}}| = |\det\tilde{\mathbf{X}}|^{-(n+1)}| = |\prod_{i=1}^{n}x_{ii}^{-(n+1)}|.$$

(b) If $\tilde{\mathbf{X}}$ has fixed diagonal elements, then

$$|J_{\tilde{\mathbf{Y}}\to\tilde{\mathbf{X}}}| = |\det\tilde{\mathbf{X}}|^{-(n-1)} = |\prod_{i=1}^{n}x_{ii}^{n-1}|.$$

18.34. Let $\tilde{\mathbf{Y}} = (\det\tilde{\mathbf{X}})\tilde{\mathbf{X}}^{-1}$, where $\tilde{\mathbf{Y}}$ and $\tilde{\mathbf{X}}$ are both lower- or both upper-triangular matrices.

(a) Then $\det\tilde{\mathbf{Y}} = (\det\tilde{\mathbf{X}})^{n-1}$.

   (i)

$$|J_{\tilde{\mathbf{Y}}\to\tilde{\mathbf{X}}}| = (n-1)|\det\tilde{\mathbf{X}}|^{(n+1)(n-2)/2}.$$

Note that $(\det\tilde{\mathbf{X}})^{(n+1)(n-2)} > 0$ as $(n+1)(n-2)$ is divisible by 2, so we take the positive square root. For the transformation to be one-to-one, we assume $\det\tilde{\mathbf{X}} > 0$ (for example, $x_{ii} > 0$ for all i) so that $\det\tilde{\mathbf{Y}} > 0$, and define $\det\tilde{\mathbf{X}} = (\det\tilde{\mathbf{Y}})^{1/(n-1)}$, the $(n-1)$th positive root of $\det\tilde{\mathbf{Y}}$. We can then write

$$|J_{\tilde{\mathbf{Y}}\to\tilde{\mathbf{X}}}| = (n-1)\prod_{i=1}^{n}x_{ii}^{(n+1)(n-2)/2}.$$

Similar comments apply to (b) and (c) below.

(ii) When $\tilde{\mathbf{X}} = \tilde{\mathbf{X}}_1 + i\tilde{\mathbf{X}}_2$ is complex and $\tilde{\mathbf{Y}} = \tilde{\mathbf{Y}}_1 + i\tilde{\mathbf{Y}}_2$, then

$$|J_{\tilde{\mathbf{Y}}_1,\tilde{\mathbf{Y}}_2 \to \tilde{\mathbf{X}}_1,\tilde{\mathbf{X}}_2}| = (n-1)|\det \tilde{\mathbf{X}}|^{(n+1)(n-2)}.$$

(b) If $\tilde{\mathbf{X}}$ has fixed diagonal elements, then

$$|J_{\tilde{\mathbf{Y}} \to \tilde{\mathbf{X}}}| = (n-1)|(\det \tilde{\mathbf{X}})^{(n-1)(n-2)/2}|.$$

(c) If $\tilde{\mathbf{Z}} = \tilde{\mathbf{Y}}^{-1} = \tilde{\mathbf{X}}/\det \tilde{\mathbf{X}}$, then

(i) $|J_{\tilde{\mathbf{Z}} \to \tilde{\mathbf{X}}}| = (n-1)|(\det \tilde{\mathbf{X}})^{-n(n+1)/2}|.$

(ii) When $\tilde{\mathbf{X}} = \tilde{\mathbf{X}}_1 + i\tilde{\mathbf{X}}_2$ is complex and $\tilde{\mathbf{Z}} = \tilde{\mathbf{Z}}_1 + i\tilde{\mathbf{Z}}_2$,

$$|J_{\tilde{\mathbf{Z}}_1,\tilde{\mathbf{Z}}_2 \to \tilde{\mathbf{X}}_1,\tilde{\mathbf{X}}_2}| = (n-1)|\det \tilde{\mathbf{X}}|^{-n(n+1)}.$$

18.35. Let $\tilde{\mathbf{Y}} = \tilde{\mathbf{X}}^k$, $k = 2, 3, \ldots$.

(a) $|J_{\tilde{\mathbf{Y}} \to \tilde{\mathbf{X}}}| = k^n |(\det \tilde{\mathbf{X}})^{k-1} \prod_{i=1}^{n} \prod_{j=i+1}^{n} \mu_{ij}|$, where

$$\mu_{ij} = \begin{cases} (x_{ii}^k - x_{jj}^k)/(x_{ii} - x_{jj}), & \text{if } x_{ii} \neq x_{jj}, \\ kx_{ii}^{k-1}, & \text{if } x_{ii} = x_{jj}. \end{cases}$$

(b) If $\tilde{\mathbf{X}}$ has fixed diagonal elements, then

$$|J_{\tilde{\mathbf{Y}} \to \tilde{\mathbf{X}}}| = |\prod_{i=1}^{n} \prod_{j=i+1}^{n} \mu_{ij}|.$$

Proofs. Section 18.9.2.

18.32a. Magnus [1988: 132].

18.32b. Magnus [1988: 137].

18.33a. Magnus [1988: 132] and Olkin and Sampson [1972: 265].

18.33b. Magnus [1988: 137].

18.34a(i). Magnus [1988: 132] and Mathai [1997: 65].

18.34a(ii). Mathai [1997: 201].

18.34b. Magnus [1988: 137].

18.34c(i). Mathai [1997: 65].

18.34c(ii). Mathai [1997: 199].

18.35a. Magnus [1988: 132].

18.35b. Magnus [1988: 137].

18.9.3 Decompositions with One Skew-Symmetric Matrix

If $\mathbf{S}$ is an $n \times n$ skew-symmetric matrix, then $\mathbf{I}_n + \mathbf{S}$ is nonsingular (cf. 5.19). Also, $\mathbf{T} = 2(\mathbf{S} + \mathbf{I}_n)^{-1} - \mathbf{I}_n$ is orthogonal and represents a one-to-one transformation as $\mathbf{S} = 2(\mathbf{T} + \mathbf{I}_n)^{-1} - \mathbf{I}_n$. (This is a special case of the Caley transformation mentioned in Section 18.12.) Any nonsingular matrix $\mathbf{Y}$ can expressed in the form $\mathbf{Y} = \tilde{\mathbf{X}}\mathbf{T}$, where $\tilde{\mathbf{X}}$ is a nonsingular lower-triangular matrix. This representation is unique under two situations (Mathai [1997: 100]):(1) $x_{ii} > 0$ for $i = 1, 2, \ldots, n-1$, and (2) the elements of $\tilde{\mathbf{X}}$ are unrestricted, but the elements of $\mathbf{S}$ are restricted in some way. For example, the elements of the first row of $(\mathbf{S} + \mathbf{I}_n)^{-1}$, except the first, being of a specific sign such as all negative (Mathai [1997: 99]) or all positive (Deemer and Olkin [1951: 361]).

18.36. Assuming that the appropriate conditions above hold so that the representation $\mathbf{Y} = \tilde{\mathbf{X}}[2(\mathbf{S} + \mathbf{I}_n)^{-1} - \mathbf{I}_n] = \tilde{\mathbf{X}}(\mathbf{I}_n - \mathbf{S})(\mathbf{I}_n + \mathbf{S})^{-1}$ is unique (i.e., one-to-one), then

$$|J_{\mathbf{Y} \to \tilde{\mathbf{X}}, \mathbf{S}}| = 2^{n(n-1)/2} \left| \left\{ \prod_{i=1}^{n} x_{ii}^{n-i} \right\} \det(\mathbf{S} + \mathbf{I}_n)^{-(n-1)} \right|.$$

(Note that $\mathbf{Y}\mathbf{Y}' = \tilde{\mathbf{X}}\tilde{\mathbf{X}}'$.)

18.37. If $\mathbf{Y} = \mathbf{T}\tilde{\mathbf{X}}\mathbf{T}' = [2(\mathbf{S} + \mathbf{I}_n)^{-1} - \mathbf{I}_n]\tilde{\mathbf{X}}[2(\mathbf{S} + \mathbf{I}_n)^{-1} - \mathbf{I}_n]'$, then

$$|J_{\mathbf{Y} \to \tilde{\mathbf{X}}, \mathbf{S}}| = 2^{n(n-1)/2} \left| \det(\mathbf{I}_n + \mathbf{S})^{-(n-1)} \prod_{i=1}^{n} \prod_{j=i+1}^{n} (x_{ii} - x_{jj}) \right|.$$

If $\mathbf{Y} = \mathbf{T}\mathbf{V}\mathbf{T}'$, where $\mathbf{V}$ is upper-triangular, then transposing we have $\mathbf{Z} = \mathbf{T}\tilde{\mathbf{X}}\mathbf{T}'$, where $\mathbf{Z} = \mathbf{Y}'$ and $\tilde{\mathbf{X}} = \mathbf{V}'$. This implies that the absolute value of the Jacobian is the same as above.

18.38. Let $\mathbf{Y} = \mathbf{T}\mathbf{D}_{\mathbf{x}}\mathbf{T}' = [2(\mathbf{S} + \mathbf{I}_n)^{-1} - \mathbf{I}_n]\mathbf{D}_{\mathbf{x}}[2(\mathbf{S} + \mathbf{I}_n)^{-1} - \mathbf{I}_n]'$ be a symmetric matrix, where $\mathbf{x} = (x_1, x_2, \ldots, x_n)'$ with $x_1 > x_2 > \ldots > x_n$, and $\mathbf{D}_{\mathbf{x}} = \mathrm{diag}(\mathbf{x})$. If the elements of the first row of $(\mathbf{S} + \mathbf{I}_n)^{-1}$ except the first are of a specific sign, then

$$|J_{\mathbf{X} \to \mathbf{S}, \mathbf{x}}| = 2^{n(n-1)/2} \left| \det(\mathbf{I}_n + \mathbf{S})^{-(n-1)} \prod_{i=1}^{n} \prod_{j=i+1}^{n} (x_i - x_j) \right|.$$

The decomposition of $\mathbf{Y}$ is unique if we add the condition that $\mathbf{Y}$ should not belong to a set of symmetric matrices that constitutes a set of measure zero in the $n(n+1)/2$-dimensional space. Olkin and Sampson [1972: 273] also quote the result, but their constant term is incorrectly inverted.

Proofs. Section 18.9.3.

18.36. Deemer and Olkin [1951: 358] and Mathai [1997: 101].

18.37. Mathai [1997: 109] and Olkin [1953: 46].

18.38. Deemer and Olkin [1951: 360–361] and Mathai [1997: 106]

18.9.4   Symmetric Y

18.39. Let $\mathbf{Y}$ be symmetric, and let $\tilde{\mathbf{P}} = (p_{ij})$ and $\tilde{\mathbf{X}}$ be $n \times n$ nonsingular lower-triangular matrices. Conditions for the following transformations to be one-to-one can be found using vec and vech as in (18.11) and (18.17). For example, in (a) below, $\text{vec}\,\mathbf{Y} = (\mathbf{I}_{n^2} + \mathbf{I}_{(n,n)})\text{vec}\,\mathbf{X}$, where $\mathbf{I}_{(n,n)}$ is the vec-permutation (commutation) matrix. We also have $\text{vech}\,\mathbf{Y} = \mathbf{H}_n\text{vec}\,\mathbf{Y}$ and $\text{vec}\,\mathbf{X} = \mathbf{G}_n\text{vech}\,\mathbf{X}$. The following matrices are all real, unless otherwise stated.

(a) If $\mathbf{Y} = \tilde{\mathbf{X}} + \tilde{\mathbf{X}}'$, then $|J_{\mathbf{Y} \to \tilde{\mathbf{X}}}| = 2^n$.

(b) Suppose $\tilde{\mathbf{X}} = \tilde{\mathbf{X}}_1 + i\tilde{\mathbf{X}}_2$ is complex.

    (i) The Jacobian is either 2^{2n} or 2^n if the x_{ii}'s are real.

    (ii) If $\mathbf{Y} = \tilde{\mathbf{X}} + \tilde{\mathbf{X}}^*$, then $\mathbf{Y}$ is now Hermitian and the transformation is no longer one-to-one unless the x_{ii} are real. In the latter case the Jacobian is 2^n.

(c) If $\mathbf{Y} = \tilde{\mathbf{X}}\tilde{\mathbf{P}} + \tilde{\mathbf{P}}'\tilde{\mathbf{X}}'$ and $\tilde{\mathbf{P}}$ is nonsingular, then

$$|J_{\mathbf{Y} \to \tilde{\mathbf{X}}}| = 2^n \prod_{i=1}^{n} |p_{ii}|^{n-i+1}.$$

(d) If $\mathbf{Y} = \tilde{\mathbf{P}}\tilde{\mathbf{X}} + \tilde{\mathbf{X}}'\tilde{\mathbf{P}}'$, then

$$|J_{\mathbf{Y} \to \tilde{\mathbf{X}}}| = 2^n \,|\prod_{i=1}^{n} |p_{ii}|^{i}.$$

(e) If $\mathbf{Y} = \tilde{\mathbf{X}}'\tilde{\mathbf{P}} + \tilde{\mathbf{P}}'\tilde{\mathbf{X}}$, then

$$|J_{\mathbf{Y} \to \tilde{\mathbf{X}}}| = 2^n \prod_{i=1}^{n} |p_{ii}|^{i}.$$

(f) If $\mathbf{Y} = \tilde{\mathbf{P}}\tilde{\mathbf{X}}' + \tilde{\mathbf{X}}\tilde{\mathbf{P}}'$, then

$$|J_{\mathbf{Y} \to \mathbf{X}}| = 2^n \prod_{i=1}^{n} |p_{ii}|^{n-i+1}.$$

18.40. Let $\mathbf{Y}$ be symmetric, and let $\tilde{\mathbf{P}}$, $\tilde{\mathbf{Q}}$, $\tilde{\mathbf{R}}$, and $\tilde{\mathbf{X}}$ be nonsingular lower-triangular matrices. Conditions for the following transformations to be one-to-one can be found using vec and vech.

(a) If $\mathbf{Y} = \tilde{\mathbf{Q}}'\tilde{\mathbf{X}}\tilde{\mathbf{P}} + \tilde{\mathbf{P}}'\tilde{\mathbf{X}}'\tilde{\mathbf{Q}}$, then

$$|J_{\mathbf{Y} \to \tilde{\mathbf{X}}}| = 2^n \,|\det \tilde{\mathbf{P}}(\det \tilde{\mathbf{Q}})^n \prod_{i=1}^{n-1} \det \mathbf{C}_{(i)}|,$$

where $\mathbf{C}_{(i)}$ is the ith $(i \times i)$ leading principal minor of $\mathbf{C} = \tilde{\mathbf{P}}\tilde{\mathbf{Q}}^{-1}$.

(b) If $\mathbf{Y} = \tilde{\mathbf{R}}'\tilde{\mathbf{Q}}\tilde{\mathbf{X}}\tilde{\mathbf{P}} + \tilde{\mathbf{P}}'\tilde{\mathbf{X}}'\tilde{\mathbf{Q}}'\tilde{\mathbf{R}}$, then

$$|J_{\mathbf{Y}\to\tilde{\mathbf{X}}}| = 2^n |\prod_{i=1}^{n} (q_{ii}r_{ii})^i p_{ii}^{n-i+1}|.$$

(c) If $\mathbf{Y} = \tilde{\mathbf{R}}\tilde{\mathbf{X}}\tilde{\mathbf{P}}\tilde{\mathbf{Q}}' + \tilde{\mathbf{Q}}\tilde{\mathbf{P}}'\tilde{\mathbf{X}}'\tilde{\mathbf{R}}'$, then

$$|J_{\mathbf{Y}\to\tilde{\mathbf{X}}}| = 2^n |\prod_{i=1}^{n} (p_{ii}q_{ii})^{n-i+1} r_{ii}^i|.$$

Proofs. Section 18.9.4.

18.39a. Mathai [1997: 28].

18.39b(i). Mathai [1997: 179].

18.39b(ii). Mathai [1997: 181].

18.39c. Mathai [1997: 32].

18.39d. Mathai [1997: 32].

18.39e. Mathai [1997: 37] and Olkin [1953: 43].

18.39f. Deemer and Olkin [1951: 349] and Mathai [1997: 37].

18.40. Magnus [1988: 133].

18.9.5 Positive Definite Y

18.41. Let $\mathbf{Y}$ be positive definite, and let $\tilde{\mathbf{X}}$ be lower-triangular and nonsingular, with positive diagonal elements (which implies the existence of a unique Cholesky decomposition).

(a) If $\mathbf{Y} = \tilde{\mathbf{X}}'\tilde{\mathbf{X}}$, then

$$|J_{\mathbf{Y}\to\tilde{\mathbf{X}}}| = 2^n \prod_{i=1}^{n} x_{ii}^i.$$

(b) If $\mathbf{Y} = \tilde{\mathbf{X}}\tilde{\mathbf{X}}'$, then

$$|J_{\mathbf{Y}\to\tilde{\mathbf{X}}}| = 2^n \prod_{i=1}^{n} x_{ii}^{n-i+1}.$$

(c) Let $\mathbf{Y} = \tilde{\mathbf{X}}\tilde{\mathbf{X}}'$, where $y_{ii} = 1$, and $\sum_{j=1}^{i} x_{ij}^2 = 1$ $(i = 1, 2, \ldots, n)$. Then

$$|J_{\mathbf{Y}\to\tilde{\mathbf{X}}}| = \prod_{i=1}^{n} x_{ii}^{n-i}.$$

Proofs. Section 18.9.5.

18.41a. Magnus [1988: 134], Mathai [1997: 56], and Olkin [1953: 43].

18.41b. Deemer and Olkin [1951: 349], Magnus [1988: 133], and Mathai [1997: 56].

18.41c. Olkin [1953: 44, theorem 5].

18.9.6   Hermitian Positive Definite Y

18.42. Suppose $\mathbf{Y}$ is an Hermitian positive definite matrix. Let $\tilde{\mathbf{X}} = \tilde{\mathbf{X}}_1 + i\tilde{\mathbf{X}}_2$ be a complex lower-triangular matrix, with $\tilde{\mathbf{X}}_i$ a real lower-triangular matrix $(i = 1, 2)$, and let $\mathbf{V}$ be a complex upper-triangular matrix. Both $\tilde{\mathbf{X}}$ and $\mathbf{V}$ are assumed to have real positive diagonal elements, which implies the existence of the unique Cholesky decompositions given below.

(a) If $\mathbf{Y} = \tilde{\mathbf{X}}\tilde{\mathbf{X}}^*$, then $|J_{\mathbf{Y}_1, \mathbf{Y}_2 \to \tilde{\mathbf{X}}_1, \tilde{\mathbf{X}}_2}| = 2^n \prod_{i=1}^{n} x_{ii}^{2(n-i)+1}$.

If, in addition, $y_{ii} = 1$ and $\sum_{k=1}^{i} x_{ik}\bar{x}_{ik} = \sum_{k=1}^{i} |x_{ik}|^2 = 1$ for $i = 1, 2, \ldots, n$, then

$$|J_{\mathbf{Y}_1, \mathbf{Y}_2 \to \tilde{\mathbf{X}}_1, \tilde{\mathbf{X}}_2}| = \prod_{i=1}^{n} x_{ii}^{2(n-i)}, \quad x_{11} = 1.$$

(b) If $\mathbf{Y} = \mathbf{V}\mathbf{V}^*$, then $|J_{\mathbf{Y}_1, \mathbf{Y}_2 \to \tilde{\mathbf{X}}_1, \tilde{\mathbf{X}}_2}| = 2^n \prod_{i=1}^{n} v_{ii}^{2(i-1)+1}$.

If, in addition, $y_{ii} = 1$ and $\sum_{k=i}^{n} |v_{ik}|^2 = 1$ for $i = 1, 2, \ldots, n$, then

$$|J_{\mathbf{Y}_1, \mathbf{Y}_2 \to \tilde{\mathbf{X}}_1, \tilde{\mathbf{X}}_2}| = \prod_{i=1}^{n} x_{ii}^{2(i-1)}, \quad x_{nn} = 1.$$

Proofs. Section 18.9.6.

18.42. Mathai [1997: 187, 194].

18.9.7   Skew-Symmetric Y

18.43. Let $\tilde{\mathbf{X}}$ be lower-triangular with fixed diagonal elements, let $\tilde{\mathbf{P}}$, $\tilde{\mathbf{Q}}$, and $\tilde{\mathbf{R}}$ be lower-triangular, and let $\mathbf{Y}$ be skew-symmetric.

(a) If $\mathbf{Y} = \mathbf{B}'\tilde{\mathbf{X}}\mathbf{A} - \mathbf{A}'\tilde{\mathbf{X}}'\mathbf{B}$, then

$$|J_{\mathbf{Y} \to \tilde{\mathbf{X}}}| = |\det \mathbf{B}^{n-1} \prod_{i=1}^{n-1} \det \mathbf{C}_{(i)}|,$$

where $\mathbf{C}_{(i)}$ is the ith $(i \times i)$ leading principal minor of $\mathbf{C} = \mathbf{A}\mathbf{B}^{-1}$.

(b) If $\mathbf{Y} = \tilde{\mathbf{R}}'\tilde{\mathbf{Q}}\tilde{\mathbf{X}}\tilde{\mathbf{P}} - \tilde{\mathbf{P}}'\tilde{\mathbf{X}}'\tilde{\mathbf{Q}}'\tilde{\mathbf{R}}$, then

$$|J_{\mathbf{Y} \to \tilde{\mathbf{X}}}| = |\prod_{i=1}^{n} (q_{ii}r_{ii})^{i-1} p_{ii}^{n-i}|.$$

(c) If $\mathbf{Y} = \tilde{\mathbf{R}}\tilde{\mathbf{X}}\tilde{\mathbf{P}}\tilde{\mathbf{Q}}' - \tilde{\mathbf{Q}}\tilde{\mathbf{P}}'\tilde{\mathbf{X}}'\tilde{\mathbf{R}}'$, then

$$|J_{\mathbf{Y} \to \tilde{\mathbf{X}}}| = |\prod_{i=1}^{n} (p_{ii}q_{ii})^{n-i} r_{ii}^{i-1}|.$$

Proofs. Section 18.9.7.

18.43. Magnus [1988: 138].

18.9.8 LU Decomposition

18.44. Let $\mathbf{Y}$ be any $n \times n$ nonsingular matrix. Then, from Section 16.4, $\mathbf{Y}$ can be expressed uniquely as a lower-triangular $\tilde{\mathbf{L}}$ with unit diagonal elements and an upper-triangular $\mathbf{U}$, that is, $\mathbf{Y} = \tilde{\mathbf{L}}\mathbf{U}$ (or $\mathbf{Y} = \mathbf{U}\tilde{\mathbf{L}}$, with different $\mathbf{U}$ and $\tilde{\mathbf{L}}$). In general, if $\tilde{\mathbf{L}}$ is lower-triangular with fixed diagonal elements (not necessarily equal to unity), then we have the following.

(a) If $\mathbf{Y} = \tilde{\mathbf{L}}\mathbf{U}$, then $|J_{\mathbf{Y} \to \tilde{\mathbf{L}}, \mathbf{U}}| = |\prod_{i=1}^{n} l_{ii}^{n-i+1} u_{ii}^{n-i}|$.

(b) If $\mathbf{Y} = \mathbf{U}\tilde{\mathbf{L}}$, then $|J_{\mathbf{Y} \to \tilde{\mathbf{L}}, \mathbf{U}}| = |\prod_{i=1}^{n} l_{ii}^{i} u_{ii}^{i-1}|$.

Proofs. Section 18.9.8.

18.44. Magnus [1988: 139]. The case when $\tilde{\mathbf{L}}$ has unit diagonal elements is proved by Mathai [1997: 92].

18.10 DECOMPOSITIONS INVOLVING DIAGONAL MATRICES

18.10.1 Square Matrices

In what follows, we define $\mathbf{D}_{\mathbf{w}} = \operatorname{diag} \mathbf{w} = \operatorname{diag}(w_1, w_2, \ldots, w_n)$, where the w_i are functionally independent, distinct, and nonzero. We can also use $|J_{\mathbf{Y} \to \mathbf{Y}'}| = 1$ for any matrix $\mathbf{Y}$. Unless stated otherwise, all matrices are real. When all the matrices are complex, we assume that $\mathbf{X} = \mathbf{X}_1 + i\mathbf{X}_2$, $\mathbf{w} = \mathbf{w}_1 + i\mathbf{w}_2$, and $\mathbf{Y} = \mathbf{Y}_1 + i\mathbf{Y}_2$, where the $\mathbf{X}_i$, $\mathbf{w}_i$, and $\mathbf{Y}_i$ are all real.

18.45. Let $\mathbf{X}$ and $\mathbf{Y}$ be $n \times n$ matrices with $\mathbf{X}$ having unit diagonal elements.

(a) Let $\mathbf{Y} = \mathbf{D}_{\mathbf{w}}\mathbf{X}$.

 (i) $|J_{\mathbf{Y} \to \mathbf{w}, \mathbf{X}}| = \prod_{i=1}^{n} |w_i|^{n-1}$. Since $x_{ij} = y_{ij}/y_{ii}$ for $i \neq j$, and $w_i = y_{ii}$ for all i, the transformation is one-to-one.

 (ii) For complex matrices, $|J_{\mathbf{Y}_1, \mathbf{Y}_2 \to \mathbf{w}_1, \mathbf{w}_2, \mathbf{X}_1, \mathbf{X}_2}| = \prod_{i=1}^{n} |w_i|^{2(n-1)}$. The result is still true if the y_{ii} and w_i are all real and positive.

(b) If $\mathbf{Y} = \mathbf{X}\mathbf{D}_{\mathbf{w}}$, we get the same answers as for (a).

(c) Let $\mathbf{Y} = \mathbf{D}_{\mathbf{w}}\mathbf{X}\mathbf{D}_{\mathbf{w}}$, with $y_{ii} > 0$ and $w_i > 0$ for $i = 1, 2, \ldots, n$.

 (i) $|J_{\mathbf{Y} \to \mathbf{w}, \mathbf{X}}| = 2^n \prod_{i=1}^{n} w_i^{2n-1}$. The transformation is one-to-one as $w_i = \sqrt{y_{ii}}$ for all i and $x_{ij} = y_{ij}/(\sqrt{y_{ii}}\sqrt{y_{jj}})$ for $i \neq j$.

 (ii) For complex $\mathbf{X}$ and $\mathbf{D}_{\mathbf{w}}$,

$$|J_{\mathbf{Y}_1, \mathbf{Y}_2 \to \mathbf{w}_1, \mathbf{w}_2, \mathbf{X}_1, \mathbf{X}_2}| = 2^{2n} \prod_{i=1}^{n} |w_i|^{2(2n-1)}.$$

 If y_{ii} and w_i are real and positive, the corresponding value is $2^n \prod_{i=1}^{n} w_i^{4n-3}$ when $\mathbf{Y} = \mathbf{D}_{\mathbf{w}}\mathbf{X}\mathbf{D}_{\mathbf{w}}$, and $2^n \prod_{i=1}^{n} w_i^{2n-1}$ when $\mathbf{Y} = \mathbf{D}_{\mathbf{w}}\mathbf{X}\mathbf{D}_{\mathbf{w}}^*$ with Hermitian $\mathbf{X}$. The transformation is no longer one-to-one.

Proofs. Section 18.10.1.

 18.45a(i). Mathai [1997: 86].

 18.45a(ii). Mathai [1997: 215].

 18.45c(i). Mathai [1997: 86].

 18.45c(ii). Mathai [1997: 215, 217].

18.10.2 One Triangular Matrix

18.46. Suppose $\tilde{\mathbf{Y}}$ is a lower-triangular matrix and $\tilde{\mathbf{X}}$ is lower-triangular with fixed diagonal elements (for example, unit elements). All matrices are real, unless otherwise stated. Note that $\det \tilde{\mathbf{X}} = \prod_{i=1}^{n} x_{ii}$.

(a) Let $\tilde{\mathbf{Y}} = \mathbf{D_w}\tilde{\mathbf{X}}$.

 (i) We have

$$|J_{\tilde{\mathbf{Y}} \to \mathbf{w}, \tilde{\mathbf{X}}}| = |\det \tilde{\mathbf{X}}| \prod_{i=1}^{n} |w_i|^{i-1}.$$

 (ii) If the matrices are complex and $\tilde{\mathbf{X}} = \tilde{\mathbf{X}}_1 + i\tilde{\mathbf{X}}_2$ has unit diagonal elements, then

$$|J_{\tilde{\mathbf{Y}}_1, \tilde{\mathbf{Y}}_2 \to \mathbf{w}_1, \mathbf{w}_2, \tilde{\mathbf{X}}_1, \tilde{\mathbf{X}}_2}| = \prod_{i=1}^{n} |w_i|^{2(i-1)}.$$

(b) Let $\tilde{\mathbf{Y}} = \tilde{\mathbf{X}}\mathbf{D_w}$.

 (i) We have

$$|J_{\tilde{\mathbf{Y}} \to \mathbf{w}, \tilde{\mathbf{X}}}| = |\det \tilde{\mathbf{X}}| \prod_{i=1}^{n} |w_i|^{n-i}.$$

 (ii) If the matrices are complex and $\tilde{\mathbf{X}}$ has unit diagonal elements, then

$$|J_{\tilde{\mathbf{Y}}_1, \tilde{\mathbf{Y}}_2 \to \mathbf{w}_1, \mathbf{w}_2, \tilde{\mathbf{X}}_1, \tilde{\mathbf{X}}_2}| = \prod_{i=1}^{n} |w_i|^{2(n-i)}.$$

The above transformations are one-to-one. The results for upper-triangular matrices are obtained by taking the transposes of the above. For example, if $\mathbf{Z} = \mathbf{U}\mathbf{D_w} = \tilde{\mathbf{X}}'\mathbf{D_w}$, where $\mathbf{Z}$ and $\mathbf{U}$ are upper-triangular, then $\mathbf{Z}' = \mathbf{D_w}\tilde{\mathbf{X}}$ and the Jacobian is given by (a). If $\mathbf{Z} = \mathbf{D_w}\mathbf{U}$, then the Jacobian is given by (b).

(c) Let $\mathbf{Y} = \tilde{\mathbf{X}}\mathbf{D_w}\tilde{\mathbf{X}}'$, then

 (i) $|J_{\mathbf{Y} \to \mathbf{w}, \tilde{\mathbf{X}}}| = (\det \tilde{\mathbf{X}})^2 \prod_{i=1}^{n} |(w_i x_{ii})|^{n-i}.$

 (ii) When $\tilde{\mathbf{X}}$ has unit diagonal elements, the Jacobian becomes $\prod_{i=1}^{n} |w_i|^{n-i}$. This case is also given below.

18.47. Let $\tilde{\mathbf{X}}$ be lower-triangular with *unit* diagonal elements, and suppose $y_{ii} > 0$ and $w_i > 0$ for $i = 1, 2, \ldots, n$.

(a) If $\tilde{\mathbf{X}}$ is real, we have the following Jacobians.

    (i) If $\mathbf{Y} = \tilde{\mathbf{X}}\mathbf{D_w}\tilde{\mathbf{X}}'$, then $|J_{\mathbf{Y}\to\mathbf{w},\tilde{\mathbf{X}}}| = \prod_{i=1}^{n} w_i^{n-i}$.

    (ii) If $\mathbf{Y} = \tilde{\mathbf{X}}'\mathbf{D_w}\tilde{\mathbf{X}}$, then $|J_{\mathbf{Y}\to\mathbf{w},\tilde{\mathbf{X}}}| = \prod_{i=1}^{n} w_i^{i-1}$.

    (iii) If $\mathbf{Y} = \mathbf{D_w}^{1/2}\tilde{\mathbf{X}}\tilde{\mathbf{X}}'\mathbf{D_w}^{1/2}$, then $|J_{\mathbf{Y}\to\mathbf{w},\tilde{\mathbf{X}}}| = \prod_{i=1}^{n} w_i^{(n-1)/2}$.

    (iv) If $\mathbf{Y} = \mathbf{D_w}^{1/2}\tilde{\mathbf{X}}'\tilde{\mathbf{X}}\mathbf{D_w}^{1/2}$, then $|J_{\mathbf{Y}\to\mathbf{w},\tilde{\mathbf{X}}}| = \prod_{i=1}^{n} w_i^{(n-1)/2}$, that is, the same as (iii).

The above transformations are one-to-one as we can express $\mathbf{Y}$ (which is positive definite) in either the form $\tilde{\mathbf{Z}}\tilde{\mathbf{Z}}'$ or $\tilde{\mathbf{Z}}'\tilde{\mathbf{Z}}$, where $\tilde{\mathbf{Z}}$ is lower-triangular with positive diagonal elements, that is, a unique Cholesky decomposition (Section 16.5).

To get the results for upper-triangular matrices, we simply write $\mathbf{U} = \tilde{\mathbf{X}}'$. For example, if $\mathbf{Y} = \mathbf{U}\mathbf{D_w}\mathbf{U}'$, the Jacobian is given by (ii) as $|J_{\tilde{\mathbf{X}}\to\tilde{\mathbf{X}}'}| = 1$. Similarly, if $\mathbf{Y} = \mathbf{U}'\mathbf{D_w}\mathbf{U}$, the Jacobian is given by (i).

(b) Suppose $\tilde{\mathbf{X}} = \tilde{\mathbf{X}}_1 + i\tilde{\mathbf{X}}_2$ is complex, but $w_i > 0$ for all i. Then:

    (i) If $\mathbf{Y} = \tilde{\mathbf{X}}\mathbf{D_w}\tilde{\mathbf{X}}^*$, $|J_{\mathbf{Y}\to\mathbf{w},\tilde{\mathbf{X}}_1,\tilde{\mathbf{X}}_2}| = \prod_{i=1}^{n} w_i^{2(n-i)}$.

    (ii) If $\mathbf{Y} = \mathbf{D_w}^{1/2}\tilde{\mathbf{X}}\tilde{\mathbf{X}}^*\mathbf{D_w}^{1/2}$, $|J_{\mathbf{Y}\to\mathbf{w},\tilde{\mathbf{X}}_1,\tilde{\mathbf{X}}_2}| = \prod_{i=1}^{n} w_i^{n-1}$.

(c) Suppose $\mathbf{U} = \mathbf{U}_1 + i\mathbf{U}_2$ is upper-triangular and complex with unit diagonal elements, and the w_i are real, positive, and distinct for all i.

    (i) If $\mathbf{Y} = \mathbf{U}\mathbf{D_w}\mathbf{U}^*$, $|J_{\mathbf{Y}\to\mathbf{w},\mathbf{U}_1,\mathbf{U}_2}| = \prod_{i=1}^{n} w_i^{2(i-1)}$.

    (ii) If $\mathbf{Y} = \mathbf{D_w}^{1/2}\mathbf{U}\mathbf{U}^*\mathbf{D_w}^{1/2}$, $|J_{\mathbf{Y}\to\mathbf{w},\mathbf{U}_1,\mathbf{U}_2}| = \prod_{i=1}^{n} w_i^{n-1}$.

18.48. Let $\tilde{\mathbf{Y}}$ and $\tilde{\mathbf{X}}$ be real nonsingular lower-triangular matrices with distinct, positive diagonal elements, and let $\mathbf{Y}$ be a positive definite matrix. Also, suppose that $\sum_{j=1}^{i} x_{ij}^2 = 1$ ($i = 1, 2, \ldots, n$), and the w_i ($i = 1, 2, \ldots, n$) are distinct and positive.

(a) If $\tilde{\mathbf{Y}} = \mathbf{D_w}\tilde{\mathbf{X}}$, then
$$|J_{\tilde{\mathbf{Y}}\to\mathbf{w},\tilde{\mathbf{X}}}| = \prod_{i=1}^{n} w_i^{i-1}x_{ii}^{-1}.$$
(For the transformation to be one-to-one we require the condition $w_i > 0$ for all i. To see this we set $x_{ii} = (1 - \sum_{j=1}^{i-1} x_{ij}^2)^{1/2}$, which leads to $w_i = (\sum_{j=1}^{i} y_{ij}^2)^{1/2}$ and $x_{ij} = y_{ij}/w_i$ for $i > j$, so that the inverse function exists.)

(b) If $\tilde{\mathbf{Y}} = \tilde{\mathbf{X}}\mathbf{D_w}$, then $|J_{\tilde{\mathbf{Y}}\to\mathbf{w},\tilde{\mathbf{X}}}| = \prod_{i=1}^{n} w_i^{n-i}x_{ii}^{-1}$.

(c) If $\tilde{\mathbf{Y}} = \mathbf{D_w}^{1/2}\tilde{\mathbf{X}}$, then $|J_{\tilde{\mathbf{Y}}\to\mathbf{w},\tilde{\mathbf{X}}}| = 2^{-n}\prod_{i=1}^{n}(w_i^{1/2})^{i-2}x_{ii}^{-1}$. (This follows from (a) by replacing w_i by $w_i^{1/2}$ and noting that $dw_i^{1/2} = \frac{1}{2}w_i^{-1/2}dw_i$.)

(d) If $\tilde{\mathbf{Y}} = \tilde{\mathbf{X}}\mathbf{D_w}^{1/2}$, then $|J_{\tilde{\mathbf{Y}}\to\mathbf{w},\tilde{\mathbf{X}}}| = 2^{-n}\prod_{i=1}^{n}(w_i^{1/2})^{n-i-1}x_{ii}^{-1}$.

(e) If $\mathbf{Y} = \mathbf{D}_{\mathbf{w}}^{1/2}\tilde{\mathbf{X}}\tilde{\mathbf{X}}'\mathbf{D}_{\mathbf{w}}^{1/2}$, then $|J_{\mathbf{Y}\to\mathbf{w},\tilde{\mathbf{X}}}| = \prod_{i=1}^{n} w_i^{(n-1)/2} x_{ii}^{n-i}$.

(f) If $\mathbf{Y} = \mathbf{D}_{\mathbf{w}}^{1/2}\tilde{\mathbf{X}}'\tilde{\mathbf{X}}\mathbf{D}_{\mathbf{w}}^{1/2}$ then, $J_{\mathbf{Y}\to\mathbf{w},\tilde{\mathbf{X}}} = \prod_{i-1}^{n} w_i^{(n-1)/2} x_{ii}^{i-1}$.

(g) If $\mathbf{Y} = \tilde{\mathbf{X}}'\mathbf{D}_{\mathbf{w}}\tilde{\mathbf{X}}$, then $|J_{\mathbf{Y}\to\mathbf{w},\tilde{\mathbf{X}}}| = \prod_{i=1}^{n} (w_i x_{ii})^{i-1}$.

(h) If $\mathbf{Y} = \tilde{\mathbf{X}}\mathbf{D}_{\mathbf{w}}\tilde{\mathbf{X}}'$, then $|J_{\mathbf{Y}\to\mathbf{w},\tilde{\mathbf{X}}}| = \prod_{i=1}^{n} (w_i x_{ii})^{n-i}$.

(i) If $\tilde{\mathbf{X}}$ is complex and $\sum_{j=1}^{i} |x_{ij}|^2 = 1$, we replace w_i by w_i^2 in the Jacobians for (a) and (b).

Proofs. Section 18.10.2.

18.46a(i). Magnus [1988: 141].

18.46a(ii). Mathai [1997: 211].

18.46b(i). Magnus [1988: 141].

18.46b(ii). Mathai [1997: 211].

18.46c(i). Magnus [1988: 141].

18.46c(ii). Olkin [1953: 45].

18.47a. Mathai [1997: 85].

18.47b. Mathai [1997: 214].

18.47c. Mathai [1997: 215].

18.48. Mathai [1997: 88–90, 218 for (i)].

18.10.3   Symmetric and Skew-Symmetric Matrices

18.49. Let $\mathbf{Y} = \mathbf{D}_{\mathbf{w}}\mathbf{X}\mathbf{D}_{\mathbf{w}}$, where $\mathbf{X}$ is symmetric with $x_{ii} = 1$ $(i = 1, 2, \ldots, n)$ and $\mathbf{D}_{\mathbf{w}} = \mathrm{diag}(\mathbf{w})$, with the w_i being distinct. Then

$$|J_{\mathbf{Y}\to\mathbf{w},\mathbf{X}}| = 2^n \prod_{i=1}^{n} |w_i|^n.$$

If we also add $w_i > 0$ for each i, then $w_i = \sqrt{y_{ii}}$ and $x_{ij} = y_{ij}/(\sqrt{y_{ii}}\sqrt{y_{jj}})$ so that the transformation is one-to-one as the inverse function exists.

18.50. Let $\mathbf{Y} = \mathbf{D}_{\mathbf{x}} + \mathbf{D}_{\mathbf{p}}\mathbf{X} - \mathbf{X}\mathbf{D}_{\mathbf{p}}$, where $\mathbf{X}$ is skew symmetric and $\mathbf{D}_{\mathbf{p}} = \mathrm{diag}(p_1, p_2, \ldots, p_n)$ is fixed. Then

$$|J_{\mathbf{Y}\to\mathbf{X},\mathbf{x}}| = \prod_{i=1}^{n} \prod_{j=i+1}^{n} |p_i - p_j|.$$

Proofs. Section 18.10.3.

18.49. Mathai [1997: 86] and Olkin [1953: 44].

18.50. Magnus [1988: 140].

18.11   POSITIVE DEFINITE MATRICES

18.51. If $\mathbf{Y} = (\det \mathbf{X})\mathbf{X}^{-1}$, where $\mathbf{X}$ (and therefore $\mathbf{Y}$), is positive definite, then

$$|J_{\mathbf{Y}\to\mathbf{X}}| = (n-1)|(\det \mathbf{X})^{(n+1)(n-2)/2}|.$$

18.52. Let $\mathbf{Y} = \mathbf{XAX}$, where all three matrices are positive definite. Then

$$|J_{\mathbf{Y}\to\mathbf{X}}| = \prod_{i=1}^{n}\prod_{j=i}^{n}(\lambda_i + \lambda_j),$$

where the λ_i are the eigenvalues of $\mathbf{XA}$, and are positive. An important special case is when $\mathbf{A} = \mathbf{I}_n$ and $\mathbf{X}$ is the positive definite square root of $\mathbf{Y}$.

18.53. If $\mathbf{Y} = \mathbf{X}^k$, where $k = 2, 3, \ldots$, and $\mathbf{X}$ is positive definite with distinct eigenvalues λ_i, then $\mathbf{Y}$ is positive definite and $|J_{\mathbf{Y}\to\mathbf{X}}|$ is given by (18.20). Let $(\mathbf{Y})^{1/k}$ denote the kth positive definite root of $\mathbf{Y}$, and μ_j $(j = 1, 2, \ldots, n)$ the eigenvalues of $\mathbf{Y}$. Then the transformation $\mathbf{Y} = \mathbf{X}^k$ is one-to-one, and $|J_{\mathbf{Y}\to\mathbf{X}}|$ can be expressed in terms of $\mathbf{Y}$ by noting that $\det \mathbf{X} = (\det \mathbf{Y})^{1/k}$ and $\lambda_j = \mu_j^{1/k}$.

Proofs. Section 18.11.

18.51. Deemer and Olkin [1951: 357] and Mathai [1997: 74].

18.52. Olkin and Sampson [1972: 269].

18.53. Mathai [1997: 98, example 2.5].

18.12   CALEY TRANSFORMATION

18.54. In this section we consider a particular transformation, called the *Caley transformation*, for nonsymmetric, symmetric, and triangular matrices, and their complex versions.

(a) Let $\mathbf{Y} = (\mathbf{A} + \mathbf{X})^{-1}(\mathbf{A} - \mathbf{X})\,[= 2(\mathbf{A} + \mathbf{X})^{-1}\mathbf{A} - \mathbf{I}_n]$, where the matrices are $n \times n$ and inverses exist so that the transformation is one-to-one.

  (i) For real matrices we have

$$|J_{\mathbf{Y}\to\mathbf{X}}| = 2^{n^2}(|\det \mathbf{A}|)^n(|\det(\mathbf{A} + \mathbf{X})|)^{-2n}.$$

  The same result holds for $\mathbf{Y} = (\mathbf{A} - \mathbf{X})(\mathbf{A} + \mathbf{X})^{-1}$.

  (ii) If the matrices are complex, we replace $\mathbf{A}$ by $\mathbf{AA}^*$, $\mathbf{A} + \mathbf{X}$ by $(\mathbf{A} + \mathbf{X})(\mathbf{A}^* + \mathbf{X}^*)$, and 2^{n^2} by 2^{2n^2} to get $|J_{\mathbf{Y}_1,\mathbf{Y}_2\to\mathbf{X}_1,\mathbf{X}_2}|$.

(b) Let $\mathbf{Y} = (\mathbf{I}_n + \mathbf{X})^{-1}(\mathbf{I}_n - \mathbf{X}) = 2(\mathbf{I}_n + \mathbf{X})^{-1} - \mathbf{I}_n$, where $\mathbf{X}$ and $\mathbf{Y}$ are symmetric.

  (i) $|J_{\mathbf{X}\to\mathbf{Y}}| = 2^{n(n+1)/2}|\det(\mathbf{I}_n + \mathbf{X})|^{-(n+1)}$.

(ii) If $\mathbf{X} = \mathbf{X}_1 + i\mathbf{X}_2$ is Hermitian and $\mathbf{Y} = \mathbf{Y}_1 + i\mathbf{Y}_2$, then

$$|J_{\mathbf{Y}_1,\mathbf{Y}_2 \to \mathbf{X}_1,\mathbf{X}_2}| = 2^{n^2} |\det\{(\mathbf{I}_n + \mathbf{X})(\mathbf{I}_n + \mathbf{X}^*)\}|^{-n}.$$

(c) Let $\tilde{\mathbf{Y}} = (\tilde{\mathbf{A}} + \tilde{\mathbf{X}})^{-1}(\tilde{\mathbf{A}} - \tilde{\mathbf{X}})$, where $\tilde{\mathbf{X}}$, $\tilde{\mathbf{A}}$, and $\tilde{\mathbf{Y}}$ are all lower-triangular, $\tilde{\mathbf{A}}$ and $\tilde{\mathbf{A}} + \tilde{\mathbf{X}}$ are nonsingular, and all matrices are real. Then

$$|J_{\tilde{\mathbf{X}} \to \tilde{\mathbf{Y}}}| = 2^{n(n+1)/2} |\det(\tilde{\mathbf{A}} + \tilde{\mathbf{X}})|^{-(n+1)} \prod_{i=1}^{n} |a_{ii}|^{n-i+1},$$

where $\det(\tilde{\mathbf{A}} + \tilde{\mathbf{X}}) = \prod_{i=1}^{n}(a_{ii} + x_{ii})$. When the matrices are upper-triangular, we see, by taking transposes, that the Jacobian is given by (d).

(d) Let $\tilde{\mathbf{Y}} = (\tilde{\mathbf{A}} - \tilde{\mathbf{X}})(\tilde{\mathbf{A}} + \tilde{\mathbf{X}})^{-1}$, where $\tilde{\mathbf{X}}$, $\tilde{\mathbf{A}}$, and $\tilde{\mathbf{Y}}$ are all lower-triangular, $\tilde{\mathbf{A}}$ and $\tilde{\mathbf{A}} + \tilde{\mathbf{X}}$ are nonsingular, and all matrices are real. Then

$$|J_{\tilde{\mathbf{X}} \to \tilde{\mathbf{Y}}}| = 2^{n(n+1)/2} |\det(\tilde{\mathbf{A}} + \tilde{\mathbf{Y}})|^{-(n+1)} \prod_{i=1}^{n} |a_{ii}|^{i}.$$

When the matrices are all upper-triangular, the equation then becomes $\tilde{\mathbf{Y}}' = (\tilde{\mathbf{A}}' - \tilde{\mathbf{X}}')(\tilde{\mathbf{A}}' + \tilde{\mathbf{X}})^{-1}$ so that, taking transposes, we get $\tilde{\mathbf{Y}} = (\tilde{\mathbf{A}} + \tilde{\mathbf{X}})^{-1}(\tilde{\mathbf{A}} - \tilde{\mathbf{X}})$, and the Jacobian is given by (c) above. We now look at the complex versions of (c) and (d).

18.55. Let $\tilde{\mathbf{X}}$, $\tilde{\mathbf{A}}$, and $\tilde{\mathbf{Y}}$ be complex lower-triangular matrices with $\tilde{\mathbf{A}}$ and $\tilde{\mathbf{A}} + \tilde{\mathbf{X}}$ nonsingular, and $\mathbf{Y} = \tilde{\mathbf{Y}}_1 + i\tilde{\mathbf{Y}}_2$ etc.

(a) Let $\tilde{\mathbf{Y}} = (\tilde{\mathbf{A}} + \tilde{\mathbf{X}})^{-1}(\tilde{\mathbf{A}} - \tilde{\mathbf{X}})$.

(i) If all the elements are complex

$$|J_{\tilde{\mathbf{Y}}_1,\tilde{\mathbf{Y}}_2 \to \tilde{\mathbf{X}}_1,\tilde{\mathbf{X}}_2}|$$
$$= 2^{n(n+1)} |\det\{(\tilde{\mathbf{A}} + \tilde{\mathbf{X}})(\tilde{\mathbf{A}} + \tilde{\mathbf{X}})^*\}|^{-(n+1)} \cdot \prod_{i=1}^{n} |a_{ii}|^{2(n-i+1)}.$$

(ii) If all the diagonal elements of $\tilde{\mathbf{X}}$ and $\tilde{\mathbf{A}}$ are real and the others complex, then

$$|J_{\tilde{\mathbf{Y}}_1,\tilde{\mathbf{Y}}_2 \to \tilde{\mathbf{X}}_1,\tilde{\mathbf{X}}_2}| = 2^{n^2} \prod_{i=1}^{n} |a_{ii} + x_{ii}|^{-2n} \cdot \prod_{i=1}^{n} |a_{ii}|^{2(n-i)+1}.$$

When the matrices are upper-triangular, we find, by taking transposes, that the Jacobians are given by (b) below.

(b) Let $\tilde{\mathbf{Y}} = (\tilde{\mathbf{A}} - \tilde{\mathbf{X}})(\tilde{\mathbf{A}} + \tilde{\mathbf{X}})^{-1}$. Then:

(i) If all the elements are complex,

$$|J_{\tilde{\mathbf{Y}}_1,\tilde{\mathbf{Y}}_2 \to \tilde{\mathbf{X}}_1,\tilde{\mathbf{X}}_2}|$$
$$= 2^{n(n+1)} |\det\{(\tilde{\mathbf{A}} + \tilde{\mathbf{X}})(\tilde{\mathbf{A}} + \tilde{\mathbf{X}})^*\}|^{-(n+1)} \cdot \prod_{i=1}^{n} |a_{ii}|^{2i}.$$

(ii) If the diagonal elements of $\tilde{\mathbf{X}}$ and $\tilde{\mathbf{A}}$ are real and the other elements complex, then

$$|J_{\tilde{\mathbf{Y}}_1,\tilde{\mathbf{Y}}_2 \to \tilde{\mathbf{X}}_1,\tilde{\mathbf{X}}_2}| = 2^{n^2} \prod_{i=1}^{n} |u_{ii} + x_{ii}|^{-2n} \cdot \prod_{i=1}^{n} |u_{ii}|^{2i-1}.$$

When the matrices are upper-triangular, we find, by taking transposes, that the Jacobians are given by (a) above.

Proofs. Section 18.12.

18.54a(i). Mathai [1997: 61] and Olkin [1953: 45].

18.54a(ii). Mathai [1997: 193].

18.54b(i). Mathai [1997: 61] and Olkin [1953: 45].

18.54b(ii). Mathai [1997: 193].

18.54c. Mathai [1997: 62–63] and Olkin [1953: 45].

18.54d. Mathai [1997: 62–63].

18.55. Mathai [1997: 195–196].

18.13 DIAGONALIZABLE MATRICES

18.56. Let $\mathbf{X}$ be a nonsingular diagonalizable matrix with real distinct eigenvalues $\lambda_1 > \lambda_2 > \ldots > \lambda_n > 0$, that is, there exists a nonsingular $\mathbf{R}$ such that $\mathbf{X} = \mathbf{R}\mathbf{D}_\lambda\mathbf{R}^{-1}$, where $\mathbf{D}_\lambda = \text{diag}(\lambda_1, \lambda_2, \ldots, \lambda_n)$. Let $\mathbf{Y} = \mathbf{F}(\mathbf{X})$, where $\mathbf{F}$ is such that $\mathbf{F}(\mathbf{X}) = \mathbf{R}\mathbf{D}_{f(\lambda)}\mathbf{R}^{-1}$, f is differentiable, and $\mathbf{D}_{f(\lambda)} = \text{diag}(f(\lambda_1), f(\lambda_2), \ldots, f(\lambda_n))$. Then, assuming $f(\lambda_i) - f(\lambda_j) \neq 0$ and $f'(\lambda_i) \neq 0$ for all $i \neq j$ $(i, j = 1, 2, \ldots, n)$,

(a) $|J_{\mathbf{Y} \to \mathbf{X}}| = |\prod_{i=1}^{n} \prod_{j=i+1}^{n} \left\{ \dfrac{f(\lambda_i) - f(\lambda_j)}{\lambda_i - \lambda_j} \right\}^2 \prod_{i=1}^{n} \dfrac{df(\lambda_i)}{d\lambda_i}|.$

(b) For example, if $\mathbf{Y} = \mathbf{X}^k$, k a positive integer, then with $f(\lambda) = \lambda^k$,

$$\begin{aligned}
\mathbf{X}^k &= (\mathbf{R}\mathbf{D}_\lambda\mathbf{R}^{-1})(\mathbf{R}\mathbf{D}_\lambda\mathbf{R}^{-1})\cdots(\mathbf{R}\mathbf{D}_\lambda\mathbf{R}^{-1}) \\
&= \mathbf{R}\mathbf{D}_{\lambda^k}\mathbf{R}^{-1} \, (= \mathbf{R}\mathbf{D}_{f(\lambda)}\mathbf{R}^{-1}).
\end{aligned}$$

Hence

$$\begin{aligned}
|J_{\mathbf{Y} \to \mathbf{X}}| &= \prod_{i=1}^{n} \prod_{j=i+1}^{n} \left\{ \frac{\lambda_i^k - \lambda_j^k}{\lambda_i - \lambda_j} \right\}^2 \prod_{i=1}^{n} k\lambda_i^{k-1} \\
&= \prod_{i=1}^{n} \prod_{j=1}^{n} (\lambda_i^{k-1} + \lambda_i^{k-2}\lambda_j + \ldots + \lambda_j^{k-1}) \\
&= \prod_{i=1}^{n} \prod_{j=1}^{n} \sum_{r=1}^{k} \lambda_i^{k-r}\lambda_j^{r-1}.
\end{aligned}$$

The reader is also refered to (18.15a).

(c) In some applications, $\mathbf{Y}$ is a random matrix whose eigenvalues are distinct with probability 1. The Jacobian $|J_{\mathbf{X}\to\mathbf{Y}}|$ is then given by $|J_{\mathbf{Y}\to\mathbf{X}}|^{-1}$, but it is expressed in terms of the eigenvalues of $\mathbf{X}$ rather than $\mathbf{Y}$, which is not so convenient in applications.

(d) If $\mathbf{X}$ is symmetric, then

$$|J_{\mathbf{Y}\to\mathbf{X}}| = |\prod_{i=1}^{n} \prod_{j=i+1}^{n} \left\{ \frac{f(\lambda_i) - f(\lambda_j)}{\lambda_i - \lambda_j} \right\} \prod_{i=1}^{n} \frac{df(\lambda_i)}{d\lambda_i}|.$$

Proofs. Section 18.13.

18.56a. Mathai [1997: 96] and Olkin and Sampson [1972: 267, lemma 9].

18.56b. Henderson and Searle [1979: 73]).

18.56d. Mathai [1997: 96] and Olkin and Sampson [1972: 268, lemma 10].

18.14   PAIRS OF MATRICES

18.57. Let $\mathbf{Y}_1$ and $\mathbf{Y}_2$ be positive definite $n \times n$ matrices. If $\det(\mathbf{Y}_1 - \lambda\mathbf{Y}_2) = 0$ has n distinct roots $\lambda_1 > \lambda_2 > \ldots > \lambda_n > 0$, there exists a unique matrix $\mathbf{W} - (w_{ij})$, with $w_{1i} > 0$ $(i = 1, 2, \ldots, n)$, such that $\mathbf{Y}_1 = \mathbf{W}\mathbf{D}_\lambda\mathbf{W}'$, $\mathbf{Y}_2 = \mathbf{W}\mathbf{W}'$, and $\mathbf{D}_\lambda = \mathrm{diag}(\lambda_1, \ldots, \lambda_n)$ (cf. 16.51c). Then

$$|J_{\mathbf{Y}_1,\mathbf{Y}_2\to\mathbf{W},\mathbf{D}_\lambda}| = 2^n |(\det \mathbf{W})^{n+2} \prod_{i=1}^{n} \prod_{j=i+1}^{n} (\lambda_i - \lambda_j)|.$$

18.58. Let $\mathbf{X}_1$ and $\mathbf{X}_2$ be positive definite.

(a) If $\mathbf{Y}_1 = \mathbf{X}_2^{-1/2}\mathbf{X}_1\mathbf{X}_2^{-1/2}$ and $\mathbf{Y}_2 = \mathbf{X}_2$, then $\mathbf{Y}_1$ and $\mathbf{Y}_2$ are positive definite and
$$|J_{\mathbf{Y}_1,\mathbf{Y}_2\to\mathbf{X}_1,\mathbf{X}_2}| = |\det \mathbf{X}_2|^{-(n+1)/2}.$$

(b) If $\mathbf{Y}_1 = (\mathbf{X}_1 + \mathbf{X}_2)^{-1/2}\mathbf{X}_1(\mathbf{X}_1 + \mathbf{X}_2)^{-1/2}$ and $\mathbf{Y}_2 = \mathbf{X}_1 + \mathbf{X}_2$, then $\mathbf{Y}_1$ and $\mathbf{Y}_2$ are positive definite and
$$|J_{\mathbf{Y}_1,\mathbf{Y}_2\to\mathbf{X}_1,\mathbf{X}_2}| = |\det(\mathbf{X}_1 + \mathbf{X}_2)|^{-(n+1)/2}.$$

18.59. Let $\mathbf{X}_1$ and $\mathbf{X}_2$ be $n \times n$ positive definite matrices. If $\mathbf{Y}_1 = \mathbf{X}_1$ and $\mathbf{Y}_2 = \mathbf{X}_1 + \mathbf{X}_2$, then $\mathbf{Y}_1$ and $\mathbf{Y}_2$ are positive definite and there exists a nonsingular $\mathbf{V}$ such that $\mathbf{Y}_1 = \mathbf{V}\mathbf{D}_\phi\mathbf{V}'$, $\mathbf{Y}_2 = \mathbf{V}\mathbf{V}'$, and $\mathbf{D}_\phi = \mathrm{diag}(\phi_1, \ldots, \phi_n)$, where $1 > \phi_1 > \phi_2 > \ldots > \phi_n > 0$ are the roots of $\det(\mathbf{X}_1 - \phi(\mathbf{X}_1 + \mathbf{X}_2)) = 0$. Then (cf. 18.57)

$$|J_{\mathbf{Y}_1,\mathbf{Y}_2\to\mathbf{V},\mathbf{D}_\phi}| = 2^n |(\det \mathbf{V})^{n+2} \prod_{i=1}^{n} \prod_{j=i+1}^{n} (\phi_i - \phi_j)|.$$

Proofs. Section 18.14.

18.57. Deemer and Olkin [1951: 350].

18.58a. Mathai [1997: 148].

18.58b. Mathai [1997: 149] and Seber [1984: 532].

18.59. Mathai [1997: 151] and Olkin and Sampson [1972: 270, lemma 14].

CHAPTER 19

MATRIX LIMITS, SEQUENCES, AND SERIES

Asymptotic theory and large sample approximations play a key role in statistical distribution theory. In this chapter we apply some of theory of limits to vectors and matrices.

19.1 LIMITS

Definition 19.1. Let $\mathbf{A}(t) = (a_{ij}(t))$. We say that $\lim_{t \to t_0} \mathbf{A}(t) = \mathbf{A}$ if $a_{ij}(t)) \to a_{ij}$ for all i, j. Of particular interest is the case when $t = \epsilon$ and $t_0 = 0$, as in the following result.

19.1. Suppose $\mathbf{A} = (a_{ij})$ is nonsingular,

(a) The elements of $\mathbf{A}^{-1}$ are continuous functions of the a_{ij}.

(b) If $\lim_{\epsilon \to 0} \mathbf{A}(\epsilon) = \mathbf{A}$, then $\lim_{\epsilon \to 0} [\mathbf{A}(\epsilon)]^{-1} = \mathbf{A}^{-1}$.

(c) $\lim_{\epsilon \to 0} (\mathbf{A} - \epsilon \mathbf{I})^{-1} = \mathbf{A}^{-1}$.

(d) If $\mathbf{A}$ is $m \times n$ and $\mathbf{B}$ is $n \times m$, both independent of ϵ, then

$$\lim_{\epsilon \to 0} \begin{pmatrix} \mathbf{I}_m & \epsilon \mathbf{A} \\ \epsilon \mathbf{B} & \mathbf{I}_n \end{pmatrix}^{-1} = \mathbf{I}_{m+n}.$$

A Matrix Handbook for Statisticians. By George A. F. Seber
Copyright © 2008 John Wiley & Sons, Inc.

19.2. (Continuity Argument) A number of matrix results can be proved by taking limits when continuity can be assumed, as was the case in (19.1) above. For example, a particular result may be true for a nonsingular matrix $\mathbf{A}$. If $\mathbf{A}$ is singular, we can choose $\epsilon > 0$ such that $\mathbf{A} + \epsilon\mathbf{I}$ is nonsingular (Abadir and Magnus [2005: 165]), set up the appropriate equation, and then let $\epsilon \to 0$. We may find that the result is then true for singular matices. For an example of this technique see Zhang [1999: 56].

Proofs. Section 19.1.

    19.1. Quoted by Zhang [1999: 58].

19.2 SEQUENCES

Sequences of vectors and matrices occur in many parts of statistics, especially in the development of asymptotic results. In particular, we are often interested in the limit of powers of matrices, as in stochastic processes where the focus is on transition matrices. We first of all consider convergence of a sequence of vectors with respect to a norm.

***Definition* 19.2.** Let $\mathcal{V}$ be a vector space over $\mathbb{F}$, and let $\|\cdot\|$ be a norm on $\mathcal{V}$. We say that the sequence of vectors $\{\mathbf{x}^{(k)}\}$ in $\mathcal{V}$ *converges with respect to the norm* to a vector $\mathbf{x} \in \mathcal{V}$ if and only if $\|\mathbf{x}^{(k)} - \mathbf{x}\| \to 0$ as $k \to \infty$. It should be noted that $\mathbf{x}$ is just an element of a vector space so that it can be regarded as either a real or complex vector or matrix, with an appropriate norm.

19.3. From (4.54) we see that if the sequence $\{\mathbf{x}^{(k)}\}$ converges to a vector $\mathbf{x}$ for one vector norm, it converges to $\mathbf{x}$ for *any* vector norm. Choosing the L_∞ norm we see that, for all vector norms on $\mathbb{R}^n$ or $\mathbb{C}^n$, $\lim_{k\to\infty} \mathbf{x}^{(k)} = \mathbf{x}$ with respect to any vector norm if and only if

$$\lim_{k\to\infty} x_i^{(k)} = x_i, \quad \text{for all} \quad i = 1, 2, \ldots, n.$$

    The extension from vectors to matrices is straightforward.

***Definition* 19.3.** Let $\{\mathbf{A}_k\}$ $(k = 1, 2, \ldots)$ be a sequence of $m \times n$ matrices, and let $a_{ij}^{(k)}$ denote the (i,j)th element of $\mathbf{A}_k$. The sequence $\{\mathbf{A}_k\}$ *converges* to $\mathbf{A} = (a_{ij})$, that is $\lim_{k\to\infty} \mathbf{A}_k = \mathbf{A}$, if

$$\lim_{k\to\infty} a_{ij}^{(k)} = a_{ij} \quad \text{for all } i, j.$$

A sequence that does not converge is said to diverge. The same definitions obviously apply to vectors as well. We shall assume that $m = n$, unless otherwise stated. If $\mathbf{A}$ is a square matrix and $\lim_{k\to\infty} \mathbf{A}^k = \mathbf{0}$, then we say that $\mathbf{A}$ is *convergent*.

19.4. Using the above notation, suppose $\lim_{k\to\infty} \mathbf{A}_k = \mathbf{A}$ and $\lim_{k\to\infty} \mathbf{B}_k = \mathbf{B}$. Let α and β be any constants, and let $\mathbf{P}$ and $\mathbf{Q}$ be any $n \times n$ matrices. From the limiting properties of scalars, the following results are straightforward.

    (a) $\lim_{k\to\infty}(\alpha\mathbf{A}_k + \beta\mathbf{B}_k) = \alpha\mathbf{A} + \beta\mathbf{B}$.

(b) $\lim_{k\to\infty} \mathbf{A}_k \mathbf{B}_k = \mathbf{AB}$.

(c) $\lim_{k\to\infty} \mathbf{PA}_k \mathbf{Q} = \mathbf{PAQ}$.

19.5. The sequence $\{\mathbf{A}^k\}$ converges if and only if the following hold.

(1) Each eigenvalue λ of $\mathbf{A}$ satisfies either $|\lambda| < 1$ or $\lambda = 1$.

(2) When $\lambda = 1$ occurs, the algebraic and geometric multiplicities of the eigenvalue 1 are the same.

19.6. If there is a matrix norm $||| \cdot |||$ such that $|||\mathbf{A}||| < 1$, then $\mathbf{A}$ is convergent.

19.7. $\mathbf{A}$ is convergent if and only if all the eigenvalues λ are less than 1 in modulus (i.e., $\rho(\mathbf{A}) < 1$, where $\rho(\mathbf{A})$ is the spectral radius of $\mathbf{A}$).

19.8. If the eigenvalue 1 occurs with algebraic and geometric multiplicity t (i.e., is semisimple), and all other eigenvalues are less than 1 in modulus, then

$$\lim_{k\to\infty} \mathbf{A}^k = \mathbf{X}(\mathbf{Y}'\mathbf{X})^{-1}\mathbf{Y}',$$

where $\mathbf{X}$ and $\mathbf{Y}$ are the $n \times t$ matrices of t linearly independent right, respectively, left eigenvectors associated with the eigenvalue 1.

19.9. Let $\mathbf{A}$ be $n \times n$, let $\{\mathbf{A}_k\}$ $(k = 1, 2, \ldots)$ be a sequence of real $n \times n$ matrices, and let $|||\cdot|||_{p,in}$ be an matrix norm induced by the L_p vector norm. If $p = 1, 2,$ or ∞ (cf. 4.74), then

(a) $\lim_{k\to\infty} \mathbf{A}_k = \mathbf{0}$ if and only if $\lim_{k\to\infty} |||\mathbf{A}_k|||_{p,in} = 0$.

(b) $\lim_{k\to\infty} \mathbf{A}_k = \mathbf{A}$ if and only if $\lim_{k\to\infty} |||\mathbf{A}_k - \mathbf{A}|||_{p,in} = 0$. If we use $\|\mathbf{A}_k - \mathbf{A}\|_F$, then this result also applies to $m \times n$ matrices; cf. Harville [1997: 431].)

(c) If $\lim_{k\to\infty} \mathbf{A}_k = \mathbf{A}$, then $\lim_{k\to\infty} |||\mathbf{A}_k|||_{p,in} = |||\mathbf{A}|||_{p,in}$. (The converse may not be true.)

19.10. The following result is useful in the context of limits. Suppose $\mathbf{C}$ is a square matrix and $(\mathbf{I} - \mathbf{C})^{-1}$ exists. If

$$\mathbf{A} = \begin{pmatrix} \mathbf{I} & \mathbf{0} \\ \mathbf{B} & \mathbf{C} \end{pmatrix}, \quad \text{then} \quad \mathbf{A}^k = \begin{pmatrix} \mathbf{I} & \mathbf{0} \\ (\mathbf{I} - \mathbf{C})^{-1}(\mathbf{I} - \mathbf{C}^k)\mathbf{B} & \mathbf{C}^k \end{pmatrix}.$$

19.11. If $||| \cdot |||$ is any matrix norm, then

$$\rho(\mathbf{A}) = \lim_{k\to\infty} |||\mathbf{A}^k|||^{1/k},$$

where ρ is the spectral radius.

Proofs. Section 19.2.

19.3. Horn and Johnson [1985: 273].

19.5. Hunter [1983a: 151–152] and Meyer [2000a: 629–630].

19.6. Harville [1997: 431–432] and Horn and Johnson [1985: 298].

19.7. Graybill [1983: 98–99], Horn and Johnson [1985: 298], and Meyer [2000a: 617].

19.8. Hunter [1983a: 153].

19.9. Graybill [1983: 96–97].

19.10. We use (19.14).

19.11. Meyer [2000a: 619].

19.3   ASYMPTOTICALLY EQUIVALENT SEQUENCES

There are situations where an $n \times n$ matrix $\mathbf{A}$ is difficult to work with, but a related matrix is easier to use that gives approximately the same result when n is large. This idea is made rigorous below.

***Definition* 19.4.** Let $\{\mathbf{A}_{(k)}\}$ and $\{\mathbf{B}_{(k)}\}$ be two sequences of real matrices, where $\mathbf{A}_{(k)}$ and $\mathbf{B}_{(k)}$ are both $k \times k$, and let $|||\mathbf{A}|||_v$ denote a matrix norm. The two sequences of matrices are defined to be *asymptotically equivalent* if and only if they satisfy the following two conditions (Graybill [1983: 101]).

(1) $|||\mathbf{A}_{(k)}|||_2 \leq c < \infty$, $|||\mathbf{B}_{(k)}|||_2 \leq c < \infty$ for $k = 1, 2, \ldots$, where c is a real number that does not depend on k.

(2) $\lim_{k \to \infty} k^{-1/2} |||(\mathbf{\Lambda}_{(k)} - \mathbf{B}_{(k)})|||_F = 0$

19.12. Let $\{\mathbf{A}_{(k)}\}$ and $\{\mathbf{B}_{(k)}\}$ be two asymptotically equivalent sequences of $k \times k$ matrices.

(a) Then $\lim_{k \to \infty} k^{-1/2} |||\mathbf{A}_k|||_F = \lim_{k \to \infty} k^{-1/2} |||\mathbf{B}_k|||_F$.

(b) Suppose $\mathbf{A}_{(k)}^{-1}$ and $\mathbf{B}_{(k)}^{-1}$ exist for each $k = 1, 2, \ldots$. If $|||\mathbf{A}_{(k)}|||_2 \leq c < \infty$ and $|||\mathbf{B}_k|||_2 \leq c < \infty$ for $k = 1, 2, \ldots$, where c is a real number that does not depend on k, then $\{\mathbf{A}_{(k)}^{-1}\}$ and $\{\mathbf{B}_{(k)}^{-1}\}$ are asymptotically equivalent.

19.13. Let $\{\mathbf{A}_{(k)}\}$, $\{\mathbf{B}_{(k)}\}$, $\{\mathbf{F}_{(k)}\}$, and $\{\mathbf{G}_{(k)}\}$ be sequences of $k \times k$ matrices.

(a) If $\{\mathbf{A}_{(k)}\}$ is asymptotically equivalent to $\{\mathbf{B}_{(k)}\}$, and $\{\mathbf{F}_{(k)}\}$ is asymptotically equivalent to $\{\mathbf{G}_{(k)}\}$, then $\{\mathbf{A}_{(k)}\mathbf{F}_{(k)}\}$ is asymptotically equivalent to $\{\mathbf{B}_{(k)}\mathbf{G}_{(k)}\}$.

(b) If, in (a,) $\{\mathbf{B}_{(k)}\}$ is asymptotically equivalent to $\{\mathbf{C}_{(k)}\}$, then $\{\mathbf{A}_{(k)}\}$ is asymptotically equivalent to $\{\mathbf{C}_{(k)}\}$.

(c) If $\{\mathbf{A}_{(k)}\mathbf{B}_{(k)}\}$ is asymptotically equivalent to $\{\mathbf{D}_{(k)}\}$, and $|||\mathbf{A}_{(k)}^{-1}|||_2 \leq c < \infty$, where c is a constant that does not depend on k, then $\{\mathbf{B}_{(k)}\}$ is asymptotically equivalent to $\{\mathbf{A}_{(k)}^{-1}\mathbf{D}_{(k)}\}$.

Proofs. Section 19.3.

19.12. Graybill [1983: 101–102].

19.13. Graybill [1983: 102].

19.4 SERIES

***Definition* 19.5.** Let $\mathbf{S}_k = \mathbf{A}_1 + \mathbf{A}_2 + \cdots + \mathbf{A}_k$, where the $\mathbf{A}_i$ are $n \times n$. The series $\mathbf{S}_k$ is said to *converge* to, or have sum, $\mathbf{S}$ if $\lim_{k \to \infty} \mathbf{S}_k = \mathbf{S}$; we write $\mathbf{S} = \sum_{k=1}^{\infty} \mathbf{A}_k$. A series that does not converge is said to diverge. (In what follows we recall that $\rho(\mathbf{A})$ is the spectral radius of $\mathbf{A}$.)

19.14. If $\mathbf{S}_k = \mathbf{I}_n + \mathbf{A} + \mathbf{A}^2 + \ldots + \mathbf{A}^k$, then

$$\mathbf{S}_k = (\mathbf{I}_n - \mathbf{A})^{-1}(\mathbf{I}_n - \mathbf{A}^{k+1}) = (\mathbf{I}_n - \mathbf{A}^{k+1})(\mathbf{I}_n - \mathbf{A})^{-1},$$

provided $(\mathbf{I}_n - \mathbf{A})^{-1}$ exists.

19.15. The following conditions are equivalent.

(1) (Newmann Series) $\mathbf{I}_n + \mathbf{A} + \mathbf{A}^2 + \cdots$ converges.

(2) $\rho(\mathbf{A}) < 1$.

(3) $\lim_{k \to \infty} \mathbf{A}^k = \mathbf{0}$, that is, $\mathbf{A}$ is convergent.

Moreover, when one (and hence all) of these conditions are satisfied, then

(a) $\mathbf{I}_n - \mathbf{A}$ is nonsingular.

(b) $\mathbf{I}_n + \mathbf{A} + \mathbf{A}^2 + \cdots$ converges to $(\mathbf{I}_n - \mathbf{A})^{-1}$.

19.16. Let $\mathbf{A} \in \mathcal{V}$ be the vector space of $n \times n$ matrices, and let $||| \cdot |||$ be any matrix norm defined on $\mathcal{V}$.

(a) If $|||\mathbf{A}||| < 1$, then $(\mathbf{I}_n - \mathbf{A})^{-1}$ exists and is given by

$$(\mathbf{I}_n - \mathbf{A})^{-1} = \sum_{k=0}^{\infty} \mathbf{A}^k.$$

We can obtain another result by setting $\mathbf{A} = \mathbf{I}_n - \mathbf{B}$.

(b) Suppose $\mathbf{B} \in \mathcal{V}$ and $\mathbf{B}$ is nonsingular. If $\mathbf{F} = \mathbf{B}^{-1}\mathbf{A}$, then the infinite series $\mathbf{B}^{-1} + \mathbf{F}\mathbf{B}^{-1} + \mathbf{F}^2\mathbf{B}^{-1} + \cdots$ converges if and only if $\lim_{k \to \infty} \mathbf{F}^k = \mathbf{0}$, in which case $\mathbf{B} - \mathbf{A}$ is nonsingular and

$$(\mathbf{B} - \mathbf{A})^{-1} = \sum_{k=0}^{\infty} \mathbf{F}^k \mathbf{B}^{-1}.$$

Replacing $\mathbf{A}$ by $-\mathbf{A}$ we get

$$(\mathbf{A} + \mathbf{B})^{-1} = \left(\sum_{k=0}^{\infty} (-\mathbf{B}^{-1}\mathbf{A})^k\right) \mathbf{B}^{-1}.$$

If $\mathbf{A}$ is small, we have the approximation $(\mathbf{A} + \mathbf{B})^{-1} \approx \mathbf{B}^{-1} - \mathbf{B}^{-1}\mathbf{A}\mathbf{B}^{-1}$.

If $\mathbf{A}$ is the matrix that is nonsingular, we interchange $\mathbf{A}$ and $\mathbf{B}$.

(c) If $\{a_k\}$, $k = 0, 1, \ldots$ is a sequence of scalars, then $\sum_{k=0}^{\infty} a_k \mathbf{A}^k$ converges if the series $\sum_{k=0}^{\infty} |a_k| \cdot |||\mathbf{A}|||^k$ of real numbers converges; $\mathbf{A}^0 = \mathbf{I}_n$.

(d) $\sum_{k=0}^{\infty} \alpha_k \mathbf{A}^k$ converges absolutely if $\sum_{k=0}^{\infty} |\alpha_k| \rho^k < \infty$, where $\rho = \rho(\mathbf{A})$ is the spectral radius of $\mathbf{A}$.

Proofs. Section 19.4.

19.14. Consider $(\mathbf{I}_n - \mathbf{A})\mathbf{S}_k$ and $\mathbf{S}_k(\mathbf{I}_n - \mathbf{A})$ and use Definition 19.3.

19.15. Graybill [1983: 100], Hunter [1983a: 154], and Meyer [2000a: 618].

19.16a. We combine (19.6) with (19.15).

19.16b. Harville [1997: 430].

19.16c. Rao and Rao [1998: 366].

19.16d. Abadir and Magnus [2005: 260].

19.5 MATRIX FUNCTIONS

Many functions $f(t)$, whether polynomial or nonpolynomial like $\exp(t)$, $\sin t$, and so on, can be generalized to have a matrix argument. Horn and Johnson [1991: chapter 6] have a good discussion on the meaning of $f(\mathbf{A})$ and associated properties. They also define a *primary matrix function* $f(\mathbf{A})$ associated with the scalar-valued *stem function* $f(t)$ using the Jordan canonical form of $\mathbf{A}$, and their book should be consulted for details. We shall only consider some nonpolynomial functions. The following theorem is helpful in this respect.

19.17. Let $f(t)$ be a scalar-valued function with power series representation $f(t) = a_0 + a_1 t + a_2 t^2 + \cdots$ that has a radius of convergence $R > 0$. If $\mathbf{A}$ is $n \times n$ and $\rho(\mathbf{A}) < R$, where $\rho(\cdot)$ is the spectral radius, the matrix power series $a_0 + a_1 \mathbf{A} + a_2 \mathbf{A}^2 + \cdots$ converges with respect to every norm on the set of $n \times n$ matrices, and its sum is denoted by the primary matrix function $f(\mathbf{A})$.

19.18. Let $\mathbf{A}$ be $n \times n$ with eigenvalues λ_i, and suppose $f(\lambda) = \sum_{k=1}^{\infty} c_k \lambda^k$ and $f(\mathbf{A}) = \sum_{k=1}^{\infty} c_k \mathbf{A}^k$.

(a) Since $\mathbf{A} = \mathbf{R}\mathbf{J}_O\mathbf{R}^{-1}$, where $\mathbf{J}_O$ is the Jordan canonical form of $\mathbf{A}$, we have $f(\mathbf{A}) = \mathbf{R}f(\mathbf{J}_O)\mathbf{R}^{-1}$.

(b) $\det f(\mathbf{A}) = \det \mathbf{R} \cdot \det f(\mathbf{J}_O) \cdot (\det \mathbf{R})^{-1} = \det(\mathbf{J}_O) = \prod_{i=1}^{n} f(\lambda_i)$.

(c) $\operatorname{trace} f(\mathbf{A}) = \operatorname{trace}(\mathbf{R}^{-1}\mathbf{R}f(\mathbf{J}_O)) = \operatorname{trace}(f(\mathbf{J}_O)) = \sum_{i=1}^{n} f(\lambda_i)$.

See also (19.21).

A number of functions with power series expansions fall into the above category, the most common being the exponential function. In fact $(\mathbf{I}_n + \mathbf{A})^{\nu}$, $\exp(\mathbf{A})$, $\log \mathbf{A}$, $\sin \mathbf{A}$, and $\cos \mathbf{A}$ can all be defined as primary matrix functions. However, using the Jordan canonical form, we find that all functions f satisfying certain derivative conditions have the property that $f(\mathbf{A})$ can be expressed as a polynomial in $\mathbf{A}$ (Meyer [2000a: 603–607]). For further details see Abadir and Magnus [2005: chapter 9] and Meyer [2000a: sections 7.3 and 7.9].

Proofs. Section 19.5.

19.17. Horn and Johnson [1991: 412].

19.18b. The determinant is the product of its eigenvalues, which are the diagonal elements of the upper-triangular matrix $\mathbf{J}_O$.

19.18c. As in (b), except that the trace is the sum of the eigenvalues.

19.6 MATRIX EXPONENTIALS

Matrix exponentials typically arise as solutions of differential equations (cf. Section 17.10).

Definition 19.6. If $\mathbf{A}$ is an $n \times n$ matrix, we define

$$\exp(\mathbf{A}t) = \mathbf{I}_n + t\mathbf{A} + \frac{t^2}{2!}\mathbf{A}^2 + \cdots + \frac{t^r}{r!}\mathbf{A}^r + \cdots, \quad -\infty < t < \infty.$$

This series converges absolutely for $\rho(\mathbf{A}) < \infty$ (by 19.16d).

19.19. Setting $t = 1$, we have the following.

(a) The eigenvalues of $\exp(\mathbf{A})$ are $\exp(\lambda_i)$ $(i = 1, 2, \ldots, n)$, where the λ_i are the eigenvalues of $\mathbf{A}$.

(b) If $\mathbf{A}$ is symmetric, then $\exp(\mathbf{A})$ is positive definite as each eigenvalue λ_i of $\mathbf{A}$ is real and $\exp(\lambda_i)$ is positive.

(c) The matrix $\exp(\mathbf{A})$ is always nonsingular (as from (a) the eigenvalues are nonzero) and
$$[\exp(\mathbf{A})]^{-1} = \exp(-\mathbf{A}).$$

(d) $\exp(k\mathbf{A}) = [\exp(\mathbf{A})]^k$ for k a positive or negative integer.

(e) $[\exp(\mathbf{A})]^* = \exp(\mathbf{A}^*)$.
It then follows that $\exp(\mathbf{A})$ is Hermitian if $\mathbf{A}$ is Hermitian, and it is unitary if $\mathbf{A}$ is skew-Hermitian.

(f) Every $n \times n$ unitary matrix $\mathbf{U}$ can expressed as $\exp(i\mathbf{A})$, where $\mathbf{A}$ is some Hermitian $n \times n$ matrix. Note that $i\mathbf{A}$ is skew-Hermitian.

(g) If $\mathbf{U}$ is an $n \times n$ symmetric unitary matrix, there exists a real symmetric matrix $\mathbf{A}$ such that $\mathbf{U} = \exp(i\mathbf{A})$.

(h) As the determinant of a matrix is the product of its eigenvalues, it follows from (a) that
$$\det[\exp(\mathbf{A})] = \exp(\text{trace } \mathbf{A}).$$
since $\prod_i \exp(\lambda_i) = \exp(\sum_i \lambda_i)$.

(i) If $\mathbf{A}$ is real skew-symmetric, then $\exp(\mathbf{A})$ is orthogonal and its determinant is 1.

(j) If $\mathbf{A}$ is skew-Hermitian, then $\mathbf{C} = \exp(\mathbf{A})$ is unitary with $|\det \mathbf{C}| = 1$.

19.20. Let $\mathbf{A}$ and $\mathbf{B}$ be $n \times n$ matrices.

(a) If $\mathbf{AB} = \mathbf{BA}$, then

$$\exp(\mathbf{A} + \mathbf{B}) = \exp(\mathbf{A})\exp(\mathbf{B}) = \exp(\mathbf{B})\exp(\mathbf{A}).$$

Although commutativity is a sufficient condition for the above to hold, it is not necessary.

(b) We can have $\exp(\mathbf{A})\exp(\mathbf{B}) = \exp(\mathbf{B})\exp(\mathbf{A})$, but $\mathbf{AB} \neq \mathbf{BA}$. For an example see Abadir and Magnus [2005: 256].

(c) $\det[\exp(\mathbf{A} + \mathbf{B})] = \det[\exp(\mathbf{A})]\det[\exp(\mathbf{B})]$.
This follows from (19.19g) above irrespective of whether (a) holds or not.

(d) (Lie Product Formula) $\lim_{n \to \infty} \left[\exp(\frac{\mathbf{A}}{n})\exp(\frac{\mathbf{B}}{n})\right]^n = \exp(\mathbf{A} + \mathbf{B})$.

19.21. Let $\mathbf{A}$ be an $n \times n$ diagonalizable matrix with eigenvalues λ_i; that is, there exists a nonsingular matrix $\mathbf{R}$ such that

$$
\begin{aligned}
\mathbf{A} &= \mathbf{R\Lambda R}^{-1} = \mathbf{R}\operatorname{diag}(\lambda_1, \lambda_2, \ldots, \lambda_n)\mathbf{R}^{-1}, \\
\mathbf{A}^k &= \mathbf{R\Lambda}^k\mathbf{R}^{-1} \\
\text{and} \quad \exp(\mathbf{A}) &= \mathbf{R}\operatorname{diag}(e^{\lambda_1}, \ldots, e^{\lambda_n})\mathbf{R}^{-1} = \mathbf{R}e^{\mathbf{\Lambda}}\mathbf{R}^{-1}, \quad \text{say.}
\end{aligned}
$$

(This method avoids using a power series expansion, and it can be generalized to any function $f(z)$ of a diagonalizable matrix by defining $f(\mathbf{\Lambda}) = \operatorname{diag}(f(\lambda_1), f(\lambda_2), \ldots, f(\lambda_n))$ and setting $f(\mathbf{A}) = \mathbf{R}f(\mathbf{\Lambda})\mathbf{R}^{-1}$.) For nondiagonalizable matrices, we can replace $\mathbf{\Lambda}$ by its Jordan form $\mathbf{J}_O$ and use (19.18).

19.22. For general t:

(a) The matrix $\exp(\mathbf{A}t)$ is nonsingular for all finite t. Noting that its eigenvalues are $\exp(t\lambda_i)$, we have from (19.19g),

$$\det(\exp(\mathbf{A}t)) = \exp(t\operatorname{trace}\mathbf{A}), \quad -\infty < t < \infty.$$

(b) Using power series expansions, $\exp(\mathbf{A}t_1)\exp(\mathbf{A}t_2) = \exp[\mathbf{A}(t_1 + t_2)]$ for all t_1 and t_2.

(c) $[\exp(\mathbf{A}t)]^{-1} = \exp(-\mathbf{A}t)$.

(d) If "$\otimes$" is the Kronecker product, we have from (11.9):

(i) $\exp(\mathbf{I}_m \otimes \mathbf{A}t) = \mathbf{I}_m \otimes \exp(\mathbf{A}t)$.
(ii) $\exp(\mathbf{A}t \otimes \mathbf{I}_m) = \exp(\mathbf{A}t) \otimes \mathbf{I}_m$.

These results hold for any primary matrix function $f(\cdot)$ and not just for $\exp(\cdot)$.

(e) $\exp(\mathbf{A}t)\exp(\mathbf{B}t) = \exp[(\mathbf{A} + \mathbf{B})t]$ for all finite $t \in \mathbb{R}$ if and only if $\mathbf{AB} = \mathbf{BA}$.

19.23. (Inequalities)

(a) Let $\mathbf{A}$ be any $n \times n$ matrix. For any matrix norm $||| \cdot |||$,

$$||| \exp(\mathbf{A}) ||| \leq \exp(||| \mathbf{A} |||).$$

(b) If $\mathbf{A}$ and $\mathbf{B}$ are $n \times n$ Hermitian matrices, then

$$\| \exp(\mathbf{A} + \mathbf{B}) \|_{ui} \leq \| \exp(\mathbf{A}) \exp(\mathbf{B}) \|_{ui},$$

for any unitarily invariant norm $\| \cdot \|_{ui}$.

19.24. Let f be a continuous function from the space of $n \times n$ complex matrices to $\mathbb{C}$ with the following properties.

(1) $f(\mathbf{XY}) = f(\mathbf{YX})$ for all $\mathbf{X}$ and $\mathbf{Y}$.

(2) $|(f(\mathbf{X}^{2k})| \leq f([\mathbf{XX}^*]^k)$ for all $\mathbf{X}$ and for all $k = 1, 2, \ldots$.

Then:

(a) $f(\mathbf{XY}) \geq 0$ for all Hermitian non-negative definite $\mathbf{X}$ and $\mathbf{Y}$. In particular,

$$f(\exp(\mathbf{A})) \geq 0$$

for all Hermitian $\mathbf{A}$.

(b) $|f(\exp(\mathbf{A}))| \leq f(\Re e \mathbf{A})$ for all $\mathbf{A}$.

(c) $|f(\exp(\mathbf{A} + \mathbf{B}))| \leq f(\Re e (\mathbf{A} + \mathbf{B})) \leq f(\Re e \, \mathbf{A}) f(\Re e \, \mathbf{B})$ for all $\mathbf{A}, \mathbf{B}$.

(d) $0 \leq f(\exp(\mathbf{A} + \mathbf{B}) \leq f(\exp(\mathbf{A}) \exp(\mathbf{B}))$ for all Hermitian $\mathbf{A}, \mathbf{B}$.

Here $\Re e$ means the "real part of." Note that (a) and (d) hold when f is the trace or the determinant. The above inequalities arise in statistical mechanics, population biology, and quantum mechanics.

Proofs. Section 19.6.

19.19a. Follows from the fact that $\mathbf{A}^k \mathbf{x} = \lambda^k \mathbf{x}$ for all $k = 1, 2, \ldots$, and for some $\mathbf{x}$. See also Meyer [2000a: : 525]

19.19c. Horn and Johnson [1991: 435].

19.19d. Abadir and Magnus [2005: 262] and Horn and Johnson [1991: 435].

19.19e. Quoted by Horn and Johnson [1991: 439, exercise 9].

19.19f. Quoted by Horn and Johnson [1991: 440, exercise 10].

19.19i. Abadir and Magnus [2005: 263].

19.19j. Abadir and Magnus [2005: 264].

19.20a. Abadir and Magnus [2005: 252–53] and Horn and Johnson [1991: 435].

19.20b. Quoted by Horn and Johnson [1991: 442, exercise 22].

19.20c. Bhatia [1997: 254] and Horn and Johnson [1997: 496].

19.21. Abadir and Magnus [2005: 260] and Meyer [2000a: 525, 601].

19.22c. Use (e) to verify that $\exp(-\mathbf{A}t)$ is the inverse.

19.22d. Quoted by Horn and Johnson [1991: 440, exercise 13].

19.22e. Abadir and Magnus [2005: 252].

19.23a. Horn and Johnson [1991: 501, equation (6.5.25)].

19.23b. Horn and Johnson [1991: 499].

19.24. Horn and Johnson [1991: 497].

CHAPTER 20

RANDOM VECTORS

20.1 NOTATION

In this chapter we do not use the convention that random variables have capital letters because of the problem of distinguishing between a random matrix and its observed value in the next chapter. As a rough rule, we generally reserve the latter part of the alphabet, $u, v, \ldots, z$ for random vectors and the rest of the alphabet for constants, including matrices of constants.

20.2 VARIANCES AND COVARIANCES

Definition 20.1. If $\mathbf{x} = (x_i)$ is a vector of random variables, then we define $\mathrm{E}(\mathbf{x}) = (\mathrm{E}(x_i))$, the vector of expected values.

20.1. For conformable vectors and matrices, $\mathrm{E}(\mathbf{Ax} + \mathbf{b}) = \mathbf{A}\mathrm{E}(\mathbf{x}) + \mathbf{b}$.

20.2. Let S be a convex subset of $\mathbb{R}^n$ and $\mathbf{x}$ an $n \times 1$ random vector with finite $\mathrm{E}(\mathbf{x})$. If $\mathrm{pr}(\mathbf{x} \in S) = 1$ then $\mathrm{E}(\mathbf{x}) \in S$.

Definition 20.2. If $\mathbf{x}$ and $\mathbf{y}$ are vectors of random variables, we define the matrix with (i, j)th elements $\mathrm{cov}(x_i, y_j)$ to be $\mathrm{cov}(\mathbf{x}, \mathbf{y})$, the *covariance matrix* of $\mathbf{x}$ and $\mathbf{y}$. When $\mathbf{x} = \mathbf{y}$, we define $\mathrm{var}(\mathbf{x}) = \mathrm{cov}(\mathbf{x}, \mathbf{x})$ to be the *variance matrix* of $\mathbf{x}$. (The terms *variance–covariance matrix*, *covariance matrix* and *dispersion matrix* are also used in the literature for var($\mathbf{x}$).) In the following, we recall that $\mathbf{A} \succeq \mathbf{B}$ implies that $\mathbf{A} - \mathbf{B}$ is non-negative definite.

20.3. Let var($\mathbf{x}$) $- \boldsymbol{\Sigma}$. Since, by (20.6b), $0 \le$ var($\mathbf{a}'\mathbf{x}$) $- \mathbf{a}'\boldsymbol{\Sigma}\mathbf{a}$, we have that $\boldsymbol{\Sigma}$ is non-negative definite. It is nonsingular (i.e., positive definite) if and only if there do not exist constants $\mathbf{a}$ ($\ne \mathbf{0}$) and b such that $\mathbf{a}'\mathbf{x} = b$ (i.e, var($\mathbf{a}'\mathbf{x}$) $= 0$).

20.4. If E($\mathbf{x}$) $= \boldsymbol{\mu}$ and var($\mathbf{x}$) $= \boldsymbol{\Sigma}$, then $\mathbf{x} - \boldsymbol{\mu} \in \mathcal{C}(\boldsymbol{\Sigma})$ or equivalently $\mathbf{x} \in \mathcal{C}(\boldsymbol{\mu}, \boldsymbol{\Sigma})$, with probability 1.

20.5. Let E($\mathbf{x}$) $= \boldsymbol{\mu}_{\mathbf{x}}$ and E($\mathbf{y}$) $= \boldsymbol{\mu}_{\mathbf{y}}$.

(a)
$$\begin{aligned}
\text{cov}(\mathbf{x} - \mathbf{a}, \mathbf{y} - \mathbf{b}) &= \text{cov}(\mathbf{x}, \mathbf{y}) \\
&= \text{E}[(\mathbf{x} - \boldsymbol{\mu}_{\mathbf{x}})(\mathbf{y} - \boldsymbol{\mu}_{\mathbf{y}})'] \\
&= \text{E}(\mathbf{x}\mathbf{y}') - \boldsymbol{\mu}_{\mathbf{x}}\boldsymbol{\mu}_{\mathbf{y}}'.
\end{aligned}$$

(b) The above result also holds if $\mathbf{x} = \mathbf{y}$ so that:

(i) var($\mathbf{x} - \mathbf{a}$) $=$ var($\mathbf{x}$).

(ii) var($\mathbf{x}$) $=$ E($\mathbf{x}\mathbf{x}'$) $- \boldsymbol{\mu}_{\mathbf{x}}\boldsymbol{\mu}_{\mathbf{x}}'$.

20.6. The following results are extremely useful.

(a) cov($\mathbf{A}\mathbf{x}, \mathbf{B}\mathbf{y}$) $= \mathbf{A}$cov($\mathbf{x}, \mathbf{y}$)$\mathbf{B}'$.

(b) From (a), var($\mathbf{A}\mathbf{x}$) $= \mathbf{A}$var($\mathbf{x}$)$\mathbf{A}'$.

(c) var($\mathbf{y} - \mathbf{A}\mathbf{x}$) $=$ var($\mathbf{y}$) $- \mathbf{A}$cov($\mathbf{x}, \mathbf{y}$) $-$ cov($\mathbf{y}, \mathbf{x}$)$\mathbf{A}'$ $+ \mathbf{A}$var($\mathbf{x}$)$\mathbf{A}'$.

20.7. If $\mathbf{x}$ and $\mathbf{y}$ are random vectors with respective means $\boldsymbol{\mu}_{\mathbf{x}}$ and $\boldsymbol{\mu}_{\mathbf{y}}$, then E($(\mathbf{y} - \mathbf{A}\mathbf{x} - \mathbf{a})(\mathbf{y} - \mathbf{A}\mathbf{x} - \mathbf{a})'$] $=$ var($\mathbf{y} - \mathbf{A}\mathbf{x}$) $+ (\boldsymbol{\mu}_{\mathbf{y}} - \mathbf{A}\boldsymbol{\mu}_{\mathbf{x}} - \mathbf{a})(\boldsymbol{\mu}_{\mathbf{y}} - \mathbf{A}\boldsymbol{\mu}_{\mathbf{x}} - \mathbf{a})'$.

20.8. If a, b, c, and d are constants, then

$$\begin{aligned}
&\text{cov}(a\mathbf{x} + b\mathbf{y}, c\mathbf{u} + d\mathbf{v}) \\
&\quad = ac\,\text{cov}(\mathbf{x}, \mathbf{u}) + ad\,\text{cov}(\mathbf{x}, \mathbf{v}) + bc\,\text{cov}(\mathbf{y}, \mathbf{u}) + bd\,\text{cov}(\mathbf{y}, \mathbf{v}).
\end{aligned}$$

In particular, var($\mathbf{u} + \mathbf{v}$) $=$ var($\mathbf{u}$) $+$ cov($\mathbf{u}, \mathbf{v}$) $+$ [cov($\mathbf{u}, \mathbf{v}$)]$'$ $+$ var($\mathbf{v}$).

20.9. (Partitioned Vector) Let $\mathbf{z} = (\mathbf{x}', \mathbf{y}')'$ be a random vector with mean $\boldsymbol{\mu}_{\mathbf{z}} = (\boldsymbol{\mu}_{\mathbf{x}}', \boldsymbol{\mu}_{\mathbf{y}}')'$, where $\mathbf{x}$ is $m \times 1$ and $\mathbf{y}$ is $p \times 1$. Then:

(a) var($\mathbf{z}$) $= \boldsymbol{\Sigma}_{zz} = \begin{pmatrix} \text{var}(\mathbf{x}) & \text{cov}(\mathbf{x}, \mathbf{y}) \\ \text{cov}(\mathbf{y}, \mathbf{x}) & \text{var}(\mathbf{y}) \end{pmatrix} = \begin{pmatrix} \boldsymbol{\Sigma}_{xx} & \boldsymbol{\Sigma}_{xy} \\ \boldsymbol{\Sigma}_{yx} & \boldsymbol{\Sigma}_{yy} \end{pmatrix}$, say,

where $\boldsymbol{\Sigma}_{yx} = \boldsymbol{\Sigma}_{xy}'$.

(b) $\mathcal{C}(\boldsymbol{\Sigma}_{xy}) \subseteq \mathcal{C}(\boldsymbol{\Sigma}_{xx})$ and $\mathcal{C}(\boldsymbol{\Sigma}_{yx}) \subseteq \mathcal{C}(\boldsymbol{\Sigma}_{yy})$.

(c) If E($\mathbf{x}$) $= \mathbf{0}$, then cov($\mathbf{y} - \mathbf{A}\mathbf{x}, \mathbf{x}$) $= \mathbf{0}$ if and only if $\mathbf{A}\mathbf{x} = \mathbf{B}\mathbf{x}$ with probability 1, where $\mathbf{B} = \boldsymbol{\Sigma}_{yx}\boldsymbol{\Sigma}_{xx}^{-}$ and $\boldsymbol{\Sigma}_{xx}^{-}$ is any weak inverse of $\boldsymbol{\Sigma}_{xx}$ (i.e., $\boldsymbol{\Sigma}_{xx}\boldsymbol{\Sigma}_{xx}^{-}\boldsymbol{\Sigma}_{xx} = \boldsymbol{\Sigma}_{xx}$).

(d) Using (20.8) and (c),

$$
\begin{aligned}
\operatorname{var}(\mathbf{y} - \mathbf{Ax}) &= \operatorname{var}[(\mathbf{y} - \mathbf{Bx}) + (\mathbf{B} - \mathbf{A})\mathbf{x}] \\
&= \operatorname{var}(\mathbf{y} - \mathbf{Bx}) + \operatorname{var}[(\mathbf{B} - \mathbf{A})\mathbf{x}] \\
&\succeq \operatorname{var}(\mathbf{y} - \mathbf{Bx}).
\end{aligned}
$$

for all $\mathbf{A}$.

(e) By (20.6b) we have $\operatorname{var}(\boldsymbol{\Sigma}_{yx}\boldsymbol{\Sigma}_{xx}^{-}\mathbf{x}) = \operatorname{var}(\mathbf{Bx}) = \boldsymbol{\Sigma}_{yx}\boldsymbol{\Sigma}_{xx}^{-}\boldsymbol{\Sigma}_{xy}$. This matrix and the generalized Schur complement $\boldsymbol{\Sigma}_{yy\cdot x} = \boldsymbol{\Sigma}_{yy} - \boldsymbol{\Sigma}_{yx}\boldsymbol{\Sigma}_{xx}^{-}\boldsymbol{\Sigma}_{xy}$ are invariant with respect to the choice of weak inverse $\boldsymbol{\Sigma}_{xx}^{-}$ (by (14.8) and (20.9b)).

(f) $\operatorname{var}(\mathbf{y} - \mathbf{Bx}) = \boldsymbol{\Sigma}_{yy\cdot x} \left(= \operatorname{var}(\mathbf{y} - \boldsymbol{\mu}_{\mathbf{y}} - \boldsymbol{\Sigma}_{yx}\boldsymbol{\Sigma}_{xx}^{-}(\mathbf{x} - \boldsymbol{\mu}_{\mathbf{x}}))\right)$, by (20.5b(i)).

(g) (Best Linear Predictor) From (20.7) and (d),

$$
\begin{aligned}
\mathrm{E}[(\mathbf{y} - \mathbf{Ax} - \mathbf{a})(\mathbf{y} - \mathbf{Ax} - \mathbf{a})'] &\succeq \operatorname{var}(\mathbf{y} - \mathbf{Ax}) \\
&\succeq \mathrm{E}[(\mathbf{y} - \widehat{\mathbf{y}}_{\mathbf{x}})(\mathbf{y} - \widehat{\mathbf{y}}_{\mathbf{x}})'] = \boldsymbol{\Sigma}_{yy\cdot x}
\end{aligned}
$$

for all conformable $\mathbf{A}$ and $\mathbf{a}$, where $\widehat{\mathbf{y}}_{\mathbf{x}} = \boldsymbol{\mu}_{\mathbf{y}} + \boldsymbol{\Sigma}_{xy}\boldsymbol{\Sigma}_{xx}^{-}(\mathbf{x} - \boldsymbol{\mu}_{\mathbf{x}})$ is called the *best linear predictor* as it minimizes the left-hand side of the above expression, the so-called *mean squared prediction error matrix*.

Proofs. Section 20.2.

20.2. Schott [2005: 377].

20.3. Seber and Lee [2003: 8].

20.4. Rao [1973a: 522] and Sengupta and Jammalamadaka [2003: 56].

20.6a. Seber and Lee [2003: 7].

20.6c. Expand $\operatorname{cov}(\mathbf{y} - \mathbf{Ax}, \mathbf{y} - \mathbf{Ax})$.

20.7. We use (20.5b(i)) with $\mathbf{x}$ replaced by $\mathbf{y} - \mathbf{Ax}$, and then use (20.5b(ii)) with $\mathbf{x}$ replaced by $\mathbf{y} - \mathbf{Ax} - \mathbf{a}$.

20.8. Seber and Lee [2003: 7].

20.9b. Sengupta and Jammalamadaka [2003: 56, with the roles of x and y interchanged].

20.9c. Sengupta and Jammalamadaka [2003: 57].

20.9d. Use (20.9c) to prove that the covariance term is zero.

20.9e. $\operatorname{var}(\mathbf{Bx}) = \mathbf{B}\boldsymbol{\Sigma}_{xx}\mathbf{B}' = \boldsymbol{\Sigma}_{yx}\boldsymbol{\Sigma}_{xx}^{-}\boldsymbol{\Sigma}_{xx}(\boldsymbol{\Sigma}_{xx}^{-})'\boldsymbol{\Sigma}_{xy} = \boldsymbol{\Sigma}_{yx}(\boldsymbol{\Sigma}_{xx}^{-}\boldsymbol{\Sigma}_{xx}\boldsymbol{\Sigma}_{xx(1)}^{-})\boldsymbol{\Sigma}_{xy}$, where $(\boldsymbol{\Sigma}_{xx}^{-})' = \boldsymbol{\Sigma}_{xx(1)}^{-}$, say, for some weak inverse $\boldsymbol{\Sigma}_{xx(1)}^{-}$ of $\boldsymbol{\Sigma}_{xx}$, as $\boldsymbol{\Sigma}_{xx}$ is symmetric. Then $\mathbf{C} = \boldsymbol{\Sigma}_{xx}^{-}\boldsymbol{\Sigma}_{xx}\boldsymbol{\Sigma}_{xx(1)}^{-}$ is a weak inverse of $\boldsymbol{\Sigma}_{xx}$ as $\boldsymbol{\Sigma}_{xx}\mathbf{C}\boldsymbol{\Sigma}_{xx} = \boldsymbol{\Sigma}_{xx}$.

20.9f. Using (c) and $\mathbf{B} = \boldsymbol{\Sigma}_{yx}\boldsymbol{\Sigma}_{xx}^{-}$, we obtain

$$
\operatorname{cov}(\mathbf{y} - \mathbf{Bx}, \mathbf{y} - \mathbf{Bx}) = \operatorname{cov}(\mathbf{y} - \mathbf{Bx}, \mathbf{y}) = \operatorname{var}(\mathbf{y}) - \mathbf{B}\operatorname{cov}(\mathbf{x}, \mathbf{y}) = \boldsymbol{\Sigma}_{yy\cdot x}.
$$

20.9g. $\mathrm{E}[(\mathbf{y} - \mathbf{Ax} - \mathbf{a})(\mathbf{y} - \mathbf{Ax} - \mathbf{a})'] \succeq \operatorname{var}(\mathbf{y} - \mathbf{Ax})$, by (20.7), and $\operatorname{var}(\mathbf{y} - \mathbf{Ax}) \succeq \operatorname{var}(\mathbf{y} - \mathbf{Bx})$, from (f).

20.3 CORRELATIONS

20.3.1 Population Correlations

Definition 20.3. If x and y are random variables, then their *population correlation coefficient* is defined to be $\rho(x, y) = \text{cov}(x, y)/[\text{var}(x)\text{var}(y)]^{1/2}(= \sigma_{xy}/(\sigma_x\sigma_y),$ say).

20.10. $\rho = \rho(x, y)$ has the following well-known properties.

  (a) $-1 \le \rho \le +1$.

  (b) $\rho^2 = 1$ if and only if x and y are linearly related.
  If $\rho = +1$, then $y - \mu_y = \frac{\sigma_y}{\sigma_x}(x - \mu_x)$.
  if $\rho = -1$, then $y - \mu_y = -\frac{\sigma_y}{\sigma_x}(x - \mu_x)$.

  (c) $\rho(ax, by) = \text{sign}(ab)\,|ab|\,\rho(x, y)$.

Definition 20.4. Suppose $\mathbf{x}$ has variance matrix $\boldsymbol{\Sigma} = (\sigma_{ij})$. If $\text{corr}(\mathbf{x}) = (\rho_{ij})$, where $\rho_{ij} = \sigma_{ij}/(\sigma_{ii}\sigma_{jj})^{1/2}$, then $\text{corr}(\mathbf{x})$ is called the *population correlation matrix* of $\mathbf{x}$.

20.11. Let $\mathbf{C} = \text{corr}(\mathbf{x}) = (\rho_{ij})$ be an $n \times n$ correlation matrix.

  (a) $\mathbf{C} = \mathbf{D}_\sigma^{-1/2}\boldsymbol{\Sigma}\mathbf{D}_\sigma^{-1/2}$, where $\mathbf{D}_\sigma$ is a diagonal matrix with positive diagonal elements $\boldsymbol{\sigma} = \text{diag}(\sigma_{11}, \sigma_{22}, \dots, \sigma_{nn})$.

  (b) $\mathbf{C}$ is non-negative definite (as $\boldsymbol{\Sigma}$ is).

  (c) $\rho_{ii} = 1$ and $|\rho_{ij}| < 1$ (for all $i, j,\ j \ne i$).

  (d) The largest eigenvalue of $\mathbf{C}$ is less than n.

  (e) $0 < \det \mathbf{C} \le 1$.

  (f) A well-known correlation matrix is of the form $\mathbf{C} = (1 - \rho)\mathbf{I}_n + \rho\mathbf{J}_n$, where $\mathbf{J}_n$ is an $n \times n$ matrix with all its elements equal to 1. For $\mathbf{C}$ to be positive definite we must have $-1/(n - 1) < \rho < 1$. For further details about this matrix see (15.18a).

Definition 20.5. Let $\mathbf{x}$ be a d-dimensional random vector with $\text{E}(\mathbf{x}) = \boldsymbol{\mu}$ and $\text{var}(\mathbf{x}) = \boldsymbol{\Sigma}$, where $\boldsymbol{\Sigma}$ is positive definite. Consider the partitions

$$\mathbf{x} = \begin{pmatrix} x_1 \\ \mathbf{x}_2 \end{pmatrix}, \quad \boldsymbol{\mu} = \begin{pmatrix} \mu_1 \\ \boldsymbol{\mu}_2 \end{pmatrix}, \quad \text{and} \quad \boldsymbol{\Sigma} = \begin{pmatrix} \sigma_{11} & \boldsymbol{\sigma}_{12}' \\ \boldsymbol{\sigma}_{12} & \boldsymbol{\Sigma}_{22} \end{pmatrix},$$

where $\mathbf{x}_2 = (x_2, \dots, x_d)'$, $\boldsymbol{\mu}_2 = (\mu_2, \dots, \mu_d)'$ and $\boldsymbol{\Sigma}_{22}$ is $(d - 1) \times (d - 1)$. Here $\text{var}(x_1) = \sigma_{11}$, $\boldsymbol{\sigma}_{12}$ is the vector of covariances betweem x_1 and each of the variables in $\mathbf{x}_2$, and $\text{var}(\mathbf{x}_2) = \boldsymbol{\Sigma}_{22}$. The *(population) multiple correlation coefficient* between x_1 and $\mathbf{x}_2$, denoted by $\rho_{1\cdot23\cdots d}$, is the maximum correlation between x_1 and any linear function $\mathbf{a}'\mathbf{x}_2$ of $x_2, \dots, x_d$. Thus

$$\mathcal{R} = \rho_{1\cdot23\cdots d} = \max_{\mathbf{a}} \frac{\text{cov}(x_1, \mathbf{a}'\mathbf{x}_2)}{[\text{var}(x_1)\text{var}(\mathbf{a}'\mathbf{x}_2)]^{1/2}} = \max_{\mathbf{a}} \frac{\mathbf{a}'\boldsymbol{\sigma}_{12}}{(\sigma_{11}\,\mathbf{a}'\boldsymbol{\Sigma}_{22}\mathbf{a})^{1/2}}.$$

Also, $\mathcal{R}^2$ is sometimes called the *(population) coefficient of multiple determination*, and $\mathcal{R}$ is the positive square root (Muirhead [1982: section 5.2] and Anderson [2003: section 2.5]).

20.12. $\mathcal{R}$ has the following properties.

(a) $\mathcal{R} = (\boldsymbol{\sigma}'_{12}\boldsymbol{\Sigma}_{22}\boldsymbol{\sigma}_{12}/\sigma_{11})^{1/2}$.

(b) $0 \leq \mathcal{R} \leq 1$.

(c) If σ^{11} is the first diagonal element of $\boldsymbol{\Sigma}^{-1}$, then, from (14.11),

$$\sigma^{11} = (\sigma_{11} - \boldsymbol{\sigma}'_{12}\boldsymbol{\Sigma}_{22}^{-1}\boldsymbol{\sigma}_{12})^{-1}$$

and $1 - \mathcal{R}^2 = 1/(\sigma^{11}\sigma_{11})$.

(d) When $\mathbf{x}$ has a nonsingular multivariate normal distribution $N_d(\boldsymbol{\mu}, \boldsymbol{\Sigma})$ we have from (20.23f) that

$$E(x_1 \mid \mathbf{x}_2) = \mu_1 + \boldsymbol{\sigma}'_{12}\boldsymbol{\Sigma}_{22}^{-1}(\mathbf{x}_2 - \boldsymbol{\mu}_2)$$

and

$$\mathrm{var}(x_1 \mid \mathbf{x}_2) = \sigma_{11\cdot23\cdots d} = \sigma_{11} - \boldsymbol{\sigma}'_{12}\boldsymbol{\Sigma}_{22}^{-1}\boldsymbol{\sigma}_{12}.$$

Then:

(i) $1 - \mathcal{R}^2 = \sigma_{11\cdot23\cdots d}/\sigma_{11}$.

(ii) $\sigma_{11\cdot23\cdots d} \leq \sigma_{11}$.

(iii) $\mathcal{R}$ is the correlation beween x_1 and $E(x_1 \mid \mathbf{x}_2)$.

(e) When $d = 2$, $\mathcal{R} = |\rho_{1\cdot2}| = |\rho(x_1, x_2)|$.

***Definition* 20.6.** The previous Definition 20.5 can be readily generalized. Let $\mathbf{x}$ be a d-dimensional random vector with $E(\mathbf{x}) = \boldsymbol{\mu}$ and positive definite variance matrix $\boldsymbol{\Sigma}$. Consider the partitions

$$\mathbf{x} = \begin{pmatrix} \mathbf{x}_1 \\ \mathbf{x}_2 \end{pmatrix}, \quad \boldsymbol{\mu} = \begin{pmatrix} \boldsymbol{\mu}_1 \\ \boldsymbol{\mu}_2 \end{pmatrix}, \quad \text{and} \quad \boldsymbol{\Sigma} = \begin{pmatrix} \boldsymbol{\Sigma}_{11} & \boldsymbol{\Sigma}_{12} \\ \boldsymbol{\Sigma}_{21} & \boldsymbol{\Sigma}_{22} \end{pmatrix},$$

where $\mathbf{x}_1$ is $k \times 1$, $\mathbf{x}_2$ is $(d - k) \times 1$, and so on, and let x_i be a variable in $\mathbf{x}_1$ ($i = 1, 2, \ldots, k$). The (population) multiple correlation coefficient between x_i and the variables $x_{k+1}, \ldots, x_d$ in $\mathbf{x}_2$, denoted by $\mathcal{R}_{i\cdot k+1, k+2,\ldots,d}$, is the maximum correlation between x_i and any linear function $\mathbf{a}'\mathbf{x}_2$ of $x_{k+1}, \ldots, x_d$. Note that $\mathcal{R}_{1\cdot2,\ldots,d} = \mathcal{R}$.

***Definition* 20.7.** Using the notation of the previous definition, let $\boldsymbol{\Sigma}_{11\cdot2} \equiv \boldsymbol{\Sigma}_{11} - \boldsymbol{\Sigma}_{12}\boldsymbol{\Sigma}_{22}^{-1}\boldsymbol{\Sigma}_{21}(= (\sigma_{ij\cdot k+1,\ldots,d}),$ say$)$. Given $\mathbf{x} \sim N_d(\boldsymbol{\mu}, \boldsymbol{\Sigma})$, then $\boldsymbol{\Sigma}_{11\cdot2} = \mathrm{var}(\mathbf{x}_1 \mid \mathbf{x}_2)$ (cf. 20.23f). We define the *(population) partial correlation coefficient* $\rho_{ij\cdot k+1,\ldots,d}$ to be the correlation coefficient between x_i and x_j, components of $\mathbf{x}_1$, in the conditional distribution of $\mathbf{x}_1$ given $\mathbf{x}_2$, that is,

$$\rho_{ij\cdot k+1,\ldots,d} = \frac{\sigma_{ij\cdot k+1,\ldots,d}}{(\sigma_{ii\cdot k+1,\ldots,d}\,\sigma_{jj\cdot k+1,\ldots,d})^{1/2}}.$$

This is the correlation beween x_i and x_j holding $\mathbf{x}_2$ fixed.

20.13. Let $\boldsymbol{\sigma}_i'$ be the ith row of $\boldsymbol{\Sigma}_{12}$.

(a) $\sigma_{ii \cdot k+1,k+2,\ldots,d} = (\sigma_{ii} - \boldsymbol{\sigma}_i' \boldsymbol{\Sigma}_{22}^{-1} \boldsymbol{\sigma}_i)$.

(b) $\mathcal{R}_{i \cdot k+1,k+2,\ldots,d} = (\boldsymbol{\sigma}_i' \boldsymbol{\Sigma}_{22}^{-1} \boldsymbol{\sigma}_i / \sigma_{ii})^{1/2}$.

(c) $1 - \mathcal{R}_{i \cdot k+1,k+2,\ldots,d}^2 = (\sigma_{ii} - \boldsymbol{\sigma}_i' \boldsymbol{\Sigma}_{22}^{-1} \boldsymbol{\sigma}_i)/\sigma_{ii}$.

(d) $\mathcal{R}_{i \cdot k+1,\ldots,d}^2 = (\sigma_{ii} - \sigma_{ii \cdot k+1,\ldots,d})/\sigma_{ii}$.

(e) $1 - \rho_{i,k+1 \cdot k+2,\ldots,d}^2 = (\sigma_{ii \cdot k+1,\ldots,d}/\sigma_{ii \cdot k+2,\ldots,d})$. In particular, with $i = 1$ and $k = 1$

$$1 - \rho_{12 \cdot 3,4,\ldots,d}^2 = (\sigma_{11 \cdot 2,\ldots,d}/\sigma_{11 \cdot 3,\ldots,d}).$$

(f) $1 - \mathcal{R}_{1 \cdot 2,3\ldots,d}^2 = (1 - \rho_{12 \cdot 3,4,\ldots,d}^2)(1 - \rho_{13 \cdot 4,5,\ldots,d}^2) \cdots (1 - \rho_{1,d-1 \cdot d}^2)(1 - \rho_{1d}^2)$.

(g) $\rho_{ij \cdot k+1,k+2,\ldots d} = \dfrac{\rho_{ij \cdot k+2,\ldots,d} - \rho_{i,k+1 \cdot k+2,\ldots,d}\, \rho_{j,k+1 \cdot k+2,\ldots,d}}{\sqrt{1 - \rho_{i,k+1 \cdot k+2,\ldots,d}^2}\sqrt{1 - \rho_{j,k+1 \cdot k+2,\ldots,d}^2}}$

(h) $\rho_{12 \cdot 3} = \dfrac{\rho_{12} - \rho_{13}\rho_{23}}{[(1 - \rho_{23}^2)(1 - \rho_{13}^2)]^{1/2}}$.

Proofs. Section 20.3.1.

20.11d. The eigenvalues λ_i of $\mathbf{C}$ are non-negative and $\sum_i \lambda_i = \text{trace } \mathbf{C} = n$.

20.11e. Follows from $(\prod_i \lambda_i)^{1/n} \leq \bar{\lambda} = 1$ (as the geometric mean does not exceed the arithmetic mean) and $\det \mathbf{C} = \prod_i \lambda_i$.

20.12. Muirhead [1982: section 5.2].

20.13. Anderson [2003: 38–41]; see also Muirhead [1982: 194].

20.3.2 Sample Correlations

Definition 20.8. If $\mathbf{x} = (x_i)$ and $\mathbf{y} = (y_i)$ are $n \times 1$ vectors representing univariate random samples of size n, then their *sample correlation coefficient* is defined to be

$$r(\mathbf{x}, \mathbf{y}) = \frac{\sum_{i=1}^{n}(x_i - \bar{x})(y_i - \bar{y})}{[\sum_{i=1}^{n}(x_i - \bar{x})^2 \sum_{i=1}^{n}(y_i - \bar{y})^2]^{1/2}}.$$

20.14. $r = r(\mathbf{x}, \mathbf{y})$ has the following properties.

(a) $r^2 \leq 1$. When $r^2 = 1$, there is a linear relation between the y_i and the x_i as in (20.10b), but with parameters replaced by their estimates.

(b) $r(a\mathbf{x}, b\mathbf{y}) = \text{sign}(ab)\,|ab|\,r(\mathbf{x}, \mathbf{y})$.

Definition 20.9. Let $\mathbf{x}_1, \mathbf{x}_2, \ldots, \mathbf{x}_n$ be a random sample from a d-dimensional distribution with mean $\boldsymbol{\mu}$ and positive definite variance matrix $\boldsymbol{\Sigma}$, and let $\mathbf{S} = (s_{ij})$ be the sample covariance matrix given by $\mathbf{S} = \sum_{i=1}^{n}(\mathbf{x}_i - \bar{\mathbf{x}})(\mathbf{x}_i - \bar{\mathbf{x}})'/(n-1)$. Let $r_{ij} = s_{ij}/(s_{ii}s_{jj})^{1/2}$, where $r_{ii} = 1$ for $i = 1, 2, \ldots, d$. Then $\mathbf{R} = (r_{ij})$ is called the *sample correlation matrix*. It does not matter if we use n instead of $(n-1)$, that is, use $\hat{\Sigma}$, the maximum likelihood estimate of Σ under normality, instead of $\mathbf{S}$ as $(n-1)$ cancels out of $\mathbf{R}$. Note that $\mathbf{S}$ is a random matrix so that it is considered again in the next chapter.

20.15. We now introduce an important device that allows the properties of population parameters to be carried over directly to their sample estimates. Consider the discrete random variable $\mathbf{y}$ with probability function

$$\text{pr}(\mathbf{y} = \mathbf{x}_i) = \frac{1}{n}, \quad i = 1, 2, \dots, n.$$

Then

$$E(\mathbf{y}) = \sum_{i=1}^{n} \mathbf{x}_i \frac{1}{n} = \overline{\mathbf{x}} \quad \text{and} \quad \text{var}(\mathbf{y}) = \sum_{i=1}^{n} (\mathbf{x}_i - \overline{\mathbf{x}})(\mathbf{x}_i - \overline{\mathbf{x}})' \frac{1}{n} = \hat{\Sigma}.$$

We can therefore "translate" sample properties into population properties using an appropriate discrete population.

Definition 20.10. Using the notation of the previous definition, consider the partition

$$\mathbf{S} = \begin{pmatrix} s_{11} & \mathbf{s}'_{12} \\ \mathbf{s}_{12} & \mathbf{S}_{22} \end{pmatrix}.$$

The *sample multiple correlation coefficient* between x_1 and $x_2, x_3, \dots, x_d$ is defined to be

$$R = (\mathbf{s}'_{12} \mathbf{S}_{22}^{-1} \mathbf{s}_{12} / s_{11})^{1/2}.$$

Under normality, R is the maximum likelihood estimate of $\mathcal{R}$, the population multiple correlation. For further details see Anderson [2003: section 4.4].

Sample versions of the other correlations and partial correlations can be defined in a similar manner. One simply replaces Σ by $\mathbf{S}$ or $\hat{\Sigma}$ in the population definitions. For example, replacing Σ by $\mathbf{S}$ in Definition 20.7, we have the sample partial correlation

$$r_{ij \cdot k+1,\dots,d} = \frac{s_{ij \cdot k+1,\dots,d}}{\left[s_{ii \cdot k+1,\dots,d} \, s_{jj \cdot k+1,\dots,d} \right]^{1/2}}.$$

For further details see Anderson [2003: section 4.3].

20.16. Using the method of (20.15), all the results and optimal properties for population parameters hold for the sample equivalents. For example, from (20.13b),

$$r_{i \cdot k+1,k+2,\dots,d} = (\mathbf{s}'_i \mathbf{S}_{22}^{-1} \mathbf{s}_i / s_{ii})^{1/2}.$$

Proofs. Section 20.3.2.

20.14a. Proved using the Cauchy–Schwarz inequality or using (20.15).

20.16. Muirhead [1982: 188].

20.4 QUADRATICS

20.17. Let $\mathbf{x}$ be an d-dimensional random vector with $E(\mathbf{x}) = \boldsymbol{\mu}$ and $\text{var}(\mathbf{x}) = \boldsymbol{\Sigma}$, a non-negative definite matrix of rank $r \le d$. Let $\mathbf{A}$ be a real $d \times d$ symmetric matrix, and let $\boldsymbol{\Sigma} = \mathbf{B}\mathbf{B}'$ (cf. 10.10), where $\mathbf{B}$ is $d \times r$ and $\mathbf{B}'\mathbf{A}\mathbf{B} \ne \mathbf{0}$. Then:

(a) With a suitable transformation of the form $\mathbf{x} = \mathbf{B}\mathbf{P}\mathbf{z}$ we find that

$$
\begin{aligned}
\mathbf{x}'\mathbf{A}\mathbf{x} &= a + 2\sum_{i=1}^{r} b_i z_i + \sum_{i=1}^{r} \lambda_i z_i^2 \\
&= \sum_{i=1}^{r} \lambda_i (z_i + \frac{b_i}{\lambda_i})^2 + (a - \sum_{i=1}^{r} \frac{b_i^2}{\lambda_i}), \quad \lambda_i \ne 0, \quad i = 1, 2, \ldots, r \\
&= \sum_{i=1}^{r} \lambda_i z_i^2 \quad \text{for} \quad \boldsymbol{\mu} = \mathbf{0},
\end{aligned}
$$

where $\mathbf{b}' = (b_1, b_2, \ldots, b_r) = \boldsymbol{\mu}'\mathbf{A}\mathbf{B}\mathbf{P}$, $\mathbf{z} = (z_1, z_2, \ldots, z_r)'$, $E(\mathbf{z}) = \mathbf{0}$, $\text{var}(\mathbf{z}) = \mathbf{I}_r$, $\mathbf{P}'\mathbf{B}'\mathbf{A}\mathbf{B}\mathbf{P} = \text{diag}(\lambda_1, \lambda_2, \ldots, \lambda_r)$, the λ_i are the eigenvalues of $\mathbf{B}'\mathbf{A}\mathbf{B}$ (i.e., of $\boldsymbol{\Sigma}\mathbf{A}$), $\mathbf{P}$ is $r \times r$ and orthogonal (i.e., $\mathbf{P}\mathbf{P}' = \mathbf{I}_r$), and $a = \boldsymbol{\mu}'\mathbf{A}\boldsymbol{\mu}$.

(b) If $\boldsymbol{\Sigma}$ is positive definite, we can choose $\mathbf{B}$ to be triangular (Cholesky decomposition) or $\boldsymbol{\Sigma}^{1/2}$ (cf. 10.32). In the latter case, we find that

$$
\mathbf{x}'\mathbf{A}\mathbf{x} = \sum_{i=1}^{d} \lambda_i (u_i + c_i)^2,
$$

where the λ_i are the eigenvalues of $\boldsymbol{\Sigma}^{1/2}\mathbf{A}\boldsymbol{\Sigma}^{1/2}$ (i.e., of $\boldsymbol{\Sigma}\mathbf{A}$), $E(\mathbf{u}) = \mathbf{0}$, $\text{var}(\mathbf{u}) = \mathbf{I}_d$, $\mathbf{c}' = (c_1, c_2, \ldots, c_d) = (\mathbf{P}'\boldsymbol{\Sigma}^{-1/2}\boldsymbol{\mu})'$, and $\mathbf{P}\mathbf{P}' = \mathbf{I}_d$.

20.18. Let $\mathbf{x}$ have mean $\boldsymbol{\mu}$ and non-negative definite variance matrix $\boldsymbol{\Sigma}$, and let $Q = E[(\mathbf{x} - \mathbf{a})'\mathbf{A}(\mathbf{x} - \mathbf{a})] \, (= E[\text{trace}\{(\mathbf{x} - \mathbf{a})'\mathbf{A}(\mathbf{x} - \mathbf{a})\}])$.

(a) Using (20.7),

$$
E(\mathbf{Q}) = \text{trace}\{\mathbf{A}E[(\mathbf{x} - \mathbf{a})(\mathbf{x} - \mathbf{a})']\} = \text{trace}(\mathbf{A}\boldsymbol{\Sigma}) + (\boldsymbol{\mu} - \mathbf{a})'\mathbf{A}(\boldsymbol{\mu} - \mathbf{a})'.
$$

(b) If $\boldsymbol{\Sigma} = \sigma^2\mathbf{I}$, we have the useful rule $E(\mathbf{Q}) = \sigma^2 \, \text{trace}\, \mathbf{A} + Q_{\mathbf{x} \to E(\mathbf{x})}$.

20.19. Let $\mathbf{x}$ be a $d \times 1$ random vector with mean $\boldsymbol{\mu}$, variance matrix $\boldsymbol{\Sigma}$, and finite fourth moments, so that $E(\mathbf{x}\mathbf{x}')$ and $E(\mathbf{x}\mathbf{x}' \otimes \mathbf{x}\mathbf{x}')$ exist, where "$\otimes$" refers to the Kronecker product. If $\mathbf{A}$ and $\mathbf{B}$ are real symmetric $d \times d$ matrices, then

$$
\begin{aligned}
\text{cov}(\mathbf{x}'\mathbf{A}\mathbf{x}, \mathbf{x}'\mathbf{B}\mathbf{x}) &= \text{trace}\{(\mathbf{A} \otimes \mathbf{B})E(\mathbf{x}\mathbf{x}' \otimes \mathbf{x}\mathbf{x}')\} \\
&\quad -\{\text{trace}(\mathbf{A}\boldsymbol{\Sigma}) + \boldsymbol{\mu}'\mathbf{A}\boldsymbol{\mu}\}\{\text{trace}(\mathbf{B}\boldsymbol{\Sigma}) + \boldsymbol{\mu}'\mathbf{B}\boldsymbol{\mu}\}.
\end{aligned}
$$

20.20. Let $\mathbf{x}$ be a random vector with elements $x_1, x_2, \ldots, x_n$ distributed as independent random variables with means $\theta_1, \theta_2, \ldots, \theta_n$, common variance μ_2, and common third and fourth moments about their means, μ_3 and μ_4, respectively (i.e., $\mu_r = E[(x_i - \theta_i)^r]$). Let $\mathbf{A}$ be any symmetric $n \times n$ matrix, and let $\mathbf{a} = \text{diag}\, \mathbf{A}$ be the column vector of the diagonal elements of $\mathbf{A}$.

(a) We have

$$\text{var}(\mathbf{x}'\mathbf{A}\mathbf{x}) = (\mu_4 - 3\mu_2^2)\mathbf{a}'\mathbf{a} + 2\mu_2^2\,\text{trace}(\mathbf{A}^2) + 4\mu_2\boldsymbol{\theta}'\mathbf{A}^2\boldsymbol{\theta} + 4\mu_3\boldsymbol{\theta}'\mathbf{A}\mathbf{a}.$$

(b) If the x_i are each normally distributed as $N(0, \sigma^2)$, then $\mu_3 = 0$, $\mu_4 = 3\mu_2^2$, $\mu_2 = \sigma^2$, and

$$\text{var}(\mathbf{x}'\mathbf{A}\mathbf{x}) = 2\sigma^4\,\text{trace}(\mathbf{A}^2).$$

If $\mathbf{B}$ is also a symmetric $n \times n$ matrix, then

$$\text{cov}(\mathbf{x}'\mathbf{A}\mathbf{x}, \mathbf{x}'\mathbf{B}\mathbf{x}) = 2\sigma^2\,\text{trace}(\mathbf{A}\mathbf{B}).$$

These results are generalized in (20.25).

(c) Let $\gamma_2 = (\mu_4 - 3\mu_2^2)/\mu_2^2$ be the common kurtosis, and let $\mathbf{P}_i$ $(i = 1, 2)$ be symmetric idempotent matrices with $\mathbf{p}_i = \text{diag}(\mathbf{P}_i)$, rank $\mathbf{P}_i = f_i\,(= \text{trace}\,\mathbf{P}_i)$, and $\mathbf{P}_1\mathbf{P}_2 = \mathbf{0}$. If $\mathbf{P}_i\boldsymbol{\theta} = \mathbf{0}$, then from (a) we have:

    (i) $\text{var}(\mathbf{y}'\mathbf{P}_i\mathbf{y}) = 2\sigma^4(f_i + \frac{1}{2}\gamma_2\mathbf{p}_i'\mathbf{p}_i)$.

    (ii) $\text{cov}(\mathbf{y}'\mathbf{P}_1\mathbf{y}, \mathbf{y}'\mathbf{P}_2\mathbf{y}) = \sigma^4\gamma_2\mathbf{p}_1'\mathbf{p}_2$.

This result is useful in examining the robustness of the F-test for a linear hypothesis and a linear model.

Proofs. Section 20.4.

    20.17a. Mathai and Provost [1992: 36].

    20.17b. Mathai and Provost [1992: 28–29].

    20.18. Schott [2005: 414] and Seber and Lee [2003: 9].

    20.19. Schott [2005: 414].

    20.20a. Quoted by Atiqullah [1962] and derived in Seber and Lee [2003: 10–11].

    20.20b. Seber and Lee [2003: 16].

    20.20c. Atiqullah [1962] and Seber and Lee [2003: 236–237]. Here (ii) is obtained from $\frac{1}{2}\{\text{var}[(\mathbf{y}'(\mathbf{P}_1 + \mathbf{P}_2)\mathbf{y}] - \sum_i \text{var}(\mathbf{y}'\mathbf{P}_i\mathbf{y})\}$.

20.5   MULTIVARIATE NORMAL DISTRIBUTION

20.5.1   Definition and Properties

Definition 20.11. Let $\mathbf{x}$ be a $d \times 1$ random vector with mean $\boldsymbol{\mu}$ and variance matrix $\boldsymbol{\Sigma}$, which is positive definite. Then $\mathbf{x}$ is said to have a (nonsingular) *multivariate normal* (or multinormal) distribution if its probability density function is given by

$$\begin{aligned}
f(\mathbf{x}) &= f(x_1, x_2, \ldots, x_d) \\
&= (2\pi)^{-d/2}(\det\boldsymbol{\Sigma})^{-1/2}\exp\{\tfrac{1}{2}(\mathbf{x} - \boldsymbol{\mu})'\boldsymbol{\Sigma}^{-1/2}(\mathbf{x} - \boldsymbol{\mu})\} \\
&\quad\quad (-\infty < x_i < \infty, \quad i = 1, 2, \ldots, d).
\end{aligned}$$

We write $x \sim N_d(\boldsymbol{\mu}, \boldsymbol{\Sigma})$. When $d = 1$, we replace N_1 by N, the univariate normal distribution. Note that $x \sim N_d(\mathbf{0}, \mathbf{I}_d)$ if and only if the x_i are independently distributed as $N(0, 1)$. If $x \sim N_d(\boldsymbol{\mu}, \boldsymbol{\Sigma})$, then $y = \boldsymbol{\Sigma}^{-1/2}(x - \boldsymbol{\mu}) \sim N_d(\mathbf{0}, \mathbf{I}_d)$. Sampling from a normal distribution is discussed in Section 21.3.

If $\boldsymbol{\Sigma}$ is positive semi-definite (i.e., singular), then the probability distribution still exists, but not the density function. However, we can extend our definition to include the so-called *singular* multivariate normal distribution using one of the two following equivalent definitions, which includes both the nonsingular and singular cases.

1. The random vector x is multivariate normal if and only if $y = \mathbf{a}'\mathbf{X}$ is univariate normal for all $\mathbf{a}$. If $y = b$ we define y to be $N(b, 0)$.

2. A random $d \times 1$ vector x with mean $\boldsymbol{\mu}$ and variance matrix $\boldsymbol{\Sigma}$ has a multivariate normal distribution if it has the same distribution as $\mathbf{Az} + \boldsymbol{\mu}$, where $\mathbf{A}$ is any $d \times m$ matrix satisfying $\boldsymbol{\Sigma} = \mathbf{AA}'$, and $z \sim N_m(\mathbf{0}, \mathbf{I}_m)$.

The singular normal distribution occurs in many places in statistics, for example the distribution of residuals from linear models (Seber and Lee [2003]) and the distribution of the estimated cell proportions from sample survey data (Rao and Scott [1984]). For a general reference on the multivariate normal see Tong [1990].

20.21. Adopting the notation of Definition 20.11, suppose that $\boldsymbol{\Sigma}$ is singular of rank r (i.e., positive semi-definite). Then, from (20.4), $x - \boldsymbol{\mu} \in \mathcal{C}(\boldsymbol{\Sigma})$ and we can express $\boldsymbol{\Sigma} = \mathbf{RR}'$, where $\mathbf{R}$ is $d \times r$ of rank r (cf. 10.10). Hence if $\mathbf{P}_{\boldsymbol{\Sigma}}$ is the orthogonal projector onto $\mathcal{C}(\boldsymbol{\Sigma})$ (cf. 2.49d), we have, for $(\mathbf{I}_d - \mathbf{P}_{\boldsymbol{\Sigma}})(x - \boldsymbol{\mu}) = \mathbf{0}$, the density function

$$f(\mathbf{x}) = (2\pi)^{r/2}[\det(\mathbf{R}'\mathbf{R})]^{-1/2} \exp[-\tfrac{1}{2}(x - \boldsymbol{\mu})\boldsymbol{\Sigma}^-(x - \boldsymbol{\mu})],$$

and 0 otherwise. Here $\boldsymbol{\Sigma}^-$ is a weak inverse of $\boldsymbol{\Sigma}$.

20.22. (A Useful Integral) If $\mathbf{A}$ and $\mathbf{B}$ are symmetric $n \times n$ matrices and $\mathbf{B}$ is positive definite, then using the multivariate normal density function we have

$$\int_{-\infty}^{+\infty} \cdots \int_{-\infty}^{+\infty} (\mathbf{x}'\mathbf{Ax} + \mathbf{a}'\mathbf{x} + a_0) \exp[-(\mathbf{x}'\mathbf{Bx} + \mathbf{b}'\mathbf{x} + b_0)] \, dx_1 \cdots dx_n$$
$$= \tfrac{1}{2}\pi^{n/2}|\mathbf{B}|^{-1/2} \exp(\tfrac{1}{4}\mathbf{b}'\mathbf{B}^{-1}\mathbf{b} - b_0)$$
$$\times [\text{trace}(\mathbf{AB}^{-1}) - \mathbf{a}'\mathbf{B}^{-1}\mathbf{b} + \tfrac{1}{2}\mathbf{b}'\mathbf{B}^{-1}\mathbf{AB}^{-1}\mathbf{b} + 2a_0].$$

20.23. Let $x \sim N_d(\boldsymbol{\mu}, \boldsymbol{\Sigma})$, where the distribution may be singular or nonsingular.

(a) The moment generating function of x is $E[\exp(\mathbf{t}'\mathbf{x})] = \exp(\mathbf{t}'\boldsymbol{\mu} + \tfrac{1}{2}\mathbf{t}'\boldsymbol{\Sigma}\mathbf{t})$. This uniquely determines the (nonsingular) distribution when $\boldsymbol{\Sigma}$ is positive definite.

(b) If $\mathbf{C}$ is $m \times d$, then $\mathbf{Cx} \sim N_m(\mathbf{C}\boldsymbol{\mu}, \mathbf{C}\boldsymbol{\Sigma}\mathbf{C}')$. The distribution is nonsingular if $\boldsymbol{\Sigma}$ is positive definite and $\mathbf{C}$ has rank m.

(c) Any subset of a multivariate normal distribution is multivariate normal.

(d) If the covariance of any two vectors that contain disjoint subsets of x is zero, then the two vectors are statistically independent.

(e) If $\text{cov}(\mathbf{Ax}, \mathbf{Bx}) = \mathbf{0}$, then $\mathbf{Ax}$ and $\mathbf{Bx}$ are statistically independent.

(f) Suppose $\boldsymbol{\Sigma}$ is positive definite, and let

$$\mathbf{x} = \begin{pmatrix} \mathbf{x}_1 \\ \mathbf{x}_2 \end{pmatrix}, \quad \boldsymbol{\mu} = \begin{pmatrix} \boldsymbol{\mu}_1 \\ \boldsymbol{\mu}_2 \end{pmatrix}, \quad \text{and} \quad \boldsymbol{\Sigma} = \begin{pmatrix} \boldsymbol{\Sigma}_{11} & \boldsymbol{\Sigma}_{12} \\ \boldsymbol{\Sigma}_{21} & \boldsymbol{\Sigma}_{22} \end{pmatrix},$$

where $\mathbf{x}_i$ and $\boldsymbol{\mu}_i$ are $d_i \times 1$, $\boldsymbol{\Sigma}_{ii}$ is $d_i \times d_i$ $(i = 1, 2)$, and $d_1 + d_2 = d$. We then have the following conditional distribution

$$\mathbf{x}_2 \mid \mathbf{x}_1 \sim N_{d_2}(\boldsymbol{\mu}_{2\cdot1}, \boldsymbol{\Sigma}_{22\cdot1}),$$

where $\boldsymbol{\mu}_{2\cdot1} = \boldsymbol{\mu}_2 + \boldsymbol{\Sigma}_{21}\boldsymbol{\Sigma}_{11}^{-1}(\mathbf{x}_1 - \boldsymbol{\mu}_1)$ and $\boldsymbol{\Sigma}_{22\cdot1} = \boldsymbol{\Sigma}_{22} - \boldsymbol{\Sigma}_{21}\boldsymbol{\Sigma}_{11}^{-1}\boldsymbol{\Sigma}_{12}$. Note that $\boldsymbol{\Sigma}_{22\cdot1}$ is the Schur complement of $\boldsymbol{\Sigma}_{11}$ in $\boldsymbol{\Sigma}$, and it is frequently expressed in the form $(\boldsymbol{\Sigma}/\boldsymbol{\Sigma}_{11})$ (cf. Section 14.1).

(g) The result (f) still holds if $\boldsymbol{\Sigma}$ is singular and we replace $\boldsymbol{\Sigma}_{11}^{-1}$ by $\boldsymbol{\Sigma}_{11}^{-}$, any weak inverse of $\boldsymbol{\Sigma}_{11}$.

20.24. (Moments) If $\mathbf{x} \sim N_d(\mathbf{0}, \boldsymbol{\Sigma})$, where $\boldsymbol{\Sigma}$ is positive definite, and $\mathbf{P}_d = \frac{1}{2}(\mathbf{I}_{d^2} + \mathbf{I}_{(d,d)})$ $(= \mathbf{N}_d$, the symmetrizer of (11.29h–i)), then:

(a) $\text{E}(\mathbf{x} \otimes \mathbf{x}) = \text{vec}\,\boldsymbol{\Sigma}$.

(b) $\text{E}(\mathbf{xx}' \otimes \mathbf{xx}') = 2\mathbf{P}_d(\boldsymbol{\Sigma} \otimes \boldsymbol{\Sigma}) + (\text{vec}\,\boldsymbol{\Sigma})(\text{vec}\,\boldsymbol{\Sigma})'$.

If just one of any of the $\mathbf{x}$'s is replaced by a constant vector, then the answer is $\mathbf{0}$.

(c) $\text{var}(\mathbf{x} \otimes \mathbf{x}) = 2\mathbf{P}_d(\boldsymbol{\Sigma} \otimes \boldsymbol{\Sigma})$.

(d) Suppose $\mathbf{x} \sim N_d(\boldsymbol{\mu}, \boldsymbol{\Sigma})$ and $\boldsymbol{\Sigma}$ is positive definite.

    (i) $\text{E}(\mathbf{x} \otimes \mathbf{x}) = \text{E}(\mathbf{z} \otimes \mathbf{z}) + \boldsymbol{\mu} \otimes \boldsymbol{\mu} = \text{vec}\,\boldsymbol{\Sigma} + (\boldsymbol{\mu} \otimes \boldsymbol{\mu})$, where $\mathbf{z} = \mathbf{x} - \boldsymbol{\mu}$.

    (ii) $\text{var}(\mathbf{x} \otimes \mathbf{x}) = 2\mathbf{P}_d(\boldsymbol{\Sigma} \otimes \boldsymbol{\Sigma} + \boldsymbol{\Sigma} \otimes \boldsymbol{\mu}\boldsymbol{\mu}' + \boldsymbol{\mu}\boldsymbol{\mu}' \otimes \boldsymbol{\Sigma})$.

(e) Higher moments are given by Graybill [1983: section 10.9] for the case $\mathbf{x} \sim N_d(\mathbf{0}, \mathbf{I}_d)$.

Proofs. Section 20.5.1.

20.21. Sengupta and Jammalamadaka [2003: 58].

20.22. Graybill [1983: 342] and Harville [1997: 322].

20.23. Anderson [2003: chapter 2] and Seber and Lee [2003: chapter 2]. For (f) see Schott [2005: 260–261] and Seber and Lee [2003: 25–26], and for (g) see Sengupta and Jammalamadaka [2003: 59].

20.24a–c. Schott [2005: 416].

20.24d(i). Schott [2005: 416].

20.24d(ii). Abadir and Magnus [2005: 310].

20.5.2 Quadratics in Normal Variables

20.25. Let $\mathbf{x} \sim N_d(\boldsymbol{\mu}, \boldsymbol{\Sigma})$, where $\boldsymbol{\Sigma}$ is positive definite.

(a) We have $(\mathbf{x} - \boldsymbol{\mu})'\boldsymbol{\Sigma}^{-1}(\mathbf{x} - \boldsymbol{\mu}) \sim \chi_d^2$ and $\mathbf{x}'\boldsymbol{\Sigma}^{-1}\mathbf{x} \sim \chi_d^2(\delta)$, the noncentral chi-squared distribution with noncentrality parameter $\delta = \boldsymbol{\mu}'\boldsymbol{\Sigma}^{-1}\boldsymbol{\mu}$.

(b) Using the notation of (20.23f), we have the following:

  (i) $Q_1 = (\mathbf{x}_1 - \boldsymbol{\mu}_1)'\boldsymbol{\Sigma}_{11}^{-1}(\mathbf{x}_1 - \boldsymbol{\mu}_1) \sim \chi_{d_1}^2$ (by 20.23c and (a) above).

  (ii) Let $Q_2 = (\mathbf{x} - \boldsymbol{\mu})'\boldsymbol{\Sigma}^{-1}(\mathbf{x} - \boldsymbol{\mu}) - (\mathbf{x}_1 - \boldsymbol{\mu}_1)'\boldsymbol{\Sigma}_{11}^{-1}(\mathbf{x}_1 - \boldsymbol{\mu}_1)$. Then, from (20.23f), $\mathbf{x}_2 \mid \mathbf{x}_1 \sim N_{d_2}(\boldsymbol{\mu}_{2\cdot1}, \boldsymbol{\Sigma}_{22\cdot1})$ and, conditional on $\mathbf{x}_1$,

$$Q_2 = (\mathbf{x}_2 - \boldsymbol{\mu}_{2\cdot1})'\boldsymbol{\Sigma}_{22\cdot1}^{-1}(\mathbf{x}_2 - \boldsymbol{\mu}_{2\cdot1}) \sim \chi_{d_2}^2.$$

Since this distribution is not a function of $\mathbf{x}_1$, it is also the unconditional distribution; (iii) below holds for the same reason.

  (iii) Q_1 and Q_2 are statistically independent.

(c) Let $\mathbf{A}$ and $\mathbf{B}$ be $d \times d$ matrices.

  (i)

$$
\begin{aligned}
\mathrm{E}(\mathbf{x}'\mathbf{A}\mathbf{x} \cdot \mathbf{x}'\mathbf{B}\mathbf{x}) \;=\;& \mathrm{trace}(\mathbf{A}\boldsymbol{\Sigma})\,\mathrm{trace}(\mathbf{B}\boldsymbol{\Sigma}) + 2\,\mathrm{trace}(\mathbf{A}\boldsymbol{\Sigma}\mathbf{B}\boldsymbol{\Sigma}) \\
&+ \mathrm{trace}(\mathbf{A}\boldsymbol{\Sigma})\boldsymbol{\mu}'\mathbf{B}\boldsymbol{\mu} + \mathrm{trace}(\mathbf{B}\boldsymbol{\Sigma})\boldsymbol{\mu}'\mathbf{A}\boldsymbol{\mu} \\
&+ 4\boldsymbol{\mu}'\mathbf{A}\boldsymbol{\Sigma}\mathbf{B}\boldsymbol{\mu} + (\boldsymbol{\mu}'\mathbf{A}\boldsymbol{\mu})(\boldsymbol{\mu}'\mathbf{B}\boldsymbol{\mu}).
\end{aligned}
$$

  (ii) $\mathrm{cov}(\mathbf{x}'\mathbf{A}\mathbf{x}, \mathbf{x}'\mathbf{B}\mathbf{x}) = 2\,\mathrm{trace}(\mathbf{A}\boldsymbol{\Sigma}\mathbf{B}\boldsymbol{\Sigma}) + 4\boldsymbol{\mu}'\mathbf{A}\boldsymbol{\Sigma}\mathbf{B}\boldsymbol{\mu}$.

  (iii) Setting $\mathbf{A} = \mathbf{B}$ in (ii), we have

$$\mathrm{var}(\mathbf{x}'\mathbf{A}\mathbf{x}) = 2\,\mathrm{trace}[(\mathbf{A}\boldsymbol{\Sigma})^2] + 4\boldsymbol{\mu}'\mathbf{A}\boldsymbol{\Sigma}\mathbf{A}\boldsymbol{\mu}.$$

(d) Let $\mathbf{x} \sim N_d(\mathbf{0}, \mathbf{I}_d)$, and let $\mathbf{A}$, $\mathbf{B}$, and $\mathbf{C}$ be all $d \times d$ symmetric matrices.

  (i)

$$
\begin{aligned}
\mathrm{E}(\mathbf{x}'\mathbf{A}\mathbf{x} &\cdot \mathbf{x}'\mathbf{B}\mathbf{x} \cdot \mathbf{x}'\mathbf{C}\mathbf{x}) \\
=\;& \mathrm{trace}\,\mathbf{A}\,\mathrm{trace}\,\mathbf{B}\,\mathrm{trace}\,\mathbf{C} + 2\,\mathrm{trace}\,\mathbf{A}\,\mathrm{trace}(\mathbf{B}\mathbf{C}) \\
&+ 2\,\mathrm{trace}\,\mathbf{B}\,\mathrm{trace}(\mathbf{A}\mathbf{C}) + 2\,\mathrm{trace}\,\mathbf{C}\,\mathrm{trace}(\mathbf{A}\mathbf{B}) \\
&+ 8\,\mathrm{trace}(\mathbf{A}\mathbf{B}\mathbf{C}).
\end{aligned}
$$

  (ii) If $\mathbf{x} \sim N_d(\mathbf{0}, \boldsymbol{\Sigma})$, we replace $\mathbf{A}$, $\mathbf{B}$, and $\mathbf{C}$ in the right-hand side of (i) by $\mathbf{A}\boldsymbol{\Sigma}$, $\mathbf{B}\boldsymbol{\Sigma}$, and $\mathbf{C}\boldsymbol{\Sigma}$, respectively.

(e) If $\mathbf{x} \sim N_d(\mathbf{0}, \boldsymbol{\Sigma})$, then:

  (i)

$$
\begin{aligned}
\mathrm{cov}[(\mathbf{x}'\mathbf{A}\mathbf{x})^2, (\mathbf{x}'\mathbf{B}\mathbf{x})] \\
=\;& 4\,\mathrm{trace}(\mathbf{A}\boldsymbol{\Sigma})\,\mathrm{trace}(\mathbf{A}\boldsymbol{\Sigma}\mathbf{B}\boldsymbol{\Sigma}) + 8\,\mathrm{trace}[(\mathbf{A}\boldsymbol{\Sigma})^2\mathbf{B}\boldsymbol{\Sigma}].
\end{aligned}
$$

(ii)

$$E[(\mathbf{x}'\mathbf{A}\mathbf{x})^3]$$
$$= \quad [\text{trace}(\mathbf{A}\boldsymbol{\Sigma})]^3 + 6\,\text{trace}(\mathbf{A}\boldsymbol{\Sigma})\,\text{trace}[(\mathbf{A}\boldsymbol{\Sigma})^2] + 8\,\text{trace}[(\mathbf{A}\boldsymbol{\Sigma})^3].$$

(f) (Moment Generating Function) If $Q = \mathbf{x}'\mathbf{A}\mathbf{x} + \mathbf{a}'\mathbf{x} + d$, where $\mathbf{A}$ is real and symmetric, then the moment generating function (m.g.f.) of Q is

$$
\begin{aligned}
M_Q(t) \quad &= \quad E[\exp(Qt)] \\
&= \quad [\det(\mathbf{I}_d - 2t\mathbf{A}\boldsymbol{\Sigma})]^{-1/2}\exp\{-\tfrac{1}{2}(\boldsymbol{\mu}'\boldsymbol{\Sigma}^{-1}\boldsymbol{\mu} - 2td) \\
&\quad + \tfrac{1}{2}(\boldsymbol{\mu} + t\boldsymbol{\Sigma}\mathbf{a})'(\mathbf{I}_d - 2t\mathbf{A}\boldsymbol{\Sigma})^{-1}\boldsymbol{\Sigma}^{-1}(\boldsymbol{\mu} + t\boldsymbol{\Sigma}\mathbf{a})\} \\
&= \quad [\det(\mathbf{I}_d - 2t\boldsymbol{\Sigma}^{1/2}\mathbf{A}\boldsymbol{\Sigma}^{1/2})]^{-1/2}\exp\{t(d + \boldsymbol{\mu}'\mathbf{A}\boldsymbol{\mu} + \mathbf{a}'\boldsymbol{\mu}) \\
&\quad + \tfrac{t^2}{2}(\boldsymbol{\Sigma}^{1/2}\mathbf{a} + 2\boldsymbol{\Sigma}^{1/2}\mathbf{A}\boldsymbol{\mu})'(\mathbf{I}_d - 2t\boldsymbol{\Sigma}^{1/2}\mathbf{A}\boldsymbol{\Sigma}^{1/2})^{-1} \\
&\quad\quad \times (\boldsymbol{\Sigma}^{1/2}\mathbf{a} + 2\boldsymbol{\Sigma}^{1/2}\mathbf{A}\boldsymbol{\mu})\} \\
&= \quad \exp\{t(d + \boldsymbol{\mu}'\mathbf{A}\boldsymbol{\mu} + \mathbf{a}'\boldsymbol{\mu}) + \tfrac{t^2}{2}\sum_{i=1}^{d}c_i^2(1 - 2t\lambda_i)^{-1}\} \\
&\quad \times \prod_{i=1}^{d}(1 - 2t\lambda_i)^{-1/2},
\end{aligned}
$$

where

$$\mathbf{T}'\boldsymbol{\Sigma}^{1/2}\mathbf{A}\boldsymbol{\Sigma}^{1/2}\mathbf{T} = \text{diag}(\lambda_1, \ldots, \lambda_d),$$

$\mathbf{T}$ is orthogonal, and

$$\mathbf{c} = (c_1, \ldots, c_d)' = \mathbf{T}'(\boldsymbol{\Sigma}^{1/2}\mathbf{a} + 2\boldsymbol{\Sigma}^{1/2}\mathbf{A}\boldsymbol{\mu}).$$

Note that

$$\prod_{i=1}^{d}(1 - 2t\lambda_i)^{-1/2} = [\det(\mathbf{I}_d - 2t\boldsymbol{\Sigma}^{1/2}\mathbf{A}\boldsymbol{\Sigma}^{1/2})]^{-1/2} = [\det(\mathbf{I}_d - 2t\mathbf{A}\boldsymbol{\Sigma})]^{-1/2}.$$

The m.g.f. can be used to obtain moments of the quadratic form—for example, (c)(iii). Thus if $Q_0 = \mathbf{x}'\mathbf{A}\mathbf{x}$ we have:

(i) $E(Q_0) = \text{trace}(\mathbf{A}\boldsymbol{\Sigma}) + \boldsymbol{\mu}'\mathbf{A}\boldsymbol{\mu}$.

(ii) $E[(Q_0)^2] = 2\,\text{trace}[(\mathbf{A}\boldsymbol{\Sigma})^2] + 4\boldsymbol{\mu}'(\mathbf{A}\boldsymbol{\Sigma})\mathbf{A}\boldsymbol{\mu} + \{\text{trace}(\mathbf{A}\boldsymbol{\Sigma}) + \boldsymbol{\mu}'\mathbf{A}\boldsymbol{\mu}\}^2$.

(iii)

$$
\begin{aligned}
E[(Q_0)^3] \quad &= \quad 8\{\text{trace}[(\mathbf{A}\boldsymbol{\Sigma})^3] + 3\boldsymbol{\mu}'(\mathbf{A}\boldsymbol{\Sigma})^2\mathbf{A}\boldsymbol{\mu}\} \\
&\quad + 6\{\text{trace}[(\mathbf{A}\boldsymbol{\Sigma})^2] + 2\boldsymbol{\mu}'\mathbf{A}\boldsymbol{\Sigma}\boldsymbol{\mu}\}\{\text{trace}(\mathbf{A}\boldsymbol{\Sigma}) + \boldsymbol{\mu}'\mathbf{A}\boldsymbol{\mu}\} \\
&\quad + \{\text{trace}(\mathbf{A}\boldsymbol{\Sigma}) + \boldsymbol{\mu}'\mathbf{A}\boldsymbol{\mu}\}^3.
\end{aligned}
$$

General expressions for $E(Q_0^r)$ and $E(Q^r)$ are given by Mathai and Provost [1992: 53–54], who also give formulae for $E(Q_0^{-h})$, where $h > 0$ and can be a fraction (Mathai and Provost [1992: 56–59]).

20.26. Let $\mathbf{x}$ have the singular normal distribution $N_d(\boldsymbol{\mu}, \boldsymbol{\Sigma})$, where $\boldsymbol{\Sigma}$ is non-negative definite of rank r $(r < d)$.

(a) $\mathbf{x}'\boldsymbol{\Sigma}^+\mathbf{x} \sim \chi_r^2(\delta)$, the noncentral chi-squared distribution with noncentrality parameter $\delta = \boldsymbol{\mu}'\boldsymbol{\Sigma}\boldsymbol{\mu}$. Here $\boldsymbol{\Sigma}^+$ is the Moore–Penrose inverse of $\boldsymbol{\Sigma}$.

(b) (Moment Generating Function) Let $\boldsymbol{\Sigma} = \mathbf{BB}'$ (cf. 10.10), where $\mathbf{B}$ is $d \times r$ of rank r and $\mathbf{B}'\mathbf{AB} \neq \mathbf{0}$. If $\mathbf{A}$ is a $d \times d$ real symmetric matrix, then the moment generating functiom (m.g.f.) of $Q = \mathbf{x}'\mathbf{Ax} + \mathbf{a}'\mathbf{x} + d$ is

$$
\begin{aligned}
M_Q(t) \;=\; & [\det(\mathbf{I}_r - 2t\mathbf{B}'\mathbf{AB})]^{-1/2} \exp\{t(\boldsymbol{\mu}'\mathbf{A}\boldsymbol{\mu} + \mathbf{a}'\boldsymbol{\mu} + d) \\
& + \tfrac{t^2}{2}(\mathbf{B}'\mathbf{a} + 2\mathbf{B}'\mathbf{A}\boldsymbol{\mu})'(\mathbf{I}_d - 2t\mathbf{B}'\mathbf{AB})^{-1}(\mathbf{B}'\mathbf{a} + 2\mathbf{B}'\mathbf{A}\boldsymbol{\mu})\}.
\end{aligned}
$$

An alternative expression in terms of eigenvalues is given by Mathai and Provost [1992: 46–47]. Positive, negative, and fractional moments of $Q_0 = \mathbf{x}'\mathbf{Ax}$ are given by Mathai and Provost [1992: 54–55, 61–65].

The characteristic function of Q is obtained from the m.g.f. by replacing t by it, where $i = \sqrt{-1}$.

20.27. Let $Q_i = \mathbf{x}'\mathbf{A}_i\mathbf{x} + \mathbf{a}_i'\mathbf{x} + d_i$, where $\mathbf{A}_i$ is a real $d \times d$ symmetric matrix $(i = 1, 2)$, and suppose $\mathbf{x} \sim N_d(\boldsymbol{\mu}, \boldsymbol{\Sigma})$.

(a) If $\boldsymbol{\Sigma}$ is positive definite, then the joint moment generating function (m.g.f.) of Q_1 and Q_2 is

$$
\begin{aligned}
M_{Q_1,Q_2}(t_1, t_2) \;=\; & [\det(\mathbf{I}_n - 2t_1\mathbf{A}_1\boldsymbol{\Sigma} - 2t_2\mathbf{A}_2\boldsymbol{\Sigma})]^{-1/2} \\
& \times \exp\{-\tfrac{1}{2}(\boldsymbol{\mu}'\boldsymbol{\Sigma}^{-1}\boldsymbol{\mu} - 2t_1 d_1 - 2t_2 d_2) \\
& + \tfrac{1}{2}(t_1\boldsymbol{\Sigma}\mathbf{a}_1 + t_2\boldsymbol{\Sigma}\mathbf{a}_2 + \boldsymbol{\mu})'(\mathbf{I}_d - 2t_1\mathbf{A}_1\boldsymbol{\Sigma} - 2t_2\mathbf{A}_2\boldsymbol{\Sigma})^{-1} \\
& \times \boldsymbol{\Sigma}^{-1}(t_1\boldsymbol{\Sigma}\mathbf{a}_1 + t_2\boldsymbol{\Sigma}\mathbf{a}_2 + \boldsymbol{\mu})\}.
\end{aligned}
$$

(b) If $\boldsymbol{\Sigma}$ is non-negative definite and $\boldsymbol{\Sigma} = \mathbf{BB}'$ (cf. 10.10), where $\mathbf{B}$ is $d \times r$ of rank r $(r < d)$, then the joint m.g.f. of Q_1 and Q_2 is

$$
\begin{aligned}
M_{Q_1,Q_2}(t_1, t_2) \;=\; & [\det(\mathbf{I}_r - 2t_1\mathbf{B}'\mathbf{A}_1\mathbf{B} - 2t_2\mathbf{B}'\mathbf{A}_2\mathbf{B})]^{-1/2} \\
& \times \exp\{t_1(\boldsymbol{\mu}'\mathbf{A}_1\boldsymbol{\mu} + \mathbf{a}_1'\boldsymbol{\mu} + d_1) + t_2(\boldsymbol{\mu}'\mathbf{A}_2\boldsymbol{\mu} + \mathbf{a}_2'\boldsymbol{\mu} + d_2) \\
& + \tfrac{1}{2}\boldsymbol{\beta}'(\mathbf{I}_r - 2t_1\mathbf{B}'\mathbf{AB} - 2t_2\mathbf{B}'\mathbf{AB})\boldsymbol{\beta}\},
\end{aligned}
$$

where

$$
\begin{aligned}
\boldsymbol{\beta} = & (\mathbf{I}_r - 2t_1\mathbf{B}'\mathbf{A}_1\mathbf{B} - 2t_2\mathbf{B}'\mathbf{A}_2\mathbf{B})^{-1} \\
& \times (t_1\mathbf{B}'\mathbf{a}_1 + 2t_2\mathbf{B}'\mathbf{A}_1\boldsymbol{\mu} + t_2\mathbf{B}'\mathbf{a}_2 + 2t_2\mathbf{B}'\mathbf{A}_2\boldsymbol{\mu}).
\end{aligned}
$$

Note that (a) follows from (b) by setting $\mathbf{B} = \mathbf{B}' = \boldsymbol{\Sigma}^{1/2}$ (cf. 10.32). We can obtain various special cases, for example: (i) if we set $\mathbf{a}_i = \mathbf{0}$ and $d_i = 0$ for $i = 1, 2$, and $\boldsymbol{\mu} = \mathbf{0}$, we get the joint m.g.f. of $\mathbf{x}'\mathbf{A}_1\mathbf{x}$ and $\mathbf{x}'\mathbf{A}_2\mathbf{x}$, or (ii) if we set $\mathbf{A}_2 = \mathbf{0}$ we get the joint m.g.f. for a quadratic and a linear form.

In (a) and (b), the joint characteristic function is obtained by replacing t_1 and t_2 by it_1 and it_2, respectively where $i = \sqrt{-1}$.

The above results were proved by Mathai and Provost [1992: 66, section 3.2c] and extended to more than two quadratics. They can also be used to obtain various product moments, for example if $\boldsymbol{\Sigma}$ is positive definite or non-negative definite we have

(i) $\text{cov}(\mathbf{x}'\mathbf{A}\mathbf{x}, \mathbf{a}'\mathbf{x}) = 2\boldsymbol{\mu}'\mathbf{A}\boldsymbol{\Sigma}\mathbf{a}$.

(ii) $\text{cov}(\mathbf{x}'\mathbf{A}_1\mathbf{x}, \mathbf{x}'\mathbf{A}_2\mathbf{x}) = 2\,\text{trace}(\mathbf{A}_1\boldsymbol{\Sigma}\mathbf{A}_2\boldsymbol{\Sigma}) + \boldsymbol{\mu}'\mathbf{A}_1\boldsymbol{\Sigma}\mathbf{A}_2\boldsymbol{\mu}$.

The m.g.f.s can also be used to obtain cumulants. The reader is referred to Mathai and Provost [1992: sections 3.2d and 3.3] for further details.

20.28. (Distribution of a Quadratic) If $\mathbf{x} \sim N_d(\boldsymbol{\mu}, \boldsymbol{\Sigma})$, where $\boldsymbol{\Sigma}$ is positive definite, and $\mathbf{A}$ is a $d \times d$ symmetric matrix, then from (20.17b) we have the representation

$$Q = \mathbf{x}'\mathbf{A}\mathbf{x} = \sum_{i=1}^{d} \lambda_j (u_i + c_i)^2,$$

where the λ_i are the eigenvalues of $\boldsymbol{\Sigma}^{1/2}\mathbf{A}\boldsymbol{\Sigma}^{1/2}$ (i.e., of $\boldsymbol{\Sigma}\mathbf{A}$), $\mathbf{u} = (u_1, u_2, \ldots, u_d)'$ is distributed as $N_d(\mathbf{0}, \mathbf{I}_d)$, $\mathbf{c} = (c_i) = \mathbf{P}'\boldsymbol{\Sigma}^{-1/2}\boldsymbol{\mu}$, and $\mathbf{P}$ is orthogonal. Here the u_i^2 are independently and identically distributed as χ_1^2, while the $(u_i + c_i)^2$ are independently distributed as non-central chi-square $\chi_1^2(c_i^2)$ with noncentrality parameter c_i^2.

(a) When $\boldsymbol{\mu} = \mathbf{0}$, then $\mathbf{c} = \mathbf{0}$ and we find that Q is a linear combination of statistically independent χ_1^2 variables. If the distinct eigenvalues are μ_j with algebraic multiplicity m_j $(j = 1, 2, \ldots, s)$, then $\mathbf{Q} \sim \sum_{j=1}^{s} \mu_j \chi_{m_j}^2$.

(b) If $\boldsymbol{\mu} \neq \mathbf{0}$, then Q is a linear combination of statistically independent non-central chi-square variables, each with one degree of freedom.

The above results can be used to find various infinite series expansions, including one in terms of chi-square densities, and some approximations for the distribution of Q. If, in (a), m_j is even $(m_j = 2\nu_j$, say), then a finite expression for the distribution of Q is available. Details of all this material including expressions for the case when $\boldsymbol{\Sigma}$ is singular and results on ratios of quadratics are given by Mathai and Provost [1992: chapter 4]. They also give extensive reference lists.

Proofs. Section 20.5.2.

20.25a. Muirhead [1982: 26–27].

20.25b. Schott [2005: 261].

20.25c. Graybill [1983: 367] and Schott [2005: 418–419].

20.25d. Graybill [1983: 368] and Schott [2005: 420, the expected value of the product of four quadratics is also given]. Magnus [1978] gives an expression for the expectation of the product of any number of quadratics.

20.25e. Graybill [1983: 368].

20.25f. Mathai and Provost [1992: 40, 42].

20.26a. Schott [2005: 405, he calls $\delta/2$ the noncentrality parameter].

20.26b. Mathai and Provost [1992: 45].

20.27a–b. Mathai and Provost [1992: 67–68].

20.5.3   Quadratics and Chi-Squared

20.29. Suppose $\mathbf{x} \sim N_d(\boldsymbol{\mu}, \boldsymbol{\Sigma})$ and $\mathbf{A}$ is a $d \times d$ real symmetric matrix.

(a) If $\boldsymbol{\Sigma}$ is positive definite, then $\mathbf{x}'\mathbf{A}\mathbf{x} \sim \chi_r^2(\delta)$, where $r = \operatorname{rank} \mathbf{A}$ $(= \operatorname{rank}(\mathbf{A}\boldsymbol{\Sigma}))$ and $\delta = \boldsymbol{\mu}'\mathbf{A}\boldsymbol{\mu}$, if and only if $\mathbf{A}\boldsymbol{\Sigma}$ is idempotent, namely $\mathbf{A}\boldsymbol{\Sigma}\mathbf{A}\boldsymbol{\Sigma} = \mathbf{A}\boldsymbol{\Sigma}$ (i.e., $\mathbf{A}\boldsymbol{\Sigma}\mathbf{A} = \mathbf{A}$). We get two special cases by setting (i) $\boldsymbol{\mu} = \mathbf{0}$ (i.e., $\delta = 0$ and the distribution is central chi-square, χ_r^2) and (ii) $\boldsymbol{\Sigma} = \mathbf{I}_d$.

(b) If $\boldsymbol{\Sigma}$ is non-negative definite, then $\mathbf{x}'\mathbf{A}\mathbf{x} \sim \chi_s^2(\delta)$, the noncentral chi-square distribution with noncentrality parameter $\delta = \boldsymbol{\mu}'\mathbf{A}\boldsymbol{\mu}$, if and only if:

   (1) $\boldsymbol{\Sigma}\mathbf{A}\boldsymbol{\Sigma}\mathbf{A}\boldsymbol{\Sigma} = \boldsymbol{\Sigma}\mathbf{A}\boldsymbol{\Sigma}$,

   (2) $\boldsymbol{\mu}'\mathbf{A}\boldsymbol{\Sigma}\mathbf{A}\boldsymbol{\Sigma} = \boldsymbol{\mu}'\mathbf{A}\boldsymbol{\Sigma}$,

   (3) $\boldsymbol{\mu}'\mathbf{A}\boldsymbol{\Sigma}\mathbf{A}\boldsymbol{\mu} = \boldsymbol{\mu}'\mathbf{A}\boldsymbol{\mu}$,

   (4) $\operatorname{trace}(\mathbf{A}\boldsymbol{\Sigma}) = s$.

Note that when $\boldsymbol{\Sigma}$ is positive definite, the four conditions reduce to (i) $\mathbf{A}\boldsymbol{\Sigma}\mathbf{A} = \mathbf{A}$ and (ii) $\operatorname{trace}(\mathbf{A}\boldsymbol{\Sigma}) = r$.

The above results for $\mathbf{x}'\mathbf{A}\mathbf{x}$ extend to $\mathbf{Q} = \mathbf{x}'\mathbf{A}\mathbf{x} + \mathbf{a}'\mathbf{x} + d$ (Mathai and Provost [1992: 201–214]).

20.30. If $\mathbf{x} \sim N_d(\mathbf{0}, \mathbf{I}_d)$, then $\mathbf{x}'\mathbf{A}\mathbf{x}$ is distributed as the difference of two independently distributed chi-squared variables if and only if $\mathbf{A}^3 = \mathbf{A}$ (i.e., $\mathbf{A}$ is tripotent). If $\mathbf{x} \sim N_d(\boldsymbol{\mu}, \mathbf{I}_n)$, then the chi-squared distributions are noncentral. This follows from (8.94b) and (20.29a) above.

Proofs. Section 20.5.3.

   20.29a. Muirhead [1982: 31] and Schott [2005: 403].

   20.29b. Christensen [2002: 10], Mathai and Provost [1992: 199], and Schott [2005: 405–406].

20.5.4   Independence and Quadratics

20.31. Let $\mathbf{x} \sim N_d(\boldsymbol{\mu}, \boldsymbol{\Sigma})$, and let $\mathbf{A}_1$ and $\mathbf{A}_2$ be $d \times d$ symmetric matrices.

(a) (Craig–Sakamoto)   If $\boldsymbol{\Sigma}$ is non-negative definite, then $\mathbf{x}'\mathbf{A}_1\mathbf{x}$ and $\mathbf{x}'\mathbf{A}_2\mathbf{x}$ are statistically independent if and only if

$$\binom{\boldsymbol{\Sigma}}{\boldsymbol{\mu}'} \mathbf{A}_1 \boldsymbol{\Sigma} \mathbf{A}_2 (\boldsymbol{\Sigma}, \boldsymbol{\mu}) = \mathbf{0},$$

or, equivalently,

   (1) $\boldsymbol{\Sigma}\mathbf{A}_1\boldsymbol{\Sigma}\mathbf{A}_2\boldsymbol{\Sigma} = \mathbf{0}$,

   (2) $\boldsymbol{\Sigma}\mathbf{A}_1\boldsymbol{\Sigma}\mathbf{A}_2\boldsymbol{\mu} = \mathbf{0}$,

   (3) $\boldsymbol{\Sigma}\mathbf{A}_2\boldsymbol{\Sigma}\mathbf{A}_1\boldsymbol{\mu} = \mathbf{0}$,

   (4) $\boldsymbol{\mu}'\mathbf{A}_1\boldsymbol{\Sigma}\mathbf{A}_2\boldsymbol{\mu} = \mathbf{0}$.

When $\boldsymbol{\mu} = \mathbf{0}$, these reduce to $\boldsymbol{\Sigma}\mathbf{A}_1\boldsymbol{\Sigma}\mathbf{A}_2\boldsymbol{\Sigma} = \mathbf{0}$, and if $\boldsymbol{\Sigma}$ is positive definite, the first equation reduces to $\mathbf{A}_1\boldsymbol{\Sigma}\mathbf{A}_2 = \mathbf{0}$ (or $\mathbf{A}_2\boldsymbol{\Sigma}\mathbf{A}_1 = \mathbf{0}$).

(b) We can extend (a) as follows. If $\boldsymbol{\Sigma}$ is non-negative definite and $Q_i = \mathbf{x}'\mathbf{A}_i\mathbf{x} + \mathbf{a}_i'\mathbf{x} + d_i$ ($i = 1, 2$), then Q_1 and Q_2 are statistically independent if and only if:

(1) $\boldsymbol{\Sigma}\mathbf{A}_1\boldsymbol{\Sigma}\mathbf{A}_2\boldsymbol{\Sigma} = \mathbf{0}$,

(2) $\boldsymbol{\Sigma}\mathbf{A}_1\boldsymbol{\Sigma}(2\mathbf{A}_2\boldsymbol{\mu} + \mathbf{a}_2) = \boldsymbol{\Sigma}\mathbf{A}_2\boldsymbol{\Sigma}(2\mathbf{A}_1\boldsymbol{\mu} + \mathbf{a}_1) = \mathbf{0}$.

(3) $(\mathbf{a}_1 + 2\mathbf{A}_1\boldsymbol{\mu})'\boldsymbol{\Sigma}(\mathbf{a}_2 + 2\mathbf{A}_2\boldsymbol{\mu}) = 0$.

These are the same as (a) when $\mathbf{a}_1 = \mathbf{a}_2 = \mathbf{0}$. The presence of the constants d_1 and d_2 do not affect independence. Note also the following.

(i) If $\mathrm{rank}(\boldsymbol{\Sigma}\mathbf{A}_1) = \mathrm{rank}\,\mathbf{A}_1$ or $\mathrm{rank}(\boldsymbol{\Sigma}\mathbf{A}_1\boldsymbol{\Sigma}) = \mathrm{rank}(\boldsymbol{\Sigma}\mathbf{A}_1)$, then $\boldsymbol{\Sigma}\mathbf{A}_1\boldsymbol{\Sigma}\mathbf{A}_2\boldsymbol{\Sigma} = \mathbf{0}$ implies that $\mathbf{A}_1\boldsymbol{\Sigma}\mathbf{A}_2\boldsymbol{\Sigma} = \mathbf{0}$.

(ii) If $\mathrm{rank}(\boldsymbol{\Sigma}\mathbf{A}_2) = \mathrm{rank}\,\mathbf{A}_2$ or $\mathrm{rank}(\boldsymbol{\Sigma}\mathbf{A}_1\boldsymbol{\Sigma}) = \mathrm{rank}(\boldsymbol{\Sigma}\mathbf{A}_2)$, then $\boldsymbol{\Sigma}\mathbf{A}_1\boldsymbol{\Sigma}\mathbf{A}_2\boldsymbol{\Sigma} = \mathbf{0}$ implies that $\boldsymbol{\Sigma}\mathbf{A}_1\boldsymbol{\Sigma}\mathbf{A}_2 = \mathbf{0}$.

(c) If $\boldsymbol{\Sigma}$ is non-negative definite and $\mathbf{C}$ is a $p \times d$ matrix, then $\mathbf{x}'\mathbf{A}\mathbf{x}$ and $\mathbf{C}\mathbf{x}$ are statistically independent if

$$\mathbf{C}\boldsymbol{\Sigma}\mathbf{A}(\boldsymbol{\Sigma}, \boldsymbol{\mu}) = \mathbf{0}.$$

If $\boldsymbol{\Sigma}$ is positive definite, then the two conditions reduce to just one condition, namely $\mathbf{C}\boldsymbol{\Sigma}\mathbf{A} = \mathbf{0}$, or $\mathbf{C}\mathbf{A} = \mathbf{0}$ when $\boldsymbol{\Sigma} = \mathbf{I}_d$.

(d) If $\boldsymbol{\Sigma}$ is non-negative definite, then setting $\mathbf{A}_2 = \mathbf{0}$ in (b), $\mathbf{x}'\mathbf{A}_1\mathbf{x} + \mathbf{a}_1'\mathbf{x} + d_1$ and $\mathbf{a}_2'\mathbf{x} + d_2$ are statistically independent if and only if

(i) $\boldsymbol{\Sigma}\mathbf{A}_1\boldsymbol{\Sigma}\mathbf{a}_2 = \mathbf{0}$.

(ii) $(\mathbf{a}_1 + 2\mathbf{A}_1\boldsymbol{\mu})'\boldsymbol{\Sigma}\mathbf{a}_2 = \mathbf{0}$.

Setting $\mathbf{A}_1 = \mathbf{0}$ as well, we see that $\mathbf{a}_1'\mathbf{x} + d_1$ and $\mathbf{a}_2'\mathbf{x} + d_2$ are statistically independent if and only if $\mathbf{a}_1'\boldsymbol{\Sigma}\mathbf{a}_2 = \mathbf{0}$.

20.32. (Bilinear Forms) Suppose $\mathbf{x}_i \sim N_{d_i}(\mathbf{0}, \boldsymbol{\Sigma}_{ii})$, where $\boldsymbol{\Sigma}_{ii}$ is positive definite ($i = 1, 2$), and $\mathbf{x}_1$ and $\mathbf{x}_2$ are statistically independent.

(a) The joint moment generating function of $Q_A = \mathbf{x}_1'\mathbf{A}\mathbf{x}_2$ and $Q_B = \mathbf{x}_1'\mathbf{B}\mathbf{x}_2$ is

$$M_{Q_A, Q_B}(t_1, t_2) = \{\det[\mathbf{I}_{d_2} - \boldsymbol{\Sigma}_{22}^{1/2}(t_1\mathbf{A} + t_2\mathbf{B})'\boldsymbol{\Sigma}_{11}(t_1\mathbf{A} + t_2\mathbf{B})\boldsymbol{\Sigma}_{22}^{1/2}]\}^{-1/2}.$$

(b) Q_A and Q_B are statistically independent if and only if $\mathbf{A}'\boldsymbol{\Sigma}_1\mathbf{B} = \mathbf{0}$ and $\mathbf{B}\boldsymbol{\Sigma}_2\mathbf{A}' = \mathbf{0}$.

20.33. (Hadamard Product) Suppose $\mathbf{x} \sim N_d(\boldsymbol{\mu}, \boldsymbol{\Sigma})$ and $\mathbf{A}$ and $\mathbf{B}$ are $m \times d$ matrices. Now $\mathbf{y} = (\mathbf{A}\mathbf{x}) \circ (\mathbf{B}\mathbf{x}) = \boldsymbol{\Phi}_m(\mathbf{A}\mathbf{x} \otimes \mathbf{B}\mathbf{x})\boldsymbol{\Phi}_1'$ (cf. 11.38a) with $\boldsymbol{\Phi}_1 = 1$, and we have the following.

(a) $\mathrm{E}(\mathbf{y}) = \mathbf{D}\mathbf{1}_m + (\mathbf{A}\boldsymbol{\mu}) \circ (\mathbf{B}\boldsymbol{\mu})$, where $\mathbf{D}$ is the diagonal matrix with diagonal elements equal to those of $\mathbf{B}\boldsymbol{\Sigma}\mathbf{A}'$.

(b)

$$\begin{aligned}
\text{var}(\mathbf{A}\mathbf{x} \circ \mathbf{B}\mathbf{x}) \;=\;& [\mathbf{A}(\boldsymbol{\Sigma} + \boldsymbol{\mu}\boldsymbol{\mu}')\mathbf{A}'] \circ [\mathbf{B}(\boldsymbol{\Sigma} + \boldsymbol{\mu}\boldsymbol{\mu}')\mathbf{B}'] \\
&+ [\mathbf{B}(\boldsymbol{\Sigma} + \boldsymbol{\mu}\boldsymbol{\mu}')\mathbf{A}'] \circ [\mathbf{A}(\boldsymbol{\Sigma} \mid \boldsymbol{\mu}\boldsymbol{\mu}')\mathbf{B}'] \\
&- [\mathbf{A}\boldsymbol{\mu}\boldsymbol{\mu}'\mathbf{A}'] \circ [\mathbf{B}\boldsymbol{\mu}\boldsymbol{\mu}'\mathbf{B}'] - [\mathbf{A}\boldsymbol{\mu}\boldsymbol{\mu}'\mathbf{B}'] \circ [\mathbf{B}\boldsymbol{\mu}\boldsymbol{\mu}'\mathbf{A}'].
\end{aligned}$$

Proofs. Section 20.5.4.

20.31a. Schott [2005: 408, 412–413], Driscoll and Krasnicka [1995], and Mathai and Provost [1992: 209–211].

20.31b. Mathai and Provost [1992: 224–225].

20.31c. Quoted by Schott [2005: 413].

20.32. Mathai and Provost [1992: 230–231].

20.33a. Quoted by Schott [2005: 439, exercise 10.44]. Using the multiplication rule for the Kronecker product,

$$\begin{aligned}
\text{E}(\mathbf{y}) \;=\;& \boldsymbol{\Psi}_m(\mathbf{A} \otimes \mathbf{B})\text{E}(\mathbf{x} \otimes \mathbf{x}) \\
=\;& \boldsymbol{\Psi}_m(\mathbf{A} \otimes \mathbf{B})(\text{vec}\,\boldsymbol{\Sigma} + \boldsymbol{\mu} \otimes \boldsymbol{\mu}) \quad \text{by (20.24d(i))} \\
=\;& \boldsymbol{\Psi}_m\text{vec}\,(\mathbf{B}'\boldsymbol{\Sigma}\mathbf{A}) + \mathbf{A}\boldsymbol{\mu} \circ \mathbf{B}\boldsymbol{\mu},
\end{aligned}$$

then use (11.38a(iii)).

20.33b. Using (20.6b), we obtain

$$\text{var}(\mathbf{y}) \;=\; \boldsymbol{\Psi}_m(\mathbf{A} \otimes \mathbf{B})\text{var}(\mathbf{x} \otimes \mathbf{x})(\mathbf{A}' \otimes \mathbf{B}')\boldsymbol{\Psi}'_m.$$

Now substitute for $\text{var}(\mathbf{x} \otimes \mathbf{x})$ using (20.24d(ii)) with $2\mathbf{P}_d = \mathbf{I}_{d^2} + \mathbf{I}_{(d,d)}$ and $\mathbf{I}_{(d,d)}$ the commutation matrix. We then have $(\mathbf{A} \otimes \mathbf{B})\mathbf{I}_{(d,d)} = \mathbf{I}_{(d,d)}(\mathbf{B} \otimes \mathbf{A})$ and, from (11.38a(iv)), $\boldsymbol{\Psi}_m\mathbf{I}_{(d,d)} = \boldsymbol{\Psi}_m$. Finally, multiply out and reintroduce "$\circ$".

20.5.5   Independence of Several Quadratics

20.34. Suppose $\mathbf{x} \sim N_d(\boldsymbol{\mu}, \boldsymbol{\Sigma})$, where $\boldsymbol{\Sigma}$ is positive definite. Let $\mathbf{A}_i$ be a $d \times d$ symmetric matrix of rank r_i, for $i = 1, 2, \ldots, k$, and let $\mathbf{A} = \mathbf{A}_1 + \cdots + \mathbf{A}_k$ be of rank r. Let $\chi_\nu^2(\delta)$ denote the noncentral chi-square distribution with ν degrees of freedom and noncentrality parameter δ. Consider the following conditions:

(1) $\mathbf{A}_i\boldsymbol{\Sigma}$ is idempotent for each i (i.e., $\mathbf{A}_i\boldsymbol{\Sigma}\mathbf{A}_i = \mathbf{A}_i$),

(2) $\mathbf{A}\boldsymbol{\Sigma}$ is idempotent. (i.e., $\mathbf{A}\boldsymbol{\Sigma}\mathbf{A} = \mathbf{A}$)

(3) $\mathbf{A}_i\boldsymbol{\Sigma}\mathbf{A}_j = \mathbf{0}$, for all $i, j, i \neq j$,

(4) $r = \sum_{i=1}^k r_i$,

If any two of (1), (2), and (3) hold, or if (2) and (4) hold, then

(a) $\mathbf{x}'\mathbf{A}_i\mathbf{x} \sim \chi^2_{r_i}(\boldsymbol{\mu}'\mathbf{A}_i\boldsymbol{\mu})$ for all i.

(b) $\mathbf{x}'\mathbf{A}\mathbf{x} \sim \chi^2_r(\boldsymbol{\mu}'\mathbf{A}\boldsymbol{\mu})$.

(c) $\mathbf{x}'\mathbf{A}_1\mathbf{x}, \ldots, \mathbf{x}'\mathbf{A}_k\mathbf{x}$ are statistically independent.

The extension of the above result to the case when $\boldsymbol{\Sigma}$ is non-negative definite is considered by Mathai and Provost [1992: 239]. When $\boldsymbol{\mu} = \mathbf{0}$ and $\boldsymbol{\Sigma}$ may be singular, further conditions for quadratics to be independently distributed as chi-square variables are given by Rao and Mitra [1971: section 9.3].

Proofs. Section 20.5.5.

20.34. Schott [2005: 413].

20.6 COMPLEX RANDOM VECTORS

Complex random vectors arise in several places in statistics, the most notable being multivariate time series (cf. Brillinger [1975: 89]) and random matrices (Mehta [2004]).

Definition 20.12. (Complex Random Vectors) Let $\mathbf{x} = \mathbf{x}_1 + i\mathbf{x}_2$ and $\mathbf{y} = \mathbf{y}_1 + i\mathbf{y}_2$ be complex random vectors, where the $\mathbf{x}_i$ and $\mathbf{y}_i$ are all real random vectors. We then define the following:

$$\begin{aligned} \mathrm{E}(\mathbf{x}) &= \mathrm{E}(\mathbf{x}_1) + i\mathrm{E}(\mathbf{x}_2), \\ \mathrm{var}(\mathbf{x}) &= \mathrm{E}[(\mathbf{x} - \mathrm{E}\mathbf{x})(\mathbf{x} - \mathrm{E}\mathbf{x})^*], \quad \text{and} \\ \mathrm{cov}(\mathbf{x}, \mathbf{y}) &= \mathrm{E}[(\mathbf{x} - \mathrm{E}\mathbf{x})(\mathbf{y} - \mathrm{E}\mathbf{y})^*], \end{aligned}$$

where $\mathbf{x}^* = \mathbf{x}_1' - i\mathbf{x}_2'$.

20.35. Using the notation in the above definition, we readily obtain:

(a) $\mathrm{var}(\mathbf{x}) = \mathbf{V}_{11} + \mathbf{V}_{22} + i(-\mathbf{V}_{12} + \mathbf{V}_{21})$, where $\mathbf{V}_{ij} = \mathrm{cov}(\mathbf{x}_i, \mathbf{x}_j)$, $i, j = 1, 2$.

(b) $\mathrm{cov}(\mathbf{x}, \mathbf{y}) = \mathrm{cov}(\mathbf{x}_1, \mathbf{y}_1) + \mathrm{cov}(\mathbf{x}_2, \mathbf{y}_2) + i[-\mathrm{cov}(\mathbf{x}_1, \mathbf{y}_2) + \mathrm{cov}(\mathbf{x}_2, \mathbf{y}_1)]$.

Definition 20.13. (Complex Normal Distribution) Let $\mathbf{x}_1$ and $\mathbf{x}_2$ be $d \times 1$ (real) random vectors such that $(\mathbf{x}_1', \mathbf{x}_2')'$ is $N_{2d}(\boldsymbol{\mu}, \boldsymbol{\Sigma})$, where $\boldsymbol{\mu} = (\boldsymbol{\mu}_1', \boldsymbol{\mu}_2')'$ and

$$\boldsymbol{\Sigma} = \begin{pmatrix} \boldsymbol{\Gamma} & -\boldsymbol{\Phi} \\ \boldsymbol{\Phi} & \boldsymbol{\Gamma} \end{pmatrix},$$

where $\boldsymbol{\Gamma}$ is non-negative definite, and $\boldsymbol{\Phi} = -\boldsymbol{\Phi}'$ (i.e., real skew-symmetric). Then $\mathbf{x} = \mathbf{x}_1 + i\mathbf{x}_2$ is said to have a *complex normal distribution* with mean $\boldsymbol{\mu}_{\mathbf{x}} = \boldsymbol{\mu}_1 + i\boldsymbol{\mu}_2$ and variance matrix $\mathrm{E}[(\mathbf{x} - \boldsymbol{\mu}_{\mathbf{x}})(\mathbf{x} - \boldsymbol{\mu}_{\mathbf{x}})^*] = \boldsymbol{\Sigma}_{\mathbf{x}}$, where $\boldsymbol{\Sigma}_{\mathbf{x}} = \boldsymbol{\Sigma}_1 + i\boldsymbol{\Sigma}_2$ is Hermitian non-negative definite. Here $\boldsymbol{\Sigma}_1 = 2\boldsymbol{\Gamma}$ and $\boldsymbol{\Sigma}_2 = 2\boldsymbol{\Phi}$ are real matrices. We say that $\mathbf{x} \sim N_d^c(\boldsymbol{\mu}_{\mathbf{x}}, \boldsymbol{\Sigma}_{\mathbf{x}})$. Thus $\mathbf{x}_1 + i\mathbf{x}_2$ is complex normal if and only if

$$\begin{pmatrix} \mathbf{x}_1 \\ \mathbf{x}_2 \end{pmatrix} \sim N_{2d}\left[\begin{pmatrix} \mu_1 \\ \mu_2 \end{pmatrix}, \frac{1}{2} \begin{pmatrix} \boldsymbol{\Sigma}_1 & -\boldsymbol{\Sigma}_2 \\ \boldsymbol{\Sigma}_2 & \boldsymbol{\Sigma}_1 \end{pmatrix} \right].$$

From (20.35) we can identify

$$\boldsymbol{\Sigma}_1 = \text{var}(\mathbf{x}_1) + \text{var}(\mathbf{x}_2) \quad \text{and} \quad \boldsymbol{\Sigma}_2 = -\text{cov}(\mathbf{x}_1, \mathbf{x}_2) + \text{cov}(\mathbf{x}_2, \mathbf{x}_1).$$

See Mathai [1997: 406–409] for further details.

20.36. Using the above notation, and assuming that $\boldsymbol{\Sigma}_\mathbf{x}$ is Hermitian positive definite (i.e., $\boldsymbol{\Sigma}_1$ is positive definite), we have:

(a) $(\det \boldsymbol{\Sigma}_\mathbf{x})^2 = \det(2\boldsymbol{\Sigma})$.

(b) $\boldsymbol{\Sigma}_\mathbf{x}^{-1} = (\boldsymbol{\Sigma}_1 + \boldsymbol{\Sigma}_2 \boldsymbol{\Sigma}_1^{-1} \boldsymbol{\Sigma}_2)^{-1} - i\boldsymbol{\Sigma}_1^{-1} \boldsymbol{\Sigma}_2 (\boldsymbol{\Sigma}_1 + \boldsymbol{\Sigma}_2 \boldsymbol{\Sigma}_1^{-1} \boldsymbol{\Sigma}_2)^{-1}$.

(c) The probability density function of $\mathbf{x}$ can be written as

$$\frac{1}{\pi^d \det \boldsymbol{\Sigma}} \exp[-(\mathbf{x} - \boldsymbol{\mu}_\mathbf{x})^* \boldsymbol{\Sigma}_\mathbf{x}^{-1} (\mathbf{x} - \boldsymbol{\mu}_\mathbf{x})].$$

(d) If $\mathbf{x} \sim N_d^c(\boldsymbol{\mu}_\mathbf{x}, \boldsymbol{\Sigma}_\mathbf{x})$ and $\mathbf{A}$ is $q \times d$, then $\mathbf{Ax} \sim N_d^c(\mathbf{A}\boldsymbol{\mu}_\mathbf{x}, \mathbf{A}\boldsymbol{\Sigma}_\mathbf{x}\mathbf{A}^*)$. It follows that the marginal distributions of a multivariate complex normal are complex normal.

(e) If $\boldsymbol{\Sigma}_2 = \mathbf{0}$, then $\mathbf{x}_1$ and $\mathbf{x}_2$ are statistically independent.

(f) The characteristic function of $\mathbf{x}$ is

$$\text{E} \exp[i \, \Re(\mathbf{t}^* \mathbf{x})] = \exp[i \, \Re(\mathbf{t}^* \boldsymbol{\mu}_\mathbf{x}) - \mathbf{t}^* \boldsymbol{\Sigma}_\mathbf{x} \mathbf{t}],$$

where $\Re e$ is the "real part."

For further background see Krishnaiah [1976]. Brillinger [1975: 313–314] gives some asymptotic results for comparing two vector times series.

Proofs. Section 20.6.

20.36. Quoted by Anderson [2003: 64–65].

20.7   REGRESSION MODELS

The study of random vectors would not complete without some discussion of regression models. I shall consider mainly linear models, because matrices play a prominent role in these models. Also, other models can sometimes be transformed into linear ones, or else, with large samples, can be approximated by linear ones. There are many good books on linear regression with several different approaches. I personally prefer a geometrical approach using orthogonal projections as developed by Seber [1977, 1980, 1984] and, to a lesser extent, by Seber and Lee [2003]. This approach is being used a lot more in texts because it avoids some of the algebraic manipulations. For the various kinds of linear model see, for example, Christensen [1997, 2001]. An extensive and detailed theoretical treatment of all aspects of linear models is given by Sengupta and Jammalamadaka [2003]. For results on modifying a linear model by, for example, adding or deleting an observation see Section 15.3.

A typical regression model takes the form $\mathbf{y} = \boldsymbol{\mu} + \boldsymbol{\epsilon}$, where $\boldsymbol{\mu} = (\mu_i)$ is an $n \times 1$ vector of unknown parameters, $\mathbf{y} = (y_i)$, and $\boldsymbol{\epsilon} = (\epsilon_i)$ are $n \times 1$ random vectors with $\mathrm{E}(\boldsymbol{\epsilon}) = \mathbf{0}$ and $\mathrm{var}(\boldsymbol{\epsilon}) = \sigma^2 \mathbf{V}$, where σ^2 is generally unknown and $n \times n$ $\mathbf{V}$ may be known. This is usually known as a *generalized (weighted) least squares* model (cf. Kariya and Kurata [2004], for example). If $\mu_i = f(\mathbf{x}_i; \boldsymbol{\theta})$, where f is a nonlinear function, $\mathbf{x}_i$ is a known observation, and $\boldsymbol{\theta}$ is unknown, we have a typical *nonlinear regression model*. The theory of such models is discussed in detail by, for example, Seber and Wild [1989]. In some models, $\mathbf{V}$ is known function of $\boldsymbol{\mu}$, and quasi-likelihood methods can be used (Seber and Wild [1989: section 2.3]). Sometimes $\mathbf{V}$ can be a function of other parameters such as autocorrelations in time series models (Seber and Wild [1989: chapter 6]) and variances in components of variance models (Sengupta and Jammalamadaka [2003: section 8.3] and Faraway [2006]). We can also have errors-in-variables models where, for example, the $\mathbf{x}_i$s are measured with error (Seber and Wild [1989: chapter 10] and Carroll et al. [2006]). Other models where $\boldsymbol{\mu}$ may contain random components are, for example, mixed models and components of variance models (e.g., P.S.R.S Rao [1997] and Searle et al. [1992]).

We get another type of nonlinear model when $\mathrm{E}(y_i) = \mu_i$, but now $g(\mu_i) = \alpha + \boldsymbol{\beta}'\mathbf{x}_i$ and the distribution of y_i belongs to the exponential family. This is called a *generalized linear model*, and such models are discussed by McCullagh and Nelder [1989] and Dobson [2001]. Other transformation methods are described by Carroll and Ruppert [1988], and another approach is via generalized additive models (Tibshirani and Hastie [1990] and Wood [2006]).

Finally, applying large sample maximum likelihood theory to very general probability distributions, we can prove the asymptotic equivalence of large sample tests for nonlinear hypotheses such as the *Likelihood ratio*, *Wald* and *Score (Lagrange multiplier)* tests by asymptotically linearizing the model and hypothesis. In the linear case, all three test statistics are equivalent (Seber and Wild [1989: section 12.4] and Seber [1980: chapter 11]). Examples using this linearization technique are given by Seber [1967, 1980] and Lee et al. [2002]. We shall now consider the linear regression model.

Definition 20.14. We call the above model $\mathbf{y} = \boldsymbol{\theta} + \boldsymbol{\epsilon}$ the *general linear model* if $\boldsymbol{\theta} = \mathbf{X}\boldsymbol{\beta}$, where $\mathbf{X}$ is a known $n \times p$ matrix of rank r ($r \leq p < n$), and $\boldsymbol{\beta}$ is a $p \times 1$ vector of unknown parameters. We also assume that $\mathbf{V}$ is non-negative definite of rank v, and we shall be interested in testing a *linear hypothesis* $\mathbf{A}\boldsymbol{\beta} = \mathbf{c}$, where $\mathbf{A}$ is $q \times p$ of rank s ($s \leq q < p$) and $\mathbf{c} \in \mathcal{C}(\mathbf{A})$. We shall refer to the general linear model as $\mathcal{M} = (\mathbf{y}, \mathbf{X}\boldsymbol{\beta}, \sigma^2\mathbf{V})$ and the linear hypothesis as H_0. For making inferences, we shall also assume that $\mathbf{y}$ is multivariate normal, namely $N_n(\mathbf{X}\boldsymbol{\beta}, \sigma^2\mathbf{V})$.

There have been a large number of theoretical results proved for the above general setup and its special cases. However, my approach to linear models when $\mathbf{V}$ is a known nonsingular matrix is somewhat pragmatic: Formulate the theoretical model and hypothesis (e.g., Seber and Lee [2003: section 6.4]) and use a statistical computer package such as R to get the required results as well as the diagnostics for validating the model.

If $\mathbf{V}$ is a known positive definite matrix, there exists a unique positive definite square root (cf. 10.32) $\mathbf{V}^{1/2}$ so that making the transformation $\mathbf{V}^{-1/2}\mathbf{y} = \mathbf{V}^{-1/2}\mathbf{X}\boldsymbol{\beta} + \mathbf{V}^{-1/2}\boldsymbol{\epsilon}$ we get the model

$$\mathbf{z} = \mathbf{W}\boldsymbol{\beta} + \boldsymbol{\eta}, \quad \text{where} \quad \mathrm{var}(\boldsymbol{\eta}) = \mathbf{V}^{-1/2}\mathbf{V}\mathbf{V}^{-1/2} = \mathbf{I}_n.$$

We therefore begin with the model $(\mathbf{y}, \mathbf{X}\boldsymbol{\beta}, \sigma^2 \mathbf{I}_n)$.

With regard to H_0, we frequently have $\mathbf{c} = \mathbf{0}$. If not we can "remove" $\mathbf{c}$ as follows. Let $\boldsymbol{\beta}_0$ satisfy $\mathbf{A}\boldsymbol{\beta}_0 = \mathbf{c}$ and consider the model $\mathbf{z} = \mathbf{X}\boldsymbol{\gamma} + \boldsymbol{\epsilon}$, where $\mathbf{z} = \mathbf{y} - \mathbf{X}\boldsymbol{\beta}_0$ and $\boldsymbol{\gamma} - \boldsymbol{\beta} - \boldsymbol{\beta}_0$. Then H_0 becomes $\mathbf{A}\boldsymbol{\gamma} = \mathbf{0}$, so for the moment we assume $\mathbf{c} = \mathbf{0}$, which fits in with the method of least squares described below. It should be noted that we can incorporate $\mathbf{c}$ if we use a different method of estimation, namely we find *affine unbiased estimators* that satisfy a minimum trace criterion (Magnus and Neudecker [1999: chapter 13]).

20.7.1 V Is the Identity Matrix

20.37. (Estimation) Consider the model $(\mathbf{y}, \mathbf{X}\boldsymbol{\beta}, \sigma^2 \mathbf{I}_n)$, where $\mathbf{X}$ is $n \times p$ of rank r and $\boldsymbol{\theta} = \mathbf{X}\boldsymbol{\beta}$.

(a) Assume $r < p$.

    (i) $\|\mathbf{y} - \boldsymbol{\theta}\|_2^2$ is minimized uniquely subject to $\boldsymbol{\theta} \in \mathcal{C}(\mathbf{X}) = \Omega$, say, at $\widehat{\boldsymbol{\theta}}$, the *least squares estimate* of $\boldsymbol{\theta}$, where

$$\widehat{\boldsymbol{\theta}} = \mathbf{P}_\Omega \mathbf{y}, \quad \mathbf{P}_\Omega = \mathbf{X}(\mathbf{X}'\mathbf{X})^- \mathbf{X}',$$

and $\mathbf{P}_\Omega$ is the unique symmetric idempotent matrix representing the orthogonal projection of $\mathbf{Y}$ onto Ω; $(\mathbf{X}'\mathbf{X})^-$ is any weak inverse of $\mathbf{X}'\mathbf{X}$. Note that $\mathbf{y} = \widehat{\boldsymbol{\theta}} + (\mathbf{y} - \widehat{\boldsymbol{\theta}})$ is an orthogonal decomposition of $\mathbf{y}$, and $\mathbf{P}_\Omega \mathbf{X} = \mathbf{X}$. The matrix $\mathbf{P}_\Omega$ is also referred to as the *hat matrix* in regression diagnostics (Seber and Lee [2003: section 10.2]).

        Other norms and measures can be used for the minimization process to produce alternative estimators to least squares (cf. Groß [2003], Rao and Tountenburg [1999], and Seber and Lee [2003: section 3.13].

    (ii) If $\widehat{\boldsymbol{\theta}} = \mathbf{X}\widehat{\boldsymbol{\beta}}$, then $\widehat{\boldsymbol{\beta}}$ is not unique. Since $\mathbf{X}'(\mathbf{y} - \widehat{\boldsymbol{\theta}}) = \mathbf{0}$, $\widehat{\boldsymbol{\beta}}$ satisfies the so-called normal equations $\mathbf{X}'\mathbf{X}\widehat{\boldsymbol{\beta}} = \mathbf{X}'\mathbf{y}$, which have *a* solution $\widehat{\boldsymbol{\beta}} = (\mathbf{X}'\mathbf{X})^- \mathbf{X}'\mathbf{y} = (\mathbf{X}'\mathbf{X})^- \mathbf{X}'\mathbf{P}_\Omega \mathbf{y} = (\mathbf{X}'\mathbf{X})^- \mathbf{X}'\widehat{\boldsymbol{\theta}}$.

    (iii) $Q = \|\mathbf{y} - \widehat{\boldsymbol{\theta}}\|_2^2 = \mathbf{y}'(\mathbf{I}_n - \mathbf{P}_\Omega)\mathbf{y} = \boldsymbol{\epsilon}'(\mathbf{I}_n - \mathbf{P}_\Omega)\boldsymbol{\epsilon}$, since $(\mathbf{I}_n - \mathbf{P}_\Omega)\boldsymbol{\theta} = \mathbf{0}$.

    (iv) σ^2 is usually estimated by its unbiased estimate $s^2 = Q/(n - r)$, which has certain optimal properties. For example, it is the MINQUE of σ^2 (Rao and Rao [1998: section 12.6]).

    (v) $\mathbf{r} = \mathbf{y} - \widehat{\boldsymbol{\theta}} = (\mathbf{I}_n - \mathbf{P}_\Omega)\mathbf{y}$ is called the *residual vector* and is used for diagnostic purposes.

        $Q = \mathbf{r}'\mathbf{r}$ is usually called the *residual sum of squares* and is often denoted by RSS.

    (vi) $\mathrm{E}(\mathbf{r}) = \mathbf{0}$ and $\mathrm{var}(\mathbf{r}) = \sigma^2(\mathbf{I}_n - \mathbf{P}_\Omega)$.

    (vii) For any $\mathbf{a}$, $\widehat{\phi} = \mathbf{a}'\widehat{\boldsymbol{\theta}} = \mathbf{a}'\mathbf{P}_\Omega \mathbf{y}$ is a linear estimate of $\phi = \mathbf{a}'\boldsymbol{\theta}$ (i.e., linear in $\mathbf{y}$) and is unbiased as $\mathbf{P}_\Omega \mathbf{X} = \mathbf{X}$ implies that $\mathrm{E}(\widehat{\phi}) = \phi$.

    (viii) (Gauss–Markov) Of all linear unbiased estimates of ϕ, $\widehat{\phi}$ is the unique estimate with minimum variance. We refer to $\widehat{\phi}$ as the *BLUE (Best Linear Unbiased Estimate)* of ϕ.

(ix) If $\mathbf{By}$ is any unbiased estimate of $\boldsymbol{\theta}$, then

$$\mathbf{D} = \text{var}(\mathbf{By}) - \text{var}(\widehat{\boldsymbol{\theta}})$$

is non-negative definite (n.n.d.) and $\mathbf{D} = \mathbf{0}$ if and only if $\mathbf{By} = \widehat{\boldsymbol{\theta}}$. We call $\widehat{\boldsymbol{\theta}}$ the *BLUE* of $\boldsymbol{\theta}$.

(x) Let $\mathbf{y} = \mathbf{X}\boldsymbol{\beta} + \sigma\mathbf{u}$, where the elements of $\mathbf{u}$ are i.i.d. with mean zero and variance 1, and density $h(\cdot)$ satisfying $h(-u) = h(u)$ for all u. Then, provided it exists, the expected information matrix for the parameter vector $(\boldsymbol{\beta}', \sigma^2)'$ is

$$\begin{pmatrix} c\mathbf{X}'\mathbf{X} & \mathbf{0} \\ \mathbf{0} & nI_{\sigma^2} \end{pmatrix}.$$

When $\mathbf{u}$ is normal, $c = \sigma^{-2}$ and $I_{\sigma^2} = 1/(2\sigma^4)$.

(b) When $\mathbf{X}$ has full rank (i.e., $r = p$) then $(\mathbf{X}'\mathbf{X})^- = (\mathbf{X}'\mathbf{X})^{-1}$ and we have the following:

(i) $\boldsymbol{\beta} = (\mathbf{X}'\mathbf{X})^{-1}\mathbf{X}'\boldsymbol{\theta}$.

(ii) $\widehat{\boldsymbol{\beta}} = (\mathbf{X}'\mathbf{X})^{-1}\mathbf{X}'\mathbf{y}$, $\text{E}(\widehat{\boldsymbol{\beta}}) = \boldsymbol{\beta}$, and $\text{var}(\widehat{\boldsymbol{\beta}}) = \sigma^2(\mathbf{X}'\mathbf{X})^{-1}$.

(iii) From (a) (viii) we find that $\mathbf{b}'\widehat{\boldsymbol{\beta}}$ is the BLUE of $\mathbf{b}'\boldsymbol{\beta}$.

(iv) Suppose the y_i are independent random variables with common variance σ^2 and common third and fourth moments, μ_3 and μ_4, respectively, about their means. Then s^2 is the unique non-negative quadratic unbiased estimate of σ^2 with minimum variance when $\mu_4 = 3\sigma^4$ or when the diagonal elements of $\mathbf{P}_\Omega = \mathbf{X}(\mathbf{X}'\mathbf{X})^{-1}\mathbf{X}'$ are all equal.

(v) If, as $n \to \infty$, $n^{-1}\mathbf{X}'\mathbf{X}$ converges to a finite positive definite matrix $\mathbf{V}_*$, and the largest diagonal element of $\mathbf{P}_\Omega$ goes to zero, then $\sqrt{n}(\widehat{\boldsymbol{\beta}} - \boldsymbol{\beta})$ converges in distribution to $N_d(\mathbf{0}, \sigma^2\mathbf{V}_*)$.

(c) If $\mathbf{X}$ has full rank and $\mathbf{y} \sim N_n(\mathbf{X}\boldsymbol{\beta}, \sigma^2\mathbf{I}_n)$, then the following hold:

(i) $\widehat{\boldsymbol{\beta}} \sim N_p(\boldsymbol{\beta}, \sigma^2(\mathbf{X}'\mathbf{X})^{-1})$.

(ii) $\widehat{\boldsymbol{\beta}}$ is statistically independent of s^2.

(iii) $Q/\sigma^2 = (n-p)s^2/\sigma^2 \sim \chi^2_{n-p}$.

(iv) $\widehat{\boldsymbol{\beta}}$ and Q/n are the maximum likelihood estimates of $\boldsymbol{\beta}$ and σ^2, respectively, and are also sufficient statistics.

(v) $\widehat{\boldsymbol{\beta}}$ is the best unbiased estimate of $\boldsymbol{\beta}$ in the sense that, for any $\mathbf{b}$, $\mathbf{b}'\widehat{\boldsymbol{\beta}}$ is the estimate of $\mathbf{b}'\boldsymbol{\beta}$ with minimum variance among all unbiased estimates, and not just among linear ones; that is, it is the *MINVUE* of $\mathbf{b}'\boldsymbol{\beta}$.

(vi) s^2 is the MINVUE of σ^2.

Definition 20.15. Let $\mathbf{X}$ be $n \times p$ of rank r. The function $\mathbf{b}'\boldsymbol{\beta}$ of $\boldsymbol{\beta}$ is said to be *estimable* if it has a linear unbiased estimate, $\mathbf{c}'\mathbf{y}$, say. Then $\mathbf{b}'\boldsymbol{\beta} = \text{E}(\mathbf{c}'\mathbf{y}) = \mathbf{c}'\mathbf{X}\boldsymbol{\beta}$ for all $\boldsymbol{\beta}$ so that $\mathbf{b}' = \mathbf{c}'\mathbf{X}$. Let $\mathbf{A}' = (\mathbf{a}_1, \mathbf{a}_2, \ldots, \mathbf{a}_q)$ be a $q \times p$ matrix. The hypothesis $H_0 : \mathbf{A}\boldsymbol{\beta} = \mathbf{0}$ is said to be *testable* if each $\mathbf{a}'_i\boldsymbol{\beta}$ is estimable for $i = 1, 2, \ldots, q$, that is $\mathbf{A}\boldsymbol{\beta}$ is estimable. If $\mathbf{X}$ has full column rank, then $\mathbf{A}\boldsymbol{\beta}$ is always estimable.

20.38. Suppose that $r < p$. The following conditions are equivalent.

(1) $\mathbf{A}\boldsymbol{\beta}$ is estimable.

(2) The rows of $\mathbf{A}$ are linearly dependent on the rows of $\mathbf{X}$, that is, there exists a $q \times n$ matrix $\mathbf{L}$ such that $\mathbf{A} = \mathbf{LX}$. If rank $\mathbf{A} = q$, then rank $\mathbf{L} = q$ (as rank $\mathbf{A} \le$ rank $\mathbf{L}$, by (3.12)). Note that $\mathbf{Ly}$ is a linear unbiased estimator of $\mathbf{A}\boldsymbol{\beta}$.

(3) $\mathcal{C}(\mathbf{A}') \subseteq \mathcal{C}(\mathbf{X}')$.

(4) $\mathbf{A}\widehat{\boldsymbol{\beta}}$ is invariant for any choice of $\widehat{\boldsymbol{\beta}} = (\mathbf{X}'\mathbf{X})^-\mathbf{X}'\mathbf{y}$.

(5) $\mathbf{A}(\mathbf{X}'\mathbf{X})^-\mathbf{X}'\mathbf{X} = \mathbf{A}$.

20.39. If $r < p$ and $\mathbf{A}\boldsymbol{\beta}$ is estimable, then from (20.38(2)) above we have:

(a) $\mathbf{A}\widehat{\boldsymbol{\beta}} = \mathbf{L}\widehat{\boldsymbol{\theta}} = \mathbf{LP}_\Omega\mathbf{y} = \mathbf{LX}(\mathbf{X}'\mathbf{X})^-\mathbf{X}'\mathbf{y} = \mathbf{A}(\mathbf{X}'\mathbf{X})^-\mathbf{X}'\mathbf{y}$.

(b) $\mathrm{E}(\mathbf{A}\widehat{\boldsymbol{\beta}}) = \mathrm{E}(\mathbf{L}\widehat{\boldsymbol{\theta}}) = \mathbf{L}\boldsymbol{\theta} = \mathbf{A}\boldsymbol{\beta}$.

(c) $\mathrm{var}(\mathbf{A}\widehat{\boldsymbol{\beta}}) = \mathrm{var}(\mathbf{L}\widehat{\boldsymbol{\theta}}) = \sigma^2\mathbf{LP}_\Omega\mathbf{L}' = \sigma^2\mathbf{A}(\mathbf{X}'\mathbf{X})^-\mathbf{A}'$. For a single estimable function, $\mathbf{A}$ reduces to $\mathbf{a}'$.

20.40. (Estimation with Constraints) Suppose $r < p$. We wish to find the least squares estimate of $\boldsymbol{\beta}$ subject to the q estimable constraints $\mathbf{A}\boldsymbol{\beta} = \mathbf{0}$, that is, subject to $\mathbf{0} = \mathbf{A}\boldsymbol{\beta} = \mathbf{LX}\boldsymbol{\beta} = \mathbf{L}\boldsymbol{\theta}$, or $\boldsymbol{\theta} \in \mathcal{N}(\mathbf{L})$.

(a) $\|\mathbf{y} - \boldsymbol{\theta}\|_2^2$ is uniquely minimized subject to $\boldsymbol{\theta} \in \mathcal{N}(\mathbf{L}) \cap \Omega = \omega$, say, when $\boldsymbol{\theta} = \widehat{\boldsymbol{\theta}}_H$, where $\widehat{\boldsymbol{\theta}}_H = \mathbf{P}_\omega\mathbf{y}$ and $\mathbf{P}_\omega$ represents the orthogonal projection onto ω.

(b) $Q_H = \|\mathbf{y} - \widehat{\boldsymbol{\theta}}_H\|_2^2 = \mathbf{y}'(\mathbf{I}_n - \mathbf{P}_\omega)\mathbf{y}$.

(c) From (2.51b) and (2.51d), $\mathbf{P}_\Omega - \mathbf{P}_\omega = \mathbf{P}_{\omega^\perp \cap \Omega}$, where $\omega^\perp \cap \Omega = \mathcal{C}(\mathbf{B})$ and $\mathbf{B} = \mathbf{P}_\Omega\mathbf{L}'$.

(d) From (c),

$$
\begin{aligned}
\widehat{\boldsymbol{\theta}}_H &= \mathbf{P}_\omega\mathbf{y} \\
&= \mathbf{P}_\Omega\mathbf{y} - \mathbf{P}_{\omega^\perp \cap \Omega}\mathbf{y} \\
&= \widehat{\boldsymbol{\theta}} - \mathbf{B}(\mathbf{B}'\mathbf{B})^-\mathbf{B}'\mathbf{y} \\
&= \widehat{\boldsymbol{\theta}} - \mathbf{P}_\Omega\mathbf{L}'(\mathbf{LP}_\Omega\mathbf{L}')^-\mathbf{LP}_\Omega\mathbf{y}.
\end{aligned}
$$

.

(e) From (d) and $\mathbf{A} = \mathbf{LX}$,

$$\mathbf{P}_{\omega^\perp \cap \Omega} = \mathbf{X}(\mathbf{X}'\mathbf{X})^-\mathbf{A}'[\mathbf{A}(\mathbf{X}'\mathbf{X})^-\mathbf{A}']^-\mathbf{A}(\mathbf{X}'\mathbf{X})^-\mathbf{X}'.$$

(f) If $\widehat{\boldsymbol{\theta}}_H = \mathbf{X}\widehat{\boldsymbol{\beta}}_H$, we have from (d) and (e)

$$\mathbf{X}\widehat{\boldsymbol{\beta}}_H = \mathbf{X}\widehat{\boldsymbol{\beta}} - \mathbf{X}(\mathbf{X}'\mathbf{X})^-\mathbf{A}'[\mathbf{A}(\mathbf{X}'\mathbf{X})^-\mathbf{A}']^-\mathbf{A}\widehat{\boldsymbol{\beta}}.$$

If $\mathbf{A}$ has full row rank, $[\mathbf{A}(\mathbf{X}'\mathbf{X})^-\mathbf{A}']^- = [\mathbf{A}(\mathbf{X}'\mathbf{X})^-\mathbf{A}']^{-1}$. If, in addition, $\mathbf{X}$ has full column rank, then multiplying by $\mathbf{X}'$,

$$\widehat{\boldsymbol{\beta}}_H = \widehat{\boldsymbol{\beta}} - (\mathbf{X}'\mathbf{X})^{-1}\mathbf{A}'[\mathbf{A}(\mathbf{X}'\mathbf{X})^{-1}\mathbf{A}']^{-1}\mathbf{A}\widehat{\boldsymbol{\beta}}.$$

We can also obtain this result using Lagrange multipliers (cf. Seber and Lee [2003: 60].

(g) If we now want to change the constraints to $\mathbf{A}\boldsymbol{\beta} = \mathbf{c}$, where $\mathbf{c} \in \mathcal{C}(\mathbf{A})$, we replace $\widehat{\boldsymbol{\beta}}$ by $\widehat{\boldsymbol{\beta}} - \boldsymbol{\beta}_0$ (where $\mathbf{A}\boldsymbol{\beta}_0 = \mathbf{c}$ for some $\boldsymbol{\beta}_0$), $\mathbf{A}\widehat{\boldsymbol{\beta}}$ by $\mathbf{A}\widehat{\boldsymbol{\beta}} - \mathbf{c}$, and $\mathbf{y}$ by $\tilde{\mathbf{y}} = \mathbf{y} - \mathbf{X}\boldsymbol{\beta}_0$. Q_H then becomes $\tilde{\mathbf{y}}'(\mathbf{I}_n - \mathbf{P}_\omega)\tilde{\mathbf{y}}$, and Q remains unchanged as $(\mathbf{I}_n - \mathbf{P}_\Omega)\tilde{\mathbf{y}} = (\mathbf{I}_n - \mathbf{P}_\Omega)\mathbf{y}$.

(i) Given the estimable constraints $\mathbf{A}\boldsymbol{\beta} = \mathbf{c}$, another approach is to note that $\mathbf{A}\boldsymbol{\beta} = \mathbf{c}$ if and only if $\boldsymbol{\beta} = \mathbf{A}^-\mathbf{c} + (\mathbf{I}_p - \mathbf{A}^-\mathbf{A})\boldsymbol{\phi}$, where (cf. 13.4) $\boldsymbol{\phi}$ is arbitrary, and $\mathbf{A}^-$ is any weak inverse of $\mathbf{A}$. Substituting for $\boldsymbol{\beta}$, the constrained model then becomes

$$(\mathbf{y} - \mathbf{X}\mathbf{A}^-\mathbf{c}, \mathbf{X}(\mathbf{I}_p - \mathbf{A}^-\mathbf{A})\boldsymbol{\phi}, \sigma^2\mathbf{I}_n).$$

A reasonable choice for $\mathbf{A}^-$ is $\mathbf{A}'(\mathbf{A}\mathbf{A}')^-$.

A second approach is use the model $(\mathbf{y}_*, \mathbf{X}_*\boldsymbol{\beta}, \sigma^2\mathbf{V})$, where

$$\mathbf{y}_* = \begin{pmatrix} \mathbf{y} \\ \mathbf{c} \end{pmatrix}, \quad \mathbf{X}_* = \begin{pmatrix} \mathbf{X} \\ \mathbf{A} \end{pmatrix}, \quad \mathbf{V} = \begin{pmatrix} \mathbf{I}_n & \mathbf{0} \\ \mathbf{0} & \mathbf{0} \end{pmatrix},$$

and $\mathbf{V}$ is singular (Sengupta and Jammalamadaka [2003: 123, 125, 244]).

20.41. (Hypothesis Testing) Suppose we wish to test $H_0 : \mathbf{A}\boldsymbol{\beta} = \mathbf{c}$ ($\mathbf{c} \neq \mathbf{0}$), where H_0 is testable, and $\mathbf{A}$ is $q \times p$ of rank s ($s \leq q$).

(a) From (20.40g) and (20.40e),

$$Q_H - Q = \tilde{\mathbf{y}}'\mathbf{P}_{\omega^\perp \cap \Omega}\tilde{\mathbf{y}} = (\mathbf{A}\widehat{\boldsymbol{\beta}} - \mathbf{c})'[\mathbf{A}(\mathbf{X}'\mathbf{X})^-\mathbf{A}']^-(\mathbf{A}\widehat{\boldsymbol{\beta}} - \mathbf{c}).$$

(b) $\mathrm{E}(Q_H - Q) = \sigma^2 s + (\mathbf{A}\boldsymbol{\beta} - \mathbf{c})'[\mathbf{A}(\mathbf{X}'\mathbf{X})^-\mathbf{A}']^-(\mathbf{A}\boldsymbol{\beta} - \mathbf{c})$.

(c) The test statistic for H_0 is

$$F = \frac{(Q_H - Q)/s}{Q/(n-r)}.$$

(d) When H_0 is true, we have the following.

(i) $\mathbf{P}_{\omega^\perp \cap \Omega}\tilde{\mathbf{y}} = \mathbf{P}_{\omega^\perp \cap \Omega}[\mathbf{y} - \mathbf{X}\boldsymbol{\beta} - (\mathbf{X}\boldsymbol{\beta} - \mathbf{X}\boldsymbol{\beta}_0)] = \mathbf{P}_{\omega^\perp \cap \omega}\boldsymbol{\epsilon}$, since from (20.40d) with $\mathbf{B}' = \mathbf{L}\mathbf{P}_\Omega$ we have

$$\mathbf{L}\mathbf{P}_\Omega\mathbf{X}(\boldsymbol{\beta} - \boldsymbol{\beta}_0) = \mathbf{L}\mathbf{X}(\boldsymbol{\beta} - \boldsymbol{\beta}_0) = \mathbf{A}\boldsymbol{\beta} - \mathbf{c} = \mathbf{0}.$$

(ii) If $\mathbf{y} \sim N_n(\mathbf{X}\boldsymbol{\beta}, \sigma^2\mathbf{I}_n)$, then from (i) the following ratio has an F-distribution, that is,

$$F = \frac{\boldsymbol{\epsilon}'\mathbf{P}_{\omega^\perp \cap \Omega}\boldsymbol{\epsilon}/s}{\boldsymbol{\epsilon}'(\mathbf{I}_n - \mathbf{P}_\Omega)\boldsymbol{\epsilon}/(n-r)} \sim F_{s,n-r}.$$

Definition 20.16. (Multiple Correlation) Consider the linear model $y_i = \beta_0 + \beta_1 x_{i1} + \beta_2 x_{i2} + \cdots + \beta_{p-1} x_{i,p-1} + \epsilon_i$ so that $\mathbf{X} = (\mathbf{1}_n, \mathbf{x}^{(1)}, \ldots, \mathbf{x}^{(p-1)})$, where rank $\mathbf{X} = p$. If we define $\widehat{\mathbf{y}} - \bar{\mathbf{0}}$, the correlation coefficient of $\mathbf{y}$ and the fitted model $\widehat{\mathbf{y}}$ is called the *multiple correlation coefficient*, and is denoted by R. Its square R^2, is called the *coefficient of (multiple) determination.*

20.42. If $\bar{y} = \sum_{i=1}^{n} y_i/n$, then:

(a) $(\mathbf{y} - \widehat{\mathbf{y}})' \mathbf{1}_n = 0$.

(b) $\sum_{i=1}^{n} (y_i - \bar{y})^2 = \sum_{i=1}^{n} (y_i - \widehat{y}_i)^2 + \sum_{i=1}^{n} (\widehat{y}_i - \bar{y})^2$.

(c) $R^2 = \dfrac{\sum_i (\widehat{y}_i - \bar{y})^2}{\sum_i (y_i - \bar{y})^2}$.

(d) $RSS = \sum_i (y_i - \widehat{y}_i)^2 = (1 - R^2) \sum_i (y_i - \bar{y})^2$.

Proofs. Section 20.7.1.

20.37a(i). Seber and Lee [2003: 36–37].

20.37a(ii). Seber and Lee [2003: 38].

20.37a(iv). Seber and Lee [2003: 44].

20.37a(vi). Use (20.6b). Seber and Lee [2003: section 10.2].

20.37a(viii). Seber and Lee [2003: 42–43].

20.37a(ix). $\mathbf{P}_\Omega$ is symmetric and idempotent. Because $\mathbf{By}$ is unbiased, $(\mathbf{I}_n - \mathbf{B})\boldsymbol{\theta} = \mathbf{0}$, so that $\mathcal{C}[(\mathbf{I}_n - \mathbf{B})'] \perp \Omega$ and $\mathbf{P}_\Omega = \mathbf{P}_\Omega \mathbf{B}' = \mathbf{B}\mathbf{P}_\Omega$. This leads to $\mathbf{D} = \mathbf{B}(\mathbf{I}_n - \mathbf{P}_\Omega)\mathbf{B}'$, which is non-negative definite (n.n.d.) as $\mathbf{I}_n - \mathbf{P}_\Omega$ is n.n.d. Finally, $\mathbf{D} = \mathbf{0}$ if and only if $(\mathbf{I}_n - \mathbf{P}_\Omega)\mathbf{B}' = \mathbf{0}$ or $\mathbf{B} = \mathbf{P}_\Omega$.

20.37a(x). Sengupta and Jammalamadaka [2003: 133; they omit the word "expected"] and Seber and Lee [2003: 49, for the normal case].

20.37b(ii). Seber and Lee [2003: 42].

20.37b(iv). Seber and Lee [2003: 45].

20.37b(v). Sen and Singer [1993: section 7.2].

20.37c. Seber and Lee [2003: 47–48 for (i)–(iii); section 3.5 for (iv); 50 for (v)] and Rao [1973a: 319 for (vi)].

20.38. Searle [1971].

20.41b. Use (20.18b) with $E(\mathbf{A}\widehat{\boldsymbol{\beta}}) = \mathbf{A}\boldsymbol{\beta}$ and trace $\mathbf{P}_{\omega^\perp \cap \Omega} = s$.

20.41d(ii). Seber [1977: section 4.5].

20.42. Seber and Lee [2003: 111–113].

20.7.2   V Is Positive Definite

20.43. The results for $\mathbf{V}$, a known positive definite matrix, follow directly from the results for $\mathbf{V} = \mathbf{I}_n$ by replacing $\mathbf{y}$ by $\mathbf{V}^{-1/2}\mathbf{y}$, $\boldsymbol{\theta}$ by $\mathbf{V}^{-1/2}\boldsymbol{\theta}$, and $\mathbf{X}$ by $\mathbf{V}^{-1/2}\mathbf{X}$ through all the previous theory. For example, we now minimize $(\mathbf{y} - \boldsymbol{\theta})'\mathbf{V}^{-1}(\mathbf{y} - \boldsymbol{\theta})$ subject to $\boldsymbol{\theta} \in \Omega$ to obtain $\hat{\boldsymbol{\theta}}$, say. We can do this by changing the inner product space or, more simply, by using the transformation $\mathbf{V}^{-1/2}$ so that $\mathbf{X}'\mathbf{X}$ becomes $\mathbf{X}'\mathbf{V}^{-1}\mathbf{X}$ and $\mathbf{X}'\mathbf{y}$ becomes $\mathbf{X}'\mathbf{V}^{-1}\mathbf{y}$, giving us

$$\mathbf{V}^{-1/2}\tilde{\boldsymbol{\theta}} = \mathbf{V}^{-1/2}\mathbf{X}(\mathbf{X}'\mathbf{V}^{-1}\mathbf{X})^{-}\mathbf{X}'\mathbf{V}^{-1}\mathbf{y}.$$

Assuming rank $\mathbf{X} = p$, and setting $\tilde{\boldsymbol{\theta}} = \mathbf{X}\tilde{\boldsymbol{\beta}}$, we now have two unbiased estimates of $\boldsymbol{\beta}$, namely

$$\begin{aligned} \widehat{\boldsymbol{\beta}} &= (\mathbf{X}'\mathbf{X})^{-1}\mathbf{X}'\mathbf{y} \quad \text{and} \\ \tilde{\boldsymbol{\beta}} &= (\mathbf{X}'\mathbf{V}^{-1}\mathbf{X})^{-1}\mathbf{X}'\mathbf{V}^{-1}\mathbf{y}. \end{aligned}$$

Then using (20.6b),

$$\text{var}(\widehat{\boldsymbol{\beta}}) = \sigma^2(\mathbf{X}'\mathbf{X})^{-1}\mathbf{X}'\mathbf{V}\mathbf{X}(\mathbf{X}'\mathbf{X})^{-1} \quad \text{and} \quad \text{var}(\tilde{\boldsymbol{\beta}}) = \sigma^2(\mathbf{X}'\mathbf{V}\mathbf{X})^{-1}.$$

The above estimators are often called the *ordinary least squares estimate* OLSE$(\boldsymbol{\beta}) = \widehat{\boldsymbol{\beta}}$ and the *generalized least squares estimate* GLSE$(\boldsymbol{\beta}) = \tilde{\boldsymbol{\beta}}$. As $\tilde{\boldsymbol{\beta}}$ is the *BLUE* of $\boldsymbol{\beta}$, we can expect this estimator to be more efficient in some sense than $\widehat{\boldsymbol{\beta}}$. Various measures of efficiency are given in (12.6), and a popular one is the Watson efficiency $\phi = 1/E_1$, where

$$\phi = \frac{\det[\text{var}(\tilde{\boldsymbol{\beta}}_{\text{GLS}})]}{\det[\text{var}(\widehat{\boldsymbol{\beta}}_{\text{OLS}})]} = \frac{[\det(\mathbf{X}'\mathbf{X})]^2}{\det(\mathbf{X}'\mathbf{V}\mathbf{X}) \cdot \det(\mathbf{X}'\mathbf{V}^{-1}\mathbf{X})}.$$

From (12.6a),

$$1 \geq \phi \geq \prod_{i=1}^{m} \frac{4\lambda_i \lambda_{n-i+1}}{(\lambda_i + \lambda_{n-i+1})^2} = \prod_{i=1}^{m}(1 - \rho_i^2),$$

where $m = \min\{p, n - p\}$ and $\lambda_1 \geq \lambda_2 \geq \cdots \geq \lambda_n > 0$ are the eigenvalues of $\mathbf{V}$. $E_1 = 1$ if and only if the two estimators are the same. The ratios $4\lambda_i\lambda_{n-i+1}/(\lambda_i + \lambda_{n-i+1})^2$ are the squared antieigenvalues of $\mathbf{V}$ (cf. Section 6.7), and the ρ_i can be taken as the canonical correlations between the OLSE $\hat{\boldsymbol{\theta}}$ and its residual $\mathbf{r} = \mathbf{y} - \hat{\boldsymbol{\theta}}$. For references on the topic see Drury et al. [2002] and the survey by Chu et al. [2005b].

The Watson efficiency has also been applied to partitioned regression models

$$\mathbf{X}\boldsymbol{\beta} = (\mathbf{X}_1, \mathbf{X}_2)\boldsymbol{\beta} = \mathbf{X}_1\boldsymbol{\beta}_1 + \mathbf{X}_2\boldsymbol{\beta}_2,$$

where $\mathbf{X}$ and $\mathbf{V}$ have full rank. A *subset Watson efficiency* can be defined for the estimate of $\boldsymbol{\beta}_i$, and the overall efficiency factorized into components. For details and examples see Chu et al. [2004; 2005a,b].

20.7.3   V Is Non-negative Definite

We now assume that $\mathbf{V}$ is a known singular matrix of rank v ($v < n$) and that rank $\mathbf{X} <$ rank $\mathbf{V}$. For a thorough review and historical summary of the topic see Puntanen and Styan [1989]. Theoretical details are given by Sengupta and Jammalamadaka [2003: chapter 7], Baksalary et al. [1990], and, more briefly, by Christensen [2002]. Singular models arise, for example, in finite population sampling, in some experimental designs, in some state-space models, and in models where some of the y-variables are virtually error-free and are effectively constants.

20.44. Consider the model $(\mathbf{y}, \mathbf{X}\boldsymbol{\beta}, \sigma^2\mathbf{V})$, where $\mathbf{V}$ is singular and rank $\mathbf{X} = r$ ($r \le p$). Let $\mathbf{H} = \mathbf{P}_\Omega$ and $\mathbf{M} = \mathbf{I}_n - \mathbf{P}_\Omega = \mathbf{I}_n - \mathbf{H}$, so that from (20.37a(i)) $\widehat{\boldsymbol{\theta}} = \mathbf{H}\mathbf{y}$.

(a) The model is consistent (i.e., the inference base is not self-contradictory) if $\mathbf{y} \in \mathcal{C}(\mathbf{X}, \mathbf{V})$ with probability 1. This follows from the fact that $\mathbf{y} - \mathbf{X}\boldsymbol{\beta} \in \mathcal{C}(\mathbf{V})$ with probability 1.

(b) One expression for the best linear unbiased estimate (BLUE) $\boldsymbol{\theta}^\dagger$ of $\boldsymbol{\theta}$ takes the form $\boldsymbol{\theta}^\dagger = \mathbf{G}\mathbf{y}$ if and only if $\mathbf{G}(\mathbf{X}, \mathbf{VM}) = (\mathbf{X}, \mathbf{0})$ (Puntanen et al. [2000]). The numerical value of $\boldsymbol{\theta}^\dagger$ is unique with probablity 1, but $\mathbf{G}$ is unique if and only if $\mathcal{C}(\mathbf{X}, \mathbf{V}) = \mathbb{R}^n$.

When $\mathbf{V}$ is nonsingular, $\boldsymbol{\theta}^\dagger = \tilde{\boldsymbol{\theta}}$ (cf. 20.43).

(c) One general solution to $\mathbf{G}(\mathbf{X}, \mathbf{VM}) = (\mathbf{X}, \mathbf{0})$ is

$$\mathbf{G} = \mathbf{I}_n - \mathbf{VM}(\mathbf{MVM})^-\mathbf{M} + \mathbf{F}[\mathbf{I}_n - \mathbf{MVM}(\mathbf{MVM})^-\mathbf{M},$$

where $\mathbf{F}$ is arbitrary.

(d) (i) Some representations of $\boldsymbol{\theta}^\dagger$ are

$$\begin{aligned}
\boldsymbol{\theta}^\dagger &= \widehat{\boldsymbol{\theta}} - \mathbf{HVM}(\mathbf{MVM})^-\mathbf{My} \\
&= \widehat{\boldsymbol{\theta}} - \mathbf{HVM}(\mathbf{MVM})^+\mathbf{My} \\
&= \widehat{\boldsymbol{\theta}} - \mathbf{HVM}(\mathbf{MVM})^+\mathbf{y} \\
&= \mathbf{y} - \mathbf{VM}(\mathbf{MVM})^-\mathbf{My}.
\end{aligned}$$

(ii) Also

$$\boldsymbol{\theta}^\dagger = \mathbf{X}(\mathbf{X}'\mathbf{W}^-\mathbf{X})^-\mathbf{X}'\mathbf{W}^-\mathbf{y},$$

where $\mathbf{W} = \mathbf{V} + \mathbf{XUX}'$ and $\mathbf{U}$ is an arbitrary matrix such that $\mathcal{C}(\mathbf{W}) = \mathcal{C}(\mathbf{X}, \mathbf{V})$.

(e) If $\mathbf{X}$ has full column rank and $\boldsymbol{\theta}^\dagger = \mathbf{X}\boldsymbol{\beta}^\dagger$, then $\boldsymbol{\beta}^\dagger = (\mathbf{X}'\mathbf{X})^{-1}\mathbf{X}'\boldsymbol{\theta}^\dagger$.

(f) $\boldsymbol{\theta}^\dagger$ is invariant to the choice of $(\mathbf{MVM})^-$.

(g) Asymptotic theory for $\boldsymbol{\theta}^\dagger$ is given by Sengupta and Jammalamadaka [2003: 522].

(h) (Mean and Variance)

(i) Since $\mathrm{E}(\mathbf{My}) = \mathbf{MX}\boldsymbol{\theta} = \mathbf{0}$, it follows from (d) that $\mathrm{E}(\boldsymbol{\theta}^\dagger) = \boldsymbol{\theta}$.

(ii) From (d) and $\mathbf{HM} = \mathbf{0}$,

$$
\begin{aligned}
\operatorname{var}(\boldsymbol{\theta}^{\dagger}) &= \sigma^2[\mathbf{HVH} - \mathbf{HVM}(\mathbf{MVM})^-\mathbf{MVH}] \\
&= \sigma^2[\mathbf{V} - \mathbf{VM}(\mathbf{MVM})^-\mathbf{MV}] \\
&= \sigma^2[\mathbf{X}(\mathbf{X}'\mathbf{W}^-\mathbf{X})^-\mathbf{X}' - \mathbf{XUX}'].
\end{aligned}
$$

(iii) If $\mathbf{X}$ has full column rank and $\boldsymbol{\theta}^{\dagger} = \mathbf{X}\boldsymbol{\beta}^{\dagger}$, then

$$
\operatorname{var}(\boldsymbol{\beta}^{\dagger}) = \operatorname{var}(\widehat{\boldsymbol{\beta}}) - (\mathbf{X}'\mathbf{X})^{-1}\mathbf{X}'\mathbf{VM}(\mathbf{MVM})^-\mathbf{MVX}(\mathbf{X}'\mathbf{X})^{-1},
$$

where $\operatorname{var}(\widehat{\boldsymbol{\beta}}) = (\mathbf{X}'\mathbf{X})^{-1}\mathbf{X}'\mathbf{VX}(\mathbf{X}'\mathbf{X})^{-1}$.

(i) The residual is

$$
\begin{aligned}
\mathbf{r} &= \mathbf{y} - \boldsymbol{\theta}^{\dagger} \\
&= \mathbf{My} + \mathbf{HVM}(\mathbf{MVM})^-\mathbf{My} \\
&= \mathbf{VM}(\mathbf{MVM})^-\mathbf{My},
\end{aligned}
$$

and

$$
\operatorname{var}(\mathbf{r}) = \sigma^2\mathbf{VM}(\mathbf{MVM})^-\mathbf{MV}.
$$

(j) Let $f = \operatorname{rank}(\mathbf{V}, \mathbf{X}) - \operatorname{rank}\mathbf{X}$. The weighted residual sum of squares is

$$
Q^{\dagger} = (\mathbf{y} - \boldsymbol{\theta}^{\dagger})'\mathbf{V}^-(\mathbf{y} - \boldsymbol{\theta}^{\dagger}) = \mathbf{y}'\mathbf{M}(\mathbf{MVM})^-\mathbf{My},
$$

and

$$
\operatorname{E}(Q^{\dagger}/f) = \sigma^2.
$$

(k) If $\mathbf{A}\boldsymbol{\beta}$ is estimable, then the BLUE of $\mathbf{A}\boldsymbol{\beta}$ is

$$
(\mathbf{A}\boldsymbol{\beta})^{\dagger} = \mathbf{AX}^-[\mathbf{I}_n - \mathbf{VM}(\mathbf{MVM})^-\mathbf{M})]\mathbf{y}.
$$

Furthermore,

$$
\operatorname{var}[(\mathbf{A}\boldsymbol{\beta})^{\dagger}] = \sigma^2\mathbf{AX}^-[\mathbf{V} - \mathbf{VM}(\mathbf{MVM})^-\mathbf{MV}](\mathbf{AX}^-)'.
$$

(l) (Inverse Partitioned Matrix Approach) Let

$$
\begin{pmatrix} \mathbf{V} & \mathbf{X} \\ \mathbf{X}' & \mathbf{0} \end{pmatrix}^- = \begin{pmatrix} \mathbf{C}_1 & \mathbf{C}_2 \\ \mathbf{C}_3 & -\mathbf{C}_4 \end{pmatrix}.
$$

(i) $\boldsymbol{\theta}^{\dagger} = \mathbf{XC}_2'\mathbf{y} = \mathbf{XC}_3\mathbf{y}$.

(ii) $\operatorname{var}(\boldsymbol{\theta}^{\dagger}) = \sigma^2\mathbf{XC}_4\mathbf{X}'$.

(iii) $\mathbf{r} = \mathbf{VC}_1\mathbf{y}$.

(iv) Referring to (j), $Q^{\dagger} = \mathbf{y}'\mathbf{C}_1\mathbf{y}$.

(m) $\boldsymbol{\theta}^{\dagger} = \widehat{\boldsymbol{\theta}}$ if and only any one of the following conditions hold.

(1) $\mathbf{HV} = \mathbf{VH}$.

(2) $\mathbf{VV} = \mathbf{HVH}$.

(3) $\mathbf{HVM} = \mathbf{0}$.

(4) $\mathcal{C}(\mathbf{VX}) \subseteq \mathcal{C}(\mathbf{X})$.

For these and further conditions see Puntanen and Styan [1989, 2006] and Isotalo et al. [2005b: chapter 6].

20.45. If $\mathbf{X}$ has full column rank p and $\mathcal{C}(\mathbf{X}) \subseteq \mathcal{C}(\mathbf{V})$, the so-called *weakly singular model*, then the Watson efficiency is

$$\phi = \frac{[\det(\mathbf{X}'\mathbf{X})]^2}{\det(\mathbf{X}'\mathbf{VX}) \cdot \det(\mathbf{X}'\mathbf{V}^+\mathbf{X})}.$$

If rank $\mathbf{V} = v$ $(p+1 \leq v \leq n)$,

$$1 \geq \phi \geq \prod_{i=1}^{m} \frac{4\lambda_i \lambda_{v-i+1}}{(\lambda_i + \lambda_{v-i+1})^2} = \prod_{i=1}^{h}(1 - \rho_i^2),$$

where $m = \min\{p, v-p\}$ and h is the number of nonzero canonical correlations between the ordinary least squares estimate and its residual (Chu et al. [2004, 2005a]). These authors have also applied the Watson efficiency to partitioned regression models for the case when $\mathbf{V}$ is positive definite. The result about the canonical correlations still applies. For other bounds on the efficiency for the singular model see Sengupta and Jammalamadaka [2003: 316–318].

20.46. Let $\mathcal{U} = \{\mathbf{U} : \mathbf{0} \preceq \mathbf{U} \preceq \mathbf{V}, \mathcal{C}(\mathbf{U}) \subseteq \mathcal{C}(\mathbf{X})\}$, where $\mathbf{A} \preceq \mathbf{B}$ means that $\mathbf{B} - \mathbf{A}$ is non-negative definite. The maximal element $\mathbf{U}$ in $\mathcal{U}$ is called the *shorted matrix (operator)* with respect to $\mathbf{X}$, and is denoted by $\mathbf{S}(\mathbf{V} \mid \mathbf{X})$. Then

$$\text{var}(\boldsymbol{\theta}^\dagger) = \mathbf{S}(\mathbf{V} \mid \mathbf{X}).$$

For further references relating to shorted matrices see Mitra and Puntanen [1991], Mitra and Puri [1979], and Mitra et al. [1995].

20.47. (Linear Restrictions) Suppose that we are interested in the linear (estimable) restrictions $\mathbf{A}\boldsymbol{\beta} = \mathbf{c}$. Let $Q_H^\dagger$ be the residual sum of squares after fitting the model subject to the constraints.

(a) $Q_H^\dagger - Q^\dagger = [(\mathbf{A}\boldsymbol{\beta})^\dagger - \mathbf{c}]'[\sigma^{-2}\text{var}(\mathbf{A}\boldsymbol{\beta})^\dagger]^-[(\mathbf{A}\boldsymbol{\beta})^\dagger - \mathbf{c}]$.

(b) $\dfrac{Q_H^\dagger - Q^\dagger}{Q^\dagger} \cdot \dfrac{f}{m} \sim F_{m,f}$,

the F-distribution, where $f = \text{rank}(\mathbf{V}, \mathbf{X}) - \text{rank}\,\mathbf{X}$, $m = \text{rank}[\text{var}(\mathbf{A}\boldsymbol{\beta})^\dagger]$, and $Q^\dagger$ is given in (20.44j).

Proofs. Section 20.7.3.

20.44a. Christensen [2002: 10] and Rao [1973a: 297].

20.44b. Groß [2004], Puntanen et al. [2000], and Rao [1973b: 282].

20.44c. Rao [1978: 1202].

20.44d. See Rao [1973b]. The last expression for $\boldsymbol{\theta}^\dagger$ is derived by Sengupta and Jammalamadaka [2003: 252, with $\mathbf{L} = \mathbf{I}$].

We use the fact that $(\mathbf{M}\mathbf{A})^{+}\mathbf{M} = (\mathbf{M}\mathbf{A})^{+}$ for any $\mathbf{A}$ such that $(\mathbf{M}\mathbf{A})^{+}$ exists, since $(\mathbf{M}\mathbf{A})^{+}\mathbf{M}$ satisfies the four conditions for it to be the Moore–Penrose inverse of $\mathbf{M}\mathbf{A}$, and $\mathbf{M}$ is idempotent.

20.44f. Sengupta and Jammalamadaka [2003: 252].

20.44h(iii). Isotalo et al. [2005b: 11].

20.44i. Sengupta and Jammalamadaka [2003: 253–255].

20.44j. Sengupta and Jammalamadaka [2003: 259–260].

20.44k. Sengupta and Jammalamadaka [2003: 252, 255].

20.44l. Rao [1973b] and Sengupta and Jammalamadaka [2003: 269].

20.47. Sengupta and Jammalamakada [2003: 277, 288].

20.8 OTHER MULTIVARIATE DISTRIBUTIONS

In this section we consider a number of continuous multivariate distributions. These distributions can be regarded as special cases of matrix variate distributions (cf. Section 21.9).

20.8.1 Multivariate t-Distribution

Definition 20.17. A $d \times 1$ random vector $\mathbf{x} = x_1, x_2, \ldots, x_d)'$ has a *multivariate t-distribution* if its probability density function is given by

$$f(\mathbf{x}) \quad = \quad \frac{\Gamma(\frac{1}{2}[\nu + d])}{(\pi\nu)^{d/2}\Gamma(\frac{1}{2}\nu)(\det \boldsymbol{\Sigma})^{1/2}}[1 + \nu^{-1}(\mathbf{x} - \boldsymbol{\mu})'\boldsymbol{\Sigma}^{-1}(\mathbf{x} - \boldsymbol{\mu})]^{-(\nu+d)/2}$$

$$(-\infty < x_i < \infty, \quad i = 1, 2, \ldots, d),$$

where $\boldsymbol{\Sigma} = (\sigma_{ij})$ is positive definite and $\Gamma(\cdot)$ is the Gamma function. We write $\mathbf{x} \sim t_d(\nu, \boldsymbol{\mu}, \boldsymbol{\Sigma})$. The distribution $t_d(1, \mathbf{0}, \mathbf{I}_d)$ is called the *multivariate Cauchy distribution*.

20.48. Suppose $\mathbf{x} \sim t_d(\nu, \boldsymbol{\mu}, \boldsymbol{\Sigma})$, then:

(a) $E(\mathbf{x}) = \boldsymbol{\mu}$ and $\text{var}(\mathbf{x}) = \nu\boldsymbol{\Sigma}/(\nu - 2)$ $(n > 2)$.

(b) $(x_i - \mu_i)/\sqrt{\sigma_{ii}} \sim t_\nu$, where t_ν is the univariate t-distribution with ν degrees of freedom.

(c) $(\mathbf{x} - \boldsymbol{\mu})'\boldsymbol{\Sigma}^{-1}(\mathbf{x} - \boldsymbol{\mu})/d \sim F_{d,\nu}$, where $F_{d,\nu}$ is the univariate F-distribution with d and ν degrees of freedom, respectively.

(d) Any subset of $\mathbf{x}$ has a multivariate t-distribution.

For further details see Kotz and Nadarajah [2004]. They also give a number of probability integrals and discuss the noncentral t-distribution.

20.8.2   Elliptical and Spherical Distributions

Definition 20.18. A $d \times 1$ random vector $\mathbf{x}$ is said to have an *elliptical distribution* with parameters $\boldsymbol{\mu}$ $(d \times 1)$ and $\mathbf{V}$ $(d \times d)$ positive definite if its density function is of the form

$$c_d (\det \mathbf{V})^{-1/2} h[(\mathbf{x} - \boldsymbol{\mu})' \mathbf{V}^{-1} (\mathbf{x} - \boldsymbol{\mu})]$$

for some function h. We will write $\mathbf{x} \sim E_d(\boldsymbol{\mu}, \mathbf{V})$ to denote that the distribution belongs to the class of elliptical distributions. The name comes from the fact that the above probability density function is constant on concentric ellipsoids

$$(\mathbf{x} - \boldsymbol{\mu})' \mathbf{V}^{-1} (\mathbf{x} - \boldsymbol{\mu}) = \mathbf{c},$$

and an alternative name is *elliptically contoured distribution*. The multivariate t, multivariate normal, the contaminated normal, and a mixture of normal distributions are examples of elliptical distributions. Fang et al. [1990], Gupta and Varga [1993], and Kollo and von Rosen [2005: section 2.3] discuss these and other examples of elliptical distributions. The kurtosis for elliptical distributions need not be zero, so that its typical bell-shaped surface can be more or less peaked than the multivariate normal. This flexibility allows one to study the robustness of statistical inference based on the normal distribution. For the theory and statistical inference based on samples from elliptical distributions see Anderson [1993; 2003, section 3.6], Fang and Anderson [1990], and Kariya and Sinha [1989]. For some asymptotic theory see Anderson [2003: 102, 158]. Matrix versions of the elliptical distribution are also available (Anderson [2003] and Girko and Gupta [1996]).

20.49. Let $\mathbf{x} \sim E_d(\boldsymbol{\mu}, \mathbf{V})$.

(a) For some function ψ, the characteristic function of $\mathbf{x}$ is

$$\phi(\mathbf{t}) = \mathrm{E}(e^{it'\mathbf{x}}) = e^{it'\boldsymbol{\mu}} \, \psi(\mathbf{t}' \mathbf{V} \mathbf{t}).$$

(b) From (a) we have that any subset of $\mathbf{x}$ has an elliptical distribution of the same form. For example, if $\mathbf{x}_1$ and $\boldsymbol{\mu}_1$ are the first k elements of $\mathbf{x}$ and $\boldsymbol{\mu}$ respectively, and $\mathbf{V}_{11}$ is the leading principal $k \times k$ submatrix of $\boldsymbol{\Sigma}$, then $\mathbf{x}_1 \sim E_k(\boldsymbol{\mu}_1, \mathbf{V}_{11})$.

(c) Provided they exist, $\mathrm{E}(\mathbf{x}) = \boldsymbol{\mu}$ and $\mathrm{var}(\mathbf{x}) = a\mathbf{V}$ for some constant a. In terms of the characteristic function $a = -2\psi'(0)$.

(d) It follows from (c) that all distributions in the class $E_d(\boldsymbol{\mu}, \mathbf{V})$ have the same mean $\boldsymbol{\mu}$ and the same correlation matrix $\mathrm{corr}(\mathbf{x}) = (\rho_{ij})$. Since a cancels out, we have $\rho_{ij} = v_{ij}/(v_{ii}v_{jj})^{1/2}$.

(e) Let $\boldsymbol{\Sigma} = -2\psi'(0)\mathbf{V} = (\sigma_{ij})$ be the variance matrix of $\mathbf{x}$, and suppose that $\mathbf{x}$ has finite fourth moments. Then:

  (i) The marginal distributions of the x_i all have zero skewness and the same kurtosis

$$\gamma_2 = \frac{\kappa_4}{(\kappa_2)^2} = \frac{3[\psi''(0) - \psi'(0)^2]}{\psi'(0)^2} = 3\kappa, \quad \text{say},$$

  where κ_r is the rth cumulant.

(ii) All fourth-order cumulants are determined by κ, namely

$$\kappa_{1111}^{ijkl} - \kappa(\sigma_{ij}\sigma_{kl} + \sigma_{ik}\sigma_{jl} + \sigma_{il}\sigma_{jk}).$$

This result is useful in asymptotic theory relating to smooth functions of elements of the sample variance matrix.

20.50. Let $\mathbf{x} \sim E_d(\boldsymbol{\mu}, \mathbf{V})$, where $\mathbf{V}$ is diagonal. If $x_1, x_2, \ldots, x_n$ are mutually independent, then $\mathbf{x}$ is multivariate normal.

20.51. Suppose $\mathbf{x} \sim E_d(\boldsymbol{\mu}, \mathbf{V})$ and $\mathbf{x}$, $\boldsymbol{\mu}$, and $\mathbf{V}$ are partitioned as follows:

$$\mathbf{x} = \begin{pmatrix} \mathbf{x}_1 \\ \mathbf{x}_2 \end{pmatrix}, \quad \boldsymbol{\mu} = \begin{pmatrix} \boldsymbol{\mu}_1 \\ \boldsymbol{\mu}_2 \end{pmatrix}, \quad \text{and} \quad \mathbf{V} = \begin{pmatrix} \mathbf{V}_{11} & \mathbf{V}_{12} \\ \mathbf{V}_{21} & \mathbf{V}_{22} \end{pmatrix},$$

where $\mathbf{x}_1$ and $\boldsymbol{\mu}_1$ are $k \times 1$ and $\mathbf{V}_{11}$ is $k \times k$. Provided the following exist, then:

(a) $E(\mathbf{x}_1 \mid \mathbf{x}_2) = \boldsymbol{\mu}_1 + \mathbf{V}_{12}\mathbf{V}_{22}^{-1}(\mathbf{x}_2 - \boldsymbol{\mu}_2)$.

(b) $\text{var}(\mathbf{x}_1 \mid \mathbf{x}_2) = g(\mathbf{x}_2)(\mathbf{V}_{11} - \mathbf{V}_{12}\mathbf{V}_{22}^{-1}\mathbf{V}_{21})$
for some function g. Moreover, the conditional distribution of $\mathbf{x}_1$ given $\mathbf{x}_2$ is k-variate elliptical. If $g(\mathbf{x}_2)$ is a constant so that $\text{var}(\mathbf{x}_1|\mathbf{x}_2)$ does not depend on $\mathbf{x}_2$, then $\mathbf{x}$ must be normal.

***Definition* 20.19.** A $d \times 1$ random vector $\mathbf{x}$ is said to have a *spherical distribution* if $\mathbf{x}$ and $\mathbf{Tx}$ have the same distribution for all $d \times d$ orthogonal matrices $\mathbf{T}$. If $\mathbf{x}$ has a density function, then this function depends on $\mathbf{x}$ only through $\mathbf{x}'\mathbf{x}$. The multivariate normal $N_d(\mathbf{0}, \sigma^2\mathbf{I}_d)$ and the multivariate t, $t_d(\nu, \mathbf{0}, \sigma^2\mathbf{I}_d)$, are spherical distributions.

20.52. If $\mathbf{x} \sim E_d(\mathbf{0}, \mathbf{I}_d)$, then $\mathbf{x}$ has a spherical distribution with a density function of the form $c_d h(\mathbf{x}'\mathbf{x})$. Let

$$
\begin{aligned}
x_1 &= r\sin\theta_1\sin\theta_2\cdots\sin\theta_{d-2}\sin\theta_{d-1} \\
x_2 &= r\sin\theta_1\sin\theta_2\cdots\sin\theta_{d-2}\cos\theta_{d-1} \\
x_3 &= r\sin\theta_1\sin\theta_2\cdots\cos\theta_{d-2} \\
&\quad \cdots \\
x_{d-1} &= r\sin\theta_1\cos\theta_2 \\
x_d &= r\cos\theta_1,
\end{aligned}
$$

where $r > 0$, $0 < \theta_i \le \pi$ ($i = 1, 2, \ldots, d-2$), and $0 < \theta_{d-1} \le 2\pi$. Then r, $\theta_1, \theta_2, \ldots, \theta_{d-1}$ are independent, and the distributions of $\theta_1, \ldots, \theta_{d-1}$ are the same for all $\mathbf{x}$, with θ_k having a probability density function proportional to $\sin^{d-1-k}\theta_k$ (so that θ_{d-1} is uniformly distributed on $(0, 2\pi)$). Also $y = \mathbf{x}'\mathbf{x} = r^2$ has probability density function

$$f_y(y) = \frac{c_d \pi^{d/2}}{\Gamma(\frac{1}{2}d)} y^{(d/2)-1} h(y), \quad (y > 0).$$

In particular, if $\mathbf{x} \sim N_d(\mathbf{0}, \mathbf{I}_d)$, we have $h(y) = e^{-y/2}$ leading to the familiar result that $y \sim \chi_d^2$.

20.53. Let $\mathbf{x}$ have a d-dimensional spherical distribution with $\mathrm{pr}(\mathbf{x} = \mathbf{0}) = 0$.

(a) If $u = \|\mathbf{x}\|_2 = (\mathbf{x}'\mathbf{x})^{1/2}$ and $\mathbf{y} = \|\mathbf{x}\|_2^{-1}\mathbf{x}$, then $\mathbf{y}$ is uniformly distributed over a d-dimensional sphere located at the origin with unit radius, and $\mathbf{y}$ and u are independent.

(b) If $v = \mathbf{a}'\mathbf{x}/\|\mathbf{x}\|_2$, where $\mathbf{a} \in \mathbb{R}^d$ and $\mathbf{a}'\mathbf{a} = 1$, then

$$w = (d-1)^{1/2}\frac{v}{(1-v^2)^{1/2}} \sim t_{d-1},$$

the t-distribution.

(c) If $\mathbf{A}$ is a $d \times d$ symmetric idempotent matrix of rank k, then $z = \mathbf{x}'\mathbf{A}\mathbf{x}/\mathbf{x}'\mathbf{x}$ has a beta distribution with parameters $\frac{1}{2}k$ and $\frac{1}{2}(d-k)$.

Proofs. Section 20.8.2.

   20.49. Muirhead [1982: 34–42].

   20.50. Muirhead [1982: 35].

   20.51. Muirhead [1982: 36].

   20.52. Anderson [2003: 47–48] and Muirhead [1982: 36–37].

   20.53. Muirhead [1982: 38–39].

20.8.3   Dirichlet Distributions

Definition 20.20. Let $\mathbf{x} = (x_1, x_2, \ldots, x_d)'$ be a $d \times 1$ random vector. Then $\mathbf{x}$ is said to have a *Type-1 Dirichlet distribution* if its density function is given by

$$f(\mathbf{x}) = \frac{\Gamma(\alpha_1 + \cdots + \alpha_{d+1})}{\Gamma(\alpha_1)\cdots\Gamma(\alpha_{d+1})}x_1^{\alpha_1 - 1}\cdots x_d^{\alpha_d - 1}(1 - x_1 - \cdots - x_d)^{\alpha_{d+1} - 1}$$

$(0 < x_i < 1, i = 1, \ldots, d, x_1 + \cdots + x_d < 1,$ and $\alpha_i > 0$ for $i = 1, \ldots, d+1$.)

We shall write $\mathbf{x} \sim D_1(d, \boldsymbol{\alpha})$, where $\boldsymbol{\alpha} = (\alpha_i)$.

   Also $\mathbf{x}$ is said to have a *Type-2 Dirichlet distribution* if its density function is given by

$$f(\mathbf{x}) = \frac{\Gamma(\alpha_1 + \cdots + \alpha_{d+1})}{\Gamma(\alpha_1)\cdots\Gamma(\alpha_{d+1})}x_1^{\alpha_1 - 1}\cdots x_d^{\alpha_d - 1}(1 + x_1 + \cdots + x_d)^{-(\alpha_1 + \cdots + \alpha_{d+1})}$$

$(0 < x_i < \infty, i = 1, \ldots, d,$ and $\alpha_i > 0$ for $i = 1, 2, \ldots, d+1.$

We shall write $\mathbf{x} \sim D_2(d, \boldsymbol{\alpha})$. The above are special cases of the matrix versions in Section 21.9.

20.54. Let $y_1, y_2, \ldots, y_m$ be independently distributed as χ^2 variables with degrees of freedom $\alpha_1, \alpha_2, \ldots \alpha_m$, respectively. If $x_i = y_i/\sum_{j=1}^m y_j$ for $i = 1, 2, \ldots, m-1$, then $\mathbf{x} \sim D_1(m-1, \boldsymbol{\alpha}/2)$.

Proofs. Section 20.8.3.

   20.54. Anderson [2003: 290, quoted in exercise 7.36].

CHAPTER 21

RANDOM MATRICES

21.1 INTRODUCTION

Matrices of random variables occur frequently in statistics, especially in the subject of multivariate analysis. Because the latter is a large subject with numerous reference books, I have had to be somewhat selective in my choice of topics. In this chapter, as in the previous one, the upper- and lowercase letters of the alphabet from a to t, excluding $\mathbf{Q}$, refer to constants, while the remainder generally refer to random variables. Unless otherwise stated, all vectors and matrices are real.

Definition 21.1. Let

$$
\mathbf{X} = (x_{ij}) = \begin{pmatrix} \mathbf{x}_1' \\ \mathbf{x}_2' \\ \vdots \\ \mathbf{x}_n' \end{pmatrix} = (\mathbf{x}^{(1)}, \mathbf{x}^{(2)}, \dots, \mathbf{x}^{(d)})
$$

be an $n \times d$ matrix of random variables with rows that all have the same variance matrix $\boldsymbol{\Sigma}$ and are uncorrelated, that is,

$$
\operatorname{cov}(\mathbf{x}_r, \mathbf{x}_s) = \delta_{rs}\boldsymbol{\Sigma},
$$

where $\delta_{rs} = 1$ for $r = s$ and $\delta_{rs} = 0$ for $r \neq s$. We shall call a matrix with the above properties a *data matrix*.

A Matrix Handbook for Statisticians. By George A. F. Seber
Copyright © 2008 John Wiley & Sons, Inc.

In practice, the x_r are usually a random sample, which implies they are independently and identically distributed, that is, i.i.d. However, this won't be assumed unless indicated.

21.1. If $\mathbf{x}_i$ defined above has mean $\boldsymbol{\mu}_i$ for each i, then $E(\mathbf{x}_i\mathbf{x}_j') = \delta_{ij}\boldsymbol{\Sigma} + \boldsymbol{\mu}_i\boldsymbol{\mu}_j'$.

21.2. (Some Vec Properties) Let $\mathbf{X}$ be a data matrix as defined above; that is, the columns of $\mathbf{X}'$ are uncorrelated and have the same variance matrix $\boldsymbol{\Sigma}$. Here "$\otimes$" is the Kronecker product.

(a) $\text{var}(\text{vec }\mathbf{X}) = \boldsymbol{\Sigma} \otimes \mathbf{I}_n$ and $\text{var}(\text{vec }\mathbf{X}') = \mathbf{I}_n \otimes \boldsymbol{\Sigma}$.

(b) Using $\text{vec}\,(\mathbf{AXB}) = (\mathbf{B}' \otimes \mathbf{A})\text{vec }\mathbf{X}$, we have, from (20.6b), (11.1e), and (11.11a),

$$\text{var}[\text{vec}\,(\mathbf{AXB})] = (\mathbf{B}' \otimes \mathbf{A})(\boldsymbol{\Sigma} \otimes \mathbf{I}_n)(\mathbf{B}' \otimes \mathbf{A})' = (\mathbf{B}'\boldsymbol{\Sigma}\mathbf{B}) \otimes (\mathbf{AA}').$$

(c) $\text{var}[\text{vec}\,(\mathbf{AX'B})] = \text{var}[(\mathbf{B}' \otimes \mathbf{A})\text{vec }\mathbf{X}'] = (\mathbf{B}'\mathbf{B}) \otimes (\mathbf{A}\boldsymbol{\Sigma}\mathbf{A}')$.

(d) $\text{cov}[\text{vec}\,(\mathbf{AXB}), \text{vec}\,(\mathbf{CXD})] = (\mathbf{B}'\boldsymbol{\Sigma}\mathbf{D}) \otimes (\mathbf{AC}')$.

(e) From (d) we see that if $\mathbf{U} = \mathbf{AXB}$ and $\mathbf{V} = \mathbf{CXD}$, then $\mathbf{U}$ and $\mathbf{V}$ are pairwise uncorrelated, that is $\text{cov}(u_{ij}, v_{rs}) = 0$ for all i, j, r, and s, if $\mathbf{AC}' = \mathbf{0}$ and/or $\mathbf{B}'\boldsymbol{\Sigma}\mathbf{D} = \mathbf{0}$.

Proofs. Section 21.1.

    21.1. Set $\mathbf{x}_i = \mathbf{x}_i - \boldsymbol{\mu}_i + \boldsymbol{\mu}_i$ and use (20.5).

    21.2. For (a), see Henderson and Searle [1979: 78, with our $\mathbf{X}$ being their $\mathbf{X}'$]; using (20.6), the proofs of (c) and (d) are similar to (b); and (e) follows from (d).

21.2   GENERALIZED QUADRATIC FORMS

21.2.1   General Results

Definition **21.2.** If $\mathbf{X} = (\mathbf{x}_1, \mathbf{x}_2, \dots, \mathbf{x}_n)'$ is an $n \times d$ data matrix and $\mathbf{A} = (a_{ij})$ is a symmetric $n \times n$ matrix, we shall call the expression $\mathbf{X}'\mathbf{AX} = \sum_{i=1}^{n} \sum_{j=1}^{n} a_{ij}\mathbf{x}_i\mathbf{x}_j'$ a *generalized quadratic*.

21.3. Using the above notation, let $\overline{\mathbf{x}} = \sum_{i=1}^{n} \mathbf{x}_i/n$ and $\widetilde{\mathbf{X}} = (\mathbf{x}_1 - \overline{\mathbf{x}}, \dots, \mathbf{x}_n - \overline{\mathbf{x}})'$. Then:

(a) $\widetilde{\mathbf{X}}'\widetilde{\mathbf{X}} = \sum_{i=1}^{n}(\mathbf{x}_i - \overline{\mathbf{x}})(\mathbf{x}_i - \overline{\mathbf{x}})' = \sum_{i=1}^{n} \mathbf{x}_i\mathbf{x}_i' - n\overline{\mathbf{x}}\,\overline{\mathbf{x}}' = \mathbf{X}'\mathbf{AX}\ (= \mathbf{Q}, \text{ say}),$
where $\mathbf{A} = (a_{ij})$ and $a_{ij} = \delta_{ij} - n^{-1}$. Thus, $\mathbf{A} = \mathbf{I}_n - \mathbf{1}_n\mathbf{1}_n'/n = \mathbf{I}_n - \mathbf{P}_1$ is the so-called centering matrix, where $\mathbf{P}_1$ is the orthogonal projection onto $\mathcal{C}(\mathbf{1}_n)$.

(b) Suppose the $\mathbf{x}_i$ are i.i.d. with mean $\boldsymbol{\mu}$ and variance matrix $\boldsymbol{\Sigma}$, and $\mathbf{S} = \mathbf{Q}/(n-1)$. Then

$$E(\mathbf{S}) = \boldsymbol{\Sigma},$$

so that **S**, the so-called *sample variance matrix*, is an unbiased estimator of Σ. Some writers define $\hat{\Sigma} = (n-1)\mathbf{S}/n$ to be the sample variance matrix; it is the maximum likelihood estimator of Σ under normality assumptions.

(c) If diag(**s**) is the diagonal matrix whose elements are the diagonal elements of **S**, then the sample correlation matrix is given by

$$\mathbf{R} = [\text{diag}(\mathbf{s})]^{-1/2}\mathbf{S}[\text{diag}(\mathbf{s})]^{-1/2}.$$

(d) The (sample) Mahalanobis distance

$$D_i^2 = (\mathbf{x}_i - \bar{\mathbf{x}})'\mathbf{S}^{-1}(\mathbf{x}_i - \bar{\mathbf{x}})$$

is the ith diagonal element of $(n-1)\tilde{\mathbf{X}}(\tilde{\mathbf{X}}'\tilde{\mathbf{X}})^{-1}\tilde{\mathbf{X}}$, and it is often used for diagnostic purposes.

(e) (i) Taking the trace of (a) and using $\text{trace}(\mathbf{cd}') = \text{trace}(\mathbf{d}'\mathbf{c}) = \mathbf{d}'\mathbf{c}$, we get

$$\sum_{i=1}^{n}(\mathbf{x}_i - \bar{\mathbf{x}})'(\mathbf{x}_i - \bar{\mathbf{x}}) = \sum_{i=1}^{n}\mathbf{x}_i'\mathbf{x}_i - n\bar{\mathbf{x}}'\bar{\mathbf{x}}.$$

(ii)

$$\sum_{r=1}^{n}\sum_{s=r+1}^{n}(\mathbf{x}_r - \mathbf{x}_s)(\mathbf{x}_r - \mathbf{x}_s)' = \frac{1}{2}\sum_{i\neq j}(\mathbf{x}_r - \mathbf{x}_s)(\mathbf{x}_r - \mathbf{x}_s)'$$

$$= n\sum_{i=1}^{n}(\mathbf{x}_i - \bar{\mathbf{x}})(\mathbf{x}_i - \bar{\mathbf{x}})'.$$

(iii) Taking traces in (ii),

$$\sum_{r=1}^{n}\sum_{s=r+1}^{n}\|\mathbf{x}_r - \mathbf{x}_s\|^2 = n\sum_{i=1}^{n}\|\mathbf{x}_i - \bar{\mathbf{x}}\|^2.$$

This result arises in cluster analysis.

We obtain the corresponding univariate cases by taking $d = 1$ in the above results.

21.4. (Asymptotic Sample Theory) Suppose that the $d \times 1$ vectors $\mathbf{x}_1, \mathbf{x}_2, \ldots, \mathbf{x}_n$ are independently and identically distributed (i.i.d.) with mean $\boldsymbol{\mu}$ and variance matrix Σ.

(a) As $n \to \infty$, $\sqrt{n}(\bar{\mathbf{x}} - \boldsymbol{\mu})$ is asymptotically distributed as $N_d(\mathbf{0}, \Sigma)$.

(b) Let $\mathbf{Q} = (n-1)\mathbf{S}$.

   (i) As $n \to \infty$, $n^{-1/2}(\text{vec}\,\mathbf{Q} - n\text{vec}\,\Sigma)$ is asymptotically $N_{d^2}(\mathbf{0}, \mathbf{V})$, where

$$\mathbf{V} = \text{var}\{\text{vec}\,[(\mathbf{x}_i - \boldsymbol{\mu})(\mathbf{x}_i - \boldsymbol{\mu})']\} = \text{var}[(\mathbf{x}_i - \boldsymbol{\mu}) \otimes (\mathbf{x}_i - \boldsymbol{\mu})],$$

   by (11.15c).

   (ii) In terms of **S**, we have $(n-1)^{1/2}(\text{vec}\,\mathbf{S} - \text{vec}\,\Sigma)$ is asymptotically $N_{d^2}(\mathbf{0}, \mathbf{V})$.

(c) If $\mathbf{A}$ is a real symmetric $n \times n$ matrix, then, under certain conditions, $\mathbf{Q} = \mathbf{X}'\mathbf{A}\mathbf{X}$ is asymptotically normal as $n \to \infty$.

For further details and references see Mathai and Provost [1992: section 4.6b].

21.5. (Asymptotic Theory) Suppose $\sqrt{n}(\mathbf{y} - \boldsymbol{\theta})$ is asymptotically $N_d(\mathbf{0}, \boldsymbol{\Sigma})$, and let $\mathbf{f}(\mathbf{u})$ be a q-dimensional vector-valued function of $\mathbf{u}$ such that each component $f_j(\mathbf{u})$ has a nonzero differential at $\mathbf{u} = \boldsymbol{\theta}$. If $\mathbf{F} = (f_{ij})$, where $f_{ij} = \partial f_j(\mathbf{u})/\partial u_i$, then

$$\sqrt{n}[\mathbf{f}(\mathbf{u}) - \mathbf{f}(\boldsymbol{\theta})] \text{ is asymptotically } N_q(\mathbf{0}, \mathbf{F}'\boldsymbol{\Sigma}\mathbf{F}).$$

21.6. Suppose $\mathbf{X}'$ has columns $\mathbf{x}_i$, then

$$E(\mathbf{X}'\mathbf{A}\mathbf{X}) = \sum_{i=1}^{n}\sum_{j=1}^{n} a_{ij}\text{cov}(\mathbf{x}_i, \mathbf{x}_j) + E(\mathbf{X}')\mathbf{A}E(\mathbf{X}).$$

21.7. Suppose the $\mathbf{x}_i$, the columns of $\mathbf{X}'$, are statistically independent and $\text{var}(\mathbf{x}_i) = \boldsymbol{\Sigma}_i$ for $i = 1, 2, \ldots, n$. Then, from (21.6),

(a) $E(\mathbf{X}'\mathbf{A}\mathbf{X}) = \sum_{i=1}^{n} a_{ii}\boldsymbol{\Sigma}_i + E(\mathbf{X}')\mathbf{A}E(\mathbf{X}).$

(b) If $\boldsymbol{\Sigma}_i = \boldsymbol{\Sigma}$ for all i, then from (a) we have

$$E(\mathbf{X}'\mathbf{A}\mathbf{X}) = (\text{trace }\mathbf{A})\boldsymbol{\Sigma} + E(\mathbf{X}')\mathbf{A}E(\mathbf{X}).$$

(c) Suppose that the $\mathbf{x}_i$ are i.i.d. with mean $\mathbf{0}$ and variance matrix $\boldsymbol{\Sigma}$. If $\mathbf{V} = \mathbf{X}'\mathbf{X}$ and $E(\mathbf{x}_i\mathbf{x}_i' \otimes \mathbf{x}_i\mathbf{x}_i') = \boldsymbol{\Psi}$, then

$$\text{var}(\text{vec }\mathbf{V}) = n[\boldsymbol{\Psi} - (\text{vec }\boldsymbol{\Sigma})(\text{vec }\boldsymbol{\Sigma})'].$$

21.8. (Independence) Let $\mathbf{X}' = (\mathbf{x}_1, \mathbf{x}_2, \ldots, \mathbf{x}_n)$, where the $\mathbf{x}_i$ are independently distributed as $N_d(\boldsymbol{\mu}_i, \boldsymbol{\Sigma})$ $(i = 1, 2, \ldots, n)$ and $\boldsymbol{\Sigma}$ is positive definite. Suppose $\mathbf{A}$ and $\mathbf{B}$ are $n \times n$ symmetric matrices and $\mathbf{C}$ is a $k \times n$ matrix. Then:

(a) $\mathbf{X}'\mathbf{A}\mathbf{X}$ and $\mathbf{X}'\mathbf{B}\mathbf{X}$ are statistically independent if and only if $\mathbf{A}\mathbf{B} = \mathbf{0}$.

(b) $\mathbf{C}\mathbf{X}$ and $\mathbf{X}'\mathbf{A}\mathbf{X}$ are statistically independent if and only if $\mathbf{C}\mathbf{A} = \mathbf{0}$.
Setting $\mathbf{C} = \mathbf{b}'$, we have that $\mathbf{X}'\mathbf{b}$ and $\mathbf{X}'\mathbf{A}\mathbf{X}$ are statistically independent if and only if $\mathbf{A}\mathbf{b} = \mathbf{0}$.

More generally:

(c) Let $\mathbf{Q}_i = \mathbf{X}'\mathbf{A}_i\mathbf{X} + \frac{1}{2}(\mathbf{L}_i\mathbf{X} + \mathbf{X}'\mathbf{L}_i') + \mathbf{C}_i$, where $\mathbf{A}_i$ and $\mathbf{C}_i$ are real symmetric matrices $(i = 1, 2)$. Then $\mathbf{Q}_1$ and $\mathbf{Q}_2$ are statistically independent if and only if

$$\mathbf{A}_1\mathbf{A}_2 = \mathbf{0}, \quad \mathbf{L}_1\mathbf{A}_2 = \mathbf{0}, \quad \mathbf{L}_2\mathbf{A}_1 = \mathbf{0}, \quad \text{and} \quad \mathbf{L}_1\mathbf{L}_2' = \mathbf{0}.$$

We can get various special cases by setting $\mathbf{A}_i = \mathbf{0}$ or $\mathbf{L}_i = \mathbf{0}$. Mathai and Provost [1992: 286–287] also give results for the case when $\boldsymbol{\Sigma}$ is singular.

Proofs. Section 21.2.1.

21.3a–b. Seber [1984: 8–9].

21.4a. Anderson [2003: 86–87] and Muirhead [1982: 15].

21.4b. Anderson [2003: 87] and Muirhead [1982: 18].

21.5. Anderson [2003: 132].

21.6. Mathai and Provost [1992: 244].

21.7a. Seber [1984: 6–7].

21.7c. Schott [2005: 424–425].

21.8a–b. Mathai and Provost [1992: 285] and Schott [2005: 422–424].

21.8c. Mathai and Provost [1992: 286–287].

21.2.2   Wishart Distribution

Definition 21.3. Let $\mathbf{X} = (\mathbf{x}_1, \mathbf{x}_2, \ldots, \mathbf{x}_n)'$ be an $m \times d$ matrix with rows which are independently and identically distributed (i.i.d.) as the multivariate normal distribution $N_d(\mathbf{0}, \boldsymbol{\Sigma})$, subject to two conditions, namely (i) $d \leq m$ and (ii) $\boldsymbol{\Sigma}$ is positive definite. If $\mathbf{W} = \mathbf{X}'\mathbf{X}$, then $\mathbf{W}$ is said to a have a (nonsingular) *Wishart disribution* with m degrees of freedom and we write $\mathbf{W} \sim W_d(m, \boldsymbol{\Sigma})$. The joint probability probability density function of the distinct elements of the symmetric matrix $\mathbf{W}$ (the $\frac{1}{2}d(d+1)$ variables in the upper triangle, say) is given in (21.67). This Wishart distribution can, of course, be defined directly in terms of its density function, though the above representation is more convenient for developing the theory, especially in the singular case. If at least one of the two conditions does not hold, then the distribution is said to be singular (cf. Srivastava [2003] for some details). We use m instead of n here as $\mathbf{X}$ is used as a "representation" rather than coming from a particular random sample of size m. If $\mathbf{W} \sim W_d(m, \boldsymbol{\Sigma})$, then we can simply choose any matrix $\mathbf{X}$ with rows which are i.i.d. $N_d(\mathbf{0}, \boldsymbol{\Sigma})$. Then $\mathbf{X}'\mathbf{X}$ has the same distribution as $\mathbf{W}$ and can be used as a "proxy" for the latter. For this reason, most authors simply set $\mathbf{W} = \mathbf{X}'\mathbf{X}$ (e.g., Seber [1984: 21]). For other general references relating to the Wishart distribution see Anderson [2003] and Muirhead [1982]; in fact most theoretical books on multivariate analysis cover the Wishart distribution in detail.

If the $\mathbf{x}_i$ are independently distributed as $N_d(\boldsymbol{\mu}_i, \boldsymbol{\Sigma})$ with $\boldsymbol{\Sigma}$ positive definite, then $\mathbf{W} = \mathbf{X}'\mathbf{X}$ has a *noncentral Wishart distribution*, generally written as $\mathbf{W}_d(m, \boldsymbol{\Sigma}; \boldsymbol{\Delta})$, where

$$\boldsymbol{\Delta} = \boldsymbol{\Sigma}^{-1/2}\mathbf{M}'\mathbf{M}\boldsymbol{\Sigma}^{-1/2}, \quad \mathbf{M} = (\boldsymbol{\mu}_1, \boldsymbol{\mu}_2, \ldots, \boldsymbol{\mu}_m)' = \mathrm{E}(\mathbf{X}),$$

and $\boldsymbol{\Sigma}^{1/2}$ is the positive square root of $\boldsymbol{\Sigma}$ (cf. 10.32). Here $\boldsymbol{\Delta}$ is called the *noncentrality matrix* and, since the distribution of $\mathbf{W}$ depends only on the eigenvalues of $\boldsymbol{\Delta}$, other expressions are used for the noncentrality parameter (Seber [1984: section 2.3.3]). Muirhead [1984: section 10.3] defines $\boldsymbol{\Delta} = \boldsymbol{\Sigma}^{-1}\mathbf{M}'\mathbf{M}$ and gives the probability density function of the noncentral distribution and its properties.

When $\mathbf{W}$ has a nonsingular distribution, $\mathbf{W}^{-1}$ is said to have an *inverted Wishart*. For some details see Anderson [2003: section 7.7] and Muirhead [1982: 113, exercise 3.6]. A generalized version also exists (Brown [2002]).

21.9. An important special case is when the $\mathbf{x}_i$ are all $N_d(\mathbf{0}, \sigma^2 \mathbf{I}_d)$. Then the elements of $\mathbf{X}$ are all i.i.d. $N(0, \sigma^2)$ and $\mathbf{X}'\mathbf{X} \sim W_d(m, \mathbf{I}_d)$.

21.10. If the $\mathbf{x}_i$ $(i = 1, 2, \ldots, m)$ are independently distributed as $N_d(\boldsymbol{\mu}_i, \boldsymbol{\Sigma})$ with $\boldsymbol{\Sigma}$ positive definite, then using (21.2a) we have

$$\text{vec}(\mathbf{X}') \sim N_{md}(\text{vec}(\mathbf{M}'), \mathbf{I}_m \otimes \boldsymbol{\Sigma}),$$

where $\mathbf{M} = (\boldsymbol{\mu}_1, \boldsymbol{\mu}_2, \ldots, \boldsymbol{\mu}_m)'$.

21.11. Suppose $\mathbf{W} = \mathbf{X}'\mathbf{X} \sim W_d(m, \boldsymbol{\Sigma}; \boldsymbol{\Delta})$, where $\mathbf{X}$ is defined above. Then:

(a) $E(\mathbf{W}) = m\boldsymbol{\Sigma} + \mathbf{M}'\mathbf{M}$, where $\mathbf{M}$ is defined above. For this reason, some authors define the noncentrality parameter to be $\boldsymbol{\Phi} = \mathbf{M}'\mathbf{M}$ or even $\frac{1}{2}\boldsymbol{\Phi}$ (e.g., Schott [2005: 422]), which have some advantages, as demonstrated in (21.12) below.

(b) $\text{var}(\text{vec}\,\mathbf{W}) = 2\mathbf{P}_m[d(\boldsymbol{\Sigma} \otimes \boldsymbol{\Sigma}) + \boldsymbol{\Sigma} \otimes \mathbf{M}'\mathbf{M} + \mathbf{M}'\mathbf{M} \otimes \boldsymbol{\Sigma}]$, where $\mathbf{P}_m$ is the symmetrizer matrix defined in Section 11.5.1.

21.12. If we redefine the parameters of a noncentral Wishart and write $\mathbf{W} \sim W_d(m, \boldsymbol{\Sigma}; \boldsymbol{\Phi})$, where $\boldsymbol{\Sigma}$ is positive definite and $\boldsymbol{\Phi} = \mathbf{M}'\mathbf{M}$, and $\mathbf{C}$ is $q \times d$ of rank q, then $\mathbf{C}\mathbf{W}\mathbf{C}' \sim W_d(m, \mathbf{C}\boldsymbol{\Sigma}\mathbf{C}'; \mathbf{C}\boldsymbol{\Phi}\mathbf{C}')$. In terms of the previous notation, we have $\boldsymbol{\Delta} = (\mathbf{C}\boldsymbol{\Sigma}\mathbf{C}')^{-1/2}\mathbf{C}\boldsymbol{\Phi}\mathbf{C}'(\mathbf{C}\boldsymbol{\Sigma}\mathbf{C}')^{-1/2}$.

21.13. Suppose $\mathbf{W} = (w_{ij})$ has a nonsingular Wishart distribution $W_d(m, \boldsymbol{\Sigma})$.

(a) $\mathbf{W}$ is positive definite with probability 1.

(b) The eigenvalues of $\mathbf{W}$ are distinct with probability 1.

(c) $E(\mathbf{W}) = m\boldsymbol{\Sigma}$, which still holds if $\mathbf{W}$ is singular.

(d) Let $\mathbf{C}$ be a $q \times d$ matrix of rank q.

   (i) $\mathbf{C}\mathbf{W}\mathbf{C}' \sim W_q(m, \mathbf{C}\boldsymbol{\Sigma}\mathbf{C}')$ and is nonsingular. We have the following special cases.

   (ii) Setting $\mathbf{C} = \mathbf{a}'$ $(\mathbf{a} \neq \mathbf{0})$, we have $\mathbf{a}'\mathbf{W}\mathbf{a} \sim \sigma_{\mathbf{a}}^2\chi_m^2$, where $\sigma_{\mathbf{a}}^2 = \mathbf{a}'\boldsymbol{\Sigma}\mathbf{a}$.

   (iii) By choosing $\mathbf{C} = (\mathbf{I}_r, \mathbf{0})$ $(r \leq d)$, or an appropriate permutation of its columns, we see that an $r \times r$ principal submatrix of $\mathbf{W}$ has the Wishart distribution $\mathbf{W}_r(m, \boldsymbol{\Sigma}_{rr})$, where $\boldsymbol{\Sigma}_{rr}$ is the corresponding $r \times r$ principal submatrix of $\boldsymbol{\Sigma}$.

   (iv) If $\mathbf{W}$ is singular and rank $\mathbf{C} \leq q$, then $\mathbf{C}\mathbf{W}\mathbf{C}'$ has a singular Wishart distribution.

(e) $w_{jj}/\sigma_{jj} \sim \chi_m^2$, $(j = 1, 2, \ldots, d)$. However, they are not statistically independent. Also w_{ij}/σ_{ij} is not chi-square for $i \neq j$.

(f) $\det\mathbf{W}/\det\boldsymbol{\Sigma}$ is distributed as a product of d independent chi-square variables with respective degrees of freedom $m, m-1, \ldots, m-d+1$.

(g) The moment generating function (m.g.f.) of $\mathbf{W} = (w_{ij})$ is

$$
\begin{aligned}
M(\mathbf{T}) &= \mathrm{E}\left[\exp\left(\sum_{j=1}^{k}\sum_{k=1}^{d} t_{jk} w_{jk}\right)\right] \\
&= \mathrm{E}\{\exp[\mathrm{trace}(\mathbf{U}\mathbf{W})]\} \\
&= [\det(\mathbf{I}_d - 2\mathbf{U}\mathbf{\Sigma})]^{-m/2},
\end{aligned}
$$

where $\mathbf{U} = \mathbf{U}'$, $u_{jj} = t_{jj}$ and $u_{jk} = u_{kj} = \frac{1}{2}t_{jk}$ ($j < k$). Since this moment generating function exists in a neighborhood of $\mathbf{T} = \mathbf{0}$, it uniquely determines the (nonsingular) probability density function of $\mathbf{W}$ and can therefore be used for deriving a number of results given below. The characteristic function is derived by Muirhead [1982: 87].

We have essentially found the m.g.f. of $\mathbf{X}'\mathbf{X}$ when the columns of $\mathbf{X}'$ are i.i.d. $N_d(\mathbf{0}, \mathbf{\Sigma})$, where $\mathbf{\Sigma}$ is positive definite. The m.g.f. of $\mathbf{X}'\mathbf{A}\mathbf{X} + \frac{1}{2}(\mathbf{L}\mathbf{X} + \mathbf{X}'\mathbf{L}') + \mathbf{C}$ when the columns of $\mathbf{X}'$ are i.i.d. $N_d(\boldsymbol{\mu}, \mathbf{\Sigma})$ is given by Mathai and Provost [1992: section 6.4]. They also give results for the cases when the $\mathbf{x}_i$ are correlated and $\mathbf{\Sigma}$ is singular.

21.14. Let $\mathbf{W} \sim W_d(m, \mathbf{\Sigma})$, where $\mathbf{\Sigma}$ is possibly singular, and let $\mathbf{A}$ be a $d \times d$ (not necessarily symmetric) matrix. Then:

(a) $\mathrm{E}(\mathbf{W}\mathbf{A}\mathbf{W}) = m[\mathbf{\Sigma}\mathbf{A}\mathbf{\Sigma} + \mathrm{trace}(\mathbf{A}\mathbf{\Sigma})\mathbf{A}] + m^2\mathbf{\Sigma}\mathbf{A}\mathbf{\Sigma}$.

(b) If $m > d + 1$, and $\mathbf{\Sigma}$ is nonsingular, then

  (i) $\mathrm{E}(\mathbf{W}\mathbf{A}\mathbf{W}^{-1}) = \dfrac{1}{m - d - 1}[m\mathbf{\Sigma}\mathbf{A}\mathbf{\Sigma}^{-1} - \mathbf{A}' - (\mathrm{trace}\,\mathbf{A})\mathbf{I}_d]$.

  (ii) $\mathrm{E}(\mathbf{W}^{-1}\mathbf{A}\mathbf{W}) = \dfrac{1}{m - d - 1}[m\mathbf{\Sigma}^{-1}\mathbf{A}\mathbf{\Sigma} - \mathbf{A}' - (\mathrm{trace}\,\mathbf{A})\mathbf{I}_d]$.

(c) If $m > d + 3$, and $\mathbf{\Sigma}$ is nonsingular, then

$$
\begin{aligned}
&\mathrm{E}(\mathbf{W}^{-1}\mathbf{A}\mathbf{W}^{-1}) \\
&= \frac{(m - d - 2)\mathbf{\Sigma}^{-1}\mathbf{A}\mathbf{\Sigma}^{-1} + \mathbf{\Sigma}^{-1}\mathbf{A}'\mathbf{\Sigma}^{-1} - \mathrm{trace}(\mathbf{A}\mathbf{\Sigma}^{-1})\mathbf{\Sigma}^{-1}}{(m - d)(m - d - 1)(m - d - 3)}.
\end{aligned}
$$

21.15. Suppose that the columns $\mathbf{x}_i$ ($i = 1, \ldots, m$) of $\mathbf{X}'$ are independently distributed as $N_d(\boldsymbol{\mu}_i, \mathbf{\Sigma})$ with $\mathbf{\Sigma}$ positive definite, and $\mathbf{A}$ is a symmetric $d \times d$ matrix of rank r. Then, if $\mathbf{A}$ is idempotent, we have

$$
\mathbf{X}'\mathbf{A}\mathbf{X} \sim W_d(m, \mathbf{\Sigma}; \mathbf{\Delta}),
$$

the noncentral Wishart distribution with

$$
\mathbf{\Delta} = \mathbf{\Sigma}^{-1/2}\mathbf{M}'\mathbf{A}\mathbf{M}\mathbf{\Sigma}^{-1/2} \quad \text{and} \quad \mathbf{M} = (\boldsymbol{\mu}_1, \boldsymbol{\mu}_2, \ldots, \boldsymbol{\mu}_m).
$$

The case when the $\mathbf{x}_i$ are not independent and $\mathbf{\Sigma}$ is non-negative definite is considered by Vaish and Chaganty [2004: 382].

21.16. Suppose that m columns of $\mathbf{X}'$ are i.i.d. as $N_d(\mathbf{0}, \mathbf{\Sigma})$, where $\mathbf{\Sigma}$ is positive definite, and let $\mathbf{A}$ and $\mathbf{B}$ be $m \times m$ symmetric matrices.

(a) Let $\mathbf{X}'\mathbf{AX}$ and $\mathbf{X}'\mathbf{BX}$ have Wishart distributions. They are statistically independent if and only if $\mathbf{AB} = \mathbf{0}$. (This result is generalized in (21.17).)

(b) Let $\mathbf{X}'\mathbf{AX}$ have a Wishart distribution. Then $\mathbf{X}'\mathbf{b}$ is statistically independent of $\mathbf{X}'\mathbf{AX}$ if and only if $\mathbf{Ab} = \mathbf{0}$.

21.17. Suppose that the columns $\mathbf{x}_i$ $(i = 1, \ldots, m)$ of $\mathbf{X}'$ are independently distributed as $N_d(\boldsymbol{\mu}_i, \boldsymbol{\Sigma})$, and let $\mathbf{A}_1, \mathbf{A}_2, \ldots, \mathbf{A}_t$ be a sequence of $n \times n$ symmetric matrices with ranks $r_1, r_2, \ldots, r_t$ such that $\sum_{i=1}^{t} \mathbf{A}_i = \mathbf{I}_n$. If one (and therefore all) of the conditions of (8.78) hold, then the $\mathbf{X}'\mathbf{A}_i\mathbf{X}$ $(i = 1, 2, \ldots, t)$ are independently distributed as the noncentral Wishart, $W_d(r_i, \boldsymbol{\Sigma}; \boldsymbol{\Delta}_i)$, where $\boldsymbol{\Delta}_i = \boldsymbol{\Sigma}^{-1/2}\mathbf{M}'\mathbf{A}_i\mathbf{M}\boldsymbol{\Sigma}^{-1/2}$ is the noncentrality parameter and $\mathbf{M} = (\boldsymbol{\mu}_1, \boldsymbol{\mu}_2, \ldots, \boldsymbol{\mu}_m)'$. An extension of this result to the case when the $\mathbf{x}_i$ are not independent and $\boldsymbol{\Sigma}$ is non-negative definite is given by Vaish and Chaganty [2004: 383] and Tian and Styan [2005: 391].

21.18. If $\mathbf{W}_i \sim W_d(m_i, \boldsymbol{\Sigma})$ $(i = 1, 2)$, and $\mathbf{W}_1$ and $\mathbf{W}_2$ are statistically independent, then $\mathbf{W}_1 + \mathbf{W}_2 \sim \mathbf{W}_d(m_1 + m_2, \boldsymbol{\Sigma})$.

21.19. Let

$$\mathbf{W} = \begin{pmatrix} \mathbf{W}_{11} & \mathbf{W}_{12} \\ \mathbf{W}_{12} & \mathbf{W}_{22} \end{pmatrix} \sim W_d(m, \boldsymbol{\Sigma}), \quad m \le d,$$

where $\boldsymbol{\Sigma}$ is positive definite, $\mathbf{W}_{ii}$ is $d_i \times d_i$, $(i = 1, 2)$, and $d_1 + d_2 = d$. Suppose $\boldsymbol{\Sigma}$ is partioned in the same way as $\mathbf{W}$ and $\boldsymbol{\Sigma}_{22\cdot1} = \boldsymbol{\Sigma}_{22} - \boldsymbol{\Sigma}_{21}\boldsymbol{\Sigma}_{11}^{-1}\boldsymbol{\Sigma}_{12}$.

(a) We have $\mathbf{W}_{22\cdot1} = \mathbf{W}_{22} - \mathbf{W}_{21}\mathbf{W}_{11}^{-1}\mathbf{W}_{12} \sim W_{d_2}(m - d_1, \boldsymbol{\Sigma}_{22\cdot1})$, and $\mathbf{W}_{22\cdot1}$ is statistically independent of $(\mathbf{W}_{11}, \mathbf{W}_{12})$. Note that $\mathbf{W}_{12} = \mathbf{W}_{21}'$.

(b) If $\boldsymbol{\Sigma}_{12} = \mathbf{0}$, then $\mathbf{Y} = \mathbf{W}_{21}\mathbf{W}_{11}^{-1}\mathbf{W}_{12} \sim W_{d_2}(d_1, \boldsymbol{\Sigma}_{22})$ and $\mathbf{Y}$ is statistically independent of $\mathbf{W}_{22\cdot1}$.

Definition 21.4. (Hotelling's Distribution) Suppose $\mathbf{y} \sim N_d(\mathbf{0}, \boldsymbol{\Sigma})$, $\mathbf{W} \sim W_d(m, \boldsymbol{\Sigma})$, $\mathbf{y}$ is statistically independent of $\mathbf{W}$, and both distributions are nonsingular. Then

$$T^2 = m\mathbf{y}'\mathbf{W}^{-1}\mathbf{y}$$

is said to have a Hotelling's distribution, and we write $\mathbf{T}^2 \sim T_{d,m}^2$.

21.20. Referring to the above definition, $F = \dfrac{T_{d,m}^2}{m} \cdot \dfrac{m - d + 1}{d} \sim F_{d,m-d+1}$, the F-distribution with d and $m - d + 1$ degrees of freedom, respectively.

If, instead, $\mathbf{y} \sim N_d(\boldsymbol{\theta}, \boldsymbol{\Sigma})$, then $F \sim F_{d,m-d+1,\delta}$, the noncentral F-distribution with noncentrality parameter $\delta = \boldsymbol{\theta}'\boldsymbol{\Sigma}\boldsymbol{\theta}$.

21.21. (Eigenvalues)

(a) If the probability density function of the $m \times d$ matrix $\mathbf{Y}$ is $f(\mathbf{Y}'\mathbf{Y})$, then the probability density of $\mathbf{B} = \mathbf{Y}'\mathbf{Y}$ is

$$\frac{\pi^{dm/2}}{\Gamma_d(\frac{1}{2}m)}(\det \mathbf{B})^{(m-d-1)/2}f(\mathbf{B}),$$

where $\Gamma_d(\cdot)$ is given by (21.67).

(b) If the real symmetric $d \times d$ matrix $\mathbf{C}$ has a probability density function of the form $g(\lambda_1, \lambda_2, \ldots, \lambda_d)$, where $\lambda_1 > \lambda_2 > \cdots > \lambda_d$ are the eigenvalues of $\mathbf{C}$, then the probability density function of the eigenvalues is

$$\frac{\pi^{d^2/2}}{\Gamma_d(\frac{1}{2}d)} g(\lambda_1, \ldots, \lambda_d) \prod_{i<j}^{d}(\lambda_i - \lambda_j).$$

(c) Suppose $\mathbf{W} \sim W_d(m, \mathbf{I}_d)$. Using (a) and (b), the probability density function of the eigenvalues of $\mathbf{W}$ is

$$\frac{\pi^{d^2/2} \prod_{i=1}^{d} \lambda_i^{(m-d-1)/2}}{2^{dm/2} \Gamma_d(\frac{1}{2}m)\Gamma_d(\frac{1}{2}d)} \exp(-\frac{1}{2}\sum_{i=1}^{d}\lambda_i) \prod_{i<j}^{d}(\lambda_i - \lambda_j).$$

(d) Suppose $\mathbf{W} \sim W_d(m, \mathbf{\Sigma})$, where $m \geq d$ and $\mathbf{\Sigma}$ is positive definite. Then the probability density function of the eigenvalues of $\mathbf{W}$ is

$$\frac{\pi^{d^2/2} \prod_{i=1}^{d} \lambda_i^{(m-d-1)/2}}{2^{dm/2} \Gamma_d(\frac{1}{2}m)\Gamma_d(\frac{1}{2}d)} \prod_{i<j}^{d}(\lambda_i - \lambda_j) \, _0F_0^d(-\frac{1}{2}\mathbf{\Lambda}, \mathbf{\Sigma}^{-1}), \quad (\lambda_1 > \cdots > \lambda_d > 0),$$

where $\mathbf{\Lambda} = \text{diag}(\lambda_1, \ldots, \lambda_d)$ and $_0F_0$ is a two-matrix hypergeometric function. When $\mathbf{\Sigma} = \mathbf{I}_n$, we have

$$_0F_0^d(-\tfrac{1}{2}\mathbf{\Lambda}, \mathbf{I}_d) = {_0F_0}(-\tfrac{1}{2}\mathbf{\Lambda},) = \exp(-\tfrac{1}{2}\text{trace}\,\mathbf{\Lambda}),$$

which gives us (c).

21.22. (Generalized Eigenvalues) Let $\mathbf{W}_i$ $(i = 1, 2)$ be independently distributed as nonsingular $W_d(m_i, \mathbf{\Sigma})$ (i.e., $m_1, m_2 \geq d$ and $\mathbf{\Sigma}$ positive definite). The probability density function of the generalized eigenvalues, namely the roots of

$$\det(\mathbf{W}_1 - \lambda\mathbf{W}_2) = 0,$$

is

$$\frac{\pi^{d^2/2}\Gamma_d(\frac{1}{2}(m_1 + m_2))}{\Gamma_d(\frac{1}{2}m_1)\Gamma_d(\frac{1}{2}m_2)\Gamma_d(\frac{1}{2}d)} \prod_{i=1}^{d} \lambda_i^{(m_1-d-1)/2} \prod_{i=1}^{d}(\lambda_i + 1)^{-(m_1+m_2)/2} \prod_{i<j}^{d}(\lambda_i - \lambda_j),$$

for $\lambda_1 \geq \cdots \geq \lambda_d \geq 0$.

Definition 21.5. (Complex Wishart Distribution) Suppose $\mathbf{x}_1, \mathbf{x}_2, \ldots, \mathbf{x}_m$ are independently and identically distributed as the complex multivariate normal distribution $N_d^c(\mathbf{0}, \mathbf{\Sigma_x})$ (cf. Section 20.6), then $\mathbf{W} = \sum_{i=1}^{m} \mathbf{x}_i\mathbf{x}_i^*$ is said to have a *complex Wishart distribution* denoted by $\mathbf{W}_d^c(m, \mathbf{\Sigma_x})$. It is used in approximating the distributions of estimates of spectral density matrices in multivariate time series and in random normal (Gaussian) processes generally. Some of the properties of the (real) Wishart distribution carry over into the complex case. For a number of references see Brillinger [1975: 90].

21.23. Suppose $\mathbf{W} = (w_{ij}) \sim \mathbf{W}_d^c(m, \mathbf{\Sigma_x})$, where $m \geq d$ and $\mathbf{\Sigma_x}$ is Hermitian positive definite. Then:

(a) The probability density function of the distinct elements of $\mathbf{W}$ is

$$\frac{(\det \mathbf{W})^{m-d} \exp[-\frac{1}{2}\operatorname{trace}(\boldsymbol{\Sigma}_{\mathbf{x}}^{-1}\mathbf{W})]}{\pi^{d(d-1)/2}(\det \boldsymbol{\Sigma}_{\mathbf{x}})^{m} \prod_{j=1}^{d}\Gamma(m-j+1)}.$$

(b) $\mathrm{E}(\mathbf{W}) = m\boldsymbol{\Sigma}_{\mathbf{x}}$.

Proofs. Section 21.2.2.

21.11a. Follows from (21.7b) with $\mathbf{A} = \mathbf{I}_{m}$.

21.11b. Schott [2005: 425].

21.12. Schott [2005: 423].

21.13. Seber [1984: 21, 27, 56].

21.14. Styan [1989].

21.15. Schott [2005: 422].

21.16. Schott [2005: 422] and Seber [1984: 24–25].

21.18. This can be readily proved using moment generating functions (cf. 21.13g).

21.19a. Seber [1984: 50–51] and Schott [2005: 423–424].

21.19b. Seber [1984: 51–52].

21.20. Seber [1984: 30-31].

21.21a. Anderson [2003: 539, with $\mathbf{Y} \to \mathbf{Y}'$ and $p \to d$].

21.21b. Anderson [2003: 538–539] proves this using (a).

21.21c. Anderson [2003: 539] and Muirhead [1982: 389, with $\lambda = 1, nl_i = \lambda_i$].

21.21d. For details see Muirhead [1982: 388–389, with $\lambda_i = nl_i$].

21.22. Anderson [2003: section 13.2, and section 13.6 for some asymptotic theory].

21.23. Srivastava [1965].

21.3 RANDOM SAMPLES

21.3.1 One Sample

21.24. Let $\mathbf{x}_i$, $i = 1, 2, \ldots, n$, be a random sample from a d-dimensional distribution with mean $\boldsymbol{\mu}$ and variance matrix $\boldsymbol{\Sigma}$. Let $\mathbf{X} = (\mathbf{x}_1, \mathbf{x}_2, \ldots, \mathbf{x}_n)'$ be the data matrix and let $\mathbf{z}_i = \mathbf{x}_i - \boldsymbol{\mu}$ $(i = 1, 2, \ldots, n)$. Suppose that the following third and fourth moment matrices exist, namely

$$\boldsymbol{\Phi} = \mathrm{E}(\mathbf{z}_i \otimes \mathbf{z}_i\mathbf{z}_i') \quad \text{and} \quad \boldsymbol{\Psi} = \mathrm{E}(\mathbf{z}_i\mathbf{z}_i' \otimes \mathbf{z}_i\mathbf{z}_i'),$$

where "$\otimes$" is the Kronecker product. Let $\mathbf{z} = (\mathbf{z}_1', \mathbf{z}_2', \ldots \mathbf{z}_n')' = \text{vec}\,(\mathbf{X}') - \mathbf{1}_n \otimes \boldsymbol{\mu}$ so that $E(\mathbf{z}) = \mathbf{0}$ and $E(\mathbf{z}\mathbf{z}') = \text{var}(\text{vec}\,\mathbf{X}') = \mathbf{I}_n \otimes \boldsymbol{\Sigma}$ (by (21.2a). Define $\boldsymbol{\Phi}_* = E(\mathbf{z} \otimes \mathbf{z}\mathbf{z}')$ and $\boldsymbol{\Psi}_* = E(\mathbf{z}\mathbf{z}' \otimes \mathbf{z}\mathbf{z}')$. Then:

(a) If $\mathbf{K}_{dn}$ $(\mathbf{I}_{(n,d)})$ is the commutation (vec-permutation) matrix, $\mathbf{E}_{ii} = \mathbf{e}_i\mathbf{e}_i'$ is an $n \times n$ matrix with 1 in the (i,i)th position and zeros elsewhere, and $\mathbf{G} = (\mathbf{E}_{11}, \mathbf{E}_{22}, \ldots, \mathbf{E}_{nn})'$, then

$$\boldsymbol{\Phi}_* = (\mathbf{I}_n \otimes \mathbf{K}_{dn} \otimes \mathbf{I}_d)(\mathbf{G} \otimes \boldsymbol{\Phi}).$$

(b) If $\widetilde{\mathbf{K}}_{nn} = \sum_{i=1}^n (\mathbf{E}_{ii} \otimes \mathbf{E}_{ii})$, then

$$\begin{aligned}
\boldsymbol{\Psi}_* &= (\mathbf{I}_{n^2 d^2} + \mathbf{K}_{nd,nd})(\mathbf{I}_n \otimes \boldsymbol{\Sigma} \otimes \mathbf{I}_n \otimes \boldsymbol{\Sigma}) + [\text{vec}\,(\mathbf{I}_n \otimes \boldsymbol{\Sigma})][\text{vec}\,(\mathbf{I}_n \otimes \boldsymbol{\Sigma})]' \\
&\quad + (\mathbf{I}_n \otimes \mathbf{K}_{dn} \otimes \mathbf{I}_d)\{\widetilde{\mathbf{K}}_{nn} \otimes [\boldsymbol{\Psi} - (\mathbf{I}_{d^2} + \mathbf{K}_{dd})(\boldsymbol{\Sigma} \otimes \boldsymbol{\Sigma}) \\
&\quad - (\text{vec}\,\boldsymbol{\Sigma})(\text{vec}\,\boldsymbol{\Sigma}')]\}(\mathbf{I}_n \otimes \mathbf{K}_{nd} \otimes \mathbf{I}_d).
\end{aligned}$$

(c) Under normality we have the following results.

  (i) $\boldsymbol{\Phi} = \boldsymbol{\Phi}_* = \mathbf{0}$.

  (ii) If $\mathbf{P}_d = \frac{1}{2}(\mathbf{I}_{d^2} + \mathbf{K}_{dd})$ (the symmetrizer matrix), then, from (20.24b),

$$\boldsymbol{\Psi} = 2\mathbf{P}_d(\boldsymbol{\Sigma} \otimes \boldsymbol{\Sigma}) + (\text{vec}\,\boldsymbol{\Sigma})(\text{vec}\,\boldsymbol{\Sigma})'.$$

  (iii) $\boldsymbol{\Psi}_* = 2\mathbf{P}_{nd}(\mathbf{I}_n \otimes \boldsymbol{\Sigma} \otimes \mathbf{I}_n \otimes \boldsymbol{\Sigma}) + [\text{vec}\,(\mathbf{I}_n \otimes \boldsymbol{\Sigma})][\text{vec}\,(\mathbf{I}_n \otimes \boldsymbol{\Sigma})]'$.

Methods for finding $E(\mathbf{x} \otimes \mathbf{x} \otimes \mathbf{x})$, $E(\mathbf{x} \otimes \mathbf{x} \otimes \mathbf{x} \otimes \mathbf{x})$, and higher moments are given by Meijer [2005].

21.25. Let $\mathbf{x}_i$, $i = 1, 2, \ldots, n$, be a random sample from a nonsingular normal distribution $N_d(\boldsymbol{\mu}, \boldsymbol{\Sigma})$. Then:

(a) $\bar{\mathbf{x}} \sim N_d(\boldsymbol{\mu}, \boldsymbol{\Sigma}/n)$.

(b) $\mathbf{Q} = (n-1)\mathbf{S} \sim W_d(n-1, \boldsymbol{\Sigma})$.

(c) From (b) we can obtain the probability density function of the eigenvalues of $\mathbf{Q}$, and therefore those of $\mathbf{S}$. As this joint distribution is rather intractable, asymptotic theory has been developed for large n for both the eigenvalues and eigenvectors of $\mathbf{S}$, especially as related to providing approximate inferential procedures for principal component analysis. The reader is referred to Seber [1984: 197-199] for a summary of the results, and to Anderson [2003: section 13.5], Muirhead [1982: chapter 9], and Schott [2005: 427–429] for further details and some derivations.

(d) We consider some properties of $\mathbf{S}$. Here $\mathbf{P}_d$ is the symmetrizer matrix (cf. 20.24b) and $\mathbf{G}_d$ is the duplication matrix.

  (i) $\text{var}(\text{vec}\,\mathbf{S}) = (n-1)^{-1}2\mathbf{P}_d(\boldsymbol{\Sigma} \otimes \boldsymbol{\Sigma}) = (n-1)^{-1}2\mathbf{P}_d(\boldsymbol{\Sigma} \otimes \boldsymbol{\Sigma})\mathbf{P}_d$.

  (ii) We note that the above matrix in (i) is singular as $\mathbf{S}$ is symmetric, which implies that $\text{vec}\,\mathbf{S}$ has repeated elements. We can get round this by using the vector $\text{vech}\,\mathbf{S}$. Then

$$\text{var}(\text{vech}\,\mathbf{S}) = \frac{2}{n-1}\mathbf{G}_d^+\mathbf{P}_d(\boldsymbol{\Sigma} \otimes \boldsymbol{\Sigma})\mathbf{P}_d\mathbf{G}_d^{+\prime}.$$

(iii) As $n \to \infty$, $(n-1)^{1/2}(\text{vec}\,\mathbf{S} - \text{vec}\,\mathbf{\Sigma})$ is asymptotically distributed as $N_{d^2}(\mathbf{0}, \mathbf{V})$, where $\mathbf{V} = 2\mathbf{P}_d(\mathbf{\Sigma} \otimes \mathbf{\Sigma})$.

(iv) From $\text{vech}\,\mathbf{S} = \mathbf{G}_d^+ \text{vec}\,\mathbf{S}$, (20.6b), and (iii), $(n-1)^{1/2}(\text{vech}\,\mathbf{S} - \text{vech}\,\mathbf{\Sigma})$ is asymptotically distributed as $N_k(\mathbf{0}, \mathbf{G}_d^+\mathbf{V}\mathbf{G}_d^{+\prime})$, where $k = d(d+1)/2$.

(v) If $\mathbf{s} = \text{diag}\,\mathbf{S}$ and $\boldsymbol{\sigma} = \text{diag}\,\mathbf{\Sigma}$, then

$$E(\mathbf{s}) = \boldsymbol{\sigma} \quad \text{and} \quad \text{var}(\mathbf{s}) = \frac{2}{n-1}(\mathbf{\Sigma} \circ \mathbf{\Sigma}),$$

where "$\circ$" is the Hadamard product. Also, as $n \to \infty$, $(n-1)^{1/2}(\mathbf{s} - \boldsymbol{\sigma})$ is asymptotically distributed as $N_d(\mathbf{0}, 2\mathbf{\Sigma} \circ \mathbf{\Sigma})$.

(vi) Schott [2005: 431–432] gives the asymptotic variance matrices for $\text{vec}\,\mathbf{R}$ and $\text{vech}\,\mathbf{R}$, where $\mathbf{R}$ is the sample correlation matrix.

(e) $\overline{\mathbf{x}}$ and $\mathbf{S}$ are statistically independent.

(f) $\overline{\mathbf{x}}$ and $\mathbf{S}$ are jointly sufficient and complete for $\boldsymbol{\mu}$ and $\mathbf{\Sigma}$.

(g) A useful statistic is

(i) $T^2 = n(\overline{\mathbf{x}} - \boldsymbol{\mu})\mathbf{S}^{-1}(\overline{\mathbf{x}} - \boldsymbol{\mu}) \sim T_{d,n-1}^2$ (cf. 21.20). This statistic can be used for testing the null hypothesis $H_0 : \boldsymbol{\mu} = \boldsymbol{\mu}_0$.

(ii) When the underlying data come from an elliptical distribution, T^2 is asymptotically χ_d^2.

(h) If $H_0 : \boldsymbol{\mu} \in \mathcal{V}$, where $\mathcal{V}$ is a p-dimensional vector subspace of $\mathbb{R}^d$, then we have the following.

(i) $T_{\min}^2 = \min_{\boldsymbol{\mu} \in \mathcal{V}} T^2 \sim T_{d-p,n-1}^2$.

(ii) If we have $H_0 : \boldsymbol{\mu} = \mathbf{K}\boldsymbol{\beta}$, where $\mathbf{K}$ is a known $d \times p$ matrix of rank p and $\boldsymbol{\beta}$ is a vector of p unknown paramters, then $\mathcal{V} = \mathcal{C}(\mathbf{K})$ and

$$T_{\min}^2 = n(\overline{\mathbf{x}}'\mathbf{S}^{-1}\overline{\mathbf{x}} - \overline{\mathbf{x}}'\mathbf{S}^{-1}\mathbf{K}\boldsymbol{\beta}^*),$$

where $\boldsymbol{\beta}^* = (\mathbf{K}'\mathbf{S}^{-1}\mathbf{K})^{-1}\mathbf{K}'\mathbf{S}^{-1}\overline{\mathbf{x}}$.

(iii) Suppose we have $H_0 : \mathbf{A}\boldsymbol{\mu} = \mathbf{0}$, where $\mathbf{A}$ is $d - p \times d$ of rank $d - p$, so that $\mathcal{V} = \mathcal{N}(\mathbf{A})$, the null space of $\mathbf{A}$. Then

$$T_{\min}^2 = n(\mathbf{A}\overline{\mathbf{x}})'(\mathbf{A}\mathbf{S}\mathbf{A}')^{-1}\mathbf{A}\overline{\mathbf{x}}.$$

A slight generalization of this is given in (i) below.

(i) Let $\mathbf{A}$ be a $q \times d$ matrix of rank q. Then:

(i) $n(\mathbf{A}\overline{\mathbf{x}} - \mathbf{A}\boldsymbol{\mu})'(\mathbf{A}\mathbf{S}\mathbf{A}')^{-1}(\mathbf{A}\overline{\mathbf{x}} - \mathbf{A}\boldsymbol{\mu}) \sim T_{q,n-1}^2$. This can be used for testing $H_0 : \mathbf{A}\boldsymbol{\mu} = \mathbf{c}$.

(ii) If $\mathbf{A}$ is a matrix of contrasts so that $\mathbf{A}\mathbf{1}_d = \mathbf{0}$, then

$$n\overline{\mathbf{x}}'\mathbf{A}'(\mathbf{A}\mathbf{S}\mathbf{A}')^{-1}\mathbf{A}\overline{\mathbf{x}} = n\overline{\mathbf{x}}'\mathbf{S}^{-1}\overline{\mathbf{x}} - \frac{n(\overline{\mathbf{x}}'\mathbf{S}^{-1}\mathbf{1}_d)^2}{\mathbf{1}_d'\mathbf{S}^{-1}\mathbf{1}_d}.$$

Proofs. Section 21.3.1.

21.24. Neudecker and Trenkler [2002].

21.25a–b. Seber [1984: 63].

21.25d(i)-(v). Schott [2005: 426–427].

21.25d(vi). Schott [2005: 431–432].

21.25e. Seber [1984: 63].

21.25f. Anderson [2003: 84].

21.25g(i). Seber [1984: 63].

21.25g(ii). Anderson [2003: 199–200].

21.25h. Seber [1984: 77–79].

21.25i(i). Seber [1984: 72].

21.25i(ii). Seber [1984: 124].

21.3.2   Two Samples

21.26. Let $\mathbf{v}_1, \mathbf{v}_2, \ldots, \mathbf{v}_{n_1}$ be a random sample from $N_d(\boldsymbol{\mu}_1, \boldsymbol{\Sigma})$, $\mathbf{w}_1, \mathbf{w}_2, \ldots, \mathbf{w}_{n_2}$ be an independent random sample from $N_d(\boldsymbol{\mu}_2, \boldsymbol{\Sigma})$, and $\boldsymbol{\theta} = \boldsymbol{\mu}_1 - \boldsymbol{\mu}_2$. Also define $\mathbf{Q}_1 = \sum_{i=1}^{n_1}(\mathbf{v}_i - \overline{\mathbf{v}})(\mathbf{v}_i - \overline{\mathbf{v}})'$ and $\mathbf{Q}_2 = \sum_{i=1}^{n_2}(\mathbf{w}_i - \overline{\mathbf{w}})(\mathbf{w}_i - \overline{\mathbf{w}})'$. Then:

(a) $\overline{\mathbf{v}} - \overline{\mathbf{w}} \sim N_d(\boldsymbol{\theta}, (n_1^{-1} + n_2^{-1})\boldsymbol{\Sigma})$.

(b) $\mathbf{Q} = \mathbf{Q}_1 + \mathbf{Q}_2 \sim W_d(n_1 + n_2 - 2, \boldsymbol{\Sigma})$.

(c) $n_1 n_2(n_1 + n_2)^{-1}(\overline{\mathbf{v}} - \overline{\mathbf{w}} - \boldsymbol{\theta})'\mathbf{S}_p^{-1}(\overline{\mathbf{v}} - \overline{\mathbf{w}} - \boldsymbol{\theta}) \sim T_{d,n_1+n_2-2}^2$, the T^2 distribution (cf. 21.20), where

$$\mathbf{S}_p = \mathbf{Q}/(n_1 + n_2 - 2).$$

We can use this statistic to test $H_0 : \boldsymbol{\theta} = \mathbf{c}$.

(d) If $\mathbf{C}$ is a $q \times d$ matrix of rank q $(q \leq d)$, then

$$\frac{n_1 n_2}{(n_1 + n_2)}[\mathbf{C}(\overline{\mathbf{v}} - \overline{\mathbf{w}}) - \mathbf{C}\boldsymbol{\theta}]'(\mathbf{C}\mathbf{S}_p\mathbf{C}')^{-1}[\mathbf{C}(\overline{\mathbf{v}} - \overline{\mathbf{w}}) - \mathbf{C}\boldsymbol{\theta}] \sim T_{q,n_1+n_2-2}^2.$$

This can be used to test $H_0 : \mathbf{C}\boldsymbol{\theta} = \mathbf{0}$. When $\mathbf{C}$ is an appropriate $d - 1 \times d$ contrast matrix then the methodology relating to H_0 is referred to as *profile analysis*.

The topic of more than two samples is best handled as a special case of the multivariate linear model described in the next section.

Proofs. Section 21.3.2.

21.26. Seber [1984: 108, 117].

21.4   MULTIVARIATE LINEAR MODEL

21.4.1   Least Squares Estimation

Definition 21.6. Let $\mathbf{Y} = \boldsymbol{\Theta} + \mathbf{U}$, where $\boldsymbol{\Theta} = \mathbf{X}\mathbb{B}$, $\mathbb{B}$ is a $p \times d$ matrix of unknown parameters, $\mathbf{X}$ is an $n \times p$ known matrix of *constants* of rank r $(r \leq p)$, $\mathbf{U} = (\mathbf{u}_1, \ldots \mathbf{u}_n)' = (\mathbf{u}^{(1)}, \ldots, \mathbf{u}^{(d)})$, and the $\mathbf{u}_i$ are a random sample from a distribution with mean $\mathbf{0}$ and variance matrix $\boldsymbol{\Sigma}$. Then $\mathbf{Y} = \mathbf{X}\mathbb{B} + \mathbf{U}$ is called a *multivariate linear model*. When $d = 1$, this reduces to the univariate linear model of Section 20.7.

We have introduced a change in notation in this section. Up till now, $\mathbf{X}$ has represented a matrix of random variables, whereas now we assume it to be a matrix of constants. This will be the case if we can carry out any analysis conditional on the observed value of $\mathbf{X}$. However, the use of $\mathbf{X}$ is traditional for linear models, and in some cases the elements of $\mathbf{X}$ take only values 0 or 1, thus representing qualitative variables. In this case, $\mathbf{X}$ is sometimes referred to as the *design matrix*, though, as Kempthorne [1980: 249] argues, a better term is perhaps *model matrix*. The matrix $\mathbf{Y}$ now takes over the role of a data matrix. In what follows we let $\Omega = \mathcal{C}(\mathbf{X})$.

Definition 21.7. If we partition $\mathbf{Y}$, $\boldsymbol{\Theta}$, and $\mathbb{B}$ in the same way that we partitioned $\mathbf{U}$, then the jth column of the multivariate linear model is the univariate model $\mathbf{y}^{(j)} = \boldsymbol{\theta}^{(j)} = \mathbf{X}\boldsymbol{\beta}^{(j)} + \mathbf{u}^{(j)}$, where $\mathbf{u}^{(j)}$ has mean $\mathbf{0}$ and variance matrix $\sigma_{jj}\mathbf{I}_n$. If $\mathbf{P}_\Omega \mathbf{y}^{(j)}$ is the (ordinary) least squares estimate of $\boldsymbol{\theta}^{(j)}$ (cf. 20.37a), where $\mathbf{P}_\Omega = \mathbf{X}(\mathbf{X}'\mathbf{X})^{-}\mathbf{X}'$, we say that $\widehat{\boldsymbol{\Theta}} = \mathbf{P}_\Omega \mathbf{Y}$ is the *least squares estimate* of $\boldsymbol{\Theta}$. When $r = p$, then setting $\widehat{\boldsymbol{\Theta}} = \mathbf{X}\widehat{\mathbb{B}}$ we have $\widehat{\mathbb{B}} = (\mathbf{X}'\mathbf{X})^{-1}\mathbf{X}'\widehat{\boldsymbol{\Theta}} = (\mathbf{X}'\mathbf{X})^{-1}\mathbf{X}'\mathbf{Y}$, called the least squares estimate of $\mathbb{B}$. If $r < p$, then $\widehat{\mathbb{B}}$ is not unique and we can use (as in the unvariate case) $\widehat{\mathbb{B}} = (\mathbf{X}'\mathbf{X})^{-}\mathbf{X}'\mathbf{Y}$, where $(\mathbf{X}'\mathbf{X})^{-}$ is any weak inverse of $(\mathbf{X}'\mathbf{X})$.

21.27. $\mathbf{y}_i = \mathbb{B}'\mathbf{x}_i + \mathbf{u}_i$, where $\mathbf{x}_i$ is the ith row of $\mathbf{X}$.

21.28. If $\mathbf{X}$ has full rank, then $\mathrm{E}(\widehat{\mathbb{B}}) = \mathbb{B}$.

21.29. We have the following covariance properties.

   (a) $\mathrm{cov}(\mathbf{y}_r, \mathbf{y}_s) = \mathrm{cov}(\mathbf{u}_r, \mathbf{u}_s) = \delta_{rs}\boldsymbol{\Sigma}$, where $\delta_{rs} = 1$ when $r = s$ and 0 otherwise.

   (b) $\mathrm{cov}(\mathbf{y}^{(j)}, \mathbf{y}^{(k)}) = \mathrm{cov}(\mathbf{u}^{(j)}, \mathbf{u}^{(k)}) = \sigma_{jk}\mathbf{I}_d$ for all $j, k = 1, \ldots, d$.

   (c) If $\mathbf{X}$ has full rank p, then $\widehat{\boldsymbol{\beta}}^{(j)} = (\mathbf{X}'\mathbf{X})^{-1}\mathbf{X}'\mathbf{y}^{(j)}$ and

$$\mathrm{cov}(\widehat{\boldsymbol{\beta}}^{(j)}, \widehat{\boldsymbol{\beta}}^{(k)}) = \sigma_{jk}(\mathbf{X}'\mathbf{X})^{-1} \quad (\text{all } j, k = 1, \ldots, d).$$

21.30. Let $\mathbf{G}(\boldsymbol{\Theta}) = (\mathbf{Y} - \boldsymbol{\Theta})'(\mathbf{Y} - \boldsymbol{\Theta})$.

   (a) (i) $\mathbf{E} = \mathbf{G}(\widehat{\boldsymbol{\Theta}}) = \mathbf{Y}'(\mathbf{I}_n - \mathbf{P}_\Omega)\mathbf{Y} = \mathbf{U}'(\mathbf{I}_n - \mathbf{P}_\Omega)\mathbf{U}$.

      (ii) $\mathrm{E}(\mathbf{E}) = (n - r)\boldsymbol{\Sigma}$.

      (iii) $\mathbf{E}$ is positive definite with probability 1.

      Here $\mathbf{E}$ is sometimes referred to as the *error matrix* or *residual matrix*.

   (b) $\mathbf{G}(\boldsymbol{\Theta}) - \mathbf{G}(\widehat{\boldsymbol{\Theta}}) = (\widehat{\boldsymbol{\Theta}} - \boldsymbol{\Theta})'(\widehat{\boldsymbol{\Theta}} - \boldsymbol{\Theta})$ is positive semidefinite for all $\boldsymbol{\Theta} = \mathbf{X}\mathbb{B}$, and equal to $\mathbf{0}$ if and only if $\boldsymbol{\Theta} = \widehat{\boldsymbol{\Theta}}$. We can say that $\widehat{\boldsymbol{\Theta}}$ is the minimum of the matrix $\mathbf{G}(\boldsymbol{\Theta})$. As a consequence we have the following properties of the least squares estimate from (10.48b,d) and (10.47a(iii)).

(i) trace $\mathbf{G}(\boldsymbol{\Theta}) \geq$ trace $\mathbf{G}(\widehat{\boldsymbol{\Theta}})$.

(ii) $\det \mathbf{G}(\boldsymbol{\Theta}) \geq \det \mathbf{G}(\widehat{\boldsymbol{\Theta}})$.

(iii) $\|\mathbf{G}(\boldsymbol{\Theta})\|_F \geq \|\mathbf{G}(\widehat{\boldsymbol{\Theta}})\|_F$, where $\|\mathbf{A}\|_F = \{\text{trace}(\mathbf{A}'\mathbf{A})\}^{1/2}$ and $\|\cdot\|_F$ is the Frobenius norm.

Any of these three results could be used as a definition of $\widehat{\boldsymbol{\Theta}}$.

21.31. (Generalized Gauss–Markov Theorem) If $\phi = \sum_{j=1}^{d} \mathbf{h}_j' \boldsymbol{\theta}^{(j)}$, a linear combination of all the elements of $\boldsymbol{\Theta}$, then $\hat{\phi} = \sum_{j=1}^{d} \mathbf{h}_j' \widehat{\boldsymbol{\theta}}^{(j)}$ is the BLUE of ϕ (i.e., the linear unbiased estimate with minimum variance).

21.32. (Two-Sample Case) Setting $\mathbf{V}' = (\mathbf{v}_1, \mathbf{v}_2, \ldots, \mathbf{v}_{n_1})$, $\mathbf{W}' = (\mathbf{w}_1, \mathbf{w}_2, \ldots, \mathbf{w}_{n_2})$, and $\mathbf{Y} = (\mathbf{V}', \mathbf{W}')'$ we see that the two-sample problem (cf. 21.26) is a special case of the multivariate model with

$$\mathbf{X}\mathbb{B} = \begin{pmatrix} \mathbf{1}_{n_1} & \mathbf{0} \\ \mathbf{0} & \mathbf{1}_{n_2} \end{pmatrix} \begin{pmatrix} \boldsymbol{\mu}_1' \\ \boldsymbol{\mu}_2' \end{pmatrix}.$$

The extension to n samples is straightforward.

Definition **21.8.** If $\mathbf{X}$ has less than full rank, then each univariate model also has less than full rank. From (20.38(2)), $\mathbf{a}_i'\boldsymbol{\beta}^{(j)}$ is estimable for each $i = 1, 2, \ldots, q$ and each model $j = 1, 2, \ldots, d$ if $\mathbf{a}_i \in \mathcal{C}(\mathbf{X}')$. Let $\mathbf{A}' = (\mathbf{a}_1, \mathbf{a}_2, \ldots, \mathbf{a}_q)$. Combining these linear combinations, we say that $\mathbf{A}\mathbb{B}$ is *estimable* if $\mathbf{A}' = \mathbf{X}'\mathbf{L}'$ or $\mathbf{A} = \mathbf{L}\mathbf{X}$ for some $q \times n$ matrix $\mathbf{L}$.

21.33. Suppose $\mathbf{A}\mathbb{B}$ is estimable.

(a) If $\mathbf{A}$ is $q \times p$ of rank q, then $\mathbf{L}$ has rank q by (20.38(2)).

(b) $\mathbf{A}\widehat{\mathbb{B}} = \mathbf{L}\mathbf{X}(\mathbf{X}'\mathbf{X})^-\mathbf{X}'\mathbf{Y} = \mathbf{L}\mathbf{P}_\Omega\mathbf{Y} = (\mathbf{P}_\Omega\mathbf{L}')'\mathbf{Y}$ is invariant for any choice of weak inverse $(\mathbf{X}\mathbf{X}')^-$ as $\mathbf{P}_\Omega$ is invariant. Here $\mathbf{P}_\Omega\mathbf{L}'$ is unique (but not $\mathbf{L}$, unless $\mathbf{X}$ has rank p) and has full row rank.

(c) $\hat{\phi} = \mathbf{a}'\mathbf{A}\widehat{\mathbb{B}}\mathbf{b} = \mathbf{a}'\mathbf{L}\widehat{\boldsymbol{\Theta}}\mathbf{b}$ is the BLUE of $\phi = \mathbf{a}'\mathbf{A}\mathbb{B}\mathbf{b} = \mathbf{a}'\mathbf{L}\boldsymbol{\Theta}\mathbf{b}$.

(d) $\mathbf{A}(\mathbf{X}'\mathbf{X})^-\mathbf{A} = \mathbf{L}\mathbf{P}_\Omega\mathbf{L}' = (\mathbf{P}_\Omega\mathbf{L}')\mathbf{P}_\Omega\mathbf{L}'$ is invariant and nonsingular by (b).

(e) $E(\mathbf{A}\widehat{\mathbb{B}}) = \mathbf{L}\mathbf{P}_\Omega\mathbf{X}\mathbb{B} = \mathbf{L}\mathbf{X}\mathbb{B} = \mathbf{A}\mathbb{B}$, since $\mathbf{P}_\Omega\mathbf{X} = \mathbf{X}$.

Proofs. Section 21.4.1.

21.27–21.29. Seber [1984: 400].

21.30a. Seber [1984: 398, 402].

21.30b. Seber [1984: 397–398].

21.31. Seber [1984: 400–401].

21.32. Seber [1984: section 8.6.4].

21.33b. $E(\mathbf{A}\widehat{\mathbb{B}}) = (\mathbf{P}_\Omega \mathbf{L}')'\boldsymbol{\Theta}$. If $(\mathbf{P}_\Omega \mathbf{M}')'\mathbf{Y}$ is another estimate, then $[\mathbf{P}_\Omega \mathbf{M}' - \mathbf{P}_\Omega \mathbf{L}']'\boldsymbol{\Theta} = \mathbf{0}$ and $\mathcal{C}[\mathbf{P}_\Omega(\mathbf{M}' - \mathbf{L}')] \perp \mathcal{C}(\mathbf{X})$ as the columns of $\boldsymbol{\Theta}$ are in $\mathcal{C}(\mathbf{X})$. Thus, $\mathbf{P}_\Omega \mathbf{M}' - \mathbf{P}_\Omega \mathbf{L}' = \mathbf{0}$ as $\mathcal{C}[\mathbf{P}_\Omega(\mathbf{M}' - \mathbf{L}'] \subseteq \mathcal{C}(\mathbf{X})$.

21.33c. The result follows from (21.31) by relabeling.

21.4.2   Statistical Inference

Let $\mathbf{Y} = \boldsymbol{\Theta} + \mathbf{U}$. In this section we now assume that the underlying distribution of the columns $\mathbf{u}_i$ of $\mathbf{U}'$ is a (nonsingular) multivariate normal distribution $N_d(\mathbf{0}, \boldsymbol{\Sigma})$. The case when $\boldsymbol{\Sigma}$ is singular is considered by, for example, Srivastava and von Rosen [2002]. The multivariate model can be expressed in terms of the univariate model vec $\mathbf{Y}$ = vec $(\mathbf{X}\mathbb{B})$ + vec $\mathbf{U}$, where from (21.2a) vec $(\mathbf{X}\mathbb{B})$ = $(\mathbf{I}_d \otimes \mathbf{X})$vec $\mathbb{B}$ and var(vec $\mathbf{U}$) = $\boldsymbol{\Sigma} \otimes \mathbf{I}_n$ (cf. Searle [1978]). A more general model in which var(vec $\mathbf{U}$) = $\boldsymbol{\Sigma} \otimes \mathbf{V}$, with $\mathbf{V}$ and $\boldsymbol{\Sigma}$ possibly singular, is considered by Sengupta and Jammalamakada [2003: chapter 10].

21.34. The likelihood function for $\mathbf{Y}$, the density function of vec $\mathbf{Y}$ (or, more conveniently, vec $(\mathbf{Y}')$) is the joint distribution of the independent $\mathbf{y}_i$, and it can be expressed in the form

$$(2\pi)^{-nd/2}(\det \boldsymbol{\Sigma})^{-n/2} \exp\{\text{trace}[-\tfrac{1}{2}(\mathbf{Y} - \boldsymbol{\Theta})'\boldsymbol{\Sigma}^{-1}(\mathbf{Y} - \boldsymbol{\Theta})]\}.$$

21.35. Suppose $\boldsymbol{\Theta} = \mathbf{X}\mathbb{B}$, where $\mathbf{X}$ has rank r, and let $\mathbf{E}$ be given by (21.30a). We assume $n - r \geq d$. Then:

(a) $\mathbf{E} \sim W_d(n - r, \boldsymbol{\Sigma})$.

(b) $\mathbf{E}$ is statistically independent of $\widehat{\boldsymbol{\Theta}}$ (and of $\widehat{\mathbb{B}}$ if $\mathbf{X}$ has full rank p).

(c) The maximum likelihood estimates of $\boldsymbol{\Sigma}$, $\boldsymbol{\Theta}$, and $\mathbb{B}$ (if $\mathbf{X}$ has full rank), are $\widehat{\boldsymbol{\Sigma}} = \mathbf{E}/n$, $\widehat{\boldsymbol{\Theta}}$, and $\widehat{\mathbb{B}}$. The maximum value of the likelihood function is $(2\pi)^{-nd/2}(\det \widehat{\boldsymbol{\Sigma}})^{-n/2}e^{-nd/2}$. (This corrects a typo in Seber [1984: 407].)

(d) If $\mathbf{X}$ has full rank, then $(\widehat{\mathbb{B}}, \widehat{\boldsymbol{\Sigma}})$ is sufficient for $(\mathbb{B}, \boldsymbol{\Sigma})$.

(e) Referring to the jth column of $\widehat{\mathbb{B}}$, if $\mathbf{X}$ has full rank (cf. 21.29c),

$$\hat{\boldsymbol{\beta}}^{(j)} \sim N_n(\boldsymbol{\beta}^{(j)}, \sigma_{jj}(\mathbf{X}'\mathbf{X})^{-1}).$$

21.36. Suppose that $\boldsymbol{\Theta} = \mathbf{X}\mathbb{B}$, where $\mathbf{X}$ has rank r. Let $\mathbf{A}$ be a known $q \times p$ matrix of rank q, and let $\mathbf{AB}$ be estimable. We are interested in testing $H_0 : \mathbf{AB} = \mathbf{C}$, where $\mathbf{C}$ is a constant matrix. Then:

(a) Referring to (21.30), the minimum $\mathbf{E}_H$, say, of $\mathbf{G}(\mathbf{X}\mathbb{B})$ subject to $\mathbf{AB} = \mathbf{C}$ occurs when $\mathbb{B}$ equals

$$\widehat{\mathbb{B}}_H = \widehat{\mathbb{B}} - (\mathbf{X}'\mathbf{X})^-\mathbf{A}'[\mathbf{A}(\mathbf{X}'\mathbf{X})^-\mathbf{A}']^{-1}(\mathbf{A}\widehat{\mathbb{B}} - \mathbf{C}).$$

Although $\widehat{\mathbb{B}}$ and $\widehat{\mathbb{B}}_H$ are not unique when $r < p$, $\widehat{\boldsymbol{\Theta}} = \mathbf{X}\widehat{\mathbb{B}}$ and $\widehat{\boldsymbol{\Theta}}_H = \mathbf{X}\widehat{\mathbb{B}}_H$ are unique. Also $\mathbf{E}_H$ is positive definite with probability one.

(b) $\mathbf{H} = \mathbf{E}_H - \mathbf{E} = (\mathbf{A}\widehat{\mathbb{B}} - \mathbf{C})'[\mathbf{A}(\mathbf{X}'\mathbf{X})^{-}\mathbf{A}']^{-1}(\mathbf{A}\widehat{\mathbb{B}} - \mathbf{C})$.

$\mathbf{H}$ is positive definite with probability one.

(c) $\mathrm{E}(\mathbf{H}) = q\boldsymbol{\Sigma} + (\mathbf{A}\mathbb{B} - \mathbf{C})'[\mathbf{A}(\mathbf{X}'\mathbf{X})^{-}\mathbf{A}']^{-1}(\mathbf{A}\mathbb{B} - \mathbf{C}) = q\boldsymbol{\Sigma} + \mathbf{D}$, say, and $\mathbf{D}$ is positive definite.

(d) $\mathbf{H}$ and $\mathbf{E}$ are statistically independent.

(e) When H_0 is true, $\mathbf{H} \sim W_d(q, \boldsymbol{\Sigma})$. When H_0 is false, $\mathbf{H}$ has a noncentral Wishart distribution $W_d(q, \boldsymbol{\Sigma}; \boldsymbol{\Delta})$, with noncentrality matrix given by $\boldsymbol{\Delta} = \boldsymbol{\Sigma}^{-1/2}\mathbf{D}\boldsymbol{\Sigma}^{-1/2}$.

(f) Let $\mathbf{E}_H^{1/2}$ be the positive definite square root of $\mathbf{E}_H$ (cf. 10.32). Then, when H_0 is true, $\mathbf{V} = \mathbf{E}_H^{-1/2}\mathbf{H}\mathbf{E}_H^{-1/2}$ has a d-dimensional matrix variate Type-1 beta distribution with degrees of freedom q and $n - r$ (cf. Section 21.9)

21.37. Four different criterion are usually computed for testing H_0, and are expressed as functions of eigenvalues of $\mathbf{V}$ given in (21.36(f)) above.

1. Roy's maximum root test ϕ_{max}, the maximum eigenvalue of $\mathbf{H}\mathbf{E}^{-1}$, based on the so-called union–intersection principle.

2. Likelihood ratio test $(\det \mathbf{E}/ \det \mathbf{E}_H)^{n/2}$.

3. The Lawley–Hotelling trace $(n - r)\operatorname{trace}(\mathbf{H}\mathbf{E}^{-1})$.

4. Pillai's trace $\operatorname{trace}(\mathbf{H}\mathbf{E}_H^{-1})$.

These tests are summarised by Seber [1984: chapter 8], but for further details and distribution theory see Muirhead [1982: chapter 10].

Proofs. Section 21.4.2.

21.34. Seber [1984: 406].

21.35. Seber [1984: section 8.4].

21.36. Seber [1984: section 8.6].

21.4.3   Two Extensions

We give two extensions to the theory, which demonstrate how matrix theory can be applied.

21.38. (Generalized Hypothesis) Suppose we want to test $H_0 : \mathbf{A}\mathbb{B}\mathbf{D} = \mathbf{0}$, where $\mathbf{A}$ is $q \times p$ of rank q ($q \leq p$) and $\mathbf{D}$ is a known $d \times v$ matrix of rank v ($v \leq d$). To do this, let $\mathbf{Y_D} = \mathbf{Y}\mathbf{D}$ so that the linear model $\mathbf{Y} = \mathbf{X}\mathbb{B} + \mathbf{U}$ is transformed to

$$\mathbf{Y_D} = \mathbf{X}\mathbb{B}\mathbf{D} + \mathbf{U}\mathbf{D} = \mathbf{X}\boldsymbol{\Lambda} + \mathbf{U}_0,$$

say, where the columns of $\mathbf{U}_0'$ are i.i.d. $N_v(\mathbf{0}, \mathbf{D}'\boldsymbol{\Sigma}\mathbf{D})$. Then H_0 becomes $\mathbf{A}\boldsymbol{\Lambda} = \mathbf{0}$ and we can apply the previous theory of (21.36) to this transformed model.

(a) $\mathbf{H}$ now becomes $\mathbf{H_D} = \mathbf{D'HD} = (\mathbf{A}\widehat{\mathbb{B}}\mathbf{D})'[\mathbf{A}(\mathbf{X'X})^{-1}\mathbf{A'}]^{-1}\mathbf{A}\widehat{\mathbb{B}}\mathbf{D}$ and $\mathbf{E}$ becomes $\mathbf{E_D} = \mathbf{D'ED} \sim W_v(n-r, \mathbf{D'\Sigma D})$.

(b) When $\mathbf{A}\mathbb{B}\mathbf{D} = \mathbf{0}$ is true, $\mathbf{H_D} \sim W_n(q, \mathbf{D'\Sigma D})$. The only change to the previous theory is to replace $\mathbf{Y}$ by $\mathbf{Y_D}$ and d by v.

(c) The above theory reduces to that of Section 21.4.2 if we set $\mathbf{D} = \mathbf{I}_d$ and $v = d$.

This hypothesis is used for carrying out a profile analysis of more than two populations (Seber [1984: section 8.7]).

21.39. (Generalized Model and Hypothesis) Consider the model

$$\mathbf{Y} = \mathbf{X\Delta K'} + \mathbf{U},$$

where $\mathbf{X}$ is a known $n \times p$ of rank p, $\mathbf{\Delta}$ is $p \times k$ matrix of unknown parameters, $\mathbf{K'}$ is a known $k \times d$ of rank k ($k < d$), and the rows of $\mathbf{U}$ are independently and identically distributed (i.i.d.) as $N_d(\mathbf{0}, \mathbf{\Sigma})$. We wish to test the hypothesis

$$H_0 : \mathbf{A\Delta D} = \mathbf{0},$$

where $\mathbf{A}$ is $q \times d$ of rank q and $\mathbf{D}$ is $k \times v$ of rank v. This model is usually called the *growth curve model* and it is considered, along with extensions, by Pan and Fang [2002] and Kollo and van Rosen [2005: chapter 4]. A brief discussion is given by Seber [1984: section 9.7].

One simple approach to the above model when there are appropriate rank conditions is to transform the model to remove $\mathbf{K'}$ using a right inverse of $\mathbf{K'}$. One method, suggested by (Potthoff and Roy [1964] and described in detail by Seber [1984: 479], is to choose a nonsingular $d \times d$ matrix $\mathbf{G}$ (usually positive definite) such that the $k \times k$ matrix $\mathbf{K'G}^{-1}\mathbf{K}$ is nonsingular, and transform $\mathbf{y}_i$ to $\mathbf{C}_1\mathbf{y}_i$, where $\mathbf{C}_1 = \mathbf{G}^{-1}\mathbf{K}(\mathbf{K'G}^{-1}\mathbf{K})^{-1}$ is $d \times k$ of rank k. Then $\mathbf{K'C}_1 = \mathbf{I}_k$ so that

$$\mathbf{Y}_1 = \mathbf{YC}_1 = \mathbf{X\Delta K'C}_1 + \mathbf{UC}_1 = \mathbf{X\Delta} + \mathbf{U}_1,$$

where the columns of $\mathbf{U}'_1$, namely $\mathbf{C}_1\mathbf{u}_i$, are i.i.d. $N_k(\mathbf{0}, \mathbf{\Sigma}_1)$ with $\mathbf{\Sigma}_1 = \mathbf{C}'_1\mathbf{\Sigma C}_1$. We have now reduced the model to the previous case, and the theory used there for testing H_0 can be applied here with $\mathbf{Y}$ replaced by $\mathbf{Y}_1$ and d by k.

21.5 DIMENSION REDUCTION TECHNIQUES

21.5.1 Principal Component Analysis (PCA)

Given a data set of interrelated variables represented by an $n \times d$ data matrix $\mathbf{X} = (\mathbf{x}_1, \mathbf{x}_2, \ldots, \mathbf{x}_n)'$, the aim of principal component analysis (PCA) is to reduce the dimensionality d of the data set, while still retaining as much of the variation present in the data set as possible. This is achieved by transforming to a new set of variables, called the principal components, which are uncorrelated and are ordered so that the first few retain most of the variation present in all of the original variables. Also, we would hope that the components may have some physical interpretation.

We shall first look at the underlying population model that generates the data, and then consider the sample estimates of various quantites. There are numerous books on multivariate analysis that contain chapters or sections on principal components, e.g., Anderson [2003], Krzanowski [1988], Muirhead [1982], and Seber [1984] (which happen to be in my office when writing this). However, more specialized books are available such as Flury [1988] and Jolliffe [2002].

Definition 21.9. Let $\mathbf{x}$ be a random d-dimensional vector with mean $\boldsymbol{\mu}$ and variance matrix $\boldsymbol{\Sigma}$. Let $\mathbf{T} = (\mathbf{t}_1, \mathbf{t}_2, \dots, \mathbf{t}_d)$ be an orthogonal matrix such that, by the spectral theorem (cf. 16.44), we have

$$\boldsymbol{\Sigma}\mathbf{t}_j = \lambda_j\mathbf{t}_j \quad \text{and} \quad \mathbf{T}'\boldsymbol{\Sigma}\mathbf{T} = \boldsymbol{\Lambda} = \mathrm{diag}(\lambda_1, \lambda_2, \dots, \lambda_d),$$

where $\lambda_1 \geq \lambda_2 \geq \cdots \geq \lambda_d$ are the ordered eigenvalues of $\boldsymbol{\Sigma}$. The sum $\mathrm{trace}\,\boldsymbol{\Sigma}$ is sometimes called the *total variance*. If $\mathbf{y} = (y_j) = \mathbf{T}'(\mathbf{x} - \boldsymbol{\mu})$, then $y_j = \mathbf{t}'_j(\mathbf{x} - \boldsymbol{\mu})$ $(j = 1, 2, \dots, d)$ is called the jth *population principal component* of $\mathbf{x}$. In developing the population theory there is no loss of generality in assuming $\boldsymbol{\mu} = \mathbf{0}$.

A major drawback to the above approach is that it can be sensitive to the units of measurement used for each x_i. For this and other reasons some authors work with the population correlation matrix $\mathrm{corr}(\mathbf{x})$ rather than the variance matrix $\boldsymbol{\Sigma}$. For a discussion of the relative merits of the two approaches see Jolliffe [2002: section 2.3]. The optimal properties described below for $\boldsymbol{\Sigma}$ also apply to $\mathrm{corr}(\mathbf{x})$ if we use the standardized vector $\mathbf{z} = (z_i)$, where $z_i = (x_i - \mu_i)/\sqrt{\sigma_{ii}}$, instead of $\mathbf{x}$.

21.40. (Population Properties)

(a) $\boldsymbol{\Sigma} = \mathbf{T}\boldsymbol{\Lambda}\mathbf{T}' = \sum_{i=1}^{d} \lambda_i\mathbf{t}_i\mathbf{t}'_i$.

(b) The principal components define the principal axes of the family of ellipsoids $(\mathbf{x} - \boldsymbol{\mu})'\boldsymbol{\Sigma}^{-1}(\mathbf{x} - \boldsymbol{\mu}) = \mathrm{const.}$

(c) Since $\mathbf{t}_j$ has unit length, y_j is the length of the orthogonal projection of $\mathbf{x} - \boldsymbol{\mu}$ in direction $\mathbf{t}_j$.

(d) As $\mathrm{var}(\mathbf{y}) = \boldsymbol{\Lambda}$, the y_j are uncorrelated and $\mathrm{var}(y_j) = \lambda_j$.

(e) $\sum_{j=1}^{d} \mathrm{var}(y_j) = \sum_{j=1}^{d} \mathrm{var}(x_j) = \mathrm{trace}\,\boldsymbol{\Sigma}$, the total variance. We can use $\lambda_j/\mathrm{trace}\,\boldsymbol{\Sigma}$ to measure the relative manitude of λ_j. If the λ_i $(i = k+1, \dots, d)$ are relatively small so that the corresponding y_i are "small" (with zero means and small variances), then $\mathbf{y}_{(k)} = (y_1, y_2, \dots, y_k)'$ can be regarded as a k dimensional approximation for $\mathbf{y}$. Thus $\mathbf{y}_{(k)}$ can be used as a "proxy" for $\mathbf{x}$ in terms of explaining a major part of the total variance.

It should be noted that the last few components are likely to be more useful than the first few in detecting outliers that are not apparent from the original variables (Jolliffe [2002: 237]).

(f) Let $\mathbf{T}_{(k)} = (\mathbf{t}_1, \dots, \mathbf{t}_k)$. Then:

   (i) $\max_{\mathbf{a}'\mathbf{a}=1} \mathrm{var}(\mathbf{a}'\mathbf{x}) = \mathrm{var}(\mathbf{t}'_1\mathbf{x}) = \mathrm{var}[\mathbf{t}'_1(\mathbf{x} - \boldsymbol{\mu})] = \mathrm{var}(y_1) = \lambda_1$,
   so that y_1 is the normalized linear combination of the elements of $\mathbf{x} - \boldsymbol{\mu}$ with maximum variance λ_1.

   (ii) $\max_{\mathbf{a}'\mathbf{a}=1, \mathbf{T}'_{(k-1)}\mathbf{a}=0} \mathrm{var}(\mathbf{a}'\mathbf{x}) = \mathrm{var}(\mathbf{t}'_k\mathbf{x}) = \mathrm{var}(y_k) = \lambda_k$, so that $\mathbf{t}'_k(\mathbf{x} - \boldsymbol{\mu})$
   is the normalized linear combination of the elements $\mathbf{x} - \boldsymbol{\mu}$ uncorrelated with $y_1, y_2, \dots, y_{k-1}$ with maximum variance λ_k.

The above results can be expressed in several different ways (e.g., Jolliffe [2002: 11–12]).

(g) (Predictive Approach) Let $\mathbf{B}$ be a $d \times k$ matrix, and consider the "best" linear predictor of $\mathbf{x} - \boldsymbol{\mu}$ on the basis of $\mathbf{B}(\mathbf{x} - \boldsymbol{\mu})$. The Frobenius norm of the variance matrix of the prediction error is

$$\|\boldsymbol{\Sigma} - \boldsymbol{\Sigma}\mathbf{B}(\mathbf{B}'\boldsymbol{\Sigma}\mathbf{B})^{-1}\mathbf{B}'\boldsymbol{\Sigma}\|_F = \|\boldsymbol{\Sigma}^{1/2}(\mathbf{I}_d - \mathbf{P}_{\boldsymbol{\Sigma}^{1/2}\mathbf{B}})\boldsymbol{\Sigma}^{1/2}\|_F,$$

where $\mathbf{P}_{\boldsymbol{\Sigma}^{1/2}\mathbf{B}}$ is a symmetric idempotent matrix representing the orthogonal projection onto $\mathcal{C}(\boldsymbol{\Sigma}^{1/2}\mathbf{B})$. The norm is a minimum when $\mathbf{B}$ is equivalent to $\mathbf{T}_{(k)}$. Moreover, minimizing the trace of the variance matrix of the prediction error—that is, maximizing $\text{trace}(\mathbf{P}_{\boldsymbol{\Sigma}^{1/2}\mathbf{B}}\boldsymbol{\Sigma})$—yields the same result (Jolliffe [2002: 17]).

The results (f) and (g) above are optimal properties shared by principal components, and (e) was used by Hotelling [1933] to define principal components. For further properties see Jolliffe [2002: section 2.1] and Seber [1984: section 5.2]. A key theorem for developing such properties is given next.

21.41. Let f be a function defined on $\mathcal{P}$, the set of all $d \times d$ non-negative definite matrices. For any $\mathbf{C} \in \mathcal{P}$, let $\lambda_1(\mathbf{C}) \geq \lambda_2(\mathbf{C}) \geq \cdots \geq \lambda_d(\mathbf{C}) \geq 0$ be the eigenvalues of $\mathbf{C}$. Then f is strictly increasing and invariant under orthogonal transformations if and only if $f(\mathbf{C}) = g[\lambda_1(\mathbf{C}), \ldots, \lambda_d(\mathbf{C})]$ for some g that is strictly increasing in each argument. This means that minimizing $f(\mathbf{C})$ with respect to $\mathbf{C}$ is equivalent to simultaneously minimizing the eigenvalues of $\mathbf{C}$. The functions $\text{trace}\,\mathbf{C}$, $\|\mathbf{C}\|_F = [\text{trace}(\mathbf{CC}')]^{1/2}$, and $\det\mathbf{C}$ satisfy the conditions on f.

21.42. Suppose f satisfies the conditions in (21.41) above and $\mathbf{v}_{(k)}$ is a k-dimensional vector. Then

$$f(\text{var}[\mathbf{x} - \boldsymbol{\mu} - \mathbf{A}\mathbf{v}_{(k)}])$$

is minimized when $\mathbf{A}\mathbf{v}_{(k)} = \mathbf{T}_{(k)}\mathbf{y}_{(k)} = \mathbf{T}_{(k)}\mathbf{T}'_{(k)}(\mathbf{x} - \boldsymbol{\mu}) = \mathbf{P}(\mathbf{x} - \boldsymbol{\mu})$, where $\mathbf{P}$ represents the orthogonal projection of $\mathbf{x} - \boldsymbol{\mu}$ onto $\mathcal{C}(\mathbf{T}_{(k)})$.

Definition 21.10. (Sample Components) In practice, $\boldsymbol{\mu}$ and $\boldsymbol{\Sigma}$ are unknown and have to be estimated from a sample $\mathbf{x}_1, \mathbf{x}_2, \ldots, \mathbf{x}_n$, that is the $\mathbf{x}_i$ are assumed to be independently and identically distributed. We can estimate $\boldsymbol{\mu}$ by $\hat{\boldsymbol{\mu}} = \overline{\mathbf{x}}$ and $\boldsymbol{\Sigma}$ by $\widehat{\boldsymbol{\Sigma}} = \widetilde{\mathbf{X}}'\widetilde{\mathbf{X}}/n$, where $\widetilde{\mathbf{X}}$ is the centered matrix $\widetilde{\mathbf{X}} = (\mathbf{x}_1 - \overline{\mathbf{x}}, \ldots, \mathbf{x}_n - \overline{\mathbf{x}})'$. Carrying out a similar factorization on $\widehat{\boldsymbol{\Sigma}}$ as we did for $\boldsymbol{\Sigma}$, we obtain the eigenvalues $\hat{\lambda}_1 \geq \hat{\lambda}_2 \geq \ldots \geq \hat{\lambda}_d > 0$ and an orthogonal matrix $\widehat{\mathbf{T}} = (\hat{\mathbf{t}}_1, \hat{\mathbf{t}}_2, \ldots, \hat{\mathbf{t}}_d)$ of corresponding eigenvectors. For each observation $\mathbf{x}_i$ we can define a vector of *sample (estimated) principal components* $\hat{\mathbf{y}}_i = \widehat{\mathbf{T}}'(\mathbf{x}_i - \overline{\mathbf{x}})$, which gives us

$$\widehat{\mathbf{Y}}' = (\hat{\mathbf{y}}_1, \hat{\mathbf{y}}_2, \ldots, \hat{\mathbf{y}}_n) = \widehat{\mathbf{T}}'\widetilde{\mathbf{X}}'.$$

Many authors prefer to use the unbiased estimator $\mathbf{S}$ of $\boldsymbol{\Sigma}$ instead of $\widehat{\boldsymbol{\Sigma}}$ in defining the sample components. In this case

$$\mathbf{S}\hat{\mathbf{t}}_j = \tfrac{n}{n-1}\widehat{\boldsymbol{\Sigma}}\hat{\mathbf{t}}_j = \left(\tfrac{n}{n-1}\hat{\lambda}_j\right)\hat{\mathbf{t}}_j,$$

and the eigenvalues of $\mathbf{S}$ are $n\hat{\lambda}_j/(n-1)$.

The question arises as to whether we should use $\mathbf{S}$ or the sample correlation matrix $\mathbf{R}$. However, it is much easier to base any inference about the population components on $\mathbf{S}$ rather than on $\mathbf{R}$ using large sample theory. A key result is that if $\mathbf{x} \sim N_d(\boldsymbol{\mu}, \boldsymbol{\Sigma})$, then from (21.25b), $(n-1)\mathbf{S} \sim W_d(n-1, \boldsymbol{\Sigma})$. For aspects of large sample theory see Seber [1984: section 5.2.5] for a brief summary. For further details see Anderson [2003: section 11.6] and Muirhead [1982: chapter 9], and see Kollo and Neudecker [1993, 1997] with regard to elliptical distributions. We note that the theory can be modified to handle dependent data such as a time series (Jolliffe [2002: chapter 12]). Also, PCA can be used in conjunction with other multivariate techniques (Jolliffe [2002: chapter 9]. With some adaption, it can be used for discrete data like contingency tables, in which case it is related to the method of *correspondence analysis* and is mentioned briefly in (21.48) below (cf. Jolliffe [2002: sections 5.4 and 13.1]).

21.43. The score of the jth element of the ith sample observation, given by $\hat{y}_{ij} = \hat{\mathbf{t}}_j'(\mathbf{x}_i - \overline{\mathbf{x}})$, is related to the orthogonal projection of $\mathbf{x}_i - \overline{\mathbf{x}}$ onto $\mathcal{C}(\hat{\mathbf{t}}_j)$, namely (cf. 2.49b)

$$\mathbf{P}_{\hat{\mathbf{t}}_j}(\mathbf{x}_i - \overline{\mathbf{x}}) = \hat{\mathbf{t}}_j \hat{\mathbf{t}}_j'(\mathbf{x}_i - \overline{\mathbf{x}}) = \hat{y}_{ij}\hat{\mathbf{t}}_j.$$

21.44. Using the result (20.15), we can show that the sample components are the population components for a discrete distribution so that all the optimal properties of population components hold for the sample components. For example, if $\mathbf{v}$ is a random vector taking the values $\mathbf{x}_i$ $(i = 1, 2, \ldots, n)$ with probability n^{-1}, then $E(\mathbf{v}) = \overline{\mathbf{x}}$ and $\mathrm{var}(\mathbf{v}) = \widehat{\boldsymbol{\Sigma}}$. Applying (20.6b), we have for $\mathbf{a}'\mathbf{a} = 1$,

$$\mathrm{var}(\mathbf{a}'\mathbf{v}) = \mathbf{a}'\mathrm{var}(\mathbf{v})\mathbf{a} = \mathbf{a}'\widehat{\boldsymbol{\Sigma}}\mathbf{a} = \frac{1}{n}\sum_{i=1}^{n}[\mathbf{a}'(\mathbf{x}_i - \overline{\mathbf{x}})]^2,$$

which takes its maximum value of $\hat{\lambda}_1$ when $\mathbf{a} = \hat{\mathbf{t}}_1$. For further details see Jolliffe [2002: section 3.7].

21.45. A sample analogue of (21.40g) can be stated as follows. Let $\mathbf{G}$ be an $n \times d$ matrix with orthonormal columns. We wish to minimize the sum of the squared distances $\mathbf{x}_i - \overline{\mathbf{x}}$ from $\mathcal{C}(\mathbf{G})$; that is, we wish to minimize $\|\widetilde{\mathbf{X}}' - \mathbf{P}_\mathbf{G}\widetilde{\mathbf{X}}'\|_F$, where $\|\cdot\|_F$ is the Frobenius norm and $\mathbf{P}_\mathbf{G}$ represents the orthogonal projection onto $\mathbf{G}$. The minimum is given by $\mathbf{G} = \widehat{\mathbf{T}}_{(k)}$.

21.46. Let $\widetilde{\mathbf{X}}$ (which has rank d with probability 1) have a singular value decomposition (thin version; cf. Section 16.3) $\widetilde{\mathbf{X}}_{n \times d} = \mathbf{U}_{n \times d}\boldsymbol{\Delta}_{d \times d}\mathbf{V}'_{d \times d}$, where $\mathbf{U}$ has orthogonal columns and $\mathbf{V}$ is an orthogonal matrix. Setting $\widehat{\mathbf{T}} = \mathbf{V}$, we have $\widehat{\mathbf{Y}} = \widetilde{\mathbf{X}}\widehat{\mathbf{T}} = \mathbf{U}\boldsymbol{\Delta}$ and

$$\boldsymbol{\Delta} = \mathrm{diag}(\sigma_1, \sigma_2, \ldots, \sigma_d) = \sqrt{n}\,\mathrm{diag}(\hat{\lambda}_1^{1/2}, \hat{\lambda}_2^{1/2}, \ldots, \hat{\lambda}_d^{1/2}) = \sqrt{n}\widehat{\boldsymbol{\Lambda}}^{1/2},$$

the diagonal matrix of singular values of $\widetilde{\mathbf{X}}$, which are the square roots of the eigenvalues of $\widetilde{\mathbf{X}}'\widetilde{\mathbf{X}}\,(= n\widehat{\boldsymbol{\Sigma}})$. For applications see Jolliffe [2002: 45].

21.47. If the $\hat{\lambda}_j$ $(j = k+1, \ldots, d)$ are small relative to trace $\widehat{\boldsymbol{\Sigma}}$ (cf. 21.40e), we can approximate $\hat{\mathbf{y}}_i$ by its first k elements $\hat{\mathbf{y}}_{i(k)}$, say.

21.48. (Contingency Tables) Consider a discrete data set of n frequency observations arranged in an $r \times c$ two-way contingency table with n_{ij} in the (i,j)th cell. Let $\mathbf{N} = (n_{ij})$ and define $\mathbf{P} = n^{-1}\mathbf{N}$, $\mathbf{D_r} = \mathrm{diag}(\mathbf{r})$, where $\mathbf{r} - \mathbf{P}\mathbf{1}_r$, $\mathbf{D_c} - \mathrm{diag}(\mathbf{c})$, where $\mathbf{c} - \mathbf{P}'\mathbf{1}_c$, and $\mathbf{X} - \mathbf{P} - \mathbf{rc}'$. If the variable defining the rows of the contingency table is independent of the variable defining the columns, then the matrix of 'expected counts' is given by $n\mathbf{rc}'$. Thus, $\mathbf{X}$ is a matrix of the residuals that remain when the 'independence' model is fitted to $\mathbf{P}$. If we apply the singular value decomposition to a redefined $\widetilde{\mathbf{X}} = \mathbf{D_r}^{-1/2}\mathbf{X}\mathbf{D_c}^{-1/2}$ in (21.46), we get the components $\widehat{\mathbf{Y}}$, which are the same as those obtained by correspondence analysis (Jolliffe [2002: sections 5.4 and 13.1]).

Proofs. Section 21.5.1.

21.40b. This follows from

$$(\mathbf{x} - \boldsymbol{\mu})'\boldsymbol{\Sigma}^{-1}(\mathbf{x} - \boldsymbol{\mu}) = \mathbf{y}'\mathbf{T}'\boldsymbol{\Sigma}^{-1}\mathbf{T}\mathbf{y} = \mathbf{y}'\boldsymbol{\Lambda}^{-1}\mathbf{y}.$$

See also Jolliffe [2002: 18].

21.40c. We use $\mathbf{t}_j'(\mathbf{x} - \boldsymbol{\mu}) = \|\mathbf{t}_j\| \cdot \|(\mathbf{x} - \boldsymbol{\mu})\| \cos\theta$.

21.40d. Seber [1984: 176].

21.40e. Seber [1984: 181–183].

21.40f. Seber [1984: 181, the inequality should be reversed in line -1].

21.41. Okamoto and Kanazawa [1968] and Seber [1984: 177–178].

21.42. Seber [1984: 179].

21.5.2 Discriminant Coordinates

Definition 21.11. Suppose we have n d-dimensional observations of which n_i belong to group i $(i = 1, 2, \ldots, g; n = \sum_{i=1}^{g} n_i)$. Let $\mathbf{x}_{ij}$ be the jth observation in group i, and define

$$\overline{\mathbf{x}}_{i\cdot} = \frac{1}{n_i}\sum_{j=1}^{n_i}\mathbf{x}_{ij} \quad \text{and} \quad \overline{\mathbf{x}}_{\cdot\cdot} = \frac{1}{n}\sum_{i=1}^{g}\sum_{j=1}^{n_i}\mathbf{x}_{ij}.$$

Let $\mathbf{W} = \sum_{i=1}^{g}\sum_{j=1}^{n_i}(\mathbf{x}_{ij} - \overline{\mathbf{x}}_{i\cdot})(\mathbf{x}_{ij} - \overline{\mathbf{x}}_{i\cdot})'$, the *within-groups* matrix, and let $\mathbf{B} = \sum_{i=1}^{g}n_i(\overline{\mathbf{x}}_{i\cdot} - \overline{\mathbf{x}}_{\cdot\cdot})(\overline{\mathbf{x}}_{i\cdot} - \overline{\mathbf{x}}_{\cdot\cdot})'$, the *between-groups* matrix. Since $\mathbf{W}$ and $\mathbf{B}$ are generally positive definite with probability 1, the eigenvalues of $\mathbf{W}^{-1}\mathbf{B}$ (which are the same as those of $\mathbf{W}^{-1/2}\mathbf{B}\mathbf{W}^{-1/2}$) are positive and distinct with probability 1, say $\lambda_1 > \lambda_2 > \cdots > \lambda_d > 0$. Let $\mathbf{W}^{-1}\mathbf{B}\mathbf{c}_r = \lambda_r\mathbf{c}_r$ $(r = 1, \ldots, d)$, where the $\mathbf{c}_r$ are suitably scaled eigenvectors, and define the $k \times d$ $(k \leq d)$ matrix $\mathbf{C} = (\mathbf{c}_1, \mathbf{c}_2, \ldots, \mathbf{c}_k)'$. If we define $\mathbf{z}_{ij} = \mathbf{C}\mathbf{x}_{ij}$, then the k elements of $\mathbf{z}_{ij}$ are called the first k *discriminant coordinates*. (Some authors have used the term *canonical variates*, which I have reserved for Section 21.5.3.) These coordinates are determined so as to emphasize group separation, but with decreasing effectiveness, so that k has to be found. The coordinates can be computed using an appropriate transformation combined with a principal component analysis. Typically, the $\mathbf{c}_i$ are scaled so that $\mathbf{CSC}' = \mathbf{I}_r$, where $\mathbf{S} = \mathbf{W}/(n-g)$. For further details see Seber [1984: section 5.8].

21.49. The above theory is based on the following results.

(a) Setting $\mathbf{x}_{ij} - \overline{\mathbf{x}}_{..} = \mathbf{x}_{ij} - \overline{\mathbf{x}}_{i.} + \overline{\mathbf{x}}_{i.} - \overline{\mathbf{x}}_{..}$, squaring, and summing over i and j, we get

$$\sum_{i=1}^{g}\sum_{j=1}^{n_i}(\mathbf{x}_{ij} - \overline{\mathbf{x}}_{..})(\mathbf{x}_{ij} - \overline{\mathbf{x}}_{..})' \;=\; \sum_{i=1}^{g} n_i(\overline{\mathbf{x}}_{i.} - \overline{\mathbf{x}}_{..})(\overline{\mathbf{x}}_{i.} - \overline{\mathbf{x}}_{..})'$$
$$+\sum_{i=1}^{g}\sum_{j=1}^{n_i}(\mathbf{x}_{ij} - \overline{\mathbf{x}}_{i.})'$$

(b) Let $\lambda = \max_{\mathbf{a}:\mathbf{a}\neq 0}(\mathbf{a}'\mathbf{Ba}/\mathbf{a}'\mathbf{Wa})$, where the maximum occurs at $\mathbf{a} = \mathbf{c}$, say. Differentiating $(\mathbf{a}'\mathbf{Ba}/\mathbf{a}'\mathbf{Wa})$ with respect to $\mathbf{a}$ we obtain $\mathbf{Bc} - \lambda\mathbf{Wc} = \mathbf{0}$ so that $\mathbf{W}^{-1}\mathbf{Bc} = \lambda\mathbf{c}$.

21.5.3 Canonical Correlations and Variates

***Definition* 21.12.** Let $\mathbf{z} = (\mathbf{x}', \mathbf{y}')'$ be a d-dimensional random vector with mean $\boldsymbol{\mu}$ and positive definite variance matrix $\boldsymbol{\Sigma}$. Let $\mathbf{x}$ and $\mathbf{y}$ have dimensions d_1 and $d_2 = d - d_1$, respectively, and consider the partition

$$\boldsymbol{\Sigma} = \begin{pmatrix} \boldsymbol{\Sigma}_{11} & \boldsymbol{\Sigma}_{12} \\ \boldsymbol{\Sigma}_{21} & \boldsymbol{\Sigma}_{22} \end{pmatrix},$$

where $\boldsymbol{\Sigma}_{ii}$ is $d_i \times d_i$ and $\boldsymbol{\Sigma}_{12} = \boldsymbol{\Sigma}_{21}'$. Let ρ_1^2 be the maximum value of the squared correlation between arbitrary linear combinations $\boldsymbol{\alpha}'\mathbf{x}$ and $\boldsymbol{\beta}'\mathbf{y}$, and let $\boldsymbol{\alpha} = \mathbf{a}_1$ and $\boldsymbol{\beta} = \mathbf{b}_1$ be the corresponding maximizing values of $\boldsymbol{\alpha}$ and $\boldsymbol{\beta}$. Then the positive square root $\sqrt{\rho_1^2}$ is called the *first (population) canonical correlation* between $\mathbf{x}$ and $\mathbf{y}$, and $u_1 = \mathbf{a}_1'\mathbf{x}$ and $v_1 = \mathbf{b}_1'\mathbf{y}$ are called the *first (population) canonical variables*. Let ρ_2^2 be the maximum value of the squared correlation between $\boldsymbol{\alpha}'\mathbf{x}$ and $\boldsymbol{\beta}'\mathbf{y}$, where $\boldsymbol{\alpha}'\mathbf{x}$ is uncorrelated with $\mathbf{a}_1'\mathbf{x}$ and $\boldsymbol{\beta}'\mathbf{y}$ is uncorrelated with $\mathbf{b}_1'\mathbf{y}$, and let $u_2 = \mathbf{a}_2'\mathbf{x}$ and $v_2 = \mathbf{b}_2'\mathbf{y}$ be the maximizing values. Then the positive square root $\sqrt{\rho_2^2}$ is called the *second canonical correlation*, and u_2 and v_2 are called the *second canonical variables*. Continuing in this manner, we obtain r pairs of canonical variables $\mathbf{u} = (u_1, u_2, \ldots, u_r)'$ and $\mathbf{v} = (v_1, v_2, \ldots, v_r)'$. We can then regard $\mathbf{u}$ and $\mathbf{v}$ as lower-dimensional "representations" of $\mathbf{x}$ and $\mathbf{y}$. We shall see below that (i) the elements of $\mathbf{u}$ are uncorrelated, (ii) the elements of $\mathbf{v}$ are uncorrelated, and (iii) the squares of the correlations between u_j and v_j $(j = 1, 2, \ldots, r)$ are collectively maximized in some sense. The mathematics is summarised in the following result.

21.50. Let $1 > \rho_1^2 \geq \rho_2^2 \geq \cdots \geq \rho_m^2 > 0$, where $m = \operatorname{rank}\boldsymbol{\Sigma}_{12}$, be the m nonzero eigenvalues of $\boldsymbol{\Sigma}_{11}^{-1}\boldsymbol{\Sigma}_{12}\boldsymbol{\Sigma}_{22}^{-1}\boldsymbol{\Sigma}_{21}$ (and of $\boldsymbol{\Sigma}_{22}^{-1}\boldsymbol{\Sigma}_{21}\boldsymbol{\Sigma}_{11}^{-1}\boldsymbol{\Sigma}_{12}$). Let the vectors $\mathbf{a}_1, \mathbf{a}_2, \ldots, \mathbf{a}_m$ and $\mathbf{b}_1, \mathbf{b}_2, \ldots, \mathbf{b}_m$ be the respective corresponding eigenvectors of $\boldsymbol{\Sigma}_{11}^{-1}\boldsymbol{\Sigma}_{12}\boldsymbol{\Sigma}_{22}^{-1}\boldsymbol{\Sigma}_{21}$ and $\boldsymbol{\Sigma}_{22}^{-1}\boldsymbol{\Sigma}_{21}\boldsymbol{\Sigma}_{11}^{-1}\boldsymbol{\Sigma}_{12}$. Suppose that $\boldsymbol{\alpha}$ and $\boldsymbol{\beta}$ are arbitrary vectors such that for $r \leq m$, $\boldsymbol{\alpha}'\mathbf{x}$ is uncorrelated with each $\mathbf{a}_j'\mathbf{x}$ $(j = 1, 2, \ldots, r-1)$, and $\boldsymbol{\beta}'\mathbf{y}$ is uncorrelated with each $\mathbf{b}_j'\mathbf{y}$ $(j = 1, 2, \ldots, r-1)$. Let $u_j = \mathbf{a}_j'\mathbf{x}$ and $v_j = \mathbf{b}_j'\mathbf{y}$, for $j = 1, 2, \ldots, r$. Then we have the following results.

(a) The maximum squared correlation between $\boldsymbol{\alpha}'\mathbf{x}$ and $\boldsymbol{\beta}'\mathbf{y}$ is given by ρ_r^2 and it occurs when $\boldsymbol{\alpha} = \mathbf{a}_r$ and $\boldsymbol{\beta} = \mathbf{b}_r$.

(b) $\operatorname{cov}(u_j, u_k) = 0$ for $j \neq k$, and $\operatorname{cov}(v_j, v_k) = 0$ for $j \neq k$.

(c) The squared (population) correlation beween u_j and v_j is ρ_j^2.

(d) $\operatorname{cov}(u_i, v_j) = 0$ for $i \neq j$.

(e) Since ρ_j^2 is independent of scale, we can scale $\mathbf{a}_j$ and $\mathbf{b}_j$ such that $\mathbf{a}_j'\boldsymbol{\Sigma}_{11}\mathbf{a}_j = 1$ and $\mathbf{b}_j'\boldsymbol{\Sigma}_{22}\mathbf{b}_j = 1$. The u_j and v_j then have unit variances. Alternatively, we can standardize so that the $\mathbf{a}_j$ and $\mathbf{b}_j$ all have unit lengths.

(f) If the $d_1 \times d_2$ matrix $\boldsymbol{\Sigma}_{12}$ has full row rank, and $d_1 < d_2$, then we have $m = d_1$. All the eigenvalues of $\boldsymbol{\Sigma}_{11}^{-1}\boldsymbol{\Sigma}_{12}\boldsymbol{\Sigma}_{22}^{-1}\boldsymbol{\Sigma}_{21}$ are therefore positive, while $\boldsymbol{\Sigma}_{22}^{-1}\boldsymbol{\Sigma}_{21}\boldsymbol{\Sigma}_{11}^{-1}\boldsymbol{\Sigma}_{12}$ has d_1 positive eigenvalues and $d_2 - d_1$ zero eigenvalues. However, the rank of $\boldsymbol{\Sigma}_{12}$ can vary as there may be constraints on $\boldsymbol{\Sigma}_{12}$ such as $\boldsymbol{\Sigma}_{12} = \mathbf{0}$ (rank 0) or $\boldsymbol{\Sigma}_{12} = \sigma^2\mathbf{1}_{d_1}\mathbf{1}_{d_2}'$ (rank 1).

21.51. Given the above notation, suppose that $\boldsymbol{\Sigma}$ is non-negative definite and singular.

(a) The key matrix is now $\mathbf{A} = \boldsymbol{\Sigma}_{11}^{-}\boldsymbol{\Sigma}_{12}\boldsymbol{\Sigma}_{22}^{-}\boldsymbol{\Sigma}_{21}$. The nonzero eigenvalues and rank of this matrix are invariant under any choices of the weak inverses $\boldsymbol{\Sigma}_{11}^{-}$ and $\boldsymbol{\Sigma}_{22}^{-}$.

(b) The eigenvalues of $\mathbf{A}$ are the squares of the canonical correlations between $\mathbf{x}$ and $\mathbf{y}$.

(c) The number of canonical correlations equal to 1 is

$$k = \operatorname{rank}\boldsymbol{\Sigma}_{11} + \operatorname{rank}\boldsymbol{\Sigma}_{22} - \operatorname{rank}\boldsymbol{\Sigma}.$$

(d) If $\boldsymbol{\Sigma}$ is positive definite, then $k = 0$.

21.52. Suppose $\mathbf{x}$ and $\mathbf{y}$ have means $\boldsymbol{\mu}_{\mathbf{x}}$ and $\boldsymbol{\mu}_{\mathbf{y}}$, respectively. Let $\mathbf{u} = \mathbf{A}(\mathbf{x} - \boldsymbol{\mu}_{\mathbf{x}})$ and $\mathbf{v} = \mathbf{B}(\mathbf{y} - \boldsymbol{\mu}_{\mathbf{y}})$, where $\mathbf{A}$ and $\mathbf{B}$ are any matrices, each with r rows that are linearly independent, satisfying $\mathbf{A}\boldsymbol{\Sigma}_{11}\mathbf{A}' = \mathbf{I}_r$ and $\mathbf{B}\boldsymbol{\Sigma}_{22}\mathbf{B}' = \mathbf{I}_r$. Then $E[(\mathbf{u} - \mathbf{v})'(\mathbf{u} - \mathbf{v})]$ is minimized when $\mathbf{u}$ and $\mathbf{v}$ are vectors of the canonical variables.

21.53. Suppose $\mathbf{z}' = (\mathbf{x}', \mathbf{y}')'$ has a positive definite variance matrix $\boldsymbol{\Sigma}$. Then $\mathbf{x}$ and $\mathbf{y}$ have the same canonical correlations as two random vectors $\mathbf{x}_0$ and $\mathbf{y}_0$ with variance matrix $\boldsymbol{\Sigma}^{-1}$, where $(\mathbf{x}_0', \mathbf{y}_0')$ is partitioned in the same way as $(\mathbf{x}', \mathbf{y}')$. This result has been extended to the case of a singular $\boldsymbol{\Sigma}$ using generalized inverses by Latour et al. [1987].

Definition 21.13. (Sample Estimates) Let $\mathbf{z}_1, \mathbf{z}_2, \ldots, \mathbf{z}_n$ be a random sample from the distribution described in Definition 21.12. Let $\bar{\mathbf{z}} = (\bar{\mathbf{x}}', \bar{\mathbf{y}}')'$ and $\widehat{\boldsymbol{\Sigma}} = \sum_{i=1}^{n}(\mathbf{z}_i - \bar{\mathbf{z}})(\mathbf{z}_i - \bar{\mathbf{z}})'/n$, where $\widehat{\boldsymbol{\Sigma}}$ is partitioned in the same way as $\boldsymbol{\Sigma}$, namely

$$n\widehat{\boldsymbol{\Sigma}} = \begin{pmatrix} \sum_{i=1}^{n}(\mathbf{x}_i - \bar{\mathbf{x}})(\mathbf{x}_i - \bar{\mathbf{x}})' & \sum_{i=1}^{n}(\mathbf{x}_i - \bar{\mathbf{x}})(\mathbf{y}_i - \bar{\mathbf{y}})' \\ \sum_{i=1}^{n}(\mathbf{y}_i - \bar{\mathbf{y}})(\mathbf{x}_i - \bar{\mathbf{x}})' & \sum_{i=1}^{n}(\mathbf{y}_i - \bar{\mathbf{y}})(\mathbf{y}_i - \bar{\mathbf{y}})' \end{pmatrix}$$

$$= \begin{pmatrix} \widetilde{\mathbf{X}}'\widetilde{\mathbf{X}} & \widetilde{\mathbf{X}}'\widetilde{\mathbf{Y}} \\ \widetilde{\mathbf{Y}}'\widetilde{\mathbf{X}} & \widetilde{\mathbf{Y}}'\widetilde{\mathbf{Y}} \end{pmatrix}$$

$$= \begin{pmatrix} \mathbf{Q}_{11} & \mathbf{Q}_{12} \\ \mathbf{Q}_{21} & \mathbf{Q}_{22} \end{pmatrix},$$

say, where $\mathbf{Q}_{ii}$ is $d_i \times d_i$ and $\mathbf{Q}_{12} = \mathbf{Q}'_{21}$ is $d_1 \times d_2$. We can assume that $d_1 \leq d_2$. Then given that $\mathbf{\Sigma}$ is positive definite and $n - 1 \geq d$, we know that, with probability 1, $n\widehat{\mathbf{\Sigma}}$ is positive definite and there are no constraints on $\mathbf{Q}_{12}$—that is, rank $\mathbf{Q}_{12} = d_1$. Let $r_1^2 > r_2^2 > \cdots > r_{d_1}^2 > 0$ be the eigenvalues of $\mathbf{Q}_{11}^{-1}\mathbf{Q}_{12}\mathbf{Q}_{22}^{-1}\mathbf{Q}_{21}$, with corresponding eigenvectors $\hat{\mathbf{a}}_1, \hat{\mathbf{a}}_2, \ldots, \hat{\mathbf{a}}_{d_1}$. We define $u_{ij} = \hat{\mathbf{a}}'_j(\mathbf{x}_i - \bar{\mathbf{x}})$, the ith element of $\mathbf{u}_j = \widetilde{\mathbf{X}}\hat{\mathbf{a}}_j$, where $\hat{\mathbf{a}}_j$ is scaled so that

$$1 = \hat{\mathbf{a}}'_j\widehat{\mathbf{\Sigma}}_{11}\hat{\mathbf{a}}_j = n^{-1}\hat{\mathbf{a}}'_j\widetilde{\mathbf{X}}'\widetilde{\mathbf{X}}\hat{\mathbf{a}}_j = \sum_{i=1}^{n} u_{ij}^2/n = \mathbf{u}'_j\mathbf{u}_j/n.$$

Then $\sqrt{r_j^2}$ is called the jth *sample canonical correlation* and these correlations are distinct with probability 1. We call u_{ij} the jth *sample canonical variable* of $\mathbf{x}_i$. In a similar fashion we define $v_{ij} = \hat{\mathbf{b}}'_j(\mathbf{y}_i - \bar{\mathbf{y}})$, the ith element of $\mathbf{v}_j = \widetilde{\mathbf{Y}}\hat{\mathbf{b}}_j$, to be the jth sample canonical variable of $\mathbf{y}_i$, where $\hat{\mathbf{b}}_1, \hat{\mathbf{b}}_2, \ldots, \hat{\mathbf{b}}_{d_1}$ are the corresponding eigenvectors of $\mathbf{Q}_{22}^{-1}\mathbf{Q}_{21}\mathbf{Q}_{11}^{-1}\mathbf{Q}_{12}$. The u_{ij} and v_{ij} are called the scores of the ith observation on the jth canonical variables. In computing the sample eigenvalues and eigenvectors we can use $\mathbf{Q}_{ab}$, $\widehat{\mathbf{\Sigma}}_{ab} = \mathbf{Q}_{ab}/n$ or $\mathbf{S}_{ab} = \mathbf{Q}_{ab}/(n-1)$ $(a, b, = 1, 2)$, as the factors n and $n - 1$ cancel out. Some computer packages use the sample correlation matrix $\mathbf{R}$ instead of $\widehat{\mathbf{\Sigma}}$. For further details see Seber [1984: section 5.7].

21.54. Using the method of (20.15), the sample canonical variables have the same optimal properties as those described for the population variables, except that population variances and covariances are replaced by their sample counterparts.

21.55. We have the following properties of r_j.

(a) The r_j^2 are distinct with probability 1.

(b) r_j^2 is the square of the sample correlation between the canonical variables whose values are in the vectors $\mathbf{u}_j$ and $\mathbf{v}_j$.

Proofs. Section 21.5.3.

    21.50. Seber [1984: section 5.7; for (d) see 278, solution to exercise 5.28].

    21.51. Rao [1981] and Styan [1985: 50–52].

    21.52. Brillinger [1975: 370].

    21.53. Jewell and Bloomfield [1983].

21.5.4   Latent Variable Methods

Latent variable methods are similar to PCA in that they endeavor to reduce the dimensionality of the data. However, they do this by imposing a model structure on the data that relates some observed variables to some underlying *latent* or hidden variables. When the latent variables are continuous or discrete, the method is called *factor analysis*, while if the latent variables are categorical, the method is usually referred to as *latent class analysis*. For general references see Bartholomew [1987] and Everitt [1984].

Definition 21.14. (Factor Analysis) Let $\mathbf{x} = (x_1, x_2, \ldots, x_d)'$ be a random vector with mean $\boldsymbol{\mu}$ and variance matrix $\boldsymbol{\Sigma}$. Let $\mathbf{f} = (f_1, f_2, \ldots, f_m)'$ be an m-dimensional random vector with mean $\mathbf{0}$ and variance matrix $\mathbf{I}_m$. The factor analysis model is defined to be

$$\mathbf{x} = \boldsymbol{\mu} + \boldsymbol{\Gamma}\mathbf{f} + \boldsymbol{\epsilon},$$

where $\boldsymbol{\epsilon}$ is assumed to be uncorrelated with $\mathbf{f}$ and has a diagonal variance matrix $\boldsymbol{\Psi} = \mathrm{diag}(\psi_1^2, \psi_2^2, \ldots, \psi_d^2)$. Here $\boldsymbol{\Gamma} = (\gamma_{jk})$ is a $d \times m$ unknown matrix of constants. The elements of $\mathbf{f}$ are called *(common) factors* or *latent variables*, the elements of $\boldsymbol{\epsilon}$ are usually called *specific* or *unique* factors, and γ_{jk} is called the weight or *factor loading* of x_j on the factor f_k.

21.56. $\boldsymbol{\Sigma} = \boldsymbol{\Gamma}\boldsymbol{\Gamma}' + \boldsymbol{\Psi}$, which leads to

$$\sigma_{jj} = \sum_{k=1}^{m} \gamma_{jk}^2 + \psi_j^2 = h_{jj}^2 + \psi_j^2,$$

say, where h_j^2 is called the *communality* or *common variance* and ψ_j^2 is called the *residual variance* or *unique variance*. The aim of factor analysis is to see if $\boldsymbol{\Sigma}$ can be expressed in the above form for a reasonably small value of m and to estimate the elements of $\boldsymbol{\Gamma}$ and $\boldsymbol{\Psi}$.

21.57. The model is not unique as $\boldsymbol{\Gamma}\mathbf{f} = (\boldsymbol{\Gamma}\mathbf{L})\mathbf{L}'\mathbf{f} = \boldsymbol{\Gamma}_0\mathbf{f}_0$ for any orthogonal $\mathbf{L}$ with $\mathrm{var}(\mathbf{f}_0) = \mathbf{L}'\mathrm{var}(\mathbf{f})\mathbf{L} = \mathbf{L}'\mathbf{I}_m\mathbf{L} = \mathbf{I}_m$. It is therefore usual to impose the constraint that $\boldsymbol{\Gamma}'\boldsymbol{\Psi}^{-1}\boldsymbol{\Gamma}$ has positive diagonal elements; under certain conditions this constraint may provide a unique $\boldsymbol{\Gamma}$. Although factor analysis is very different from PCA, it is often confused with PCA (Jolliffe [2002: chapter 7] and Srivastava [2002: chapter 12]).

21.58. Let $\hat{\mathbf{f}} = \mathbf{A}(\mathbf{x} - \boldsymbol{\mu}) = \mathbf{A}\mathbf{y}$ be a linear "estimate" of $\mathbf{f}$. Then the mean square error is

$$\mathrm{E}(\|\hat{\mathbf{f}} - \mathbf{f}\|_2^2) = \mathrm{trace}(\mathbf{A}'\mathbf{A}\boldsymbol{\Sigma}) - 2\,\mathrm{trace}(\mathbf{A}\boldsymbol{\Gamma}) + m.$$

This is minimzed when

$$\mathbf{A} = \boldsymbol{\Gamma}'\boldsymbol{\Sigma}^{-1} = (\mathbf{I}_m + \boldsymbol{\Gamma}'\boldsymbol{\Psi}^{-1}\boldsymbol{\Gamma})^{-1}\boldsymbol{\Gamma}'\boldsymbol{\Psi}^{-1}.$$

Proofs. Section 21.5.4.

21.58. Seber [1984: 221].

21.5.5   Classical (Metric) Scaling

Definition 21.15. Given a set of n objects, a *proximity measure* δ_{rs} is a measure of the "closeness" of objects r and s; here closeness does not necessarily refer to physical distance. A proximity δ_{rs} is called a *dissimilarity* if $\delta_{rr} = 0$, $\delta_{rs} \geq 0$, and $\delta_{rs} = \delta_{sr}$, for all $r, s = 1, 2, \ldots, n$; the matrix $\mathbf{D} = (\delta_{rs})$ is called a *dissimilarity matrix*. We say that $\mathbf{D}$ is *Euclidean* if there exists a p-dimensional configuration of points $\mathbf{y}_1, \mathbf{y}_2, \ldots, \mathbf{y}_n$ for some p such that the interpoint Euclidean distance $d_{rs} = \|\mathbf{y}_r - \mathbf{y}_s\|_2 = \delta_{rs}$.

21.59. Let $\mathbf{A} = (a_{ij})$ be a symmetric $n \times n$ matrix, where $a_{rs} = -\frac{1}{2}\delta_{rs}^2$. Define $b_{rs} = a_{rs} - \bar{a}_{r\cdot} - \bar{a}_{\cdot s} + \bar{a}_{\cdot\cdot}$ so that

$$\mathbf{B} = (b_{rs}) = \mathbf{CAC},$$

where $\mathbf{C} = (\mathbf{I}_n - n^{-1}\mathbf{1}_n\mathbf{1}_n')$, the usual centering matrix.

(a) $\mathbf{D}$ is Euclidean if and only if $\mathbf{B}$ is non-negative definite.

(b) When $\delta_{rs}^2 = \|\mathbf{x}_r - \mathbf{x}_s\|_2^2$, $\widetilde{\mathbf{X}} = (\mathbf{x}_1 - \bar{\mathbf{x}}, \mathbf{x}_2 - \bar{\mathbf{x}}, \ldots, \mathbf{x}_n - \bar{\mathbf{x}})'$, and $\boldsymbol{\Delta} = (\delta_{rs}^2)$, we find that

$$\boldsymbol{\Delta} = \mathbf{1}_n\mathbf{1}_n' \operatorname{diag}(\widetilde{\mathbf{X}}\widetilde{\mathbf{X}}') - 2\widetilde{\mathbf{X}}\widetilde{\mathbf{X}}' + \operatorname{diag}(\widetilde{\mathbf{X}}\widetilde{\mathbf{X}}')\mathbf{1}_n\mathbf{1}_n'.$$

Then $\mathbf{B} = -\frac{1}{2}\mathbf{C}\boldsymbol{\Delta}\mathbf{C} = \widetilde{\mathbf{X}}\widetilde{\mathbf{X}}'$, where $\widetilde{\mathbf{X}}'\mathbf{1}_n = \mathbf{0}$. For further details and extensions (e.g., using weights), see Takane [2004]. The next result looks at the reverse of the above process.

21.60. If $\mathbf{B}$ of (21.59) is non-negative definite, then we can find the $\mathbf{y}_i$ as follows. There exists an orthogonal matrix $\mathbf{V} = (\mathbf{v}_1, \mathbf{v}_2, \ldots, \mathbf{v}_n)$ such that

$$\mathbf{V}'\mathbf{B}\mathbf{V} = \begin{pmatrix} \boldsymbol{\Gamma} & \mathbf{0} \\ \mathbf{0} & \mathbf{0} \end{pmatrix} \quad (= \boldsymbol{\Lambda}, \text{ say}),$$

where $\boldsymbol{\Gamma} = \operatorname{diag}(\gamma_1, \gamma_2, \ldots, \gamma_p)$ and $\gamma_1 \geq \gamma_2 \geq \ldots \geq \gamma_p > 0$ are the positive eigenvalues of $\mathbf{B}$. Let $\mathbf{V}_1 = (\mathbf{v}_1, \mathbf{v}_2, \ldots, \mathbf{v}_p)$ and

$$\begin{aligned} \mathbf{Y} &= (\sqrt{\gamma_1}\mathbf{v}_1, \sqrt{\gamma_2}\mathbf{v}_2, \ldots, \sqrt{\gamma_p}\mathbf{v}_p) \\ &= (\mathbf{y}^{(1)}, \mathbf{y}^{(2)}, \ldots, \mathbf{y}^{(p)}) \\ &= (\mathbf{y}_1, \mathbf{y}_2, \ldots, \mathbf{y}_n)', \quad \text{say.} \end{aligned}$$

Then:

(a) $\mathbf{B}\mathbf{1}_n = \mathbf{0}$, since $\mathbf{C}\mathbf{1}_n = \mathbf{0}$.

(b) $\mathbf{B} = \mathbf{V}\boldsymbol{\Lambda}\mathbf{V}' = \mathbf{Y}\mathbf{Y}'$.

(c) $n^2\bar{\mathbf{y}}'\bar{\mathbf{y}} = (\mathbf{Y}'\mathbf{1}_n)'(\mathbf{Y}'\mathbf{1}_n) = \mathbf{1}_n'\mathbf{B}\mathbf{1}_n = \mathbf{0}$, so that $\bar{\mathbf{y}} = \mathbf{0}$.

(d) $\|\mathbf{y}_r - \mathbf{y}_s\|_2^2 = \delta_{rs}^2$.

21.61. If $\mathbf{D}$ is not Euclidean, then some of the eigenvalues of $\mathbf{B}$ will be negative. However, if the first k eigenvalues are comparatively large and positive, and the remaining positive or negative eigenvalues are near zero, then the rows of $\mathbf{Y}_k = (\mathbf{y}^{(1)}, \mathbf{y}^{(2)} \ldots, \mathbf{y}^{(k)})$ will give a reasonable k-dimensional configuration. If the original objects are d-dimensional points $\mathbf{x}_i$ ($i = 1, 2, \ldots, n$) so that $\|\mathbf{x}_r - \mathbf{x}_s\|^2 = \delta_{rs}^2$, then the n rows of $\mathbf{Y}_k$ will give an approximate k-dimensional reduction of a d-dimensional system of points. The above procedure is often referred to as *classical scaling* or *principal coordinate analysis* (Jolliffe [2002: section 5.2] and Seber [1984: section 5.5.1]). Jolliffe [2002: section 5.5] notes that principal coordinate analysis is similar to principal component analysis for certain types of similarity matrix.

Proofs. Section 21.5.5.

21.59a. Seber [1984: 236].

21.59b. Takane [2004].

21.60. Seber [1984: 237].

21.6 PROCRUSTES ANALYSIS (MATCHING CONFIGURATIONS)

Classical multidimensional scaling of Section 21.5 can be regarded as a technique for trying to match one set of n points in d-dimensional space by another set in a lower dimensional space. A related technique, commonly known as *procrustes analysis*, refers to the problem of matching two configurations of n points in d-dimensional space where there is a preassigned correspondence between the points of one configuration and the points of the other.

21.62. Let $\mathbf{A}$ be a real $d \times d$ matrix with a singular value decomposition $\mathbf{A} = \mathbf{P\Sigma Q}'$, where $\mathbf{P}$ and $\mathbf{Q}$ are $d \times d$ orthogonal matrices and $\mathbf{\Sigma} = \mathrm{diag}(\sigma_1, \sigma_2, \ldots, \sigma_d)$, where the σ_i are the singular values. Then, for all orthogonal $\mathbf{T}$,

$$\mathrm{trace}(\mathbf{AT}) = \mathrm{trace}(\mathbf{TA}) \le \mathrm{trace}[(\mathbf{A'A})^{1/2}],$$

with equality if $\mathbf{T} = \widehat{\mathbf{T}} = \mathbf{QP}'$. At the maximum,

$$\mathbf{AT} = \mathbf{P\Sigma Q'QP'} = \mathbf{P\Sigma P'},$$

which is non-negative definite. In fact $\mathrm{trace}(\mathbf{AT})$ is maximized if and only if $\mathbf{AT}$ is non-negative definite.

If $\mathbf{A}$ is nonsingular, then $\mathbf{TA} = (\mathbf{A'A})^{1/2}$ has a unique solution

$$\widehat{\mathbf{T}} = (\mathbf{A'A})^{-1/2}\mathbf{A}'.$$

21.63. Given two sets of d-dimensional points $\mathbf{x}_i$ and $\mathbf{y}_i$ $(i = 1, 2, \ldots, n)$, we wish to move the $\mathbf{y}_i$ relative to the $\mathbf{x}_i$ through rotation, reflection and translation, i.e., by the linear transformation $\mathbf{T'y}_i + \mathbf{c}$, where $\mathbf{T}$ is an orthogonal matrix, such that $\sum_{i=1}^{n} \|\mathbf{x}_i - \mathbf{T'y}_i - \mathbf{c}\|_2^2$ is minimised. The answer is $\mathbf{c} = \overline{\mathbf{x}} - \overline{\mathbf{y}}$ together with the minimum of $\sum_{i=1}^{n} \|\mathbf{x}_i - \overline{\mathbf{x}} - \mathbf{T'}(\mathbf{y}_i - \overline{\mathbf{y}})\|_2^2 = \|\widetilde{\mathbf{X}} - \widetilde{\mathbf{Y}}\mathbf{T}\|_F^2$ with respect to orthogonal $\mathbf{T}$, where $\|\cdot\|_F$ is the Frobenius norm. Here

$$\begin{aligned} \|\widetilde{\mathbf{X}} - \widetilde{\mathbf{Y}}\mathbf{T}\|_F^2 &= \mathrm{trace}[(\widetilde{\mathbf{X}} - \widetilde{\mathbf{Y}}\mathbf{T})'(\widetilde{\mathbf{X}} - \widetilde{\mathbf{Y}}\mathbf{T})] \\ &= \mathrm{trace}(\widetilde{\mathbf{X}}'\widetilde{\mathbf{X}}) + \mathrm{trace}(\widetilde{\mathbf{Y}}'\widetilde{\mathbf{Y}}) - 2\,\mathrm{trace}(\widetilde{\mathbf{X}}'\widetilde{\mathbf{Y}}\mathbf{T}), \end{aligned}$$

where $\widetilde{\mathbf{X}}' = (\mathbf{x}_1 - \overline{\mathbf{x}}, \ldots, \mathbf{x}_n - \overline{\mathbf{x}})$ and $\widetilde{\mathbf{Y}}$ is similarly defined. We have to maximize $\mathrm{trace}(\mathbf{T}\widetilde{\mathbf{X}}'\widetilde{\mathbf{Y}}) = \mathrm{trace}(\mathbf{TA})$, where $\mathbf{A} = \widetilde{\mathbf{X}}'\widetilde{\mathbf{Y}}$, with respect to $\mathbf{T}$. From (21.62) the answer is $\widehat{\mathbf{T}} = \mathbf{QP}'$. If $\mathbf{A}$ is nonsingular, we also have that the minimizing $\mathbf{T}$ for our original problem is

$$\widehat{\mathbf{T}} = (\widetilde{\mathbf{Y}}'\widetilde{\mathbf{X}}\widetilde{\mathbf{X}}'\widetilde{\mathbf{Y}})^{-1/2}(\widetilde{\mathbf{Y}}'\widetilde{\mathbf{X}}).$$

For further details concerning various aspects of procrustes analysis such as scaling, rotations and/or reflections, projections, and nonorthogonal transformations, see Gower and Dijksterhuis [2004].

Proofs. Section 21.6.

21.62. Gower and Dijksterhuis [2004: section 4.1] and Seber [1984: 254–255].

21.63. Seber [1984: 253].

21.7 SOME SPECIFIC RANDOM MATRICES

21.64. Let $\mathbf{A}(x)$ be a matrix whose elements are function of a random variable x. If $\mathbf{A}$ is positive definite for all x, then, provided that the expectations exist,

$$\mathrm{E}(\mathbf{A}^{-1}) - [\mathrm{E}(\mathbf{A})]^{-1} \succeq \mathbf{0},$$

that is, is non-negative definite.

21.65. (Generalized Quadratics)

(a) (Positive Definite) Suppose that the columns of $\mathbf{X}' = (\mathbf{x}_1, \mathbf{x}_2, \ldots, \mathbf{x}_n)$ are statistically independent and $\mathbf{A}$ is an $n \times n$ non-negative definite matrix of rank r $(r \geq d)$. If for each $\mathbf{x}_i$ and all $\mathbf{b}$ $(\neq \mathbf{0})$ and c, $\mathrm{pr}(\mathbf{b}'\mathbf{x}_i = c) = 0$, then

$$\mathbf{X}'\mathbf{A}\mathbf{X} \text{ is positive definite with probability 1.}$$

(b) Let $\mathbf{X}'$ be defined as in (a), and let $\mathbf{A}$ be a symmetric matrix of rank r. If the joint distribution of the elements of $\mathbf{X}$ is absolutely continuous with respect to the nd-dimensional Lebesque measure, then the following statements hold with probability 1:

$$\mathrm{rank}(\mathbf{X}'\mathbf{A}\mathbf{X}) = \min\{d, r\}$$

and the nonzero eigenvalues of $\mathbf{X}'\mathbf{A}\mathbf{X}$ are distinct.

Proofs. Section 21.7.

21.64. Groves and Rothenberg [1969].

21.65a. DasGupta [1971: theorem 5] and Eaton and Perlman [1973: theorem 2.3].

21.65b. Okamoto [1973].

21.8 ALLOCATION PROBLEMS

There is a subject area, which is mentioned for completeness, that sometimes uses dimension reducing techniques. This might be described generally as allocation, and includes two topics, discriminant analysis and cluster analysis, for which there are very extensive literatures. The emphasis tends to be on vectors rather than matrices. In essence, discriminant analysis is the problem of allocating an observation to one of two (or more) multivariate distributions, given samples from each distribution. Cluster analysis is a method of partitioning a cluster of observations into "sensible" groupings or classes (e.g., classifying psychiatric illnesses). Both topics are discussed in Seber [1984: chapters 6 and 7]. For further practical overviews of cluster analysis see Everitt [1993], Gordon [1999], and Kaufmann and Rousseeuw [1990]. Discriminant analysis is considered in detail by McLachlan [1992].

21.9   MATRIX-VARIATE DISTRIBUTIONS

In this section we give the density functions of some well known matrix distributions.

Definition 21.16. (Matrix-Variate Normal)   A random matrix $p \times n$ matrix $\mathbf{Y}$ with $E(\mathbf{Y}) = \mathbf{M}$ is said to have a *matrix-variate normal* distribution if $\mathbf{y} = \text{vec}\,\mathbf{Y} \sim N_{pn}(\text{vec}\,\mathbf{M}, \mathbf{\Psi} \otimes \mathbf{\Sigma})$. Following Kollo and von Rosen [2005: section 2.2], we say that $\mathbf{Y} \sim N_{p,n}(\mathbf{M}, \mathbf{\Sigma}, \mathbf{\Psi})$. These authors show in detail that many of the properties of the multivariate normal carry over to the matrix normal distribution. They also give moments for generalized quadratics and describe *matrix-variate elliptical* distributions (see also Gupta and Varga [1993]). If $\boldsymbol{\mu}_i$ is the ith column of $\mathbf{M}$ and $\mathbf{\Psi} = \mathbf{I}_n$, then the columns of $\mathbf{Y}$ are independently distributed as $N_p(\boldsymbol{\mu}_i, \mathbf{\Sigma})$ and $\mathbf{Y}'$ is now the data matrix. In this case we can identify $\mathbf{Y} = \mathbf{X}'$ and $p = d$, where $\mathbf{X}$ is the data matrix.

21.66. Using the above notation, if $\mathbf{\Psi}$ and $\mathbf{\Sigma}$ are positive definite so that $\mathbf{\Psi} \otimes \mathbf{\Sigma}$ is positive definite, and $\mathbf{m} = \text{vec}\,\mathbf{M}$, then the probability density function of $\mathbf{y} = \text{vec}\,\mathbf{Y}$ is

$$
\begin{aligned}
f(\mathbf{y}) &= (2\pi)^{-pn/2}[\det(\mathbf{\Psi} \otimes \mathbf{\Sigma})]^{-1/2} \exp[-\tfrac{1}{2}(\mathbf{y} - \mathbf{m})'(\mathbf{\Psi} \otimes \mathbf{\Sigma})^{-1}(\mathbf{y} - \mathbf{m})] \\
&= (2\pi)^{-pn/2}(\det \mathbf{\Psi})^{-p/2}(\det \mathbf{\Sigma})^{-n/2}\text{etr}[-\tfrac{1}{2}\mathbf{\Sigma}^{-1}(\mathbf{Y} - \mathbf{M})\mathbf{\Psi}^{-1}(\mathbf{Y} - \mathbf{M})'],
\end{aligned}
$$

where $\text{etr} = e^{\text{trace}}$.

21.67. (Wishart Distribution)   In Section 21.2.2 we introduced the random symmetric $d \times d$ matrix $\mathbf{W}$, which has a distribution $\mathbf{W}_d(m, \mathbf{\Sigma})$. When $\mathbf{\Sigma}$ is positive definite and $m \geq d$, we can obtain the probability density function of $\text{vech}\,\mathbf{W}$ (the distinct elements of $\mathbf{W}$) as

$$
f(\text{vech}\,\mathbf{W}) = c^{-1}(\det \mathbf{W})^{(m-d-1)/2}\text{etr}(-\tfrac{1}{2}\mathbf{\Sigma}^{-1}\mathbf{W}),
$$

where $c = 2^{md/2}(\det \mathbf{\Sigma})^{m/2}\Gamma_d(\tfrac{1}{2}m)$, "etr " is defined in (21.66) above, and

$$
\Gamma_d(a) = \pi^{d(d-1)/4} \prod_{j=1}^{d} \Gamma[a - \tfrac{1}{2}(j - 1)].
$$

Definition 21.17. (Matrix-Variate Gamma Distribution)   Let $\mathbf{X}$ be a positive definite $d \times d$ random matrix and $\mathbf{B}$ a positive definite $d \times d$ matrix of constants. Then $\mathbf{X}$ is said to have a *matrix-variate gamma distribution* if the probability density function of $\mathbf{x} = \text{vech}\,\mathbf{X}$ is

$$
f(\mathbf{x}) = \frac{1}{(\det \mathbf{B})^{-a}\Gamma_d(a)}(\det \mathbf{X})^{a-(d+1)/2}\text{etr}(-\mathbf{BX}),
$$

where $a > (d - 1)/2$. For some applications see Mathai [1991].

Definition 21.18. (Matrix-Variate Beta Distributions) A $d \times d$ positive definite random matrix $\mathbf{U}$ such that $\mathbf{V} = \mathbf{I}_d - \mathbf{U}$ is positive definite is said to have a *matrix-variate Type-1 beta distribution* with a and b degrees of freedom $(a, b \geq (d - 1)/2)$ if the density function of $\mathbf{u} = \text{vech}\,\mathbf{U}$ is

$$
f(\mathbf{u}) = \frac{1}{B_d(a, b)}(\det \mathbf{U})^{a-(d+1)/2}(\det[\mathbf{I}_d - \mathbf{U}])^{b-(d+1)/2},
$$

where

$$B_d(a,b) = \frac{\Gamma_d(a)\Gamma_d(b)}{\Gamma_d(a+b)}, \quad d \le 2a, 2b$$

and $\Gamma_d(a)$ is given in (21.67). Note that $\mathbf{V} = \mathbf{I} \quad \mathbf{U}$ also has a matrix-variate beta Type-1 distribution with b and a degrees of freedom, respectively. Mathai [1997: 259–260] proves that $f(\mathbf{u})$ is a density function.

The positive definite random $d \times d$ matrix $\mathbf{Y}$ is said to have a *matrix-variate Type-2 beta distribution* with a and b degrees of freedom $(a, b \ge (d-1)/2)$ if the density function of $\mathbf{y} = \text{vech}\,\mathbf{Y}$ is

$$f(\mathbf{y}) = \frac{1}{B_d(a,b)} (\det \mathbf{Y})^{a-(d+1)/2} [\det[\mathbf{I}_d + \mathbf{Y}]]^{-(a+b)}.$$

For further details see Mathai [1997: 262–264].

21.68. Suppose, for $i = 1, 2$, that $\mathbf{W}_i$ has a nonsingular Wishart distribution $W_d(m_i, \mathbf{\Sigma})$ ($\mathbf{\Sigma}$ positive definite, $m_1, m_2 \ge d$) and $\mathbf{W}_1$ and $\mathbf{W}_2$ are statistically independent. Since, by (21.61a), $\mathbf{W}_i$ is positive definite (with probability 1), then so is $\mathbf{W}_1 + \mathbf{W}_2$. Let $\mathbf{V} = (\mathbf{W}_1 + \mathbf{W}_2)^{-1/2} \mathbf{W}_1 (\mathbf{W}_1 + \mathbf{W}_2)^{-1/2}$, where $(\mathbf{W}_1 + \mathbf{W}_2)^{1/2}$ is the positive definite square root of $\mathbf{W}_1 + \mathbf{W}_2$ (cf. 10.32). Then:

(a) $\mathbf{V}$ has a matrix-variate Type-1 beta distribution defined above with $\frac{1}{2}m_1$ and $\frac{1}{2}m_2$ degrees of freedom, respectively.

(b) The eigenvalues λ_i of $\mathbf{V}$ are distinct with probablity 1 and can be ordered $1 > \lambda_1 > \lambda_2 \cdots > \lambda_d > 0$ (cf. 21.65b)

(c) The joint probability density function of the λ_i is

$$f(\boldsymbol{\lambda}) = c^{-1} \left(\prod_{i=1}^{d} \theta_i \right)^{(m_1-d-1)/2} \left[\prod_{i=1}^{d} (1 - \theta_j) \right]^{(m_2-d-1)/2} \prod_{i<j}^{d} (\theta_i - \theta_j),$$

where $c = \pi^{-d^2/2} B_d(\frac{1}{2}m_1, \frac{1}{2}m_2) \Gamma_d(\frac{1}{2}d)$.

(d) $\mathbf{Y} = \mathbf{W}_2^{-1/2} \mathbf{W}_1 \mathbf{W}_2^{-1/2}$ has a matrix-variate Type-2 beta distribution with $\frac{1}{2}m_1$ and $\frac{1}{2}m_2$ degrees of freedom, respectively.

Definition 21.19. (Matrix-Variate Dirichlet Distributions) A set of positive definite $p \times p$ random matrices $\mathbf{X}_1, \mathbf{X}_2, \ldots, \mathbf{X}_k$ (i.e., each $\mathbf{X}_i \succ \mathbf{0}$) is said to have a *matrix-variate Type-1 Dirichlet distribution* with parameter $\boldsymbol{\alpha} = (\alpha_1, \ldots, \alpha_{k+1})'$, where $\alpha_i > \frac{p-1}{2}$ for $i = 1, 2, \ldots, k+1$, if their joint density function is

$$f(\mathbf{X}_1, \ldots, \mathbf{X}_k) = \frac{\Gamma_p(\alpha_1 + \alpha_2 + \cdots + \alpha_{k+1})}{\Gamma_p(\alpha_1)\Gamma_p(\alpha_2) \cdots \Gamma_p(\alpha_{k+1})} \cdot \left\{ \prod_{i=1}^{k} (\det \mathbf{X}_i)^{\alpha_i - \frac{p+1}{2}} \right\}$$

$$\times \{\det(\mathbf{I}_p - \mathbf{X}_1 - \cdots - \mathbf{X}_k)\}^{\alpha_{k+1} - \frac{p+1}{2}},$$

$$\mathbf{0} \prec \mathbf{X}_i \prec \mathbf{I}_p, \quad (i = 1, 2, \ldots, k), \quad \mathbf{0} \prec \sum_{i=1}^{k} \mathbf{X}_i \prec \mathbf{I}_p.$$

The $\mathbf{X}_i$ are said to have a *matrix-variate Type-2 Dirichlet distribution* with parameter $\boldsymbol{\alpha}$ if their joint density function is

$$f(\mathbf{X}_1, \ldots, \mathbf{X}_k) = \frac{\Gamma_p(\alpha_1 + \alpha_2 + \cdots + \alpha_{k+1})}{\Gamma_p(\alpha_1)\Gamma_p(\alpha_2) \cdots \Gamma_p(\alpha_{k+1})} \cdot \left\{ \prod_{i=1}^k (\det \mathbf{X}_i)^{\alpha_i - \frac{\mu+1}{2}} \right\}$$

$$\times [\det(\mathbf{I}_p + \mathbf{X}_1 + \cdots + \mathbf{X}_k)]^{-(\alpha_1 + \cdots + \alpha_{k+1})}, \quad \text{each } \mathbf{X}_i \succ \mathbf{0},$$

where $\alpha_i > \frac{p-1}{2}$ for $i = 1, 2, \ldots, k+1$ (Mathai [1997: section 5.1.8]).

21.69. Let the $p \times p$ random matrices $\mathbf{X}_1, \ldots, \mathbf{X}_k$ have a joint matrix-variate Type-1 Dirichlet distribution.

(a) Any subset of the k matrices also has a joint matrix-variate Type-1 Dirichlet distribution.

(b) $\mathbf{U} = \mathbf{X}_1 + \cdots + \mathbf{X}_k$ has a matrix-variate Type-1 beta distribution with degrees of freedom $\alpha_1 + \cdots + \alpha_k$ and α_{k+1}, respectively.

21.70. Let $\mathbf{X}_1, \ldots, \mathbf{X}_k$ have a joint matrix-variate Type-2 Dirichlet distribution with parameter $\boldsymbol{\alpha}$, and let $\mathbf{X}_0 = \mathbf{X}_1 + \cdots + \mathbf{X}_k$. Then the $\mathbf{Y}_i = (\mathbf{I} + \mathbf{X}_0)^{-1/2} \mathbf{X}_i (\mathbf{I} + \mathbf{X}_0)^{-1/2}$ $(i = 1, 2, \ldots, k)$ are jointly distributed as a matrix-variate Type-1 Dirichlet distribution with parameter $\boldsymbol{\alpha}$.

Proofs. Section 21.9.

21.66. The second equation follows from $(\boldsymbol{\Psi} \otimes \boldsymbol{\Sigma})^{-1} = \boldsymbol{\Psi}^{-1} \otimes \boldsymbol{\Sigma}^{-1}$ and applying (11.17d(ii)).

21.68. Mathai and Provost [1992: 256–257] and Seber [1984: 33–36].

21.69. Mathai [1997: 276–277].

21.70. Mathai [1997: 278].

21.10   MATRIX ENSEMBLES

In some situations an $n \times d$ matrix $\mathbf{X}$ is simply a matrix of random variables rather than a data matrix involving random vectors. In the former case, some distribution theory for sucn a random matrix, including $\mathbf{X}'\mathbf{X}$, is given by Olkin [2002]. However, random matrices have seen an upsurge of interest in nuclear physics and related topics. Random matrix ensembles were first introduced in physics by Wigner to describe the correlations of nuclear spectra. Underlying the subject is the idea that the characteristic energies of chaotic systems behave locally as if they were the eigenvalues of a very large matrix with randomly distributed elements. The dynamical systems considered are characterized by their Hamiltonians, which in turn are represented by Hermitian matrices. There are also some curious links such as that between certain zeros of the Riemann zeta function and eigenvalues of a random matrix. For an introduction to these ideas see Mehta [2004: chapter 1]. The reader should also refer to Section 5.7 for the definition of terms.

Definition 21.20. A *Gausssian orthogonal ensemble* is a set of real symmetric $n \times n$ matrices of random variables, $\mathbf{H} = (h_{ij})$ say, where $\mathbf{H}$

(1) has a probability distribution that is invariant under transformations $\mathbf{T}^{-1}\mathbf{H}\mathbf{T}$, where $\mathbf{T}$ is a real orthogonal matrix (i.e., $\mathbf{T}^{-1} = \mathbf{T}'$),

(2) and all the h_{ij} $(i \leq j)$ are statistically independent.

This model applies when the dynamical system is "symmetric under time reversal".

When there is no time reversal symmetry, we can have a *Gaussian unitary ensemble* with $\mathbf{H}$ a Hermitian matrix and $\mathbf{T}$ replaced by $\mathbf{U}$, a unitary matrix (with $\mathbf{U}^{-1} = \mathbf{U}^*$). There is also a *Gaussian symplectic ensemble* with $\mathbf{H}$ a self-dual Hermitian matrix and invariance with respect to the transformation $\mathbf{W}^R\mathbf{H}\mathbf{W}$, where $\mathbf{W}$ is any symplectic matrix and $\mathbf{W}^R$ is its dual (i.e., $\mathbf{W}^R\mathbf{W} = \mathbf{I}$). This ensemble arises when there is time reversal symmetry and the total spin is a half-integer.

We define β to be the number of variables representing the number of components making up the particular entity under consideration. Thus $\beta = 1$ for real numbers, $\beta = 2$ for complex numbers and $\beta = 4$ for quaternions.

21.71. Given any one of the three Gausssian ensembles above, then the probability density function of $\mathbf{H}$ satisfies

$$f(\text{vech }\mathbf{Z}) = \exp(-a\,\text{trace}(\mathbf{H}^2) + b\,\text{trace }\mathbf{H} + c),$$

where a is real and positive, b and c are real, and b is usually zero. In each of the three cases, the eigenvalues of $\mathbf{H}$ are real. The total number of real variables in $\mathbf{H}$ consists of n diagonal elements and $\frac{1}{2}n(n-1)\beta$ off-diagonal elements. Also

$$\text{trace}(\mathbf{H}^2) = \sum_{i=1}^{n} \lambda_i^2 \quad \text{and} \quad \text{trace }\mathbf{H} = \sum_{i=1}^{n} \lambda_i,$$

where the λ_i are the eigenvalues of $\mathbf{H}$. By choosing $\frac{1}{2}n(n-1)\beta$ certain angular parameters together with the λ_i, and making the transformation to these new parameters, the Jacobian can be found. This leads to the density function of the λ_i as

$$c(\beta, n)\exp(-\sum_{i=1}^{n} \lambda_i^2) \prod_{1 \leq i < j \leq n} |\lambda_j - \lambda_k|^{\beta}.$$

It is these eigenvalues that are of interest in nuclear physics (Mehta [2004: 53, 56, 58]).

CHAPTER 22

INEQUALITIES FOR PROBABILITIES AND RANDOM VARIABLES

Inequalities arise in many places in probability and statistics. For example, Tong [1980: chapter 8] gives a number of applications of probability inequalities to simultaneous confidence regions, hypothesis testing and simultaneous comparisons, ranking and selection problems, and reliability and life testing. Some of the results in this chapter can be proved using the concept of majorization and Schur convexity, discussed in Chapter 23.

22.1 GENERAL PROBABILITIES

Let E_i $(i = 1, 2, \ldots, n)$ be any events.

22.1. (Boole's Formula)

$$
\begin{aligned}
\mathrm{pr}(\cup_{i=1}^n E_i) &= \sum_{i=1}^n \mathrm{pr}(E_i) - \sum_{i=1}^{n-1} \sum_{j=i+1}^n \mathrm{pr}(E_i \cap E_j) \\
&\quad + \sum_{i=1}^{n-2} \sum_{j=i+1}^{n-1} \sum_{k=j+1}^n \mathrm{pr}(E_i \cap E_j \cap E_k) - \cdots + (-1)^{n+1} \mathrm{pr}[(\cap_{i=1}^n E_i)].
\end{aligned}
$$

From this we can derive the following inequalities:

$$
\mathrm{pr}(\cup_{i=1}^n E_i) \leq \sum_{i=1}^n \mathrm{pr}(E_i),
$$

A Matrix Handbook for Statisticians. By George A. F. Seber
Copyright © 2008 John Wiley & Sons, Inc.

$$\operatorname{pr}(\cup_{i=1}^{n} E_i) \geq \sum_{i=1}^{n} \operatorname{pr}(E_i) - \sum_{i<j} \operatorname{pr}(E_i \cap E_j),$$

$$\operatorname{pr}(\cup_{i=1}^{n} E_i) \leq \sum_{i=1}^{n} \operatorname{pr}(E_i) - \sum_{i<j} \operatorname{pr}(E_i \cap E_j) + \sum_{i<j<k} \operatorname{pr}(E_i \cap E_j \cap E_k),$$

and so on

22.2. Let $\overline{E}_i$ be the complement of E_i.

(a) $\operatorname{pr}(\cap_{i=1}^{n} E_i) = 1 - \operatorname{pr}(\overline{\cap_{i=1}^{n} E_i}) = 1 - \operatorname{pr}(\cup_{i=1}^{n} \overline{E}_i).$

(b) This leads to the *Kounias inequality*

$$\operatorname{pr}(\cap_{i=1}^{n} E_i) \geq 1 - \sum_{i=1}^{n} \operatorname{pr}(\overline{E}_i) + \max_{j} \sum_{i \neq j} \operatorname{pr}(\overline{E}_i \cap \overline{E}_j).$$

22.3. Since the probability of the union of disjoint events is the sum of the individual probabilities, we have

(a)

$$\begin{aligned} \operatorname{pr}(\cup_{i=1}^{n} \overline{E}_i) &= \operatorname{pr}(\overline{E}_1) + \operatorname{pr}(\overline{E}_2 \cap E_1) + \cdots + \operatorname{pr}(\overline{E}_n \cap E_{n-1} \cap \cdots \cap E_1) \\ &= \operatorname{pr}(\overline{E}_1) + \sum_{i=2}^{n} \operatorname{pr}(\overline{E}_i \cap E_{i-1} \cap \cdots \cap E_1). \end{aligned}$$

(b) If (i) denotes an arbitray index in the set $\{1, 2, \ldots, i-1\}$ $(i > 1)$, then

$$\begin{aligned} \operatorname{pr}(\cup_{i=1}^{n} \overline{E}_i) &\leq \operatorname{pr}[\overline{E}_1] + \sum_{i=2}^{n} \operatorname{pr}(\overline{E}_i \cap E_{(i)}) \\ &= \sum_{i=1}^{n} \operatorname{pr}(\overline{E}_i) - \sum_{i=2}^{n} \operatorname{pr}(\overline{E}_i \cap \overline{E}_{(i)}). \end{aligned}$$

Since the labeling of the E_i is arbitrary we have the following generalization.

22.4. (Hunter–Worsley Inequality) Let G be a graph representing events $\overline{E}_1, \ldots, \overline{E}_n$ as vertices with $\overline{E}_i$ and $\overline{E}_j$ joined by an edge e_{ij} if and only if $\overline{E}_i \cap \overline{E}_j \neq \phi$. Then, for any spanning tree T of G,

$$\operatorname{pr}(\cap_{i=1}^{n} E_i) \geq 1 - \sum_{i=1}^{n} \operatorname{pr}(\overline{E}_i) + \sum_{(i,j):e_{ij} \in T} \operatorname{pr}[\overline{E}_i \cap \overline{E}_j].$$

In the class of the above bounds, the sharpest bound is obtained by finding the spanning tree T^* for which the term

$$\sum_{(i,j):e_{ij} \in T} \operatorname{pr}(\overline{E}_i \cap \overline{E}_j)$$

is maximum (cf. Hochberg and Tamhane [1987: 364 for further details]). The Kounias inequality (cf. 22.2b) is never sharper as it uses the maximum only over a subset of all spanning trees.

Proofs. Section 22.1.

22.1. This result can be readily proved by induction.

22.2. Hochberg and Tamhane [1987: 363].

22.3a. We take the union of the events on the right-hand side and use results like $\overline{E}_1 \cup (\overline{E}_2 \cap E_1) = \overline{E}_1 \cup \overline{E}_2$ and $\overline{E}_1 \cap (\overline{E}_2 \cap \overline{E}_1) = \phi$, and so on, to show that the events are disjoint.

22.3b. Follows from $\mathrm{pr}(\overline{E}_i) = \mathrm{pr}(\overline{E}_i \cap E_{(i)}) + \mathrm{pr}(\overline{E}_i \cap \overline{E}_{(i)})$.

22.2   BONFERRONI-TYPE INEQUALITIES

22.5. We have the following results.

(a) (Degree-One Inequality) If $p_i = \mathrm{pr}(E_i)$, $i = 1, 2, \ldots, k$, then

$$\mathrm{pr}(\cap_{i=1}^k E_i) \geq 1 - \sum_{i=1}^{k}(1 - p_i).$$

(b) (Degree-Two Inequality)   Let $q_i = \mathrm{pr}(\overline{E}_i) = 1 - p_i$ and $q_{ij} = q_{ji} = \mathrm{pr}(\overline{E}_i \cap \overline{E}_j)$ for $i, j = 1, 2, \ldots, k$. Then

   (i)

$$1 - Q_1 \leq \mathrm{pr}(\cap_{i=1}^k E_i) \leq 1 - \frac{Q_1^2}{Q_1 + 2Q_2},$$

where

$$Q_1 = \sum_{i=1}^n q_i \quad \text{and} \quad Q_2 = \sum_{i=2}^n \sum_{j=1}^{i-1} q_{ij} = \sum_{j<i} q_{ij}.$$

Note that $Q_1 + 2Q_2$ is simply $\sum_{i=1}^n \sum_{j=1}^n q_{ij}$, where $q_{ii} = q_i$ for all i. Also, from (22.2a), we obtain

$$\mathrm{pr}\left[\cup_{i=1}^n \overline{E}_i\right] \geq \frac{Q_1^2}{Q_1 + 2Q_2}.$$

   (ii) If $\mathbf{q} = (q_1, q_2, \ldots, q_n)'$ and $\mathbf{Q} = (q_{ij})$ is nonsingular, then

$$\mathrm{pr}(\cup_{i=1}^n \overline{E}_i) \geq \mathbf{q}'\mathbf{Q}^{-1}\mathbf{q},$$

by (12.1d). The nonsingularity condition for $\mathbf{Q}$ was removed as follows.

   (iii) $\mathrm{pr}(\cup_{i=1}^n \overline{E}_i) \geq \mathbf{q}'\mathbf{Q}^-\mathbf{q}$,   where $\mathbf{Q}^-$ is a weak inverse of $\mathbf{Q}$.

(iv) The lower bound given in (a) is sharpened in the following result.

$$1 - Q_1 + \max_{1 \le j \le n} \sum_{i:i \neq j} q_{ij} \le \mathrm{pr}(\cap_{i=1}^n E_i) \le 1 - Q_1 + Q_2.$$

This is equivalent to

$$Q_1 - Q_2 \le \mathrm{pr}(\cup_{i=1}^n \overline{E}_i) \le Q_1 - \max_{1 \le j \le n} \sum_{i:i \neq j} q_{ij}.$$

The above results are called second degree because they require only knowledge of pairwise intersections of events. For further information on higher degree inequalities see Tong[1980: 147-148]. Some statistical applications of Bonferroni inequalities to simultaneous confidence inervals are given by Galambos and Simonelli [1996: chapter 8].

Proofs. Section 22.1.

22.5a. Tong [1980: 143, theorem 7.1.1]

22.5b(i). Tong [1980: 143, theorem 7.1.2].

22.5b(ii). Tong [1980: 145, lemma 7.1.1].

22.5b(iii). Kounias [1968] and quoted by Tong [1980: 146, theorem 7.1.3].

22.5b(iv). Tong [1980: 147, theorem 7.4.1].

22.3 DISTRIBUTION-FREE PROBABILITY INEQUALITIES

22.3.1 Chebyshev-Type Inequalities

If x is a random variable with mean μ and variance σ^2, then for $a > 0$,

$$\mathrm{pr}[|x - \mu| \le a\sigma] \ge 1 - 1/a^2.$$

This is known as the univariate *Chebyshev inequality*. A one-sided version is given by

$$\mathrm{pr}[x - \mu \le a\sigma] \ge 1 - 1/(1 + a^2),$$

and a multivariate version (with equal variances and common correlation) is considered by Tong [1980: 155, lemma 7.2.1]. We now consider further generalizations of these from Tong [1980: section 7.2].

22.6. Let $\mathbf{x} = (x_1, x_2)'$ be a random vector with mean $\boldsymbol{\mu} = (\mu_1, \mu_2)'$, variances σ_1^2 and σ_2^2, and correlation ρ. Then:

(a) For all $a_i > 0$,

$$\mathrm{pr}[\cap_{i=1}^2 (|x_i - \mu_i| \le a_i \sigma_i)] \ge 1 - \{(a_1^2 + a_2^2) + [(a_1^2 + a_2^2)^2 - 4\rho^2 a_1^2 a_2^2]^{1/2}\}/(2a_1^2 a_2^2).$$

(b) When $a_1 = a_2 = a$, $\mathrm{pr}[\cap_{i=1}^2 (|x_i - \mu_i| \le a\sigma_i)] \ge 1 - [1 + (1 - \rho^2)^{1/2}]/a^2$.
The equality is attainable.

22.7. Let $\mathbf{x} = (x_1, x_2, \ldots, x_n)'$ be a random vector with mean vector $\boldsymbol{\mu}$ and variance matrix $\boldsymbol{\Sigma} = (\sigma_{ij})$. If $\sigma_i^2 = \sigma_{ii}$ and $a_i > 0$ for all i, then

$$\mathrm{pr}\left[\cap_{i=1}^{n}(|x_i - \mu_i| \leq a_i\sigma_i)\right] \geq 1 - \sum_{i=1}^{n} \frac{1}{a_i^2}.$$

Tong [1980: 153] described a more general result that gives the sharpest lower bound.

22.8. Let $\mathbf{x} = (x_1, x_2, \ldots, x_n)'$ be a random vector with mean $\boldsymbol{\mu}$, variances σ_i^2 $(i = 1, 2, \ldots, n)$, and common correlation ρ, where $\rho \in [-1/(n-1), 1]$ (to ensure that the variance matrix is positive definite; cf. 15.18a(iv)). Then, for a satisfying

$$a > (n-1)[(1-\rho)/n]^{1/2},$$

we have

$$\mathrm{pr}[\cap_{i=1}^{n}(x_i \leq \mu_i + a\sigma_i)]$$
$$\geq 1 - \frac{\{[(1 + (n-1)\rho)(a^2 - u)]^{1/2} + a(n-1)(1-\rho)^{1/2}\}^2}{n\{a^2 + [1 + (n-1)\rho]/n\}^2},$$

where $u = (n-1)(1-\rho) - 1$.

22.9. Let $\mathbf{x} = (x_1, x_2, \ldots, x_n)'$ be a random vector with mean $\boldsymbol{\mu}$.

(a) Let ϕ be a concave function from $\mathbb{R}^n$ to $[0, \infty)$. For fixed $a > 0$, define $A = \{\mathbf{x} \mid \phi(\mathbf{x}) \leq a\}$. If $E[\phi(\mathbf{x})]$ exists, then

$$\mathrm{pr}(\mathbf{x} \in A) \geq 1 - \phi(\boldsymbol{\mu})/a.$$

We now give several applications of the above result.

(b) If y is a non-negative random variable with $E(y) < \infty$, then setting $A = \{y : y \leq \delta\}$ we have

$$\mathrm{pr}(y \geq \delta) \leq \frac{E(y)}{\delta}, \quad \delta > 0.$$

(c) Suppose that the x_i are all non-negative, and let $x_{(1)} \leq \cdots \leq x_{(n)}$ denote the order statistics. Then, for $c_1 \geq \cdots \geq c_n \geq 0$, the function $\phi(\mathbf{x}) = \sum_{i=1}^{n} c_i x_{(i)}$ is concave in $\mathbf{x}$. If $E(x_{(i)})$ exists, then

$$\mathrm{pr}(\sum_{i=1}^{n} c_i x_{(i)} \leq a) \geq 1 - \sum_{i=1}^{n} c_i \mu_{(i)}/a, \quad a > 0,$$

where $\mu_{(1)} \leq \cdots \leq \mu_{(n)}$ are the ordered means.

For the special case $c_1 = 1$ and $c_2 = \cdots = c_n = 0$, this reduces to

$$\mathrm{pr}(\min_{1 \leq i \leq n} x_i \leq a) \geq 1 - \frac{1}{a} \min_{1 \leq i \leq n} \mu_i,$$

or equivalently

$$\mathrm{pr}[\cap_{i=1}^{n}(x_i > a)] \leq \frac{1}{a} \min_{1 \leq i \leq n} \mu_i.$$

Proofs. Section 22.3.1.

22.6a. Quoted by Tong [1980: 152].

22.6b. Tong [1980: 150, theorem 7.2.1].

22.8. Tong [1980: 156, theorem 7.2.3].

22.9a. Tong [1980: 157, theorem 7.2.4].

22.9b. Mathai and Provost [1992: 188].

22.9c. Quoted by Tong [1980: 158].

22.3.2   Kolmogorov-Type Inequalities

22.10. Let $y_1, y_2 \ldots, y_n$ be n independent random variables with means η_i and variances τ_i^2 $(i = 1, 2, \ldots, n)$, and let $v_n = (\sum_{i=1}^n \tau_i^2)^{1/2}$. Then, for every fixed $a > 0$, we have the following:

(a) $\text{pr} \left[\bigcap_{r=1}^n \left(\left| \sum_{i=1}^r (y_i - \eta_i) \right| \leq av_n \right) \right] \geq 1 - 1/a^2.$

(b) $\text{pr} \left[\bigcap_{r=1}^n \left(\sum_{i=1}^r (y_i - \eta_i) \leq av_n \right) \right] \geq a^2/(1 + a^2).$

Proofs. Section 22.3.2.

22.10a. Tong [1980: 158, theorem 7.3.1].

22.10b. Tong [1980: 159, theorem 7.3.2].

22.3.3   Quadratics and Inequalities

22.11. Let $\mathbf{x}$ be a $n \times 1$ random vector with $E(\mathbf{x}) = \boldsymbol{\mu}$ and $\text{var}(\mathbf{x}) = \boldsymbol{\Sigma}$, and consider the quadratics $Q_i = (\mathbf{x} - \mathbf{a})' \mathbf{A}_i (\mathbf{x} - \mathbf{a})$, where $\mathbf{A}_i$ is non-negative definite $(i = 1, 2, \ldots, k)$ and $\mathbf{a}$ is an arbitrary constant vector.

(a) $\text{pr}[\cap_{i=1}^k (Q_i \leq \delta_i)] \geq 1 - \left(\dfrac{\gamma_1}{\delta_1} + \dfrac{\gamma_2}{\delta_2} + \cdots + \dfrac{\gamma_k}{\delta_k} \right),$

where, for $i = 1, 2, \ldots, k$, we have $\delta_i > 0$ and $\gamma_i = \text{trace}(\mathbf{A}_i \boldsymbol{\Sigma}) + (\boldsymbol{\mu} - \mathbf{a})' \mathbf{A}_i (\boldsymbol{\mu} - \mathbf{a}).$

(b) $\text{pr}[\cap_{i=1}^k (Q_i \geq \delta_i)] \leq \dfrac{\gamma_1 + \gamma_2 + \cdots + \gamma_k}{\delta_1 + \delta_2 + \cdots + \delta_k}.$

Proofs. Section 22.3.3.

22.11. Mathai and Provost [1992: 188–189] and the references therein.

22.4   DATA INEQUALITIES

22.12. The following inequalities hold for any numbers, but the main application is to random observations.

(a) Let $x_1, x_2, \ldots, x_n$ be n observations and define $\overline{x} = \sum_{i=1}^{n} x_i/n$ and $\widehat{\sigma}^2 = \sum_{i=1}^{n}(x_i - \overline{x})^2/n$. Then

$$(x_i - \overline{x})^2 \leq (n-1)\widehat{\sigma}^2, \quad i = 1, 2, \ldots, n.$$

Equality holds if all the other x_j's are equal except x_i. For an extension of the above see Kabe [1980].

(b) Let $x_1 \geq x_2 \geq \cdots \geq x_n$. Then

$$\overline{x} - \widehat{\sigma}\sqrt{(k-1))/(n-k+1)} \leq x_k \leq \overline{x} + \widehat{\sigma}\sqrt{(n-k)/k}.$$

Equality occurs on the left-hand side if and only if

$$x_1 = x_2 = \cdots = x_{k-1} \quad \text{and} \quad x_k = x_{k+1} = \cdots = x_n,$$

and on the right-hand side if and only if

$$x_1 = x_2 = \cdots = x_k \quad \text{and} \quad x_{k+1} = x_{k+2} = \cdots = x_n.$$

(c) Suppose $\mathbf{x}_1, \mathbf{x}_2, \ldots, \mathbf{x}_n$ are d-dimensional observations. Let $\overline{\mathbf{x}}_n = \sum_{i=1}^{n} \mathbf{x}_i/n$ and $\mathbf{S}_n = \sum_{i=1}^{n}(\mathbf{x}_i - \overline{\mathbf{x}}_n)(\mathbf{x}_i - \overline{\mathbf{x}}_n)'/n$. Then

$$(n-1)\mathbf{S}_n - (\mathbf{x}_j - \overline{\mathbf{x}}_n)(\mathbf{x}_j - \overline{\mathbf{x}}_n)' \quad \text{is non-negative definite,}$$

or equivalently,

$$(\mathbf{x}_j - \overline{\mathbf{x}}_n)'\mathbf{S}_n^{-1}(\mathbf{x}_j - \overline{\mathbf{x}}_n) \leq n - 1, \quad j = 1, 2, \ldots, n.$$

Thus each $\mathbf{x}_j$ lies in the interior or on the surface of the ellipsoid $(\mathbf{x} - \overline{\mathbf{x}}_n)'\mathbf{S}_n^{-1}(\mathbf{x} - \overline{\mathbf{x}}_n) = n - 1$. If $\mathbf{S}_n$ is singular, we can replace $\mathbf{S}_n^{-1}$ by $\mathbf{S}_n^{-}$, a weak inverse of $\mathbf{S}_n$.

(d) If $\overline{\mathbf{x}}_{(j)} = \frac{1}{n-1}\sum_{i:i\neq j} \mathbf{x}_i$, then

$$\frac{n^2}{n-1}\mathbf{S}_n - (\mathbf{x}_j - \overline{\mathbf{x}}_{(j)})(\mathbf{x}_j - \overline{\mathbf{x}}_{(j)})' \quad \text{is non-negative definite,}$$

or equivalently,

$$(\mathbf{x}_j - \overline{\mathbf{x}}_{(j)})'\mathbf{S}_n^{-1}(\mathbf{x}_j - \overline{\mathbf{x}}_{(j)}) \leq \frac{n^2}{n-1}, \quad j = 1, 2, \ldots, n.$$

Proofs. Section 22.4.

22.12a. Isotalo et al. [2005b: 176] and Samuelson [1968].

22.12b. Farnum [1989] and Wolkowicz and Styan [1979].

22.12c. Trenkler and Puntanen [2005]

22.12d. Trenkler and Puntanen [2005].

22.5   INEQUALITIES FOR EXPECTATIONS

22.13. (Multivariate Jensen's Inequality) Let $\mathbf{x}$ be an $n \times 1$ random vector with finite expectation $E(\mathbf{x}) = \boldsymbol{\mu}$.

(a) Let ϕ be a real-valued convex function defined on S, where S is a convex subset of $\mathbb{R}^n$. If $\operatorname{pr}(\mathbf{x} \in S) = 1$, then

$$E[\phi(\mathbf{x})] \geq \phi(\boldsymbol{\mu}).$$

(b) If ϕ is a symmetric (cf. Definition 23.6 above (23.14)) and continuous function on $\mathbb{R}^n$, then

$$\phi(\mu_{(1)}, \mu_{(2)}, \dots, \mu_{(n)}) \leq E[\phi(x_{(1)}, x_{(2)}, \dots, x_{(n)})],$$

where $\mu_{(1)} \geq \mu_{(2)} \geq \cdots \geq \mu_{(n)}$ and $x_{(1)} \geq x_{(2)} \geq \cdots \geq x_{(n)}$.

22.14. (Finite Population) Suppose that $x_1, x_2, \dots, x_n$ are obtained by sampling without replacement from a finite population, and $y_1, y_2, \dots, y_n$ are obtained by sampling with replacement from the same population. Then if g is continuous and convex,

$$E\left[\left(\sum_{i=1}^{n} x_i\right)\right] \leq E\left[\left(\sum_{i=1}^{n} y_i\right)\right].$$

Proofs. Section 22.5.

22.13. Schott [2005: 378].

22.14. Hoeffding [1963] and quoted by Marshall and Olkin [1979: 331–343].

22.6   MULTIVARIATE INEQUALITIES

22.6.1   Convex Subsets

22.15. Let $\mathbf{x} \in \mathbb{R}^d$ be a random vector with symmetric probability density function $f(\mathbf{x})$, that is, $f(-\mathbf{x}) = f(\mathbf{x})$, such that the set $\{\mathbf{x} : f(\mathbf{x}) \geq \alpha\}$ is convex for all $\alpha > 0$. Suppose that S is a convex subset of $\mathbb{R}^d$ and is symmetric about $\mathbf{0}$ (i.e., if $\mathbf{x} \in S$ then $-\mathbf{x} \in S$ also). Then:

(a) $\operatorname{pr}(\mathbf{x} + c\mathbf{b} \in S) \geq \operatorname{pr}(\mathbf{x} + \mathbf{b} \in S)$ for any constant $\mathbf{b} \in S$ and $0 \leq c \leq 1$.

(b) The result (a) still hold if $\mathbf{b}$ is replaced by $\mathbf{y}$, a random vector distributed independently of $\mathbf{x}$.

(c) If $\mathbf{x} \sim N_d(\mathbf{0}, \boldsymbol{\Sigma})$, then its probability density function satisfies the above conditions.

(d) If $\mathbf{x} \sim N_d(\mathbf{0}, \boldsymbol{\Sigma}_1)$ and $\mathbf{y} \sim N_d(\mathbf{0}, \boldsymbol{\Sigma}_2)$, where $\boldsymbol{\Sigma}_1 - \boldsymbol{\Sigma}_2$ is non-negative definite, then $\operatorname{pr}(\mathbf{x} \in S) \leq \operatorname{pr}(\mathbf{y} \in S)$. This type of result has been extended to elliptically contoured distributions by Perlman [1993].

Many of the unimodal symmetric multivariate distributions centered at the origin, like the multivariate normal and multivariate t-distribution, satisfy the conditions of this theorem. For further background see Anderson [1996].

22.6.2   Multivariate Normal

22.16. (Slepian Inequality) Suppose $\mathbf{x} \sim N_d(\mathbf{0}, \mathbf{\Sigma})$, where $\mathbf{\Sigma} = (\sigma_{ij})$ is non-negative definite. Let $y_i = x_i/\sqrt{\sigma_{ii}}$, $i = 1, 2, \ldots, d$ so that $\mathbf{y} \sim N_d(\mathbf{0}, \mathbf{R})$, where $\mathbf{R} = (\rho_{ij})$ is the population correlation matrix. Then, for any constants $c_1, c_2, \ldots, c_d$,

$$\mathrm{pr}\left[\cap_{i=1}^{d}(y_i \le c_i)\right]$$

is an increasing function for each ρ_{ij}, $i \ne j$. If $\mathbf{R}$ is positive definite, then the above is a strictly increasing function of ρ_{ij}, $i \ne j$.

Replacing c_i by $\sqrt{\sigma_{ii}}c_i$, we see that the result still holds if use the x_i instead of the y_i.

If all the $\rho_{ij} \ge 0$ $(i \ne j)$, then

$$\mathrm{pr}\left[\cap_{i=1}^{d}(y_i \le c_i)\right] \ge \prod_{i=1}^{d} \mathrm{pr}(y_i \le c_i).$$

If $\rho_{ij} \le 0$ for all $i, j, i \ne j$, the above inequality is reversed.

Because we can transform from x_i to y_i, researchers have focused, without any loss of generality, on deriving results for $\mathbf{y}$, where $\mathbf{y} \sim N_d(\mathbf{0}, \mathbf{R})$ and $\mathbf{R}$ is the correlation matrix.

22.17. (Khatri)

(a) Suppose $\mathbf{x} = (\mathbf{x}'_{(1)}, \mathbf{x}'_{(2)})' \sim N_d(\mathbf{0}, \mathbf{\Sigma})$, where $\mathbf{x}_{(k)}$ is $d_k \times 1$ $(k = 1, 2)$ and $d = d_1 + d_2$. Let

$$\mathbf{\Sigma} = \begin{pmatrix} \mathbf{\Sigma}_{11} & \mathbf{\Sigma}_{12} \\ \mathbf{\Sigma}'_{12} & \mathbf{\Sigma}_{22} \end{pmatrix},$$

where $\mathbf{\Sigma}_{kk}$ is $d_k \times d_k$. Let $A_1 \subset \mathbb{R}^{d_1}$ and $A_2 \subset \mathbb{R}^{d_2}$ be two convex regions that are symmetric about the origin (cf. 22.15). If $\mathbf{\Sigma}_{12}$ has rank zero (i.e., $\mathbf{\Sigma}_{12} = \mathbf{0}$) or has rank one, then

$$\mathrm{pr}\left[\cap_{k=1}^{2}(\mathbf{x}_{(k)} \in A_k)\right] \ge \prod_{k=1}^{2} \mathrm{pr}(\mathbf{x}_{(k)} \in A_k).$$

Setting $\mathbf{x}_{(1)} = x_1$, we have

$$\mathrm{pr}(|x_1| \le a_1, \mathbf{x}_{(2)} \in A_2) \ge \mathrm{pr}(|x_1| \le a_1) \, \mathrm{pr}(\mathbf{x}_{(2)} \in A_2),$$

and repeatedly applying this result to each element of $\mathbf{x}_{(2)}$, we obtain

$$\mathrm{pr}\left[\cap_{i=1}^{d}(|x_i| \le a_i)\right] \ge \prod_{i=1}^{d} \mathrm{pr}(|x_i| \le a_i).$$

This inequality is strict if $\mathbf{\Sigma}$ is positive definite, $\mathbf{\Sigma}$ is not a diagonal matrix, and all the a_is are positive.

The above results have been generalized to the case when the mean of $\mathbf{x}$ is not necessarily zero. For details see Tong [1980: theorem 2.2.3].

If the correlation matrix $\mathbf{R}$ of $\mathbf{x}$ has structure l (see (b) below) and is positive definite, then $\mathbf{\Sigma}_{12}$ has rank 0 or 1 (quoted by Tong [1980: 33, example 11]).

(b) Suppose $\mathbf{x} = (\mathbf{x}'_{(1)}, \mathbf{x}'_{(2)}, \ldots, \mathbf{x}'_{(r)})'$, where $\mathbf{x} \sim N_d(\mathbf{0}, \mathbf{R})$ and $\mathbf{R} = (\rho_{ij})$ is the correlation matrix. Here $\mathbf{x}_{(k)}$ is $d_k \times 1$ $(k = 1, 2, \ldots, r)$, where $\sum_k^r d_k = d$. For each k, let $A_k \subset \mathbb{R}^{d_k}$ be a convex region symmetric about the origin. Suppose we have the *product structure* $\rho_{ij} = \lambda_i \lambda_j$, $\lambda_i \in (-1, 1)$, for all i, j $(j \neq i)$, called *structure l* by Tong, then the following inequalities hold.

(i) Firstly,

$$\text{pr}\left[\cap_{k=1}^r (\mathbf{x}_{(k)} \in A_k)\right] \geq \text{pr}\left[\cap_{k \in C}(\mathbf{x}_{(k)} \in A_k)\right]$$
$$\times \text{pr}\left[\cap_{k \notin C}(\mathbf{x}_{(k)} \in A_k)\right]$$
$$\geq \prod_{k=1}^r \text{pr}(\mathbf{x}_{(k)} \in A_k)$$

and

$$\text{pr}\left[\cap_{k=1}^r (\mathbf{x}_{(k)} \notin A_k)\right] \geq \text{pr}\left[\cap_{k \in C}(\mathbf{x}_{(k)} \notin A_k)\right]$$
$$\times \text{pr}\left[\cap_{k \notin C}(\mathbf{x}_{(k)} \notin A_k)\right]$$
$$\geq \prod_{k=1}^r \text{pr}(\mathbf{x}_{(k)} \notin A_k)$$

holds for every subset C of the integers $\{1, 2, \ldots, r\}$. The inequalities are strict if the A_i's are bounded sets with positive probabilities and the λ_i's are nonzero.

(ii) We then have the following special case of the above.

$$\text{pr}\left[\cap_{i=1}^d (|x_i| \geq a_i)\right] \geq \prod_{i=1}^d \text{pr}(|x_i| \geq a_i).$$

22.18. (Šidák's Theorem) Let $\mathbf{x} \sim N_d(\mathbf{0}, \mathbf{R}(\lambda))$, where $\mathbf{R}(\lambda) = (\rho_{ij}(\lambda))$ is a correlation matrix depending on $\lambda \in [0, 1]$ in the following way: For a fixed non-negative definite correlation matrix $\mathbf{T} = (\tau_{ij})$, we define $\rho_{ij}(\lambda) = \tau_{ij}$ for all $i, j = 2, 3, \ldots, d$ and $\rho_{1j}(\lambda) = \rho_{j1}(\lambda) = \lambda \tau_{1j}$ for $j = 2, \ldots, d$. Then $\mathbf{R}(\lambda)$ is non-negative definite for $\lambda \in [0, 1]$, and

$$\text{pr}\left[\cap_{i=1}^d (|x_i| \leq a_i)\right]$$

is monotonically nondecreasing in $\lambda \in [0, 1]$ for every a_i $(i = 1, 2, \ldots, d)$. If $\mathbf{T}$ is positive definite (which implies $\mathbf{R}(\lambda)$ is positive definite), $\tau_{1j} \neq 0$ for some $j > 1$, and all the a_is are positive, then the above probability is strictly increasing in λ.

A consequence of the above result is that if we now have correlation matrix $\mathbf{R}(\boldsymbol{\lambda})$, where $\boldsymbol{\lambda} = (\lambda_1, \lambda_2, \ldots, \lambda_d)'$, $\rho_{ij}(\boldsymbol{\lambda}) = \lambda_i \lambda_j \tau_{ij}$ for all $i \neq j$, and each $\lambda_i \in [0, 1]$, then the above probability is monotonically nondecreasing in each $\lambda_i \in [0, 1]$. It is strictly increasing in λ_i if $\mathbf{T} = (\tau_{ij})$ is positive definite, $\lambda_i \tau_{ij} \neq 0$ for some $j \neq i$, and all the a_is are positive.

22.19. Suppose $\mathbf{x} \sim N_d(\mathbf{0}, \boldsymbol{\Sigma})$, where $\sigma_{ij} = \sigma^2$ for $i = j$, $\sigma_{ij} = \rho\sigma^2$ for $i \neq j$, and $\rho \in [0, 1]$. Define for $k = 1, 2, \ldots, d$ and $a > 0$,

$$\text{pr}_1(k) = \text{pr}\left[\cap_{i=1}^k (x_i \leq a)\right]$$
$$\text{pr}_2(k) = \text{pr}\left[\cap_{i=1}^k (|x_i| \leq a)\right]$$

and

$$\mathrm{pr}_3(k) \;=\; \mathrm{pr}\left[\cap_{i=1}^k(|x_i| \geq a)\right].$$

Then, for $m = 1, 2, 3$, we have

$$\mathrm{pr}_m(d) \geq [\mathrm{pr}_m(k)]^{d/k} \geq [\mathrm{pr}_m(1)]^d, \quad d > k \geq 2.$$

The inequalities are strict if $\rho > 0$.

***Definition* 22.1.** Let $\mathbf{x} = (x_1, x_2, \ldots, x_d)'$ be a random vector. We say that the elements of $\mathbf{x}$ are *associated random variables* if for any two univariate functions $g_i(\mathbf{x})$ of $\mathbf{x}$ such that $\mathrm{E}[g_i(\mathbf{x})]$ exists $(i = 1, 2)$ that are nondecreasing in each argument, we have $\mathrm{cov}[g_1(\mathbf{x}), g_2(\mathbf{x})] \geq 0$. Esary et al. [1967], who introduced the concept, have given a number of results that can be used to readily verify that a given set of random variables is associated.

22.20. The following statements are true.

(a) Any subset of associated random variables is a set of associated random variables.

(b) The set consisting of a single random variable is associated.

(c) If two sets of associated random variables are independent, then their union is a set of associated random variables.

(d) Independent random variables are associated.

(e) Nondecreasing functions of associated random variables are associated random variables.

22.21. If the elements of the $d \times 1$ vector $\mathbf{x}$ are associated random variables, then

$$\mathrm{pr}\left[\cap_{i=1}^d(x_i \leq a_i)\right] \;\geq\; \mathrm{pr}\left[\cap_{i\in C}(x_i \leq a_i)\right]\mathrm{pr}\left[\cap_{i\notin C}(x_i \leq a_i)\right]$$

$$\geq\; \prod_{i=1}^d \mathrm{pr}(x_i \leq a_i)$$

holds for all a_i and all subsets C of $\{1, 2, \ldots, d\}$.

Proofs. Section 22.6.2.

22.15. Results quoted by Schott [2005: 83, exercise 2.61] and follow from (2.65e).

22.16. Tong [1980: section 2.1].

22.17a. Tong [1980: 16–19].

22.17b(i). Quoted by Hochberg and Tamhane [1987: 367] and proved by Tong [1980: theorems 2.2.4 and 2.3.2].

22.17b(ii). Tong [1980: 28, theorem 2.3.3].

22.18. Tong [1980: 21, theorem 2.2.5, corollary 1].

22.19. Tong [1980: 30, theorem 2.3.4].

22.20. Tong [1980: 87, theorem 5.2.2].

22.21. Tong [1980: 89, theorem 5.2.4].

22.6.3 Inequalities For Other Distributions

22.22. (Multivariate t-Distribution) Let $\mathbf{x} = (x_1, x_2, \ldots, x_d)' \sim N_d(\mathbf{0}, \mathbf{R})$, where we can assume without loss of generality that $\mathbf{R} = (\rho_{ij})$ is the correlation matrix with $\rho_{ii} = 1$ (cf. 22.16). Let $z_i = x_i/u$, where u is distributed as a $\sqrt{\chi_\nu^2/\nu}$ random variable independent of $\mathbf{x}$. Then $\mathbf{z} = (z_1, z_2, \ldots, z_d)'$ has a multivariate t-distribution with ν degrees of freedom and associated correlation matrix $\mathbf{R}$. In terms of the notation of Section 20.8.1, $\mathbf{y} \sim t_d(\nu, \mathbf{0}, \mathbf{R})$.

By conditioning on u, the following inequalities from above still hold with x_i or y_i replaced by z_i.

(a) If all the $\rho_{ij} \geq 0$, then

$$\mathrm{pr}\left[\cap_{i=1}^d (z_i \leq c_i)\right] \geq \prod_{i=1}^d \mathrm{pr}(z_i \leq c_i).$$

(b) $\mathrm{pr}\left[\cap_{i=1}^d (|z_i| \leq a_i)\right] \geq \prod_{i=1}^d \mathrm{pr}(|z_i| \leq a_i).$

This inequality is strict if $\mathbf{R}$ is positive definite, $\mathbf{R}$ is not a diagonal matrix, and all the a_is are positive.

(c) If $\rho_{ij} = \lambda_i \lambda_j$ (all i, j, $j \neq i$), where each $\lambda_i \in (-1, +1)$, then

$$\mathrm{pr}\left[\cap_{i=1}^d (|z_i| \geq a_i)\right] \geq \prod_{i=1}^d \mathrm{pr}(|z_i| \geq a_i).$$

22.23. (Correlated F-ratios) Let $\chi_0^2, \chi_1^2, \ldots, \chi_k^2$ be independent chi-square variables with degrees of freedoom $\nu_0, \nu_1, \ldots \nu_k$, and let $F_i = (\chi_i^2/\nu_i)/(\chi_0^2/\nu_0)$, $i = 1, 2, \ldots, k$. Then

(a) $\mathrm{pr}\left[\cap_{i=1}^k (F_i \leq a_i)\right] > \prod_{i=1}^k \mathrm{pr}(F_i \leq a_i).$

(b) $\mathrm{pr}\left[\cap_{i=1}^k (F_i > a_i)\right] > \prod_{i=1}^k \mathrm{pr}(F_i > a_i).$

Proofs. Section 22.6.3.

22.22. Hochberg and Tamhane [1987: 369].

22.23. Tong [1980: 43, theorem 3.2.2].

CHAPTER 23

MAJORIZATION

Majorization does not seem to be a topic very well known in statistical circles. However, majorization can be used to prove a number of inequalities. A key result is (23.7), from which we may assume without any loss of generality (Tong [1980: 105]) that only two coordinates need be different when proving inequalities. Two applications are, for example, species-diversity indices (Tong [1983]) and optimal design theory (Bhaumik [1995]). The topic is also relevant to the finding of optimal statistical tests (Anderson [2003: section 8.10]).

23.1 GENERAL PROPERTIES

***Definition* 23.1.** Let $\mathbf{x} = (x_1, x_2, \ldots, x_n)'$ and $\mathbf{y} = (y_i, y_2, \ldots, y_n)'$ be vectors in $\mathbb{R}^n$. Suppose the x_i are ordered in decreasing order of magnitude as $x_{(1)} \geq x_{(2)} \geq \cdots \geq x_{(n)}$, with the y_i ordered in a similar fashion. We say that $\mathbf{x}$ is *(strongly) majorized* by $\mathbf{y}$ (or $\mathbf{y}$ *majorizes* $\mathbf{x}$), and use the symbol $\mathbf{x} \ll \mathbf{y}$ (or $\mathbf{y} \gg \mathbf{x}$), if

$$x_{(1)} + x_{(2)} + \cdots + x_{(i)} \leq y_{(1)} + y_{(2)} + \cdots + y_{(i)}, \quad i = 1, 2, \ldots, n-1,$$
$$x_1 + x_2 + \cdots + x_n = y_1 + y_2 + \cdots + y_n.$$

This definition is given by Marshall and Olkin [1979] and Horn and Johnson [1991], except they use $x_{[i]}$ instead of $x_{(i)}$. They and most other authors, except Rao and Rao [1998: chapter 9], use $\mathbf{x} \prec \mathbf{y}$ instead of $\mathbf{x} \ll \mathbf{y}$; however, I have reserved the former symbol in the form of $\mathbf{A} \succ \mathbf{0}$ for positive definite matrices and used $\mathbf{A} > \mathbf{0}$ for matrices with all positive elements.

The above definition has also been exended to infinite sequences (Marshall and Olkin [1979: 16]) and to random vectors (Marshall and Olkin [1979: chapter 11]).

23.1. If the x_i (and the y_i) are ordered in increasing magnitude, say $x_{\{1\}} \leq x_{\{2\}} \leq \cdots \leq x_{\{n\}}$, then since $x_{\{i\}} = x_{(n-i+1)}$, $\mathbf{x} \ll \mathbf{y}$ if and only if

$$
\begin{aligned}
x_{\{1\}} + x_{\{2\}} + \cdots + x_{\{i\}} &\geq y_{\{1\}} + y_{\{2\}} + \cdots + y_{\{i\}}, \quad i = 1, 2, \ldots, n-1, \\
x_1 + x_2 + \cdots + x_n &= y_1 + y_2 + \cdots + y_n.
\end{aligned}
$$

Some authors use this result as their definition, which I find confusing because of the direction of the inequalities. Using the notation $x_{[i]}$ instead of $x_{\{i\}}$, Rao and Rao [1998: 304] proved the equivalence of the two definitions.

23.2. Let π be a permutation of $\{1, 2, \ldots, n\}$. If $\mathbf{x}$, $\mathbf{y}$, and $\mathbf{z}$ are in $\mathbb{R}^n$, and $\mathbf{x}_\pi$ is the vector whose components are a permutation of the elements of $\mathbf{x}$, then the following hold.

(a) $\mathbf{x} \ll \mathbf{x}$.

(b) If $\bar{x} = \sum_{i=1}^n x_i/n$, then $\bar{x}\mathbf{1}_n \ll \mathbf{x}$.

(c) $\mathbf{x} \ll \mathbf{x}_\pi$ for every permutation π.

(d) If $\mathbf{x} \ll \mathbf{y}$ and $\mathbf{y} \ll \mathbf{z}$, then $\mathbf{x} \ll \mathbf{z}$.

(c) If $\mathbf{x} \ll \mathbf{y}$ and $\mathbf{y} \ll \mathbf{x}$, then $\mathbf{y} = \mathbf{x}_\pi$ for some permutation π.

(e) If $\mathbf{x} \ll \mathbf{z}$, $\mathbf{y} \ll \mathbf{z}$, and $0 \leq \alpha \leq 1$, then $\alpha\mathbf{x} + (1-\alpha)\mathbf{y} \ll \mathbf{z}$.

(f) If $\mathbf{x} \ll \mathbf{y}$, $\mathbf{x} \ll \mathbf{z}$, and $0 \leq \alpha \leq 1$, then $\mathbf{x} \ll \alpha\mathbf{y} + (1-\alpha)\mathbf{z}$.

(g) If $\mathbf{z}$ is any vector, then

$$
\begin{pmatrix} \mathbf{x} \\ \mathbf{z} \end{pmatrix} \ll \begin{pmatrix} \mathbf{y} \\ \mathbf{z} \end{pmatrix} \text{ if and only if } \mathbf{x} \ll \mathbf{y}.
$$

(h)

$$
\left(\frac{1}{n}, \frac{1}{n}, \ldots, \frac{1}{n} \right)' \ll \left(\frac{1}{n-1}, \frac{1}{n-1}, \ldots, \frac{1}{n-1}, 0 \right)'
$$
$$
\ll \cdots \ll \left(\frac{1}{2}, \frac{1}{2}, 0, \ldots, 0 \right)' \ll (1, 0, \ldots, 0)'.
$$

23.3. The following conditions are equivalent.

(1) $\mathbf{x} \ll \mathbf{y}$.

(2) $\mathbf{x} = \mathbf{A}\mathbf{y}$ for some doubly stochastic matrix $\mathbf{A}$.

(3) $\mathbf{x} = \mathbf{B}\mathbf{y}$ for some orthostochastic matrix $\mathbf{B}$.

Note that $\mathbf{A}$ and $\mathbf{B}$ are not unique.

23.4. If $\mathbf{x} \geq \mathbf{0}$, $\mathbf{y} \geq \mathbf{0}$, and $\mathbf{x} \ll \mathbf{y}$, then $\prod_{i=1}^{n} x_i \geq \prod_{i=1}^{n} y_i$.

23.5. (Schur) If $\mathbf{H}$ be an $n \times n$ Hermitian matrix with (real) eigenvalues given by the vector $\boldsymbol{\lambda}(\mathbf{H}) = (\lambda_1, \lambda_2, \ldots, \lambda_n)'$, where $\lambda_1 > \ldots > \lambda_n$, and (real) diagonal elements $\mathbf{h} = (h_{11}, h_{22}, \ldots h_{nn})'$, then

$$\mathbf{h} \ll \boldsymbol{\lambda}(\mathbf{H})$$

on $\mathbb{R}^n$. This result is an example of a number of inequalities involving the eigenvalues and singular values of a matrix (cf. Marshall and Olkin [1979: chapter 9]), some of which are quoted elsewhere.

23.6. (Fan) If $\mathbf{A}$ and $\mathbf{B}$ are $n \times n$ Hermitian matrices, then

$$\boldsymbol{\lambda}(\mathbf{A} + \mathbf{B}) \ll \boldsymbol{\lambda}(\mathbf{A}) + \boldsymbol{\lambda}(\mathbf{B}).$$

***Definition* 23.2.** If $\mathbf{Q}$ is a permutation matrix that interchanges just two coordinates of a vector $\mathbf{x}$ (say x_j and x_k), then a *T-transform*, denoted by $\mathbf{Tx}$, has $\mathbf{T}$ of the form $\mathbf{T} = \lambda \mathbf{I} + (1 - \lambda)\mathbf{Q}$, where $0 \leq \lambda \leq 1$. What $\mathbf{Tx}$ does is transform x_j into $\lambda x_j + (1 - \lambda)x_k$ and x_k into $\lambda x_k + (1 - \lambda)x_j$, leaving the other elements of $\mathbf{x}$ unchanged. For example, if $\mathbf{Q}$ interchanges the first two elements of $\mathbf{x} \in \mathbb{R}^n$, then $\mathbf{Q} = (\mathbf{e}_2, \mathbf{e}_1, \ldots, \mathbf{e}_n)$ is $\mathbf{I}_n$ with its first two columns interchanged.

23.7. We have $\mathbf{x} \ll \mathbf{y}$ if and only if there exists a finite number of real vectors $\mathbf{c}_1, \mathbf{c}_2, \ldots \mathbf{c}_m$ such that $\mathbf{x} = \mathbf{c}_1 \ll \mathbf{c}_2 \ll \cdots \ll \mathbf{c}_{m-1} \ll \mathbf{c}_m = \mathbf{y}$, where, for all i, $\mathbf{c}_i$ and $\mathbf{c}_{i+1}$ differ in two coordinates only. Thus if $\mathbf{x} \ll \mathbf{y}$, then $\mathbf{y}$ can be derived from $\mathbf{x}$ by successive applications of a finite number of T-transforms.

***Definition* 23.3.** (Weak Majorization) We now generalize the Definition 23.1 above. A vector $\mathbf{x}$ is said to be *weakly (sub)majorized* by $\mathbf{y}$, and we denote the relationship by $\mathbf{x} \ll_w \mathbf{y}$, if

$$x_{(1)} + x_{(2)} + \cdots + x_{(i)} \leq y_{(1)} + y_{(2)} + \cdots + y_{(i)}, \quad i = 1, 2, \ldots, n.$$

Marshall and Olkin [1979] use the notation $\mathbf{x} \prec_w \mathbf{y}$ for weak (sub)majorization. Some authors omit the prefix "sub".

We say that $\mathbf{x}$ is *weakly (super)majorized* by $\mathbf{y}$ and denote the relationship by $\mathbf{x} \ll^w \mathbf{y}$ if

$$x_{\{1\}} + x_{\{2\}} + \cdots + x_{\{i\}} \geq y_{\{1\}} + y_{\{2\}} + \cdots + y_{\{i\}}, \quad i = 1, 2, \ldots, n.$$

23.8. The results below follow directly from the previous definitions.

(a) $\mathbf{x} \ll_w \mathbf{y}$ if and only if $-\mathbf{x} \ll^w -\mathbf{y}$.

(b) $\mathbf{x} \ll_w \mathbf{y}$ and $\mathbf{x} \ll^w \mathbf{y}$ if and only if $\mathbf{x} \ll \mathbf{y}$.

(c) $\mathbf{x} \leq \mathbf{y}$ (i.e., $x_i \leq y_i$ for all i) implies that $\mathbf{x} \ll_w \mathbf{y}$ and $\mathbf{x} \gg^w \mathbf{y}$.

(d) $\mathbf{x} \ll_w \mathbf{y}$ if and only if for some $\mathbf{u}$, $\mathbf{x} \leq \mathbf{u}$ and $\mathbf{u} \ll \mathbf{y}$.

(e) $\mathbf{x} \ll^w \mathbf{y}$ if and only if for some $\mathbf{v}$, $\mathbf{x} \ll \mathbf{v}$ and $\mathbf{v} \geq \mathbf{y}$.

23.9. Let $\mathbb{R}_+ = [0, \infty)$. Then $\mathbf{x} \ll_w \mathbf{y}$ on $\mathbb{R}_+^n$ if and only if there exists a doubly stochastic matrix $\mathbf{A}$ such that $\mathbf{x} = \mathbf{A}\mathbf{y}$.

23.10. $\mathbf{x} \ll_w \mathbf{y}$ if and only if $\sum_{i=1}^n \phi(x_i) \leq \phi(y_i)$ for all continuous monotonically increasing convex functions ϕ (cf. Definition 23.5 below).

23.11. Let $\mathbf{A}$ and $\mathbf{B}$ be $n \times n$ symmetric matrices, and suppose that $\lambda_1(\mathbf{C}) \geq \lambda_2(\mathbf{C}) \geq \cdots \geq \lambda_n(\mathbf{C})$ with $\boldsymbol{\lambda}(\mathbf{C})$ the corresponding vector, where $\mathbf{C} = \mathbf{A}$ or $\mathbf{B}$. Then

$$\boldsymbol{\lambda}(\mathbf{A} + \mathbf{B})) \ll_w \boldsymbol{\lambda}(\mathbf{A}) + \boldsymbol{\lambda}(\mathbf{B}).$$

23.12. If $\mathbf{A}$ and $\mathbf{B}$ are $n \times n$ matrices and $\mathbf{A} \leq \mathbf{B}$ (i.e., $a_{ij} \leq b_{ij}$ for all $i.j$), then using the notation of (23.11) above we have

$$\boldsymbol{\lambda}(\mathbf{A}) \ll_w \boldsymbol{\lambda}(\mathbf{B}).$$

23.13. Let $\mathbf{A}$ and $\mathbf{B}$ be $n \times n$ real or complex matrices, and let $\sigma_1(\mathbf{C}) \geq \sigma_2(\mathbf{C}) \geq \cdots \geq \sigma_k(\mathbf{C})$ $(k = \min\{m, n\})$, with $\boldsymbol{\sigma}(\mathbf{C})$ the corresponding vector. Then:

(a) $\boldsymbol{\sigma}(\mathbf{A} + \mathbf{B}) \ll_w \boldsymbol{\sigma}(\mathbf{A}) + \boldsymbol{\sigma}(\mathbf{B})$.

(b) If $m = n$,

  (i) $\boldsymbol{\sigma}(\mathbf{A}\mathbf{B}) \ll_w \boldsymbol{\sigma}(\mathbf{A}) \circ \boldsymbol{\sigma}(\mathbf{B})$.

  (ii) $\boldsymbol{\sigma}(\mathbf{A} \circ \mathbf{B}) \ll \boldsymbol{\sigma}(\mathbf{A}) \circ \boldsymbol{\sigma}(\mathbf{B})$.

Here "$\circ$" is the Hadamard product.

Proofs. Section 23.1.

23.2. Rao and Rao [1998: 303–307, for (a)–(g)] and Marshall and Olkin [1979: 7, for (h)].

23.3. Marshall and Olkin [1979: 21–24].

23.4. Quoted by Rao and Rao [1998: 320].

23.5. Marshall and Olkin and Olkin [1979: 218] and Zhang [1998: 230].

23.6. Marshall and Olkin [1979: 241] and Zhang [1999: 231].

23.7. Marshall and Olkin [1979: 21] and Rao and Rao [1998: 316].

23.8. Marshall and Olkin [1979: 11].

23.9. Horn and Johnson [1991: 166-167].

23.11. Anderson [2003: 357].

23.12. Anderson [2003: 359].

23.13. Zhang [1998: 232].

23.2   SCHUR CONVEXITY

Definition 23.4. Let $\mathbf{f}$ be a function from $\mathbb{R}^n$ to $\mathbb{R}^m$ $(m > 1)$ defined on $\mathcal{A} \subset \mathbb{R}^n$. Then $\mathbf{f}$ is said to be *Schur-convex on* $\mathcal{A}$ if

$$\mathbf{x}, \mathbf{y} \in \mathbb{R}^n \text{ and } \mathbf{x} \ll \mathbf{y} \text{ on } \mathcal{A} \Rightarrow \mathbf{f}(\mathbf{x}) \ll_w \mathbf{f}(\mathbf{y}).$$

Also $\mathbf{f}$ is said to be *strongly Schur-convex on* $\mathcal{A}$ if

$$\mathbf{x}, \mathbf{y} \in \mathbb{R}^n \text{ and } \mathbf{x} \ll_w \mathbf{y} \text{ on } \mathcal{A} \Rightarrow \mathbf{f}(\mathbf{x}) \ll_w \mathbf{f}(\mathbf{y}),$$

and $\mathbf{f}$ is said to be *strictly Schur-convex on* $\mathcal{A}$ if

$$\mathbf{x}, \mathbf{y} \in \mathbb{R}^n \text{ and } \mathbf{x} \ll \mathbf{y} \text{ on } \mathcal{A} \Rightarrow \mathbf{f}(\mathbf{x}) \ll \mathbf{f}(\mathbf{y}).$$

If $\mathcal{A} = \mathbb{R}^n$, we drop the words "on $\mathcal{A}$". Note that the label "schur-convex" is a bit misleading as such a function is not necessarily convex.

A function $\mathbf{f}$ is *Schur-concave* if $(-\mathbf{f})$ is Schur-convex. (The above definitions come from Rao and Rao [1998: 307], except m and n are interchanged.)

The above definitions need to be clarified as follows when $m = 1$ and $\mathbf{f}$ is no longer a vector (say ϕ). The function ϕ is Schur-convex (respectively concave) on $\mathcal{A}$ if $\mathbf{x}, \mathbf{y} \in \mathbb{R}^n$ and $\mathbf{x} \ll \mathbf{y}$ on $\mathcal{A} \Rightarrow \phi(\mathbf{x}) \le \phi(\mathbf{y})$ (respectively $\phi(\mathbf{x}) \ge \phi(\mathbf{y})$). If $\mathbf{y}$ is not a permutation of $\mathbf{x}$, then ϕ is said to be strictly Schur-convex (respectively concave) on $\mathcal{A}$ if $\mathbf{x}, \mathbf{y} \in \mathbb{R}^n$ and $\mathbf{x} \ll \mathbf{y}$ on $\mathcal{A} \Rightarrow \phi(\mathbf{x}) < \phi(\mathbf{y})$ (respectively $\phi(\mathbf{x}) > \phi(\mathbf{y})$).

Schur convexity or concavity can be used to prove many inequalities. For example, Marshall and Olkin [1979: chapter 8] list a number of inequalities including those relating to the angles or sides of various geometrical figures such as triangles and polygons. Schur convexity also arises in combinatorial analysis, particularly with respect to graph theory, the theory of network flows, and the study of incidence matrices (Marshall and Olkin [1979: chapter 7]).

Definition 23.5. Let $\mathbf{f}$ be a function from $\mathbb{R}^n$ to $\mathbb{R}^m$ $(m > 1)$ defined on $\mathcal{A} \subset \mathbb{R}^n$, and let $\mathbf{x} \le \mathbf{y}$ (i.e., $x_i \le y_i$ for each i). Then $\mathbf{f}$ is said to be *montonically increasing* on $\mathcal{A}$ if $\mathbf{x}, \mathbf{y} \in \mathbb{R}^n$ and $\mathbf{x} \le \mathbf{y}$ on $\mathcal{A} \Rightarrow \mathbf{f}(\mathbf{x}) \le \mathbf{f}(\mathbf{y})$, *monotonically decreasing* if $-\mathbf{f}$ is monotonically increasing, and *monotone* if it is either monotonically increasing or decreasing.

We recall from Section 2.5 that $\mathbf{f}$ is *convex* if

$$\mathbf{f}(\alpha \mathbf{x} + (1 - \alpha)\mathbf{y}) \le \alpha \mathbf{f}(\mathbf{x}) + (1 - \alpha)\mathbf{f}(\mathbf{y}),$$

for every $0 \le \alpha \le 1$ and $\mathbf{x}, \mathbf{y} \in \mathbb{R}^n$; $\mathbf{f}$ is *concave* if $-\mathbf{f}$ is convex.

Definition 23.6. Let $\mathbf{f}$ be a function from $\mathbb{R}^n$ to $\mathbb{R}^m$, and let $\mathbf{y} = \mathbf{f}(\mathbf{x})$. Then $\mathbf{f}$ is said to be *symmetric* if, for every permutation π of $\{1, 2, \ldots, n\}$, there exists a permutation π' of $\{1, 2, \ldots, m\}$ such that $\mathbf{f}(\mathbf{x}_\pi) = \mathbf{y}_{\pi'}$ for all $\mathbf{x} \in \mathbb{R}^n$.

If $m = 1$, so that $\mathbf{f} = \phi$, say, then ϕ is symmetric if $\phi(\mathbf{x}_\pi) = \phi(\mathbf{x})$ for all π. Examples of symmetric functions for $n = 3$ are $\phi(\mathbf{x}) = x_1 + x_2 + x_3$, $\phi(\mathbf{x}) = x_1 x_2 + x_2 x_3 + x_3 x_1$, and $\phi(\mathbf{x}) = x_1 x_2 x_3$.

23.14. If a function $\mathbf{f}$ from $\mathbb{R}^n$ to $\mathbb{R}^m$ is convex and symmetric, then it is Schur-convex. In addition, if $\mathbf{f}$ is monotonically increasing, then $\mathbf{f}$ is strongly Schur-convex.

23.15. Let g be a function from $\mathbb{R}$ to $\mathbb{R}$, and define

$$\mathbf{f}(\mathbf{x}) = (g(x_1), g(x_2), \dots, g(x_n))'.$$

Then, from the previous result, we have the following.

(a) If g is convex, then $\mathbf{f}$ is Schur-convex.

(b) If g is a convex monotonically increasing function, then $\mathbf{f}$ is strongly Schur-convex.

(c) Taking $g(x) = |x|$, $g(x) = x^2$, and $g(x) = \max\{x, 0\} = x^+$, respectively, we have

    (i) $\mathbf{x} \ll \mathbf{y}$ implies that $|\mathbf{x}| \ll_w |\mathbf{y}|$.

    (ii) $\mathbf{x} \ll \mathbf{y}$, $u_i = x_i^2$ and $v_i = y_i^2$ implies that $\mathbf{u} \ll_w \mathbf{v}$.

    (iii) $\mathbf{x} \ll \mathbf{y}$ implies that $\mathbf{x}^+ \ll_w \mathbf{y}^+$.

23.16. The following symmetric convex functions are Schur-convex.

(a) $\phi(\mathbf{x}) = \max_i |x_i|$.

(b) $\phi(\mathbf{x}) = (\sum_{i=1}^n |x_i|^r)^{1/r}$, $r \geq 1$.

23.17. (Sum)

(a) Let $\mathcal{I}$ be an interval of $\mathbb{R}$, and let g be a function from $\mathcal{I}$ to $\mathbb{R}$. If y is (strictly) convex on $\mathcal{I}$, then $\phi(\mathbf{x}) = \sum_{i=1}^n g(x_i)$ is (strictly) Schur-convex on $\mathcal{I}^n$, as ϕ is symmetric. In this case, $\mathbf{x} \ll \mathbf{y}$ on $\mathcal{I}$ implies that $\phi(\mathbf{x}) \leq \phi(\mathbf{y})$ (or $\phi(\mathbf{x}) < \phi(\mathbf{y})$ for strict convexity).

There is also a converse result. Suppose g is continuous on $\mathcal{I}$. If ϕ is (strictly) Schur-convex on $\mathcal{I}^n$, then g is (strictly) convex on $\mathcal{I}$.

(b) Combining the above, the inequality

$$\sum_{i=1}^n g(x_i) \leq \sum_{i=1}^n g(y_i)$$

holds for all continuous convex functions g from $\mathbb{R}$ to $\mathbb{R}$ if and only if $\mathbf{x} \ll \mathbf{y}$. Also, the same inequality holds for all continuous increasing convex functions g if and only if $\mathbf{x} \ll_w \mathbf{y}$. It holds for continuous decreasing convex functions g if and only if $\mathbf{x} \ll^w \mathbf{y}$.

23.18. The following are examples of strictly convex functions.

(a) For $a > 0$, $g(x) = [x + (1/x)]^a$ is strictly convex on $(0, 1]$. For $a \geq 1$, g is strictly convex on $(0, \infty)$.

(b) $-\log x$ is strictly convex on $(0, \infty)$.

(c) $g(x) = 1/x$ is strictly convex on $(0, \infty)$.

In each case $\phi(\mathbf{x}) = \sum_{i=1}^{n} g(x_i)$ is strictly Schur-convex. If $x_i > 0$ and $\sum_{i=1}^{n} x_i = 1$, then $n^{-1}\mathbf{1}_n \ll \mathbf{x}$ (by 23.2b) and $\phi(n^{-1}\mathbf{1}_n) \leq \phi(\mathbf{x})$. We can use this result to set up inequalities. For example, using (a),

$$\sum_{i=1}^{n} \left(x_i + \frac{1}{x_i}\right)^a \geq \frac{(n^2 + 1)^a}{n^{a-1}}.$$

23.19. (Product) Let g be a continuous non-negative function defined on an interval $\mathcal{I}$. Then $\phi(\mathbf{x}) = \prod_{i=1}^{n} g(x_i)$ is (strictly) Schur-convex on $\mathcal{I}^n$ if and only if $\log g$ is (strictly) convex on $\mathcal{I}$. Since, by (23.2b), $\overline{x}\mathbf{1}_n \ll \mathbf{x}$, we can use this result to obtain various inequalities. For example, $\log \Gamma(x)$ is strictly convex on $\mathbb{R}_{++} = (0, \infty)$ so that $\phi(\mathbf{x}) = \prod_{i=1}^{n} \Gamma(x_i)$ is strictly Schur-convex on $\mathbb{R}_{++}^n$. Hence

$$[\Gamma(\overline{x})]^n \leq \prod_{i=1}^{n} \Gamma(x_i).$$

For further details about Schur-convex or Schur-concave functions see Marshall and Olkin [1979: chapter 3].
Proofs. Section 23.2.

23.14. Rao and Rao [1998: 318].

23.15a. Marshall and Olkin [1979: 115] and Rao and Rao [1998: 319].

23.15b. Marshall Olkin [1979: 116] and Rao and Rao [1998: 319].

23.15c. Rao and Rao [1998: 319].

23.16. Marshall and Olkin [1979: 96].

23.17a. Marshall and Olkin [1979: 64, 67].

23.17b. Marshall and Olkin [1979: 108–109].

23.18–23.19. Marshall and Olkin [1979: 70–73, 75].

23.3 PROBABILITIES AND RANDOM VARIABLES

23.20. (Probabilities) For $i = 1, 2, \ldots, n$, let $p_i = \mathrm{pr}(E_i)$, and let q_i be the probability that at least k of the events $E_1, E_2, \ldots, E_n$ occurs.

(a) If $\mathbf{p} = (p_1, \ldots, p_n)'$ and $\mathbf{q} = (q_1, \ldots, q_n)'$, then $\mathbf{p} \ll \mathbf{q}$.

(b) From (a) we have $\sum_{i=1}^{n} p_i = \sum_{i=1}^{n} q_i$ and $\prod_{i=1}^{n} p_i \geq \prod_{i=1}^{n} q_i$.

23.21. (Expectations) Let $\mathbf{x} = (x_1, x_2, \ldots, x_n)'$ be a random vector with finite expectation, and define $a_i = \mathrm{E}(x_i)$, $\tilde{a}_i = a_{(i)}$, and $b_i = \mathrm{E}(x_{(i)})$, where $a_{(1)} \geq a_{(2)} \geq \cdots \geq a_{(n)}$ and $x_{(1)} \geq x_{(2)} \geq \cdots \geq x_{(n)}$. Then:

(a) $\mathbf{a} \ll \mathbf{b}$.

(b) $\tilde{\mathbf{a}} \ll \mathbf{b}$.

23.22. (Eigenvalues) Let $\mathbf{Z}$ be a random Hermitian $n \times n$ matrix with eigenvalues $\lambda_1(\mathbf{Z}) \geq \lambda_2(\mathbf{Z}) \geq \ldots \geq \lambda_n(\mathbf{Z})$ (which are all real). Then

$$(\lambda_1(\mathrm{F}[\mathbf{Z}]), \ldots, \lambda_n(\mathrm{E}[\mathbf{Z}]))' \ll (\mathrm{E}[\lambda_1(\mathbf{Z})], \ldots, \mathrm{E}[\lambda_n(\mathbf{Z})])',$$

where the $\lambda_i(\mathrm{E}[\mathbf{Z}])$ are the eigenvalues of $\mathrm{E}[\mathbf{Z}]$, the expectation of $\mathbf{Z}$.

23.23. (Singular Values) Let $\mathbf{W}$ be an $m \times n$ random complex matrix with singular values $\sigma_1(\mathbf{W}) \geq \ldots \geq \sigma_t(\mathbf{W})$, where $t = \min(m, n)$. Then

$$(\sigma_1(\mathrm{E}[\mathbf{Z}]), \ldots, \sigma_n(\mathrm{E}[\mathbf{Z}]))' \ll_w (\mathrm{E}[\sigma_1(\mathbf{Z})], \ldots, \mathrm{E}[\sigma_n(\mathbf{Z})])'.$$

Proofs. Section 23.2.

23.20. Marshall and Olkin [1979: 345–347].

23.21a. Rao and Rao [1998: 305].

23.21b. Marshall and Olkin [1979: 348].

23.22. Marshall and Olkin [1979: 355].

23.23. Marshall and Olkin [1979: 357].

CHAPTER 24

OPTIMIZATION AND MATRIX APPROXIMATION

The subject of finding unconstrained or constrained maxima and minima of functions is an extensive one. Schott [2005: section 9.7] gives a helpful summary. We consider only a few basic results in this chapter.

24.1 STATIONARY VALUES

Definition 24.1. Let $f : \mathbf{x} \to f(\mathbf{x})$ be a real-valued function defined on S, a subset of $\mathbb{R}^n$. Then f has a *local maximum* at $\mathbf{c}$ if, for some $\delta > 0$, $f(\mathbf{c}) \geq f(\mathbf{x})$ for all $\mathbf{x}$ such that $\|\mathbf{x} - \mathbf{c}\|_2 < \delta$. It has a *strict local maximum* if $f(\mathbf{c}) > f(\mathbf{x})$ for all $\mathbf{x} \neq \mathbf{c}$ such that $\|\mathbf{x} - \mathbf{c}\|_2 < \delta$. Also, f has a *global (absolute) maximum* at $\mathbf{c}$ if $f(\mathbf{c}) \geq f(\mathbf{x})$ for all $\mathbf{x} \in S$. The function f has a *local minimum* at $\mathbf{c}$ if $-f$ has a local maximum at $\mathbf{c}$, and a *global (absolute) minimum* at $\mathbf{c}$ if $-f$ has a global maximum at $\mathbf{c}$.

Let $\mathbf{c}$ be an interior point of S (cf. Definition 2.29 below (2.63)). Then there exists a $\delta > 0$ such that $\mathbf{x} \in S$ for all $\mathbf{x}$ satisfying $\|\mathbf{x} - \mathbf{c}\|_2 < \delta$. Suppose f is differentiable at $\mathbf{c}$. If

$$\frac{\partial f(\mathbf{c})}{\partial \mathbf{c}'} = \left.\frac{\partial f(\mathbf{x})}{\partial \mathbf{x}'}\right|_{\mathbf{x}=\mathbf{c}} - \mathbf{0},$$

then any point $\mathbf{c}$ satisfying the above equation is called a *stationary point*. (Note that $f(\mathbf{c})$ is also called a *critical value* of f at $\mathbf{c}$.) Such a point can be a local maximum, a local minimum, or a *saddle point*.

24.1. (Unconstrained Local Optimization) Let f be defined as in Definition 24.1 above, and suppose that f is twice differentiable at $\mathbf{c}$, where $\mathbf{c}$ is an interior point of S. If $\mathbf{c}$ is a stationary value of f and $\nabla^2 f(\mathbf{c})$ is the Hessian of f at $\mathbf{c}$ (cf. Section 17.11), then:

(a) f has a strict local minimum at $\mathbf{c}$ if $\nabla^2 f(\mathbf{c})$ is positive definite.

(b) f has a strict local maximum at $\mathbf{c}$ if $-\nabla^2 f(\mathbf{c})$ is positive definite (i.e., $\nabla^2 f(\mathbf{c})$ is negative definite).

(c) f has a saddle point at $\mathbf{c}$ if $\nabla^2 f(\mathbf{c})$ is neither positive definite nor negative definite, but is nonsingular.

(c) f may have a local minimum, a local maximum, or a saddlepoint at $\mathbf{c}$ if $\nabla^2 f(\mathbf{c})$ is singular.

24.2. Minimizing a function is equivalent to minimizing a monotonically increasing transformation of that function. This result is particularly useful in maximum likelihood estimation, which is discussed in Section 24.3.1 below.

24.3. (Method of Lagrange Multipliers for Constrained Optimization) We now give sufficient conditions for finding a strict local maximum or minimum of a real-valued function f defined on $S \subseteq \mathbb{R}^n$ subject to the vector of constraints $\mathbf{g}(\mathbf{x}) = \mathbf{0}$, where $\mathbf{g} = (g_1, g_2, \ldots, g_m)'$ is $m \times 1$ $(m < n)$.

Let $\mathbf{c}$ be an interior point of S, let $F(\mathbf{x}, \boldsymbol{\lambda}) = f(\mathbf{x}) + \boldsymbol{\lambda}' \mathbf{g}(\mathbf{x})$, where $\boldsymbol{\lambda} \in \mathbb{R}^m$ (called the *Lagrange multiplier*; some use $-\boldsymbol{\lambda}$), and suppose that the following conditions hold:

(1) f and $\mathbf{g}$ are twice differentiable at $\mathbf{c}$.

(2) $\mathbf{B} = \partial \mathbf{g}(\mathbf{x})/\partial \mathbf{x}' = (\partial g_i(\mathbf{x})/\partial x_j)$ has full row rank m at $\mathbf{x} = \mathbf{c}$.

(3) $\mathbf{x} = \mathbf{c}$ is a solution of $\mathbf{g}(\mathbf{x}) = \mathbf{0}$ and $\partial F(\mathbf{x}, \boldsymbol{\lambda})/\partial \mathbf{x} = \mathbf{0}$ (for some $\boldsymbol{\lambda}$).

If $\mathbf{A}$ is $\nabla^2 f(\mathbf{x}) - \sum_{i=1}^{m} \lambda_i \nabla^2 g_i(\mathbf{x})$ evaluated at $\mathbf{x} = \mathbf{c}$, then f has a strict local maximum at $\mathbf{x} = \mathbf{c}$, subject to $\mathbf{g}(\mathbf{x}) = \mathbf{0}$, if

$$\mathbf{x}' \mathbf{A} \mathbf{x} < 0, \quad \text{for all } \mathbf{x} \neq \mathbf{0} \text{ for which } \mathbf{B} \mathbf{x} = \mathbf{0}.$$

A similar result holds for a strict local minimum with the inequality $\mathbf{x}' \mathbf{A} \mathbf{x} > 0$ replacing $\mathbf{x}' \mathbf{A} \mathbf{x} < 0$. In practice, one can often just simply solve the equations $\partial F(\mathbf{x}, \boldsymbol{\lambda})/\partial \mathbf{x} = \mathbf{0}$ and $\mathbf{g}(\mathbf{x}) = \mathbf{0}$ for $\mathbf{x}$ and $\boldsymbol{\lambda}$, and then use ad hoc methods to check the nature of the constrained stationary value without having to investigate $\mathbf{A}$.

24.4. Assuming that the conditions of the previous result (24.3) hold, we now give some equivalent sufficient conditions for a strict local maximum or a strict local minimum to exist. Let $\mathbf{A}$ be a symmetric $n \times n$ matrix and $\mathbf{B}$ an $m \times n$ matrix of rank m. Let $\mathbf{A}_{rr}$ be the leading principal $r \times r$ submatrix of $\mathbf{A}$, and let $\mathbf{B}_r$ be the $m \times r$ matrix obtained by deleting the last $n - r$ columns of $\mathbf{B}$. For $r = 1, 2, \ldots, n$, define the $(m + r) \times (m + r)$ matrix $\boldsymbol{\Delta}_r$ as

$$\boldsymbol{\Delta}_r = \begin{pmatrix} \mathbf{0} & \mathbf{B}_r \\ \mathbf{B}'_r & \mathbf{A}_{rr} \end{pmatrix}.$$

If $\mathbf{B}_m$ is nonsingular (which can be achieved by rearranging the x_i variables in (24.3)), then $\mathbf{x}'\mathbf{A}\mathbf{x} > 0$ holds for all $\mathbf{x} \neq \mathbf{0}$ satisfying $\mathbf{B}\mathbf{x} = \mathbf{0}$ if and only if

$$(-1)^m \det \mathbf{\Delta}_r > 0 \quad \text{for all } r = m+1, \ldots, n.$$

Also, $\mathbf{x}'\mathbf{A}\mathbf{x} < 0$ holds for all $\mathbf{x} \neq \mathbf{0}$ satisfying $\mathbf{B}\mathbf{x} = \mathbf{0}$ if and only if

$$(-1)^r \det \mathbf{\Delta}_r > 0 \quad \text{for all } r = m+1, \ldots, n.$$

Examples using the above theory are given by Schott [2005: 380–381] and Magnus and Neudecker [1999: 138].

24.5. (Global Optimization)   Finding the global maximum or minimum is sometimes best achieved by using the ideas of convex sets and functions, as seen in the following results.

(a) On a convex set, the set of points at which the minimum of a convex function is attained is convex, and any local minimum is a global minimum. The same is true for a concave function, except replacing minimum by maximum.

A strictly convex function attains a minimum at no more than one point of a convex set, and a stationary (critical) point is necessarily a minimum.

(b) On a compact convex set, the maximum of a convex function occurs at an extreme point. The same is true for the minimum of a concave function.

We now focus on convex and concave functions.

Proofs. Section 24.1.

24.1. Schott [2005: 371–372; he omits the word "strict"] and Magnus and Neudecker [1999: 122–123].

24.2. Magnus and Neudecker [1999: 129].

24.3. Magnus and Neudecker [1999: 135–138] and quoted by Schott [2005: 379–380].

24.4. Magnus and Neudecker [1999: 53–54, 136].

24.5. Quoted by Horn and Johnson [1985: 535].

24.2   USING CONVEX AND CONCAVE FUNCTIONS

24.6. Let $f : \mathbf{x} \to f(\mathbf{x})$ be a real-valued convex function defined on S, a convex subset of $\mathbb{R}^n$.

(a) Corresponding to each interior point $\mathbf{a} \in S$, an $n \times 1$ vector $\mathbf{t}$ exists, such that

$$f(\mathbf{x}) \geq f(\mathbf{a}) + \mathbf{t}'(\mathbf{x} - \mathbf{a})$$

for all $\mathbf{x} \in S$.

(b) If S is an open convex set, f is differentiable, and $\mathbf{a} \in S$, then

$$f(\mathbf{x}) \geq f(\mathbf{a}) + \frac{\partial f(\mathbf{a})}{\partial \mathbf{a}'}(\mathbf{x} - \mathbf{a})$$

for all $\mathbf{x} \in S$.

24.7. (Global Minimum or Maximum) Let $f(\mathbf{x})$ be a real-valued convex (respectively concave) function defined for all $\mathbf{x} \in S$, an open convex subset of $\mathbb{R}^n$. If f is differentiable and $\mathbf{c} \in S$ is a stationary point of f, then f has a global minimum (respectively maximum) at $\mathbf{c}$. If f is strictly convex or strictly concave, then $\mathbf{c}$ is unique.

24.8. Let $y = f(\mathbf{X})$. If $\mathrm{d}^2 f \geq 0$, then f is convex and f has a global minimum at $\mathrm{d}f = 0$. However, if $\mathrm{d}^2 f > 0$ for all $\mathrm{d}\mathbf{X} \neq \mathbf{0}$, then f is strictly convex and f has a strict global minimum at $\mathrm{d}f = 0$. For second-order differentials see Section 17.11.

24.9. (Constrained Global Minimum) Let f be a real-valued function defined and differentiable on an open convex set S in $\mathbb{R}^n$, and let $\mathbf{g}$ be an $m \times 1$ vector function ($m < n$) defined and differentiable on S. Let $\mathbf{c}$ be a point of S, and let $F(\mathbf{x}) = f(\mathbf{x}) + \boldsymbol{\lambda}'\mathbf{g}(\mathbf{x})$, where $\boldsymbol{\lambda} \in \mathbb{R}^m$. Assume that $\mathbf{x} = \mathbf{c}$ is a solution of $\mathbf{g}(\mathbf{x}) = \mathbf{0}$ and $\partial F(\mathbf{x})/\partial \mathbf{x} = \mathbf{0}$. If F is convex (respectively strictly convex) on S, then f has an absolute minimum (respectively unique absolute minimum) at $\mathbf{c}$ under the constraint $\mathbf{g}(\mathbf{c}) = \mathbf{0}$. Under the same conditions, if F is (strictly) concave, then f has a (unique) absolute maximum at $\mathbf{c}$ under the constraint $\mathbf{g}(\mathbf{x}) = \mathbf{0}$.

24.10. Suppose we wish to minimize $y = f(\mathbf{X})$ subject to the constraints $\mathbf{G}(\mathbf{X}) = \mathbf{0}$, where $\mathbf{G}$ is a matrix function of $\mathbf{X}$. Define the Lagrangian function $\psi(\mathbf{X}) = f(\mathbf{X}) - \mathrm{trace}[\mathbf{L}'\mathbf{G}(\mathbf{X})]$, where $\mathbf{L}$ is a matrix of Langrange multipliers. (If $\mathbf{G}$ happens to be symmetric, then we can take $\mathbf{L}$ to be symmetric also.) If ψ is (strictly) convex, then ψ has a (strict) global minimum at the point where $\mathrm{d}\psi = 0$ under the constraint $\mathbf{G}(\mathbf{X}) = \mathbf{0}$.

Proofs. Section 24.2.

24.6. Schott [2005: section 9.8].

24.7. Magnus and Neudecker [1999: 128–129].

24.8. Abadir and Magnus [2005: 354].

24.9. Magnus and Neudecker [1999: 139].

24.10. Abadir and Magnus [2005: 354].

24.3 TWO GENERAL METHODS

24.3.1 Maximum Likelihood

Definition **24.2.** Suppose we have a set of random variables denoted by $\mathbf{x}$ with continuous probability density function or discrete probability function $f(\mathbf{x} \mid \boldsymbol{\theta})$ depending on d unknown parameters $\boldsymbol{\theta} = (\theta_1, \theta_2, \ldots, \theta_d)'$, where $\boldsymbol{\theta} \in \Omega$ (often

$\Omega = \mathbb{R}^d$). We now express this function as a function of $\boldsymbol{\theta}$, namely $l(\boldsymbol{\theta})$, called the *likelihood function*. (Any constants or unknown functions of $\mathbf{x}$ are sometimes supressed.)

A value of $\boldsymbol{\theta}$, $\widehat{\boldsymbol{\theta}}$ say, that maximizes $l(\boldsymbol{\theta})$ or equivalently $L(\boldsymbol{\theta}) = \log l(\boldsymbol{0})$ for $\boldsymbol{\theta} \in \Omega$, is called a *maximum likelihood estimate* of $\boldsymbol{\theta}$. There is no guarantee that such an estimate exists for (almost) every $\mathbf{y}$, nor that is unique if it exists. If L is based on a set of n independent observations, we denote the estimate by $\widehat{\boldsymbol{\theta}}_n$ to emphasize its dependence on n.

The vector $\mathbf{u}(\boldsymbol{\theta}) = \partial L(\boldsymbol{\theta})/\partial \boldsymbol{\theta}$, or more briefly $\partial L/\partial \boldsymbol{\theta}$, is usually referred to as the *score vector*. The equations $\mathbf{u}(\boldsymbol{\theta}) = \mathbf{0}$ are called the *likelihood equations*.

24.11. Under fairly general conditions (e.g., Cox and Hinkley [1974] and Mäkeläinen et al. [1981]), we have the following results.

(a) $\mathrm{E}(\mathbf{u}) = \mathbf{0}$.

(b) $\mathrm{E}(\mathbf{u}\mathbf{u}') = \mathrm{var}(\mathbf{u}) = -\mathrm{E}(\partial \mathbf{u}/\partial \boldsymbol{\theta}') = \mathrm{E}(\mathcal{I}) = \mathbf{I}_\theta$, where $\mathcal{I} = -\partial^2 L/\partial \boldsymbol{\theta}\partial \boldsymbol{\theta}'$ is usually called the *(observed) information matrix* and $\mathbf{I}_\theta$ the *expected (Fisher) information matrix*.

(c) As $n \to \infty$, $\mathbf{u}(\boldsymbol{\theta})$ is approximately distributed as the multivariate normal distribution $N_d(\mathbf{0}, \mathbf{I}_\theta)$.

(d) $\widehat{\boldsymbol{\theta}}$ is the unique solution of $\mathbf{u}(\boldsymbol{\theta}) = \mathbf{0}$.

(e) If L is based on a set of n independent observations and $\boldsymbol{\theta}_0$ is the true value of $\boldsymbol{\theta}$, then as $n \to \infty$,

   (i) $(\widehat{\boldsymbol{\theta}}_n - \boldsymbol{\theta}_0)$ is approximately distributed as $N_n(\mathbf{0}, \mathbf{I}_{\theta_0}^{-1})$, and

   (ii) $-2[L(\widehat{\boldsymbol{\theta}}_n) - L(\boldsymbol{\theta}_0)]$ is approximately distributed as χ_d^2, the chi-square distribution.

With additional assumptions, the above theory extends to mutually independent non-identically distributed random variables, and even to dependent variables.

24.12. (Constrained Maximization) Recalling that $\boldsymbol{\theta} \in \Omega \subseteq \mathbb{R}^d$, sometimes Ω is restricted in some way. For example, in multivariate analysis (cf. Chapter 21) a matrix of parameters may be symmetric or even positive definite, as in the case of a variance matrix, so that technically these constraints should be built into the optimization process. For example, if we wish to maximize an expression subject to a matrix restricted to being positive definite, what frequently happens is that the unrestricted maximum turns out to be positive definite with probability 1. This unrestricted maximum is then also the restricted maximum (e.g., Calvert and Seber [1978: 274–276]). Alternatively, we can express the positive definite matrix in the form $\mathbf{A}'\mathbf{A}$ (cf. 10.32), where $\mathbf{A}$ is unknown. For a selection of examples and proofs see Abadir and Magnus [2005: section 13.12] and Magnus and Neudecker [1999: chapters 15 and 16].

A major problem with maximum likelihood is showing that the estimate obtained is actually a maximum. As a result, various ad hoc methods are used such as convexity arguments.

24.13. Let $\boldsymbol{\Sigma}$ and $\mathbf{A}$ be $n \times n$ matrices. Consider the matrix function f, where

$$f(\boldsymbol{\Sigma}) = \log(\det \mathbf{A}) + \text{trace}(\boldsymbol{\Sigma}^{-1}\mathbf{A}).$$

If $\mathbf{A}$ is positive definite, then, subject to $\boldsymbol{\Sigma}$ being positive definite, $f(\boldsymbol{\Sigma})$ is minimized uniquely at $\boldsymbol{\Sigma} = \mathbf{A}$.

24.14. Let $\mathbf{A}$ be $n \times n$, then:

(a) $\text{trace}(\mathbf{A}\mathbf{A}')$ is a strictly convex function of $\mathbf{A}$.

(b) If $\mathbf{A}$ is positive definite then $-\log(\det \mathbf{A})$ is a convex function.

Proofs. Section 24.3.1.

24.13. Seber [1984: 523].

24.14. Calvert and Seber [1978: 280].

24.3.2   Least Squares

***Definition* 24.3.** Let $\mathbf{y}$ be an $n \times 1$ random vector with mean $\text{E}(\mathbf{y}) = \mathbf{f}(\boldsymbol{\theta})$, where $\boldsymbol{\theta}$ is a $d \times 1$ vector of parameters and $\boldsymbol{\theta} \in \Omega$. Then $\tilde{\boldsymbol{\theta}}$ is a *least squares estimate* of $\boldsymbol{\theta}$ if $\tilde{\boldsymbol{\theta}}$ minimizes $[\mathbf{y} - \mathbf{f}(\boldsymbol{\theta})]'[\mathbf{y} - \mathbf{f}(\boldsymbol{\theta})]$ with respect to $\boldsymbol{\theta} \in \Omega$. In practice, $\mathbf{f}(\boldsymbol{\theta})$ will also depend on some data observations.

If a weight matrix function $\mathbf{W}(\boldsymbol{\theta})$ (generally a positive definite matrix) is included, then a minimum of $[\mathbf{y} - \mathbf{f}(\boldsymbol{\theta})]'\mathbf{W}(\boldsymbol{\theta})[\mathbf{y} - \mathbf{f}(\boldsymbol{\theta})]$, $\tilde{\boldsymbol{\theta}}_w$ say, is called a *generalized or weighted least squares estimate*. Various iterative methods such as *Iteratively Reweighted Least Squares (IRLS)* (e.g., Seber and Wild [1989: 37]) are available. In some applications $\mathbf{W}$ does not depend on $\boldsymbol{\theta}$.

Under certain general conditions, least squares and generalized least squares estimates are unique and have certain optimal properties. They generally have some useful asymptotic properties as well, which do not depend on normality assumptions. However, under normality assumptions, such estimates may be the same as the maximum likelihood estimates, for example in univariate (Seber and Lee [2003]) and multivariate (Seber [1984]) linear models, and nonlinear models (Seber and Wild [1989]). For a further discussion of least squares with respect to regression models see Section 20.7.

24.4   OPTIMIZING A FUNCTION OF A MATRIX

24.4.1   Trace

24.15. Let $\mathbf{A}$ be a real $n \times n$ matrix with singular values $\sigma_1(\mathbf{A}) \geq \cdots \geq \sigma_n(\mathbf{A})$ and singular value decomposition $\mathbf{A} = \mathbf{P}\boldsymbol{\Sigma}\mathbf{Q}'$, where $\boldsymbol{\Sigma} = \text{diag}(\sigma_1(\mathbf{A}), \ldots, \sigma_n(\mathbf{A}))$ and $\mathbf{P}$ and $\mathbf{Q}$ are $n \times n$ orthogonal matrices. Let $\mathcal{T}_n$ be the collection of all $n \times n$ orthogonal matrices. Then

$$\max_{\mathbf{T} \in \mathcal{T}} \text{trace}(\mathbf{A}\mathbf{T}) = \sum_{i=1}^{n} \sigma_i(\mathbf{A}),$$

and the maximum is attained at $\mathbf{T}_0 = \mathbf{QP}'$, where $\mathbf{T}_0$ is not necessarily unique. Also $\mathbf{AT}_0$ is non-negative definite. Furthermore, if $\mathbf{T}_1$ is an orthogonal matrix such that $\mathbf{AT}_1$ is non-negative definite, then $\mathbf{T}_1$ is the maximizer.

24.16. Let $\mathbf{A}$ be an $m \times n$ real matrix with singular value decomposition $\mathbf{P}\boldsymbol{\Sigma}_1\mathbf{Q}'$, let $\mathbf{B}$ be a real $n \times m$ matrix with singular value decomposition $\mathbf{B} = \mathbf{R}\boldsymbol{\Sigma}_2\mathbf{S}'$, $p = \min\{m, n\}$, and let $\mathcal{T}_p$ be the set of all $p \times p$ orthogonal matrices. Here $\mathbf{P}$, $\mathbf{Q}$, $\mathbf{R}$, and $\mathbf{S}$ are conformable orthogonal matrices. Then

$$\max_{\mathbf{T} \in \mathcal{T}_n, \mathbf{U} \in \mathcal{T}_m} \mathrm{trace}(\mathbf{ATBU}) = \sum_{i=1}^{p} \sigma_i(\mathbf{A})\sigma_i(\mathbf{B}) \quad (= \mathrm{trace}(\boldsymbol{\Sigma}_1\boldsymbol{\Sigma}_2)).$$

By substitution we see that equality occurs when $\mathbf{T} = \mathbf{QR}'$ and $\mathbf{U} = \mathbf{SP}'$. The above holds for complex matrices if we replace orthogonal matrices by unitary matrices and the trace by its real part.

24.17. Let $\mathcal{V}$ be the set of all real $m \times n$ matrices $\mathbf{C}$. Suppose $\mathbf{X}$ is a given $n \times r$ matrix, $\mathbf{V}_1, \mathbf{V}_2, \ldots, \mathbf{V}_k$ are given $m \times n$ matrices, and $a_1, a_2, \ldots, a_k$ are given real scalars. Let

$$\mathcal{V}_1 = \{\mathbf{C} : \mathbf{C} \in \mathcal{V}, \mathbf{CX} = \mathbf{0}, \mathrm{trace}(\mathbf{CV}_i') = a_i \text{ for each } i\},$$

and let $\mathbf{Q} = \mathbf{I}_n - \mathbf{X}(\mathbf{X}'\mathbf{X})^-\mathbf{X}$ represent the orthogonal projection perpendicular to $\mathcal{C}(\mathbf{X})$. Then:

(a) $\min_{\mathbf{C} \in \mathcal{V}_1} \mathrm{trace}(\mathbf{CC}') = \mathrm{trace}(\mathbf{C}_0\mathbf{C}_0')$,

where $\mathbf{C}_0 = \sum_{i=1}^{k} \alpha_i \mathbf{V}_i\mathbf{Q}$, and $(\alpha_1, \alpha_2, \ldots, \alpha_k)$ is a solution to the following system of linear equations:

$$\sum_{i=1}^{k}[\mathrm{trace}(\mathbf{V}_i\mathbf{Q}\mathbf{V}_j')]\alpha_i = a_j, \quad j = 1, 2, \ldots, k.$$

(b) Suppose now that $m = n$, $\mathcal{V}_1$ is the set of all symmetric $m \times m$ matrices, and the $\mathbf{V}_i$ are now all symmetric $m \times m$ matrices. Then

$$\min_{\mathbf{C} \in \mathcal{V}_1} \mathrm{trace}(\mathbf{C}^2) = \mathrm{trace}(\mathbf{C}_1^2),$$

where $\mathbf{C}_1 = \sum_{i=1}^{k} \alpha_i \mathbf{QV}_i\mathbf{Q}$, and $(\alpha_1, \alpha_2, \ldots, \alpha_k)$ is a solution to the following system of linear equations:

$$\sum_{i=1}^{k}[\mathrm{trace}(\mathbf{QV}_i\mathbf{QV}_j)]\alpha_i = a_j, \quad j = 1, 2, \ldots, k.$$

The above solutions are not necessarily unique if the matrices in the linear equations are singular. This theory can be applied to variance estimation (Rao and Rao [1998: sections 12.5–12.10]).

24.18. Let $\mathbf{X}$ be an $n \times p$ matrix of rank p, $\mathbf{V}$ be an $n \times n$ positive definite matrix, and $\mathbf{W}$ be an $m \times p$ matrix. Then $\mathrm{trace}(\mathbf{GVG}')$ is minimized with respect to $\mathbf{G}$, subject to $\mathbf{GX} = \mathbf{W}$, when $\mathbf{G} = \mathbf{G}_0$ and

$$\mathbf{G}_0 = \mathbf{W}(\mathbf{X}'\mathbf{V}^{-1}\mathbf{X})^{-1}\mathbf{X}'\mathbf{V}^{-1}.$$

If we drop the assumption that rank $\mathbf{X} = p$, then trace($\mathbf{GVG}'$) is minimized with respect to $\mathbf{G}$, subject to $\mathbf{GX} = \mathbf{X}$, when $\mathbf{G} = \mathbf{G}_1$ and

$$\mathbf{G}_1 = \mathbf{X}(\mathbf{X}'\mathbf{V}^{-1}\mathbf{X})^+\mathbf{X}'\mathbf{V}^{-1},$$

where $(\mathbf{X}'\mathbf{V}^{-1}\mathbf{X})^+$ is the Moore–Penrose inverse.

24.19. Let $\mathbf{X}$ be an $n \times p$ matrix of rank r and $\mathbf{V}$ be a non-negative definite $n \times n$ matrix. Then the minimum of $[\text{trace}(\mathbf{V}^2)/n]$ subject to $\mathbf{VX} = \mathbf{0}$ and trace $\mathbf{V} = 1$ is given by $\mathbf{V}_0$, where

$$\mathbf{V}_0 = \frac{1}{n-r}(\mathbf{I}_n - \mathbf{XX}^+),$$

and $\mathbf{X}^+$ is the Moore–Penrose inverse of $\mathbf{X}$.

Proofs. Section 24.4.1.

> 24.15. Horn and Johnson [1985: 432] and Rao and Rao [1998: 347–348] for the complex case.
>
> 24.16. Horn and Johnson [1985: 436, with some matrices replaced by their complex conjugates] and Rao and Rao [1998: 357–359, complex case].
>
> 24.17. Rao and Rao [1998: 410–413].
>
> 24.18. Abadir and Magnus [2005: 384–386].
>
> 24.19. Abadir and Magnus [2005: 386].

24.4.2   Norm

In this section we are also involved with matrix approximation as well as optimization.

Definition 24.4. Let $\mathcal{U}$ be the vector space of all $m \times n$ real matrices, and let $\mathcal{V}$ be a subspace. Given $\mathbf{A} \in \mathcal{U}$ and $\mathbf{B} \in \mathcal{V}$, we say that $\mathbf{B}$ is the *closest* to $\mathbf{A}$ with respect to a given norm $\| \cdot \|$ if $\mathbf{B}$ minimizes $\|\mathbf{A} - \mathbf{B}\|$. Note that $\mathbf{B}$ may not be unique.

24.20. (Eckart–Young) Some of the dimension reduction techniques of Section 21.5 may be described as approximating an $n \times d$ data matrix $\mathbf{X}$ by another $n \times r$ matrix, with $r < d$. We now consider the broader problem of approximating one matrix by another of lower rank.

Let $\mathbf{A}$ be an $m \times n$ real matrix of rank r with singular value decomposition

$$\mathbf{A} = \mathbf{P}\mathbf{\Sigma}\mathbf{Q}' = \sum_{i=1}^{r} \sigma_i \mathbf{p}_i \mathbf{q}_i',$$

where $\mathbf{P} = (\mathbf{p}_1, \ldots, \mathbf{p}_m)$, $\mathbf{Q} = (\mathbf{q}_1, \ldots, \mathbf{q}_n)$, and $\sigma_i = \sigma_i(\mathbf{A})$, the ordered singular values of $\mathbf{A}$. Let $\mathcal{V}$ be the set of $m \times n$ matrices of rank s $(s < r)$.

(a) Then

$$\min_{\mathbf{B} \in \mathcal{V}} \|\mathbf{A} - \mathbf{B}\|_{oi} = \|\mathbf{A} - \mathbf{B}_0\|_{oi},$$

for all orthogonally invariant norms $\| \cdot \|_{oi}$, where

$$\mathbf{B}_0 = \sum_{i=1}^{s} \sigma_i \mathbf{p}_i \mathbf{q}_i' \quad (= \mathbf{P}\boldsymbol{\Sigma}_1 \mathbf{Q}', \ \text{say}).$$

(b) We have from (a) the special case

$$\begin{aligned}
\|\mathbf{A} - \mathbf{B}_0\|_F^2 &= \text{trace}[(\mathbf{A} - \mathbf{B}_0)(\mathbf{A} - \mathbf{B}_0)'] \\
&= \sum_{i=s+1}^{r} \sigma_i^2 \, \text{trace}(\mathbf{p}_i \mathbf{q}_i' \mathbf{q}_i \mathbf{p}_i') \\
&= \sigma_{s+1}^2 + \sigma_{s+2}^2 \cdots + \sigma_r^2,
\end{aligned}$$

where $\| \cdot \|_F$ is the Frobenius norm.

(c) If the rows of $\mathbf{A}$ sum to zero (i.e., $\mathbf{A}'\mathbf{1}_m = \mathbf{0}$), then, in terms of the Frobenius norm, $\mathbf{B}$ is the rank s matrix whose column differences best approximate the column differences of $\mathbf{A}$. The same result applies to row differences if $\mathbf{A}\mathbf{1}_n = \mathbf{0}$.

The above results are used in many places in statistics such as the biplot (Jolliffe [2002: section 5.3] and Seber [1984: section 5.3]), classical multidimensional scaling (Seber [1984: 240]), sample principal components (Jolliffe [2002: 36–38]), and procrustes analysis (Gower and Dijksterhuis [2004] and Seber [1984: 252]).

24.21. Let $\mathbf{A}$ and $\mathbf{B}$ be $m \times n$ real matrices, $p = \min\{m, n\}$, and let $\mathcal{T}_k$ be the set of $k \times k$ real orthogonal matrices. Then, if $\| \cdot \|_F$ is the Frobenius norm,

$$\|\mathbf{A} - \mathbf{U}\mathbf{B}\mathbf{V}\|_F^2 = \|\mathbf{A}\|_F^2 - 2\,\text{trace}(\mathbf{A}\mathbf{V}'\mathbf{B}'\mathbf{U}') + \|\mathbf{B}\|_F^2$$

and

$$\min_{\mathbf{U} \in \mathcal{T}_m, \mathbf{V} \in \mathcal{T}_n} \|\mathbf{A} - \mathbf{U}\mathbf{B}\mathbf{V}\|_F = \left\{ \sum_{i=1}^{p} [\sigma_i(\mathbf{A}) - \sigma_i(\mathbf{B})]^2 \right\}^{1/2}.$$

The minimizing values of $\mathbf{U}$ and $\mathbf{V}$ follow from (24.16) above with appropriate substitutions.

24.22. Let $\mathbf{A}$ and $\mathbf{B}$ be $m \times n$ real matrices with respective singular value decompositions $\mathbf{A} = \mathbf{P}\boldsymbol{\Sigma}_1 \mathbf{Q}'$ and $\mathbf{B} = \mathbf{R}\boldsymbol{\Sigma}_2 \mathbf{S}'$. Let $\mathcal{T}_p$ be the set of all $p \times p$ real orthogonal matrices. Then, if $\| \cdot \|_F$ is the Frobenius norm,

$$\min_{\mathbf{U} \in \mathcal{T}_m, \mathbf{T} \in \mathcal{T}_n} \|\mathbf{U}\mathbf{A} - \mathbf{B}\mathbf{T}\|_F = \|\mathbf{U}_0\mathbf{A} - \mathbf{B}\mathbf{T}_0\|_F,$$

where $\mathbf{U}_0 = \mathbf{R}\mathbf{P}'$ and $\mathbf{T}_0 = \mathbf{S}\mathbf{Q}'$.

24.23. If $\mathbf{A}$ is an $n \times n$ symmetric matrix, $\mathbf{Q}_r$ is an $n \times r$ matrix with orthonormal columns, and $\mathbf{S}$ is any $r \times r$ matrix, then

$$\min_{\mathbf{S}} \|\mathbf{A}\mathbf{Q}_r - \mathbf{Q}_r\mathbf{S}\|_F = \|(\mathbf{I}_n - \mathbf{Q}_r\mathbf{Q}_r')\mathbf{A}\mathbf{Q}_r\|_F$$

and $\mathbf{S} = \mathbf{Q}_r'\mathbf{A}\mathbf{Q}_r$ is the minimizer.

24.24. We now find nearest approximations for several matrices.

(a) (Symmetric Matrix)   Let $\mathbf{A}$ be $n \times n$ real matrix, and define $\mathbf{B} = \frac{1}{2}(\mathbf{A} + \mathbf{A}')$. Then $\mathbf{B}$ is a symmetric matrix closest to $\mathbf{A}$ with respect to any orthogonally invariant norm $\| \cdot \|_{oi}$. Thus if $\mathbf{C}$ is any $n \times n$ real symmetric matrix, then

$$\|\mathbf{A} - \mathbf{B}\|_{oi} \leq \|\mathbf{A} - \mathbf{C}\|_{oi}.$$

(b) (Skew-Symmetric Matrix)   Referring to (a), if we now have $\mathbf{B} = \frac{1}{2}(\mathbf{A} - \mathbf{A}')$, then $\mathbf{B}$ is a skew-symmetric matrix closest to $\mathbf{A}$ with respect to any orthogonally invariant norm.

(c) (Orthogonal Matrix)   Let $\mathbf{A}$ be a real $n \times n$ matrix with singular value decomposition $\mathbf{P}\mathbf{\Sigma}\mathbf{Q}'$, and let $\mathcal{T}_n$ be the set of $n \times n$ real orthogonal matrices. Then, if $\| \cdot \|_F$ is the Frobenius norm,

$$\min_{\mathbf{T} \in \mathcal{T}_n} \|\mathbf{A} - \mathbf{T}\|_F = \|\mathbf{A} - \mathbf{T}_0\|_F,$$

with $\mathbf{T}_0 = \mathbf{P}\mathbf{Q}'$.

(d) (Non-negative Definite Matrix)   Let $\mathbf{A}$ be a real $n \times n$ matrix, $\mathbf{B} = \frac{1}{2}(\mathbf{A} + \mathbf{A}')$, and $\mathbf{B} = \mathbf{Q}\mathbf{H}$ be a polar decomposition with $\mathbf{Q}$ orthogonal and $\mathbf{H}$ non-negative definite. If $\mathcal{N}$ is the set of all non-negative definite matrices, then

$$\min_{\mathbf{C} \in \mathcal{N}} \|\mathbf{A} - \mathbf{C}\|_F = \|\mathbf{A} - \mathbf{C}_0\|_F,$$

where $\mathbf{C}_0 = \frac{1}{2}(\mathbf{B} + \mathbf{H})$ is non-negative definite and unique. Rao and Rao [1998: Sections 11.6 and 11.7] give a number of approximations like the above based on the M, N-invariant generalized matrix norm.

24.25. Suppose $\mathbf{X} = (\mathbf{x}_1, \ldots, \mathbf{x}_n)$ is an $m \times n$ matrix, $\mathbf{B} = (\mathbf{b}_1, \ldots, \mathbf{b}_k)$ is an $m \times k$ matrix of rank k $(k \leq m)$, $\mathbf{Z} = (\mathbf{z}_1, \ldots, \mathbf{z}_n)$ is $k \times n$ matrix, and $\mathbf{a} \in \mathbb{R}^m$. If $\| \cdot \|_{ui}$ is any unitarily invariant norm, then $\|\mathbf{X} - \mathbf{a}\mathbf{1}_n' - \mathbf{B}\mathbf{Z}\|_{ui}$ is minimized with respect to $\mathbf{a}$, $\mathbf{B}$, and $\mathbf{Z}$ when

$$\mathbf{b} = n^{-1}\mathbf{X}\mathbf{1}_n \ (= \bar{\mathbf{x}}, \text{ say}), \quad \mathbf{B} = (\mathbf{u}_1, \ldots, \mathbf{u}_k), \quad \text{and} \quad \mathbf{Z}' = (\sigma_1 \mathbf{v}_1, \ldots, \sigma_k \mathbf{v}_k),$$

where σ_i, $\mathbf{u}_i$, and $\mathbf{v}_i$ are defined by the singular value decomposition

$$\mathbf{X} - \bar{\mathbf{x}}\mathbf{1}_n' = \mathbf{U}\mathbf{\Sigma}\mathbf{V}' = \sigma_1\mathbf{u}_1\mathbf{v}_1' + \cdots + \sigma_r\mathbf{u}_r\mathbf{v}_r'.$$

In particular,

$$\min_{\mathbf{a},\mathbf{B},\mathbf{Z}} \|\mathbf{X} - \mathbf{a}\mathbf{1}_n' - \mathbf{B}\mathbf{Z}\|_F = \sigma_{k+1}^2 + \cdots + \sigma_r^2.$$

This problem is related to that of finding a hyperplane that is "nearest" to a set of points (Rao and Rao [1998: 399–400]).

Proofs. Section 24.4.2.

24.20a. Rao and Rao [1998: 392].

24.20c. Harville [1997: 556–559] and Seber [1984: 206–207].

24.21. Horn and Johnson [1985: 435–436, complex case].

24.22. Rao and Rao [1998: 389].

24.23. Golub and Van Loan [1996: 401].

24.24a. Rao and Rao [1998: 388].

24.24c. Quoted by Rao and Rao [1998: 393].

24.24d. Rao and Rao [1998: 389–391].

24.4.3 Quadratics

24.26. Suppose $q(\mathbf{x}) = \mathbf{x}'\mathbf{A}\mathbf{x} + \mathbf{b}'\mathbf{x} + c$, where $\mathbf{A}$ is a real symmetric matrix. Then:

(a) $q(\mathbf{x}) = (\mathbf{x} + \frac{1}{2}\mathbf{A}^-\mathbf{b})'\mathbf{A}(\mathbf{x} + \frac{1}{2}\mathbf{A}^-\mathbf{b}) + (c - \frac{1}{4}\mathbf{b}'\mathbf{A}^-\mathbf{b})$.

(b) $q(\mathbf{x})$ has a maximum if and only if $\mathbf{b} \in \mathcal{C}(\mathbf{A})$ and $-\mathbf{A}$ is non-negative definite. If such is the case, a maximizer of $q(\mathbf{x})$ is of the form

$$\mathbf{x}_{\max} = -\tfrac{1}{2}\mathbf{A}^-\mathbf{b} + (\mathbf{I} - \mathbf{A}^-\mathbf{A})\mathbf{x}_0,$$

where $\mathbf{x}_0$ is arbitrary.

(c) $q(\mathbf{x})$ has a minimum if and only if $\mathbf{b} \in \mathcal{C}(\mathbf{A})$ and $\mathbf{A}$ is non-negative definite. If such is the case, a minimizer of $q(\mathbf{x})$ is of the same form

$$\mathbf{x}_{\min} = -\tfrac{1}{2}\mathbf{A}^-\mathbf{b} + (\mathbf{I} - \mathbf{A}^-\mathbf{A})\mathbf{x}_0,$$

where $\mathbf{x}_0$ is arbitrary.

24.27. Suppose $\mathbf{x}, \mathbf{a} \in \mathbb{R}^m$, $\mathbf{c} \in \mathbb{R}^k$, and $\mathbf{B}$ is $m \times k$ of rank k. If $\mathbf{\Sigma}$ is a positive definite $m \times m$ matrix, then

$$\min_{\mathbf{c}}(\mathbf{x} - \mathbf{a} - \mathbf{B}\mathbf{c})'\mathbf{\Sigma}^{-1}(\mathbf{x} - \mathbf{a} - \mathbf{B}\mathbf{c}) = (\mathbf{x} - \mathbf{a})'(\mathbf{\Sigma}^{-1} - \mathbf{\Sigma}^{-1}\mathbf{P_B})(\mathbf{x} - \mathbf{a}),$$

occurs at

$$\mathbf{c} = (\mathbf{B}'\mathbf{\Sigma}\mathbf{B})^{-1}\mathbf{B}'\mathbf{\Sigma}^{-1}(\mathbf{x} - \mathbf{a}),$$

where

$$\mathbf{P_B} = \mathbf{B}(\mathbf{B}'\mathbf{\Sigma}^{-1}\mathbf{B})^{-1}\mathbf{B}'\mathbf{\Sigma}^{-1}.$$

If we now have $\mathbf{x}_i$, $\mathbf{c}_i$, and $w_i > 0$ for $i = 1, 2, \ldots, n$, then

$$\min_{\mathbf{a}, \mathbf{B}, \mathbf{c}_1, \ldots, \mathbf{c}_n} \sum_{i=1}^{n} w_i(\mathbf{x}_i - \mathbf{a} - \mathbf{B}\mathbf{c}_i)'\mathbf{\Sigma}^{-1}(\mathbf{x}_i - \mathbf{a} - \mathbf{B}\mathbf{c}_i)$$

$$= \sum_{i=1}^{n} w_i(\mathbf{x}_i - \overline{\mathbf{x}})'\mathbf{\Sigma}^{-1}(\mathbf{x}_i - \overline{\mathbf{x}}) - \operatorname{trace}(\mathbf{\Sigma}^{-1}\mathbf{P}_{\mathbf{B}_0}\mathbf{S}),$$

where $\mathbf{S} = \sum_{i=1}^{n} w_i(\mathbf{x}_i - \overline{\mathbf{x}})'\mathbf{\Sigma}^{-1}(\mathbf{x}_i - \overline{\mathbf{x}})$, $\mathbf{B}_0 = \mathbf{\Sigma}^{1/2}\mathbf{Q}_*$, and $\mathbf{Q}_*$ is the matrix of the first k eigenvectors of $\mathbf{\Sigma}^{-1/2}\mathbf{S}\mathbf{\Sigma}^{-1/2}$.

24.28. Let $\mathbf{A}$ be a real symmetric matrix.

(a) If $r(\mathbf{x}) = \mathbf{x}'\mathbf{A}\mathbf{x}/\mathbf{x}'\mathbf{x}$, then by differentiation the stationary values of $r(\mathbf{x})$ occur when $\mathbf{x}$ is an eigenvector of $\mathbf{A}$ and are equal to the eigenvalues of $\mathbf{A}$. Note that we can set $\mathbf{x}'\mathbf{x} = 1$ without any change in the result. We then have that $r(\mathbf{x})$ is maximized with respect to $\mathbf{x}$ when $\mathbf{x}$ is a unit-norm eigenvector of $\mathbf{A}$ corresponding to its largest eigenvalue. The minimum relates to the minimum eigenvalue (see 6.58a). If we also have $\mathbf{C}'\mathbf{x} = \mathbf{0}$, where $\mathbf{C}$ is $n \times p$ ($p \leq n$), then Golub and Van Loan [1996: 621] give a method for finding the stationary values of $r(\mathbf{x})$ subject to this constraint.

(b) Suppose, in addition to $\mathbf{C}'\mathbf{x} = \mathbf{0}$, we also have $\mathbf{x}'\mathbf{B}\mathbf{x} = 1$, where $\mathbf{B}$ is positive definite.

   (i) The stationary values of $\mathbf{x}'\mathbf{A}\mathbf{x}$ subject to these constraints are attained at the eigenvectors of $\mathbf{B}^{-1}(\mathbf{I}_n - \mathbf{P})\mathbf{A}$, where $\mathbf{P}$ is the projection matrix

$$\mathbf{P} = \mathbf{C}(\mathbf{C}'\mathbf{B}^{-1}\mathbf{C})^{-}\mathbf{C}'\mathbf{B}^{-1},$$

   that is, $\mathbf{x}$ satisfies $(\mathbf{I}_n - \mathbf{P})\mathbf{A}\mathbf{x} = \lambda\mathbf{B}\mathbf{x}$. Setting $\mathbf{B} = \mathbf{I}_n$ gives a solution to the second part of (a).

   (ii) If $\mathbf{A} = \mathbf{a}\mathbf{a}'$, then $\mathbf{x}'\mathbf{A}\mathbf{x}$ has a maximum value when $\mathbf{x} \propto \mathbf{B}^{-1}(\mathbf{I}_n - \mathbf{P})\mathbf{a}$. This result occurs in problems of genetic selection. Rao and Rao [1998: 507] give references to two extensions of the above.

24.29. Let $\mathbf{A}$, $\mathbf{B}$, and $\mathbf{C}$ be $n \times n$ matrices with $\mathbf{A}$ and $\mathbf{B}$ positive definite and $\mathbf{C}$ symmetric. The stationary values of

$$\frac{\mathbf{x}'\mathbf{C}\mathbf{x}}{(\mathbf{x}'\mathbf{A}\mathbf{x})^{1/2}(\mathbf{x}'\mathbf{B}\mathbf{x})^{1/2}}$$

are $\lambda_i^2 \nu_i$ ($i = 1, 2, \ldots$), where λ_i and ν_i are solutions of the equations

$$\begin{aligned} 2\mathbf{C}\mathbf{x} &= \lambda(\mathbf{A} + \nu\mathbf{B})\mathbf{x}, \\ \nu &= \mathbf{x}'\mathbf{A}\mathbf{x}/(\mathbf{x}'\mathbf{B}\mathbf{x}). \end{aligned}$$

This result occurs in the study of canonical variates (Rao and Rao [1987]).

24.30. Let $\mathbf{A}$ be a positive definite $n \times n$ matrix, let $\mathbf{B}$ be an $n \times k$ matrix, and let $\mathbf{c}$ be a given $k \times 1$ vector.

(a) If $\mathbf{S}^{-}$ is any weak inverse of $\mathbf{B}'\mathbf{A}^{-1}\mathbf{B}$, then for $n \times 1$ $\mathbf{x}$,

$$\min_{\mathbf{B}'\mathbf{x}=\mathbf{c}} \mathbf{x}'\mathbf{A}\mathbf{x} = \mathbf{c}'\mathbf{S}^{-}\mathbf{c},$$

   where the minimum attained at $\mathbf{x} = \mathbf{A}^{-1}\mathbf{B}\mathbf{S}^{-}\mathbf{c}$.

(b) If rank $\mathbf{B} = k$, then from (a) we have

$$\min_{\mathbf{B}'\mathbf{x}=\mathbf{c}} \mathbf{x}'\mathbf{x} = \mathbf{c}'(\mathbf{B}'\mathbf{B})^{-1}\mathbf{c},$$

   where the minimum is attained at $\mathbf{x} = \mathbf{B}(\mathbf{B}'\mathbf{B})^{-1}\mathbf{c}$.

(c) Suppose $\mathbf{x}' = (\mathbf{x}_1', \mathbf{x}_2')$ and $(\mathbf{B}_1', \mathbf{B}_2')\begin{pmatrix} \mathbf{x}_1 \\ \mathbf{x}_2 \end{pmatrix} = \mathbf{c}$. If

$$\begin{pmatrix} \mathbf{B}_1'\mathbf{B}_1 & \mathbf{B}_2' \\ \mathbf{B}_2 & \mathbf{0} \end{pmatrix}^{-} = \begin{pmatrix} \mathbf{C}_1 & \mathbf{C}_2 \\ \mathbf{C}_3 & \mathbf{C}_4 \end{pmatrix},$$

then

$$\min_{\mathbf{B}'\mathbf{x}=\mathbf{c}} \mathbf{x}_1'\mathbf{x}_1 = \mathbf{c}'\mathbf{C}_1\mathbf{c},$$

where the minimum is attained at $\mathbf{x}_1 = \mathbf{B}_1\mathbf{C}_1'\mathbf{c}$.

(d) Suppose $\mathbf{A}$ is now non-negative definite and $\mathbf{c} \in \mathcal{C}(\mathbf{B}')$. Let

$$\begin{pmatrix} \mathbf{A} & \mathbf{B} \\ \mathbf{B}' & \mathbf{0} \end{pmatrix}^{-} = \begin{pmatrix} \mathbf{C}_1 & \mathbf{C}_2 \\ \mathbf{C}_3 & -\mathbf{C}_4 \end{pmatrix},$$

then

$$\min_{\mathbf{B}'\mathbf{x}=\mathbf{c}} \mathbf{x}'\mathbf{A}\mathbf{x} = \mathbf{c}'\mathbf{C}_4\mathbf{c}.$$

24.31. Let $0 < p_0 < 1$ be given, and let $\mathbf{p} = (p_1, p_2, \ldots, p_m)'$, where $0 < p_0 \leq p_i \leq 1$ for $i = 1, 2, \ldots, m$. Let $\mathcal{R}$ be the region

$$\mathcal{R} = \{\mathbf{p} : 0 < p_0 \leq p_i \leq 1, i = 1, \ldots, m, p_0 < 1\},$$

and define

$$f(\mathbf{p}) = \frac{1 + p_0 + \mathbf{1}_m'\mathbf{p}}{(1 + p_0^2 + \mathbf{p}'\mathbf{p})^{1/2}}.$$

Then

$$\max_{\mathbf{p} \in \mathcal{R}} f(\mathbf{p}) = \left(m + \frac{(1 + p_0)^2}{1 + p_0^2} \right)^{1/2},$$

and it occurs in the interior of $\mathcal{R}$ at the point given by

$$p_i = \frac{1 + p_0^2}{1 + p_0}, \quad i = 1, 2, \ldots, m.$$

The minimum occurs at one or more of the extreme points of $\mathcal{R}$. For a proof and further details see Thibaudeau and Styan [1985]. They point out that their above result applies to a measure of imbalance for experimental designs introduced by Chakabarti [1963].

Proofs. Section 24.4.3.

24.26. Sengupta and Jammalamadaka [2003: 49–50].

24.27. Rao and Rao [1998: 400–401].

24.28b(i). Rao and Rao [1998: 507]. Note that $\mathbf{x}'(\mathbf{I}_n - \mathbf{P})\mathbf{A}\mathbf{x} = \lambda$.

24.28b(ii). When $\mathbf{A} = \mathbf{a}\mathbf{a}'$, $\mathbf{a}$ is an eigenvector of $\mathbf{A}$. We then set $\lambda = \mathbf{a}'\mathbf{B}^{-1}(\mathbf{I}_n - \mathbf{P})\mathbf{a}$ in (i) and substitute for $\mathbf{x}$.

24.29. Rao and Rao [1998: 507–508].

24.30. Rao [1973a: 60–61].

24.5   OPTIMAL DESIGNS

In fitting the linear model $\mathbf{y} = \mathbf{X}\boldsymbol{\beta} + \boldsymbol{\epsilon}$, where $\mathbf{X}$ is $n \times p$ of rank p (cf. Section 20.7), we may wish to find the best design for minimizing some function of $\mathrm{var}(\widehat{\boldsymbol{\beta}}) = \sigma^2(\mathbf{X}'\mathbf{X})^{-1}$, where $\widehat{\boldsymbol{\beta}} = (\mathbf{X}'\mathbf{X})^{-1}\mathbf{X}'\mathbf{y}$, the least squares estimate of $\boldsymbol{\beta}$. Depending on which function is chosen, there are three main critera, namely:

(1) *A-optimality*: minimize $\mathrm{trace}[(\mathbf{X}'\mathbf{X})^{-1}]$.

(2) *E-optimality*: minimize the largest eigenvalue (i.e., the spectral radius $\rho[(\mathbf{X}'\mathbf{X})^{-1}]$) of $(\mathbf{X}'\mathbf{X})^{-1}$.

(3) *D-optimality*: minimize $\det[(\mathbf{X}'\mathbf{X})^{-1}]$ or maximize $\det(\mathbf{X}'\mathbf{X})$.

For general references to optimal designs see Atkinson and Donev [1992], Druilhet [2004], and Melas [2006].

24.32. (D-Optimality) This is probably the most commonly used criterion for two reasons. Firstly, when $\boldsymbol{\epsilon}$ in the above linear model is multivariate normal $N_n(\mathbf{0}, \sigma^2\mathbf{I}_n)$, the D-optimal design gives the smallest volume of the confidence ellipsoid for $\boldsymbol{\beta}$. Secondly, the computations are the simplest. To find the optimal $\mathbf{X}$ with a given number n rows from a set of N potential rows, one begins with an initial choice of n rows, for example at random, and then determines the effect on the determinant by exchanging a deleted row with a different row from the set of potential rows using a result like (15.13b). For further references and details see Gentle [1998: 190]

REFERENCES

Abadir, K. M. and Magnus, J. R. (2005). *Matrix Algebra.* Cambridge University Press, New York.

Agaian, S. S. (1985). *Hadamard Matrices and Their Applications.* Springer-Verlag, New York.

Anderson, T. W. (1955). The integral of a symmetric unimodal function over a symmetric convex set and some probability inequalities. *Proceedings of the American Mathematical Society,* **6**, 170–176.

Anderson, T. W. (1993). Non-normal multivariate distributions: Inference based on elliptically contoured distributions. In C. R. Rao (Ed.), *Multivariate Analysis: Future Directions,* Vol. 1, 1–24. North-Holland, Amsterdam.

Anderson, T. W. (1996). Some inequalities for symmetric convex sets with applications. *Annals of Statistics,* **24**, 753–762.

Anderson, T. W. (2003). *An Introduction to Multivariate Statistical Analysis,* 3rd ed. John Wiley, New York.

Anderson, T. W. and Styan, G. P. H. (1982). Cochran's theorem, rank additivity and tripotent matrices. In G. Kallianpur, P. R. Krishnaiah, and J. K. Ghosh (Eds.), *Statistics and Probability: Essays in Honor of C. R. Rao,* 1–23. North-Holland, New York.

Anderson, W. N., Jr. and Duffin, R. J. (1969). Series and parallel addition of matrices. *SIAM Journal of Applied Mathematics,* **26**, 576–594.

Ansley, C. F. (1985). Quick proofs of some regression theorems via the QR algorithm. *The American Statistician,* **39** (1), 57–59.

Atiqullah, M. (1962). The estimation of residual variance in quadratically balanced least squares problems and the robustness of the F-test. *Biometrika,* **49**, 83–91.

Atkinson, A. C. and Donev, A. N. (1992). *Optimum Experimental Designs.* Oxford University Press, Oxford, UK.

A Matrix Handbook for Statisticians. By George A. F. Seber
Copyright © 2008 John Wiley & Sons, Inc.

Bacharach, M. (1965). Estimating non-negative matrices from marginal data. *International Economic Review*, **6**, 294–310.

Baksalary, J. K. and Baksalary, O. M. (2004a). On linear combinations of generalized projectors. *Linear Algebra and Its Applications*, **388**, 17–24.

Baksalary, J. K. and Baksalary, O. M. (2004b). Nonsingularity of linear combinations of idempotent matrices. *Linear Algebra and Its Applications*, **388**, 25–29.

Baksalary, J. K. and Baksalary, O. M. (2004c). Relationships between generalized inverses of a matrix and generalized inverses of its rank-one-modifications. *Linear Algebra and Its Applications*, **388**, 31–44.

Baksalary, J. K. and Puntanen, S. (1990). Spectrum and trace invariance criterion and its statistical applications. *Linear Algebra and Its Applications*, **142**, 121–128.

Baksalary, J. K. and Puntanen, S. (1991). Generalized matrix versions of the Cauchy–Schwarz and Kantorivich inequalities. *Aequationes Mathematicae*, **41**, 103–110.

Baksalary, J. K. and Styan, G. P. H. (2002). Generalized inverses of partitioned matrices in Banachiewicz–Schur form. *Linear Algebra and Its Applications*, **354**, 41–47.

Baksalary, J. K., Baksalary, O. M., and Styan, G. P. H. (2002). Idempotency of linear combinations of an idempotent matrix and a tripotent matrix. *Linear Algebra and Its Applications*, **354**, 21–34.

Baksalary, J. K., Puntanen, S., and Styan, G. P. H. (1990). A property of the dispersion matrix of the best linear unbiased estimator in the general Gauss–Markov model. *Sankhyā Series A*, **52** (3), 279–296.

Ballantyne, C. S. (1978). Products of idempotent matrices. *Linear Algebra and Its Applications*, **19**, 81–86.

Bapat, R. B. (1990). Permanents in probability and statistics. *Linear Algebra and Its Applications*. **127**, 3–25.

Bapat, R. B. and Raghavan, T. E. S. (1997). *Nonnegative Matrices and Applications*. Cambridge University Press, Cambridge.

Bartholomew, D. J. (1987). *Latent Variable Models and Factor Analysis*. Oxford University Press, New York.

Basilevsky A. (1983). *Applied Matrix Algebra in the Statistical Sciences*. North-Holland, Elsevier Science Publishing, New York.

Bates, D. M. (1983). The derivative of the determinant of $X'X$ and its uses. *Technometrics*, **25**, 373–376.

Bates, D. M. and Watts, D. G. (1985). Multiresponse estimation with special application to linear systems of differential equations (with Discussion). *Technometrics*, **27**, 329–360.

Bates, D. M. and Watts, D. G. (1987). A generalized Gauss–Newton procedure for multiresponse parameter estimation. *SIAM Journal of Science and Statistical Computing*, **8**, 49–55.

Bates, D. M. and Watts, D. G. (1988). *Nonlinear Regression Analysis*. John Wiley, New York.

Bellman, R. (1970). *Introduction to Marix Analysis*, 2nd ed. McGraw-Hill, New York.

Bénasséni, J. (2002). A complementary proof of an eigenvalue property in correspondence analysis. *Linear Algebra and Its Applications*, **354**, 49–51.

Ben-Israel, A. and Greville, T. N. E. (2003). *Generalized Inverses: Theory and Applications*, 2nd ed. Springer-Verlag, New York.

Berkovitz, L. D. (2002). *Convexity and Optimization in R^n*. John Wiley, New York.

Berman, A. and Plemmons, R. J. (1994). *Non-negative Matrices in the Mathematical Sciences*. SIAM, Philadelphia. Originally published in 1979 by Academic Press.

Berman, A. and Shaked-Monderer, N. (2003). *Completely Positive Matrices.* World Scientific, Hackensack, NJ.

Bhatia, R. (1997). *Matrix Analysis.* Springer, New York.

Bhaumik, D. K. (1995). Majorization and D-optimality for complete block designs under correlations. *Journal of the Royal Statistical Society, Series B,* **57**, 139–143.

Bini, D., Tyrtyshnikov, E., and Yalamov, P. (2001). *Structured Matrices: Recent Developments in Theory and Computation.* Nova Science Publishers, Inc., New York.

Birkhoff, G. and Varga, R. S. (1958). Reactor criticality and non-negative matrices. *Journal of the Society for Industrial and Applied Mathematics,* **6**, 354–377.

Bloomfield, P. and Watson, G. S. (1975). The inefficiency of least squares. *Biometrika,* **62**, 121–128.

Boshnakov, G. N. (2002). Multi-companion matrices. *Linear Algebra and Its Applications,* **354**, 53–83.

Böttcher, A. and Silbermann, B. (1999). *Introduction to Large Truncated Toeplitz Matrices.* Springer, New York.

Boullion, T. L. and Odell, P. L. (1971). *Generalized Inverse Matrices.* John Wiley, New York.

Brillinger, D. R. (1975). *Times Series : Data Analysis and Theory.* Holt, Rinehart and Winston, New York.

Brown, P. J. (2002). Inverted Wishart distribution, generalized. In A. H. El-Shaarwi and W. W. Piegorsch (Eds), *Encyclopedia of Environmetrics.* John Wiley, New York

Bullen, P. S. (2003). *Handbook of Means and Their Inequalities.* Kluwer, Netherlands.

Calvert, B. and Seber, G. A. F. (1978). Minimisation of functions of a positive semidefinite matrix A subject to $AX = 0$. *Journal of Multivariate Analysis,* **8**, 274–281.

Campbell, S. L. and Meyer, C. D. (1979). *Generalized Inverses of Linear Transformations.* Pitman, London.

Carlitz, L. (1959). Some cyclotomic matrices. *Acta Arithmetica* **5**, 293–308.

Carmeli, M. (1983). *Statistical Theory and Random Matrices.* Marcel Dekker, New York.

Carroll, R. J. and Ruppert, D. (1988). *Transformation and Weighting in Regression.* Chapman and Hall, London.

Carroll, R. J., Ruppert, D., and Stefanski, L. A. (2006), *Measurement Error in Nonlinear Models: A Modern Perspective,* 2nd ed. Chapman and Hall/CRC, Boca Raton, FL.

Caswell, H. (2001). *Matrix Population Models,* 2nd ed. Sinauer Associates, Sunderland, MA.

Caswell, H. (2006). Applications of Markov chains in demography. In A. N. Langville and W. J. Stewart (Eds), *MAM2006: Markov Anniversary Meeting,* 319-334. Boson Books, Raleigh, NC.

Caswell, H. (2007). Sensitivity analysis of transient population dynamics. *Ecology Letters,* **10**, in press.

Chakrabarti, M. C. (1963). On the C-matrix in design of experiments. *Journal of the Indian Statistical Association,* **1**, 8–23.

Ching, W.-K. (2006). *Markov Chains: Models, Algorithms and Applications.* Springer, New York.

Christensen, R. (1997). *Log-linear Models and Logistic Regression.* Springer, New York.

Christensen, R. (2001). *Advanced Linear Modeling: Multivariate, Times Series, and Spatial Data; Nonparametric Regression and Response Surface Maximization,* 2nd ed. Springer, New York.

Christensen, R. (2002). *Plane Answers to Complex Questions: The Theory of Linear Models,* 3rd ed. Springer, New York.

Chu, K. L., Isotalo, J., Puntanen, S., and Styan, G. P. H. (2004). On decomposing the Watson efficiency of ordinary least squares in a partitioned weakly singular model. *Sankhā: The Indian Jounal of Statistics*, **66** (4), 634–651.

Chu, K. L., Isotalo, J., Puntanen, S., and Styan, G. P. H. (2005a). Some further results concerning the decomposition of the Watson efficiency in partitioned linear models. *Sankhā: The Indian Jounal of Statistics*, **67** (1), 74–89.

Chu, K. L., Isotalo, J., Puntanen, S., and Styan, G. P. H. (2005b). The efficiency factorization multiplier for the Watson efficiency in partitioned linear models; Some examples and a literature review. *Research Letters in the Information and Mathematical Sciences*, **8**, 165–187. Available online at http://iims.massey.ac.nz/research/letters/.

Clarke, M. R. B. (1971). Algorithm AS41: Updating the sample mean and dispersion matrix. *Applied Statistics*, **20**, 206–209.

Cox, D. R. and Hinkley, D. V. (1974). *Theoretical Statistics*. Chapman and Hall, London.

Cullen, C. G. (1997). *Linear Algebra with Applications*, 2nd ed. Addison-Wesley, Reading, MA.

Dagum, E. B. and Luati, A. (2004). A linear transformation and its properties with special application in times series filtering. *Linear Algebra and Its Applications*, **388**, 107–117.

Darroch, J. N. and Silvey, S. D. (1963). On testing more than one hypothesis. *Annals of Mathematical Statistics*, **34**, 555–567.

Davis, P. J. (1979). *Circulant Matrices*. John Wiley, New York.

Debreu, G. and Herstein, I. N. (1953). Nonnegative square matrices. *Econometrica*, **21**, 597–607.

Deemer, W. L. and Olkin, I. (1951). The Jacobians of certain matrix transformations useful in multivariate analysis. *Biometrika*, **38**, 345–367.

Demetrius, L. (1971). Primitivity conditions for growth matrices. *Mathematical Biosciences*, **12**, 53–58.

Dobson, A. J. (2001). *An Introduction to Generalized Linear Models*, 2nd ed. Chapman and Hall, London.

Dhrymes, P. J. (2000). *Mathematics for Econometrics*, 3rd ed. Springer-Verlag, New York.

Dragomir, S. S. (2004). *Discrete Inequalities of the Cauchy-Bunyakovsky-Schwarz Type*. Nova Science Publishers, New York.

Dress, A. W. M. and Wenzel, W. (1995). A simple proof of an identity concerning pfaffians of skew symmetric matrices. *Advances in Mathematics*, **112**, 120–134.

Driscoll, M. F. and Krasnicka, B. (1995). An accessible proof of Craig's theorem in the general case. *The American Statistician*, **49** (1), 59–62.

Druilhet, P. (2004). Conditions for optimality in experimental designs. *Linear Algebra and Its Applications*, **388**, 147–157.

Drury, S. W., Liu, S., Lu, C.-Y., Puntanen, S., and Styan, G. P. H. (2002). Some comments on several matrix inequalities with applications to canonical correlations; historical background and recent developments. *Sankhā A*, **62**, 453–507.

Drygas, H. (1970). *The Coordinate Free Approach to Gauss-Markov Estimation. Lecture Notes in Operations Research and Mathematical Systems*, **40**. Springer-Verlag, Berlin.

Duff, I. S., Erisman, A. M., and Reid, J. K. (1986). *Direct Methods for Sparse Matrices*. Clarendon Press, Oxford.

Dümbgen, L. (1995). A simple proof and refinement of Wielandt's eigenvalues inequality. *Statistics & Probability Letters*, **25**, 113–115.

Dwyer, P. S. (1967). Some applications of matrix derivatives in multivariate analysis. *Journal of the American Statistical Association*, **62**, 607–625.

Eaton, M. L. and Perlman, M. D. (1973). The non-singularity of generalized sample covariance matrices. *Annals of Statistics*, **1**, 710–717.

Eaton, M. L. and Tyler, D. E. (1991). On Wielandt's inequality and its application to the asymptotic distribution of the eigenvalues of a random symmetric matrix. *Annals of Statistics*, **19**, 260–271.

Edelman, A. (1997). The probability that a random real Gaussian matrix has k real eigenvalues, related distributions and the circular law. *Journal of Multivariate Analysis*, **60**, 203-232.

Elsner, L. (1982)). On the variation of the spectra of matrices. *Linear Algebra and Its Applications*, **7**, 127–138.

Esary, J. D., Proschan, F., and Walkup, D. W. (1967). Associated random variables, with applications. *The Annals of Mathematical Statistics*, **38**, 1466–1474.

Escobar, L. A. and Moser, E. B. (1993). A note on the updating of regression estimates. *The American Statistician*, **47**, 192–194.

Eubank, R. L. and Webster, J. T. (1985). The singular-value decomposition as a tool for solving estimability problems. *The American Statistician*, **39** (1), 64–66.

Everitt, B. S. (1984). *An Introduction to Latent Variable Models*. Chapman and Hall, London.

Everitt, B. S. (1993). *Cluster Analysis*, 3rd ed. Hodder and Stoughton, London.

Fan, K. (1949). On a theorem of Weyl concerning eigenvalues of linear transformations I. *Proceedings of the National Academy of Sciences of the USA*, **35**, 652–655.

Fan, K. (1951). Maximum properties and inequalities for eigenvalues of completely continuous operators. *Proceedings of the National Academy of Sciences of the USA*, **37**, 760–766.

Fang, K.-T. and Anderson, T. W. (Eds.) (1990). *Statistical Inference in Elliptically Contoured and Related Distributions*. Allerton Press, New York.

Fang, K.-T., Kotz, S., and Ng, K.-W. (1990). *Symmetric Multivariate and Related Distributions*. Chapman and Hall, New York.

Faraway, J. J. (2006). *Extending the Linear Model with R: Generalized Linear, Mixed Effects and Nonparametric Models*. Chapman and Hall/CRC, FA, USA.

Farebrother, W. (1976). Further results on the mean square error of ridge regression. *Journal of the Royal Statistical Society B*, **38**, 248–250.

Farnum, N. R. (1989). An alternative proof of Samuelson's inequality and its extensions. *The American Statistician*, **43** (1), 46–47.

Flury, B. (1988). *Common Principal Components and Related Multivariate models*. John Wiley, New York.

Friedberg, S. H., Insel, A. J., and Spence, L. E. (2003). *Linear Algebra*, 3rd ed. Pearson Education, Upper Saddle River, NJ.

Galambos, J. and Simonelli, I. (1996). *Bonferroni-type Inequalities with Applications*. Springer-Verlag, New York.

Gantmacher, F. R. (1959). *The Theory of Matrices*, Vol. 1. Chelsea Publishing Co., New York.

Gentle, J. E. (1998). *Numerical Linear Algebra for Applications in Statistics*. Springer, New York.

Girko, V. L. and Gupta, A. K. (1996). Multivariate elliptically contoured linear models and some aspects of the theory of random matrices. In A. K. Gupta and V. C. Girko (Eds.), *Proceedings of the Sixth Lukacs Symposium*, Analysis and Theory of Random Matrices, 327–386. VSL, The Netherlands.

Golub, G. H and Van Loan, C. F. (1996). *Matrix Computations*, 3rd ed. John Hopkins University Press, Baltimore.

534 REFERENCES

Golyandina, N., Nekrutkin, V., and Zhigljavsky, A. (2001). *Analysis of Time Series Structure: SSA and Related Techniques*. Monographs on Statistics and Applied Probability, **90**. Chapman and Hall/CRC, Boca Raton, FL.

Goodman, N. R. (1963). Statistical analysis based upon a certain multivariate complex Gaussian distribution (an introduction). *Annals of Mathematical Statistics*, **34**, 152–177.

Gordon, A. D. (1999). *Classification*, 2nd ed. Chapman and Hall, Boca Raton, FL.

Gower, J. C. and Dijksterhuis, G. B. (2004). *Procrustes Problems*. Oxford Statistical Science Series **30**, Oxford University Press.

Graham, A. (1981). *Kronecker Products and Matrix Calculus*. Ellis Horwood, Chichester, UK.

Graham, A. (1987). *Nonnegative Matrices and Applicable Topics in Linear Algebra*. Ellis Horwood, Chichester, UK.

Graybill, F. A. (1983). *Matrices With Applications in Statistics*, 2nd ed. Wadsworth, Belmont, CA.

Greenbaum, A. (2007). Iterative solution methods for linear systems. In L. Hogben (Ed.), *Handbook of Linear Algebra*, Chapter 41. Chapman and Hall/CRC, Taylor and Francis Group, Boca Raton, FL.

Greenberg, B. G. and Sarhan, A. E. (1959). Matrix inversion, its interest and application in data analysis. *Journal of the American Statistical Association*, **54**, 755–766.

Grenander, U. and Szegő, G. (1958). *Toeplitz Forms and their Applications*. University of California Press, Berkeley, CA.

Greub, W. and Rheinboldt, W. (1959). On a generalization of an inequality of L. V. Kantorovich. *Proceedings of the American Mathematical Society*, **10**, 407–415.

Groß, J. (1999). Solution to a rank equation. *Linear Algebra and Its Applications*, **289**, 127–130.

Groß, J. (2000). The Moore–Penrose inverse of a partitioned nonnegative definite matrix. *Linear Algebra and Its applications*, **321**, 113–121.

Groß, J. (2003). *Linear Regression*. Springer, New York.

Groß, J. (2004). The general Gauss-Markov model with possibly singular covariance matrix. *Statistical Papers*, **45**, 311–336.

Groves, T. and Rothenberg, T. (1969). A note on the expected value of an inverse matrix. *Biometrika*, **56**, 690–691.

Gupta, A. K. and Varga, R. (1993). *Elliptically Contoured Models in Statistics*. Kluwer Academic, Dordrecht.

Gustafson, K. E. (1968). The angle of an operator and positive operator products. *Bulletin of the American Mathematical Society*, **74**, 488–492.

Gustafson, K. E. (2000). An extended operator trigonometry. *Linear Algebra and Its Applications*, **319**, 117–135.

Gustafson, K. E. (2002). Operator trigonometry of statistics and economics. *Linear Algebra and its Applications*, **354**, 141–158.

Gustafson, K. E. (2005). The geometry of statistical efficiency. *Research Letters in the Information and Mathematical Sciences*, **8**, 105–121. Available online at http://iims.massey.ac.nz/research/letters/.

Gustafson, K. E. and Rao, D. K. M. (1997). *Numerical Range: The Field of Values of Linear Operators and Matrices*. Springer, New York.

Guttman, L. (1955). A generalized simplex for factor analysis. *Psychometrika*, **20**, 173–192.

Hager, W. W. (1989). Updating the inverse of a matrix. *SIAM REview*, **11** (2), 221–239.

Halmos, P. R. (1958). *Finite-Dimensional Vector Spaces.* Van Nostrand, New York.

Halton, J. H. (1966a). A combinatorial proof of Cayley's theorem on Pfaffians. *Journal of Combinatorial Theory,* **1**, 224–232.

Halton, J. H. (1966b). An identity of the Jacobi type for Pfaffians. *Journal of Combinatorial Theory,* **1**, 333–337.

Hanson, A. T. (2006). *Visualizing Quaternions.* Elsevier, Morgan Kaufmann Publishers, San Francisco, CA.

Hardy, G. H., Littlewood, J. E., and Pólya, G. (1952). *Inequalities.* Cambridge University Press, Cambridge, UK.

Harville, D. A. (1997). *Matrix Algebra from a Statistician's Perspective.* Springer, New York.

Harville, D. A. (2001). *Matrix Algebra: Exercises and Solutions.* Springer-Verlag, New York.

Henderson H. V. and Searle, S. R. (1979). Vec and vech operators for matrices, with some uses in Jacobians and multivariate statistics. *Canadian Journal of Statistics,* **7**, 65–81.

Henderson H. V. and Searle, S. R. (1981a). The vec-permutation matrix, the vec operator and Kronecker products. *Linear and Multilinear Algebra,* **9**, 271–288.

Henderson H. V. and Searle, S. R. (1981b). On deriving the inverse of a sum of matrices. *SIAM Review,* **23**, 53–60.

Henderson H. V., Pukelsheim, F., and Searle, S. R. (1983). On the history of the Konecker product. *Linear and Multilinear Algebra,* **14**, 113–120.

Hernández-Lerma, O. and Lasserre, J. B. (2003). *Markov Chains and Invariant Probabilities.* Birkhäuser, Boston.

Hochberg, Y. and Tamhane, A. C. (1987). *Multiple Comparison Procedures.* John Wiley, New York.

Hoeffding, W. (1963). Probability inequalities for sums of bounded random variables. *Journal of the American Statistical Association,* **58**, 13–30.

Hoppensteadt, F., Salehi, H., and Skorokhod, A. (1996). On the asymptotic behaviour of Markov chains with small random perturbations of transition probabilities. In A. K. Gupta and V. C. Girko (Eds.), *Proceedings of the Sixth Lukacs Symposium, Analysis and Theory of Random Matrices,* 93–100. VSL, The Netherlands.

Horn R. and Johnson, C.A. (1985). *Matrix Analysis.* Cambridge University Press, Cambridge, UK. Corrected reprint edition (1990), which I was unaware of in time to incorporate any changes there may have been in the theorem statements.

Horn R. and Johnson, C.A. (1991). *Topics in Matrix Analysis.* Cambridge University Press, Cambridge, UK.

Hotelling, H. (1933). Analysis of a complex of statistical variables into principal components. *Journal of Educational Psychology,* **24**, 417–441.

Hsu, J. C. (1996). *Multiple Comparisons: Theory and Methods.* Chapman and Hall, London.

Hsu, P. L. (1953). On symmetric, orthogonal and skew-symmetric matrices. *Proceedings of the Royal Society of Edinburgh,* **10**, 37–44.

Hunter, C. M. and Caswell, H. (2005). The use of the vec-permutation matrix in spatial matrix population models. *Ecological Modelling,* **188**, 15-21.

Hunter, J. J. (1983a). *Mathematical Techniques of Applied Probability,* Vol. 1. Academic Press, New York.

Hunter, J. J. (1983b). *Mathematical Techniques of Applied Probability,* Vol. 2. Academic Press, New York.

Hunter, J. J. (1988). Characterizations of generalized inverses associated with Markovian kernels. *Linear Algebra and Its Applications,* **102**, 121–142.

Hunter, J. J. (1990). Parametric forms for generalized inverses of Markovian kernels and their applications. *Linear Algebra and Its Applications*, **127**, 71–84.

Hunter, J. J. (1991). The computation of stationary distributions of Markov chains through perturbations. *Journal of Applied Mathematics and Stochastic Analysis*, **4** (1), 29–46.

Hunter, J. J. (1992). Stationary distributions and mean first passage times in Markov chains using generalised inverses. *Asia-Pacific Journal of Operational Research*, **9**, 145–153.

Hunter, J. J. (2005). Stationary distributions and mean first passage times of perturbed Markov chains. *Linear Algebra and Its Applications*, **410**, 217–243.

Isotalo, J., Puntanen, S., and Styan, G. P. H. (2005a). *Formulas Useful for Linear Regression Analysis and Related Matrix Theory*, 3rd ed. Report A 350, Dept. of Mathematics, Statistics, and Philosophy, University of Tampere, Tampere, Finland, 85 pp.

Isotalo, J., Puntanen, S., and Styan, G. P. H. (2005b). *Matrix Tricks for Linear Statistical Models: Our Personal Top Sixteen*, 3rd ed. Report A 363, Dept. of Mathematics, Statistics, and Philosophy, University of Tampere, Tampere, Finland, 207 pp.

Jeffrey, D. J. and Corless, R. M. (2007). Linear algebra in Maple. In L. Hogben (Ed.), *Handbook of Linear Algebra*, Chapter 72. Chapman and Hall/CRC, Taylor and Francis Group, Boca Raton, FA.

Jewell, N. P. and Bloomfield, P. (1983). Canonical correlations of past and future for times series. *Annals of Statistics*, **11**, 837–847.

John, J. A. (2001). Updating formula in an analysis of variance model. *Biometrika*, **88**, 1175–1178.

John, J. A. and Williams, E. R. (1995). *Cyclic and Computer Generated Designs*, 2nd ed. Monographs on Statistics and Applied probability, **38**. Chapman and Hall, London, UK.

Jolliffe, I. T. (2002). *Principal Component Analysis*, 2nd ed. Springer, New York.

Kabe, D. G. (1980). On extensions of Samuelson's inequality (Letter to the Editor). *The American Statistician*, **34** (4), 249.

Kantor, I. L. and Solodovnikov, A. S. (1989). *Hypercomplex Numbers: An Elementary Introduction to Algebras*. Springer–Verlag, Berlin.

Kariya, T. and Kurata, H. (2004). *Generalized Least Squares*. Wiley, New York.

Kariya, T. and Sinha, B. K. (1989). *Robustness of Statistical Tests*. Academic Press, Boston.

Kaufman, L. and Rousseeuw, P. J. (1990). *Finding Groups in Data: An Introduction to Cluster Analysis*. John Wiley, New York.

Kelly, P. J. and Weiss, M. L. (1979). *Geometry and Convexity*. John Wiley, New York.

Kempthorne, O. (1980). The term design matrix. *The American Statistician*, **34** (4), 249.

Khatri, C. G. and Rao, C. R. (1981). Some extensions of the Kantorovich inequality and statistical applications. *Journal of Multivariate Analysis*, **11**, 498–505.

Khatri, C. G. and Rao, C. R. (1982). Some generalizations of the Kantorovich inequality. *Sankhyā A*, **44**, 91–102.

Khattree, R. (1996). Multivariate statistical inferences involving circulant matrices. In A. K. Gupta and V. C. Girko (Eds.), *Proceedings of the Sixth Lukacs Symposium, Analysis and Theory of Random Matrices*, 101–110. VSL, The Netherlands.

Klein, T. (2004). Invariant symmetric block matrices for the design of mixture experiments. *Linear Algebra and Its Applications*, **388**, 261–278.

Knott, M. (1975). On the minimum efficiency of least squares. *Biometrika*, **62**, 129–132.

Koliha, J. J., Rakočević, V., and Straškraba, I. (2004). The difference and sum of projectors. *Linear Algebra and Its Applications*, **388**, 279–288.

Kollo, T. and Neudecker, H. (1993). Asymptotics of eigenvalues and unit-length eigen-vectors of sample variance and correlation matrices. *Journal of Multivariate Analysis*, **47**, 283–300. Corrigendum: 1994, **51**, 210.

Kollo, T. and Neudecker, H. (1997). Asymptotics of Pearson-Hotelling principal-component vectors of sample variance and correlation matrices. *Behaviormetrika*, **24** (1), 51–69.

Kollo, T. and von Rosen, D. (2005). *Advanced Multivariate Statistics with Matrices.* Springer, New York.

Kotz, S. and Nadarajah, S. (2004). *Multivariate t Distributions and Their Applications.* Cambridge University Press, New York.

Kounias, E. (1968). Bounds for the probability of a union of events, with applications. *Annals of Statistics*, **39**, 2154–2158.

Krishnaiah, P. R. (1976). Some recent developments on complex multivariate distribution. *Journal of Multivariate Analysis*, **6**, 1–30.

Krzanowski, W. J. (1988). *Principles of Multivariate Analysis.* Clarendon Press, Oxford, UK.

Lancaster, P. (1964). On eigenvalues of matrices dependent on a parameter. *Numerische Mathematik*, **6**, 337–346.

Latour, D., Puntanen, S., and Styan, G. P. H. (1987). Equalities and inequalities for the canonical correlations associated with some partitioned generalized inverses of a covariance matrix. *Proceedings of the Second International Tampere Seminar on Linear Statistical Models and Their Applications*, 541–553. Dept. of Mathematical Sciences, University of Tampere, Finland.

Lay, D. C. (2003). *Linear Algebra and Its Applications*, 3rd ed. Addison Wesley, Boston.

Lay, S. R. (1982). *Convex Sets and Their Applications.* John Wiley, New York.

Lee, A. J., Nyangoma, S. O., and Seber, G. A. F. (2002). Confidence regions for multino-mial parameters. *Computational Statistics & Data Analysis*, **39**, 329–342.

Leon, S. J. (2007). Matlab. In L. Hogben (Ed.), *Handbook of Linear Algebra*, Chapter 71. Chapman and Hall/CRC, Taylor and Francis Group, Boca Raton, FL.

Lidskiĭ, V. (1950). The proper values of the sum and product of symmetric matrices. *Dokl. Akad. Nauk. SSSR*, **75**, 769-772 (in Russian). (Translated by C. D. Benster, U.S. Department of Commerce, National Bureau of Standards, Washingon, D.C., N.B.S. Rep. 2248, 1953).

Liski, E. P. and Puntanen, S. (1989). A further note on a theorem on the difference of the generalized inverses of two non-negative definite matrices. *Communications in Statistics—Theory and Methods*, **18** (5), 1747–1751.

Liu, S. (2002a). Several inequalities involving Khatri–Rao products of positive semidefinite matrices. *Linear Algebra and Its Applications*, **354**, 175–186.

Liu, S. (2002b). Local influence in multivariate elliptical linear regression models. *Linear Algebra and Its Applications*, **354**, 159–174.

Liu, S. and Neudecker, H. (1996). Several matrix Kantorovich-type inequalities. *Journal of Mathematical Analysis and Applications*, **197**, 23–26.

McCullagh, P. and Nelder, J. A. (1989). *Generalized Linear Models*, 2nd ed. Chapman and Hall, New York.

McLachlan, G. J. (1992). *Discriminant Analysis and Statistical Pattern Recognition.* John Wiley, New York. Reproduced in paperback, Wiley-Interscience Series (2004), Hobo-ken, NJ.

MacRae, E. C. (1974). Matrix derivatives with an application to an adaptive decision problem. *Annals of Statistics*, **2** (2), 337-346.

Magnus, J. R. (1978). The moments of products of quadratic forms in normal variables. *Statistica Neerlandica*, **32**, 201–210.

Magnus, J. R. (1982). Multivariate error components analysis of linear and nonlinear regression models by maximum likelihood. *Journal of Econometrics*, **19**, 239–285.

Magnus, J. R. (1988). *Linear Structures*. Charles Griffin, London and Oxford University Press, New York.

Magnus, J. R. and Neudecker, H. (1979). The commutation matrix: Some properties and applications. *Annals of Statistics*, **7**, 381–394.

Magnus, J. R. and Neudecker, H. (1999). *Matrix Differential Calculus: With Applications in Statistics and Econometrics*, revised ed. John Wiley, Chichester, UK.

Maindonald, J. H. (1984). *Statistical Computation*. John Wiley, New York.

Mäkeläinen, T. (1970). Extrema for characterisic roots of product matrices. *Soc. Sci. Fenn. Commentationes Physico-Mathematicae*, **38** (4), 27–53.

Mäkeläinen, T., Schmidt, K., and Styan, G. P. H. (1981). On the existence and uniqueness of the maximum likelihood estimate of a vector-valued parameter in fixed-size samples. *Annals of Statistics*, **9**, 758–767.

Mandel, J. (1982). Use of the singular value decomposition in regression analysis. *The American Statistician*, **36** (1), 15–24.

Marcus, M. and Minc, H. (1964). *A Survey of Matrix theory and Matrix Inequalities*. Allyn and Bacon, Boston.

Marsaglia, G. and Styan, G. P. H. (1974a). Equalities and inequalities for ranks of matrices. *Linear and Multilinear Algebra*, **2**, 269–292.

Marsaglia, G. and Styan, G. P. H. (1974b). Rank conditions for generalized inverses of partitioned matrices. *Sankhyā*, **36**, 437–442.

Marshall, A. W. and Olkin, I. (1979). *Inequalities: Theory of Majorization and Its Applications*. Mathematics in Science and Engineering, **143**. Academic Press, New York.

Marshall, A. W. and Olkin, I. (1990). Matrix versions of the Cauchy and Kantorovich inequalities. *Aequationes Mathematicae*, **40**, 89–93.

Mathai, A. M. (1997). *Jacobians of Matrix Transformations and Functions of Matrix Argument*. World Scientific, Singapore.

Mathai, A. M. and Provost, S. B. (1992). *Quadratic Forms in Random Variables: Theory and Applications*. Marcel Dekker, New York.

Mathai, A. M. and Provost, S. B. (2005). Some complex matrix-variate statistical distributions on rectangular matrices. *Linear Algebra and Its Applications*, **410**, 198–216.

Mehta, M. L. (1989). *Matrix Theory*. Les Editions de Physique, 91944 Les Ulis Cedex, France.

Mehta, M. L. (2004). *Random Matrices*, 3rd ed. Elsevier Academic Press, New York.

Meijer, E. (2005). Matrix algebra for higher order moments. *Linear Algebra and Its Applications*, **410**, 112-134.

Melas, V. B. (2006). *Functional Approach to Optimal Design*. Springer, New York.

Merikoski, J. K. and Virtanen, A. (2004). The best possible lower bound for the Perron root using traces. *Linear Algebra and Its Applications*, **388**, 301–313.

Meyer, C. D. (1975). The role of the group generalized inverse in the theory of finite Markov chains. *SIAM Review*, **17** (3), 443–464.

Meyer, C. D. (2000). *Matrix Analysis and Applied Linear Algebra*. SIAM, Philadelphia. See www.matrixanalysis.com/Errata.html for a list of errata (which do not affect my book).

Miller, R. G. (1981). *Simultaneous Statistical Inference*, 2nd edit. McGraw-Hill, New York.

Minc, H. (1978). *Permanents*. Addison-Wesley, Reading, MA.

Minc, H. (1987). Theory of Permanents 1982–1985. *Linear and Multilinear Algebra*, **21**, 109–148.

Mitra, S. K. and Puntanen, S. (1991). The shorted operator and generalized inverses of matrices. *Calcutta Statistical Association Bulletin*, **40**, 97–102.

Mitra, S. K., Puntanen, S., and Styan, G. P. H. (1995). Shorted matrices and their applications in linear statistical models: A review. *Multivariate Statistics and Matrices in Statistics: Proceedings of the 5th Tartu Conference, Tartu-Pühajärve, Estonia, 23–28 May 1994.* In E.-M. Tiit, T. Kollo, and H. Niemi (Eds.), *New Trends in Probability and Statistics*, Vol. 3, 289–311. VSP, Utrecht, Netherlands, and TEV, Vilnius, Lithuania.

Mitra, S. K. and Puri, M. L. (1979). Shorted operators and generalized inverses of matrices. *Linear Algebra and Its Applications*, **25**, 45–56.

Moore, E. H (1935). General analysis. *Memoirs of the American Philosophical Society*, **1**, 147–209.

Mudholkar, G. S. (1965). A class of tests with monotone power functions for two problems in multivariate statistical analysis. *Annals of Mathematical Statistics*, **36**, 1794–1801.

Mudholkar, G. S. (1966). On confidence bounds associated with multivariate analysis and non-independence between two sets of variates. *Annals of Mathematical Statistics*, **37**, 1736–1746.

Muirhead, R. J. (1982) *Aspects of Multivariate Statistical Theory.* John Wiley, New York.

Mulholland, H. P. and Smith, C. A. B. (1959). An inequality arising in genetical theory. *American Mathematical Monthly*, **66**, 673–683.

Nel, D. G. (1980). On matrix differentiation in statistics. *South African Statistical Journal*, **14**, 137–193.

Nelder J. A. (1985). An alternative interpretation of the singular-value decomposition in regression. *The American Statistician*, **39** (1), 63–64.

Neudecker, H. (2003). On two matrix derivatives by Kollo and von Rosen. *SORT*, **27** (2), 153–164.

Neudecker, H. and Liu, S. (1994). Letters to the editor. *The American Statistician*, **48** (4), 351.

Neudecker, H. and Trenkler, G (2002). Third and fourth moment matrices of vec $\mathbf{X}'$ in multivariate analysis. *Linear Algebra and Its Applications*, **354**, 223–229.

Norris, J. R. (1997). *Markov Chains.* Cambridge University Press, New York.

Northcott, D. G. (1984). *Multilinear Algebra.* Cambridge University Press, Cambridge.

Noumann, M. and Xu, J. (2005). A parallel algorithm for computing the group inverse via Perron complementation. *Electronic Journal of Linear Algebra*, **13**, 131–145.

Okamoto, M. (1973). Distinctness of the eigenvalues of a quadratic form in a multivariate sample. *Annals of Statistics*, **1**, 763–765.

Okamoto, M. and Kanazawa, M. (1968). Minimization of eigenvalues of a matrix and optimality of principal components. *Annals of Mathematical Statistics*, **39**, 859–863.

Olkin, I. (1953). Note on the Jacobians of certain matrix transformations useful in multivariate analysis. *Biometrika*, **40**, 43–46.

Olkin, I. (2002). The 70th anniversary of the distribution of random matrices. *Linear Algebra and Its Applications*, **354**, 231–243.

Olkin, I. and Sampson, A. R. (1972). Jacobians of matrix transformations and induced functional equations. *Linear Algebra and Its Applications*, **5**, 257–276.

Ostrowski, A. M. (1973). *Solutions of Equations in Euclidean and Banach Spaces.* Academic Press, New York.

Ouellette, D. V. (1981). Schur complements and statistics. *Linear Algebra and Its Applications*, **36**, 187–195.

Pan, J.-X. and Fang, K.-T. (2002). *Growth Curve Models and Statistical Diagnostics.* Springer, New York.

Parameswaran, S. (1954). Skew-symmetric determinants. *The American Mathematical Monthly*, **61** (2), 116.

Parthsarathy, T. and Raghaven, T. E. S. (1971). *Some Topics in Two Person Games*. Elsevier, New York.

Penrose, R. A. (1955). A generalised inverse for matrices. *Proceedings of the Cambridge Philosophical Society*, **51**, 406–413.

Penrose, R. (1956). On best approximate solutions of linear matrix equations. *Proceedings of the Cambridge Philosophical Society*, **52**, 17–19.

Perlman, M. D. (1990). T.W. Anderson's theorem on the integral of a symmetric function over a symmetric convex set and its applications in probability and statistics. In G. P. H. Styan (Ed.), *The Collected Papers of T. W. Anderson, 1943–1985*, **2**, 1627–1641. John Wiley, New York.

Perlman, M. D. (1993). Concentration inequalities for multivariate distributions: II. Elliptically contoured distributions. In M. Shaked and Y. L. Tong (Eds.), *Stochastic Inequalities*, IMS Lecture Notes-Monograph Series, **22**, 284–308. Institute of Mathematical Statistics, Hayward, Californina.

Potthoff,, R. F. and Roy, S. N. (1964). A generalized multivariate analysis of variance model useful especially for growth curve models. *Biometrika*, **51**, 313–326.

Pronzato, L, Wynn, H. P., and Zhigljavsky, A. (2005). Kantorovich-type inequalities for operators via D-optimal design theory. *Linear Algebra and Its Applications*, **410**, 160–169.

Puntanen, S. and Styan, G. P. H. (1989). The equality of the ordinary least squares estimator and the best linear unbiased estimator (with comments by Oscar Kempthorne and by Shayle Searle, and with "Reply" by the authors). *The American Statistician*, **43** (3), 153–164. See also further comments in **44** (2), 191–192.

Puntanen, S. and Styan, G. P. H. (2005a). Historical introduction: Issai Schur and the early development of the Schur complement. In F. Zhang (Ed.), *The Schur Complement and Its Applications*, 1–16, Springer, New York.

Puntanen, S. and Styan, G. P. H. (2005b). Schur complements in statistics and probability. In F. Zhang (Ed.), *The Schur Complement and Its Applications*, 163–226. Springer, New York.

Puntanen, S. and Styan, G. P. H. (2006). Chapter 52: Random Vectors and Linear Statistical Models. In L. Hogben (Ed.), *Handbook of Linear Algebra*, 52.1–52.17. Taylor and Francis Group, Boca Raton, FL.

Puntanen, S., Styan, G. P. H., and Jensen, S. T. (1998). *Three Biblographies and a Guide: A third Guide on Books on Matrices and Books on Inequalities with Statistical and Other Applications*, The Seventh Inernational Workshop on Matrices and Statistics, in Celebration of T. W. Anderson's 80th Birthday, Fort Lauderdale, Florida, 27–100. Nova Southern University, Fort Lauderdale, FL.

Puntanen, S., Styan, G. P. H., and Werner, H. J. (2000). Two matrix-based proofs that the linear estimator Gy is the best linear unbiased estimator. *Journal of Statistical Planning and Inference*, **88**, 173–179.

Rao, A. R. and Bhimasankaram, P. (2000). *Linear Algebra*, 2nd ed. Hindustan Book Agency, New Delhi.

Rao, C. R. (1973a). *Linear Statistical Inference and Its Applications*, 2nd ed. John Wiley, New York.

Rao, C. R. (1973b). Representations of best linear unbiased estimators in the Gauss–Markoff model with a singular dispersion matrix. *Journal of Multivariate Analysis*, **3**, 276–292.

Rao, C. R. (1978). Choice of best linear estimators in the Gauss–Markoff model with a singular dispersion matrix. *Communications in Statistics–Theory and Methods*, **7**, 1199–1208.

Rao, C. R. (1980). Matrix approximations and reduction of dimensionality in multivariate statistical analysis. In P. R Krishnaiah (Ed.), *Multivariate Analysis–V*, 3–22. North-Holland Publishing Co.

Rao, C. R. (1981). A lemma on g-inverse of a matrix and computation of correlation coefficients in the singular case. *Communications in Statistics–Theory and Methods*, **10**, 1–10.

Rao, C. R. (1985). The inefficiency of least squares: Extension of Kantorovich inequality. *Linear Algebra and Its Applications*, **70**, 249–255.

Rao, C. R. (2000). Statistical proofs of some matrix inequalities. *Linear Algebra and Its Applications*, **321**, 307–320.

Rao, C. R. (2005). Antieigenvalues and antisingular values of a matrix and applications to problems in statistics. *Research Letters in the Information and Mathematical Sciences*, **8**, 53–76. Available online at http://iims.massey.ac.nz/research/letters/.

Rao, C. R. and Mitra, S. K. (1971). *Generalized Inverse of Matrices and its Applications*. John Wiley, New York.

Rao, C. R. and Rao, C. V. (1987). Stationary values of the product of two Raleigh coefficients: Homologous canonical variates. *Sankhyā B*, **49**, 113–125.

Rao, C. R. and Rao, M. B. (1998). *Matrix Alebra and Its Applications to Statistics and Econometrics*. World Scientific, Singapore.

Rao, C. R. and Toutenburg, H. (1999). *Linear models: Least Squares and Alternatives*. Springer-Verlag, New York.

Rao, J. N. K. and Scott, A. J. (1984). On chi-squared tests for multiway contingency tables with cell proportions estimated from survey data. *Annals of Statistics*, **12**, 46–60.

Rao, P. S. R. S. (1997). *Variance Components Estimation: Mixed Models, Methodologies and Applications*, Monographs on Statistics and Applied Probability, **48**. Chapman and Hall, London.

Richards, D. St. P. (1996). New classes of probability inequalities for some classical distributions. In A. K. Gupta and V. C. Girko (Eds.), *Proceedings of the Sixth Lukacs Symposium*, Analysis and Theory of Random Matrices, 217–224. VSL, The Netherlands.

Rockafellar, R. T. (1970). *Convex Analysis*. Princeton University Press, Princeton, NJ.

Rogers, G. S. (1980). *Matrix Derivatives*. Lecture Notes in Statistics, Vol. 2. Marcel Dekker, New York.

Rogers, G. S. (1984). Kronecker products in ANOV – a first step. *The American Statistician*, **38** (3), 197–202.

Rogers, G. S. and Young, D. L. (1978). On testing a multivariate linear hypothesis when the covariance matrix and its inverse have the same pattern. *Journal of the American Statistical Association*, **73**, 203–207.

Rothblum, U. G. (2007). Non-negative matrices and stochastic matrices. In L. Hogben (Ed.), *Handbook of Linear Algebra*, Chapter 9. Chapman and Hall/CRC, Taylor and Francis Group, Boca Raton, FL.

Roy, S. N. (1954). A useful theorem in matrix theory. *Proceedings of the American Mathematical Society*, **54** (4), 635–638.

Roy, S. N. and Sarhan, A. E. (1956). On inverting a class of patterned matrices. *Biometrika*, **43**, 227–231.

Ruskeepää, H. (2007). Mathematica. In L. Hogben (Ed.), *Handbook of Linear Algebra*, Chapter 73. Chapman and Hall/CRC, Taylor and Francis Group, Boca Raton, FL.

Ryan, T. P. (1996). *Modern Regression Methods*. John Wiley, New York.

Samuelson, P. A. (1968). How deviant can you be? *Journal of the American Statistical Association*, **63**, 1522–1525.

Scheffé, H. (1953). A method of judging all contrasts in analysis of variance. *Annals of Mathematical Statistics*, **40**, 87–104.

Schott, J. R. (2005). *Matrix Analysis for Statistics*, 2nd ed. John Wiley, New York.

Scott, A. J. and Styan, G. P. H. (1985). On a separation theorem for generalized eigenvalues and a problem in the analysis of sample surveys. *Linear Algebra and Its Applications*, **70**, 209–224.

Searle, S. R. (1971). *Linear Models*. John Wiley, New York. Reprinted in the Wiley Classics Library Reprint Edition (1997).

Searle, S. R. (1978). A univariate formulation of the multivariate linear model. In H. A. David (Ed.), *Contributions to Survey Sampling and Applied Statistics, Papers in Honor of H. O. Hartley*, 181-189. Academic Press, New York

Searle, S. R. (1982). *Matrix Algebra for Statistics*. John Wiley, New York.

Searle, S. R., McCulloch, C. E., and Casella, G. (1992). *Variance Components*. John Wiley, New York.

Seber, G. A. F. (1964). The linear hypothesis and large sample theory. *Annals of Mathematical Statistics*, **35**, 773–779.

Seber, G. A. F. (1967). Asymptotic linearisation of non-linear hypotheses. *Sankhyā, Series A*, **29** (2), 183–190.

Seber, G. A. F. (1977). *Linear Regression Analysis*. John Wiley, New York. For the second edition see Seber and Lee (2003) below.

Seber, G. A. F. (1980). *The Linear Hypothesis: A General Theory*, 2nd ed. Griffin's Statistical Monographs No. 19. Griffin, High Wycombe, UK.

Seber, G. A. F. (1982). *The Estimation of Animal Abundance and Related Parameters*, 2nd ed. Griffin, London. Also reprinted in paperback by Blackburn Press, Caldwell, NJ (2002).

Seber, G. A. F. (1984). *Multivariate Observations*. John Wiley, New York. Also reprinted in the Wiley Interscience Paperback Series (2004).

Seber, G. A. F. and Lee, A. J. (2003). *Linear Regression Analysis*, 2nd ed. John Wiley, New York.

Seber, G. A. F. and Nyangoma, S. O. (2000). Residuals for multinomial models. *Biometrika*, **87**, 183–191.

Seber, G. A. F. and Wild, C. J. (1989). *Nonlinear Regression*. Wiley: New York. Also reprinted in the Wiley Interscience Paperback Series (2003).

Seeley, J. F. (1971). Quadratic subspaces and completeness. *Annals of Mathematical Statistics*, **42**, 710–721.

Sen, P. K. and Singer, J. M. (1993). *Large Sample Methods in Statistics: An Introduction with Applications*. Chapman and Hall, New York.

Seneta, E. (1981). *Non-negative Matrices and Markov Chains*, 2nd ed. Springer–Verlag, New York.

Sengupta, D. and Jammalamadaka, S. R. (2003). *Linear Models: An Integrated Approach*. World Scientific, Singapore.

Shisha, O. and Mond, B. (1967). Bounds on the difference of means. *Inequalities: Proceedings of a Symposium held at Wright-Patterson Air Force Base, Ohio, August 19–27, 1965*, 293–308. Academic Press, New York.

Slapničar, I. (2007). Numerical methods for eigenvalues. In L. Hogben (ed.), *Handbook of Linear Algebra*, Chapter 42. Chapman and Hall/CRC, Taylor and Francis Group, Boca Raton, FL.

Smith, S. P. (2001). The factorability of symmetric matrices and some implications for statistical linear models. *Linear Algebra and Its Applications*, **335**, 63–80.

Srivastava, M. S. 1965). On the complex Wishart. *Annals of Mathematical Statistics*, **36**, 313–315.

Srivastava, M. S. (2002). *Methods of Multivariate Statistics*. Wiley, New York.

Srivastava, M. S. (2003). Singular Wishart and multivariate beta distributions. *Annals of Statistics*, **31**, 1537–1560.

Srivastava, M. S. and von Rosen D. (2002). Regression models with unknown singular covariance matrix. *Linear Algebra and Its Applications*, **354**, 255–273.

Steerneman, T. and van Perlo-ten Kleij, F. (2005). Properties of the matrix $\mathbf{A} - \mathbf{XY}^*$. *Linear Algebra and Its Applications*, **410**, 70–86.

Stefanski, L. A. (1996). A note on the arithmetic–geometric–harmonic mean inequalities. *The American Statistician*, **50** (3), 246–247.

Stewart, G. W. (1972). On the sensitivity of the eigenvalue problem $Ax = \lambda Bx$. *SIAM Journal of Numerical Analysis*, **9**, 669–696.

Stewart, G. W. (1998). *Matrix Algorithms. Volume 1: Basic Decompositions*. SIAM, Philadelphia.

Stewart, G. W. (2001). *Matrix Algorithms. Volume 2: Eigen Systems*. SIAM, Philadelphia.

Stigler, S. M. (1999). *Statistics on the Table: The History of Statistical Concepts and Methods*. Harvard University Press.

Strang, W. C. (1960). On the Kantorovich inequality. *Proceedings of the American Mathematical Society*, **11**, 468.

Styan, G. P. H. (1983). On some inequalities associated with ordinary least squares and the Kantorovich inequality. *Festscrift for Eino Haikala on his Seventieth birthday, Acta Universitatis Tamperensis, Series A*, **153**, 158–166. University of Tampere.

Styan, G. P. H. (1985). Schur complements and linear statistical models. In T. Pukkila and S. Putanen (Eds.), *Proceedings of the First Tampere Seminar on Linear Statistical Models and Their Applications*, 37–35. Department of Mathematical Sciences, University of Tampere, Finland.

Styan, G. P. H. (1989). Three useful expressions for expectations involving a Wishart matrix and its inverse. In Y. Dodge (Ed.), *Statistical Data Analysis and Inference*, 283–296. North-Holland, Amsterdam.

Takane, Y. (2004). Matrices with special reference to applications in psychometrics. *Linear Algebra and Its Applications*, **388**, 341–361.

Takane, Y. and Yanai, H. (2005). On the Wedderburn–Guttman theorem. *Linear Algebra and Its Applications*, **410**, 267–278.

Tee, G. J. (2005). Eigenvectors of block circulant and alternating circulant matrices. *Research Letters in the Information and Mathematical Sciences*, **8**, 123–142. Available online at http://iims.massey.ac.nz/research/letters/.

Thibaudeau, Y. and Styan, G. P. H. (1985). Bounds for Chakrabarti's measure of imbalance in experimental design. In T. Pukkila and S. Putanen (Eds.), *Proceedings of the First Tampere Seminar on Linear Statistical Models and Their Applications*, 323–347. Department of Mathematical Sciences, University of Tampere, Finland.

Tian, Y. (2000). Completing block matrices with maximal and minimal ranks. *Linear Algebra and Its Applications*, **321**, 327–345.

Tian, Y. (2002). Upper and lower bounds of matrix expressions using generalized inverses. *Linear Algebra and Its Applications*, **355**, 187–214.

Tian,Y. (2004). Rank equalities for block matrices and their Moore–Penrose inverses. *Houston Journal of Mathematics*, **30** (2), 483–510.

Tian, Y. (2005a). The reverse-order law $(AB)^+ = B^+(A^+BB^+)^+A^+$ and its equivalent equalities. *Journal of Mathematics of Kyoto University*, **45**, 841–850.

Tian, Y. (2005b). Special forms of generalized inverses of row block matrices. *Electronic Journal of Linear Algebra*, **13**, 249–261.

Tian, Y. (2006a). Invariance properties of a triple matrix product involving generalized inverses. *Linear Algebra and Its Applications*, **417**, 94–107.

Tian, Y. (2006b). Ranks and independence of solutions of the matrix $AXB + CYD = M$. *Acta Math. Univ. Comeniannae*, **LXXV** (1), 1–10.

Tian, Y. (2006c). The equivalence between $(\mathbf{AB})^+ = \mathbf{B}^+\mathbf{A}^+$ and other mixed-type reverse-order laws. *International Journal of Mathematical Education in Science and Technology*, **37** (3), 331–339.

Tian, Y. and Styan, G. P. H. (2001). Some rank inequalities for idempotent and involutionary matrices. *Linear Algebra and Its Applications*, **335**, 101–117.

Tian, Y. and Styan, G. P. H. (2005). Cochran's theorem for outer inverses of matrices and matrix quadratic forms. *Linear and Multilinear Algebra*, **53** (5), 387–392.

Tian, Y. and Styan, G. P. H. (2006). Cochran's statistical theorem revisited. *Journal of Statistical Planning and Inference*, **136**, 2659–2667.

Tian, Y. and Takane, Y. (2005). Schur complements and Banachiewicz-Schur forms. *Electronic Journal of Linear Algebra*, **13**, 405–418.

Tibshirani, R. and Hastie, T. J. (1990). *Generalized Additive Models*. Chapman and Hall, New York.

Tong, Y. L. (1980). *Probability Inequalities in Multivariate Distributions*. Academic Press, New York.

Tong, Y. L. (1983). Some distribution properties of the sample species-diversity indices and their applications. *Biometrics*, **39**, 999–1008.

Tong, Y. L. (1990). *Mulitvariate Normal Distribution*. Springer-Verlag, New York.

Tracey, D. S. and Dwyer, P. S. (1969). Multivariate maxima and minima with matrix derivatives. *Journal of the American Statistical Association*, **64**, 1576–1594.

Trenkler, G. (1994). Characterizations of oblique and orthogonal projectors. In T. Caliński and R. Kala (Eds.), *Proceedings of the International Conference on Linear Statistical Inference LINSTAT'93*, 255–270. Kluwer Academic Publishers, Netherlands.

Trenkler, G. and Puntanen, S. (2005). A multivariate version of Samuelson's inequality. *Linear Algebra and Its Applications*, **410**, 143–149.

Tsatsomeros, M. (2007). Matrix equalities and inequalities. In L. Hogben (Ed.), *Handbook of Linear Algebra*, Chapter 14. Chapman and Hall/CRC, Taylor and Francis Group, Boca Raton, FL.

Turkington, D. A. (2002). *Matrix Calculas & Zero-One Matrices: Statistical and Econometric Applications*. Cambridge University Press, Cambridge, UK.

Ukita, Y. (1955). Characterization of 2-type diagonal matrices with an application to order statistics. *Journal of Hokkaido College of Art and Literature*, **6**, 66–75.

Vaish, A. K. and Chaganty, N. R. (2004). Wishartness and independence of matrix quadratric forms for Kronecker product covariance structures. *Linear Algebra and Its Applications*, **388**, 379–388.

Varga, R. S. (1962). *Matrix Iterative Analysis*. Prentice-Hall, Englewood Cliffs, NJ.

Vetter, W. J. (1970). Derivative operations on matrices. *IEEE Transactions Automatic Control*, **AC–15**, 241–244.

Wanless, I. M. (2007). Permanents. In L. Hogben (Ed.), *Handbook of Linear Algebra*, Chapter 31. Chapman and Hall/CRC, Taylor and Francis Group, Boca Raton, FL.

Wang, S. G. and Ip, W. C. (2000). A matrix version of the Wielandt inequality and its application to statistics. *Linear Algebra and Its Applications*, **296**, 171–181.

Wei, B.-C. (1997). *Exponential Family Nonlinear Models*. Springer-Verlag, Singapore.

Widom, H. (1965). Toeplitz matrices. In I.I. Hirschmann, Jr. (Ed.), *Studies in Real and Complex Analysis*, MAA Studies in Mathematics, Prentice–Hall, Englewood Cliffs, NJ.

Wielandt, H. (1955). An extremum property of sums of eigenvalues. *Proceedings of the American Mathematical Society*, **6**, 106-110.

Wijsmann, R. A. (1979). Constructing all simultaneous confidence sets in a given class, with applications to MANOVA. *Annals of Statistics*, **7**, 1003–1018.

Williams, E. R. and John, J. A. (2000). Updating the average efficiency factor in α-designs. *Biometrika*, **87**, 695–699.

Wolkowicz, H. and Styan, G. P. H. (1979). Extensions of Samuelson's inequality. *The American Statistician*, **33**, 143–144. See also comments and a reply in **34** (4), 249–250.

Wolkowicz, H. and Styan, G. P. H. (1980). Bounds for eigenvalues using traces. *Linear Algebra and Its Applications*, **29**, 471–506.

Wood, S. D. (2006). *Generalized Additive models: An Introduction with R*. Chapman and Hall/CRC, Taylor and Francis Group, Boca Raton, FL.

Zhang, Fuzhen (1997). Quaternions and matrices of quaternions. *Linear Algebra and Its Applications*, **251**, 21–57.

Zhang, Fuzhen (1999). *Matrix Theory: Basic Results and Techniques*. Springer-Verlag, New York.

INDEX

WILEY SERIES IN PROBABILITY AND STATISTICS
ESTABLISHED BY WALTER A. SHEWHART AND SAMUEL S. WILKS

Editors: *David J. Balding, Noel A. C. Cressie, Nicholas I. Fisher,*
Iain M. Johnstone, J. B. Kadane, Geert Molenberghs, David W. Scott,
Adrian F. M. Smith, Sanford Weisberg
Editors Emeriti: *Vic Barnett, J. Stuart Hunter, David G. Kendall,*
Jozef L. Teugels

The **Wiley Series in Probability and Statistics** is well established and authoritative. It covers many topics of current research interest in both pure and applied statistics and probability theory. Written by leading statisticians and institutions, the titles span both state-of-the-art developments in the field and classical methods.

Reflecting the wide range of current research in statistics, the series encompasses applied, methodological and theoretical statistics, ranging from applications and new techniques made possible by advances in computerized practice to rigorous treatment of theoretical approaches.

This series provides essential and invaluable reading for all statisticians, whether in academia, industry, government, or research.

*Now available in a lower priced paperback edition in the Wiley Classics Library.
†Now available in a lower priced paperback edition in the Wiley–Interscience Paperback Series.

*Now available in a lower priced paperback edition in the Wiley Classics Library.

†Now available in a lower priced paperback edition in the Wiley–Interscience Paperback Series.

*Now available in a lower priced paperback edition in the Wiley Classics Library.
†Now available in a lower priced paperback edition in the Wiley–Interscience Paperback Series.

*Now available in a lower priced paperback edition in the Wiley Classics Library.
†Now available in a lower priced paperback edition in the Wiley–Interscience Paperback Series.

*Now available in a lower priced paperback edition in the Wiley Classics Library.
†Now available in a lower priced paperback edition in the Wiley–Interscience Paperback Series.

---

*Now available in a lower priced paperback edition in the Wiley Classics Library.

†Now available in a lower priced paperback edition in the Wiley–Interscience Paperback Series.

*Now available in a lower priced paperback edition in the Wiley Classics Library.
†Now available in a lower priced paperback edition in the Wiley–Interscience Paperback Series.